ELECTRICAL ENERGY CONVERSION AND TRANSPORT

IEEE Press
445 Hoes Lane
Piscataway, NJ 08854

IEEE Press Editorial Board 2013
John Anderson, *Editor in Chief*

Linda Shafer	Saeid Nahavandi	George Zobrist
George W. Arnold	David Jacobson	Tariq Samad
Ekram Hossain	Mary Lanzerotti	Dmitry Goldgof
Om P. Malik		

Kenneth Moore, *Director of IEEE Book and Information Services (BIS)*

A complete list of titles in the IEEE Press Series on Power Engineering appears at the end of this book.

ELECTRICAL ENERGY CONVERSION AND TRANSPORT

An Interactive Computer-Based Approach

SECOND EDITION

George G. Karady
Keith E. Holbert

Mohamed E. El-Hawary, *Series Editor*

IEEE PRESS

WILEY

Cover Design: John Wiley & Sons, Inc.
Cover Illustration: Courtesy of Siemens AG; Power Lines © Corbis Super Royalty Free/Alamy

Copyright © 2013 by the Institute of Electrical and Electronics Engineers, Inc. All rights reserved.

Published by John Wiley & Sons, Inc., Hoboken, New Jersey.
Published simultaneously in Canada.

No part of this publication may be reproduced, stored in a retrieval system, or transmitted in any form or by any means, electronic, mechanical, photocopying, recording, scanning, or otherwise, except as permitted under Section 107 or 108 of the 1976 United States Copyright Act, without either the prior written permission of the Publisher, or authorization through payment of the appropriate per-copy fee to the Copyright Clearance Center, Inc., 222 Rosewood Drive, Danvers, MA 01923, (978) 750-8400, fax (978) 750-4470, or on the web at www.copyright.com. Requests to the Publisher for permission should be addressed to the Permissions Department, John Wiley & Sons, Inc., 111 River Street, Hoboken, NJ 07030, (201) 748-6011, fax (201) 748-6008, or online at http://www.wiley.com/go/permissions.

Limit of Liability/Disclaimer of Warranty: While the publisher and author have used their best efforts in preparing this book, they make no representations or warranties with respect to the accuracy or completeness of the contents of this book and specifically disclaim any implied warranties of merchantability or fitness for a particular purpose. No warranty may be created or extended by sales representatives or written sales materials. The advice and strategies contained herein may not be suitable for your situation. You should consult with a professional where appropriate. Neither the publisher nor author shall be liable for any loss of profit or any other commercial damages, including but not limited to special, incidental, consequential, or other damages.

For general information on our other products and services or for technical support, please contact our Customer Care Department within the United States at (800) 762-2974, outside the United States at (317) 572-3993 or fax (317) 572-4002.

Wiley also publishes its books in a variety of electronic formats. Some content that appears in print may not be available in electronic formats. For more information about Wiley products, visit our web site at www.wiley.com.

Library of Congress Cataloging-in-Publication Data:

Karady, George G.
   Electrical energy conversion and transport : an interactive computer-based approach / George G. Karady, Keith E. Holbert. – Second edition.
      pages cm
   Includes bibliographical references and index.
   ISBN 978-0-470-93699-3 (cloth)
   1. Electric power distribution. 2. Electric current converters. 3. Electric power production–Data processing. I. Holbert, Keith E. II. Title.
   TK1001.K36 2012
   621.31–dc23

2012029241

Printed in the United States of America.

10 9 8 7 6 5 4 3 2 1

CONTENTS

Preface and Acknowledgments — xv

1 ELECTRIC POWER SYSTEMS — 1
  1.1. Electric Networks — 2
      1.1.1. Transmission Systems — 4
      1.1.2. Distribution Systems — 6
  1.2. Traditional Transmission Systems — 6
      1.2.1. Substation Components — 8
      1.2.2. Substations and Equipment — 9
      1.2.3. Gas Insulated Switchgear — 17
      1.2.4. Power System Operation in Steady-State Conditions — 18
      1.2.5. Network Dynamic Operation (Transient Condition) — 20
  1.3. Traditional Distribution Systems — 20
      1.3.1. Distribution Feeder — 21
      1.3.2. Residential Electrical Connection — 24
  1.4. Intelligent Electrical Grids — 26
      1.4.1. Intelligent High-Voltage Transmission Systems — 26
      1.4.2. Intelligent Distribution Networks — 28
  1.5. Exercises — 28
  1.6. Problems — 29

2 ELECTRIC GENERATING STATIONS — 30
  2.1. Fossil Power Plants — 34
      2.1.1. Fuel Storage and Handling — 34
      2.1.2. Boiler — 35
      2.1.3. Turbine — 41
      2.1.4. Generator and Electrical System — 43

	2.1.5.	Combustion Turbine	47
	2.1.6.	Combined Cycle Plants	48
2.2.	Nuclear Power Plants		49
	2.2.1.	Nuclear Reactor	50
	2.2.2.	Pressurized Water Reactor	53
	2.2.3.	Boiling Water Reactor	55
2.3.	Hydroelectric Power Plants		56
	2.3.1.	Low Head Hydroplants	59
	2.3.2.	Medium- and High-Head Hydroplants	60
	2.3.3.	Pumped Storage Facility	62
2.4.	Wind Farms		63
2.5.	Solar Power Plants		66
	2.5.1.	Photovoltaics	66
	2.5.2.	Solar Thermal Plants	70
2.6.	Geothermal Power Plants		72
2.7.	Ocean Power		73
	2.7.1.	Ocean Tidal	74
	2.7.2.	Ocean Current	75
	2.7.3.	Ocean Wave	75
	2.7.4.	Ocean Thermal	76
2.8.	Other Generation Schemes		76
2.9.	Electricity Generation Economics		77
	2.9.1.	O&M Cost	79
	2.9.2.	Fuel Cost	79
	2.9.3.	Capital Cost	80
	2.9.4.	Overall Generation Costs	81
2.10.	Load Characteristics and Forecasting		81
2.11.	Environmental Impact		85
2.12.	Exercises		86
2.13.	Problems		86

3 SINGLE-PHASE CIRCUITS   89

3.1.	Circuit Analysis Fundamentals		90
	3.1.1.	Basic Definitions and Nomenclature	90
	3.1.2.	Voltage and Current Phasors	91
	3.1.3.	Power	92
3.2.	AC Circuits		94

3.3.	Impedance	96
	3.3.1.	Series Connection	100
	3.3.2.	Parallel Connection	100
	3.3.3.	Impedance Examples	104
3.4.	Loads	109
	3.4.1.	Power Factor	111
	3.4.2.	Voltage Regulation	116
3.5.	Basic Laws and Circuit Analysis Techniques	116
	3.5.1.	Kirchhoff's Current Law	117
	3.5.2.	Kirchhoff's Voltage Law	123
	3.5.3.	Thévenin's and Norton's Theorems	127
3.6.	Applications of Single-Phase Circuit Analysis	128
3.7.	Summary	140
3.8.	Exercises	141
3.9.	Problems	141

4 THREE-PHASE CIRCUITS — 145

4.1.	Three-Phase Quantities	146
4.2.	Wye-Connected Generator	151
4.3.	Wye-Connected Loads	155
	4.3.1.	Balanced Wye Load (Four-Wire System)	156
	4.3.2.	Unbalanced Wye Load (Four-Wire System)	158
	4.3.3.	Wye-Connected Three-Wire System	160
4.4.	Delta-Connected System	162
	4.4.1.	Delta-Connected Generator	162
	4.4.2.	Balanced Delta Load	163
	4.4.3.	Unbalanced Delta Load	166
4.5.	Summary	168
4.6.	Three-Phase Power Measurement	174
	4.6.1.	Four-Wire System	175
	4.6.2.	Three-Wire System	175
4.7.	Per-Unit System	177
4.8.	Symmetrical Components	182
	4.8.1.	Calculation of Phase Voltages from Sequential Components	182
	4.8.2.	Calculation of Sequential Components from Phase Voltages	183
	4.8.3.	Sequential Components of Impedance Loads	184

4.9.	Application Examples	188
4.10.	Exercises	203
4.11.	Problems	204

5 TRANSMISSION LINES AND CABLES — 207

5.1.	Construction	208
5.2.	Components of the Transmission Lines	215
	5.2.1. Towers and Foundations	215
	5.2.2. Conductors	216
	5.2.3. Insulators	218
5.3.	Cables	223
5.4.	Transmission Line Electrical Parameters	224
5.5.	Magnetic Field Generated by Transmission Lines	225
	5.5.1. Magnetic Field Energy Content	229
	5.5.2. Single Conductor Generated Magnetic Field	230
	5.5.3. Complex Spatial Vector Mathematics	233
	5.5.4. Three-Phase Transmission Line-Generated Magnetic Field	234
5.6.	Transmission Line Inductance	239
	5.6.1. External Magnetic Flux	240
	5.6.2. Internal Magnetic Flux	241
	5.6.3. Total Conductor Magnetic Flux	243
	5.6.4. Three-Phase Line Inductance	244
5.7.	Transmission Line Capacitance	249
	5.7.1. Electric Field Generation	249
	5.7.2. Electrical Field around a Conductor	250
	5.7.3. Three-Phase Transmission Line Generated Electric Field	256
	5.7.4. Three-Phase Line Capacitance	271
5.8.	Transmission Line Networks	273
	5.8.1. Equivalent Circuit for a Balanced System	273
	5.8.2. Long Transmission Lines	277
5.9.	Concept of Transmission Line Protection	282
	5.9.1. Transmission Line Faults	282
	5.9.2. Protection Methods	285
	5.9.3. Fuse Protection	285
	5.9.4. Overcurrent Protection	285
	5.9.5. Distance Protection	288

5.10.	Application Examples	289
	5.10.1. Mathcad® Examples	289
	5.10.2. PSpice®: Transient Short-Circuit Current in Transmission Lines	302
	5.10.3. PSpice: Transmission Line Energization	304
5.11.	Exercises	307
5.12.	Problems	308

6 ELECTROMECHANICAL ENERGY CONVERSION 313

6.1.	Magnetic Circuits	314
	6.1.1. Magnetic Circuit Theory	315
	6.1.2. Magnetic Circuit Analysis	317
	6.1.3. Magnetic Energy	323
	6.1.4. Magnetization Curve	324
	6.1.5. Magnetization Curve Modeling	329
6.2.	Magnetic and Electric Field Generated Forces	336
	6.2.1. Electric Field-Generated Force	336
	6.2.2. Magnetic Field-Generated Force	337
6.3.	Electromechanical System	343
	6.3.1. Electric Field	344
	6.3.2. Magnetic Field	345
6.4.	Calculation of Electromagnetic Forces	347
6.5.	Applications	352
	6.5.1. Actuators	353
	6.5.2. Transducers	356
	6.5.3. Permanent Magnet Motors and Generators	362
	6.5.4. Microelectromechanical Systems	365
6.6.	Summary	368
6.7.	Exercises	368
6.8.	Problems	369

7 TRANSFORMERS 375

7.1.	Construction	376
7.2.	Single-Phase Transformers	381
	7.2.1. Ideal Transformer	382
	7.2.2. Real Transformer	391
	7.2.3. Determination of Equivalent Transformer Circuit Parameters	399

7.3.		Three-Phase Transformers	408
	7.3.1.	Wye–Wye Connection	410
	7.3.2.	Wye–Delta Connection	415
	7.3.3.	Delta–Wye Connection	418
	7.3.4.	Delta–Delta Connection	420
	7.3.5.	Summary	420
	7.3.6.	Analysis of Three-Phase Transformer Configurations	421
	7.3.7.	Equivalent Circuit Parameters of a Three-Phase Transformer	429
	7.3.8.	General Program for Computing Transformer Parameters	432
	7.3.9.	Application Examples	435
	7.3.10.	Concept of Transformer Protection	447
7.4.		Exercises	450
7.5.		Problems	451

8 SYNCHRONOUS MACHINES 456

8.1.		Construction	456
	8.1.1.	Round Rotor Generator	457
	8.1.2.	Salient Pole Generator	459
	8.1.3.	Exciter	462
8.2.		Operating Concept	465
	8.2.1.	Main Rotating Flux	465
	8.2.2.	Armature Flux	468
8.3.		Generator Application	472
	8.3.1.	Loading	472
	8.3.2.	Reactive Power Regulation	472
	8.3.3.	Synchronization	473
	8.3.4.	Static Stability	474
8.4.		Induced Voltage and Armature Reactance Calculation	487
	8.4.1.	Induced Voltage Calculation	488
	8.4.2.	Armature Reactance Calculation	496
8.5.		Concept of Generator Protection	507
8.6.		Application Examples	511
8.7.		Exercises	535
8.8.		Problems	536

9 INDUCTION MACHINES — 541

- 9.1. Introduction — 541
- 9.2. Construction — 543
  - 9.2.1. Stator — 543
  - 9.2.2. Rotor — 546
- 9.3. Three-Phase Induction Motor — 547
  - 9.3.1. Operating Principle — 547
  - 9.3.2. Equivalent Circuit — 553
  - 9.3.3. Motor Performance — 556
  - 9.3.4. Motor Maximum Output — 557
  - 9.3.5. Performance Analyses — 560
  - 9.3.6. Determination of Motor Parameters by Measurement — 570
- 9.4. Single-Phase Induction Motor — 591
  - 9.4.1. Operating Principle — 592
  - 9.4.2. Single-Phase Induction Motor Performance Analysis — 595
- 9.5. Induction Generators — 603
  - 9.5.1. Induction Generator Analysis — 603
  - 9.5.2. Doubly Fed Induction Generator — 606
- 9.6. Concept of Motor Protection — 608
- 9.7. Exercises — 610
- 9.8. Problems — 611

10 DC MACHINES — 616

- 10.1. Construction — 616
- 10.2. Operating Principle — 620
  - 10.2.1. DC Motor — 620
  - 10.2.2. DC Generator — 623
  - 10.2.3. Equivalent Circuit — 625
  - 10.2.4. Excitation Methods — 628
- 10.3. Operation Analyses — 629
  - 10.3.1. Separately Excited Machine — 630
  - 10.3.2. Shunt Machine — 637
  - 10.3.3. Series Motor — 645
  - 10.3.4. Summary — 651
- 10.4. Application Examples — 652
- 10.5. Exercises — 669
- 10.6. Problems — 669

11 INTRODUCTION TO POWER ELECTRONICS AND MOTOR CONTROL — 673

- 11.1. Concept of DC Motor Control — 674
- 11.2. Concept of AC Induction Motor Control — 678
- 11.3. Semiconductor Switches — 685
    - 11.3.1. Diode — 685
    - 11.3.2. Thyristor — 687
    - 11.3.3. Gate Turn-Off Thyristor — 692
    - 11.3.4. Metal–Oxide–Semiconductor Field-Effect Transistor — 693
    - 11.3.5. Insulated Gate Bipolar Transistor — 695
    - 11.3.6. Summary — 696
- 11.4. Rectifiers — 697
    - 11.4.1. Simple Passive Diode Rectifiers — 697
    - 11.4.2. Single-Phase Controllable Rectifiers — 709
    - 11.4.3. Firing and Snubber Circuits — 726
    - 11.4.4. Three-Phase Rectifiers — 728
- 11.5. Inverters — 729
    - 11.5.1. Voltage Source Inverter with Pulse Width Modulation — 732
    - 11.5.2. Line-Commutated Thyristor-Controlled Inverter — 735
    - 11.5.3. High-Voltage DC Transmission — 738
- 11.6. Flexible AC Transmission — 739
    - 11.6.1. Static VAR Compensator — 740
    - 11.6.2. Static Synchronous Compensator — 744
    - 11.6.3. Thyristor-Controlled Series Capacitor — 744
    - 11.6.4. Unified Power Controller — 747
- 11.7. DC-to-DC Converters — 747
    - 11.7.1. Boost Converter — 748
    - 11.7.2. Buck Converter — 754
- 11.8. Application Examples — 757
- 11.9. Exercises — 773
- 11.10. Problems — 774

Appendix A Introduction to Mathcad® — 777

- A.1. Worksheet and Toolbars — 777
    - A.1.1. Text Regions — 780
    - A.1.2. Calculations — 780
- A.2. Functions — 783
    - A.2.1. Repetitive Calculations — 784
    - A.2.2. Defining a Function — 785

	A.2.3. Plotting a Function	786
	A.2.4. Minimum and Maximum Function Values	788
A.3.	Equation Solvers	788
	A.3.1. Root Equation Solver	789
	A.3.2. Find Equation Solver	789
A.4.	Vectors and Matrices	790

Appendix B Introduction to MATLAB® — 794

B.1.	Desktop Tools	794
B.2.	Operators, Variables, and Functions	796
B.3.	Vectors and Matrices	797
B.4.	Colon Operator	799
B.5.	Repeated Evaluation of an Equation	799
B.6.	Plotting	800
B.7.	Basic Programming	803

Appendix C Fundamental Units and Constants — 805

| C.1. | Fundamental Units | 805 |
| C.2. | Fundamental Physical Constants | 809 |

Appendix D Introduction to PSpice® — 810

D.1.	Obtaining and Installing PSpice	810
D.2.	Using PSpice	811
	D.2.1. Creating a Circuit	811
	D.2.2. Simulating a Circuit	812
	D.2.3. Analyzing Simulation Results	813

Problem Solution Key — 815

Bibliography — 822

Index — 824

PREFACE AND ACKNOWLEDGMENTS

This book provides material essential for an undergraduate course covering the fundamental concepts of electric energy conversion and transport—a key branch of electrical engineering. Every electrical engineer should know why a motor rotates and how electric energy is generated and transported. Moreover, the electric power grid is a critical part of any national infrastructure. The maintenance and development of this vital industry requires well-trained engineers who are able to use modern computation techniques to analyze electric systems and understand the theory of electrical energy conversion.

Engineering education has improved significantly during the last decade due to advancements in technology and the widespread use of personal computers. Engineering educators have also recognized the need to transform students from passive listeners in the classroom to active learners. The paradigm shift is from a teacher-centered delivery approach to that of a learner-centered environment.

Computer-equipped classrooms and the computer aptitude of students open up new possibilities to improve engineering education by changing the delivery method. We advocate an interactive presentation of the subject matter, in which the students are intimately engaged in the lectures. This book is designed to support active learning, especially in a computer-based classroom environment. The computer-assisted teaching method increases student mastery of the course material as a result of their participation in its development. The primary goal of this approach is to increase student learning through their dynamic involvement; secondarily, students' interest in power engineering is enhanced through their own attraction to computer technologies. This interactive approach provides students with a better understanding of the theory and the development of solid problem-solving skills.

As many universities and instructors firmly favor the use of one software package versus another, we leave the instructor to freely choose the software employed. This book applies Mathcad®, MATLAB®, and PSpice® throughout, and as such appendices introduce the basic use of these three programs. Less emphasis is paid on dedicated power engineering simulation tools due to the extended time and effort needed to learn such specialized software. In contrast, general-purpose programs permit students to focus more on the connection between the theory and computational analysis.

The extensive computer use permits analyzing complex problems that are not easily solvable by hand computations with calculators. In fact, the experienced instructor will find that their students are able to work complicated problems that were previously too difficult at this level. This is a significant modernization of the classical topic of electric energy conversion. Students familiar with the application of modern computational techniques to electrical power applications are better prepared to meet the needs of industry.

This textbook facilitates interactive teaching of the subject material. Through the students' active participation, learning is enhanced. The advantages of this method include:

1. Better understanding of the subject because the students participate in its development;
2. Development and advancement of problem-solving skills;
3. Simultaneously learning the practical engineering application of the material using computerized methods accepted by industry;
4. Extending the students' attention span and maintaining their interest during the lecture—this method eliminates boredom that inhibits students toward the end of most lectures;
5. The students analyze the results and draw the conclusions, thus enhancing learning; and
6. The students gain experience with general-purpose mathematical and scientific computing programs frequently utilized by industry.

The authors recommend the textbook to faculty who want to modernize their electric power curriculum. The book is also intended for engineers interested in increasing their knowledge of electrical power and computer-based problem-solving skills. Such knowledge may open up or expand career opportunities in the electric power industry.

This second edition has inserted an additional chapter by moving and substantially expanding the treatment of electric power generation. Furthermore, the technical coverage of all the chapters has been expanded with the addition of material in areas such as the intelligent (smart) grid, symmetrical components, long transmission lines, induction generators, flexible alternating current (ac) transmission systems, buck and boost converters, and the protection of transformers, generators, motors, and transmission lines.

HOW TO USE THIS BOOK EFFECTIVELY

This textbook differs noticeably from others in that classical derivations are combined with numerical examples. In doing so, the reader is not only provided with the general analytical expressions as the theoretical development proceeds, but additionally, the concurrent numerical results assist the student in developing a sense for the correct magnitude of various parameters and variables. The authors have found Mathcad particularly well suited to this approach. Regardless of which software the reader chooses to use, we recommend that the reader first familiarize himself or herself with the information in Appendix A ("Introduction to Mathcad"), since Mathcad expressions are utilized throughout the text. This will allow the reader to reap the full benefits of this delivery method. Although this book employs Mathcad, MATLAB, and PSpice, other

PREFACE AND ACKNOWLEDGMENTS

computational software can also be utilized effectively—this includes HSpice, Maple, Mathematica, and even spreadsheet packages such as Excel.

The authors suggest a course syllabus ordering that parallels the textbook. The textbook may be used for either a single semester or a two-semester course. For instance, Chapter 2 ("Electric Generating Stations") can be skipped without significant loss of continuity for those instructors and readers who wish to do so. Similarly, Chapter 3 ("Single-Phase Circuits") represents a review of basic circuit analysis, albeit in the context of computer-based analysis, which is generally a prerequisite to a course such as this. Suggested timelines for one and two three-semester-hour courses are outlined in the tables below.

One-Semester Course		Two-Semester Course	
Week	Topic	Week	Topic
1	Chapter 1: Electric Power Systems	1	Chapter 1: Electric Power Systems
2	Chapter 3: Single-Phase Circuits (emphasizing Section 3.4 and Section 3.5)	2–3	Chapter 2: Electric Generating Stations
		4–5	Chapter 3: Single-Phase Circuits
3–4	Chapter 4: Three-Phase Circuits (omit Section 4.6, Section 4.7, and Section 4.8)	6–8	Chapter 4: Three-Phase Circuits
		9–12	Chapter 5: Transmission Lines
		13–15	Chapter 6: Electromechanical Energy Conversion
5–6	Chapter 5: Transmission Lines (omit Section 5.5.3, Section 5.5.4, Section 5.7.3, Section 5.8.2, and Section 5.9)	16–18	Chapter 7: Transformers
		19–21	Chapter 8: Synchronous Machines
		22–24	Chapter 9: Induction Machines
7–8	Chapter 6: Electromechanical Energy Conversion (omit Section 6.1.5, Section 6.4, and Section 6.5)	25–26	Chapter 10: DC Machines
		27–30	Chapter 11: Introduction to Power Electronics and Motor Control
9–10	Chapter 7: Transformers (omit Section 7.2.3, Section 7.3.7, Section 7.3.8, and Section 7.3.10)		
11–12	Chapter 8: Synchronous Machines (omit Section 8.3.4 and Section 8.5)		
13–14	Chapter 9: Induction Machines (omit Section 9.3.6, Section 9.5, and Section 9.6)		
15	Chapter 10: DC Machines (omit Section 10.3)		

Here, we present a brief overview of the suggested instructional technique for a representative class period. The basis of the approach is that after introducing the hardware and theory, the basic formulae and their practical application are developed jointly with the students using computers. Having divided the particular topic into sections, the instructor outlines each step of the analysis, and students then proceed to develop the equation(s) using his or her computer. While students are working together, the instructor is free to move about the classroom, answer student questions, and assess their understanding. After allowing students sufficient time to complete the process and reach conclusions, the instructor confirms the results and the students make corrections as needed. This procedure leads to student theory development and analysis of performance—*learner-centered education*.

Through computer utilization, a seamless integration of theory and application is achieved, thereby increasing student interest in the subject. The textbook derivation of the system equations and the operational analyses are presented using numerical examples. The numerical examples reinforce the theory and provide deeper understanding of the physical phenomena. In addition, computer utilization provides immediate feedback to the student.

Again, paralleling the classroom activities, each chapter first describes the hardware associated with that topic; for example, the construction and components are presented using drawings and photographs. This is followed by the theory and physics of the chapter material together with the development of an equivalent circuit. The major emphasis of the chapters is operational analysis. The questions at the end of each chapter are open ended to promote deeper investigation by the reader.

The interactive method is also applicable in a self-learning environment. In this case, the text outlines each step. The reader is encouraged to initially ignore the solution given in the text, but instead derive the equations and calculate the value using his or her computer. The reader then compares his or her results with the correct answers. This process is continued until the completion of the instructional unit.

ACKNOWLEDGMENTS

The second edition of this textbook has benefited from the constructive criticism of others. The authors would like to express their sincere gratitude to the late Professor Richard Farmer, who was a member of the National Academy of Engineering, for his thorough review of both the first and second editions of the book manuscript. We also humbly thank the Institute of Electrical and Electronics Engineers (IEEE) Education Society for its recognition of the merits of computer-based active learning through the IEEE Transactions on Education Best Paper[1] award to us.

GEORGE G. KARADY
KEITH E. HOLBERT
Tempe, AZ
April 2013

---

[1] Holbert, K.E. and Karady, G.G., "Strategies, challenges and prospects for active learning in the computer-based classroom," *IEEE Transactions on Education*, 52(1), 31–38, 2009.

1

ELECTRIC POWER SYSTEMS

The purpose of the electric power system is to generate, transmit, and distribute electrical energy. Usually, a three-phase alternating current (ac) system is used for generation and transmission of the electric power. The frequency of the voltage and current is 60 Hz in the United States and some Asian countries, and is 50 Hz in Europe, Australia, and parts of Asia. Sometimes, exceptions are the rule, as in the case of Japan for which the western portion of the country is served by 60 Hz, whereas the eastern side operates at 50 Hz.

In the 1880s, during the development of electricity distribution, the pioneers' choice as to whether to use direct current (dc) or ac was contested. In particular, Thomas Edison favored dc, whereas both George Westinghouse and Nikola Tesla supported ac. AC transmission won this so-called War of the Currents due to the ability to convert ac voltages from higher to lower voltages using transformers and vice versa. This increased ac voltage permitted electric energy transport over longer distances with less power line losses than with dc.

The ac electrical system development started in the end of the 19th century, when the system frequency varied between 16.66 and 133 Hz. A large German company introduced 50 Hz frequency around 1891, after flickering was observed in systems

---

Electrical Energy Conversion and Transport: An Interactive Computer-Based Approach, Second Edition. George G. Karady and Keith E. Holbert.
© 2013 Institute of Electrical and Electronics Engineers, Inc. Published 2013 by John Wiley & Sons, Inc.

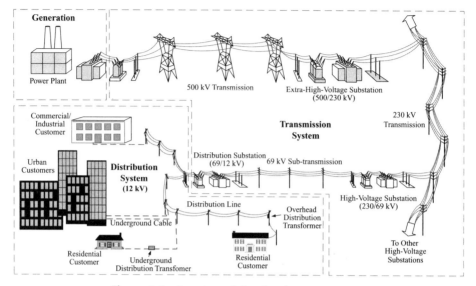

Figure 1.1. Overview of the electric power system.

operating at 40 Hz. In 1890, the leading U.S. electric company, Westinghouse Electric, introduced the 60 Hz frequency to avoid arc light flickering at lower frequencies.

The major components of the power system are:

- power plants, which produce electric energy,
- transmission and distribution lines, which transport the electric energy,
- substations with switchgear, which transform voltages, provide protection, and form node points, and
- loads, which consume the energy.

Figure 1.1 shows the major components of the electric power system.

This chapter describes the construction of the electric transmission and distribution system; discusses the substation equipment, including circuit breakers (CBs), disconnect switches, and protection; and describes the low voltage distribution system, including residential electric connections.

1.1. ELECTRIC NETWORKS

Power plants convert the chemical energy in coal, oil, or natural gas, or the potential energy of water, or nuclear energy into electric energy. In fossil nuclear power plants, the thermal energy is converted to high-pressure, high-temperature steam that drives a turbine which is mechanically connected to an electric generator. In a hydroelectric plant, the water falling to a lower elevation drives the turbine-generator set. The

ELECTRIC NETWORKS

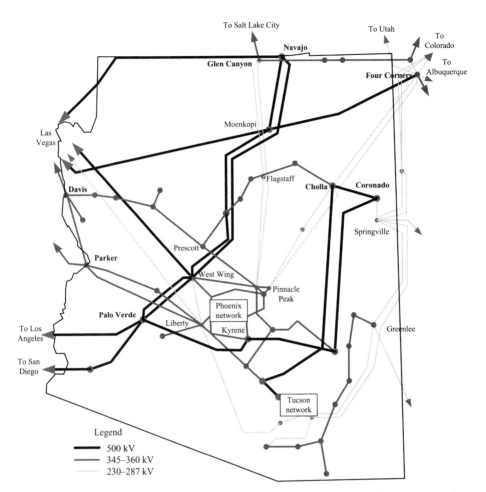

Figure 1.2. High- and extra-high-voltage transmission system in Arizona (power generation sites are shown in bold letters). (Data are from Western Systems Coordinating Council, 1999).

generator produces electric energy in the form of voltage and current. The generator voltage is around 15–25 kV, which is insufficient for long-distance transmission of the energy. To permit long-distance energy transportation, the voltage is increased and, simultaneously, the current is reduced by a transformer at the generation station. In Figure 1.1, the voltage is raised to 500 kV, and an extra-high-voltage (EHV) line carries the energy to a faraway substation, which is usually located in the outskirts of a large town or in the center of several large loads. For example, in Arizona, a 500 kV transmission line connects the Palo Verde Nuclear Generating Station to the Kyrene and Westwing substations, which supply a large part of Phoenix (see Fig. 1.2).

The electric power network is divided into separate transmission and distribution systems based on the voltage level. The system voltage is described by the

TABLE 1.1. Standard System Voltages (ANSI C84.1-1995[a] and C92.2-1987[b])

Name or Category	Nominal Voltage (kV)
Subtransmission	34.5
	46
	69
High voltage	115
	138
	161
	230
EHV	345
	400 (Europe)
	500
	765
Ultra-high voltage	1000 (China)

[a]ANSI C84.1-1995, *Voltage ratings for electric power systems and equipment (60 Hz)*.
[b]ANSI C92.2-1987, *Alternating-current electrical systems and equipment operating at voltages above 230 kV nominal—preferred voltage ratings*.

root-mean-square (rms) value of the *line-to-line* voltage, which is the voltage between phase conductors. Table 1.1 lists the standard transmission line and the subtransmission voltages. The line voltage of the transmission systems in the United States is between 115 and 765 kV. The ultra-high-voltage lines are generally not in commercial use; although in 2011 China started the operation of a 392 miles (630 km) long 1000 kV ultra-high-voltage ac line with a maximum capacity of 3000 MVA. The 345–765 kV transmission lines are the EHV lines, with a maximum length of 400–500 miles. The 115–230 kV lines are the high-voltage lines with a maximum length of 100–200 miles. The high-voltage lines are terminated at substations, which form the node points on the network. The substations supply the loads through transformers and switchgear. The transformer changes the voltage and current. The *switchgear* protects the system. The most important part of the switchgear is the *circuit breaker*, which automatically switches off (opens) the line in the event of a fault. Distribution line lengths are around 5–30 miles (8–48 km) with voltages at or below 46 kV.

1.1.1. Transmission Systems

The *transmission system* transfers three-phase power from the electric generating stations to the load centers. As an example, Figure 1.2 sketches a typical electrical network that supplies the metropolitan areas in Arizona and interconnects to the power systems of neighboring states. In this system 500, 345, 230, and 115 kV lines connect the loads and power plants. Note that the system in Figure 1.2 is a loop network, where at least two lines supply each load and the generating stations are connected to the network with three, or even four, lines. This arrangement assures that the failure of one line does not produce an outage. The electric system in the United States must withstand at least a single contingency, which means that loads and generators at a specific node are connected by at least two independent power system paths (e.g., power lines).

In addition, the map shows that 500, 345, and 230 kV lines interconnect the Arizona (AZ) system with California, Nevada, Utah, and New Mexico. These interconnections provide instantaneous assistance in cases of lost generation and line outages in the AZ system. Interconnection also permits the export or import of energy depending on the need of the area.

In open areas, overhead transmission lines are used. Typical examples are the interconnection between towns or a line running along a road within a city. In large, congested cities, underground cables are frequently used for electric energy transmission. An underground system has significantly higher costs but is environmentally and aesthetically preferable. Typically, the cost per mile of the overhead transmission lines is 6–10 times less than the underground cables.

At an EHV substation, transformers reduce the voltage to 230 or 345 kV. In Figure 1.1, a 230 kV high-voltage transmission line transports the energy to a high-voltage substation, typically located on the outskirts of the town. The voltage is further reduced at the high-voltage substation. Typically, 69 kV subtransmission lines connect the high-voltage substation to local distribution stations, which are located in the town. The subtransmission lines are built along larger streets.

In addition to the ac transmission system, high-voltage dc (HVDC) lines are used for long-distance, large energy transmission. Figure 1.3 depicts the main components of an HVDC system. The HVDC link contains two converters interconnected by a dc transmission line. The converters are electronic devices able to operate as a rectifier or as an inverter. Figure 1.3 shows that both converters are divided into two units connected in series. The middle point of the series-connected units is grounded. If the power is transferred from Converter 1 to Converter 2, then Converter 1 functions as a rectifier and Converter 2 acts as an inverter. The rectifier mode converts ac voltage to dc, and the inverter mode changes the dc voltage to ac. The dc transmission line typically has only two conductors, a positive (+) conductor and a negative (−) conductor.

HVDC is used to transport large amounts of energy over a long distance; typically, a dc line is not economical for less than around 300 miles (~500 km). A representative

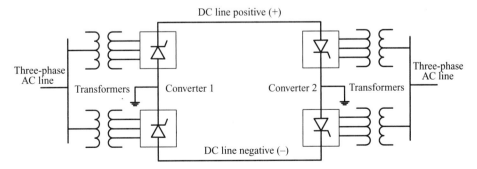

Figure 1.3. Main components of a high-voltage dc (HVDC) transmission system.

example for HVDC transmission is the Pacific DC Intertie, which is an 846 miles (1362 km) long HVDC transmission line between the Celilo Converter Station at The Dalles, Oregon and the Sylmar Converter Station north of Los Angeles, California. The line has two conductors with a maximum operating voltage of ±500 kV between the conductors and the ground; the maximum capacity of the Intertie is 3100 MW.

The large capacitance of the ac cables limits the power transfer through the cable, because the cable must carry both the load and the capacitive current. Using dc eliminates the capacitive current, which justified building HVDC underwater cable systems all over the world. One of the frequently discussed systems is the HVDC cable interconnection between the United Kingdom (UK) and France. This system is capable of transporting 2000 MW through a 45-km long HVDC underwater cable. Another advantage of the HVDC system is the elimination of the inductive voltage drop.

1.1.2. Distribution Systems

The *distribution system* uses both three-phase and single-phase networks. The larger industrial loads require a three-phase supply. A subtransmission line or a dedicated distribution line directly supplies large industrial plants and factories. A single-phase system delivers power to ordinary residences.

The voltage is reduced at the distribution substation, which supplies several distribution lines that deliver the energy along streets. The distribution system voltage is less than or equal to 46 kV. The most popular distribution voltage in the United States is the 15 kV class, but the actual voltage varies. Typical examples for the 15 kV class are 12.47 and 13.8 kV. As an example, in Figure 1.1, a 12 kV distribution line is connected to a 12 kV cable, which supplies commercial or industrial customers. The graphic also illustrates that 12 kV cables supply the downtown area in a large city.

A 12 kV cable can also supply the residential areas through *step-down transformers*, as shown in Figure 1.1. Each distribution line supplies several step-down transformers distributed along the line. The distribution transformer, frequently mounted on a pole or placed in the yard of a house, reduces the voltage to 240/120 V. Short-length low-voltage lines power the homes, shopping centers, and other local loads. One distribution transformer can serve six to eight residential customers.

1.2. TRADITIONAL TRANSMISSION SYSTEMS

The North American electric power system is presently divided into four isolated systems referred to as interconnections. The interconnections, as indicated in Figure 1.4, are:

1. the Eastern Interconnection,
2. the Electric Reliability Council of Texas (ERCOT) Interconnection,
3. the Western Interconnection, and
4. the Québec Interconnection.

TRADITIONAL TRANSMISSION SYSTEMS

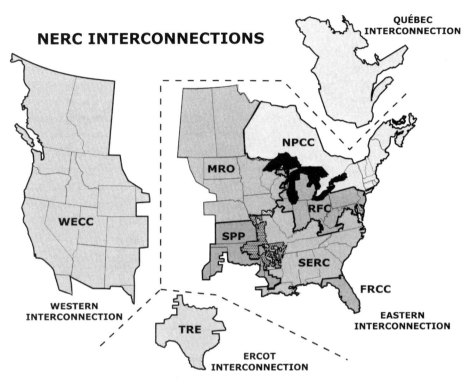

Figure 1.4. North American Electric Reliability Corporation (NERC) interconnections. FRCC, Florida Reliability Coordinating Council; MRO, Midwest Reliability Organization; NPCC, Northeast Power Coordinating Council; RFC, Reliability First Corporation; SERC, SERC Reliability Corporation; SPP, Southwest Power Pool, RE; TRE, Texas Reliability Entity (TRE). This image from the North American Electric Reliability Corporation's website is the property of the North American Electric Reliability Corporation and is available at http://www.nerc.com/page.php?cid=1%7C9%7C119. This content may not be reproduced in whole or any part without the prior express written permission of the North American Electric Reliability Corporation.

The four systems are connected through regulated back-to-back HVDC links, HVDC transmission lines, and regulated ac tie lines. A back-to-back HVDC link is an HVDC system without a transmission line, that is, it contains two directly interconnected converters. High-power electronic devices can regulate the power flow through the ac line. In the last two decades, the industry developed the flexible ac transmission system (FACTS), which is able to electronically control the operation of a high-voltage ac line. Chapter 11 discusses both HVDC and FACTS systems.

The regulated connections permit energy transfer in normal operation and in case of an emergency. They block system oscillations and cascading outages. As examples, ERCOT in Texas uses back-to-back HVDC links, and the Western Electricity Coordinating Council (WECC) connects to the Eastern Interconnection through powerful HVDC transmission ties.

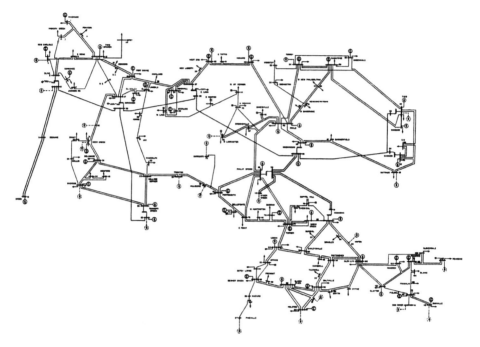

Figure 1.5. IEEE 118 bus power flow test case network.

1.2.1. Substation Components

The connection diagrams for actual power networks are confidential material because of security concerns. Figure 1.5 presents the Institute of Electrical and Electronics Engineers (IEEE) published 118 bus power flow test case network, which is the one-line diagram of a typical three-phase system illustrating the nature of an actual power network. The diagram shows a loop network that should withstand at least a single contingency, but in most cases will withstand multiple contingencies. This implies that at least two transmission lines supply each bus.

Figure 1.6 details a portion of the system, where each transmission line is connected to a substation bus, which is a node point of the system. There are simple load buses, like the Pokagon bus, which is supplied by only two lines (i.e., meets single-contingency requirement). Other buses have both load and generation like Twin Branch with seven connecting lines—it may withstand six outages. A third type of bus has load, generation, and parallel connected capacitor, or synchronous condenser, for example, New Carlisle. The capacitor is a switched unit, which is used at high load to produce reactive power and reduce voltage drop. Similarly, switched inductive load is connected in parallel to selected buses to reduce overvoltages in case of light loading. A *synchronous condenser* is a rotating device, like a generator, which produces or absorbs reactive power (vars). It can be permanently connected to the system and regulates voltage by producing or absorbing vars. A synchronous condenser can be used instead of a capacitor. At the Olive substation in Figure 1.6, a regulating autotransformer

TRADITIONAL TRANSMISSION SYSTEMS

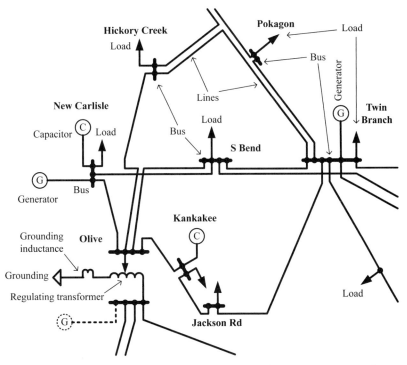

Figure 1.6. Segment of the IEEE 118 bus power flow test case network.

interconnects the substation with the lower portion of the network. This transformer regulates the voltage within a ±10% range. The transformer neutral point may be grounded though a reactance to reduce the ground fault-produced short circuit current.

Other components not shown include:

- switched or electronically controlled series capacitors that are inserted in selected transmission lines to compensate for the line inductance and reduce voltage drop, and
- CBs, which are protecting the system and switch off the line in case of short circuit.

1.2.2. Substations and Equipment

Substations form the node points of the electric system. Figure 1.7 pictures a typical distribution substation. The major role of substations is to distribute the electric energy and provide protection against faults on the lines and other equipment. Figure 1.1 reveals three types of substations that are used:

1. EHV substations (500/230 kV);
2. high voltage substations (230/69 kV); and
3. distribution substations (69/12 kV).

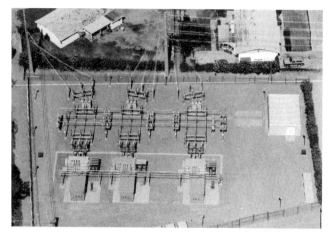

Figure 1.7. Aerial view of a three-bay distribution substation (courtesy of Salt River Project).

Although the circuit diagrams of these substations are different, the general circuit concept and major components are the same. Figure 1.8 presents a conceptual diagram for an EHV substation. That circuit is frequently called the "breaker-and-a-half bus scheme." The rationale behind the name is that two lines have three CBs.

The primary substation equipment is as follows:

The CB is a large switch that interrupts load and fault currents. The fault current automatically triggers the CB, but the CB can also be operated manually. A CB has a fixed contact and a moving contact placed in a housing that is filled with gas or oil. Sulfur hexafluoride (SF_6) gas is the most common. Figure 1.9 illustrates a simplified contact arrangement for a typical breaker. In the closed position, the moving contact is inside the tubular fixed contact. Strong spring loading assures low contact resistance in the closed position. The switch is operated by pulling the moving contact out of the tubular fixed contact. The opening of the switch generates arcing between the contacts. The simultaneous injection of high-pressure SF_6 blows out the arc. Figure 1.10 demonstrates the operating principle for an actual CB. The CB has two tubes serving as fixed contacts (marked 1, 2, and 9) placed in a porcelain housing and a moving part with sliding contacts (3, 8, and 5), which connect the two fixed parts when the breaker is closed (Scene 1). The breaker is filled with SF_6 gas, which has high dielectric strength. The opening of the breaker drives the moving part downward (Scene 2). First, contact 3 separates and the moving contact compresses the SF_6 gas in chamber 7. This is followed by the separation of the main contact 5. The opening of contact 5 produces arcing between 4 and 5, and simultaneously initiates the fast, jet-like flow of the compressed SF_6, as portrayed by the small arrows in Scene 3. The SF_6 jet blows out the arc and interrupts the current (Scene 4).

The industry uses two types of CBs: live-tank and dead-tank breakers. In a live-tank breaker, insulators support the breaker, and the breaker is placed in a horizontal porcelain housing and insulated from the ground. Figure 1.11 presents a live-tank breaker. The switch is in the crossarm. The vertical porcelain column insulates the

TRADITIONAL TRANSMISSION SYSTEMS

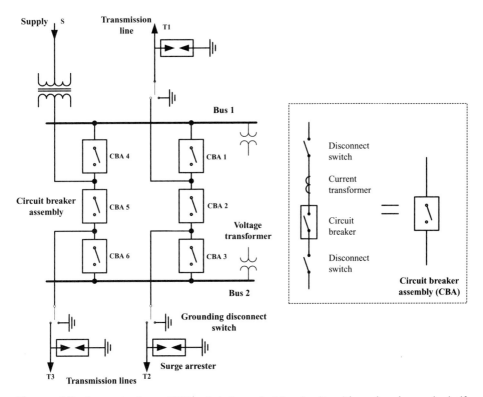

Figure 1.8. Concept of an EHV substation electric circuit with a breaker-and-a-half configuration.

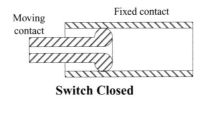

Switch Closed

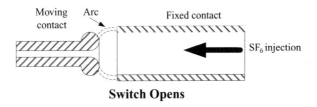

Switch Opens

Figure 1.9. Simplified CB operation.

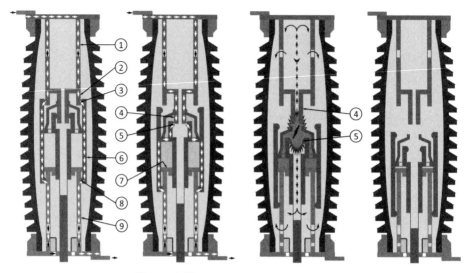

Figure 1.10. SF$_6$ CB operation sequence.

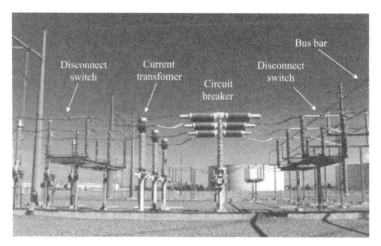

Figure 1.11. CBA with live-tank breaker in a 69 kV substation.

switch and houses the control rods. The dead-tank breaker has a grounded metal housing. The switch is placed in this grounded (dead) tank and insulated by oil or SF$_6$. Large bushings isolate the circuit conductor from the tank. Figure 1.12 pictures a 500 kV SF$_6$ dead-tank CB.

The *disconnect switch* provides circuit separation and facilitates CB maintenance. The CB position cannot be determined by observation. Nevertheless, the lineman needs to know that the breaker is open for safety reasons. Furthermore, in the event that CB

TRADITIONAL TRANSMISSION SYSTEMS

Figure 1.12. Dead-tank 500 kV SF_6 CB.

maintenance is required, a disconnect switch is required on each side of the CB to completely isolate the CB. A disconnect switch is a large device that provides visible evidence that the circuit is open, and it can be operated only when the CB is open. Figure 1.13 provides a typical disconnect switch with a vertically rotating bar that opens the switch. Figure 1.11 shows disconnect switches with horizontally moving bars.

The *current and voltage transformers* reduce the current to 5 A or less and the voltage to about 120 V, respectively. The current transformers (CTs) and potential transformers (PTs) insulate the instrumentation circuits from the high voltage and current, and as such, CTs and PTs are collectively known as *instrument transformers*. These signals trigger the protection relays, which operate the CB in the event of a fault. In addition, the low power quantities are used for metering and system control.

The *surge arresters* are used for protection against lightning and switching overvoltages. Figure 1.14 presents a surge arrester. The surge arrester contains a nonlinear resistor housed in a porcelain tube. The nonlinear resistor has very high resistance at normal voltage, but the resistance is greatly reduced when the voltage exceeds a specified level. This diverts high lightning or switching current to ground and protects the substation from *overvoltage*.

The major component of the substation is the circuit breaker assembly (CBA), which requires two disconnect switches and one or more CTs for proper operation. The right side of Figure 1.8 illustrates a CBA with a single CT. In the main diagram, the simplified box is used. The two disconnect switches in the CBA permit maintenance of any CB. In case of a CB failure, other breakers will provide backup to clear the fault.

Figure 1.13. Disconnect switch, 500 kV.

Figure 1.14. Surge arrester, 69 kV.

TRADITIONAL TRANSMISSION SYSTEMS   15

The opening of the two disconnect switches, after deenergization, permits breaker maintenance to be performed. The CT is used to measure the line current and activates protection in case of a line fault. The protection triggers the CB, which opens the line to stop current flow. Figure 1.11 shows the CBA on a 69 kV substation.

The breaker-and-a-half bus scheme is a redundant system where a fault of any of the components does not jeopardize operation. Figure 1.8 reveals that power entering through the supply transformer may flow directly through CBA 5 and supply transmission line T3. However, a part of the power can flow through CBA 4, Bus 1, and CBA 1 to supply T1. Transmission line T2 is supplied through CBA 5, CBA 6, Bus 2, and CBA 3, and/or through CBA 4, Bus 1, CBA 1, and CBA 2.

EXAMPLE 1.1: Failure analysis of the breaker-and-a-half substation configuration

It is an interesting exercise to analyze the operation when one of the components fails. It can be seen that any CBA can be removed without affecting service integrity.

Normal Operation: Referring to Figure 1.8, there are two independent current paths between the supply (S) and each of the transmission lines (T1, T2, and T3). For instance, the supply S can feed T3 directly through CBA 5, or through the series combination of CBA 4, Bus 1, CBA 1, 2, 3, Bus 2, and CBA 6.

Fault Operation: The substation electric circuit of Figure 1.8 is analyzed here for three cases: (a) short circuit on transmission line T1, (b) short circuit on Bus 1, and (c) CB failure of CBA 5.

(a) *Short Circuit on Transmission Line T1.* The protective response to a short circuit on a transmission line is to isolate the affected line using the adjoining CBs. For instance, a short circuit on line T1 triggers the protection that opens the two CBs, CBA 1 and CBA 2, thus separating T1 from the substation. In this case, the supply S feeds T3 directly through CBA 5, and S serves T2 through CBA 5, 6, Bus 2, and CBA 3. Figure 1.15 shows the resultant current paths.

(b) *Short Circuit on Bus 1.* Likewise, a short circuit on a bus initiates isolation of the bus from the remainder of the circuit. A short circuit on Bus 1 triggers the opening of both CBA 4 and CBA 1. In this case, T1 is supplied through CBA 5, 6, Bus 2, and CBA 3 and 2; T2 is powered through CBA 5, 6, Bus 2, and CBA 3; and T3 is supplied directly through CBA 5. Figure 1.16 indicates the current pathways.

(c) *CB Failure.* The two CB failure modes are *fail open* and *fail close*. If a CB cannot be closed, it fails in the open position. Similarly, if the CB cannot be opened (i.e., current switched off), then it has failed closed. In the following, we analyze CBA 5 for each of the two failure modes.

   Case 1: CBA 5 Fails Open. If CBA 5 cannot be closed, it has failed in the open position. Consequently, T1 is supplied through CBA 4, Bus 1, and CBA 1. T2 is powered through CBA 4, Bus 1, and CBA 1 and 2; and T3 is supplied through CBA 4, Bus 1, CBA 1, 2, 3, Bus 2, and CBA 6. Figure 1.17

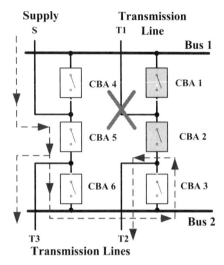

Figure 1.15. Current paths in the case of a short circuit on transmission line T1.

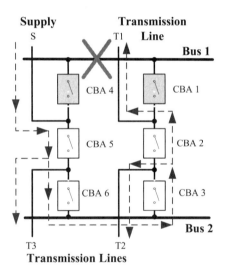

Figure 1.16. Current paths in the case of Bus 1 fault.

provides the current paths. It is observed from this scenario that the CBs and buses must be specified to carry all three load currents simultaneously. Specifically in this case the full supply current passes through CBA 4 and 1, and Bus 1.

Case 2: CBA 5 Fails Closed. If CBA 5 fails in the closed position (that is, the CB cannot be opened), then a short circuit in T3 cannot be isolated

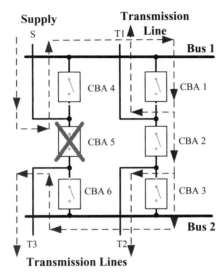

Figure 1.17. CB CBA 5 fails in open position.

locally because the faulty CBA 5 directly connects the source to the shorted line. Backup protection at the sources (not shown) is required to switch off the supply. Similarly, if a CBA that is directly connected to a bus fails closed, then a short circuit on that bus cannot be locally isolated.

1.2.3. Gas Insulated Switchgear

Most high-voltage substations use open-air switchgear as seen in Figure 1.7. However, modern cities with high-rise buildings consume large amounts of power, requiring high-voltage supplies and a substation located in an area with limited space. The industry developed the SF_6 gas insulated switchgear (GIS) that can be placed in constrained spaces, even underground. Figure 1.18 shows a typical GIS.

All components are placed in an aluminum tube that is filled with SF_6 gas. The high dielectric strength of the gas permits the use of short distances between the conductors, which in turn reduces the size of the switchgear. Figure 1.19a exhibits the cross section of a GIS unit, which contains the bus bars, a CB, disconnect switch, grounding switch, and current and voltage transformers. Figure 1.19b provides the connection diagram of the GIS unit.

Although GIS significantly reduces the size of the high-voltage switchgear, its high price and the adverse environmental effects of SF_6 gas limit the use of this switchgear. The International Electrotechnical Commission (IEC) 60694 standard permits 1–3% in SF_6 gas emission per year. SF_6 is an anthropogenically produced compound, which, in addition to being a greenhouse gas, can also decompose under electrical stress, forming toxic by-products that are a health threat for workers in the event of exposure.

Figure 1.18. GIS placed inside building (courtesy of Siemens, Erlangen, Germany).

1.2.4. Power System Operation in Steady-State Conditions

Synchronous generators supply the power system. All synchronous generators rotate with the same speed and produce 60 Hz voltage in the United States. An increase in load changes the angle between the induced voltage and terminal voltage. In steady-state conditions, this power angle must be much less than 90°. The operation with constant synchronous speed maintains the frequency at a constant level. The permitted frequency deviation is less than ±0.5 Hz.

The electric power system practically has no storage capacity, which implies that the generated power must equal the power consumed by the loads plus the system losses. The power system load is continuously changing. Typically, the load is very low at night, higher during the day, and the maximum load occurs in the early evening or late afternoon. Most of the time, the load changes gradually.

The system must be able to supply the load at any instant and simultaneously maintain the system voltage on each bus near the rated value. Standards require that the voltage on each bus must be within the range of ±5–8%.

The load is forecast and most of the generators follow a predetermined schedule. However, selected generators provide the necessary power to balance the system. The system power limitations are based on keeping voltages within range and equipment loading within the current carrying capability. The corrective measures are to shift generation between power plants to unload the heavy loaded equipment. The system voltage, in practice, is maximized to reduce transmission line losses. System voltage is controlled by producing or absorbing reactive power.

Utilities perform studies to anticipate the power system conditions at specific load levels. This is usually done with the use of a power flow program that simulates the actual power system. The generator terminal voltages in the power flow study are selected to minimize losses in the system and have voltages within equipment rating.

TRADITIONAL TRANSMISSION SYSTEMS

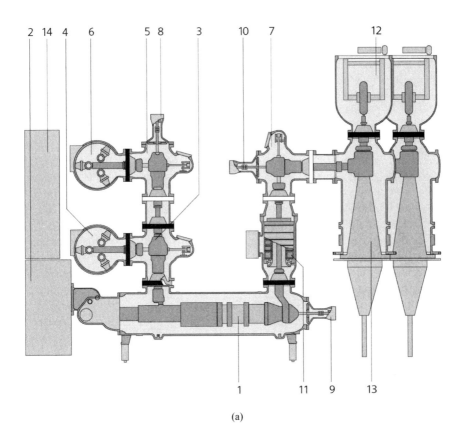

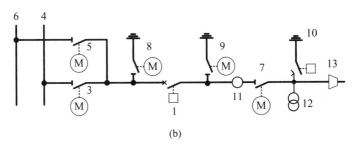

Figure 1.19. Gas insulated switchgear. (a) Cross section; (b) Electrical connection diagram. (1) Circuit breaker (CB) interrupter unit; (2) spring-stored energy mechanism with CB control unit; (3) and (5) busbar disconnector; (4) and (6) busbar; (7) outgoing feeder disconnector; (8), (9), and (10) earthing switch; (11) CT; (12) voltage transformer; (13) cable sealing end; (14) integrated local control cubicle. Courtesy of Siemens, Erlangen, Germany.

Generator power levels are defined at each generator bus except one. This one is called the slack bus. The power flow program causes the real power at the slack bus to be adjusted to balance generation, load, and losses.

The power flow program is a powerful tool for planning the future power system and anticipating operating problems by studying contingencies.

1.2.5. Network Dynamic Operation (Transient Condition)

A short circuit produces large current, typically 5–10 times the rated current. The system protection detects the short circuit and triggers the operation of the CB, which opens and interrupts the current after a few cycles. The large short circuit current produces severe voltage drop. Actually, the voltage at the fault location is practically zero.

This short duration voltage dip produces a sudden reduction of load near the fault. However, the generator input power remains practically constant, but the output power is reduced significantly. The larger input power than output power accelerates the generators near the fault. If the fault clears in a short time (the critical clearing time), the system restores voltage and loads, which stops the generator acceleration and the generator returns to normal operation.

However, if the fault clearing is delayed, the generator acceleration causes the affected generators to fall out of synchronism with the other generators of the system. The outage of several generators can collapse the system and require lengthy restart procedures. The described event is called a transient stability-caused outage. In addition to the fault transient instability, an outage can be initiated by switching operations and other disturbances. There have been cases where switching created a system configuration, resulting in undamped oscillations at a frequency around 0.5 Hz. These oscillations can cause lines to open and generators to trip. This can result in a system blackout. This is referred to as steady-state instability due to insufficient damping.

1.3. TRADITIONAL DISTRIBUTION SYSTEMS

Figure 1.20 depicts the structure of the electrical system. The transmission system contains three looped networks: (1) EHV network, (2) high-voltage network, and (3) subtransmission line network. These electrical networks connect the power plants and load centers together and thus bring electricity to towns and other loads. However, the medium voltage distribution systems, which supply the residential and industrial customers, are radial networks which may not withstand a contingency.

The American National Standards Institute (ANSI) standard C84.1 for 60 Hz electric power systems and equipment in North America limits the voltage magnitude deviation at the service entrance to ±5% for normal conditions and from −8.3% to +5.8% for short durations or unusual conditions. There is not an established standard concerning the frequency deviation from nominal, but the frequency does not deviate more than 0.1 Hz, from 60 Hz, 99% of the time. Continuity of service is another important consideration since an unacceptable number of interruptions prompts customers to complain to the utility and the regulatory agency. The SAIFI (System Average

TRADITIONAL DISTRIBUTION SYSTEMS

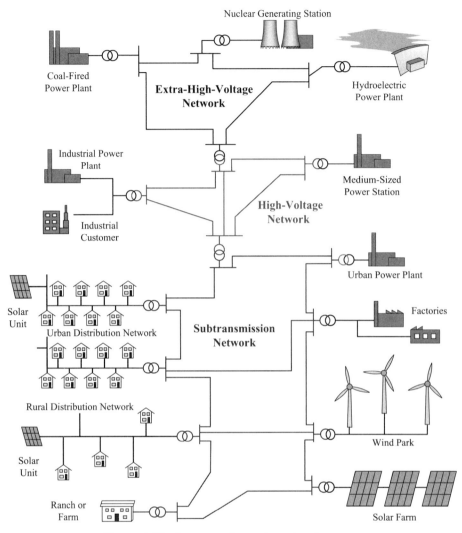

Figure 1.20. Structure of a modern electric system.

Interruption Frequency Index) is a measure of system reliability—the SAIFI is generally in the range of one to five interruptions per year.

1.3.1. Distribution Feeder

The distribution system is a radial system, without loops. Figure 1.21 illustrates the concept of a typical distribution system, where a main three-phase feeder is positioned along a major thoroughfare. The voltage of this primary distribution system is around 15 kV. In Arizona, the nominal voltage for most urban distribution is 12.47 or 13.8 kV.

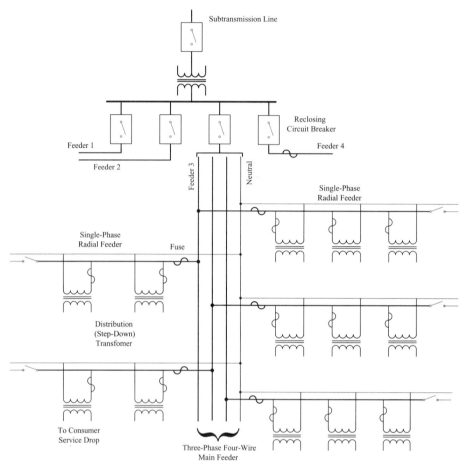

Figure 1.21. Concept of radial distribution system.

The main feeder is protected by a reclosing CB that switches off the feeder in case of a fault, and after a few cycles, the breaker recloses and restores the energy supply. This is an effective way of protection for overhead distribution circuits because most faults on an overhead line are temporary—originating from a weather-related event. However, if the reclosing is unsuccessful, the breaker opens the line permanently. For an underground system, most faults are permanent, so reclosing is not used.

Many commercial customers (e.g., grocery stores, office buildings, and schools) are supplied with three-phase power due to heavy loads such as fan motors and air conditioning. Residential and light commercial customers are powered by single-phase subfeeders, which are protected by fuses. Close to the residential and light commercial loads, distribution transformers are connected to the single-phase subfeeders. Low voltage (120/240 V) secondary circuits, called *consumer service drops*, supply the

Figure 1.22. Overhead line and cable connections.

individual customers. The distribution transformers are protected on the primary side by fuses. This fuse operates in case of transformer fault or short on the service drop cable. The consumer is protected by CBs on the service panel.

Figure 1.22 shows the interconnection of a distribution line with a distribution cable. This connection is used to supply residential areas with underground distribution. The figure shows the cable termination, the surge arrester that is used for overvoltage protection, and the fuse cutout, which is used for *overcurrent* protection. The fuse cutout contains a fuse mounted on a pivoted insulator and it serves as a disconnect switch that can be opened using an insulated rod—commonly known as a *hot stick*. A metal conduit attached to the wooden pole protects the cable.

Figure 1.23 presents a typical consumer service drop, where a step-down transformer mounted on the distribution pole supplies an individual house or a group of homes. For aesthetic purposes, underground cables have replaced overhead distribution lines in some residential areas. In such cases, the transformer is housed in a ground-level metal casing, which is placed on a concrete slab, as seen in Figure 1.24. Figure 1.23 shows a typical single-phase pole-mounted transformer that supplies a few (one to eight) houses. A surge arrester and a fuse cutout protect the pole-mounted oil-insulated transformer. The secondary of the transformer supplies a 240/120 V insulated conductor that is attached to a carrier steel wire, leading the electricity to the houses. Such transformers may supply a 240/120 V low voltage line.

Figure 1.23. Consumer service drop.

1.3.2. Residential Electrical Connection

A low-voltage secondary feeder supplies individual houses. A distribution transformer has a three-wire electrical system, which accommodates both 120 and 240 V loads. Figure 1.25 draws the electrical connection of a typical household supply.

The step-down transformer has one neutral and two phase wires. The neutral is grounded at the transformer secondary side. The three wires are connected to the kilowatt and kilowatt-hour meter at the house. The residence has a four-wire system, which consists of three insulated wires and one bare ground wire. The ground wire and insulated neutral wire are grounded at the house service entrance.

The kilowatt and kilowatt-hour meters measure and record the kilowatt-hour energy consumption, and in some cases the maximum 15-minute kilowatt demand. Figure 1.26 pictures a typical kilowatt-hour meter and service panel with the main CBs.

The power company is responsible for the system up to the secondary terminals of the meter. The service panel and the house wiring are the homeowner's responsibility. The service panel is equipped with the main CB, which protects the house against short circuit and overload. Lighting and small appliances are supplied by the 120 V lines, which are also protected by CBs. In older installations, fuses have been used instead of CBs. The load is connected between the phase and neutral conductors. The

TRADITIONAL DISTRIBUTION SYSTEMS

Figure 1.24. Ground-level transformer in a residential area.

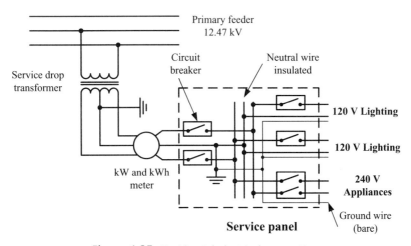

Figure 1.25. Residential electrical connection.

Figure 1.26. Typical residential kilowatt-hour meter and service panel.

housing of appliances and lamps is grounded by the ground wire. Larger appliances, such as the cooking stove and clothes dryer, are powered by 240 V circuits.

1.4. INTELLIGENT ELECTRICAL GRIDS

The U.S. Energy Independence and Security Act of 2007 describes the smart grid as the modernization of the electricity infrastructure to maintain a reliable and secure system that can meet future growth. Figure 1.27 shows a conceptual model of the intelligent grid. The diagram depicts the electricity flow within the electric power system (generation, transmission, distribution, and customer) and the communication links between the individual components and the operations, service provider, and market domains. These require monitoring the electric system operation parameters at each level using an advanced sensor system, the communication of the measured data through a secure and fast system, and evaluation of the obtained information using advanced computer software. The goal is a more reliable electrical power system.

1.4.1. Intelligent High-Voltage Transmission Systems

The cascading outages observed in the last decade demonstrated that the electrical power grid remains vulnerable. Unexpected small faults can initiate a large-scale

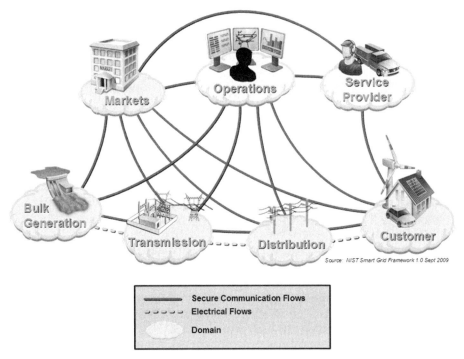

Figure 1.27. Smart grid conceptual model. Source: National Institute of Standards and Technology.

network outage. The electrical network did not keep pace with the advancements in digital technology and communication. Because of the importance of the power grid to the national infrastructure, the modernization of the power grid became a high priority. The smart grid concept emerged at the end of the 1990s, when the application of digital processing and communications to the power grid began.

One of the first and most important innovations was the phasor measurement unit (PMU). Using PMUs, Bonneville Power Administration introduced the wide area measurement system (WAMS) in 2000. In this system, the Global Positioning System (GPS) time signal synchronizes a large number of sensors distributed throughout the network. The PMU sensors measure the magnitude and phase angle of the current or both current and voltage. The phasor data (magnitude, phase angle, and time stamp) obtained at different locations of the power system are sent to a computer called a phasor data concentrator, which compares and evaluates the data. The power system condition is assessed using the data. The combination of phasor measurement data with dynamic line ratings has emerged to monitor more accurately the transmission system operation and integrate renewable energy sources.

Large-scale solar- and wind generation-produced power is variable, depending upon meteorological conditions. Application of these electric power generation methods requires real-time monitoring of operational parameters and continuous analysis of the obtained data to assure reliable operation and foresee emerging operation problems as well as initiate remedial actions in case of equipment failure.

1.4.2. Intelligent Distribution Networks

Some of the renewable generation, like residential and commercial solar generation, is connected directly to the distribution system, and is referred to as distributed generation. The increasing use of electrical cars will increase the load on the distribution system. The present radial distribution system must be upgraded to an electric network, with at least a single contingency to handle distributed generation and increased loading.

The introduction of advanced metering infrastructure permits two-way communication between the individual customers and the utility. This permits dynamic electricity pricing. The utility will be able to control the voltage to minimize losses, and curtail customer usage in the event of generation shortage. The customer will be able to reduce costs by using appliances when the price of electricity is low. Furthermore, smart appliances, which acquire pricing signals from a smart meter, are programmed to reduce energy use during peak periods.

Simultaneously, the utility will instantaneously detect customer outages and initiate remedial actions. Smart meters will report voltage reduction or zero voltage, which is a sign of an outage. The low voltage network will automatically reconfigure the system to reduce the number of customers affected by a short circuit. The smart distribution grid was in an early stage of development in 2011, but most utilities in the United States are working toward its deployment.

1.5. EXERCISES

1. Draw a sketch and explain the concept of electric energy transmission. What are the advantages of the multilevel voltages?
2. What are the typical voltages for subtransmission, high-voltage, extra-high-voltage, and ultra-high-voltage systems?
3. What are the typical voltages for the distribution system?
4. Draw the connection diagram of a high-voltage substation using the breaker-and-a-half configuration. Identify the components.
5. What is a circuit breaker? Discuss its role and operating principle.
6. What are current and voltage transformers?
7. What is the disconnect switch and its purpose?
8. What is a surge arrester? Why is it important?
9. Draw a diagram of a typical radial distribution system. Describe briefly its operation.
10. Describe the residential electric connection. Draw the connection diagram.

1.6. PROBLEMS

Problem 1.1

Using the one-line diagram of Figure 1.8, determine the protective response of the circuit breakers due to a short circuit (a) solely on transmission line T2 and (b) only on transmission line T3.

Problem 1.2

Substations frequently use the connections shown in the one-line diagram of Figure 1.8. Analyze the circuit operation for the cases of when CBA 4 (a) fails open and (b) fails close.

Problem 1.3

For the breaker-and-a-half configuration of Figure 1.8, determine the circuit breaker positions (open or closed) needed to supply power to the three lines, if a short circuit occurs on Bus 2.

2

ELECTRIC GENERATING STATIONS

Electric generating stations are known by several names, including power plants and powerhouses. This chapter describes the various types of electric power stations including fossil, nuclear, hydropower, and renewable generation. An assortment of energy sources, including both renewable and nonrenewable, may be employed to produce electricity. Most nonrenewable electricity generating schemes today require a thermal-to-mechanical-to-electrical energy conversion process, as illustrated in Figure 2.1. In contrast, the renewable forms, including solar, wind, hydroelectric, and ocean-based energy, take a variety of energy conversion paths to produce electric power.

A thermal electric generating station converts the chemical energy of natural gas, oil, coal, biomass, or the nuclear energy of uranium to electric energy. In the late 1800s, reciprocating steam engines were used to drive generators and produce electricity. The more efficient steam boiler–turbine system replaced the steam engine around the turn of the 19th century. The steam boiler burns fuel in a furnace. The heat generates steam that drives the turbine–generator set. Typically, the steam turbine and electric generator are mounted on a common platform/foundation and the shafts are connected together. The turbine-driven generator converts the mechanical rotational

---

Electrical Energy Conversion and Transport: An Interactive Computer-Based Approach, Second Edition.
George G. Karady and Keith E. Holbert.
© 2013 Institute of Electrical and Electronics Engineers, Inc. Published 2013 by John Wiley & Sons, Inc.

ELECTRIC GENERATING STATIONS

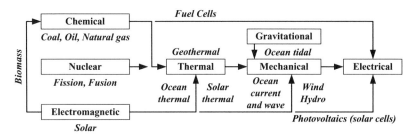

Figure 2.1. Energy sources and corresponding conversion processes to generate electricity.

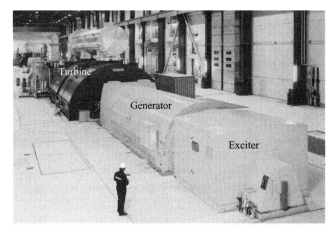

Figure 2.2. Steam turbine and electric generator with exciter (courtesy of Siemens, Erlangen, Germany).

energy to electrical energy. Figure 2.2 shows a steam turbine and electric generator with its exciter unit.

In the beginning, oil was the most frequently used fuel. Increased petroleum costs, owing to increased gasoline consumption by automobiles, elevated coal to the primary fuel for electricity generation. However, environmental concerns (e.g., sulfur dioxide generation, acid rain, coal dust pollution, and ash-handling problems) of coal burning have curtailed the building of new coal-fired power plants.

Recently, natural gas has emerged as the power plant fuel of choice due to three factors. First, natural gas burns cleaner, making plant siting and adherence to environmental regulations easier. Second, natural gas is available in large quantities at a reasonable price. Third, significant increases in plant thermal efficiency have been achieved using combined cycle plants that utilize gas turbine technology advances from the aerospace industry.

Most fossil and nuclear plants utilize a Rankine thermal cycle employing steam. The *thermal efficiency* is defined as

$$\eta_{th} = \frac{P_e}{\dot{Q}_{th}} = \frac{\text{plant net electric power (energy) output}}{\text{plant thermal power (energy) input}}. \tag{2.1}$$

Utility personnel often use the terminology *heat rate*, which is the number of Btus (British thermal units) needed to produce 1 kWh of electricity; thus, the heat rate is $3412/\eta_{th}$. The maximum thermal efficiency is defined by the Carnot cycle as:

$$\eta_{\text{Carnot}} = 1 - \frac{T_L}{T_H}, \tag{2.2}$$

where T_L and T_H are the minimum and maximum absolute temperatures, respectively, for heat rejection and addition to the thermal cycle. For instance, in a coal-fired unit, T_L is the condensate temperature and T_H is the superheated steam temperature.

EXAMPLE 2.1: Plot the maximum thermal efficiency as a function of the steam temperature in degrees Fahrenheit for a condenser temperature of 100°F.

Solution: This problem is readily solved using a variety of software including MATLAB, Mathcad, or a simple spreadsheet. The MATLAB code below uses Equation (2.2), while adjusting degrees Fahrenheit to degrees Rankine with °R = °F + 460°, to determine the Carnot (i.e., maximum theoretical) thermal efficiency. Noteworthy is that this code employs the element-by-element division operator (./) to accomplish this. An introduction to MATLAB is provided in Appendix B.

```
% Example 2.1: Compute Carnot cycle efficiency

clear all

CONDtemp = 100;   % condensate temperature (°F)

STEAMtemp = 250 : 1 : 1000;   % steam temperature (°F)

Carnot = 1 - (CONDtemp + 460) ./ (STEAMtemp + 460);

plot(STEAMtemp, Carnot, 'LineWidth',2.5)
set(gca,'fontname','Times','fontsize',12);
title('Example 2.1')
xlabel('Steam Temperature (°F)')
ylabel('Maximum Thermal Efficiency')
legend(num2str(CONDtemp,'Condensate Temperature = %3.0f °F'), ...
    'Location','NorthWest')
```

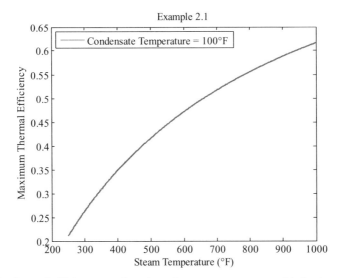

Figure 2.3. Thermal efficiency as a function of steam temperature with feedwater temperature held constant.

The results, plotted in Figure 2.3, prove clearly that increased steam temperature leads to higher possible thermal efficiency in converting the heat energy from fuel burning to electricity. It is not difficult to surmise that higher thermal efficiency can lead to lower electricity generation costs and reduced emissions on a per-kilowatt-hour basis.

Historically, the hydroelectric power plants developed nearly simultaneously with the thermal power plants. The river water level is increased by a dam, which creates a head. The head-generated pressure difference produces fast-flowing water that drives a hydraulic turbine, which turns the generator. The generator converts the mechanical energy to electrical energy.

After World War II, electricity generation from nuclear power plants emerged. More than 450 nuclear plants operate worldwide. In these plants, nuclear fission is created using enriched uranium. The fission chain reaction heats water and produces steam. The steam drives a traditional turbine–generator set. In the last few decades, environmental considerations and nuclear power plant capital costs halted the building of new plants in the United States and curtailed the operation of some existing plants. Presently, concerns about climate change (i.e., global warming due to greenhouse gas emissions) are encouraging new nuclear power plant construction despite the events at Fukushima, Japan.

Coal, nuclear, natural gas, and hydro power plants generate most of the electricity in the United States. A breakdown of electricity generation in the United States is given in Figure 2.4. At the present time, other energy sources such as wind, solar, geothermal, and biomass produce little of the electricity consumed in the United States.

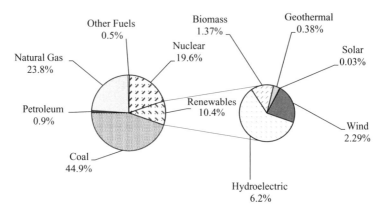

Figure 2.4. Net electricity generation in the United States for 2010. (Data are from *Annual Energy Review 2010*, Energy Information Administration.)

2.1. FOSSIL POWER PLANTS

Fossil power plants include coal-fired, oil-fired, and natural gas fueled power plants. Figure 2.5 shows an aerial view of a small generating station. The major components of a thermal power plant are:

- fuel storage and handling,
- boiler,
- turbine, and
- generator and electrical system.

2.1.1. Fuel Storage and Handling

The fossil fuels are characterized according to various properties. A key parameter is the fuel energy content, referred to as the heating value (HV), which is expressed on a per-unit mass basis for coal, and on a per-unit volume basis for oil and natural gas. Other important fuel characteristics include coal grindability and weatherability (which are the resistance to crushing and outdoor environmental conditions, respectively), oil specific gravity and viscosity, and sulfur content (for all).

Coal is transported by long coal trains with special railcars or by barge if the power plant is at a river or seaside. The railcars are tipped and the coal dropped into a dumper. Conveyer belts carry the coal to an open-air coal yard. The coal yard stores several weeks of supply. Additional conveyer belts move the coal into the power plant where the coal is fed through hoppers to large mills. The mills pulverize the coal to a fine powder. The coal powder is mixed with air and injected into the boiler through burners. The mixture is ignited as it enters the furnace.

FOSSIL POWER PLANTS

Figure 2.5. Aerial view of a small generating station (Kyrene Generating Station). Courtesy of the Salt River Project.

Oil and liquefied natural gas are also transported by rail or pipelines. The power plant stores this fuel in large steel tanks, holding several days of supply. The oil is pumped to the burners, which atomize the oil and mix the small oil particles with air. The mixture is injected into the furnace and ignited. The natural gas, also mixed with air, is fed to the boiler through the burners, which ignite the mixture as it enters the furnace. Natural gas is the easiest of the fossil fuels to burn as it mixes well with air and it burns cleanly with little ash.

2.1.2. Boiler

Figure 2.6 exhibits the flow diagram of a boiler. The boiler is an inverted U-shaped steel structure. Water tubes cover the walls of the boiler. The major systems of a boiler are the following, and are discussed in the succeeding sections:

- fuel injection system,
- water–steam system,
- air–flue gas system, and
- ash handling system.

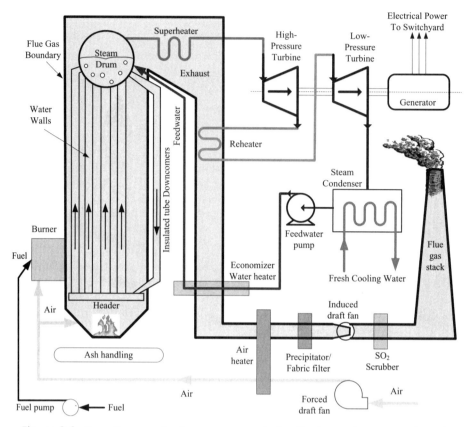

Figure 2.6. Flow diagram of a drum type steam boiler (feedwater heaters not shown).

2.1.2.1. Fuel Injection System. Natural gas, atomized oil, or pulverized coal is mixed with primary air by the nozzles in the burners and injected into the furnace. The mixture is ignited either by the high heat in the boiler or by an oil or gas torch.

Secondary air is pumped into the boiler to assure complete burning of the fuel. The burning fuel produces a high-temperature (around 3000°F, 1650°C) combustion gas in the boiler, which heats the water in the tubes covering the walls by convection and radiative heat transfer. The high heat evaporates the water and produces steam, which is collected in the steam drum.

2.1.2.2. Water–Steam System. A large water pump drives the feedwater through a high-pressure water heater (not shown in Fig. 2.6) and the economizer. Steam extracted from the turbine heats the high-pressure feedwater heater. The hot flue gas heats the economizer. The water is preheated to 400–500°F (200–260°C).

The warmed high-pressure feedwater is pumped into the steam drum. Insulated tubes (called downcomers) located outside the boiler connect the steam drum to the header at the bottom of the boiler. The water flows through the downcomer tubes to the header. The header distributes the water among the riser tubes covering the furnace walls. The water circulation is assured by the density difference between the water in the downcomer and riser tubes.

Heat from the burning fuel evaporates the water to produce steam. The liquid and vapor components of the saturated steam are separated in the steam drum. The superheater dries the saturated vapor and increases its temperature to around 1000°F (540°C). The superheated steam drives the high-pressure turbine. The steam exhausted from the high-pressure turbine is reheated by the flue gas heated reheater. The intermediate-pressure and/or low-pressure turbines are driven by the reheated steam. The steam exhausted from the turbine is condensed into liquid in the condenser. The condensation generates a vacuum, which extracts the steam from the turbine.

A deaerator is built in the condenser to remove the air from the condensed water. This is necessary because the air (oxygen) in the water causes corrosion of the pipes. The power plant loses a small fraction of water during operation. This necessitates that the gas-free condensed water is mixed with purified feedwater that replaces the water loss. The mixture is pumped back to the boiler through the high-pressure feedwater heaters and economizer. It can be seen that the boiler has a closed water circuit. The replacement feedwater is highly purified and chemically treated. The use of highly purified water reduces corrosion in the system.

The condenser is a shell-and-tube heat exchanger, where the steam condenses on the tubes, which are cooled by water from a nearby source. Heat dissipation techniques include:

- once-through cooling to a river, lake, or ocean;
- cooling ponds including spray ponds; and
- cooling towers.

The former method is the least expensive but can result in thermal pollution. *Thermal pollution* is the introduction of waste heat into natural bodies of water supporting aquatic life. The addition of heat reduces the water's ability to hold dissolved gases, including oxygen, which aquatic life requires, although aquatic life growth is usually enhanced by warm water. To avoid thermal pollution, the latter two methods have been employed. A cooling pond and its smaller, specialized version—the spray pond—are artificial lakes. In a spray pond, the water is pumped through nozzles to generate fine spray. The evaporation cools the water as it falls back to the pond.

Most cooling towers are of the wet (vs. dry) variety that employs direct water-to-air contact which can cool more efficiently from the evaporation process but which suffers water loss. Another classification of cooling towers is the air draft mechanism: natural versus mechanical draft. The natural draft towers are large, tall, hyperbolic-shaped structures, whereas mechanical draft towers are shorter and employ either forced or induced draft fans. Most cooling towers are filled with a latticework of horizontal bars (see Fig. 2.7). This baffling within the tower increases the water surface area for more efficient cooling. The warm water is sprayed on the latticework at the top of the tower.

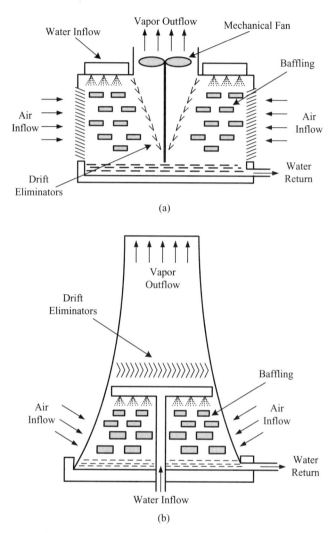

Figure 2.7. Wet cooling towers. (a) Cross-flow induced draft tower; (b) hyperbolic natural draft tower.

The water slowly drifts from the top of the tower to the bottom through the bars. Simultaneously, fans and/or the natural draft drive air from the bottom of the tower to the top. The evaporation cools the water efficiently.

In contrast, dry cooling towers retain the cooling water within tubes that are air cooled. Use of dry cooling towers reduces water consumption by the power plant at a cost of reduced plant thermal efficiency due to increased turbine backpressure in the condenser. The water–energy nexus (e.g., energy is needed to extract, process, and transport water, while at the same time water is required in many energy conversion processes) and the public need for clean water are motivating greater utilization of dry and hybrid wet–dry cooling towers.

2.1.2.3. Air–Flue Gas System. The forced draft fan drives the fresh ambient air through the air heater, which increases the air temperature to 500–600°F (260–315°C). The warm flue gas heats the air heater. The preheated primary air, mixed with the fuel, supports the burning in the boiler. The secondary air assures complete burning of all the injected fuel by meeting or exceeding the required stoichiometric air–fuel ratio.

The hot combustion gas flows through the boiler, generates steam, and heats the superheater, reheater, economizer, and air heater. An induced draft fan drives the flue gas into the atmosphere through the stack. The exhaust gas temperature is around 300°F (150°C). The stack must disperse the flue gases into the atmosphere without disturbing the environment. This requires filters to remove harmful chemicals and particles, and sufficient stack height that assures that the residual pollution in the flue gas is distributed over a large area without causing dangerous concentrations of pollutants downwind.

Particulate emissions (e.g., fly ash) from burning have historically received the greatest attention since they are easily seen leaving smokestacks. For a pulverized coal unit, 60–80% of the ash leaves the furnace with the flue gas. Two emission control devices for fly ash are the traditional fabric filters and the more recent electrostatic precipitators. The fabric filters are large baghouse filters having a high maintenance cost, since the cloth bags have a life of only 18–36 months, although the bags can be temporarily cleaned by shaking or back-flushing with air. These fabric filters are inherently large structures, resulting in a large pressure drop, which reduces the plant efficiency and net electric output due to the need for larger fans.

Electrostatic precipitators have a collection efficiency of 99%, but do not work well for fly ash with a high electrical resistivity (as commonly results from combustion of low-sulfur coal). In addition, the designer must avoid allowing unburned gas to enter the electrostatic precipitator since the gas could be ignited. A side view of an electrostatic precipitator is provided in Figure 2.8. The flue gas, laden with fly ash, is sent through channels having negatively charged electrodes that give the particles a negative charge. The particles are then routed past positively charged plates, or grounded plates, which attract the now negatively charged ash particles. The particles stick to the metal plates until they are collected. Mechanical rappers are activated to shake the particles loose so the ash exits through the hoppers at the base of the unit. The air that leaves the plates is then nearly free of particulates.

Ideally, the hydrocarbon combustion produces water and carbon dioxide gases; however, without sufficient oxygen (air), incomplete combustion may occur, leaving the intermediate product of carbon monoxide (CO). Carbon monoxide production is generally reduced by providing excess air (oxygen) to the furnace:

$$\begin{aligned} 2H_2 + O_2 &\to 2H_2O, \\ C + O_2 &\to CO_2. \end{aligned} \quad (2.3)$$

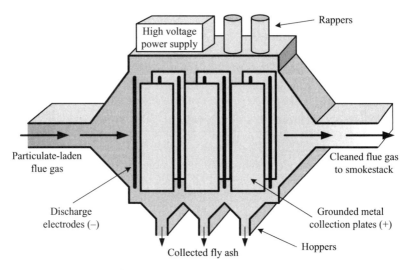

Figure 2.8. Electrostatic precipitator.

EXAMPLE 2.2: A 350 MWe fossil unit utilizes a fuel with an HV of 28,000 kJ/kg. If the plant has a thermal efficiency of 38% and the fuel is composed of 80% carbon by weight, determine the CO_2 production rate at full power.

Solution: Not every problem in engineering requires a computer solution, and knowing the right tool(s) to utilize is an important skill; in this case, an electronic calculator suffices. At 100% power, the fuel use rate is.

$$\dot{m}_{fuel} = \frac{\dot{Q}_{th}}{HV} = \frac{P_e}{HV\, \eta_{th}} = \frac{350 \times 10^3 \text{ kW}}{(28,000 \text{ kJ/kg})(0.38)} = 32.9 \text{ kg/s}.$$

The chemical balance for the combustion of carbon (i.e., $C + O_2 \rightarrow CO_2$) discloses that for every 12 kg of carbon burned, 44 kg of carbon dioxide is produced. Therefore, the CO_2 generation rate is:

$$(32.9 \text{ kg}_{fuel}/s)(0.80 \text{ kg}_C/\text{kg}_{fuel})(44 \text{ kg}_{CO_2}/12 \text{ kg}_C) = 96.5 \text{ kg}_{CO_2}/s.$$

On an annual basis, this would be 3 million metric tons of CO_2 generated per year.

The main gaseous pollutants from combustion include sulfur oxides (SO_X), nitrous oxides (NO_X), and CO. Both SO_X and NO_X can create *acid rain* composed of sulfuric acid (H_2SO_4) and nitric acid (HNO_3), respectively. The SO_X, mostly SO_2 and some SO_3, can cause respiratory irritation. The NO_X contributes to smog and ozone formation and vegetation damage. The CO reduces the oxygen-carrying capability of the blood, referred to as carbon monoxide poisoning. In the United States, the Clean Air Act sets the federal standards for plant emissions, although individual states may establish limits that are more stringent.

To reduce SO_X emissions, many power plants have chosen to use low-sulfur fuel. Coal from the western United States is typically low in sulfur content, whereas high-sulfur coal dominates from eastern states. Natural gas that is high in hydrogen sulfide (H_2S) is known as sour gas, in contrast to sweet gas, which is low in sulfur. Some coal cleaning can be accomplished to remove impurities before combustion. A flue–gas desulfurization system is often employed to remove the SO_2. Both wet and dry sulfur scrubbing processes are in use today. The dry process, in which a lime (CaO) or limestone ($CaCO_3$) solution is sprayed into the flue gas, is the most economical. The wet scrubbing process is more efficient and may be implemented as either a throwaway or a recovery method. Although the recovery products of sulfur and sulfuric acid can be sold, the more popular is the wet, throwaway lime/limestone process, which utilizes the chemical reaction:

$$SO_2 + CaCO_3 \rightarrow CaSO_3 + CO_2. \quad (2.4)$$

Most NO_X originates from the nitrogen in the fuel rather than the N_2 in the air. To reduce NO_X emissions, tighter control of the combustion process is employed through combustion temperature (reduction) or lowering the air–fuel ratio. Similar to automobiles, exhaust gas recirculation can be used to reduce combustion temperature.

2.1.2.4. Ash Handling System. The coal-fired power plant produces large amounts of ash. Ash is mineral matter present in the fuel. The larger ash particles are collected at the bottom of the furnace and mixed with water. The fly ash is extracted by the bag filters and mixed with water. The produced slurry is pumped into a clay-lined pond, from which the water evaporates, without polluting the groundwater of any nearby community. The evaporation of the water produces an environmentally undesirable deposit, which could form a cement-like hard surface. The utilities cover the ash fields with soil and restart the vegetation to minimize the adverse environmental effects. Some utilities utilize the ash as an aggregate in concrete.

2.1.3. Turbine

The high-pressure, high-temperature steam drives the turbine. The heat energy in the steam is converted to mechanical energy. The turbine has a stationary part and a rotating shaft. Both are equipped with blades. The length of the turbine blades decreases from the exhaust to the steam entrance, as the steam density increases from the turbine inlet to exit. The shaft is supported by bearings, which are supplied with high-pressure oil for lubrication. In the event of loss of lubrication, the turbine must be shut down quickly to avoid permanent damage. The steam is supplied through the stationary part of the turbine. Figure 2.9 shows the stationary part with the stationary blades and the rotor with the moving blades. The change in blade length is clearly visible. Figure 2.10 shows the turbine blades from a much larger steam turbine in common use today.

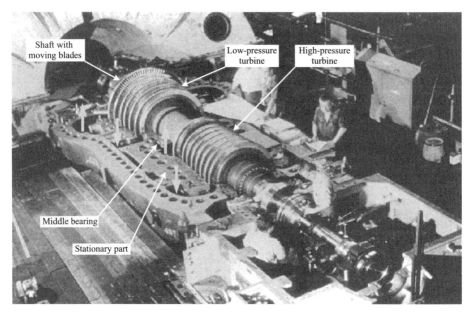

Figure 2.9. Steam turbine internals (courtesy of the Salt River Project).

Figure 2.10. Turbine blades in a large unit (courtesy of Siemens, Erlangen, Germany).

FOSSIL POWER PLANTS

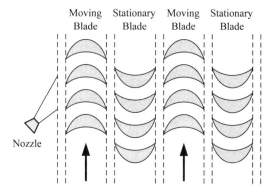

Figure 2.11. Operating concept of a steam turbine.

The turbine operating concept is that the high-pressure, high-temperature steam is injected into the turbine through nozzles. The nozzles increase the steam velocity. The high-speed steam flows through a set of blades placed on the turbine rotor. The direction of the flow changes in the moving blades and the pressure drops. The direction change-caused impact and pressure drop drive the moving blades and the rotor. The concept is illustrated in Figure 2.11. The efficiency of the process is improved by alternately using several sets of moving and stationary blades. The steam flows through the moving blades and drives the rotor shaft. After that, the stationary blades change the direction of the flow and direct the steam into the next set of moving blades. The pressure drops and the impact of the steam drives the rotor.

Modern power plants have one high-pressure and one low-pressure turbine, and, in some cases, an intermediate pressure turbine. Figure 2.9 presents a typical unit with two turbines. The right-hand side is the high-pressure turbine; the left-hand side is the low-pressure turbine. A bearing is placed between the two units. The steam enters at the right-hand side, drives the high-pressure turbine, and exhausts before the middle bearing. The exhausted steam is reheated and fed to the low-pressure turbine. The reheated steam enters just behind the middle bearing and drives the low-pressure turbine. The steam exhausts at the end of the turbine. The arrows indicate the steam entrance and exhaust points.

2.1.4. Generator and Electrical System

The generator and turbine are mounted on the same pedestal and the shafts are directly connected. A condenser, turbine, generator, and main transformer of the Kyrene Generating Station are seen in Figure 2.12.

The generator stator has a laminated iron core with slots. Three-phase windings are placed in the slots. The large generators are wye (Y) connected. The winding is made of mica insulated copper bars. Figure 2.13 exhibits the *stator* of a large generator under construction.

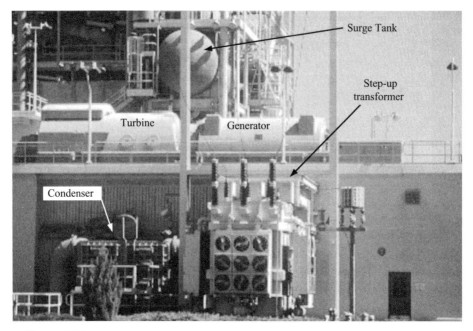

Figure 2.12. Turbine, generator, and main transformer of Kyrene Generating Station.

Figure 2.13. Synchronous generator stator (copyright by Siemens AG).

Figure 2.14. Iron core of synchronous generator rotor (© Brush Electrical Machines, reproduced by permission).

Figure 2.15. Synchronous generator rotor (courtesy of Siemens, Erlangen, Germany).

Typically, the high-speed generator used in steam power plants has a round *rotor*. The round rotor is a solid iron cylinder with slots. Insulated copper bars are placed in the slots to form a coil, which is supplied by direct current (dc) excitation current. Figure 2.14 displays the massive iron core with the slots, but without the winding. It can be seen that the slots do not cover the entire surface of the rotor. The area without the slots forms the magnetic poles. Figure 2.15 shows the completed rotor where the windings are covered with wedges and retention rings are incorporated to reinforce the structure. The figure also shows the fans at each end of the windings.

The generator operating principle is that the dc current in the rotor generates a magnetic field. The turbine spins the rotor and the magnetic field. The rotating field induces voltage in the stator three-phase windings.

Figure 2.16. Salient pole rotor (courtesy of Siemens, Erlangen, Germany).

A two-pole generator rotating at 3600 rpm produces a 60 Hz voltage. Most large generators rotate at 3600 rpm. The nuclear stations use four-pole generators spinning at 1800 rpm. Smaller generators and low speed generators will be equipped with salient poles, as seen in Figure 2.16. Most hydrogenerators operate at low speed and have more salient poles. A low-speed hydrogenerator rotating at 360 rpm would have 20 poles. The poles are supplied by dc current through brushes connected to the slip rings.

High-pressure, oil-lubricated bearings support the rotor at both ends. The bearings are insulated at the opposite side of the turbine connection to avoid shaft currents generated by stray magnetic fields.

The small generators are cooled by air, which is circulated by fans attached to the rotor. The large generators may be hydrogen cooled. The hydrogen is circulated in a closed loop and cooled by a hydrogen-to-water heat exchanger. The very large generators are water cooled. The cooling water is circulated through the special hollow conductors of the windings.

The dc excitation current may be produced by rectifiers and connected to the rotor through slip rings and brushes. Some units are equipped with a shaft-mounted brushless excitation system.

The generator operation and construction will be discussed in greater detail in Chapter 8, which addresses synchronous machines.

The generator converts the turbine mechanical energy to electrical energy. The electricity generated by the power plant supplies loads through transmission lines. The motors, mills, and pumps in a power plant require auxiliary electrical energy, which amounts to 10–15% of the power plant capacity.

Figure 2.17 shows a simplified connection diagram for a generating station. The generator is connected directly to the main transformer. The main transformer supplies the high-voltage bus through a circuit breaker, disconnect switches, and current transformer. An auxiliary transformer is also connected directly to the generator. The

FOSSIL POWER PLANTS

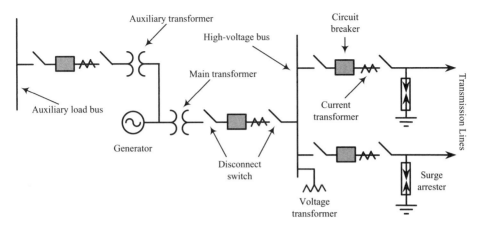

Figure 2.17. Simplified connection diagram of a generating station.

generating station auxiliary power is supplied through this transformer. A circuit breaker, disconnect switches, and current transformer protect the auxiliary transformer at the secondary side of the auxiliary transformer. The use of a circuit breaker at the generator side of the main transformer is uneconomical in the case of large generators.

The high-voltage bus forms a node point and distributes the generator power among the transmission lines. The voltage of the bus is monitored using a voltage transformer, which is also termed a potential transformer.

The two outgoing transmission lines are connected to the bus. The lines are protected against lightning and switching surge by a surge arrester. Each line is also protected by a circuit breaker. Two disconnect switches permit the separation of the circuit breaker in case of circuit breaker maintenance. The current transformer measures the line current and activates the protection in the event of a line fault. The protection relay triggers the circuit breaker, which switches off (opens) the line. The high-voltage bus configuration in Figure 2.17 is not typical; the operation of a circuit breaker and other protection components was discussed in detail within Chapter 1.

2.1.5. Combustion Turbine

Petroleum and natural gas can be burned in a furnace boiler-based plant to produce steam for input to a turbine generator. Alternatively, light fuel oil and natural gas can undergo combustion in a conventional gas turbine that utilizes the Brayton thermal cycle. Figure 2.18 shows that a gas turbine plant is significantly simpler in structure than the coal-fired unit described earlier. Fuel is added to compressed air, which upon combustion creates expansive gases that drive the gas turbine. The compressor and generator are shaft-connected to the turbine. The simplicity and size of these plants permit locating them on small sites, but the thermal efficiency of these simple cycles is low. With the higher cost of fossil fuels and the low thermal efficiency, these

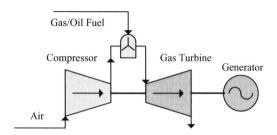

Figure 2.18. Conventional gas turbine power cycle.

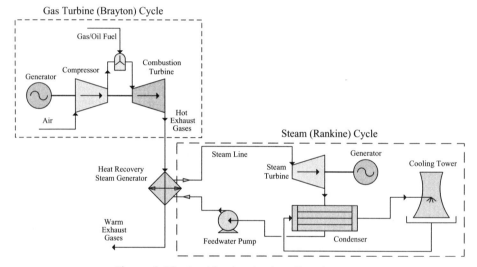

Figure 2.19. Combined cycle plant flow diagram.

conventional gas turbine plants are generally restricted to serving as peaking or reserve units. Fortunately, gas turbines can be taken from cold start to 100% power within minutes.

2.1.6. Combined Cycle Plants

Combined cycle plants have become a popular generation scheme in recent years. A combined cycle unit utilizes a gas turbine (Brayton) top cycle with the excess heat going to a steam turbine (Rankine) bottom cycle, as diagrammed in Figure 2.19. Air is compressed before injecting fuel for ignition in the gas turbine. The resulting combustion gases are first used to drive the gas turbine, then the hot exhaust gases are sent to a heat recovery steam generator (HRSG), before release through the stack. The heat transferred to the HRSG produces steam, which is used to drive a steam turbine generator set. Some combined cycle plants incorporate burners to increase (i.e., augment) the

Rankine cycle steam quality. The bottom steam cycle employs condenser cooling, as is normally found in a thermal power plant. The efficiency of a combined cycle plant without heat augmentation is determined from the efficiencies of the Brayton (η_{Bray}) and Rankine (η_{Rank}) cycles using:

$$\eta_{comb} = \eta_{Bray} + \eta_{Rank} - \eta_{Bray}\eta_{Rank}. \qquad (2.5)$$

The overall thermal efficiency of combined cycle plants built today is a remarkable 60%. Combined cycle plants are designed for intermediate load since they are relatively quick to start. Additional advantages of these plants are that they can be constructed in a relatively short period (about 2 years) and their use of natural gas, which is an environmentally good choice (except for greenhouse gas emission) and a reasonably priced fuel. Combined cycle plants should not be confused with *cogeneration*, also known as combined heat and power (CHP), which is the production of electricity in house along with industrial process steam.

2.2. NUCLEAR POWER PLANTS

Nuclear power plants are a major part of the electrical energy generation industry. More than 450 plants are in operation worldwide. The most popular plants (close to 300) are the pressurized water reactors (PWRs). In addition, there are approximately 100 boiling water reactors (BWRs) and 20 gas-cooled reactors. Further, close to 50 heavy-water reactors and a few liquid metal-cooled reactors operate around the world.

A nuclear power plant generates electricity in a manner very similar to a fossil power plant. Typically, nuclear plants provide base energy, running at practically constant load. Their electric output is around 1000 MW. Concern with thermal pollution increased with the construction of nuclear plants due to their large size and lower thermal efficiency ($\eta_{th} \approx 33\%$) as compared with coal-fired units ($\eta_{th} \approx 40\%$), and the fact that all the heat rejection from a nuclear plant is via the condenser cooling water, whereas a fossil unit also releases excess heat through the stack. For this reason, nuclear plants sought alternative heat-dissipation techniques such as the tall natural draft cooling towers, with which nuclear units are so commonly associated.

The advantages of nuclear power are the abundant and relatively cheap fuel and the pollution-free operation in normal conditions. However, leaks or equipment failure could allow radioactive gas or liquid (water) discharge that might pose a health hazard to the surrounding communities. An additional unanswered political question in the United States is the final storage of the spent fuel, which is radioactive and hazardous. A similar concern is the decommissioning of old and obsolete plants.

Decreased energy consumption in the United States after the energy crises of the 1970s, along with the listed environmental and health concerns, stopped or slowed the building of new nuclear plants and curtailed the operation of several existing plants. These actions resulted in severe financial losses for several utilities. Nevertheless, several hundred nuclear plants are in operation and generating large amounts of energy worldwide.

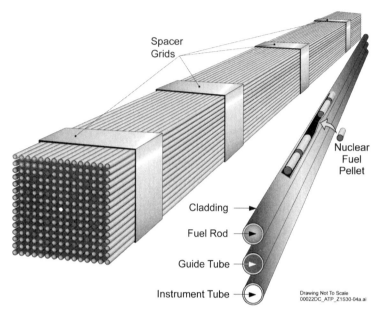

Figure 2.20. Nuclear reactor fuel pellet, rod, and assembly. Source: Department of Energy, Office of Civilian Radioactive Waste Management (DOE OCRWM).

2.2.1. Nuclear Reactor

Most power reactors use enriched uranium as a fuel. The uranium dioxide (UO_2) is pressed into pellets and the pellets are stacked in a Zircaloy-clad rod. These rods are the fuel elements used in a reactor. Many fuel rods are placed in a square lattice to construct a fuel assembly, as illustrated in Figure 2.20. A couple hundred fuel assemblies are generally needed to fuel the entire reactor core. This reactor core is housed in a reactor pressure vessel that is composed of steel 8–10 inches (20–25 cm) thick. The reactor core is populated with fuel and control rods. The nuclear fission reaction is regulated by the position of the control rods. Figure 2.21 shows the nuclear reactor vessel where the core is located. The control and fuel rods are arranged in a pattern carefully calculated during the reactor design.

Most reactors use neutrons in thermal equilibrium (<0.1 eV) with a moderator to sustain the chain reaction, and hence are called thermal reactors. The fission reaction emits neutrons at fast energy levels (>1 MeV). Thermal reactors employ a moderator such as light water, heavy water, or graphite to slow down the neutrons. Such thermal reactors are easier to control than fast (breeder) reactors and some designs can use natural uranium. In the United States, most nuclear plants utilize light water reactors (LWRs), which include the PWRs and BWRs.

The nuclear reaction starts if sufficient numbers (a critical mass) of fuel rods are placed in a confined space. Natural uranium consists of about 0.7% of the isotope uranium-235 (^{235}U) and the remainder (99.3%) is ^{238}U. Uranium-235 readily undergoes

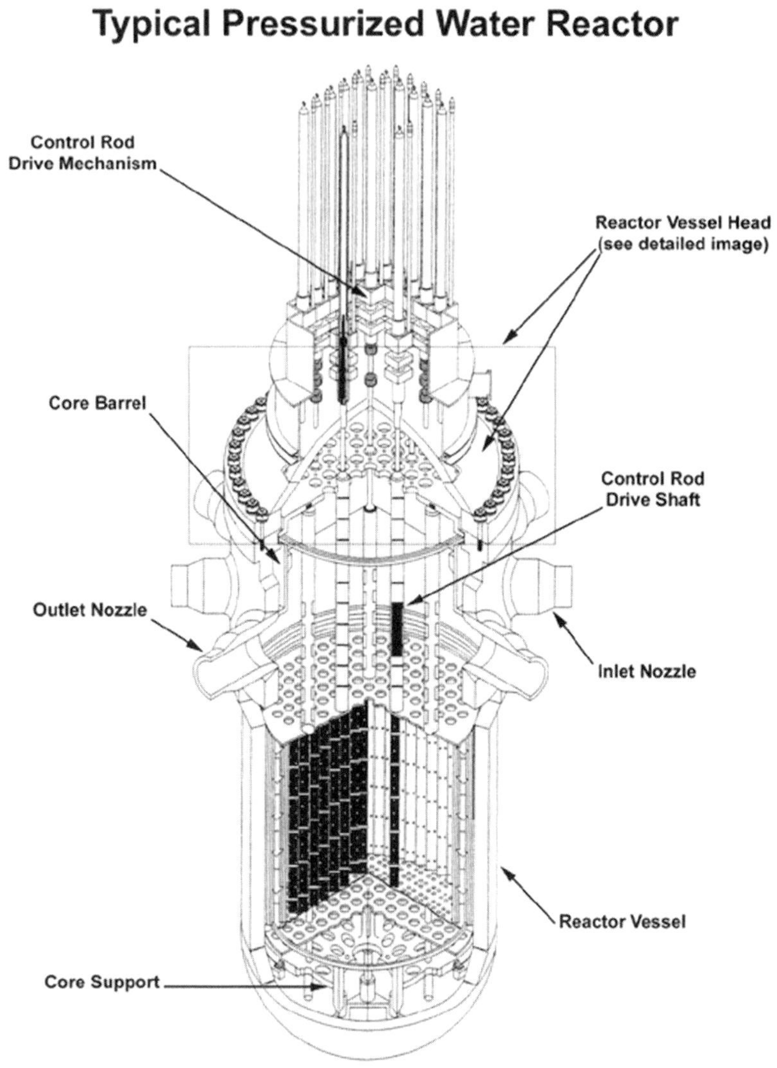

Typical Pressurized Water Reactor

Figure 2.21. Nuclear reactor vessel. Source: Department of Energy, Energy Information Administration (DOE EIA).

fission by thermal neutrons, whereas ^{238}U does not. In most cases, the fuel is enriched to about 3% ^{235}U to achieve a sustained reaction.

The neutron absorption by U-235 can initiate atomic fission. The fission of the uranium expels more neutrons and releases heat energy (Q):

$$^{235}_{92}U + ^{1}_{0}n \rightarrow \left(^{236}_{92}U\right)^* \rightarrow ^{A_1}_{Z_1}X + ^{A_2}_{Z_2}X + v\,^{1}_{0}n + Q, \tag{2.6}$$

where ν is the number of emitted neutrons per fission ($\nu \approx 2$–3). The freed neutrons continue the chain reaction; and the generated heat ($Q \approx 200$ MeV/fission) is utilized to produce steam. In addition to neutrons and heat, the nuclear fission produces two to three fission fragments (X). These fission products are radioactive and have decay times on the order of a thousand years.

The cooling water enters the reactor, flows up through the core, and removes the heat generated by nuclear fission. Safe reactor operation basically requires that heat is adequately removed from the core in order to avoid release of radioactivity from the plant. This can be ensured by maintaining the UO_2 fuel temperature below its melting temperature of about 5000°F (2760°C), and by keeping the cladding temperature below the point ($\approx$2200°F, 1200°C) at which the exothermic zirconium–water reaction is significant. This reaction produces potentially explosive hydrogen gas via:

$$Zr + 2H_2O \rightarrow ZrO_2 + 2H_2. \tag{2.7}$$

The nuclear reaction is controlled to maintain proper heat generation. The reaction is regulated using control rods, which can be constructed of neutron-absorbing material such as boron, silver, cadmium, and indium. The withdrawal of the control rods increases the reaction rate and heat generation. The insertion of the control rods reduces the power generation. The reactor is shut down by inserting all the control rods fully into the core. Even though the chain reaction ceases, some heat continues to be produced from the decay of the radioactive fission products. The plant incorporates cooling systems to remove this decay heat after shutdown. Loss of that decay heat removal capability can lead to reactor fuel melting.

In addition to causing fission, some neutrons undergo parasitic capture by the uranium-238. The added neutron increases the atomic mass of the ^{238}U, which can be transformed to plutonium-239 via the following reaction and radioactive decays:

$$\begin{aligned}^{238}_{92}U + ^{1}_{0}n &\longrightarrow {}^{239}_{92}U \xrightarrow{23.5 \text{ min}} {}^{239}_{93}Np + {}^{0}_{-1}e \\ {}^{239}_{93}Np &\xrightarrow{2.355 \text{ day}} {}^{239}_{94}Pu + {}^{0}_{-1}e.\end{aligned} \tag{2.8}$$

Pu-239 and other nuclides of higher atomic number than uranium are termed the *transuranics*. The transuranics are radioactive and long lived, having half-lives of thousands of years.

Nuclear waste is classified as either low-level or high-level waste. Low-level waste includes clothing, rags, and tools, which are sealed in a drum for ultimate placement in a dedicated landfill. The high-level waste includes fission products and transuranic isotopes, and is highly radioactive and must be stored for long periods. Approximately once every 18–24 months, the reactor is shut down for refueling, at which point about one-third of the (spent) fuel assemblies are removed. At present, U.S. government policy prohibits chemical reprocessing of spent fuel rods from commercial nuclear power plants. Instead, the used fuel assemblies will likely be placed in corrosion-resistant metal canisters, which will be housed in an underground repository. For 20 years, studies were conducted to determine whether Yucca Mountain (located 100 miles northwest of Las Vegas, Nevada) could serve as a suitable geologic repository in the United States. Political issues have led to uncertainty regarding the final decision.

EXAMPLE 2.3: Use the 200 MeV of heat released per fission to determine the number of uranium atoms required to produce a joule of heat.

Solution: First, recall that 1 eV = 1.602×10^{-19} J. Therefore, the number of ^{235}U atoms that must be fissioned to generate 1 J of heat is simply:

$$\left(\frac{1\ ^{235}\text{U atom}}{\text{fission}}\right)\left(\frac{\text{fission}}{200\ \text{MeV}}\right)\left(\frac{1\ \text{MeV}}{1.602 \times 10^{-13}\ \text{J}}\right) = 3.12 \times 10^{10}\ \text{atoms/J}.$$

Using Avogadro's number (N_A), this number of atoms (n) is equivalent to a ^{235}U mass of:

$$m = \frac{nM}{N_A} = \frac{(3.12 \times 10^{10}\ \text{atoms})(235\ \text{g/mol})}{6.022 \times 10^{23}\ \text{atoms/mol}} = 1.22 \times 10^{-11}\ \text{g},$$

where M is the atomic mass.

2.2.2. Pressurized Water Reactor

PWRs are the dominant reactor type for electric power plants and are also the basis of naval reactors. The flow diagram for a PWR is presented in Figure 2.22. The reactor

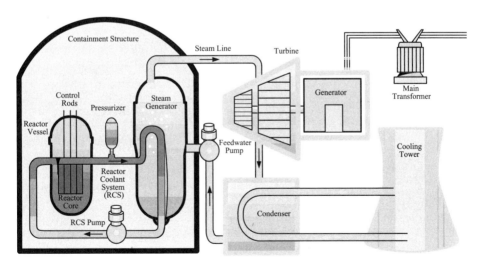

Figure 2.22. Pressurized water reactor (PWR) nuclear power plant (courtesy of the Tennessee Valley Authority).

has two water loops: a primary (potentially radioactive) water loop and a secondary water (steam) loop. This two-loop system separates the reactor cooling fluid from the turbine steam loop. The entire reactor coolant system is housed in a concrete containment building designed to prevent the release of radioactivity to the environment.

A coolant pump circulates the water in the primary loop through the reactor and steam generator (heat exchanger). The reactor heats the primary coolant system water to about 550–600°F (290–315°C). The pressurizer maintains a water pressure around 2200 psia (150 bar). This high pressure prevents water boiling and steam generation in the reactor core.

In the secondary water–steam loop, pumps drive the feedwater into the steam generator. Because of the relatively low secondary side pressure ($\approx$1000 psia, 70 bar), the heat exchanger evaporates the water and produces steam. The produced steam drives the turbine. This system is similar to that of the conventional thermal plants described previously. However, the temperatures and pressures in the PWR are significantly less than those in a coal-fired unit, thus leading to the lower thermal efficiency in the nuclear plant. The condenser produces a vacuum and extracts the steam from the turbine when the water is condensed. The condensed water is reheated by the high-pressure feedwater heaters and fed back to the steam generator. The feedwater is heated by steam extracted from the turbine.

Figure 2.23 provides an aerial view of the Palo Verde Nuclear Generating Station. The plant has three reactors that are housed in dome-shaped concrete structures

Figure 2.23. Aerial view of the Palo Verde Nuclear Generating Station (photo courtesy of the Palo Verde Nuclear Generating Station and Arizona Public Service).

(containment buildings). The turbine and generator units are placed in separate buildings (see Fig. 2.2). The nuclear fuel is stored in bunkers just in front of the reactors. The condensers are cooled by mechanical draft cooling towers. Each reactor has three cooling towers and a cooling pond. A large 500 kV switchyard is in the front of the PWR power plant.

2.2.3. Boiling Water Reactor

BWRs are another reactor type commonly used for electric power plants. The BWR has a single water–steam loop. The reactor heats the water and generates steam. Unlike a PWR, the water heated by the BWR nuclear core is sent directly to the turbine. The steam temperature is around 545°F (285°C) and the pressure is 1000 psia (70 bar). Located in Japan, the Fukushima Daiichi plant, which experienced an accident due to an earthquake and the ensuing tsunami in March 2011, utilized BWR technology, whereas the Three Mile Island plant in the United States employed PWRs.

Steam separators and dryers located within the top of the reactor vessel partition the water liquid and vapor. The liquid water flows downward, mixes with the feedwater, and returns to the reactor core inlet. The steam drives the turbine, which typically rotates with a speed of 1800 rpm. Figure 2.24 shows that the feedwater pumps drive the condensed water back to the reactor. The remainder of the plant is similar to a conventional thermal power plant system.

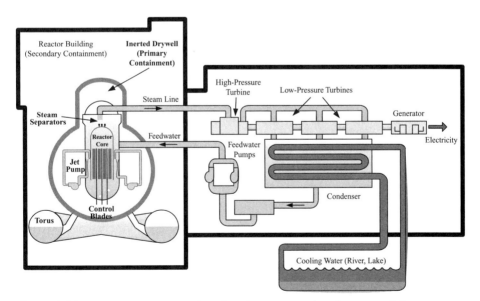

Figure 2.24. Boiling water reactor (BWR) nuclear power plant (courtesy of the Tennessee Valley Authority).

Supplied by recirculation pumps, jet pumps maintain water circulation through the reactor core. The control rods (blades) of the BWR enter the reactor from the bottom since steam separators are located above the reactor core and because the boiling of water near the top of the reactor core decreases the nuclear fission rate.

2.3. HYDROELECTRIC POWER PLANTS

Hydropower has played an important historical role in the industrialization of society, from grinding flour to powering industry. Hydroelectric power plants transform the potential energy of the water head to mechanical energy using a hydraulic turbine, and the generator then converts the mechanical energy to electrical energy. There are two categories of hydroelectric plants (dams):

1. run-of-the-river (diversion) plants, in which water is continuously passed with limited reservoir storage (e.g., the Bonneville Dam on the lower Columbia River); and
2. storage dams, in which water is released as needed and available (e.g., Hoover Dam on the Colorado River).

Although storage dams are used for peak power production, the large capital costs incurred from dam construction are nevertheless justifiable since a dam may have several purposes, including electricity generation, flood control, river navigation, irrigation, public water supply, and recreation.

Figure 2.25 depicts the general concept of a hydroplant using the medium or low head plant as an example. A dam built across a river produces an upper-level reservoir and tail water. The difference between the water level at the reservoir side and at the tail waterside of the dam is the *head*. Built into the dam is a powerhouse with a turbine hall such as that photographed in Figure 2.26. The powerhouse has a hydroturbine, generator, and control gates. The generator and turbine have vertical shafts, which are directly connected. The head generates fast-flowing water through the turbine. The water drives the turbine–generator set. The rotating generator produces the electricity. The power obtained from a hydropower plant is the product of the head (H), water density (ρ), and volumetric flow rate ($\dot{V}$) through the penstock:

$$P = H\rho g \dot{V}, \qquad (2.9)$$

where g is gravitational acceleration (9.81 m/s^2 = 32.2 ft/s^2). The water discharged from the turbine flows to the tail water reservoir, which frequently is the continuation of the original river. Control gates regulate the flow through the turbine. In the event of a flood, the spillway gates open, sending overflow across the dam, or diversion gates at the bottom of the dam may be opened. Both actions allow the direct flow of the excess water to the tail water reservoir, which eliminates the overloading of the dam. Additional gates at the water intake and draft tube permit the isolation and removal of water from the turbine during maintenance.

HYDROELECTRIC POWER PLANTS

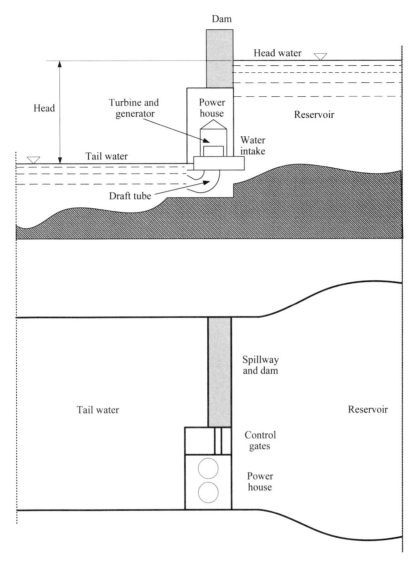

Figure 2.25. Hydroelectric power plant (medium or low head).

Figure 2.26. Turbine hall at a hydropower plant.

EXAMPLE 2.4: The Bonneville Dam on the Columbia River has eight large generators rated at 54 MWe each under a 59-ft head (H_L), and two smaller generators rated at 43.2 MWe each under a 49-ft head (H_S). Each turbine discharges water at a rate of 13,300 ft³/s. Determine the overall efficiency of the hydroplant.

Solution: First, determine the theoretical power for the large and small generator types for a nominal freshwater density of 62.4 lbm/ft³:

$$P_L = \frac{\rho g \dot{V} H_L}{g_c} = \frac{\left(62.4 \frac{\text{lbm}}{\text{ft}^3}\right)\left(32.2 \frac{\text{ft}}{\text{s}^2}\right)\left(13{,}300 \frac{\text{ft}^3}{\text{s}}\right)(59 \text{ ft})}{\left(32.2 \frac{\text{lbm} \cdot \text{ft}}{\text{lbf} \cdot \text{s}^2}\right)\left(0.7376 \frac{\text{ft} \cdot \text{lbf}}{\text{W} \cdot \text{s}}\right)} = 66.38 \text{ MW},$$

$$P_S = \frac{\rho g \dot{V} H_S}{g_c} = \frac{\left(62.4 \frac{\text{lbm}}{\text{ft}^3}\right)\left(32.2 \frac{\text{ft}}{\text{s}^2}\right)\left(13{,}300 \frac{\text{ft}^3}{\text{s}}\right)(49 \text{ ft})}{\left(32.2 \frac{\text{lbm} \cdot \text{ft}}{\text{lbf} \cdot \text{s}^2}\right)\left(0.7376 \frac{\text{ft} \cdot \text{lbf}}{\text{W} \cdot \text{s}}\right)} = 55.13 \text{ MW}.$$

Note the need for dividing by g_c because of the use of English units; this is unnecessary when performing the analysis using the International System of Units (SI) units. Finally, the overall efficiency is the ratio of total power output to input:

$$\eta = \frac{P_{\text{out}}}{P_{\text{in}}} = \frac{(8)(54 \text{ MWe}) + (2)(43.3 \text{ MWe})}{(8)(66.39 \text{ MW}) + (2)(55.14 \text{ MW})} = 0.809.$$

The initial part of this problem could have been solved with Mathcad using Equation (2.9) without having to look up any required unit conversions or potentially making a mistake by forgetting to introduce g_c into the denominator. Further, the base formula may be reused as demonstrated in the following Mathcad coding:

$$\rho := 62.4 \frac{\text{lb}}{\text{ft}^3} \quad \text{Vdot} := 13300 \frac{\text{ft}^3}{\text{s}}$$
$$\text{Pmech}(\text{head}) := \rho \cdot g \cdot \text{Vdot} \cdot \text{head}$$
$$\text{Pmech}(59 \text{ ft}) = 6.639 \times 10^7 \text{ W}$$
$$\text{Pmech}(49 \text{ ft}) = 5.514 \times 10^7 \text{ W}.$$

Note that Mathcad uses a built-in value of $g = 32.17$ ft/s² for the acceleration due to gravity. An introduction to Mathcad is provided in Appendix A.

2.3.1. Low Head Hydroplants

Figure 2.27 presents the cross section of a low head powerhouse, with a Kaplan reaction turbine. A large oil-immersed truss bearing supports both the generator and turbine.

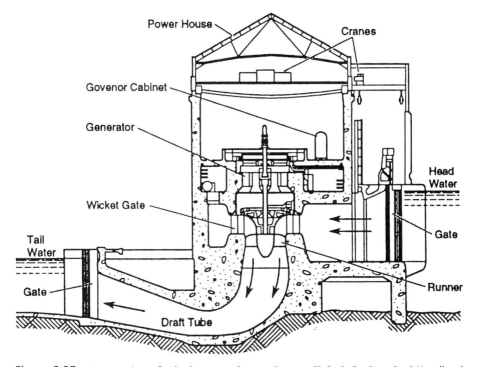

Figure 2.27. Cross-section of a hydropower house. Source: Fink, D.G., *Standard Handbook for Electrical Engineers*, McGraw-Hill, New York, 1978, with permission.

Figure 2.28. Large hydrogenerator under construction (photo by Hydro-Québec).

The upper, watertight chamber houses the vertical shaft generator. The vertical shaft Kaplan turbine is like a large propeller with 4–10 blades. The pitch of the blades is adjustable between 5° and 35° by a hydraulic servomechanism.

The water enters the turbine through gates and is evenly distributed by a spiral casing surrounding the turbine. The flow is regulated by "wicket gates" and by the adjustment of the pitch. The water speed in the turbine is around 10–30 ft/s (3–9 m/s). The water is discharged from the turbine through an elbow-shaped draft tube that reduces the water velocity to 1–2 ft/s (0.3–0.6 m/s).

The hydrogenerator is a salient pole machine. Typically, the machine has 20–72 poles. These poles are supplied by dc current and they produce the magnetic field that induces the voltage in the generator. The shaft speed is 100–300 rpm. Figure 2.28 shows a large hydro generator in the construction phase.

A welded spoke wheel supports the pole spider. The poles have a laminated iron core with dc windings using stranded copper conductors. Short-circuited damper bars are built in each pole face.

The stator is laminated iron with slots. A welded steel frame holds the iron core. A three-phase winding is placed in the stator slots. The turn-to-turn insulation is fiberglass or Dacron glass (DuPont, Kinston, NC). The turn-to-ground insulation is epoxy or polyester resin-impregnated mica tape. The larger machines have a braking system that rapidly stops the machine when removed from service.

2.3.2. Medium- and High-Head Hydroplants

In addition to the described low-head hydro, medium- and high-head hydroplants are in operation. The medium-head hydro has similar construction, but uses the Francis reaction turbine, which has a different blade arrangement, as pictured in Figure 2.29.

Figure 2.30 shows the arrangement of a high-head hydroplant. The high-head hydro uses an impulse turbine, such as a Pelton wheel. The large head-produced,

HYDROELECTRIC POWER PLANTS

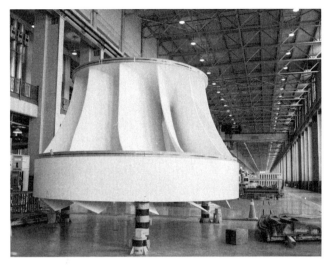

Figure 2.29. Wheel of a medium head Francis turbine (photo by Hydro-Québec).

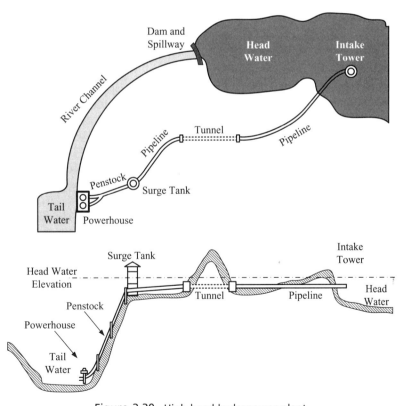

Figure 2.30. High-head hydropower plant.

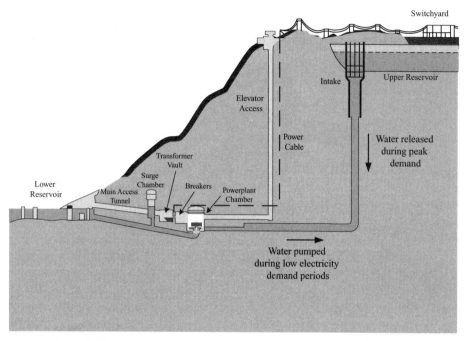

Figure 2.31. Pumped storage (courtesy of the Tennessee Valley Authority).

high-water pressure is converted to a high-velocity water jet, which drives the turbine. The rating of a high-head hydro plant is generally less than 100 MW.

2.3.3. Pumped Storage Facility

Also related to hydroelectric power is the practice of using pumped storage facilities as an electric energy storage device. Pumped storage facilities consist of both high-elevation and low-elevation reservoirs, as sketched in Figure 2.31. The powerplant chamber of the pumped storage facility houses a reversible hydraulic turbine similar to that used in a dam. The direction of this turbine can be reversed by supplying the special motor–generator with electricity, such that it becomes a pump to transfer water to the upper reservoir. Such pumped storage units, like hydroplants, can provide power on very short notice.

The pumped storage plant consumes electricity during low demand (e.g., nighttime) to pump water from the low-elevation body of water to a high-elevation reservoir. Then, during peak power demands (e.g., daytime), the water is allowed to flow back down and generate electricity before returning to the low-elevation lake. Of course, more electricity is required to pump the water uphill than is subsequently generated in the return downhill. However, overall, the pumped storage is economical because it generates high-cost, on-peak electricity while consuming low-cost, off-peak energy. Additionally, the upper reservoir experiences some evaporative water loss.

Although the first pumped storage facilities were built for load-balancing purposes as previously described, the incorporation of significant amounts of intermittent power sources such as photovoltaics (PVs) and wind into the electric grid requires electric energy storage, which can be achieved with approaches like pumped hydro, compressed air, flywheel, batteries, and cryogenic magnetic energy storage.

2.4. WIND FARMS

Similar to hydropower, wind power has enjoyed a rich history from providing the motive force for sailboats to pumping water using windmills. Winds primarily originate from uneven heating of the Earth's surface, although local conditions (e.g., mountains) affect the wind. As such, wind speed generally increases with daylight heating and then decreases in the later hours after dusk. This behavior leads to unpredictability and lower capacity factors (CFs).

Today's wind turbines convert the kinetic energy of the wind to mechanical shaft power and, ultimately, electrical power. The total power associated with the wind kinetic energy is:

$$P_W = \frac{1}{2} \dot{m} v^2 = \frac{1}{2} (\rho A v) v^2 = \frac{1}{2} \rho A v^3, \tag{2.10}$$

where $\dot{m}$ is the air mass flow rate; A is the swept area traversed by the turbine blades, and ρ and v are the air density and velocity, respectively. In 1920, Albert Betz determined that the theoretical limit for extracting the energy is 16/27 (59.3%) of the total power. Even so, a well-designed wind turbine does not achieve Betz efficiency.

Figure 2.32 shows a representative wind turbine. The tower provides a pedestal for the nacelle, which is the housing for the generator and other equipment behind the

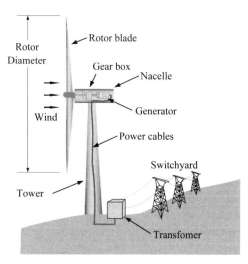

Figure 2.32. Horizontal axis wind turbine (courtesy of the Tennessee Valley Authority).

Figure 2.33. Wind farm (photo credit Joshua Winchell/U.S. Fish and Wildlife Service).

rotor blades. Since wind speed is lowest at ground level, increased tower height serves to place the rotor blades into levels of higher wind velocity. The importance of wind speed to the generation potential is obvious from Equation (2.10). Yaw control rotates the nacelle to direct the blades toward the wind. The electric outputs from multiple wind turbines in a wind farm are brought together in a collector substation prior to transferring power to the grid. Typically, an underground medium voltage network is utilized to interconnect the turbines.

Older wind turbines were equipped with induction generators to produce alternating current (ac) power directly at the system frequency. This mandated that the turbines rotate at a speed that produced the requisite frequency. Present wind turbines may utilize either induction or synchronous generators. Appropriate power conversion electronics may be employed in which the generated variable frequency current is converted to dc and then back to system frequency ac.

A wind farm is presented in Figure 2.33. Onshore wind parks benefit from reduced construction, operation, and maintenance costs as well as ease of connection to existing electric transmission. However, offshore wind farms enjoy higher and more constant wind speeds, but may encounter more adverse environmental conditions (e.g., saltwater and severe storms).

EXAMPLE 2.5: Plot the total wind power, theoretical maximum, and actual power for a wind turbine with $D = 50$ m diameter blades for wind speeds from 0 to 50 km/h. Assume that the actual wind turbine efficiency is 70% of the Betz efficiency.

Solution: The nominal density of dry air is $\rho_{air} = 1.2$ kg/m^3. The turbine blade area is simply $A = \pi D^2/4 = \pi(50 \text{ m})^2/4 = 1963.5$ m^2. The following MATLAB code computes the total wind power using Equation (2.10), and then reduces this total by the Betz efficiency to find the theoretical maximum power. The factor 1000/3600 converts kilometers per hour to meters per second:

```
% Example 2.5: Compute wind power
clear all

rhoair = 1.2;     % air density (kg/m^3)
area = 1963.5;    % turbine blade area (m^2)

speed = 0 : 1 : 50;    % wind velocity (km/hr)

% compute the total wind power (kW) from its kinetic energy
windpower = 0.5 * rhoair * area * speed.^3 * (1000/3600)^3 / 1000;
Betzpower = 16/27 * windpower;    % power based on Betz efficiency
actualpower = 0.7 * Betzpower;    % more realistic power estimate

plot(speed, windpower, 'b-.', speed, Betzpower, 'm--',...
     speed, actualpower, 'g-', 'LineWidth',2.5)
set(gca,'fontname','Times','fontsize',12);
title('Example 2.5')
xlabel('Wind Speed (km/hr)')
ylabel('Power (kW)')
legend('Total Wind Power', 'Betz Power', 'Actual Power',...
     'Location','NorthWest')
```

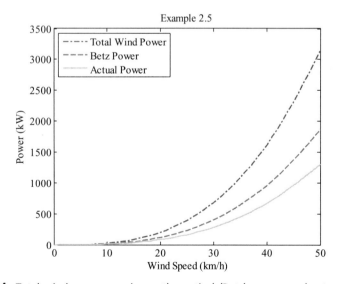

Figure 2.34. Total wind power, maximum theoretical (Betz) power, and actual power for 70% Betz efficiency for a 50 m diameter wind turbine.

The resultant plot with the three curves is provided in Figure 2.34. As can be seen from the actual power curve, a single wind turbine of this size can reasonably be expected to produce 1 MWe. This example further illustrates the importance of siting wind farms at locations with high (average) wind speed.

2.5. SOLAR POWER PLANTS

There are three energy conversion processes normally associated with the sun:

1. heliochemical, which is principally the photosynthesis process,
2. helioelectrical, which is commonly exploited in solar cells (i.e., PVs), and
3. heliothermal, which is a conversion of sunlight into thermal heat as employed within concentrating solar power (CSP) plants.

Like many renewables, solar energy boasts free fuel, but the capital costs are presently large. Although both wind and solar are intermittent power sources, sunlight availability is more predictable. Solar energy, however, is dilute with the incident power being less than 1 kW/m^2. The specific solar irradiance received depends on latitude, season, time of day, and atmospheric conditions. Practical direct normal solar radiation is on the order of 4–8 kWh/m^2/day, but these values are prior to the energy conversion process.

2.5.1. Photovoltaics

Solar cells provide a direct means to convert sunlight to electrical energy. Depending on the technology, PV conversion efficiency is in the range of 10–20%. PV installations can increase their overall electric energy output using single- or dual-axis tracking systems. A drawback of solar cells compared to CSP is that PV generates dc power that must typically be converted to ac. A 14 MW PV power system is pictured in Figure 2.35.

A diagram of a solar cell is shown in Figure 2.36. Photons from the sun enter the semiconductor structure, depositing their energy, thereby causing the creation of electron–hole pairs. The top emitter region is thin and heavily doped; the base region, where most light is absorbed, is lightly doped. A silicon solar cell produces an output voltage of about 0.6 V and tens of milliamps per square centimeter. In order to raise the current and voltage, multiple cells are connected in parallel and series, respectively, to create a PV module. Several modules are joined to form a solar panel; and multiple panels are combined to create a solar array. This modularity makes PV output scalable, and therefore suitable for distributed power generation.

An ideal solar cell may be represented simply by a current source and diode in parallel. Figure 2.37 presents an equivalent circuit model for a solar cell, including resistances characterizing the cell material resistance (R_S) and leakage current through

Figure 2.35. PV installation using single-axis tracking (photo credit Nellis Air Force Base).

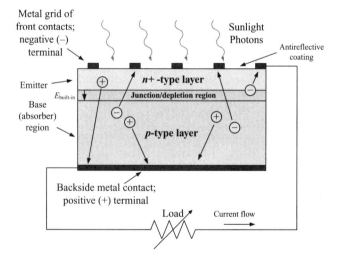

Figure 2.36. Solar cell operation principle.

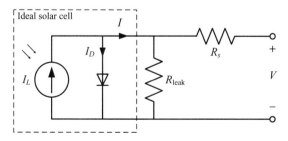

Figure 2.37. Equivalent circuit for a solar cell.

the cell (R_{leak}). PV manufacturers seek to reduce R_S to as small a value as possible while ensuring that current leakage is minimal (i.e., R_{leak} is large).

The net current (I) from an ideal solar cell is the difference between the light-induced (I_L) and dark (I_D) currents. The inherent dark current can be expressed in terms of the reverse saturation current (I_0) using the Shockley diode equation:

$$I_D = I_0 \left[\exp\left(\frac{qV}{nkT}\right) - 1 \right], \tag{2.11}$$

where V is the device voltage; q is the fundamental charge constant (1.6×10^{-19} J/V); k is the Boltzmann constant (1.38×10^{-23} J/K); and T is the absolute temperature. In full sun conditions, the ideality factor n is approximately unity, denoting ideal diode behavior. The net PV cell current is:

$$I = I_L - I_D = I_L - I_0 \left[\exp\left(\frac{qV}{nkT}\right) - 1 \right]. \tag{2.12}$$

At short-circuit conditions, $V = 0$ and hence, the short-circuit current is $I_{SC} = I_L$. Similarly, at open-circuit conditions, $I = 0$ and the open-circuit voltage is found from Equation (2.12) to be:

$$V_{OC} = \frac{nkT}{q} \ln\left(1 + \frac{I_L}{I_0}\right). \tag{2.13}$$

Of interest is the maximum power output from a solar cell. Taking the derivative of the power ($P = IV$) and setting it to zero yields the following transcendental formula for the voltage (V_{Pmax}) at maximum power output:

$$\exp\left(\frac{qV_{Pmax}}{nkT}\right) = \frac{1 + I_L/I_0}{1 + qV_{Pmax}/(nkT)}. \tag{2.14}$$

Once the previous expression is solved for V_{Pmax}, the maximum power output is seen to be:

$$P_{max} = \frac{(I_L + I_0)V_{Pmax}}{1 + nkT/(qV_{Pmax})}. \tag{2.15}$$

This equation reveals that increased cell temperatures leads to reduced power output. The conversion efficiency of a solar cell is the ratio of the electric power output to the radiant solar power input. For a single-junction silicon solar cell, the Shockley–Queisser limit for the maximum theoretical conversion efficiency is ~30%.

SOLAR POWER PLANTS

EXAMPLE 2.6: Using Mathcad, plot the output current and power versus voltage of a solar cell at 25°C with the following characteristics: short-circuit current $I_{SC} = 8.3$ A and open-circuit voltage $V_{OC} = 0.61$ V. Then determine the maximum power output (P_{max}) and the corresponding voltage and current at P_{max}.

Solution: First, the constants and PV characteristics are established:

$$I_{SC} := 8.3 \text{ A} \qquad V_{OC} := 0.61 \text{ V} \qquad T_{cell} := (25 + 273) \text{ K}$$

$$q := 1.602 \cdot 10^{-19} \frac{J}{V} \qquad n := 1 \qquad k := 1.38 \cdot 10^{-23} \frac{J}{K}$$

Knowledge of I_{SC} immediately identifies the value of I_L:

$$I_L := I_{SC}.$$

Using I_L and V_{OC}, Equation (2.13) allows the extraction of the reverse saturation current:

$$I_0 := \frac{I_L}{\exp\left(\frac{q \cdot V_{OC}}{n \cdot k \cdot T_{cell}}\right) - 1}.$$

Now an array to vary the solar cell voltage is defined, and the current and power are computed:

$$V_{cell} := 0 \text{ V}, .01 \text{ V} .. V_{OC}$$

$$I_{cell}(V_{cell}) := I_L - I_0 \cdot \left(\exp\left(\frac{q \cdot V_{cell}}{n \cdot k \cdot T_{cell}}\right) - 1\right)$$

$$P_{cell}(V_{cell}) := V_{cell} \cdot I_{cell}(V_{cell}).$$

Figure 2.38 presents the single graph with both the current and power curves. Rather than use the transcendental expression of Equation (2.14), the Mathcad Maximize function in conjunction with a solve block is used to find the maximum power and both the corresponding voltage and current at that point:

$$V_{guess} := 0.5 \text{ V}$$
Given
$$V_{guess} \leq V_{OC}$$
$$V_{Pmax} := \text{Maximize}(P_{cell}, V_{guess}) \qquad V_{Pmax} = 0.53 \text{ V}$$
$$P_{max} := P_{cell}(V_{Pmax}) \qquad P_{max} = 4.2 \text{ W}$$
$$I_{Pmax} := I_{cell}(V_{Pmax}) \qquad I_{Pmax} = 7.9 \text{ A}$$

These values concur visually with the curves in Figure 2.38.

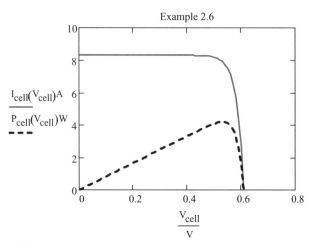

Figure 2.38. Solar cell current and power output as a function of voltage.

2.5.2. Solar Thermal Plants

In terms of large-scale electric utility power generation, CSP plants heat a fluid and eventually make steam to drive a traditional turbine–generator set. Since solar thermal power plants are Carnot limited, they employ concentrators to achieve high temperatures and thereby higher plant thermal efficiency. The two main approaches to heating the working fluid are:

1. parabolic trough collectors and
2. central receiver systems (power towers).

The power tower approach utilizes heliostats (mirrors) to focus sunlight on a "single point" atop a tower, as exemplified in Figure 2.39. The heliostats require dual-axis tracking, whereas single-axis tracking suffices for the trough collectors, as pictured in Figure 2.40. The line focus system of the parabolic trough can heat the working fluid from 150 to 350°C, whereas the point focus approach of the power tower is capable of producing temperatures from 250 to 1500°C.

A noteworthy advantage of solar thermal plants is the ability to integrate a thermal energy storage (TES) facility within the power plant, thereby giving CSP the capability of generating power for a while after sunset and during periods of cloudiness. Another feature of these two solar thermal facilities is the option to employ auxiliary burners (using biomass, hydrogen, or fossil fuels) during sunlight unavailability. For example, the Solar Electric Generating Systems (SEGS) units built at Kramer Junction, California, in the 1980s employ natural gas to provide up to 25% of the thermal energy for steam production. A third CSP plant type—the dish–Stirling engine—does not readily permit such TES.

SOLAR POWER PLANTS

Figure 2.39. Solar One central receiver concentrated solar power facility (courtesy of Department of Energy/National Renewable Energy Laboratory [DOE/NREL]).

Figure 2.40. Solar trough collectors at Kramer Junction (courtesy of DOE/NREL).

In the case of trough collectors, an oil-based working fluid is pumped through a tube positioned along the focal line of the parabolic trough. The receiver tube is encased within a glass shell that is evacuated of air to reduce convection heat losses from the pipe. Using a heat exchanger, the hot working fluid heats water to produce steam, which is then piped to a steam turbine–electric generator. As such, CSP plants require condenser cooling, typically using evaporative cooling methods—this can be problematic in arid regions where sunlight is correspondingly plentiful.

Power tower facilities utilize molten salt as the heat transfer fluid. The higher heat capacity of the molten salt permits its direct use as a thermal energy medium, but heat tracing is required to melt the salt for operational use. The 10 MW Solar One demonstration plant built near Barstow, California, operated from 1982 to 1988 using a 95 m tall tower; after upgrading its working fluid, TES, and reflector field in the early 1990s, it was designated Solar Two.

2.6. GEOTHERMAL POWER PLANTS

Unlike the majority of electric generating stations utilizing renewable energy, geothermal facilities can be operated as base load plants. Interestingly, geothermal power is the only renewable energy source of terrestrial origin. Most geothermal sites are near rifts between tectonic plates. These cracks in the earth's crust, which is 30–50 km thick, allow molten material to seep closer to the surface. Earthquake and volcanic activity are also common in these areas, for example, the Pacific Ring of Fire stretching from New Zealand northward to Japan and Alaska and then south along the western coast of the Americas.

Two categories of geothermal resources are:

1. *Hydrothermal.* The traditional geothermal energy source consisting of hot water and steam, and
2. *Petrothermal.* Hot dry rock, which uses water injection to produce hot water or steam.

Hydrothermal can be further subdivided into vapor- and liquid-dominated systems, where the former is the exception. The well-produced geothermal hot water or steam is used on either an indirect or a direct basis, for input to a heat exchanger or a turbine, respectively. Geothermal water temperatures are lower than those temperatures in fossil and nuclear plants, which consequently results in lower geothermal power plant efficiencies ($\eta_{th} \sim 20\%$).

A geothermal power plant is shown in Figure 2.41. Estimated land use including the well field is 800–1000 acres (3–4 km^2) for 100 MWe for 30 years of production. To access geothermal sources requires drilling both extraction and injection wells between 1 and 2 miles (1–3 km) deep, and after a few years, new wells must be drilled. Geothermal wells tap into and release pollutants, including radon gas, hydrogen sulfide (H_2S), CO_2, methane (CH_4), and ammonia (NH_3). The presence of the impurities in the

Figure 2.41. A geothermal power plant at The Geysers near Santa Rosa, California (photo credit: Julie Donnelly-Nolan, USGS).

extracted brine motivates the use of flash steam and binary systems. The brine (saline water) can cause mineralization problems in the power plant.

2.7. OCEAN POWER

Ocean power sources can be categorized according to the fundamental energy type being harnessed. The four types of ocean energy addressed in the succeeding sections are:

1. ocean tidal, which can be considered similar to hydropower;
2. ocean current, which can be likened to an underwater sea "wind";
3. ocean wave, which exploits the kinetic and potential energies of the waves; and
4. ocean thermal, which is Carnot limited and has a very low thermal efficiency.

These ocean energy sources are of solar origins, except for ocean tidal, which is mostly caused by lunar gravity. Hydrokinetic conversion techniques harness the kinetic energy from ocean tidal, wave, and marine current sources, and even river water flow. Ocean power facilities generally face issues with saltwater corrosion, potential collisions with shipping traffic, and the effects of large storms such as tropical cyclones. Like offshore wind, most oceanic power stations require submarine cables to transmit electricity back to the land-based grid.

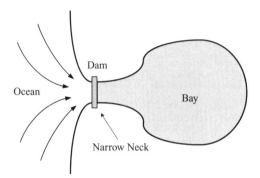

Figure 2.42. Schematic of ocean tidal barrage site.

2.7.1. Ocean Tidal

Tidal power results from changing gravitational effects due to the relative positions of the earth, sun, and moon. Extraction of potential energy from the movement of the tides is akin to hydroelectric generation. In fact, the simplest implementation of tidal power would be to construct a dam, referred to as a tidal barrage, across the mouth of a bay, as depicted in Figure 2.42, with electricity being generated both when the tide comes in and when it recedes. Although tides are predictable, the electricity generation does not match with load patterns. The instantaneous flow rate and the head are directly affected by the tidal cycle—they follow a sinusoidal behavior. The most famous tidal power facility is located at La Rance, France. In operation since 1966, this 240 MWe power plant employs 24 reversible direction turbines driven by a tidal range of 9–14 m.

Consider the potential energy held in a bay after high tide. The water mass stored in the bay is $m = RS\rho$, where S is the bay surface area and R is the tidal range, which is the height difference between high and low tides. The head (H) is not the tidal range since as water exits the bay, the level of the reservoir decreases. The actual head is obtained as half the range, that is, $H = R/2$. The potential energy from the movement of water in one-half tidal cycle is:

$$PE = mgH = (RS\rho)g\left(\frac{R}{2}\right) = \frac{R^2 S \rho g}{2}. \qquad (2.16)$$

The time period over which this energy is extracted requires some elaboration since the tidal cycle (T) is 12 hours and 24.6 minutes. For systems constructed to utilize both the incoming and outgoing tides, the potential energy of Equation (2.16) is doubled over the 12.4 hours tidal cycle. The maximum average power for a plant using both tidal directions is therefore:

$$P = \frac{PE}{T} = \frac{R^2 S \rho g}{T}. \qquad (2.17)$$

EXAMPLE 2.7: The Passamaquoddy Bay site on the Canadian–U.S. border has an average tidal range of 5.5 m and a basin area of 262 km². Determine both the maximum potential energy stored in the bay and the maximum average power that could be produced at the site.

Solution: Seawater density is approximately 1025 kg/m³. The maximum potential energy storage for one tidal cycle would be a scheme that relies on both incoming and outgoing tides, so using the doubled form of Equation (2.16):

$$E_{max} = R^2 S \rho g = (5.5 \text{ m})^2 (262 \text{ km}^2) \left(1025 \frac{\text{kg}}{\text{m}^3}\right) \left(9.81 \frac{\text{m}}{\text{s}^2}\right) \left(1000 \frac{\text{m}}{\text{km}}\right)^2 \left(\frac{1 \text{h}}{3600 \text{s}}\right),$$

$$= 2.2 \times 10^7 \frac{\text{kW} \cdot \text{h}}{\text{cycle}},$$

the average power per cycle is then:

$$P = \frac{PE}{T} = \frac{2.2 \times 10^7 \text{ kW} \cdot \text{h}}{12.41 \text{ h}} = 1,770 \text{ MWe}.$$

Annually, this would be (1770 MWe)(8760 hours) = 1.55 × 10¹⁰ kWh. The enormous capital costs of the dam make harnessing this tidal energy source impractical.

2.7.2. Ocean Current

Ocean current power generation involves the extraction of kinetic energy from marine water flow. One source of underwater currents is pinch points, in which the tides cause higher than normal velocities, known as tidal streams. Other sources of marine currents include the wind, and temperature and salinity gradients. The equations quantifying the ocean current power are identical to those for wind power, including the Betz efficiency. The difference being that ocean flow rates are about one-fifth that of the wind, but the seawater density is about 800 times that of air. An ocean current turbine is depicted in Figure 2.43.

2.7.3. Ocean Wave

An ocean wave consists of both potential and kinetic energy. The total energy of a wave per unit surface area (A) is:

$$\frac{E}{A} = \frac{1}{2} \rho g a^2, \qquad (2.18)$$

where a is the wave amplitude (a is half the wave height). The wave period is the division of the wavelength by the wave propagation velocity ($\tau = \lambda/c$). Various techniques utilizing floating buoys connected to or incorporating a generator have been proposed

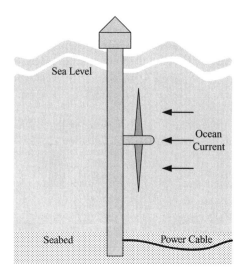

Figure 2.43. Ocean current turbine.

to produce electricity offshore from the bobbing motion of the waves. Like wind farms, ocean current and wave facilities would consist of multiple individual generators. In 1910, ocean wave power was used to light lamps on the Huntington Beach Wharf in California—until a storm carried the apparatus out to sea.

2.7.4. Ocean Thermal

Ocean thermal energy conversion (OTEC) relies on temperature gradients between warm ocean surface water and cold seawater near the ocean floor. Because the high and low temperature difference is small ($\Delta T \sim 22°C$), the thermal cycle efficiency is abysmal ($\eta_{th} \sim$ 2–3%). Both closed (Anderson) and open (Claude) thermal cycles have been designed. The closed-cycle approach transfers heat from warm saltwater to vaporize a working fluid with a low boiling point (e.g., ammonia), which drives a turbine–generator. The open cycle passes the warm seawater into an evaporator below atmospheric pressure, which causes some of the water to vaporize; the steam produced is then used directly in a low-pressure turbine. In both cycles, deep cold seawater is transported to the surface to provide condenser cooling. A few OTEC pilot plants with net power outputs in the tens of kilowatt electric have been constructed.

2.8. OTHER GENERATION SCHEMES

Other electricity generation schemes include biomass, fuel cells, and diesel generators. The cost of petroleum products has made the latter technique increasingly unpopular; however, remote or isolated locations (e.g., an island) and specific applications (e.g., emergency power) continue to keep diesel generators in use worldwide.

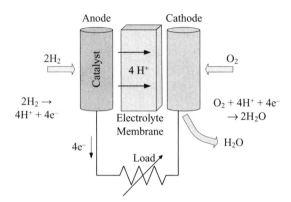

Figure 2.44. Fuel cell principle of operation.

Biomass fuel is defined in several ways. A restricted definition of biomass includes only fuels from recent photosynthesis and, as such, biomass could be considered as carbon neutral. An expanded definition of biomass also encompasses municipal waste (residential and commercial garbage) that is incinerated. In both cases, the biomass energy source is the hydrocarbon fuel that is typically burned. The biomass fuels, however, possess less chemical energy per unit mass than do traditional fossil fuels.

Fuel cells provide a direct chemical to electrical energy conversion path for fuels such as hydrogen and hydrocarbons. Because of the direct energy conversion process, fuel cell performance is independent of the Carnot cycle efficiency. Fuel cell operation is achieved by passing H_2 and O_2 (air) gases through the porous anode and cathode, respectively, as illustrated in Figure 2.44. A catalyst located at the anode oxides the fuel by stripping the electrons from the hydrogen atoms and allows the protons to pass through an electrolyte membrane separating the anode and cathode. A fuel cell is generally identified in terms of its electrolyte, for instance the polymer electrolyte membrane (PEM) fuel cell. The electrons travel through (i.e., power) an external circuit before reaching the cathode, where they combine with the protons and oxygen to form the normal water reaction product. Hence, fuel cells clearly produce dc rather than ac power.

2.9. ELECTRICITY GENERATION ECONOMICS

Economics drives the selection of an appropriate power generation scheme for the given situation. A utility may need additional generation during high electricity demand hours (peak load) or the new power may be needed 24 hours a day (base load). *Base load* is that load below which the demand never falls, that is, the base load must be supplied 100% of the time. The *peaking load* occurs less than about 15% of the time; the intermediate load is between roughly 15% and 85% of the time.

The plant operational use is quantified using the capacity factor (CF). The CF is the ratio of electric energy produced during some time interval to the energy that could have been produced at net rated power (P_e) during the same period (T), that is,

$$CF = \frac{\int_0^T P(t)dt}{P_e T}. \qquad (2.19)$$

Another plant performance indicator is the *availability factor* (*AF*), which is the time period that the plant is operable divided by the total period considered (AF ≥ CF):

$$AF = \frac{\text{Time period plant is operable}}{\text{Total time period of interest}}. \qquad (2.20)$$

The AF does not directly appear in the economics equations below; however, a peaking plant must maintain a high AF compared to its CF.

The *spinning reserve* is excess capacity that is running and synchronized with the system, and should be equal to or greater than the power the largest unit is producing in the power network in the event that unit should trip. Another consideration is the *reserve capacity* (historically maintained at about 20%), which is the difference between the total rated capacity of all the units in the grid and the expected peak demand on the system divided by the total rated capacity.

In calculating the cost of electricity production (cents per kilowatt-hour, ¢/kWh), the energy cost is broken into three categories:

1. *Capital*: land, equipment, construction, interest;
2. *Operational and Maintenance (O&M)*: wages, maintenance, some taxes, and insurance; and
3. *Fuel Costs*: coal, natural gas, oil, uranium.

Historically, costs were often expressed in mills per kilowatt-hour where 1000 mills equal $1. Since costs are expressed on a per-kilowatt-hour basis, a high CF is desired so the capital and O&M costs are spread out. As shown in Table 2.1, the capital and

TABLE 2.1. Power Plant Selection according to Load Demand

Loading	Capital Costs	Fuel Costs	CF	Example Power Plants
Base	High	Low	High	Coal and nuclear
Intermediate	Mid	Mid	Mid	Combined cycle natural gas, and older coal units
Peak	Low	High	Low	Conventional combustion turbines using oil and natural gas

ELECTRICITY GENERATION ECONOMICS

fuel costs generally dictate how a plant is used on the grid. Exceptions to these categorizations do exist, for example, hydroelectric units have high capital costs but may be used for peak power. Furthermore, small islands may utilize diesel generators to generate electricity. For some electricity generation schemes (e.g., solar and wind), it may be necessary to include some type of energy storage device or mechanism. The cost of such energy storage facilities must be considered in the overall costs.

The overall electricity generation cost is determined from the following:

$$e = \frac{\text{Capital} + \text{O\&M} + \text{Fuel}}{\text{Electric Energy Produced}} = e_C + e_{OM} + e_F. \quad (2.21)$$

Let us consider calculating the electricity cost for a single year. The electric energy (E) produced in a given year can be determined using the nominal (net) plant power rating (P_e) and the CF for that year via:

$$E = P_e \text{CF}(365 \text{ days/year}). \quad (2.22)$$

2.9.1. O&M Cost

O&M costs may be divided into (1) fixed O&M costs (dollars per kilowatt-year, $/kW·year) that vary with the plant size, and (2) variable O&M costs (¢/kWh), which are proportional to the unit energy production. The fixed costs comprise wages and overhead for permanent employees, routine maintenance, and other fees, whereas variable costs consist of equipment outage maintenance, utilities, and consumables such as chemicals. If the total annual O&M costs are known, then:

$$e_{OM} = \frac{\text{O\&M [\$/year]}}{P_e \text{CF}(8760 \text{ h/year})}. \quad (2.23)$$

2.9.2. Fuel Cost

Fuel costs are generally proportional to the plant output so the related energy cost is constant. The annual thermal heat required to produce a given amount of electricity can also be determined from the plant thermal efficiency. The thermal heat produced is the product of the fuel input rate ($\dot{m}_{fuel}$) and the fuel heat content (i.e., the HV), that is, $\dot{Q}_{th} = \dot{m}_{fuel} HV$. The annual electric energy produced is thus:

$$E = \dot{m}_{fuel} HV \eta_{th}. \quad (2.24)$$

The annual fuel cost is determined from the amount (e.g., mass or volume) of fuel used and its cost per unit mass or volume (F_C):

$$\text{Fuel} = \dot{m}_{fuel} F_C. \quad (2.25)$$

For a fossil-fired unit, F_C is commonly expressed in terms of dollars per ton or gallon, and the HV is given in British thermal units per pound-mass (kilojoules per kilogram, kJ/kg) or British thermal units per gallon (kilojoules per liter, kJ/L). For a nuclear power plant, F_C is typically quoted in dollars per kilogram of uranium, and the *burnup*, B, in megawatts thermal-day per metric ton of uranium (MTU), is the terminology employed for expressing fuel heat content. By combining the two expressions previously mentioned, the fuel cost of electricity is found to be:

$$e_F = \frac{\text{Fuel}}{E} = \frac{F_C}{HV\eta_{th}}. \tag{2.26}$$

As should be expected, the fuel cost is independent of the plant CF.

EXAMPLE 2.8: An 800 MWe coal-fired unit has a thermal efficiency of 38% and a CF of 82%. The electric utility is in negotiations for a long-term contract for coal having a HV of 36,500 kJ/kg. To hold the fuel cost at or below 1 ¢/kWh, what is the maximum price, in dollars per metric ton, that the utility should be willing to pay?

Solution: Rearranging Equation (2.26) provides the defining relation:

$$F_C \leq e_F HV\eta_{th} = \left(0.01 \frac{\$}{\text{kWh}}\right)\left(36{,}500 \frac{\text{kJ}}{\text{kg}}\right)(0.38)\left(\frac{\text{kW}}{\text{kJ/s}}\right)\left(1000 \frac{\text{kg}}{\text{tonne}}\right)\left(\frac{1\,\text{h}}{3600\,\text{s}}\right)$$

$$= \$38.53/\text{tonne}.$$

The given CF was extraneous information.

2.9.3. Capital Cost

As a substantial simplification, the capital cost can be likened to a home mortgage payment, which mostly consists of principal and interest. Significant differences between the home mortgage and power plant cases do exist, however. For example, power plants take years (vs. months) to build, hence, utilities must borrow money during the plant construction period (granted that individuals obtain home construction loans, too). In addition, the time value of money should be considered, that is, because of inflation, the value of 1 dollar 10 years from now will likely be different than today, and depreciation should also be accounted for.

The capital (or fixed) cost must be properly distributed throughout the expected operating lifetime of the plant. For financial analyses, plants are generally assumed to have a lifetime of 30–40 years, although they may operate for either a shorter or a longer period. If the utility spreads the construction costs and interest over 40 years with equal annual (or monthly) payments, then the objective is to find the annual payment amount. To do so, the payment can be expressed as a percentage of the construction cost (F_B), that is, a levelized annual fixed-charge rate (I) is established such that:

LOAD CHARACTERISTICS AND FORECASTING

$$e_C = \frac{CAP}{E} = \frac{IF_B}{P_e CF(8760 \text{ h/year})} = \left(\frac{F_B}{P_e}\right)\frac{I}{CF8760}, \qquad (2.27)$$

where CAP is the annualized amount of the entire capital costs including interest. The term F_B/P_e is the cost of building a power plant, which is generally expressed in terms of dollars per installed kilowatt-electric ($/kWe). As the electric rating is increased, the ratio F_B/P_e generally decreases—this is known as *economy of scale*. The levelized fixed-charge rate can be computed directly from:

$$I = \frac{d}{1-(1+d)^{-N}}, \qquad (2.28)$$

where d is the discount rate and N is the number of years for payment. The *discount rate* is the sum of the actual interest rate and the inflation (escalation) rate.

2.9.4. Overall Generation Costs

Substituting Equation (2.23), Equation (2.26), and Equation (2.27) into Equation (2.21) yields an overall formula for the electric generation costs:

$$e = e_C + e_{OM} + e_F = \frac{IF_B + O\&M}{P_e CF(8760 \text{ h/year})} + \frac{F_C}{HV\eta_{th}}. \qquad (2.29)$$

From the earlier expression, the parameters that should be minimized as well as those that should be maximized to achieve reduced electricity costs are recognized readily. In particular, a high CF (base load plant) means that high capital cost can be accommodated; conversely, low CF (peak plants) permits high fuel costs in conjunction with low capital cost. Table 2.2 provides representative production costs for electric generating stations in the United States for 2010.

2.10. LOAD CHARACTERISTICS AND FORECASTING

The electrical system load and the system losses equal the generation, because the system has very little storage capacity. For the analysis of load variation, the load is typically averaged for a time interval of 15 minutes, 30 minutes, or 1 hour. The short-term averaged load is called the *demand*.

Figure 2.45 plots the daily variation of the load demand curves for commercial, industrial, and residential loads as well as their combined values. The graph demonstrates that on a winter morning, the load rises and reaches a maximum around 8:00 A.M. After a short afternoon reduction, the load again increases, attaining a second maximum around 8:00 P.M. After that, the load decreases until a minimum is realized at approximately 3:00 A.M. The system peak is used to determine the necessary generation capacity. The integral of the curve gives the daily energy demand.

TABLE 2.2. Costs of New Central Electric Generating Stations

Technology	Capacity (MW)	Heat Rate (Btu/kWh)	Overnight Capital Cost ($/kW)	Annual Fixed O&M Cost ($/kW)	Variable O&M Cost ($/MWh)
Advanced pulverized coal	650	8,800	3,167	35.97	4.25
Integrated coal-gasification combined cycle (IGCC)	600	8,700	3,565	59.23	6.87
IGCC with carbon capture and sequestration (CCS)	520	10,700	5,348	69.30	8.04
Conventional natural gas combined cycle (NGCC)	540	7,050	978	14.39	3.43
Advanced NGCC with CCS	340	7,525	2,060	30.25	6.45
Conventional combustion turbine (natural gas)	85	10,850	974	6.98	14.70
Fuel cells (natural gas)	10	9,500	6,835	350	0
Advanced nuclear (dual units)	2,236	–	5,335	88.75	2.04
Biomass combined cycle (wood)	20	12,350	7,894	338.79	16.64
Wind—onshore	100	–	2,438	28.07	0
Wind—offshore	400	–	5,975	53.33	0
Solar thermal (power tower)	100	–	4,692	64.00	0
Photovoltaic—small	7	–	6,050	26.04	0
Photovoltaic—large	150	–	4,755	16.70	0
Geothermal binary cycle	50	–	4,141	84.27	9.64
Municipal solid waste	50	18,000	8,232	373.76	8.33
Hydroelectric	500	–	3,076	13.44	0

Source: U.S. Energy Information Administration, "Updated Capital Cost Estimates for Electricity Generation Plants," November 2010, p. 7. Additional generation schemes appear in this Department of Energy publication.

The load also varies with the season of the year. Figure 2.46 graphs the seasonal variation of the daily demand in the southwestern United States The plot reveals that maximum demand is in the summertime because of the air conditioning use. In the northern part of the United States, the maximum load occurs in the wintertime due to electric space heating. Furthermore, the demand varies through the week; typically, the maximum load during the weekend is only around 60% of the demand during the 5-day

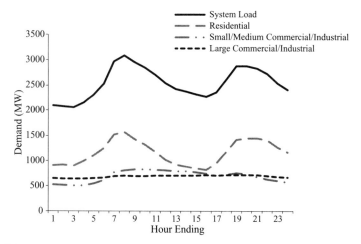

Figure 2.45. Commercial, industrial, residential, and combined demand daily variation for a winter day.

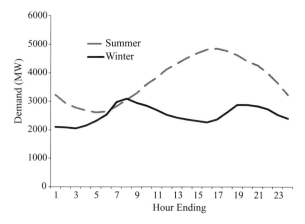

Figure 2.46. Seasonal variation of the daily demand in the southwestern United States

workweek. The load or the demand is affected by the weather. Public events (e.g., sporting contests and political speeches) watched by large numbers of people also have an impact on the loads.

Figure 2.47 exhibits a typical diurnal load curve and generation mix. The graph shows that large nuclear and coal-fired plants generate the bulk power. These large power plants operate at practically constant load. Smaller coal-fired power plants and large numbers of gas-fired power plants are producing the energy needed during the load variation. In the morning, gas-fired plants pick up the load. The older plants have gas-fired boilers and steam turbines; the newer plants are combined cycle units.

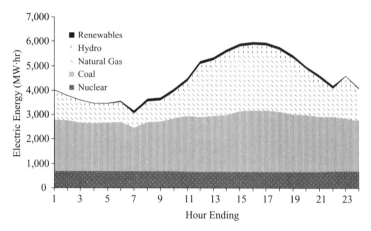

Figure 2.47. Representative generation mix and daily demand for a summer day.

Gas turbines, which can be started within a half hour, supply the short duration peak load. The wind- and solar-generated powers are fully used when they are available. Pumped storage plants also generate electricity during the peak hours and consume electricity during the nighttime.

Electrical Load Forecasting

Electrical load forecasting is important for utilities for planning the day-to-day operation, including generation and purchase of electrical power as well as long-term planning for construction of new power plants and transmission lines. The load forecasting is divided into three categories:

1. short-term forecasting, covering a 1-hour to 1-week period,
2. medium-term forecasting, encompassing 1 week to 1 year, and
3. long-term forecasting, involving more than 1 year.

Short-term forecasting is used to schedule generation and purchase of electric power for the next day or for the week. Medium-term forecasting with a 1-year horizon is employed to schedule generator and other equipment maintenance. Long-term forecasting with a 10- or 20-year horizon is utilized for generation and transmission planning.

Several mathematical methods have been developed for short-term load forecasting. The most popular methods are the similar-day approach, the regression technique, and the time series method. Typically, the forecasters use several years of historical data for the day of the week, the specific day of the year, and the weather. In the simplest case, the forecast is the demand of a similar day.

The medium-term forecasts also use long-term historical data such as the day of the week, hour of the day, and weather data like temperature and humidity. The long-term forecast, in addition to the historical data, includes projected economic and population growth in a selected location.

TABLE 2.3. Potential Impact of Using Various Renewables to Generate Electricity

Generation Scheme	Possible Negative Impacts
Geothermal	• Increased seismic activity
	• Depletion of underground water sources
	• Land subsidence (falling, sagging)
Hydroelectric	• Fish (e.g., salmon) migration inhibited
	• Land intensive (dislocation of people and wildlife)
	• Silting of the river channel behind the dam
Ocean	• Effects on aquatic life
	• Disturbance of the seabed
Solar	• Land intensive
	• Effects on indigenous plant and animal species (e.g., desert tortoise)
Wind	• Aesthetically unpleasing to some people
	• Avian (bird) and bat kills
	• Land use
	• Low frequency noise

2.11. ENVIRONMENTAL IMPACT

Each and all electricity generation approaches have impacts on the environment. For example, addressed earlier in this chapter were thermal pollution, nuclear waste, and SO_2 and NO_X production. Another impact of most thermal plants is the need to withdraw water for cooling purposes. Even the so-called green energy forms are not without their environmental effects (see Table 2.3). In all cases, the environmental impact should include the effects from the complete fuel cycle associated with the particular electricity generation scheme, for instance, coal mining and petroleum shipment (e.g., oil spills).

The *greenhouse effect* on global warming (i.e., climate change) has received significant attention in recent years. Many scientists believe that the increased emission of greenhouse gases, such as water vapor, carbon dioxide, nitrous oxide, and methane, is causing the temperature of the planet to rise. Short-wavelength radiation from the sun passes through the atmosphere without interference from the greenhouse gases. The transmitted sunlight is then absorbed by the earth's surface. Later, the absorbed energy is reemitted by the earth as long-wavelength radiation. This long-wavelength radiation (unlike the short-wavelength sunlight) is absorbed by the greenhouse gases (such as CO_2, which can originate from combustion), thus heating the earth's atmosphere.

Energy and water are inextricably linked in what is referred to as the energy–water nexus. For instance, water treatment processes (e.g., desalination) as well as routine pumping and distribution operations for water supply systems require energy input. Water utilization for condenser cooling in thermal power plants is receiving increasing attention. According to the U.S. Geological Survey (USGS), 49% of all water withdrawals in the United States during 2005 were for thermoelectric power. The reader should be cognizant of a terminology distinction: *withdrawal* is water diverted from a surface or groundwater source, while *consumption* is water use that permanently removes water from its source (e.g., from evaporation).

2.12. EXERCISES

1. What is the role of electric generating stations?
2. List some types of electric generating stations.
3. What are the base load and peak load? Illustrate them with a sketch.
4. List the components of a fossil power plant.
5. Describe typical fuel storage, handling, and injection systems used in fossil power plants.
6. Describe the boiler used in a fossil power plant. Draw a sketch showing the major subsystems.
7. Describe typical water–steam and ash-handling systems used in fossil power plants.
8. What is the condenser? Draw a sketch.
9. What is an electrostatic precipitator?
10. Describe a typical wet cooling tower. Draw a sketch. What is the function of the cooling tower?
11. Describe the operating concept of a steam turbine.
12. Draw a simplified connection diagram of a power plant and identify the components and their roles.
13. Describe the combined cycle power plant. Draw a sketch.
14. What is the operating principle of nuclear power generation?
15. Describe the concept of a boiling water reactor. Draw a sketch.
16. Describe the concept of a pressurized water reactor. Draw a sketch.
17. Describe the types of hydroelectric power plants and their operating principles.
18. What is a low-head hydroplant? Draw a sketch.
19. Describe a high-head hydropower plant. Draw a sketch.
20. What is a pumped storage facility?

2.13. PROBLEMS

Problem 2.1

A 600 MWe coal-fired power plant has a coal pile with 2.5 million metric tons of fuel available. Historically, the plant has operated with a thermal efficiency of 36% and a capacity factor of 80%. If the coal has a heating value of 25,000 kJ/kg, determine the number of days the coal reserve will last.

Problem 2.2

An electric power plant is built with a rated output of 800 MWe. In the first year of operation, the plant operates at an average of 85% of its full power for 10 months, and then is shut down for the remaining 2 months. In the second year, the plant generates an average of 700 MWe over the entire year. (a) Determine the plant capacity factor for each year. (b) If the total operating and maintenance (O&M) cost to the utility was equal in the first and second years, and the electricity-averaged O&M cost in the first year was 1.5 ¢/kWh, then what was the O&M cost in cents per kilowatt-hour for the second year?

Problem 2.3

Determine the condenser heat dissipation requirements in MWt (megawatts thermal) for two different 1000 MWe power plants. The first plant is a coal-fired unit that has a thermal efficiency of 40% and releases 15% of the heat produced in the furnace up the stack. The second plant is a nuclear unit with a thermal efficiency of 33%.

Problem 2.4

A 400 MWe power station with a thermal efficiency of 40% incurs a fuel cost of $1/GJ. If the capacity factor is 65%, determine the annual fuel cost.

Problem 2.5

Glen Canyon Dam, the fourth highest dam in the United States, has a rated head of 510 ft and a maximum flow rate of 33,200 ft^3/s. The original capacity was 950 MWe in 1964, but generator uprating has since increased the capacity to 1320 MWe. Determine the original efficiency and the present efficiency.

Problem 2.6

A particular solar module consists of thirty-six 100 cm^2 solar cells in series. Each cell produces 0.5 V and 30 mA/cm^2. Determine the nominal module output voltage, current, and power.

Problem 2.7

Complete the analytical steps to derive (a) Equation (2.14) from $P = IV$, and (b) Equation (2.15) from Equation (2.14) and $P_{max} = I V_{Pmax}$.

Problem 2.8

The short-circuit current and open-circuit voltage of a solar cell are measured as 7.6 A and 0.58 V, respectively. (a) Plot the output current and power versus voltage of the solar cell at 40°C. (b) Determine the maximum output power and the corresponding voltage.

Problem 2.9

Compare the maximum thermal efficiency of trough and power tower CSP plants having working fluid temperatures of 250°C and 1000°C, respectively. Assume that the condenser rejects heat at 40°C.

Problem 2.10

A gas turbine power plant presently has a thermal efficiency of 30%. It is proposed to add a Rankine bottom cycle to increase the overall output of the plant. If the Rankine cycle efficiency will be 25%, determine the overall efficiency of the upgraded power station.

Problem 2.11

(a) If combustion of one atom of carbon to CO_2 yields 4.08 eV of heat, calculate the number of carbon atoms required to produce 1 joule of heat. (b) Compute the equivalent mass of carbon from the number of atoms found in part (a). (c) Determine the ratio of carbon to U-235 mass required to liberate the same amount of heat energy.

Problem 2.12

A photovoltaic power plant is to be built in a location where the daily incident sunlight provides 6 kWh/m². If the solar cell and power electronic components yield an overall conversion efficiency of 12%, determine the minimum surface area of land (in km²) needed to generate 20 MWh of electricity each day.

Problem 2.13

Coal from a particular mine contains 2% sulfur and 5% mineral matter. If the fuel is being burned at a rate of 2000 kg/min, determine the annual generation of SO_2 and ash for a power plant capacity factor of 60%. Note that $S + O_2 \rightarrow SO_2$.

Problem 2.14

An ocean-based turbine with 25 m diameter blades is subjected to a 3 m/s marine current. Determine the maximum theoretical electric power output of the turbine if the mechanical-to-electrical conversion efficiency is 95%.

Problem 2.15

A pumped storage facility has upper and lower reservoirs with an average height difference of 1000 ft and capacities of 125,000 acre-ft of water each (an acre-foot of water is the water volume needed to cover an acre of land to a depth of 1 ft, and is equal to 43,560 ft³). From midnight to 4:00 A.M. (off-peak), the contents of the lower reservoir are pumped to the upper reservoir with an efficiency of 80%. During the daytime, the water in the upper reservoir is emptied into the lower reservoir with a generation efficiency of 95%. (a) Determine the electric energy input and output during the off-peak and on-peak periods, respectively. (b) What is the overall efficiency, referred to as the *turnaround efficiency*, of this facility?

3

SINGLE-PHASE CIRCUITS

An electric power system is usually a three-phase system that contains generators, transformers, transmission and distribution lines, and loads. Power lines, which are linear elements, can be overhead lines or underground cables. Generators and transformers are also linear devices if iron saturation is neglected. Loads are generally nonlinear. A typical load in the power system draws constant power at a constant power factor. The proper operation of electric power devices requires that the load voltage be kept within ±5% of the rated voltage.

Most three-phase loads are balanced, which permits the representation of the three-phase system by a single-phase circuit. This emphasizes the importance of the discussion of single-phase circuits in an electric power textbook. Although it is assumed that the reader is familiar with single-phase circuit analysis techniques, in this chapter we review the fundamental concepts and establish the nomenclature and terminology utilized in this book. After an overview of the basic principles, we demonstrate the use of PSpice, Mathcad, and MATLAB programs for circuit analysis. The principal analysis concepts are presented through interactive derivations and numerical solutions using computer tools.

Electrical Energy Conversion and Transport: An Interactive Computer-Based Approach, Second Edition. George G. Karady and Keith E. Holbert.
© 2013 Institute of Electrical and Electronics Engineers, Inc. Published 2013 by John Wiley & Sons, Inc.

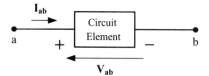

Figure 3.1. Passive sign convention; the current enters the circuit element at the positive voltage terminal.

3.1. CIRCUIT ANALYSIS FUNDAMENTALS

3.1.1. Basic Definitions and Nomenclature

The basic electrical quantities include current (I), voltage (V), and power (P). Voltage is an electromotive force or potential difference between two nodes (points). *Voltage is the difference in energy level of a unit charge located at each of two points in a circuit, and therefore represents the energy required to move the unit positive charge from one point to the other*. *Current* is time rate of change of electric charge (q) at a given location, that is, $i(t) = dq/dt$, and is measured in amperes (A). Normally we talk about the movement of positive charges (conventionally positive flow), although we know that, in general, for metallic conductors, current results from electron motion. The sign of the current indicates its direction of flow relative to some prechosen reference direction. In selecting the current direction, it is useful to follow the *passive sign convention*, which designates that current should enter the positive voltage terminal of a circuit element (see Fig. 3.1). If the passive sign convention is obeyed, then determining whether an element supplies or absorbs *power* is made easier. In particular, if power (which is the product of the current and voltage) is positive, then the element is absorbing power; conversely, if the power value is negative, the circuit element is supplying power; in summary:

$$P = IV \begin{cases} <0 & \text{element } supplies \text{ power} \\ >0 & \text{element } absorbs \text{ power} \end{cases}. \quad (3.1)$$

The conventions prescribed by the Institute for Electrical and Electronics Engineers (IEEE) are employed in this book (American National Standards Institute [ANSI]/IEEE Std 280-1985). The instantaneous values of time-varying quantities such as voltage and current are represented by lowercase letters, for example, $v(t)$ and $i(t)$. Uppercase letters characterize values associated with time-varying quantities such as the root-mean-square (rms) voltage (V_{rms}) and average power (P). Bold uppercase letters denote frequency-dependent variables such as complex power (**S**) and phasor voltages and currents (e.g., **V**$_{rms}$ and **I**). Figure 3.1 shows the equivalency of three designations for voltage polarity: (1) the use of explicit +/− labeling, (2) an arrow that points toward the positive terminal, and (3) the notation **V**$_{ab}$, for which the first letter of the subscript indicates the positive terminal and the second letter denotes the negative terminal. **V**$_{ab}$ can also be read as the voltage at node *a* relative to point *b* (often *b* is the ground).

CIRCUIT ANALYSIS FUNDAMENTALS

3.1.2. Voltage and Current Phasors

The time-domain functions are used for transient analysis. A sinusoidal voltage waveform of period T can be written as:

$$v(t) = V_M \cos(\omega t + \delta), \tag{3.2}$$

where V_M is the voltage magnitude or amplitude, $\omega = 2\pi f = 2\pi/T$ is the angular frequency in radians per second, and δ is the phase angle. The cyclic frequency, f, is 60 Hz in the United States, and 50 Hz in Europe and other parts of the world. However, specialized applications such as aircraft, spacecraft, and submarines may use a 400 Hz power system. Older industrial sites and mines use 25 Hz, although this implementation is gradually disappearing.

Power engineering calculations utilize the rms value and phase angle of both voltage and current. The rms or effective value is calculated from:

$$V_{rms} = \sqrt{\frac{1}{T}\int_0^T v(t)^2 \, dt}. \tag{3.3}$$

For a sinusoidal signal, the magnitude and the rms values are related by:

$$V_M = V_{rms}\sqrt{2}. \tag{3.4}$$

Hence, the time-domain voltage equation becomes:

$$v(t) = \sqrt{2}V_{rms}\cos(\omega t + \delta). \tag{3.5}$$

Single- and three-phase alternating current (ac) steady-state analyses are more easily performed in the frequency domain using phasor representation. A *phasor* is a complex representation of a time-invariant sinusoidal waveform. A general sinusoidal voltage may be transformed from its time-domain expression to an equivalent phasor expression by merely considering the magnitude (V_M) and phase shift (δ) of the signal relative to a chosen reference. The frequency-domain phasor representation of the above voltage is:

$$\mathbf{V} = \frac{V_M}{\sqrt{2}}\angle\delta = V_{rms}\angle\delta = V_{rms}e^{j\delta}. \tag{3.6}$$

This transformation is based on Euler's formula, $e^{\pm j\alpha} = \cos(\alpha) \pm j\sin(\alpha)$, while assuming that a specific ac frequency is analyzed which permits the removal of ω. In particular, $v(t)$ may be written in terms of the real part of a complex exponential expression:

$$v(t) = V_M \cos(\omega t + \delta) = \mathrm{Re}\left[V_M e^{j(\omega t + \delta)}\right]. \tag{3.7}$$

The supply voltage is often selected as the reference with a phase angle of δ = 0. Using phasors, the voltage amplitude and phase angle can be calculated using a polar (or complex) notation since the power system operates at a particular frequency.

Similarly, the current may be represented as:

$$i(t) = I_M \cos(\omega t + \phi) = \sqrt{2} I_{rms} \cos(\omega t + \phi) \quad \text{or} \quad \mathbf{I} = \frac{I_M}{\sqrt{2}} \angle \phi = I_{rms} \angle \phi = I_{rms} e^{j\phi}. \quad (3.8)$$

Often it is necessary to describe the phase shift between current and voltage. For ac waveforms at the same frequency, $v(t)$ is said to *lead* $i(t)$ by δ − φ, or equivalently $i(t)$ *lags* $v(t)$ by the same amount. If δ = φ, then the two waveforms are said to be *in phase*; if δ ≠ φ, then the two waveforms are *out of phase*.

EXAMPLE 3.1: The current drawn by an ac-powered hairdryer is listed as 12 A (rms). Determine the peak-to-peak current and the current magnitude.

Solution: The current magnitude is simply $I_M = \sqrt{2} I_{rms} = \sqrt{2}(12\,\text{A}) = 16.97\,\text{A}$. The peak-to-peak value is the difference between the maximum positive and maximum negative values of the waveform. Thus, the peak-to-peak current is (2)(16.97 A) = 33.94 A.

3.1.3. Power

The *instantaneous power*, $p(t)$, is the product of the instantaneous values of the voltage and current:

$$p(t) = v(t) i(t) = \sqrt{2} V_{rms} \cos(\omega t + \delta) \sqrt{2} I_{rms} \cos(\omega t + \phi). \quad (3.9)$$

Using the trigonometric identity:

$$\cos(\alpha)\cos(\beta) = \tfrac{1}{2}\cos(\alpha - \beta) + \tfrac{1}{2}\cos(\alpha + \beta),$$

the expression for instantaneous power may be rewritten as:

$$p(t) = V_{rms} I_{rms} [\cos(\delta - \phi) + \cos(2\omega t + \delta + \phi)]. \quad (3.10)$$

The first term is a constant, and the second term has a cyclic frequency at twice the applied frequency. These facts make calculation of the average power, P, easier, since we simply integrate a constant and a cosine wave over exactly twice its period (for which the latter is equal to zero):

$$P = \frac{1}{T} \int_0^T p(t)\,dt = V_{rms} I_{rms} \cos(\delta - \phi). \quad (3.11)$$

CIRCUIT ANALYSIS FUNDAMENTALS

The average power, P, is the *real power* transferred from the generator to the load impedance. The product $V_{rms}I_{rms}$ is termed the *apparent power* and is expressed in units of volt-amperes (VA) as compared with average power, which is expressed in watts (W). The real power is also known as the *active power*.

Another related quantity whose units are volt-amperes is the *complex power*, S, which is defined as:

$$S = \mathbf{V}_{rms}\mathbf{I}^*_{rms} = V_{rms}I_{rms}\angle\delta - \phi, \qquad (3.12)$$

where * denotes the complex conjugate operation. Hence, apparent power is the magnitude of the complex power, $|S| = V_{rms}I_{rms}$. The total complex power, like energy, in a system is conserved, although S has no physical significance. The conjugate of a phasor is formed by merely changing the sign of the phase angle:

$$\mathbf{I}^*_{rms} = \left(I_{rms}e^{j\phi}\right)^* = I_{rms}e^{-j\phi} = I_{rms}\angle -\phi. \qquad (3.13)$$

Expanding the expression for complex power yields:

$$\begin{aligned}\mathbf{S} = \mathbf{V}_{rms}\mathbf{I}^*_{rms} &= V_{rms}e^{j\delta}I_{rms}e^{-j\phi} = V_{rms}I_{rms}e^{j(\delta-\phi)} \\ &= V_{rms}I_{rms}\cos(\delta-\phi) + jV_{rms}I_{rms}\sin(\delta-\phi) \\ &= P + jQ.\end{aligned} \qquad (3.14)$$

The derivation shows that the real part of the complex power is the average power. The imaginary part of S is called the *reactive power*, Q. The reactive or *quadrature power* indicates temporary energy storage rather than any real power loss in an element. To distinguish Q from P and S, reactive power is expressed in units of volts-ampere reactive (VAR).

For ac steady-state circuits, each of the following quantities is conserved, that is:

$$\begin{aligned}\text{instantaneous power: } & \sum_k p_k(t) = 0, \\ \text{real power: } & \sum_k P_k = 0, \\ \text{reactive power: } & \sum_k Q_k = 0, \\ \text{complex power: } & \sum_k S_k = 0,\end{aligned} \qquad (3.15)$$

where the summations are over all the circuit elements in the network. Apparent power is not conserved.

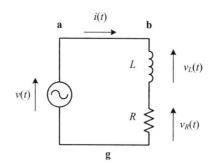

Figure 3.2. Example of a single-phase circuit.

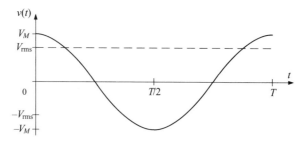

Figure 3.3. Sinusoidal source voltage, $V_M\cos(\omega t)$.

3.2. AC CIRCUITS

The typical linear single-phase ac circuit contains voltage sources and impedances (resistance, inductance, and capacitance). The components may be connected in series or parallel, or combinations of the two can occur. As an example, Figure 3.2 presents a simple single-phase circuit. In this circuit, an ac voltage source generates a sinusoidal ac voltage that drives current through an inductor (L) and a resistor (R). An actual generator used in a power system is more complex than a simple voltage source; however, to a first approximation, a generator can be represented as a simple voltage source.

Figure 3.3 plots the cosinusoidal voltage waveform generated by the voltage source. The voltage direction (or polarity) is indicated on the circuit diagram of Figure 3.2 by an arrow from **g** to **a**. This means that during the positive half-cycle from $t = -T/4$ to $t = T/4$, the potential of point **a** is larger than **g**. The generator current flows from **g** to **a** in the *positive half-cycle*; the generator current and voltage are in the same direction. The load current in the *positive half-cycle* flows from **b** to **g**; the load current and voltages are in opposite directions.

AC CIRCUITS

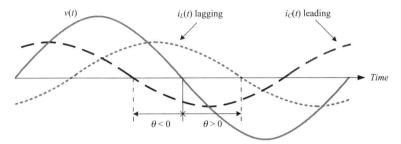

Figure 3.4. Illustration of leading and lagging current phase shifts.

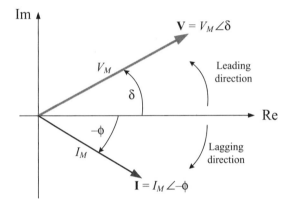

Figure 3.5. Phasor diagram of current and voltage.

The sinusoidal voltage drives a sinusoidal current through the circuit. For a voltage source with a reference phase shift of zero ($\delta = 0°$), the current equation is:

$$i(t) = \sqrt{2} I_{rms} \cos(\omega t - \theta), \qquad (3.16)$$

where $\theta = \delta - \phi$ is the phase shift between the voltage and current waveforms. The current can be *lagging* or *leading* the voltage. The current lags the applied voltage ($\theta > 0$) in an inductive circuit and leads ($\theta < 0$) in a capacitive circuit. Figure 3.4 graphs the voltage with lagging and leading currents. As an example, the inductive circuit in Figure 3.2 generates lagging current. The current will lead the voltage if we replace the inductor with a capacitor in that circuit.

A phasor diagram can be employed to graphically illustrate the relationship between current and voltage. In particular, the current and/or voltage phasor is drawn on a complex plane. The vector, whose length represents the phasor magnitude, is drawn using the positive real axis as reference with positive phase angles in the counterclockwise direction. A voltage phasor of $\mathbf{V} = V_M \angle \delta$ and a current phasor of $\mathbf{I} = I_M \angle -\phi$ are plotted in Figure 3.5. The lead/lag relationship becomes visually apparent from the

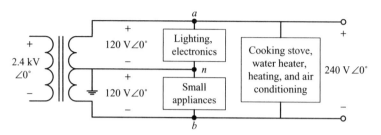

Figure 3.6. Residential single- (split-) phase power derived from center-tapped utility step-down transformer.

relative displacement of the two vectors; specifically, the voltage leads the current (or equivalently, the current lags the voltage) by $\delta + \phi$ in this case. A phasor diagram also provides a graphical analysis platform since phasors are readily added and subtracted on the complex plane.

Although the electric power network is predominantly a three-phase system, Chapter 1 related that a typical house is powered as a single-phase installation. Figure 3.6 exhibits the single- (split-) phase configuration for a North American residence that derives its power from a center-tapped utility step-down transformer. This ac power circuit provides 120 Vrms to two branches that serve smaller loads, and 240 Vrms supplies larger loads between nodes a and b. The split-phase approach aims to permit higher voltages to the larger loads for increased system efficiency while applying lower voltages to those loads that people are more likely to be near. The latter item is an electrical safety consideration.

3.3. IMPEDANCE

Ohm's law provides a linear relationship between voltage across and current through a circuit element:

$$\mathbf{V} = \mathbf{IZ}, \qquad (3.17)$$

where $\mathbf{Z}$ is the *impedance* of the element, having units of ohms, Ω. Unlike voltage and current, impedance is a truly complex quantity that can be expressed in rectangular, polar, or exponential notation. Impedance consists of a resistive term, R, and an imaginary term, the *reactance*, X. The reactance is a function of frequency, which generally makes the impedance a function of frequency:

$$\mathbf{Z}(\omega) = R + jX(\omega). \qquad (3.18)$$

The relation between the magnitude of the impedance, and resistance and reactance, can be visualized by the *impedance triangle* drawn in Figure 3.7. The polar representation of the impedance is:

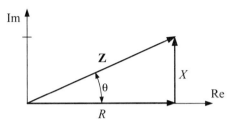

Figure 3.7. Impedance triangle.

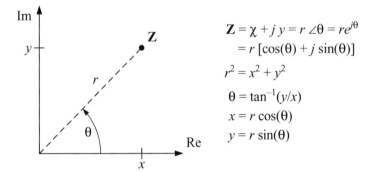

$$\mathbf{Z} = \chi + jy = r\angle\theta = re^{j\theta}$$
$$= r[\cos(\theta) + j\sin(\theta)]$$
$$r^2 = x^2 + y^2$$
$$\theta = \tan^{-1}(y/x)$$
$$x = r\cos(\theta)$$
$$y = r\sin(\theta)$$

Figure 3.8. Equivalency of rectangular, polar (phasor) and exponential complex number forms.

$$\mathbf{Z} = |\mathbf{Z}|\angle\theta = |\mathbf{Z}|e^{j\theta} = |\mathbf{Z}|[\cos(\theta) + j\sin(\theta)] = R + jX$$
$$\text{and} \quad |\mathbf{Z}| = \sqrt{R^2 + X^2} \quad \theta = \arctan\left(\frac{X}{R}\right). \tag{3.19}$$

General interrelationships between complex numbers in rectangular, polar, and exponential forms are summarized in Figure 3.8. Addition and subtraction of complex numbers is easiest using rectangular representations, whereas phasor (or polar) notation leads to simplified multiplication and division of complex quantities. Table 3.1 lists some general mathematical operations applied to phasors and complex numbers. For instance, using Ohm's law to compute a phasor voltage and current is readily accomplished using:

$$\mathbf{V} = \mathbf{IZ} = (I\angle\phi)(Z\angle\theta) = (IZ)\angle(\phi + \theta),$$
$$\mathbf{I} = \frac{\mathbf{V}}{\mathbf{Z}} = \frac{V\angle\delta}{Z\angle\theta} = \left(\frac{V}{Z}\right)\angle(\delta - \theta). \tag{3.20}$$

The reciprocal of the impedance is the *admittance*, $\mathbf{Y} = 1/\mathbf{Z}$, which is measured in siemens (S). The admittance can be expressed in terms of the conductance (G) and

TABLE 3.1. Mathematical Operations for Complex Numbers and Phasors

Operation	Procedure
Equivalence	$\mathbf{X} = a + jb = r\angle\theta = re^{j\theta} = r(\cos(\theta) + j\sin(\theta))$; $r^2 = a^2 + b^2$; $\tan(\theta) = b/a$
Addition	$\mathbf{X} + \mathbf{Y} = (a + jb) + (c + jd) = (a + c) + j(b + d)$
Subtraction	$\mathbf{X} - \mathbf{Y} = (a + jb) - (c + jd) = (a - c) + j(b - d)$
Negative value	$-\mathbf{X} = -(a - jb) = -(r\angle\theta) = r\angle\theta \pm 180° = re^{j(\theta \pm \pi)}$
Multiplication	$\mathbf{XY} = (r\angle\theta)(s\angle\phi) = rs\angle\theta + \phi = rse^{j(\theta+\phi)}$
Division	$\dfrac{\mathbf{X}}{\mathbf{Y}} = \dfrac{r\angle\theta}{s\angle\phi} = \dfrac{r}{s}\angle\theta - \phi = \dfrac{r}{s}e^{j(\theta-\phi)}$
Reciprocal	$\dfrac{1}{\mathbf{X}} = \dfrac{1}{r\angle\theta} = \dfrac{1}{r}\angle-\theta = \dfrac{e^{-j\theta}}{r}$
Square root	$\sqrt{\mathbf{X}} = \sqrt{r\angle\theta} = \sqrt{r}\angle\theta/2 = \sqrt{r}e^{j\theta/2}$
Complex conjugation	$\mathbf{X}^* = (a + jb)^* = a - jb = (r\angle\theta)^* = r\angle-\theta = re^{-j\theta}$
Conjugate multiplication	$(\mathbf{XY})^* = \mathbf{X}^*\mathbf{Y}^*$
Conjugate division	$(\mathbf{X}/\mathbf{Y})^* = \mathbf{X}^*/\mathbf{Y}^*$

TABLE 3.2. Passive Circuit Elements

Element	Impedance	Admittance
Capacitor (C)	$\mathbf{Z}_C = 1/(j\omega C) = -1/(\omega C)\angle 90°$	$\mathbf{Y}_C = j\omega C$
Inductor (L)	$\mathbf{Z}_L = j\omega L = \omega L \angle 90°$	$\mathbf{Y}_L = 1/(j\omega L)$
Resistor (R)	$\mathbf{Z}_R = R = R\angle 0°$	$\mathbf{Y}_R = 1/R$

susceptance (B) components via $\mathbf{Y} = G + jB$. The impedance and admittance relations for the resistor, capacitor, and inductor are given in Table 3.2. The reactance of an inductor and a capacitor is $X_{ind} = \omega L$ and $X_{cap} = -1/(\omega C)$, respectively.

EXAMPLE 3.2: Prove that, in general, resistance and conductance are not reciprocals of one another, and likewise, that reactance and susceptance are not reciprocals.

Solution: First, equate the impedance to the reciprocal of the admittance

$$\mathbf{Z} = \dfrac{1}{\mathbf{Y}},$$

$$R + jX = \dfrac{1}{G + jB}.$$

IMPEDANCE

Now multiply the right-hand side by unity in the form the complex conjugate of the admittance to obtain:

$$R + jX = \frac{1}{G + jB}\left(\frac{G - jB}{G - jB}\right) = \frac{G - jB}{G^2 + B^2}.$$

From equating the real and imaginary components, the relationships between the four parameters emerge:

$$R = \frac{G}{G^2 + B^2} \quad X = \frac{-B}{G^2 + B^2}.$$

Only in the case of a purely resistive network are the resistance and conductance reciprocals, that is, if $X = 0$, then $G = 1/R$.

EXAMPLE 3.3: Most calculators and computer software have built-in functions for performing the complex math associated with phasor and impedances; however, the results depend upon the user's proper utilization of such engineering tools. As a simple example, consider the conversion of the impedance $\mathbf{Z} = -3 - j4\Omega$ to a polar representation using the MATLAB command line. The magnitude is easily found using the absolute value function (*abs*):

```
>> Z = -3-4j;
>> abs (Z)
ans =
     5
```

However, an inexperienced user might simply use the *atan* function to find the angle:

```
>> atan(imag(Z)/real(Z)) *180/pi
ans =
    53.1301
```

But the previous answer is incorrect, in that the impedance lies in the third quadrant, not the first. MATLAB provides a different function, *angle*, for properly calculating the angle of a complex variable, but like *atan*, the resultant angle is in radians:

```
>> angle (Z) *180/pi
ans =
   -126.8699
```

Reducing the individual impedances in a circuit into a single equivalent impedance is the focus of the following subsections.

3.3.1. Series Connection

When M impedances are connected in series, the resulting equivalent impedance is:

$$\mathbf{Z}_e = \sum_{k=1}^{M} \mathbf{Z}_k. \qquad (3.21)$$

EXAMPLE 3.4: The equivalent impedance of a circuit with a resistor, inductor, and capacitor connected in series is:

$$\mathbf{Z}_e = \mathbf{Z}_R + \mathbf{Z}_L + \mathbf{Z}_C = R + j\omega L_{ind} + \frac{1}{j\omega C_{cap}}$$
$$= |\mathbf{Z}|e^{j\theta} = |\mathbf{Z}|[\cos(\theta) + j\sin(\theta)],$$

where

$$|\mathbf{Z}| = \sqrt{R^2 + \left(\omega L_{ind} - \frac{1}{\omega C_{cap}}\right)^2} \qquad \theta = \arctan\left(\frac{\omega L_{ind} - \frac{1}{\omega C_{cap}}}{R}\right).$$

The phase angle (θ) is positive in an inductive circuit and negative in a capacitive circuit.

3.3.2. Parallel Connection

Circuit elements that share the same two end nodes are said to be in *parallel*. When N impedances are connected in parallel (see Fig. 3.9), the resulting equivalent impedance is:

$$\frac{1}{\mathbf{Z}_e} = \sum_{k=1}^{N} \frac{1}{\mathbf{Z}_k}. \qquad (3.22)$$

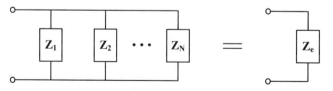

Figure 3.9. Parallel impedances with equivalent.

IMPEDANCE

In the case of two impedances connected in parallel, the previous equation can be simplified as follows:

$$Z_e = \frac{1}{\dfrac{1}{Z_1} + \dfrac{1}{Z_2}} = \frac{Z_1 Z_2}{Z_1 + Z_2}.$$

Alternatively, the admittance of a circuit with N components connected in parallel can be calculated by adding the admittances of the components together:

$$Y_e = \sum_{k=1}^{N} Y_k. \quad (3.23)$$

EXAMPLE 3.5: Determine the impedance of a circuit with a resistor, inductor, and capacitor connected in parallel.

Solution: The admittance of a network with resistance, inductance, and capacitance connected in parallel is:

$$Y = \frac{1}{R} + \frac{1}{j\omega L_{ind}} + j\omega C_{cap}.$$

The impedance of this circuit is:

$$Z = \frac{1}{Y} = \frac{1}{\dfrac{1}{R} + \dfrac{1}{j\omega L_{ind}} + j\omega C_{cap}} = \frac{1}{\dfrac{1}{R} + \dfrac{1}{jX_{ind}} + \dfrac{1}{jX_{cap}}}.$$

3.3.2.1. One-Line Diagram and System Efficiency.
A *one-line diagram* is often used to provide a concise representation of the network under study. The one-line or single-line diagram is a simplified schematic that portrays only the major components and major interconnection(s) between components. As an example, Figure 3.10 depicts a three-phase generator supplying a network through a transformer. The shorthand notation of the one-line diagram is then expanded into a circuit schematic for analysis. The one-line connection diagram of the system introduced in Figure 3.10 can be converted to the three-phase circuit given in Figure 3.11, which explicitly displays all the

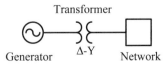

Figure 3.10. One-line diagram of a simple three-phase system.

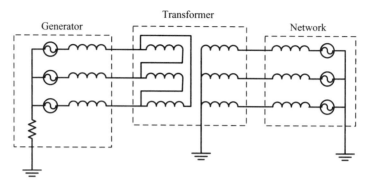

Figure 3.11. Three-phase connection diagram.

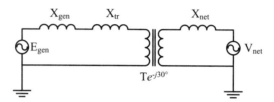

Figure 3.12. Single-phase equivalent circuit for balanced load.

components in each phase. Further, in the case of a balanced load, the single-phase equivalent ac circuit diagram, presented in Figure 3.12, could be used for analysis.

The performance of a power system can be quantified in terms of its efficiency and the voltage drop through the system. The *system efficiency* is the ratio of the output, in most cases the active load power, and the input, in most cases the supply real power. For the system of Figure 3.10, the system efficiency can be expressed as:

$$\text{efficiency} = \frac{P_{net}}{P_{gen}} = \frac{P_{gen} - P_{loss}}{P_{gen}}, \qquad (3.24)$$

where P_{gen} and P_{net} are the supplied generator power and consumed network power, respectively, and P_{loss} is the power loss in the system. Efficiency for an individual component, such as a transformer or a transmission line, can be defined in a similar manner.

EXAMPLE 3.6: A resistor, inductor, and capacitor are connected in series with a voltage source as depicted in Figure 3.13. Determine the real and reactive power absorbed or supplied by each circuit element.

Solution: The voltage across each of the passive circuit elements is found readily using Ohm's law:

IMPEDANCE

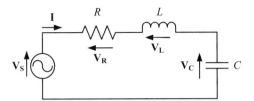

Figure 3.13. Simple series resistor-inductor-capacitor (RLC) circuit.

$$V_R = IZ_R = IR,$$
$$V_L = IZ_L = Ij\omega L,$$
$$V_C = IZ_C = I/(j\omega C).$$

From the equivalent impedance of the network, $Z_{eq} = Z_R + Z_L + Z_C$, the supply voltage is:

$$V_S = IZ_{eq} = I(R + j\omega L - j/(\omega C)).$$

The complex power in the R, L, and C components is obtained from:

$$S_R = V_R I^* = (IR)I^* = |I|^2 R,$$
$$S_L = V_L I^* = (Ij\omega L)I^* = j|I|^2 \omega L,$$
$$S_C = V_C I^* = (I/(j\omega C))I^* = -j|I|^2/(\omega C).$$

In the case of calculating the voltage source complex power, a negative sign must be inserted into Equation (3.12) to make V_S conform to the passive sign convention:

$$S_S = -V_S I^* = -I(R + j\omega L - j/(\omega C))I^*$$
$$= -|I|^2[R + j(\omega L - 1/(\omega C))].$$

The complex power results are summarized in tabular form where a positive or negative sign for the real and reactive power indicates whether power is absorbed or supplied, respectively, by the element:

Circuit Element	Real Power (P)	Reactive Power (Q)
Resistor	$|I|^2 R$ (absorbed)	0
Inductor	0	$|I|^2 \omega L$ (absorbed)
Capacitor	0	$-|I|^2/(\omega C)$ (supplied)
Voltage source	$-|I|^2 R$ (supplied)	$-|I|^2(\omega L - 1/(\omega C))$ (see later text)
Total	0	0

With respect to the reactive power of the voltage source:

- if $\omega L > 1/(\omega C)$, then the voltage source is supplying reactive power in the amount of $|I|^2(\omega L - 1/(\omega C))$; but
- if $\omega L < 1/(\omega C)$, then the voltage source is absorbing reactive power in the amount of $|I|^2(1/(\omega C) - \omega L)$.

In compliance with Equation (3.15), the totals of the real and reactive (and therefore complex) powers are zero for the circuit.

3.3.3. Impedance Examples

The practical aspects of series and parallel circuit impedance calculations are presented through numerical examples. The analysis of a simple circuit with complex impedances using a hand calculator is a complicated and lengthy procedure. The circuit analysis process can be simplified by using general-purpose calculation programs like MATLAB, Mathcad, or Mathematica. In this book, Mathcad and MATLAB are used.

This first example is performed using Mathcad and the second using MATLAB. Particular advantages of Mathcad are that the equations appear as the handwritten equations, and the program calculates and converts the units automatically. In Appendix A and Appendix B, respectively, short tutorials for Mathcad and MATLAB are presented. The tutorials present the basic knowledge necessary to get started.

EXAMPLE 3.7: Mathcad impedance examples

Series Connection of Impedances

Utilities can connect a capacitor in series with a transmission line to compensate for the line impedance. This reduces the voltage drop in cases of heavy load on the line. Figure 3.14 presents the single-line circuit diagram, and Figure 3.15 provides the equivalent circuit.

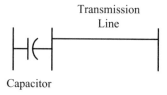

Figure 3.14. One-line diagram of a transmission line compensated by a capacitor connected in series.

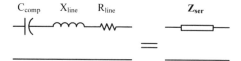

Figure 3.15. Equivalent circuit of a transmission line compensated with a series capacitor.

IMPEDANCE

The system frequency data are: $f := 60 \text{ Hz} \quad \omega := 2 \cdot \pi \cdot f$.
The transmission line resistance and reactance per mile are:

$$R_{mi} := 0.32 \frac{\Omega}{mi} \quad X_{mi} := 0.75 \frac{\Omega}{mi}.$$

The transmission line length is $d_{line} := 3 \text{ mi}$.
The transmission line resistance and reactance are:

$$R_{line} := R_{mi} \cdot d_{line} \quad R_{line} = 0.96 \, \Omega$$
$$X_{line} := X_{mi} \cdot d_{line} \quad X_{line} = 2.25 \, \Omega.$$

Note that Mathcad has two equal signs: the first equal sign (:=) is used when a variable is defined, and the second equal sign (=) is used to obtain the value of a variable. The capacitor value is $C_{comp} := 1572 \, \mu F$.
Using Table 3.2, the reactance of the capacitance is:

$$X_{line} := \frac{-1}{\omega \cdot C_{comp}} \quad X_{comp} = -1.69 \, \Omega.$$

The line resistance and reactance, and the reactance of the capacitor are connected in series. The equivalent impedance of the compensated line is:

$$Z_{ser} := j \cdot X_{comp} + R_{line} + j \cdot X_{line} \quad Z_{ser} = (0.960 + 0.563 j) \, \Omega.$$

With the compensation capacitor, the overall line reactance has been reduced from 2.25 to 0.56 Ω.

Parallel Connection of Impedances

Operators connect a capacitor in parallel with motors to improve the power factor. Figure 3.16 shows the single-line diagram for a motor and capacitor connected in parallel. The motor is represented by an inductance and resistance connected in parallel, as drawn in Figure 3.17. The motor values are:

$$R_{mot} := 20 \, \Omega \quad X_{mot} := 23 \, \Omega.$$

Figure 3.16. One-line diagram of a motor and capacitor connected in parallel.

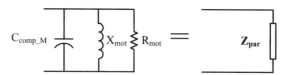

Figure 3.17. Equivalent circuit for a motor and capacitor connected in parallel.

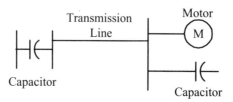

Figure 3.18. Compensated transmission line supplies a motor with capacitor load.

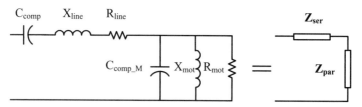

Figure 3.19. Equivalent circuit of the system in Figure 3.18.

The compensating capacitor is $C_{comp\_M} := 500$ μF.
The reactance of the capacitor is:

$$X_{comp\_M} := \frac{-1}{\omega \cdot C_{comp\_M}} \qquad X_{comp\_M} = -5.305 \; \Omega.$$

The three impedances are connected in parallel. The equivalent impedance is:

$$Z_{par} := \frac{1}{\dfrac{1}{R_{mot}} + \dfrac{1}{j \cdot X_{mot}} + \dfrac{1}{j \cdot X_{comp\_M}}} \qquad Z_{par} = -2.125 - 6.163j \; \Omega.$$

Combined Series and Parallel Connection of Impedances

Let us assume that the compensated transmission line supplies the motor and capacitor load previously described. Figure 3.18 shows the one-line circuit diagram of the combined system, and Figure 3.19 illustrates the equivalent circuit, which is a series combination of the compensated line and motor.

IMPEDANCE

The total impedance of the system can be obtained immediately by adding the two equivalent circuit impedances together:

$$Z_{equ} := Z_{ser} + Z_{par} \quad Z_{equ} = (3.085 - 5.600j) \, \Omega.$$

EXAMPLE 3.8: MATLAB impedance examples

The impedance calculation of the three series and/or parallel networks used earlier in the Mathcad examples is repeated here using MATLAB. For all three examples, the system cyclic and angular frequency must first be established as shown in **Impedance.m**:

```
%   Impedance.m
f = 60;                       % system frequency (Hz)
wfreq = 2 * pi * f;           % angular frequency (rad/sec)
```

Series Connection of Impedances

The equivalent impedance of the series circuit of Figure 3.15 is readily seen to be the summation of the individual component impedances:

$$Z_{ser} = Z_{line} + Z_{comp} = R_{line} + jX_{line} + \frac{1}{j\omega C_{comp}}.$$

The uncompensated line impedance formed by the series resistor and inductor is first computed:

```
%   1. Series connection of impedances
%   establish transmission line parameters
Rmi = 0.32;              % line resistance (ohm/mile)
Xmi = 0.75;              % line reactance (ohm/mile)
length = 3;              % line length (miles)
Zline = (Rmi + j*Xmi) * length   % uncompensated line
impedance (ohms)
```

Next, the reactance (impedance) of the compensating capacitor is calculated:

```
%   compute impedance of compensating capacitor for
%   transmission line
Ccomp = 1572e-6;              % farads
Zcomp = 1 / (j*wfreq*Ccomp)   % ohms
```

The capacitor impedance is added to the line impedance:

```
%   compute series equivalent impedance of compensated line
Zser = Zline + Zcomp
```

The results as computed by MATLAB are:

```
Zline =
    0.9600 + 2.2500i
Zcomp =
         0 - 1.6874i
Zser =
    0.9600 + 0.5626i
```

Parallel Connection of Impedances

The equivalent impedance of the parallel circuit introduced in Figure 3.17 is also calculated in a straightforward manner:

$$\frac{1}{Z_{par}} = \frac{1}{R_{mot}} + \frac{1}{jX_{mot}} + j\omega C_{comp\_M}.$$

After setting the motor resistance and reactance, the impedance of the parallel capacitor is computed:

```
%    2. Parallel connection of impedances
%    establish motor resistance and reactance
Rmot = 20;             % ohms
Xmot = 23;             % ohms
%    compute capacitor reactance
CcompM = 500e-6;                        % farads
ZcompM = 1 / (j*wfreq*CcompM)           % ohms
```

The equivalent parallel impedance of the capacitor, inductor and resistor is calculated:

```
%    combine the parallel motor elements and capacitor
Zpar = 1 / (1/Rmot + 1/(j*Xmot) + 1/ZcompM)     % ohms
```

The MATLAB computed results are:

```
ZcompM =
         0 - 5.3052i

Zpar =
    2.1249 - 6.1631i
```

Combined Series and Parallel Connection of Impedances

The overall equivalent impedance from combining the compensated transmission line and motor in series (see Fig. 3.19) is simply the sum of the their impedances:

```
% 3. Combined series and parallel connection of impedances
Zequ = Zser + Zpar      % ohms
```

The following MATLAB result matches the answer found using Mathcad:

```
Zequ =
    3.0849 - 5.6005i
```

3.4. LOADS

The loads in a power network range in size from small appliances and consumer electronics to large motors and industrial furnaces. In general, each of these loads may be represented by an impedance as depicted in Figure 3.20. If $\mathbf{V}_{load}$ and $\mathbf{I}_{load}$ are the rms voltage and current at the load impedance $\mathbf{Z}$, then:

$$\mathbf{Z} = \frac{\mathbf{V}_{load}}{\mathbf{I}_{load}} = \frac{V_{load}\angle\delta}{I_{load}\angle\phi} = \frac{V_{load}}{I_{load}}\angle\delta-\phi = |\mathbf{Z}|\angle\theta = R + jX. \quad (3.25)$$

The load phase angle $\theta = \delta - \phi$ then describes the amount by which the voltage leads the current. In particular, for an inductive load, the load phase angle is positive since:

$$\mathbf{Z}_{ind} = R + j\omega L = \sqrt{R^2 + (\omega L)^2}\angle\tan^{-1}\left(\frac{\omega L}{R}\right), \quad (3.26)$$

and the current is lagging the voltage by θ. In contrast, for a capacitive load, the load phase angle is negative because:

$$\mathbf{Z}_{cap} = R - \frac{j}{\omega C} = \sqrt{R^2 + \frac{1}{(\omega C)^2}}\angle\tan^{-1}\left(\frac{-1}{\omega RC}\right). \quad (3.27)$$

This means that in the latter case, the current is actually leading the voltage by $\phi - \delta$.

Using Equation (3.14), the complex power to the load is:

$$\mathbf{S}_{load} = \mathbf{V}_{load}\mathbf{I}^*_{load} = P_{load} + jQ_{load}. \quad (3.28)$$

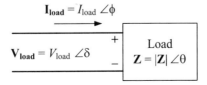

Figure 3.20. General load.

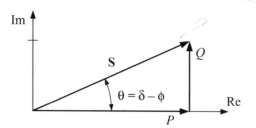

Figure 3.21. Power triangle.

Using Ohm's law, the complex power can be expressed in terms of the load impedance and either the load rms current or voltage:

$$\mathbf{S}_{load} = (\mathbf{I}_{load}\mathbf{Z})\mathbf{I}^*_{load} = I^2_{load}\mathbf{Z} = I^2_{load}(R+jX), \tag{3.29}$$

$$\mathbf{S}_{load} = \mathbf{V}_{load}\left(\frac{\mathbf{V}_{load}}{\mathbf{Z}}\right)^* = \frac{V^2_{load}}{\mathbf{Z}^*} = V^2_{load}\mathbf{Y}^* = V^2_{load}(G-jB). \tag{3.30}$$

Since the load is connected to a near-constant voltage, the load current becomes the dependent variable. The load current can be computed directly from the load voltage and impedance using Ohm's law or from the complex power formulae previously mentioned.

The complex power is also indicative of whether the load is inductive or capacitive since:

$$\begin{aligned}\mathbf{S}_{load} &= \mathbf{V}_{load}\mathbf{I}^*_{load} = I^2_{load}(R+jX) = P_{load} + jQ_{load}, \\ P_{load} &= I^2_{load}R, \\ Q_{load} &= I^2_{load}X.\end{aligned} \tag{3.31}$$

Therefore,

- the reactive power is *positive in an inductive load circuit*, where the load phase angle is positive and the current is lagging the voltage; and
- the reactive power is *negative in a capacitive load circuit*, where the load phase angle is negative and the current is leading the voltage.

The relation between the magnitude of the complex or apparent power and the real power and reactive power can be visualized by the *power triangle* exhibited in Figure 3.21.

EXAMPLE 3.9: The real power and apparent power of a compact fluorescent lamp (CFL) is measured as 25 W and 52 VA, respectively, at 120 Vrms with current lagging voltage. Determine the impedance of the CFL.

LOADS

Solution: From Equation (3.14), we note that:

$$P = V_{rms} I_{rms} \cos(\delta - \phi),$$

from which the load phase angle can be determined:

$$\theta = \delta - \phi = \arccos\left(\frac{P}{V_{rms} I_{rms}}\right) = \arccos\left(\frac{25 \text{ W}}{52 \text{ VA}}\right) = 61.3°.$$

Therefore, the complex power at the load is:

$$\mathbf{S} = V_{rms} I_{rms} \angle \theta = 52 \text{ VA} \angle 61.3°.$$

This type of load is known to be inductive, such that with the use of Equation (3.30), the CFL impedance is found to be:

$$\mathbf{Z} = \left(\frac{V_{load}^2}{\mathbf{S}}\right)^* = \left[\frac{(120 \text{ V})^2}{52 \text{ VA} \angle 61.3°}\right]^* = 276.9 \text{ }\Omega \angle 61.3° = 133 + j243 \text{ }\Omega.$$

Unlike incandescent light bulbs, CFLs may have a sizable reactive component.

3.4.1. Power Factor

The *power factor*, *pf*, is defined as the ratio of average power to apparent power at a specified point in the system:

$$pf = \frac{P}{V_{rms} I_{rms}} = \frac{P}{|\mathbf{S}|} = \cos(\delta - \phi). \quad (3.32)$$

The *power factor angle* is $\theta = \delta - \phi$, and its relationship to some other parameters is summarized in Table 3.3. In this case, the power factor angle is equivalent to the load

TABLE 3.3. Power Factor Angle Relations

Power Factor Angle	Current–Voltage (*pf*)	Load Equivalent	Reactive Power
$\theta = -90°$	$i(t)$ **leading** $v(t)$	Purely capacitive	$Q = -V_{load} I_{load}$
$-90° < \theta < 0°$		Equivalent *RC*	$Q < 0$
$\theta = 0°$	$i(t)$ **in phase** with $v(t)$	Purely resistive	$Q = 0$
$0° < \theta < 90°$	$i(t)$ **lagging** $v(t)$	Equivalent *RL*	$Q > 0$
$\theta = 90°$		Purely inductive	$Q = V_{load} I_{load}$

angle. Because the cosine of an angle between −90° and +90° ranges from zero to unity, it is necessary to supplement the actual *pf* value ($0 \leq pf \leq 1$) with a designation as to whether the current is *leading* or *lagging* the voltage, as illustrated in Table 3.3. Thus, in a purely resistive load, the power factor is unity ($pf = 1$), whereas in a purely reactive (capacitive or inductive) load, the power factor is zero ($pf = 0$).

Industrial, commercial, and domestic loads are typically motor and lighting loads. Motors are a major component of these loads. Consequently, most loads are inductive and the power factor lagging. However, the use of compensating capacitors may produce a leading power factor at night when the loads are light. Another source of leading power factors are lightly loaded, long transmission lines, where the line capacitance overcompensates for the load and line inductance.

In an electric power network, the typical load consumes constant power at a constant power factor if the load voltage is within a ±5% limit. The quality of an electrical load is judged by the power factor. A low power factor requires more rms current for the same load power, which results in greater utility transmission losses in the power lines. As an example, the power companies require most customers to maintain a power factor greater than 0.8. The definition of power factor is fortuitous since the real power can be measured readily with a wattmeter and the rms current and voltage are obtained with conventional instruments.

If the power factor is unity, then the power is simply the product of the rms current and voltage:

$$P = I_{rms} V_{rms} = \frac{I_M V_M}{2}. \tag{3.33}$$

The previous formula demonstrates one reason for utilizing the effective (or rms) values, since they allow us to write an equation for the average power like those used in direct current (dc) circuits (i.e., $P = I^2 R$); in fact, that relation is ultimately the basis for defining the rms value. For a resistor, Ohm's law can be combined with the prevoius formula to find the power dissipated in a resistor:

$$P = I_{rms}^2 R = \frac{V_{rms}^2}{R}. \tag{3.34}$$

Note that since the resistance is always positive, then the resultant power value in the previous expression must also always be positive. Recall that for $P > 0$ that power is absorbed by an element, which means that the resistor always absorbs power.

A load may be simply specified in terms of its real power and power factor. For such circumstances, a relationship between those parameters and the load current is useful. The load current calculation requires the load voltage. The first step is the calculation of the load phase angle from the power factor:

$$\theta_{load} = \pm \arccos(pf_{load}). \tag{3.35}$$

LOADS

TABLE 3.4. Sign Convention of Exponential

Power Factor	Current, **I**	Complex Power, **S**
Leading (capacitive)	$|I|e^{+j\theta}$ positive exponent	$|S|e^{-j\theta}$ negative exponent
Lagging (inductive)	$|I|e^{-j\theta}$ negative exponent	$|S|e^{+j\theta}$ positive exponent

Where $\theta = \arccos(pf)$ is the load power factor phase angle.

The phase angle (of the complex power) is *positive* if the load power factor is *lagging*, and *negative* if the power factor is *leading*. The complex power is:

$$\mathbf{S}_{\text{load}} = \mathbf{V}_{\text{load}} \mathbf{I}^*_{\text{load}} = V_{\text{load}} I_{\text{load}} e^{j\theta_{\text{load}}}$$
$$= |\mathbf{S}_{\text{load}}| e^{j\theta_{\text{load}}} = \frac{P_{\text{load}}}{pf_{\text{load}}} e^{j\theta_{\text{load}}}. \quad (3.36)$$

The real part of the complex power is the real power; the imaginary part is the reactive power. From Table 3.3, the reactive power is *positive* if the power factor is *lagging (inductive load)* and *negative* if the power factor is *leading (capacitive load)*.

The load current is the conjugate of the complex power and load voltage ratio:

$$\mathbf{I}_{\text{load}} = \left[\frac{\mathbf{S}_{\text{load}}}{\mathbf{V}_{\text{load}}}\right]^*, \quad (3.37)$$

The three equations previously cited can be combined to obtain an expression for the direct calculation of the load current. The result is:

$$\mathbf{I}_{\text{load}} = \frac{P_{\text{load}}}{\mathbf{V}_{\text{load}} pf_{\text{load}}} e^{\mp j \arccos(pf_{\text{load}})}. \quad (3.38)$$

In this equation the *plus* (+) *sign* is used if the power factor is *leading* and the *minus* (−) *sign* is used if *pf* is *lagging*. In other words, the reactive component of the current is negative if the power factor is lagging (the load is inductive). The reactive component of the current is positive if the power factor is leading (the load is capacitive). The sign of the exponential is summarized in Table 3.4.

EXAMPLE 3.10: MATLAB power factor calculation

The electrical network of a utility supplies an industrial load of 100 kW, 60 Hz, and 480 V(rms) through a feeder (see Fig. 3.22). The feeder impedance is $0.15 + j0.6$ Ω. Vary the load *pf* from 0.6 to 0.9 lagging, and plot the supply real power versus *pf* to determine the generation requirements (i.e., the "cost") for the utility to serve this industrial customer.

Figure 3.22. One-line diagram for load served via transmission or feeder line.

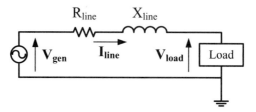

Figure 3.23. Circuit schematic for load served via transmission or feeder line.

Solution: Referring to Figure 3.23 and Equation (3.32), the rms current through the line to the load can be calculated using:

$$I_{line} = \frac{P_{load}}{pf\, V_{load}}.$$

The real power supplied by the utility is that necessary to serve the load as well as the real power loss in the transmission line:

$$P_{gen} = P_{load} + R_{line} I_{line}^2.$$

Note that this calculation is independent of the system frequency and whether the load power factor is leading or lagging.

Begin the MATLAB script file by establishing the system circuit data:

```
%   PowerFactor.m

%   Establish the circuit parameters
Pload = 100e3;        % watts
Vload = 480;          % volts-rms

Rline = 0.15;         % ohms
```

Next, we vary the power factor as requested using a step size of 0.001, which will eventually generate about 300 points to be plotted:

```
%   Vary the power factor from 0.6 to 0.9, with steps of 0.001
pf = 0.6 : 0.001 : 0.9;
```

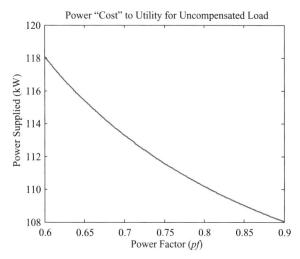

Figure 3.24. MATLAB plot of utility-supplied power as a function of power factor.

The line–load current and power supplied by the utility are then computed using the previously mentioned equations. Note in the following the use of the MATLAB "./" and ".^" element-by-element division and power operators, respectively:

```
% Compute a vector of line-load currents for the pf values
Iline = Pload ./ (pf * Vload);

% Compute the power supplied by utility for each pf value
Pgen = Pload + Rline * Iline.^2;
```

MATLAB can then be used to easily plot the results, which are given in Figure 3.24:

```
% Plot the power supplied v. the power factor
plot(pf, Pgen/1000, 'LineWidth',2.5);
set(gca,'fontname','Times','fontsize',12);
xlabel('Power Factor (pf)');
ylabel('Power Supplied (kW)');
title('Power "Cost" to Utility for Uncompensated Load');
xlim([pf(1) pf(length(pf))]);
```

The graph reveals that if the load has a power factor of 0.6, then the utility must generate about 118 kW to supply the 100 kW load. Whereas if the load power factor is raised to 0.9, then the utility needs only generate 108 kW to supply the same load. Power factor correction is the subject of Example 3.14.

3.4.2. Voltage Regulation

The power engineer is also concerned with the reduction in voltage through the system, and ultimately the voltage provided to the customer. Consider the one-line diagram of Figure 3.22, which depicts a generator supplying a load through a transmission line. The equivalent single-phase circuit is provided in Figure 3.23. The *voltage regulation* is defined based upon the input voltage (V_{gen} in the case of the network of Fig. 3.23) being held constant; the voltage regulation is the difference between the magnitudes of the voltage at the load terminals for the no-load ($V_{no\text{-}load}$) and (full) load conditions divided by the load voltage (V_{load}) magnitude:

$$\text{Voltage Regulation} = \frac{|V_{no\text{-}load}| - |V_{load}|}{|V_{load}|} \times 100\%. \tag{3.39}$$

In the case of multiple loads or a looped network, the supply voltage is not equal to the no-load voltage. Consequently, the no-load voltage must be calculated using the Kirchhoff equations when only the load under study is removed.

The no-load voltage is equal to the generator supply voltage (V_{gen}) if the load current is zero and line capacitance is small. The line capacitance can generally be neglected for distribution lines and transmission lines less than 50 mi in length. In this case, Figure 3.23 depicts a single load; if this load is switched off, the no-load current is zero and the load voltage is equal to the supply voltage. Hence, the formula for the voltage regulation can be expressed simply as the difference between the magnitudes of the supply (sending end) and load (receiving end) voltages divided by the load voltage magnitude:

$$\text{Voltage Regulation} = \frac{|V_{gen}| - |V_{load}|}{|V_{load}|} \times 100\%. \tag{3.40}$$

This latter expression for the voltage regulation is frequently applicable in the examples of this book, and requires only one circuit analysis, in contrast to Equation (3.39), which requires analyses at both loaded and unloaded conditions.

In an electric network or transmission system, the voltage must be between the standard specified limits, which is close to ±10%. In this case, the maximum voltage drop can be calculated for evaluation of system performance. The *voltage drop* is equivalent to Equation (3.40).

3.5. BASIC LAWS AND CIRCUIT ANALYSIS TECHNIQUES

Kirchhoff's circuit laws and their extension to formalized circuit analysis techniques such as mesh and nodal analysis are addressed in this section. As the reader is assumed to already be skilled in general network analysis, other approaches such as the application of the superposition theorem and source transformation are not detailed here,

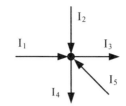

Figure 3.25. Illustration of KCL.

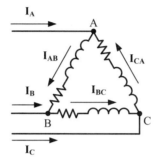

Figure 3.26. Example network for applying KCL nodal equations.

even though they are valid for sinusoidal steady-state analyses. Most of the circuits analyzed in this textbook are linear, meaning that the network output is directly proportional to its input. A linear circuit is exclusively composed of independent sources, linear-dependent sources and linear circuit elements, having a linear voltage–current relationship.

3.5.1. Kirchhoff's Current Law

Kirchhoff's current law (KCL) states that the sum of the currents entering any node is zero. The other KCL formulation is that the current flowing toward the node is equal to the current flowing away from the node (see Fig. 3.25). For N currents entering a node, KCL is written as:

$$\sum_{k=1}^{N} i_k(t) = 0 \quad \text{or} \quad \sum_{k=1}^{N} I_k = 0. \tag{3.41}$$

A network with three nodes is presented in Figure 3.26. The nodal (KCL) equations for nodes **A**, **B**, and **C** are:

$$\text{node A:} \quad I_A + I_{CA} - I_{AB} = 0,$$
$$\text{node B:} \quad I_B + I_{AB} - I_{BC} = 0,$$
$$\text{node C:} \quad I_C + I_{BC} - I_{CA} = 0.$$

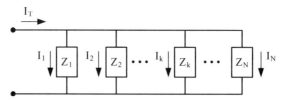

Figure 3.27. Network for applying current division.

3.5.1.1. Current Division. The combination of KCL and Ohm's law to circuit elements that are in parallel results in the well-known *current divider* equation. Referring to Figure 3.27, the current ($\mathbf{I_k}$) through the parallel impedance $\mathbf{Z_k}$ is:

$$\mathbf{I_k} = \mathbf{I_T} \frac{\mathbf{Z_{par}}}{\mathbf{Z_k}}, \qquad (3.42)$$

where $\mathbf{I_T}$ is the total current through all the parallel elements and $\mathbf{Z_{par}}$ is the equivalent impedance for the N parallel elements. For the special case of only two elements in parallel ($\mathbf{Z_1}$ and $\mathbf{Z_2}$), the previous formula reduces to:

$$\mathbf{I_1} = \mathbf{I_T} \frac{\mathbf{Z_2}}{\mathbf{Z_1} + \mathbf{Z_2}}, \qquad (3.43)$$

where this expression is normally read as the current through one branch is the total current times the impedance in the other branch divided by the sum of the two branch impedances.

3.5.1.2. Nodal Analysis. *Nodal analysis* systematically applies KCL to the nodes of a circuit in order to solve for the voltages at each node. There are three basic steps followed in nodal analysis:

1. Choose a reference (ground or neutral) node where the voltage is deemed zero, and then assign (label) unique voltages to the other nodes.
2. Apply KCL to each node other than the reference node while expressing the currents in terms of the node voltages using Ohm's law ($\mathbf{I} = \Delta \mathbf{V}/\mathbf{Z}$).
3. Solve the resulting system of KCL equations for the nodal voltages.

EXAMPLE 3.11: Nodal analysis example

The example here presents the practical application of the KCL equations method, referred to as nodal analysis. First, a solution is obtained by hand. This is followed by a computer solution using Mathcad.

BASIC LAWS AND CIRCUIT ANALYSIS TECHNIQUES 119

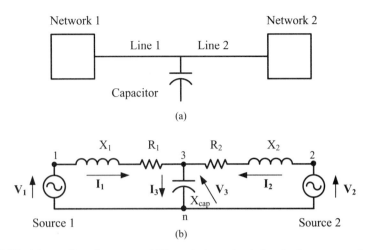

Figure 3.28. (a) One-line diagram and (b) equivalent circuit for the interconnection of two electric distribution networks.

Two electric distribution networks are interconnected through a transmission line. Figure 3.28a displays the one-line diagram (arrangement) of the system. A shunt capacitor is placed in the middle of the line to improve voltage regulation. Figure 3.28b provides the equivalent circuit diagram. Each of the distribution systems is represented by a voltage source. The magnitudes of the two voltage sources are equal, but there is a 60° phase difference between them. The voltage of Source 1 is the reference voltage, with a zero-degree (0°) phase angle. Source 2 has a phase shift of 60° with respect to Source 1, that is, $V_2 = V_1 \angle -60°$. The transmission line is represented with a resistance and reactance connected in series. A capacitance is placed at the middle of the line. This requires the division of the line into two parts: Line 1 and Line 2. It is desired to find the voltage at node 3 and the three currents entering or leaving node 3.

Method 1: Hand Solution
The hand solution follows the nodal analysis procedure as previously enumerated.

Step 1. The nodes have already been labeled in Figure 3.28b. The nodal voltages for nodes 1 and 2 are simply the source voltages since each source is directly connected to the neutral (ground) node. The remaining node 3 is assigned the voltage, V_3.

Step 2. Since the voltages at nodes 1 and 2 are already known, it is only necessary to write a single KCL expression, specifically at node 3:

$$I_1 - I_3 + I_2 = 0.$$

For each current in the KCL expression we now substitute an equivalent Ohm's law expression in terms of the node voltages:

$$\frac{V_1 - V_3}{R_1 + jX_1} - \frac{V_3}{jX_{cap}} + \frac{V_2 - V_3}{R_2 + jX_2} = 0.$$

Step 3. Here there is a single equation to be solved. Algebraic manipulation leads to an expression for the only unknown nodal voltage, that of V_3:

$$V_3 = \frac{\dfrac{V_1}{R_1 + jX_1} + \dfrac{V_2}{R_2 + jX_2}}{\dfrac{1}{R_1 + jX_1} + \dfrac{1}{jX_{cap}} + \dfrac{1}{R_2 + jX_2}}.$$

Substituting in the circuit parameter values yields a value for V_3. The values of the nodal voltages can then be used to find the three currents: I_1, I_2, and I_3:

$$I_1 = \frac{V_1 - V_3}{R_1 + jX_1} \quad I_2 = \frac{V_2 - V_3}{R_2 + jX_2} \quad I_3 = \frac{V_3}{jX_{cap}}.$$

Method 2: Mathcad Solution

Here we again solve the circuit of Figure 3.28, but using computer-based nodal analyses. Two approaches are taken using Mathcad: (1) a process following the earlier hand solution and (2) a procedure using the Mathcad *Find* equation solver. The reader is encouraged to utilize his/her computer and follow the calculation using the Mathcad program.

> **Mathcad Hint**
> Electrical engineers prefer to use a lowercase *j* to denote the imaginary number rather than a lowercase *i* to avoid confusion with current *i(t)*. However, Mathcad defaults to using the letter *i* for the imaginary number. This can be changed to *j* in the **Display Options** tab of the **Result Format** dialog box. In this book, we have chosen to follow the standard convention and use *j* for the imaginary unit.

The physical circuit data are first defined. The $-60°$ phase angle for voltage source V_2 can be incorporated using the complex exponential, that is, $\angle\theta = e^{j\theta}$:

$$V_1 := 7.2 \text{ kV} \quad V_2 := 7.2 \cdot e^{-j \cdot 60 \text{ deg}} \cdot \text{kV} \quad \omega = 2 \cdot \pi \cdot 60 \text{ Hz}$$
$$X_1 := 11\,\Omega \quad X_2 := X_1$$
$$R_1 := 4\,\Omega \quad R_2 := R_1$$
$$C_{cap} := 100\,\mu F \quad X_{cap} := \frac{-1}{\omega \cdot C_{cap}}.$$

BASIC LAWS AND CIRCUIT ANALYSIS TECHNIQUES   121

Immediate results are computed for:

$$X_{cap} = -26.53\ \Omega \quad |V_2| = 7.2\ \text{kV} \quad \arg(V_2) = -60\ \text{deg}.$$

Mathcad has an equation solver that simplifies the circuit analysis. However, for comparison we will first solve this problem using the equations derived in the previously mentioned hand calculation. After this, we use the Mathcad equation solver to demonstrate the advantages of the computer use for elementary circuit calculations.

Typing and dimensional errors may prohibit Mathcad from performing the calculations. To reduce such errors, we display the results of intermediate calculations as we progress through the problem. To do so, we select a guess (seed) value for the variable: this causes Mathcad to compute a numerical result as soon as each new equation is entered. The reasonableness of the numerical values obtained at each step helps to assure us that the equations are solvable and that we have not erred in typing the formula. In order to verify the equations in this problem, we select a guess value of $V_3 := 7\ \text{kV}$ with an implicit phase angle of zero.

Approach #1
The nodal current equation (KCL) for node 3 is:

$$I_1 - I_3 + I_2 = 0.$$

The currents can be calculated by dividing the voltage difference by the impedance, that is, Ohm's law. The voltage difference between node 1 and 3 is $V_1 - V_3$. The impedance is $R_1 + jX_1$. The current is:

$$I_1 := \frac{V_1 - V_3}{R_1 + jX_1} \quad |I_1| = 17.1\ \text{A} \quad \arg(I_1) = -70.0\ \text{deg}.$$

Current I_2 is calculated similarly. The result is:

$$I_2 := \frac{V_2 - V_3}{R_2 + jX_2} \quad |I_2| = 606.8\ \text{A} \quad \arg(I_2) = 171.4\ \text{deg}.$$

The capacitance current I_3 is the voltage V_3 divided by the capacitive impedance:

$$I_3 := \frac{V_3}{jX_{cap}} \quad |I_3| = 263.9\ \text{A} \quad \arg(I_3) = 90.0\ \text{deg}.$$

The numerical values help to assure the validity of the equations, but these are not the actual current values since V_3 is a guess value. The substitution of the current values in the nodal equation results in the following equation:

$$\frac{V_1 - V_3}{R_1 + jX_1} - \frac{V_3}{jX_{cap}} + \frac{V_2 - V_3}{R_2 + jX_2} = 0,$$

where V_3 is the unknown in this equation. The rearrangement of the equation results in:

$$\frac{V_1}{R_1 + jX_1} - V_3 \cdot \left[\frac{1}{jX_{cap}} + \frac{1}{(R_1 + jX_1)} + \frac{1}{(R_2 + jX_2)} \right] + \frac{V_2}{R_2 + jX_2} = 0.$$

The equation is solved for V_3. The result is:

$$V_3 := \frac{\dfrac{V_1}{R_1 + jX_1} + \dfrac{V_2}{R_2 + jX_2}}{\left[\dfrac{1}{jX_{cap}} + \dfrac{1}{(R_1 + jX_1)} + \dfrac{1}{(R_2 + jX_2)} \right]}$$

$V_3 = (6.38 - 4.54j) \cdot kV \quad V_3 = 7.831 \cdot kV \quad \arg(V_3) = -35.4 \cdot \deg.$

Approach #2
Another way to solve the nodal equation is by using the Mathcad *Find* equation solver. The Mathcad equation solvers are more fully explained in Appendix A. The use of the *Find* equation solver requires a guess value for V_3, which was selected before:

Given

$$\frac{V_1 - V_3}{R_1 + jX_1} - \frac{V_3}{jX_{cap}} + \frac{V_2 - V_3}{R_2 + jX_2} = 0,$$

the equation-solver immediately computes the voltage:

$$V_{3F} := \text{Find}(V_3) \quad V_{3F} = (6.38 - 4.54j) \, kV.$$

We can see that the two approaches result in the same voltage, V_3. The magnitude operator (| |) and *arg* function were utilized earlier to change the complex number expression for V_3 into a phasor quantity. The corresponding values of the currents are:

$$I_1 := \frac{V_1 - V_3}{R_1 + jX_1} \quad |I_1| = 394.2 \, A \quad \arg(I_1) = 9.8 \cdot \deg$$

$$I_2 := \frac{V_2 - V_3}{R_2 + jX_2} \quad |I_2| = 278.2 \, A \quad \arg(I_2) = 141.4 \cdot \deg$$

$$I_3 := \frac{V_3}{jX_{cap}} \quad |I_3| = 295.2 \, A \quad \arg(I_3) = 54.6 \cdot \deg$$

3.5.2. Kirchhoff's Voltage Law

Kirchhoff's voltage law (KVL) states that the sum of the voltages around a loop is zero. The other formulation is that the supply voltage is equal with the load voltages or voltage drops in a closed loop. KVL is written as:

$$\sum_k v_k(t) = 0 \quad or \quad \sum_k \mathbf{V}_k = 0. \tag{3.44}$$

It is imperative that the correct voltage polarity be used in the earlier formulae as the loop is traversed. Often the engineer prefers to begin at a voltage source, which is taken to be a voltage increase (positive voltage), and then follow the loop using that source as the reference polarity.

EXAMPLE 3.12: Figure 3.29 presents a simple circuit to demonstrate application of KVL loop equations. In this circuit, the voltage source $\mathbf{V}_S$ drives the current $\mathbf{I}$ through the reactance and resistance. The current produces a voltage drop $\mathbf{V}_R$ across the resistance and a voltage drop $\mathbf{V}_X$ across the inductance. KVL states that the sum of the three voltages is zero. The loop equation is:

$$\mathbf{V}_S - \mathbf{V}_X - \mathbf{V}_R = 0 \quad or \quad \mathbf{V}_S = \mathbf{I}R + j X,$$

where $\mathbf{V}_R = \mathbf{I}R$ and $\mathbf{V}_X = \mathbf{I}jX$ are the voltage drops across the resistor and inductor, respectively.

3.5.2.1. Voltage Division.
The combination of KVL and Ohm's law to circuit elements that are in series and have the same current flowing through them (i.e., in a single loop) results in the well-known *voltage divider* equation. Referring to Figure 3.30, the voltage ($\mathbf{V}_k$) across the series impedance $\mathbf{Z}_k$ is:

$$\mathbf{V}_k = \mathbf{V}_T \frac{\mathbf{Z}_k}{\mathbf{Z}_{ser}}, \tag{3.45}$$

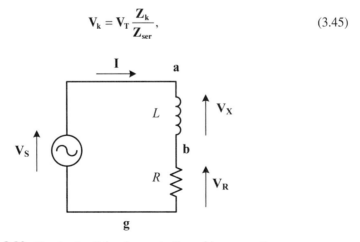

Figure 3.29. Simple circuit for demonstration of loop equation.

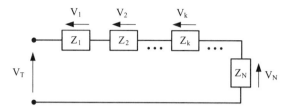

Figure 3.30. Network for applying voltage division.

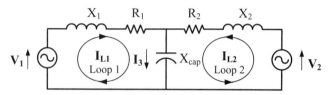

Figure 3.31. Circuit for using loop equations.

where V_T is the total voltage across all the series impedance elements and Z_{ser} is the equivalent impedance for the N series elements.

3.5.2.2. Loop Analysis. *Loop* (or *mesh*) *analysis* systematically applies KVL to the loops of a circuit in order to solve for the currents around each loop. Similar to nodal analysis, there are three basic steps followed in loop analysis:

1. Identify the loops and assign a current to each mesh or loop.
2. Apply KVL around each loop to obtain an equation in terms of the loop currents. Note that a loop/mesh current is not a branch current; specifically, a branch current may be made up of one or more loop currents.
3. Solve the resulting system of KVL equations for the loop currents.

EXAMPLE 3.13: Loop analysis example

The following examples repeat the analysis of the circuit of Figure 3.28 except that loop analysis is used here. Similar to Example 3.11, a solution is first obtained by hand, and then Mathcad is used.

Method 1: Hand Solution
The hand solution follows the loop analysis procedure previously described.

Step 1. Figure 3.31 shows that this circuit contains two independent meshes. We assume two loop currents: the current in Loop 1 is I_{L1} and the current in Loop 2 is I_{L2}. The directions of the loop currents are marked on Figure 3.31. The chosen directions are advantageous since the loop current directions match the polarity of the voltage sources.

Step 2. KVL is first applied to the left-hand loop (Loop 1):

$$\mathbf{V}_1 - \mathbf{I}_{L1}(R_1 + jX_1) - (\mathbf{I}_{L1} + \mathbf{I}_{L2})jX_{cap} = 0.$$

A KVL expression is next written for Loop 2:

$$\mathbf{V}_2 - \mathbf{I}_{L2}(R_2 + jX_2) - (\mathbf{I}_{L2} + \mathbf{I}_{L1})jX_{cap} = 0.$$

Step 3. In this example, the application of loop analysis results in two coupled equations, which may be written in a matrix formulation:

$$\begin{bmatrix} R_1 + j(X_1 + X_{cap}) & jX_{cap} \\ jX_{cap} & R_2 + j(X_2 + X_{cap}) \end{bmatrix} \begin{bmatrix} \mathbf{I}_{L1} \\ \mathbf{I}_{L2} \end{bmatrix} = \begin{bmatrix} \mathbf{V}_1 \\ \mathbf{V}_2 \end{bmatrix}.$$

This matrix equation is of the form $\underline{\underline{Z}}\,\underline{I} = \underline{V}$, which has a solution of $\underline{I} = \underline{\underline{Z}}^{-1}\underline{V}$. For a 2 × 2 matrix, the inverse may be determined using:

$$\begin{bmatrix} a & b \\ c & d \end{bmatrix}^{-1} = \frac{1}{ad - bc}\begin{bmatrix} d & -b \\ -c & a \end{bmatrix}. \tag{3.46}$$

Method 2: Mathcad Solution
Here we solve the circuit of Figure 3.31 using a Mathcad-based loop analysis. Again, two approaches are taken using Mathcad: (1) following the earlier hand solution and (2) using the Mathcad *Find* equation solver.

Using KVL, the loop equations are:

$$\mathbf{V}_1 - \mathbf{I}_{L1} \cdot (R_1 + jX_1 + jX_{cap}) - \mathbf{I}_{L2} \cdot jX_{cap} = 0$$
$$\mathbf{V}_2 - \mathbf{I}_{L2} \cdot (R_2 + jX_2 + jX_{cap}) - \mathbf{I}_{L1} \cdot jX_{cap} = 0$$

The verification of the validity of the equations requires the selection of guess values, which are:

$$\mathbf{I}_{L1} := 300 \text{ A} \quad \mathbf{I}_{L2} := 300 \text{ A}.$$

Approach #1
First, rearrange the KVL equation for Loop 1 to solve for $\mathbf{I}_{L2}$:

$$\mathbf{I}_{L2} := \frac{\mathbf{V}_1 - \mathbf{I}_{L1} \cdot (R_1 + jX_1 + jX_{cap})}{jX_{cap}}.$$

The numerical result using the guess values is $I_{L2} = -175.6 + 226.2j$ A.
The $\mathbf{I_{L2}}$ is substituted in the second KVL equation:

$$V_2 - \frac{V_1 - I_{L1} \cdot (R_1 + jX_1 + jX_{cap})}{jX_{cap}} \cdot (R_2 + jX_2 + jX_{cap}) - I_{L1} \cdot jX_{cap} = 0.$$

The further rearrangement of that equation results in:

$$V_2 - V_1 \frac{(R_2 + jX_2 + jX_{cap})}{jX_{cap}}$$
$$+ \left[\frac{I_{L1} \cdot (R_1 + jX_1 + jX_{cap}) \cdot (R_2 + jX_2 + jX_{cap})}{jX_{cap}} - I_{L1} \cdot jX_{cap} \right] = 0.$$

Solving for $\mathbf{I_{L1}}$ yields:

$$I_{L1} := \frac{V_2 - V_1 \cdot \frac{(R_2 + jX_2 + jX_{cap})}{jX_{cap}}}{jX_{cap} - \frac{(R_1 + jX_1 + jX_{cap}) \cdot (R_2 + jX_2 + jX_{cap})}{jX_{cap}}}$$

$I_{L1} = (388.5 + 66.8j)$ A $|I_{L1}| = 394.2$ A $\arg(I_{L1}) = 9.75$ deg.

The Loop 2 current is:

$$I_{L2} := \frac{V_1 - I_{L1} \cdot (R_1 + jX_1 + jX_{cap})}{jX_{cap}}$$

$I_{L2} = (-217.3 + 173.8j)$ A $|I_{L2}| = 278.2$ A $\arg(I_{L2}) = 141.4 \cdot$ deg.

The capacitance (branch) current $\mathbf{I_3}$ is the sum of the two loop currents:

$$I_3 := I_{L1} + I_{L2} = (171.2 + 240.5j) \text{ A}$$
$$|I_3| = 295.2 \text{ A} \quad \arg(I_3) = 54.6 \text{ deg.}$$

Approach #2
The Mathcad equation solver can solve the loop equations directly. The method is illustrated in the following paragraphs. The guess values are:

$$I_{L1} := 300 \text{ A} \quad L_{L2} := 250 \text{ A}.$$

Using the two KVL expressions already developed:

Given
$$V_1 - I_{L1} \cdot (R_1 + jX_1 + jX_{cap}) - I_{L2} \cdot jX_{cap} = 0$$
$$V_2 - I_{L2} \cdot (R_2 + jX_2 + jX_{cap}) - I_{L1} \cdot jX_{cap} = 0,$$

the Mathcad *Find* equation solver immediately computes:

$$\text{Find}(I_{L1}, I_{L2}) = \begin{pmatrix} 388.5 + 66.8j \\ -217.3 + 173.8j \end{pmatrix} \text{A}.$$

The obtained results are identical to those from the first approach.

3.5.3. Thévenin's and Norton's Theorems

Thévenin's theorem states that an electric network can be represented by a voltage source and an impedance (Z_{Th}) connected in series (see Fig. 3.32). The voltage of the source is the open-circuit voltage, V_{oc}, of the network. Norton's theorem is similar except that the network is modeled by Z_{Th} in parallel with a current source equal to the short-circuit current, I_{sc}. The Thévenin and Norton networks are interchangeable using source transformation with $V_{oc} = I_{sc}Z_{Th}$. These theorems imply that complicated networks can be replaced with simple networks for analysis purposes.

Application of these theorems to circuit analysis generally involves splitting the circuit into two parts: (1) the network to be replaced by an equivalent and (2) the portion of the circuit that is to remain unchanged. The network for which an equivalent is to be found must be linear, whereas the remainder (typically the load) may be either linear or nonlinear. The open-circuit voltage, V_{oc}, is the voltage at the terminals of the network when the load is removed. The short-circuit current, I_{sc}, is the current that flows at the network terminals when the network is short-circuited there. The Thévenin impedance at the terminals is the ratio of the open-circuit voltage to the short-circuit current

$$Z_{Th} = V_{oc}/I_{sc}. \tag{3.47}$$

Alternatively, after short-circuiting the voltage sources and open-circuiting the current sources in the network, Z_{Th} may be determined by combining the remaining impedances

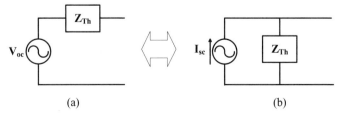

Figure 3.32. (a) Thévenin and (b) Norton equivalent circuits.

using the series and parallel impedance reduction techniques described earlier in this chapter.

For power systems analyses, the short-circuit current is computed by the operating utility and is an available datum. In a power network, the short-circuit current is usually inductive. Practically speaking, the resistance is often negligible; consequently, the network can generally be represented by an inductive reactance. The open-circuit voltage of a power network is normally within 5% of the rated line-to-ground (neutral) voltage, V_{ln}. Therefore, in this case, the Thévenin reactance of the network is estimated as:

$$X_{net} = \frac{|V_{ln}|}{|I_{sc}|}. \qquad (3.48)$$

3.6. APPLICATIONS OF SINGLE-PHASE CIRCUIT ANALYSIS

The practical application of the complex phasor analysis of a single-phase circuit is presented through three numerical examples:

1. power factor correction,
2. transmission line operation analysis, and
3. generator supplying a constant impedance network through a line.

The reader is encouraged to use a computer and follow the calculation by entering the equations and evaluating the results. This interactive method of learning will lead to a better understanding of the concepts.

EXAMPLE 3.14: Power factor improvement

Consider a factory that operates with a poor (low) power factor. The utility penalizes the factory operator with a higher electricity rate. The operator can install a capacitor connected in parallel with the load to improve the power factor and reduce the power cost. Figure 3.33 gives the system one-line diagram.

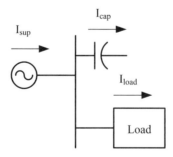

Figure 3.33. Power factor improvement by capacitor.

APPLICATIONS OF SINGLE-PHASE CIRCUIT ANALYSIS

The factory load is:

$$P_{load} := 150 \text{ kW} \quad pf_{load} := 0.65 \text{ inductive} \quad V_{load} := 7.2 \text{ kV}.$$

The desired power factor is: $pf_{sup} := 0.9$ (lagging).

The capacitor does not affect the real power but reduces the reactive power. Consequently, the load current and the desired supply current are calculated using the given load value and the appropriate power factor. Moreover, the load and supply voltages are equal. Using Equation (3.38) for a lagging power factor, the load current is:

$$I_{load} := \frac{P_{load}}{pf_{load} \cdot V_{load}} \cdot e^{-j \cdot acos(pf_{load})}$$

$$|I_{load}| = 32.1 \text{ A} \quad arg(I_{load}) = -49.5 \text{ deg.}$$

The desired supply current is:

$$I_{sup} := \frac{P_{load}}{pf_{sup} \cdot V_{load}} \cdot e^{-j \cdot acos(pf_{sup})}$$

$$|I_{sup}| = 23.1 \text{ A} \quad arg(I_{sup}) = -25.8 \text{ deg.}$$

The capacitance current is the difference of the supply and load currents. The capacitance current is:

$$I_{cap} := I_{sup} - I_{load} \quad |I_{cap}| = 14.3 \text{ A} \quad arg(I_{cap}) = 90.0 \text{ deg.}$$

The capacitor-provided reactive power is the imaginary part of the complex power:

$$Q_{cap} := Im(V_{load} \cdot \overline{I_{cap}}) \quad Q_{cap} = -102.7 \text{ kV} \cdot \text{A},$$

where the complex conjugate operation is denoted by Mathcad using the overline mark. The impedance of the capacitor is:

$$Z_{cap} := \frac{V_{load}}{I_{cap}} \quad Z_{cap} = -504.7 j \text{ }\Omega.$$

Assuming 50 Hz frequency, the capacitor needed to improve the power factor is:

$$\omega := 2 \cdot \pi \cdot 50 \text{ Hz}$$

$$C_{cap} := \frac{1}{j \cdot \omega \cdot Z_{cap}} \quad C_{cap} = 6.307 \text{ }\mu\text{F}.$$

This capacitor is connected in parallel with the load to correct the power factor without changing the load current or voltage.

EXAMPLE 3.15: Analysis of transmission line operation

In this example, the operation of a transmission line is analyzed in three different operating conditions:

1. No-load (open-circuit) condition;
2. Short-circuit condition; and
3. Loaded condition.

The input current, real and reactive powers, and the voltage and current at the receiving end are calculated for each condition.

The transmission line is represented by a Π (pi) circuit and supplied by a voltage source (V_S).

The system frequency data are $f := 60$ Hz $\quad \omega := 2 \cdot \pi \cdot f$.
The M is introduced into Mathcad to represent the mega prefix: $M := 10^6$.
The supply voltage is: $V_S := 75$ kV.
The line impedance is: $Z_{line} := (10 + j \cdot 73) \Omega$.
The sending end and receiving end capacitances are:

$$C_s := 1.6 \, \mu F \quad C_r := C_s = 1.6 \, \mu F.$$

No-Load Condition

Figure 3.34 provides the equivalent circuit for an open-circuit (unloaded) condition. Inspection of the equivalent circuit shows that the voltage source supplies two parallel current paths. The first (right side) path contains X_{line}, R_{line} and C_r connected in series. The second (left) path contains the line capacitance C_s.

The impedance of the first path is

$$Z_{circuit} := Z_{line} + \frac{1}{j \cdot \omega \cdot C_r} = (10.0 - 1584.9j) \, \Omega.$$

The voltage source supplies the first path directly. The current through the first path is calculated by Ohm's law:

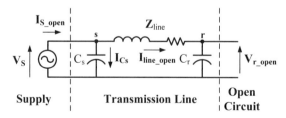

Figure 3.34. Transmission line in an open-circuit condition.

APPLICATIONS OF SINGLE-PHASE CIRCUIT ANALYSIS

$$I_{line\_open} := \frac{V_S}{Z_{circuit}} = (0.30 + 47.32j) \text{ A}$$

$$|I_{line\_open}| = 47.32 \text{ A} \quad \arg(I_{line\_open}) = 89.6 \cdot \deg.$$

The current through the second path is:

$$I_{Cs} := \frac{V_S}{\frac{1}{j \cdot \omega \cdot C_s}} = 45.2j \text{ A} \quad |I_{Cs}| = 45.2 \text{ A} \quad \arg(I_{Cs}) = 90.0 \cdot \deg.$$

The total source current in the open-circuit condition is the sum of the previously mentioned two currents:

$$I_{S\_open} := I_{Cs} + I_{line\_open} = (0.3 + 92.6j) \text{ A}$$

$$|I_{S\_open}| = 92.6 \text{ A} \quad \arg(I_{S\_open}) = 89.8 \cdot \deg.$$

The generator complex power is the product of the supply voltage and the source current conjugate:

$$S_{open} := V_S \cdot \overline{I_{S\_open}} = (0.02 - 6.94j) \cdot \text{M} \cdot \text{V} \cdot \text{A}.$$

The result shows that the real power is negligible. The negative reactive power indicates that the voltage source absorbs reactive power. The passive sign convention is not obeyed at the supply of Figure 3.34 since the current I_{S_open} exits (rather than enters) from the positive voltage terminal; hence, a negative power value indicates reactive power is absorbed rather than supplied.

The open-circuit voltage on the line end is the product of the line current and the impedance of the receiving capacitance (C_r):

$$V_{r\_open} := I_{line\_open} \cdot \frac{1}{j \cdot \omega \cdot C_r} = (78.45 - 0.50j) \cdot \text{kV}$$

$$|V_{r\_open}| = 78.45 \cdot \text{kV} \quad \arg(V_{r\_open}) = -0.36 \cdot \deg.$$

The supply voltage, V_S, is 75 kV. The open-circuit voltage is slightly higher than the supply voltage. In fact, the open-circuit voltage can be dangerously high in a long line, where the high voltage can endanger the transmission line insulation. This is referred to as the *Ferranti effect*. Power companies load the line with an inductance, when required, to eliminate the overvoltage in an open-circuit condition.

Short-Circuit Condition

Figure 3.35 presents the equivalent circuit in the short-circuit condition. It is assumed that the short circuit occurs at the receiving end of the line. The equivalent circuit

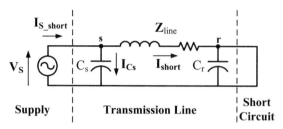

Figure 3.35. Transmission line short-circuited.

reveals that the short circuit effectively eliminates the receiving capacitance C_r from the network analysis.

The line current is the ratio of the supply voltage and line impedance:

$$I_{short} := \frac{V_S}{Z_{line}} = (0.14 - 1.01j) \cdot kA$$

$$|I_{short}| = 1.02 \cdot kA \quad \arg(I_{short}) = -82.2 \cdot \deg.$$

The active (real) part of the short-circuit current is small. The large reactive part is inductive.

Large short-circuit currents, if sustained, will damage line conductors. Utilities protect the transmission lines with circuit breakers that open the line automatically when a short circuit occurs. Typically, the short-circuit current is interrupted within 50–200 ms (3–12 cycles).

Using KCL, the supply current is:

$$I_{S\_short} := I_{short} + I_{Cs} = (0.14 - 0.96j) \cdot kA$$

$$|I_{S\_short}| = 0.97 \cdot kA \quad \arg(I_{S\_short}) = -81.8 \cdot \deg.$$

Calculating the complex input power results in:

$$S_{S\_short} := V_S \cdot \overline{I_{S\_short}} = (10.4 + 72.2j) \, M \cdot V \cdot A.$$

Both the real and the reactive powers are significant for the short-circuit condition.

Loaded Condition

Figure 3.36 exhibits the equivalent circuit of the loaded line. A lagging power factor loading produces a voltage drop, which varies with the load and power factor. To evaluate the load effect, we shall load the line and determine the load voltage for a given load power when the supply voltage is held constant. Two approaches to this calculation are demonstrated.

APPLICATIONS OF SINGLE-PHASE CIRCUIT ANALYSIS

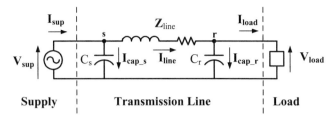

Figure 3.36. Equivalent circuit of the loaded line.

Approach #1
To begin, the calculations are tested using a load voltage of 75 kV, and the *pf* is set at 0.75 lagging. Hence, the load data are:

$$V_{load} := 75 \text{ kV} \quad pf_{load} := 0.75.$$

The load is variable, but we test the calculation by using $P_{load} := 10$ MW as a guess value.

Referring to Equation (3.38) and Table 3.4 for a lagging load, the load current as a function of the variable load power is:

$$I_{load}(V_{load}, P_{load}) := \frac{P_{load}}{V_{load} \cdot pf_{load}} \cdot e^{-j \cdot acos(pf_{load})}$$

$$|I_{load}(V_{load}, P_{load})| = 177.8 \text{ A} \quad arg(I_{load}(V_{load}, P_{load})) = -41.4 \cdot deg.$$

From Ohm's law, the capacitive current at the receiving end is:

$$I_{cap\_r}(V_{load}) := \frac{V_{load}}{\frac{1}{j \cdot \omega \cdot C_r}} \quad I_{cap\_r}(V_{load}) = 45.24j \text{ A}$$

$$|I_{cap\_r}(V_{load})| = 45.2 \text{ A} \quad arg(I_{cap\_r}(V_{load})) = 90.0 \cdot deg.$$

The line current is the sum of the load and receiving-end capacitance currents:

$$I_{line}(V_{load}, P_{load}) := I_{cap\_r}(V_{load}) + I_{load}(V_{load}, P_{load})$$
$$|I_{line}(V_{load}, P_{load})| = 151.7 \text{ A} \quad arg(I_{line}(V_{load}, P_{load})) = -28.5 \cdot deg.$$

The required supply voltage is the sum of the load voltage and the voltage drop on the line impedance:

$$V_{sup}(V_{load}, P_{load}) := V_{load} + Z_{line} \cdot I_{line}(V_{load}, P_{load})$$
$$|V_{sup}(V_{load}, P_{load})| = 82.1 \cdot kV \quad arg(V_{sup}(V_{load}, P_{load})) = 6.3 \cdot deg.$$

The capacitive current at the supply side is the supply voltage divided by the sending-end capacitive impedance:

$$I_{cap\_s}(V_{load}, P_{load}) := \frac{V_{sup}(V_{load}, P_{load})}{\frac{1}{j \cdot \omega \cdot C_s}} \quad I_{cap\_s}(V_{load}, P_{load}) = (-5.4 + 49.2j) \text{ A}$$

$$|I_{cap\_s}(V_{load}, P_{load})| = 49.5 \text{ A} \quad \arg(I_{cap\_s}(V_{load}, P_{load})) = 96.3 \cdot \deg.$$

The supply current is the sum of the line current and sending-end capacitive current:

$$I_{sup}(V_{load}, P_{load}) := I_{cap\_s}(V_{load}, P_{load}) + I_{line}(V_{load}, P_{load})$$

$$|I_{sup}(V_{load}, P_{load})| = 130.0 \text{ A} \quad \arg(I_{sup}(V_{load}, P_{load})) = -10.2 \cdot \deg.$$

The supply complex power is:

$$S_{sup}(V_{load}, P_{load}) := V_{sup}(V_{load}, P_{load}) + \overline{I_{sup}(V_{load}, P_{load})}$$

$$S_{sup}(V_{load}, P_{load}) = (10.2 + 3.0j) \cdot M \cdot V \cdot A.$$

Generally, in the operation of the power system, the supply voltage is kept near the rated value, and the load voltage depends on the load. The actual load voltage corresponding to a given load power and supply voltage is obtained using the Mathcad *root* function, which is in essence solving the above set of equations to find the load voltage that satisfies the relation for V_{sup}. For the test load of 10 MW, the load voltage is:

$$V_{load\_e}(P_{load}) := \text{root}(V_{sup}(V_{load}, P_{load}) - 75 \text{ kV}, V_{load})$$

$$|V_{load\_e}(P_{load})| = 70.03 \cdot \text{kV} \quad \arg(V_{load\_e}(P_{load})) = -9.7 \cdot \deg.$$

More information on the use of the Mathcad *root* equation solver is given in Appendix A.

As a verification of this solution, we can compare the open-circuit voltage computed earlier to the load voltage calculated with the previous equation when the load power is zero:

$$|V_{load\_e}(0 \text{ MW})| = 78.45 \cdot \text{kV} \quad \arg(V_{load\_e}(0 \text{ MW})) = -0.36 \deg$$

$$|V_{r\_open}| = 78.45 \cdot \text{kV} \quad \arg(V_{r\_open}) = -0.36 \deg.$$

This confirms that the two results are identical.

To plot the load voltage versus the loading, we vary the load power according to:

$$P_{load} := 0 \cdot \text{MW}, 1 \text{ MW} \ldots 15 \text{ MW}.$$

APPLICATIONS OF SINGLE-PHASE CIRCUIT ANALYSIS

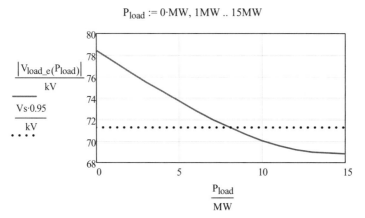

Figure 3.37. Load voltage corresponding to a given load power with a V_s = 75 kV supply voltage.

Figure 3.37 graphs the load voltage versus the load power. For the Mathcad plots presented in this book, the units are given as the denominator of a fraction having the particular variable as the numerator. Specifically in Figure 3.37, the units of P_{load} and V_{load_e} are megawatt and kilovolt, respectively. The graph discloses that the load voltage versus power is nearly linear, while the load voltage is within about 5% of the supply voltage. For load voltages less than $0.95 V_{sup}$, the relationship is nonlinear.

This load voltage calculation approximates true power system behavior well in the ±5% rated voltage range because the calculation assumes that the load is constant and independent from the load voltage. For an actual load, the load power decreases as the load voltage is reduced since the load impedance depends on the load voltage. In fact, the power companies eliminate overload by reducing the load voltage. This technique creates what is termed *brownout*, which is an approach to avert a blackout.

The proper operation of the power system requires that the load voltage be within ±5% limits. This entails the calculation of the *regulation*, which is the percentage value of the voltage drop caused by the load as compared to the unloaded condition. The voltage regulation versus load is:

$$\text{Reg}(P_{load}) := \frac{|V_{load\_e}(0\text{ MW})| - |V_{load\_e}(P_{load})|}{|V_{load\_e}(0\text{ MW})|}.$$

For example, at a 5 MW load, the voltage regulation is:

$$\text{Reg}(5\text{ MW}) = 6.1\%,$$

which is slightly elevated from the acceptable value.

Finally, another practical analysis is the determination of the load limit to maintain a specified voltage regulation. The load power for a 5% regulation is calculated by solving:

$$|\text{Reg}(P_{\text{load}})| = 5\%.$$

This can be accomplished numerically using the Mathcad *root* equation solver. A guess value for the equation solver is $P_{\text{load}} := 10$ MW and the result is:

$$P_{\text{load}5\%} := \text{root}(\text{Reg}(P_{\text{load}}) - 5\%, P_{\text{load}}) \quad P_{\text{load}5\%} = 4.07 \cdot \text{MW}.$$

For this system, the load should remain below 4 MW.
The load voltage corresponding to this power level is:

$$V_{\text{loade}} := V_{\text{load\_e}}(P_{\text{load}5\%}) \quad |V_{\text{loade}}| = 74.5 \text{ kV}.$$

Approach #2
Alternatively, the network analysis can be performed using nodal analysis. KCL is first applied to the node labeled "r":

$$I_{\text{line}} = I_{\text{load}} + I_{\text{cap\_r}}.$$

Except for the load current which is expressed using Equation (3.38), the currents are written in terms of nodal voltages:

$$\frac{V_{\text{sup}} - V_{\text{Load}}}{Z_{\text{line}}} = \frac{P_{\text{load}}}{\text{pf}_{\text{load}} \cdot V_{\text{Load}}} \cdot e^{-j \cdot \text{acos}(\text{pf}_{\text{load}})} + \frac{V_{\text{Load}}}{\dfrac{1}{j \cdot \omega \cdot C_r}}.$$

This node point equation can be solved either symbolically or numerically in Mathcad. A numerical solution can be obtained using the Mathcad *Find* function:

Given

$$\frac{V_{\text{sup}} - V_{\text{Load}}}{Z_{\text{line}}} = \frac{P_{\text{load}}}{\text{pf}_{\text{load}} \cdot V_{\text{Load}}} \cdot e^{-j \cdot \text{acos}(\text{pf}_{\text{load}})} + \frac{V_{\text{Load}}}{\dfrac{1}{j \cdot \omega \cdot C_r}}$$

$$V_{\text{load\_a}}(P_{\text{load}}) := \text{Find}(V_{\text{Load}}).$$

For the 10 MW test value and the open-circuit conditions, the load voltages are:

$$|V_{\text{load\_a}}(0 \text{ MW})| = 78.45 \cdot \text{kV} \quad \arg(V_{\text{load\_a}}(0 \text{ MW})) = -0.36 \cdot \text{deg}$$
$$|V_{\text{load\_a}}(10 \text{ MW})| = 70.03 \cdot \text{kV} \quad \arg(V_{\text{load\_a}}(10 \text{ MW})) = -9.7 \cdot \text{deg}.$$

These results agree with those found using the first approach.

APPLICATIONS OF SINGLE-PHASE CIRCUIT ANALYSIS

EXAMPLE 3.16: Generator supplies a constant impedance network through a line

Figure 3.38 presents the one-line diagram of a system where a generator supplies a constant impedance load through a transmission line. The generator can be represented by a voltage source and impedance. In this example, we neglect the generator impedance and represent the generator with a voltage source only. A Π circuit represents the line. The load is represented by a resistance and inductive reactance connected in parallel. The equivalent circuit is drawn in Figure 3.39. The circuit data are given in the succeeding paragraphs.

The generator voltage rating and frequency are:

$$V_g := 15 \text{ kV} \quad f_g := 60 \text{ Hz} \quad \omega := 2 \cdot \pi \cdot f_g.$$

The transmission line parameters are:

$$R_{line} := 1.1 \,\Omega \quad X_{line} := 6 \,\Omega$$
$$C_s := 30 \,\mu\text{F} \quad C_r := C_s.$$

The load impedance is: $R_L := 45 \,\Omega \quad X_L := 60 \,\Omega$.

Equivalent Impedance Calculation

If the system frequency is 60 Hz, the capacitive impedances at the sending and receiving ends of the line are:

$$Z_{Cs} := \frac{1}{j \cdot \omega \cdot C_s} = -88.4j \,\Omega \quad Z_{Cr} := Z_{Cs}.$$

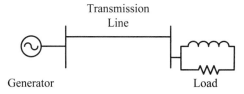

Figure 3.38. Generator supplies an impedance load through a line.

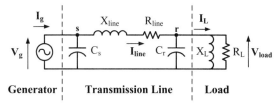

Figure 3.39. Equivalent circuit of the system in Figure 3.38.

The load is represented by a resistance and reactance connected in parallel. The equivalent load impedance is:

$$Z_L := \frac{1}{\frac{1}{R_L} + \frac{1}{j \cdot X_L}} \quad Z_L = 28.8 + 21.6j \; \Omega.$$

The load impedance and the capacitance at the line end are also connected in parallel. The equivalent impedance is:

$$Z_{L\_Cr} := \frac{Z_{Cr} \cdot Z_L}{Z_{Cr} + Z_L} \quad Z_{L\_Cr} = (42.5 + 10.3j) \; \Omega.$$

The transmission line impedance is connected in series with the combined impedance of the load and capacitor. The equivalent impedance is:

$$Z_{line\_L\_Cr} := R_{line} + j \cdot X_{line} + Z_{L\_Cr} \quad Z_{line\_L\_Cr} = (43.6 + 16.3j) \; \Omega.$$

The combination of the impedances results in a simplified circuit, shown in Figure 3.40.

Current Calculation
The currents in this circuit can be calculated directly using Ohm's law. Further circuit simplification is unnecessary. The sending-end capacitor current is:

$$I_{Cs} := \frac{V_g}{Z_{Cs}} \quad |I_{Cs}| = 169.6 \text{ A} \quad \arg(I_{Cs}) = 90 \text{ deg}.$$

The line current is:

$$I_{line} := \frac{V_g}{Z_{line\_L\_Cr}} \quad |I_{line}| = 322.2 \text{ A} \quad \arg(I_{line}) = -20.4 \text{ deg}.$$

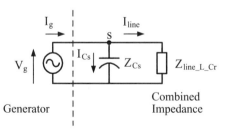

Figure 3.40. Simplified equivalent circuit of Figure 3.39.

APPLICATIONS OF SINGLE-PHASE CIRCUIT ANALYSIS

The generator current is the sum of the capacitor and line currents. The result is:

$$I_g := I_{Cs} + I_{line} \quad I_g = (301.9 + 57.2j) \text{ A}$$
$$|I_g| = 307.3 \text{ A} \quad \phi_g := \arg(I_g) = 10.7 \cdot \deg.$$

Voltage and Voltage Drop Calculation

The next step is the calculation of the load voltage, which is the product of the line current and the combined impedance of the load and receiving-end capacitor:

$$V_{load} := I_{line} \cdot Z_{L\_Cr} \quad V_{load} = (13.99 - 1.69j) \cdot \text{kV}$$
$$|V_{load}| = 14.09 \cdot \text{kV} \quad \arg(V_{load}) = -6.9 \cdot \deg.$$

An alternative method for calculating the load voltage is the subtraction of the voltage drop on the line impedance from the generator voltage:

$$V_{load\_alt} := V_g - I_{line} \cdot (R_{line} + j \cdot X_{line}) \quad V_{load\_alt} = 13.99 - 1.69j \text{ kV}.$$

The load current is the ratio of the load voltage and load impedance:

$$I_L := \frac{V_{load}}{Z_L} \quad I_L = (282.8 - 270.7j) \text{ A}$$
$$|I_L| = 391.5 \text{ A} \quad \arg(I_L) = -43.7 \cdot \deg.$$

An alternate method for load current calculation is the use of the current division equation:

$$I_{L\_alt} := I_{line} \cdot \frac{Z_{Cr}}{Z_{Cr} + Z_L} \quad I_{L\_alt} = 282.8 - 270.7j \text{ A}.$$

The proper operation of household and other equipment requires the load voltage to be within about ±5% of the rated voltage. The standards for electric utilities specify that the voltage be within ±5% for 95% of the time, and within ±8% for 99% of the time.

The most frequently used method to evaluate the operating conditions is the calculation of the percentage voltage drop or voltage regulation. The percentage voltage drop is defined as:

$$\text{Voltage\_drop} := \frac{|V_g| - |V_{load}|}{|V_{load}|} \quad \text{Voltage\_drop} = 6.4\%.$$

Note that Mathcad automatically converts the fractional voltage drop to a percentage value; therefore, the usual 100% multiplier is not needed in the previous formula. The voltage drop in the investigated circuit is slightly more than 5%—it is not acceptable.

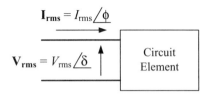

Figure 3.41. General circuit element obeying passive sign convention.

TABLE 3.5. Important Interrelations

Load	Power Factor Angle	Reactive Power	Complex Power
Capacitive	θ < 0° (leading)	$Q < 0$	$|S|e^{-j\theta}$
Inductive	θ > 0° (lagging)	$Q > 0$	$|S|e^{+j\theta}$
Resistive	θ < 0°	$Q = 0$	P

3.7. SUMMARY

It has been assumed that the reader was already familiar with much of the basic circuit analysis laws, conventions, and techniques presented in this chapter. The goal of this chapter has been to review those basics while preparing the reader to apply computer tools, such as Mathcad and MATLAB, to network analysis. In addition, several quantities that are possibly new to the reader have been introduced; for example, system efficiency and voltage regulation, which are defined in Equation (3.24) and Equation (3.39), respectively. Further, we note that the engineer should be prepared to utilize interrelations between real power, reactive power, complex power, voltage, current, and power factor to solve for the unknown variable(s) based on those quantities that are known. Specifically, the reader will find the following relations to be particularly useful throughout the remainder of this textbook. For $\mathbf{V_{rms}}$ and $\mathbf{I_{rms}}$, obeying the passive sign convention, as depicted in Figure 3.41:

$$\begin{aligned} \mathbf{S} = \mathbf{V_{rms}}\mathbf{I^*_{rms}} &= V_{rms}e^{j\delta}I_{rms}e^{-j\phi} \\ &= V_{rms}I_{rms}\cos(\delta - \phi) + jV_{rms}I_{rms}\sin(\delta - \phi) \\ &= P + jQ. \end{aligned} \quad (3.49)$$

If that circuit element is an impedance, then:

$$\mathbf{Z} = \frac{\mathbf{V_{rms}}}{\mathbf{I_{rms}}} = \frac{V_{rms}\angle\delta}{I_{rms}\angle\phi} = \frac{V_{rms}}{I_{rms}}\angle\delta - \phi = |\mathbf{Z}|\angle\theta, \quad (3.50)$$

$$\mathbf{S_Z} = \mathbf{V_{rms}}\mathbf{I^*_{rms}} = V_{rms}I_{rms}e^{\pm j\theta} = |\mathbf{S_Z}|e^{\pm j\theta} = \frac{P}{pf}e^{\pm j\theta}, \quad (3.51)$$

where $\theta = \pm \arccos(pf)$ and the proper sign can be determined by referring to Table 3.5.

3.8. EXERCISES

1. What is the rms value of a sinusoidal voltage or current? Present the equation.
2. Describe the impedance triangle and present the equations for complex and polar representation of an impedance consisting of a resistance and reactance connected in series.
3. Describe the rules for the calculation of an equivalent impedance when impedances are connected in series and in parallel.
4. How do you calculate the instantaneous power?
5. What is the complex power? What is the unit of measure for complex power?
6. Present the equations for the complex power calculation.
7. What is the active power? What is the unit of real power?
8. What is the reactive power? What is the unit of reactive power?
9. What is the power triangle?
10. What is the power factor? How it is used? Give the basic equations for the power factor calculation.
11. What is the sign of the reactive power if the current is leading the voltage? Draw a sketch of the current and voltage waveforms.
12. What is the sign of the reactive power if the current is lagging the voltage? Draw a sketch of the current and voltage waveforms.
13. Describe Kirchhoff's current law.
14. Describe Kirchhoff's voltage law.
15. Discuss the Thévenin equivalent of a circuit. Illustrate the concept with a diagram.
16. Present the equations for the current calculation in the case of loads with lagging and leading power factors.
17. Draw an equivalent circuit for a transmission line. Identify the components.
18. What is the definition of voltage regulation? Present the equation and describe the meaning of the variables.
19. Discuss the concept of power factor correction.

3.9. PROBLEMS

Problem 3.1

Determine the period (in milliseconds) of an ac sinusoidal waveform in a power system operating at (a) 50 Hz and (b) 60 Hz.

Problem 3.2

Household lighting voltage is 120 Vrms. (a) Determine the voltage magnitude, V_M, of the sinusoidal waveform. (b) What is the range of the rms voltage if the voltage is within ±5% of the nominal value?

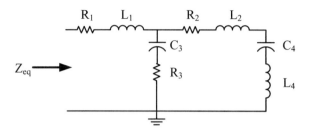

Figure 3.42. Network for Problem 3.5.

Problem 3.3

Use phasors to prove that:

$$\left(\frac{\mathbf{X}}{\mathbf{Y}}\right)^* = \frac{\mathbf{X}^*}{\mathbf{Y}^*}.$$

Problem 3.4

Draw a phasor diagram of the following currents and voltages: (a) $i(t) = 12\cos(377t + 18°)$ A, (b) $v(t) = 17.7\cos(377t - 36°)$ kV, (c) $\mathbf{V} = 69$ kV$\angle 130°$, and (d) $\mathbf{I} = 100$ A$\angle -120°$.

Problem 3.5

Calculate the impedance of the network ($\mathbf{Z}_{eq}$) given in Figure 3.42 at (a) 50 Hz and (b) 60 Hz using the following network circuit element values:

Branch 1:	$R_1 = 25\ \Omega$	$L_1 = 0.1$ H
Branch 2:	$R_2 = 25\ \Omega$	$L_2 = 0.1$ H
Branch 3:	$R_3 = 48\ \Omega$	$C_3 = 0.1\ \mu$F
Branch 4:	$C_4 = 25\ \mu$F	$L_4 = 0.3$ H

Problem 3.6

The 240 V air conditioning motor at a residence is rated at 1 hp with a power factor of 0.65 (lagging). Determine the equivalent (a) impedance and (b) admittance of the ac motor.

Problem 3.7

A single-phase 3 hp, 240 V, 60 Hz pool pump (motor) runs at full load with 92% efficiency and a power factor of 0.73 lagging. Calculate the capacitor needed to improve the power factor to 0.98 lagging.

PROBLEMS   143

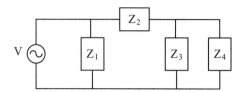

Figure 3.43. Electric network for Problem 3.9.

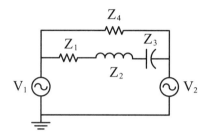

Figure 3.44. Circuit for Problem 3.11.

Problem 3.8

Two voltage sources are connected together through an impedance of $\mathbf{Z} = 12 + j21\ \Omega$. Determine which source operates as the load and which is the generator. The voltage sources are: $\mathbf{V}_1 = 14\ \text{kV} \angle 0°$ and $\mathbf{V}_2 = 13\ \text{kV} \angle 12°$.

Problem 3.9

The electrical network shown in Figure 3.43 has a voltage source of 140 V and the values of the impedances are as follows: $\mathbf{Z}_1 = 5 - j8\ \Omega$, $\mathbf{Z}_2 = 10 + j5\ \Omega$, $\mathbf{Z}_3 = 5 - j5\ \Omega$, and $\mathbf{Z}_4 = 15 + j10\ \Omega$. Determine: (a) the real power absorbed by each impedance and (b) the reactive power of each impedance. Clearly state whether reactive power is delivered to or absorbed by each impedance.

Problem 3.10

A single-phase voltage source of $\mathbf{V}_S = 220\ \text{V} \angle 0°$ supplies a load impedance of $\mathbf{Z} = 100\ \Omega \angle 50°$ Determine: (a) the resistance and reactance of the load, (b) the real and reactive power absorbed by the load, and (c) the power factor of the circuit as viewed from the source, and state whether the power factor is lagging or leading.

Problem 3.11

Two ideal voltage sources are connected to each other through a feeder with a combination of impedances, as illustrated in Figure 3.44. The voltage source values are $\mathbf{V}_1 = 100\ \text{V}$ and $\mathbf{V}_2 = 120\ \text{V} \angle -25°$, and the impedances are $\mathbf{Z}_1 = 10\ \Omega$, $\mathbf{Z}_2 = 5j\ \Omega$, $\mathbf{Z}_3 = -25j\ \Omega$, and $\mathbf{Z}_4 = 3\ \Omega$. Determine: (a) the real power of each voltage source, and state whether the source is supplying or absorbing real power; (b) the reactive power of each source, and state whether the source is supplying or absorbing reactive power; and (c) the real and reactive power of the impedances, and state whether the power is supplied or absorbed. (d) Plot the real power consumed by the feeder against the power angle, assuming that the angle varies between 0° and 360°. Find the maximum power consumed. Explain the trend of the power consumed versus the power angle. *Note*: The *power angle* is the phase angle difference between phasors V_1 and V_2. In this part of the problem, the phase angle of V_1 is zero and replace −25° by a variable angle for V_2.

Problem 3.12

An industrial plant consists of several 60 Hz single-phase motors with low-power factor. The plant absorbs 600 kW with a power factor of 0.75 lagging from the substation bus. The supply voltage is 12.47 kV. The power factor can be improved by connecting a capacitor in parallel with the supply or by using a synchronous motor, which generates reactive power. Analyze both of these cases independently:

(a) Find the required kilovolt-amperes reactive rating of a capacitor connected across the load to raise the power factor to 0.95 lagging.

(b) Assuming a synchronous motor rated at 250 hp with an 80% efficiency is operated from the same bus at rated conditions and a power factor of 0.85 leading, calculate the resultant supply power factor.

Problem 3.13

A 220 V bus supplies a motor and capacitor. The 60 Hz single-phase motor has the following rating:

$$\mathbf{V} = 220 \text{ V} \quad \mathbf{S} = 65 \text{ kVA} \quad pf = 0.8 \text{ (lagging)}.$$

The motor is loaded to its rated capability. The capacitor is connected in parallel with the motor to obtain a total power factor of 0.99 leading. (a) Draw the equivalent circuit. (b) Determine the required kilovolt-ampere rating of the capacitor and the capacitance value. (c) Plot the *pf* as a function of capacitance. (d) Find the capacitance needed to obtain a final *pf* = 1.

4

THREE-PHASE CIRCUITS

Practically all electrical energy generation and transmission systems use a three-phase circuit. The three-phase energy is transported through three or four conductors to large customers. Only the small household and light commercial loads are supplied by a single phase. The major advantage of the three-phase system is the efficiency of power transmission. Another advantage is the generation of constant torque, which reduces vibration, for rotating machines. This is particularly important for industries with large motors. In addition, the three-phase generator efficiency is higher than a single-phase unit. A three-phase transmission line carries three times the power of a single-phase line while requiring practically the same right-of-way. These advantages resulted in the worldwide use of three-phase circuits for electric power generation, transmission, and distribution.

A polyphase system or circuit utilizes alternating current (ac) sources generating at the same frequency but with different phases. An example includes the use of six-phase and 12-phase systems to supply power to large rectifiers. This chapter presents the basis of the three-phase electric transmission system; explains the wye and delta connections; and introduces the per-unit system frequently used in power engineering. This chapter also details the calculation of voltage, current, and power in three-phase wye and delta systems, and discusses three-phase power measurements.

Electrical Energy Conversion and Transport: An Interactive Computer-Based Approach, Second Edition. George G. Karady and Keith E. Holbert.
© 2013 Institute of Electrical and Electronics Engineers, Inc. Published 2013 by John Wiley & Sons, Inc.

4.1. THREE-PHASE QUANTITIES

This book only addresses *balanced three-phase systems* unless stated otherwise. A balanced system is one in which the three sinusoidal voltages have the same magnitude and frequency, and each is 120° out of phase with the other two; that is,

$$\begin{aligned} v_{an}(t) &= V_M \cos(\omega t) & \mathbf{V}_{an} &= V_M \angle 0°, \\ v_{bn}(t) &= V_M \cos(\omega t - 120°) & \mathbf{V}_{bn} &= V_M \angle -120°, \\ v_{cn}(t) &= V_M \cos(\omega t - 240°) & \mathbf{V}_{cn} &= V_M \angle -240°. \end{aligned} \quad (4.1)$$

When $\mathbf{V}_{bn}$ lags $\mathbf{V}_{an}$ by 120° and $\mathbf{V}_{cn}$ lags $\mathbf{V}_{bn}$ by 120°, then the system is said to have a *positive phase sequence*, also called an *abc* phase sequence. The phase sequence describes the order in which the phase voltages reach their maximum (peak) values with respect to time. If $\mathbf{V}_{cn}$ and $\mathbf{V}_{bn}$ lag $\mathbf{V}_{an}$ by 120° and 240°, respectively, the system has a *negative* (or *acb*) *phase sequence*. A positive phase sequence as exemplified in Equation (4.1) is used throughout this book.

A *balanced three-phase circuit* is one in which the loads are such that the currents produced by the voltages are also balanced:

$$\begin{aligned} i_a(t) &= I_M \cos(\omega t - \theta) & \mathbf{I_a} &= I_M \angle -\theta, \\ i_b(t) &= I_M \cos(\omega t - \theta - 120°) & \mathbf{I_b} &= I_M \angle -\theta - 120°, \\ i_c(t) &= I_M \cos(\omega t - \theta - 240°) & \mathbf{I_c} &= I_M \angle -\theta - 240°, \end{aligned} \quad (4.2)$$

where the voltage of each phase leads its corresponding current by an angle of θ. The trigonometric identity:

$$\cos(\alpha) + \cos(\alpha - 120°) + \cos(\alpha + 120°) = 0 \quad (4.3)$$

can be used with Equation (4.1) and Equation (4.2) to find that both the sum of the balanced voltages and the sum of the balanced currents equal zero (i.e., $v_{an}(t) + v_{bn}(t) + v_{cn}(t) = 0$ and $i_a(t) + i_b(t) + i_c(t) = 0$). This fact can also be seen graphically in the phasor diagram of Figure 4.1.

Recall that the instantaneous power for a single phase is $p(t) = i(t)v(t)$. The total instantaneous power for three balanced phases is then:

$$p_T(t) = p_a(t) + p_b(t) + p_c(t) = 3 \frac{V_M I_M}{2} \cos(\theta) = 3 V_{rms} I_{rms} \cos(\theta) \quad (4.4)$$

The instantaneous three-phase power is a constant over time! In similar manner, the total three-phase complex power is:

$$\mathbf{S}_T = \mathbf{S}_A + \mathbf{S}_B + \mathbf{S}_C = 3\mathbf{S}_1, \quad (4.5)$$

where $\mathbf{S}_1$ is the complex power of any one the three phases in the balanced circuit.

THREE-PHASE QUANTITIES

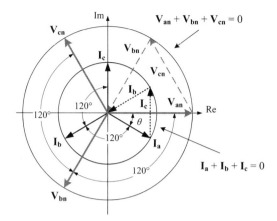

Figure 4.1. Phasor diagram for a balanced three-phase circuit with a positive phase sequence. The graphical additions of both the voltages and currents equal zero.

EXAMPLE 4.1: The following MATLAB script is used to visualize the time-varying voltages and currents of a balanced three-phase 60 Hz circuit. Consider the case where the currents lag their corresponding voltages by an angle, θ:

$$v_a(t) = \sqrt{2}V_{rms}\cos(\omega t),$$
$$v_b(t) = \sqrt{2}V_{rms}\cos(\omega t - 120°),$$
$$v_c(t) = \sqrt{2}V_{rms}\cos(\omega t - 240°),$$
$$i_a(t) = \sqrt{2}I_{rms}\cos(\omega t - \theta),$$
$$i_b(t) = \sqrt{2}I_{rms}\cos(\omega t - 120° - \theta),$$
$$i_c(t) = \sqrt{2}I_{rms}\cos(\omega t - 240° - \theta).$$

We select a root-mean-square (rms) voltage (V_{rms}) of 230 kV, an rms current (I_{rms}) of 500 A, and a phase angle of $\theta = 45°$. The vector functions available in MATLAB make such computations and graphing straightforward:

```
% ThreePhasePower.m
clear all

% consider a balanced three-phase system at 60 Hz
freq = 60;   w = 2*pi*freq;      % Hz and rad/sec

% establish time vector in ms (about 2 cycles long)
t = 0 : 0.5 : 40;

% use high voltage ac values
Vrms = 230;      % rms voltage (kV)
```

```
Irms = 0.5;          % rms current (kA)
angle = pi/4;        % angle V leads I
%    need to change rms to a magnitude
Va = Vrms * sqrt(2) * cos(w*t/1000);
Vb = Vrms * sqrt(2) * cos(w*t/1000 - 2*pi/3);
Vc = Vrms * sqrt(2) * cos(w*t/1000 - 4*pi/3);
% plot the voltage for each phase
subplot(3,1,1);
plot(t,Va,'b-.',t,Vb,'m--',t,Vc,'g-','LineWidth',2.5);
set(gca,'fontname','Times','fontsize',12);
ylabel('Voltage (kV)');
legend('V_a','V_b','V_c');
title(['Three-Phase Balanced System at ',...
    num2str(freq),' Hz, and V Leads I by ',...
    num2str(angle*180/pi),'º']);
%   change rms to a magnitude, and shift angle
Ia = Irms * sqrt(2) * cos(w*t/1000 - angle);
Ib = Irms * sqrt(2) * cos(w*t/1000 - angle - 2*pi/3);
Ic = Irms * sqrt(2) * cos(w*t/1000 - angle - 4*pi/3);
% plot the current for each phase
subplot(3,1,2);
plot(t,Ia,'b-.',t,Ib,'m--',t,Ic,'g-','LineWidth',2.5)
set(gca,'fontname','Times','fontsize',12);
ylabel('Current (kA)');
legend('I_a','I_b','I_c');
```

The instantaneous real power in each phase and the total real power are calculated from:

$$p_a(t) = i_a(t)v_a(t),$$
$$p_b(t) = i_b(t)v_b(t),$$
$$p_c(t) = i_c(t)v_c(t),$$
$$p_T(t) = p_a(t) + p_b(t) + p_c(t).$$

using MATLAB:

```
% compute instantaneous power in each phase
pa = Va .* Ia;       % MW
pb = Vb .* Ib;
pc = Vc .* Ic;
% compute total power in all three phases
pt = pa + pb + pc;

% plot the power for each phase, and total power
subplot(3,1,3);
```

THREE-PHASE QUANTITIES

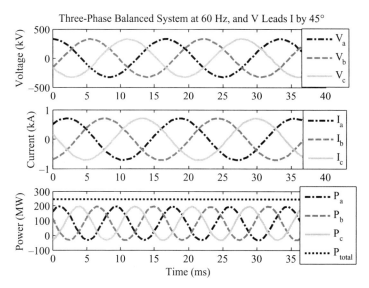

Figure 4.2. Time-varying voltage, current, and power in a balanced three-phase circuit.

```
plot(t,pa,'b-.',  t,pb,'m--',  t,pc,'g-',  t,pt,...
    'k:','LineWidth',2.5)
set(gca,'fontname','Times','fontsize',12);
ylim([-100 300]);
xlabel('Time (ms)');
ylabel('Power (MW)');
legend('P_a','P_b','P_c','P_{total}');
```

The resulting plots from MATLAB are given in Figure 4.2. Note that the instantaneous power of each phase varies at a frequency of 120 Hz, thus verifying Equation (3.10). The instantaneous total power is seen to be a constant as expected. The total power shown in the graph can be verified using Equation (4.4):

$$p_{total} = 3V_{rms}I_{rms}\cos(\theta) = 3(230\text{ kV})(0.5\text{ kA})\cos(45°) = 244\text{ MW},$$

where V_{rms} is the line-to-neutral voltage. The reader is encouraged to vary the phase angle (θ) between the current and voltage from +90° to −90° to study its effect on the instantaneous real powers.

Delta–Wye Connections

Three-phase circuits employ wye- (Y) and delta- (Δ) connected networks. Figure 4.3 and Figure 4.4 illustrate the physical arrangements for wye and delta loads, respectively, to demonstrate the basis for their names. The delta configuration does not lend itself to using series and parallel impedance combination techniques. In such cases, the delta (Δ) connection is converted to a wye (Y) configuration, as depicted in Figure 4.5, using:

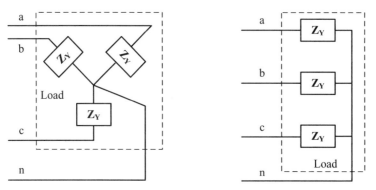

Figure 4.3. Wye-connected loads.

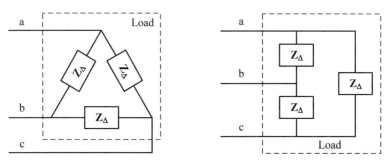

Figure 4.4. Delta-connected loads.

$$\mathbf{Z}_a = \frac{\mathbf{Z}_1 \mathbf{Z}_2}{\mathbf{Z}_1 + \mathbf{Z}_2 + \mathbf{Z}_3},$$
$$\mathbf{Z}_b = \frac{\mathbf{Z}_1 \mathbf{Z}_3}{\mathbf{Z}_1 + \mathbf{Z}_2 + \mathbf{Z}_3}, \quad (4.6)$$
$$\mathbf{Z}_c = \frac{\mathbf{Z}_2 \mathbf{Z}_3}{\mathbf{Z}_1 + \mathbf{Z}_2 + \mathbf{Z}_3}.$$

The previous formula is more easily remembered as the following: the impedance next to a particular Y node is the product of the two delta impedances connected to that node divided by the sum of the three delta impedances. The reverse transformation (Y to Δ) can also be performed:

$$\mathbf{Z}_1 = \frac{\mathbf{Z}_a \mathbf{Z}_b + \mathbf{Z}_b \mathbf{Z}_c + \mathbf{Z}_c \mathbf{Z}_a}{\mathbf{Z}_c},$$
$$\mathbf{Z}_2 = \frac{\mathbf{Z}_a \mathbf{Z}_b + \mathbf{Z}_b \mathbf{Z}_c + \mathbf{Z}_c \mathbf{Z}_a}{\mathbf{Z}_b}, \quad (4.7)$$
$$\mathbf{Z}_3 = \frac{\mathbf{Z}_a \mathbf{Z}_b + \mathbf{Z}_b \mathbf{Z}_c + \mathbf{Z}_c \mathbf{Z}_a}{\mathbf{Z}_a}.$$

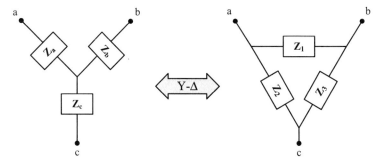

Figure 4.5. Delta–wye impedance transformation.

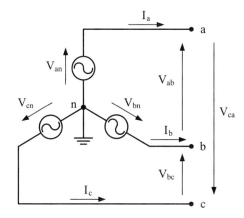

Figure 4.6. Three-phase wye-connected generator.

For the balanced case in which all the delta ($\mathbf{Z}_\Delta = \mathbf{Z}_1 = \mathbf{Z}_2 = \mathbf{Z}_3$) and wye ($\mathbf{Z}_Y = \mathbf{Z}_a = \mathbf{Z}_b = \mathbf{Z}_c$) impedances are equal, the transformation formula reduces to simply:

$$\mathbf{Z}_\Delta = 3\mathbf{Z}_Y. \tag{4.8}$$

Generators and loads connected in wye and delta configurations are presented in the following sections. There are four combinations: (1) Y-Y, (2) Y-Δ, (3) Δ-Y, and (4) Δ-Δ. While analyzing these connections, we shall determine phase and line values for both voltage and current, where *phase* quantities describe the values across or through each phase of the source or load.

4.2. WYE-CONNECTED GENERATOR

Figure 4.6 exhibits a three-phase generator connected in a wye (Y) configuration. The generator can be represented by three voltage sources where the magnitudes of the ac source voltages ($|\mathbf{V}_{an}|, |\mathbf{V}_{bn}|, |\mathbf{V}_{cn}|$) are the same. The phase shift between the voltages is 120°. This system has three phase conductors (**a**, **b**, and **c**) and a grounded neutral (**n**).

The neutral point of a real generator is not directly grounded. An impedance to ground is utilized to limit current if a short circuit occurs. The impedance is sized to limit fault current to 5 A, which is the threshold for iron damage.

The voltage of phase **a** (**V**$_{an}$) is selected as the reference with a phase angle of δ. Similar to Equation (4.1), the phasor voltage expressions are:

$$\begin{aligned} \mathbf{V}_{an} &= V_P \angle \delta, \\ \mathbf{V}_{bn} &= V_P \angle \delta - 120°, \\ \mathbf{V}_{cn} &= V_P \angle \delta - 240°, \end{aligned} \quad (4.9)$$

where V_P is the phase voltage magnitude between the phase conductors and neutral, and is called the *line-to-neutral voltage*. An important property of this balanced voltage set is:

$$\mathbf{V}_{an} + \mathbf{V}_{bn} + \mathbf{V}_{cn} = 0. \quad (4.10)$$

As noted in Chapter 1, the power system voltage is described by the rms value of the *line-to-line* voltage. The *line-to-line voltage* is also termed simply the *line voltage*. To find general expressions for the three line-to-line voltages, we employ Kirchhoff's voltage law (KVL). Referring to Figure 4.6, the KVL relation for the **n** → **a** → **b** → **n** loop is:

$$\mathbf{V}_{an} - \mathbf{V}_{ab} - \mathbf{V}_{bn} = 0. \quad (4.11)$$

We rearrange the previous equation and then substitute for **V**$_{bn}$ in terms of **V**$_{an}$ to find that:

$$\begin{aligned} \mathbf{V}_{ab} &= \mathbf{V}_{an} - \mathbf{V}_{bn} \\ &= \mathbf{V}_{an} - \mathbf{V}_{an} \angle -120° \\ &= \mathbf{V}_{an} \left(1 - e^{-j120°}\right) \\ &= \mathbf{V}_{an} \sqrt{3} \angle 30°. \end{aligned} \quad (4.12)$$

Consider the line voltage **V**$_{ab}$ to have magnitude V_L and phase angle ψ (i.e., **V**$_{ab}$ = $V_L \angle \psi$). The relationship between the line-to-neutral and line-to-line voltages is then:

$$\begin{aligned} \mathbf{V}_{ab} &= \mathbf{V}_{an} \sqrt{3} \angle 30°, \\ V_L \angle \psi &= V_P \angle \delta \sqrt{3} \angle 30°, \\ V_L &= V_P \sqrt{3} \quad \text{and} \quad \psi = \delta + 30°. \end{aligned} \quad (4.13)$$

WYE-CONNECTED GENERATOR

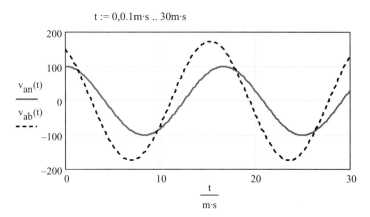

Figure 4.7. Comparison of the relative magnitudes of and the phase shift between the line, $v_{ab}(t)$, and phase, $v_{an}(t)$, voltages.

The line-to-line voltage magnitude is $\sqrt{3}$ times larger than the line-to-neutral (phase) voltage, and the line voltage leads the corresponding phase voltage by $\psi - \delta = 30°$. Figure 4.7 is a comparison of the relative magnitudes of the line and phase voltages and the phase shift between them.

Similar KVL expressions can be developed for the other two line voltages. The three line voltages are:

$$\mathbf{V}_{ab} = \mathbf{V}_{an}\sqrt{3}\angle 30°,$$
$$\mathbf{V}_{bc} = \mathbf{V}_{bn}\sqrt{3}\angle 30° = \mathbf{V}_{an}\sqrt{3}\angle -90°, \qquad (4.14)$$
$$\mathbf{V}_{ca} = \mathbf{V}_{cn}\sqrt{3}\angle 30° = \mathbf{V}_{an}\sqrt{3}\angle -210°.$$

We will find that the wye-connected generator with its three line-to-neutral voltage sources can be equivalently represented by a delta-connected generator having three sources equal to the above line voltages, and vice versa.

EXAMPLE 4.2: Mathcad implementation

A useful exercise is to numerically verify the previous equations using Mathcad. The convention is that the three-phase system is defined by the line-to-line voltage with **ab** being the reference. An rms line-to-line voltage of 69 kV is selected as the reference voltage with a zero-phase angle ($\delta = 0°$):

$$\mathbf{V}_{ab} := 69 \text{ kV} \cdot e^{j0} \quad \mathbf{V}_{bc} := \mathbf{V}_{ab} \cdot e^{-j120 \text{ deg}} \quad \mathbf{V}_{ca} := \mathbf{V}_{ab} \cdot e^{-j240 \text{ deg}}$$

$$\mathbf{V}_{ab} = 69 \cdot \text{kV} \qquad |\mathbf{V}_{ab}| = 69 \cdot \text{kV} \quad \arg(\mathbf{V}_{ab}) = 0 \cdot \text{deg}$$
$$\mathbf{V}_{bc} = (-34.5 - 59.8j) \cdot \text{kV} \quad |\mathbf{V}_{bc}| = 69 \cdot \text{kV} \quad \arg(\mathbf{V}_{bc}) = -120 \cdot \text{deg}$$
$$\mathbf{V}_{ca} = (-34.5 + 59.8j) \cdot \text{kV} \quad |\mathbf{V}_{ca}| = 69 \cdot \text{kV} \quad \arg(\mathbf{V}_{ca}) = 120 \cdot \text{deg}$$

Note that Mathcad has two equal signs: the first equal sign (:=) is used when a variable is defined; the second equal sign (=) is used to obtain the value of a variable. The line-to-neutral voltage of phase **a** (V_{an}) is found using Equation (4.14):

$$V_{an} := \frac{V_{ab}}{\sqrt{3} \cdot e^{j30\,deg}}.$$

The other two phase voltages are defined in reference to the phase **a** voltage:

$$V_{bn} := V_{an} \cdot e^{-j120\,deg} \quad V_{cn} := V_{an} \cdot e^{-j240\,deg}.$$

The numerical values of the phase voltages are:

$$|V_{an}| = 39.8 \cdot kV \qquad |V_{bn}| = 39.8 \cdot kV \qquad |V_{cn}| = 39.8 \cdot kV$$
$$\arg(V_{an}) = -30.0 \cdot deg \quad \arg(V_{bn}) = -150.0 \cdot deg \quad \arg(V_{cn}) = 90.0 \cdot deg.$$

The sum of the three phase voltages is zero, thereby confirming Equation (4.10):

$$V_{an} + V_{bn} + V_{cn} = 0 \text{ kV}.$$

The obtained numerical values demonstrate that the phase angle between the line-to-line voltages is 120°, and the phase angle between the line-to-neutral and the line-to-line voltage is 30° (e.g., $\angle V_{ab} - \angle V_{an} = 30°$). The line-to-line voltage is $\sqrt{3}$ times larger than the line-to-neutral voltage (e.g., $|V_{ab}| = \sqrt{3}|V_{an}|$). This relationship is illustrated on the phasor diagram of Figure 4.8. The phasor diagram clearly shows the 30° phase shift between the line-to-line and line-to-neutral voltages, and the construction of the line-to-line voltages from the difference of the two line-to-neutral voltages.

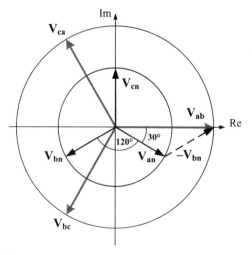

Figure 4.8. Phasor diagram of a wye-connected balanced system.

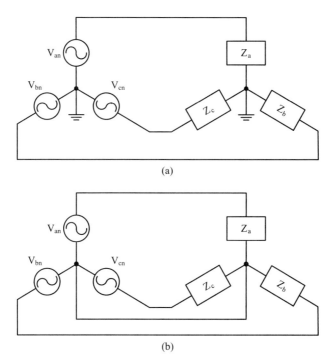

Figure 4.9. (a) Three- and (b) four-wire wye-connected systems.

4.3. WYE-CONNECTED LOADS

A wye-connected load can be connected to a wye generator via either a three-wire or a four-wire system, as drawn in Figure 4.9. Most high-voltage transmission lines use a three-wire grounded system. The loads are connected between the phases through transformers. The ground current is minimal, preferably zero. However, if the neutral point at both the supply and the load are grounded, the earth acts as a conductor and interconnects the two neutral points. This effectively transforms the three-wire system to act similar to a four-wire system.

In the distribution level, the four-wire system is frequently used. The neutral wire is insulated, although it is grounded at the transformer sites.

At the low voltage level, both the three-wire and four-wire systems are used by industry. The larger industrial complexes use a three-wire 460 V system for larger motor loads. The smaller loads are supplied by a 208 V four-wire system, where lighting and small loads are connected between the phases and the neutral wire at 120 V. The larger loads are connected between the phases and supplied by 208 V. The motors are supplied by the 208 V three-phase system.

A wye-connected generator can be loaded by three impedances connected in a wye configuration, as seen in Figure 4.10. Three phase conductors (**a**, **b**, **c**) connect the load impedances (Z_a, Z_b, Z_c) to the sources, and the neutral conductor connects the neutral

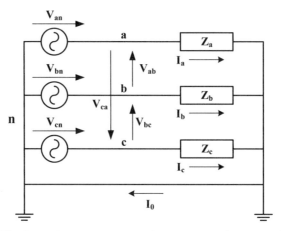

Figure 4.10. Three-phase wye-connected generator with an impedance load.

point of the sources to the neutral point of the load. The neutral point of the sources is grounded to assure that the potential of the phase conductors to the ground remains constant. This is an important safety consideration. In actual applications, generators are not generally connected directly to the load; an exception is shipboard power systems. Loads are connected to the generators through transformers and transmission lines, which are the subjects of later chapters.

For the four-wire circuit, the source voltages are connected directly to the load impedances and the neutral conductor provides a return path for each phase current. Each source line-to-neutral voltage drives a current through the corresponding impedance. However, in general, for a wye-connected load, the terminology "load voltage" refers to the line-to-line voltage. Figure 4.10 reveals that the current in each phase can be calculated by dividing the appropriate line-to-neutral voltage by the corresponding load impedance in that phase (e.g., $\mathbf{I_b} = \mathbf{V_{bn}}/\mathbf{Z_b}$).

4.3.1. Balanced Wye Load (Four-Wire System)

Most three-phase systems operate with a (near-) balanced load, where each phase carries the same load current or is loaded with the same impedance. Because the phases do not affect one another, it is sufficient to calculate the current in phase **a**. The currents in phase **b** and **c** have the same amplitude and relative phase angle, but the phase shift between the currents is 120°.

Referring to Figure 4.10 and using KVL applied to phase **a** yields:

$$\mathbf{V_{an}} - \mathbf{I_a}\mathbf{Z_a} = 0. \qquad (4.15)$$

The line current in phase **a** is:

$$\mathbf{I_a} = \frac{\mathbf{V_{an}}}{\mathbf{Z_a}}. \qquad (4.16)$$

WYE-CONNECTED LOADS

Since $Z_a = Z_b = Z_c = Z_Y$, the line currents in phases **b** and **c** are identical in magnitude to I_a, but lagging by 120° and 240°, respectively. Using this fact and applying Kirchhoff's current law (KCL) at the neutral node yields:

$$I_0 = I_a + I_b + I_c = 0. \tag{4.17}$$

The complex power in the load of phase **a** is:

$$S_a = V_{an} I_a^*. \tag{4.18}$$

The total three-phase complex power is three times the single-phase power above:

$$S_T = 3 S_a. \tag{4.19}$$

EXAMPLE 4.3: The balanced three-phase Y-Y system is represented by a single-phase equivalent circuit. This circuit carries one-third of the total three-phase power. The circuit is energized by the line-to-neutral voltage. To illustrate the calculation of the current in a balanced three-phase system, we assume that the generator of Figure 4.6 is loaded by a three-phase load of 3 kW with a power factor (*pf*) of 0.8 lagging and line-to-line voltage of 480 V:

$$P_{3ph\_load} := 3000 \text{ W} \quad pf_{load} := 0.8 \quad V_{an} := \frac{480 \text{ V}}{\sqrt{3}} = 277.1 \text{ V}.$$

The system is represented by the single-phase equivalent diagram provided in Figure 4.11. The load in this circuit is one-third of the three-phase load:

$$P_{1ph} := \frac{P_{3ph\_load}}{3} \quad P_{1ph} = 1000 \text{ W}.$$

The line current in phase **a** is:

$$I_a := \frac{P_{1ph}}{V_{an} \cdot pf_{load}} \cdot e^{-j \cdot acos(pf_{load})}.$$

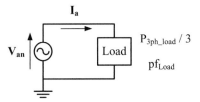

Figure 4.11. Single-phase equivalent of a three-phase system with a balanced load.

The exponential is negative because *pf* is lagging. The currents of phases **b** and **c** are shifted by 120° and 240°, respectively:

$$I_b := I_a \cdot e^{-j120 \text{ deg}} \quad I_c := I_a \cdot e^{-j240 \text{ deg}}.$$

The numerical current values are:

$$|I_a| = 4.51 \text{ A} \quad \arg(I_a) = -36.9 \cdot \text{deg}$$
$$|I_b| = 4.51 \text{ A} \quad \arg(I_b) = -156.9 \cdot \text{deg}$$
$$|I_c| = 4.51 \text{ A} \quad \arg(I_c) = 83.1 \cdot \text{deg}.$$

4.3.2. Unbalanced Wye Load (Four-Wire System)

An analytical derivation of the line currents in the unbalanced case proceeds in the same manner as that for the balanced case, except that each phase must be separately analyzed. Assuming that the neutral return line has a zero impedance, as is implied in Figure 4.10, means that each line current can be determined using Ohm's law:

$$I_a = \frac{V_{an}}{Z_a} \quad I_b = \frac{V_{bn}}{Z_b} \quad I_c = \frac{V_{cn}}{Z_c}. \tag{4.20}$$

For the unbalanced case:

$$I_0 = I_a + I_b + I_c \neq 0, \tag{4.21}$$

and the total complex power supplied by the generator is:

$$S_T = S_a + S_b + S_c = V_{an}I_a^* + V_{bn}I_b^* + V_{cn}I_c^*. \tag{4.22}$$

The unbalanced case requires the analysis of all three phases or the application of symmetrical components. In either case, use of a computer program is highly advantageous to perform such repetitive complex-valued calculations. The symmetrical components technique is presented later in this chapter.

EXAMPLE 4.4: This Mathcad example analyzes an unbalanced four-wire Y-Y system to find the line currents and the total three-phase power. To illustrate this load current calculation, the load in phase **a** is selected as a resistance and inductance connected in series:

$$Z_a := (10 + j \cdot 5) \cdot \Omega.$$

The load in phase **b** is set as a resistance and capacitance connected in series:

$$Z_b := (12 - j \cdot 7) \cdot \Omega.$$

WYE-CONNECTED LOADS

The load in phase **c** is chosen as a pure resistance: $Z_c := 13 \; \Omega$.
The phase currents are calculated by Ohm's law:

$$I_a := \frac{V_{an}}{Z_a} \quad I_b := \frac{V_{bn}}{Z_b} \quad I_c := \frac{V_{cn}}{Z_c}.$$

By KCL, the current in the neutral conductor is the sum of the phase currents:

$$I_0 := I_a + I_b + I_c.$$

For line-to-neutral voltages of:

$$V_{an} := 120 \; V \quad V_{bn} := 120 \; V \cdot e^{-j \cdot 120 \; deg} \quad V_{cn} := 120 \; V \cdot e^{-j \cdot 240 \; deg},$$

the numerical results for the line currents are:

$$|I_a| = 10.7 \; A \quad \arg(I_a) = -26.6 \cdot deg \quad |I_b| = 8.6 \; A \quad \arg(I_b) = -89.7 \cdot deg$$
$$|I_c| = 9.2 \; A \quad \arg(I_c) = 120 \cdot deg \quad |I_0| = 7.4 \; A \quad \arg(I_0) = -47.3 \cdot deg.$$

Figure 4.10 and this calculation illustrate that the phase currents do not affect each other if the neutral conductor impedance is negligible.

The analysis reveals that a wye-connected system has two voltages (line-to-line and line-to-neutral) and a single line–load current (because the line and load currents are identical) in each phase. In contrast, for a delta-connected system, there is only one voltage (i.e., the line voltage) and two currents (line and load), as will be demonstrated in Section 4.4.

Power Calculation

The generator complex power can be calculated by computing the complex power of each phase and adding them together to obtain the total three-phase power. The complex power of each phase is calculated by multiplying the appropriate source voltage with the corresponding current conjugate:

$$S_a := V_{an} \cdot \overline{I_a} \quad S_b := V_{bn} \cdot \overline{I_b} \quad S_c := V_{cn} \cdot \overline{I_c}.$$

As a reminder, Mathcad signifies the complex conjugate operation using an overline notation. The three-phase complex power is:

$$S_{3\_phase} := S_a + S_b + S_c.$$

The corresponding numerical values are:

$$S_a = 1152 + 576j \; V \cdot A \quad S_b = 895 - 522j \; V \cdot A \quad S_c = 1108 \; V \cdot A,$$
$$S_{3\_phase} = 3155 + 53.7j \; V \cdot A,$$

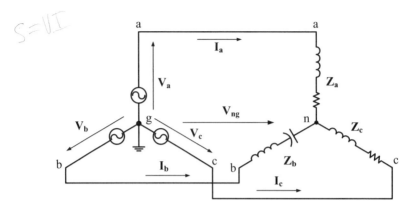

Figure 4.12. Three-wire wye-connected system.

where the first number is the real power and the second number is the reactive power. It can be seen that phase **a** carries inductive reactive power (lagging *pf*). Phase **b** carries capacitive reactive power, and in phase **c**, the reactive power is zero.

4.3.3. Wye-Connected Three-Wire System

Figure 4.12 presents a wye-connected three-wire system. Three wye-connected sources supply three impedances also connected in a wye. The neutral point (**g**) of the generator is grounded. But the supply and load neutral points are not connected together. This eliminates the direct return path. The unequal line currents generate a voltage difference (V_{gn}) between the neutral point of the generator and the neutral point (**n**) of the load impedances.

EXAMPLE 4.5: The system operation is demonstrated through a numerical example. The balanced supply line-to-line voltage is: $V_{L\_L} := 480$ V.

The three load impedances are:

$$Z_a := (70 + j \cdot 60)\,\Omega \quad Z_b := (40 - j \cdot 50)\,\Omega \quad Z_c := (80 + j \cdot 30)\,\Omega.$$

The reference voltage is the line-to-neutral voltage of phase **a**. The line-to-neutral voltages are:

$$V_a := \frac{480\text{ V}}{\sqrt{3}} \quad V_b := V_a \cdot e^{-j120\text{ deg}} \quad V_c := V_a \cdot e^{-j240\text{ deg}},$$

$|V_a| = 277.1$ V $\quad |V_b| = 277.1$ V $\quad |V_c| = 277.1$ V
$\arg(V_a) = 0$ deg $\arg(V_b) = -120$ deg $\arg(V_c) = 120$ deg.

WYE-CONNECTED LOADS

Voltage Difference between the Neutral Points

The system contains three lines, which are $g \to a \to n$, $g \to b \to n$, and $g \to c \to n$. The currents are calculated from the loop voltage equations (KVL). As an example in loop $g \to a \to n \to g$, the difference of the line-to-neutral voltage and the voltage between the neutral points (V_{ng}) drives the current though impedance Z_a. The line currents are:

$$I_a := \frac{V_a - V_{ng}}{Z_a} \quad I_b := \frac{V_b - V_{ng}}{Z_b} \quad I_c := \frac{V_c - V_{ng}}{Z_c}. \tag{4.23}$$

The sum of the currents in node point **n** is zero. The node point equation for node **n** is:

$$\frac{V_a - V_{ng}}{Z_a} + \frac{V_b - V_{ng}}{Z_b} + \frac{V_c - V_{ng}}{Z_c} = 0.$$

The voltage difference between the neutral points is calculated from this equation. The voltage difference between the neutral points is:

$$V_{ng} := \frac{\dfrac{V_a}{Z_a} + \dfrac{V_b}{Z_b} + \dfrac{V_c}{Z_c}}{\dfrac{1}{Z_a} + \dfrac{1}{Z_b} + \dfrac{1}{Z_c}}. \tag{4.24}$$

The numerical result discloses that the voltage difference is significant:

$$V_{ng} = 111.3 - 100.2j \text{ V} \quad |V_{ng}| = 149.8 \text{ V}.$$

The currents are calculated by substituting the obtained voltage difference in Equation (4.23). The currents are:

$$I_a := \frac{V_a - V_{ng}}{Z_a} \quad |I_a| = 2.10 \text{ A} \quad \arg(I_a) = -9.4 \text{ deg}$$

$$I_b := \frac{V_b - V_{ng}}{Z_b} \quad |I_b| = 4.47 \text{ A} \quad \arg(I_b) = -99.4 \text{ deg}$$

$$I_c := \frac{V_c - V_{ng}}{Z_c} \quad |I_c| = 4.94 \text{ A} \quad \arg(I_c) = 105.7 \text{ deg}$$

The calculation of the power is similar to Example 4.4.

The earlier analysis is for an unbalanced load. We can extend the results to the case of a balanced load by considering Equation (4.24) when $\mathbf{Z_a} = \mathbf{Z_b} = \mathbf{Z_c} = \mathbf{Z_Y}$:

$$\mathbf{V}_{ng} = \frac{\dfrac{\mathbf{V}_a}{\mathbf{Z}_a} + \dfrac{\mathbf{V}_b}{\mathbf{Z}_b} + \dfrac{\mathbf{V}_c}{\mathbf{Z}_c}}{\dfrac{1}{\mathbf{Z}_a} + \dfrac{1}{\mathbf{Z}_b} + \dfrac{1}{\mathbf{Z}_c}} = \frac{\dfrac{1}{\mathbf{Z}_Y}(\mathbf{V}_a + \mathbf{V}_b + \mathbf{V}_c)}{\dfrac{3}{\mathbf{Z}_Y}} = \frac{\mathbf{V}_a + \mathbf{V}_b + \mathbf{V}_c}{3}.$$

In addition, we know from Equation (4.10) that the sum of the balanced voltage sources is zero; therefore $\mathbf{V}_{ng} = 0$ for a balanced three-phase circuit.

4.4. DELTA-CONNECTED SYSTEM

Again, in a delta system, we examine both Δ-connected generators and loads.

4.4.1. Delta-Connected Generator

Figure 4.13 depicts a three-phase generator connected in a delta (Δ). The voltage between terminals **ab**, **bc**, and **ca** are the line-to-line voltages. For a balanced three-phase system, the magnitudes of these voltages are equal, and the phase shift between them is 120°. The voltage between phase **ab** is typically selected as the reference with a zero phase angle ($\psi = 0$). The delta source (line) voltages are:

$$\begin{aligned}\mathbf{V}_{ab} &= V_L \angle \psi, \\ \mathbf{V}_{bc} &= V_L \angle \psi - 120°, \\ \mathbf{V}_{ca} &= V_L \angle \psi - 240°.\end{aligned} \quad (4.25)$$

The delta-connected generator can be converted to an equivalent wye-connected supply consisting of three voltage sources, as illustrated in Figure 4.13. The voltages of the

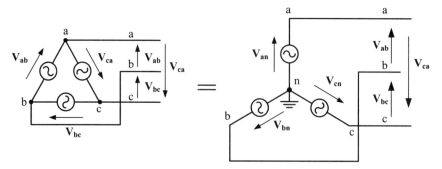

Figure 4.13. Delta-connected generators and the equivalent wye-connected system.

equivalent wye-connected sources are the line-to-neutral voltages, which are calculated from the line-to-line voltages using the following relations, which are essentially a restatement of Equation (4.14):

$$\mathbf{V}_{an} = \frac{\mathbf{V}_{ab}}{\sqrt{3}} e^{-j30°} \quad \mathbf{V}_{bn} = \frac{\mathbf{V}_{bc}}{\sqrt{3}} e^{-j30°} \quad \mathbf{V}_{cn} = \frac{\mathbf{V}_{ca}}{\sqrt{3}} e^{-j30°}. \quad (4.26)$$

The practical use of this conversion is that we can assume either a wye or a delta configuration for the supply depending on the load. In the case of a delta load, the delta supply connection is more advantageous, and likewise, in the case of a wye load, the assumption of a wye-connected source simplifies the calculation. Most of the time, we assume that the supply is arranged in a wye. A delta-connected generator would lead to a circulating current within the delta mesh if the three phase voltages are unbalanced.

In general, the line-to-line voltage is stated as the rated voltage of a three-phase system. As an example, the 500 kV system, which is used for bulk energy transmission in most parts of the United States, has a line-to-ground voltage of 500 kV/$\sqrt{3}$ = 288.67 kV, since the line-to-line voltage is the 500 kV.

4.4.2. Balanced Delta Load

Figure 4.14 diagrams a delta-connected load in which the line-to-line (source) voltages are directly applied to the load if the line impedance is negligible. This implies that the applied voltages are independent of each other. Consequently, the calculation of the load current **ab** is sufficient in the case of a balanced load. The current in phases **bc** and **ca** will have the same amplitude and relative phase angle as $\mathbf{I}_{ab}$, but will be shifted by ±120°.

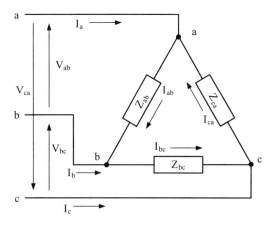

Figure 4.14. Delta-connected load.

The current through impedance Z_{ab} is computed using Ohm's law:

$$\mathbf{I}_{ab} = \frac{\mathbf{V}_{ab}}{\mathbf{Z}_{ab}}. \tag{4.27}$$

To determine the line current $\mathbf{I}_a$, we employ KCL at node **a**:

$$\mathbf{I}_a = \mathbf{I}_{ab} - \mathbf{I}_{ca}. \tag{4.28}$$

Substituting the Ohm's law expressions for the phase currents yields:

$$\mathbf{I}_a = \frac{\mathbf{V}_{ab}}{\mathbf{Z}_{ab}} - \frac{\mathbf{V}_{ca}}{\mathbf{Z}_{ca}}. \tag{4.29}$$

For a balanced delta load, $\mathbf{Z}_{ab} = \mathbf{Z}_{bc} = \mathbf{Z}_{ca} = \mathbf{Z}_\Delta$, and for a balanced source, $\mathbf{V}_{ca} = \mathbf{V}_{ab} \angle -240°$, such that:

$$\begin{aligned}
\mathbf{I}_a &= \frac{\mathbf{V}_{ab}}{\mathbf{Z}_\Delta} - \frac{\mathbf{V}_{ab} \angle -240°}{\mathbf{Z}_\Delta} \\
&= \frac{\mathbf{V}_{ab}}{\mathbf{Z}_\Delta}(1 - \angle -240°) \\
&= \mathbf{I}_{ab}(1 - e^{-j240°}) \\
&= \mathbf{I}_{ab}\sqrt{3}\angle -30°.
\end{aligned} \tag{4.30}$$

Thus, the line current is $\sqrt{3}$ times larger than the delta load (phase) current.

The voltage $\mathbf{V}_{ab}$ and current $\mathbf{I}_{ab}$ for $\mathbf{Z}_{ab}$ obey the passive sign convention, and the complex power is:

$$\mathbf{S}_{ab} = \mathbf{V}_{ab}\mathbf{I}_{ab}^*. \tag{4.31}$$

The total three-phase complex power is three times the single-phase power previously mentioned ($\mathbf{S}_T = 3\,\mathbf{S}_{ab}$).

It is important to note that the earlier analysis is applicable to a delta load being supplied by either (1) a delta generator with line voltages $\mathbf{V}_{ab}$, $\mathbf{V}_{bc}$, and $\mathbf{V}_{ca}$, or (2) a wye generator with line-to-neutral voltages $\mathbf{V}_{an}$, $\mathbf{V}_{bn}$, and $\mathbf{V}_{cn}$.

EXAMPLE 4.6: The analysis of the delta-connected system is presented through a numerical example. The line-to-line voltages with $\mathbf{V}_{ab}$ as the reference are:

$$\mathbf{V}_{ab} := 208\text{ V} \quad \mathbf{V}_{bc} := \mathbf{V}_{ab} \cdot e^{-j120\text{ deg}} \quad \mathbf{V}_{ca} := \mathbf{V}_{ab} \cdot e^{-j240\text{ deg}}.$$

DELTA-CONNECTED SYSTEM

The numerical values are:

$$|V_{bc}| = 208 \text{ V} \quad \arg(V_{bc}) = -120 \cdot \deg$$
$$|V_{ca}| = 208 \text{ V} \quad \arg(V_{ca}) = 120 \cdot \deg.$$

This delta-connected generator will be connected to a balanced load in this example and then to an unbalanced load in the example in the next subsection below.

To illustrate the calculation of the line and phase currents in a balanced three-phase system, we assume that the sources in Figure 4.13 are loaded by a three-phase load of 3 kW with a power factor of 0.8 lagging:

$$P_{3ph\_load} := 3000 \text{ W} \quad pf_{load} := 0.8.$$

The line-to-line phasor voltage is set to $V_{ab} := 208 \cdot e^{-j \cdot 0 \deg}$ V.

Phase **ab** carries one-third of the total three-phase load and the line-to-line voltage supplies this load:

$$P_{1ph} := \frac{P_{3ph\_load}}{3} \quad P_{1ph} = 1000 \text{ W}.$$

The current of phase **ab** is:

$$I_{ab} := \frac{P_{1ph}}{V_{ab} \cdot pf_{load}} \cdot e^{-j \cdot a\cos(pf_{load})}.$$

The numerical values for the phase current **ab** are:

$$I_{ab} = 4.81 - 3.61j \text{ A} \quad |I_{ab}| = 6.01 \text{ A} \quad \arg(I_{ab}) = -36.9 \deg.$$

The currents of phases **bc** and **ca** are shifted by 120° and 240°, respectively:

$$I_{bc} := I_{ab} \cdot e^{-j \cdot 120 \deg} \quad |I_{bc}| = 6.01 \text{ A} \quad \arg(I_{bc}) = -157 \cdot \deg$$
$$I_{ca} := I_{ab} \cdot e^{-j \cdot 240 \deg} \quad |I_{ca}| = 6.01 \text{ A} \quad \arg(I_{ca}) = 83 \cdot \deg.$$

From KCL and referring to Figure 4.14, the line currents are:

$$I_a := I_{ab} - I_{ca} \quad I_b := I_{bc} - I_{ab} \quad I_c := I_{ca} - I_{bc}.$$

The numerical values for the line currents are:

$$I_a = (4.09 - 9.57j) \text{ A} \quad |I_a| = 10.41 \text{ A} \quad \arg(I_a) = -66.9 \cdot \deg$$
$$I_b = (-10.33 + 1.25j) \text{ A} \quad |I_b| = 10.41 \text{ A} \quad \arg(I_b) = 173.1 \cdot \deg$$
$$I_c = (6.25 + 8.33j) \text{ A} \quad |I_c| = 10.41 \text{ A} \quad \arg(I_c) = 53.1 \cdot \deg$$

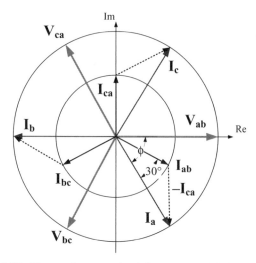

Figure 4.15. Phasor diagram of a delta-connected balanced system.

Comparison of the line currents reveals that their amplitudes are the same in each phase and the phase shift between them is 120°. Comparison of the line and phase currents shows that the line current is $\sqrt{3}$ times the phase current and the phase shift is −30°, as annotated in Figure 4.15. That phasor diagram is, in general, applicable to all delta-connected balanced systems. This example also points out that in a delta system there are two currents (phase and line) and one voltage (the line-to-line voltage).

An alternative method for calculating the line current in a balanced system is to use Equation (4.30):

$$I_a := \sqrt{3} \cdot I_{ab} \cdot e^{-j \cdot 30 \text{ deg}}.$$

The numerical values are identical to those previously mentioned:

$$I_a = 4.09 - 9.57j \text{ A} \quad |I_a| = 10.41 \text{ A} \quad \arg(I_a) = -66.9 \text{ deg}.$$

Similar formulas are used to calculate the currents in phases **b** and **c**.

4.4.3. Unbalanced Delta Load

Derivation of the line and phase currents in the case of an unbalanced delta load proceeds in a similar manner as that of the balanced delta load. Referring to Figure 4.14, each load (phase) current is directly computed from Ohm's law:

$$I_{ab} = \frac{V_{ab}}{Z_{ab}} \quad I_{bc} = \frac{V_{bc}}{Z_{bc}} \quad I_{ca} = \frac{V_{ca}}{Z_{ca}}. \tag{4.32}$$

DELTA-CONNECTED SYSTEM

The line currents must be determined from applying KCL at each node of the delta load:

$$\mathbf{I}_a = \mathbf{I}_{ab} - \mathbf{I}_{ca},$$
$$\mathbf{I}_b = \mathbf{I}_{bc} - \mathbf{I}_{ab}, \quad (4.33)$$
$$\mathbf{I}_c = \mathbf{I}_{ca} - \mathbf{I}_{bc}.$$

For an unbalanced load, we will find that the line currents and load currents through each phase are different, and that they are not evenly separated by 120°.

EXAMPLE 4.7: The delta- or wye-connected source of the prior example is loaded with three impedances connected in a delta. The impedances are:

$$Z_{ab} := 23 \, \Omega \quad Z_{bc} := (22 + j \cdot 15) \cdot \Omega \quad Z_{ca} := (25 - j \cdot 22) \cdot \Omega.$$

The three-phase equivalent circuit of the load in Figure 4.14 shows that each line-to-line voltage is directly applied to the corresponding impedance, which implies that the current through each impedance can be calculated by dividing the appropriate line-to-line voltage by the corresponding impedance. In other words, the currents through the impedances are calculated using Ohm's law. The phase currents are:

$$\mathbf{I}_{ab} := \frac{\mathbf{V}_{ab}}{Z_{ab}} \quad \mathbf{I}_{bc} := \frac{\mathbf{V}_{bc}}{Z_{bc}} \quad \mathbf{I}_{ca} := \frac{\mathbf{V}_{ca}}{Z_{ca}}.$$

The numerical values are:

$$\mathbf{I}_{ab} = 9.04 \, \text{A} \quad |\mathbf{I}_{ab}| = 9.04 \, \text{A} \quad \arg(\mathbf{I}_{ab}) = 0 \cdot \deg$$
$$\mathbf{I}_{bc} = (-7.04 - 3.39 j) \, \text{A} \quad |\mathbf{I}_{bc}| = 7.81 \, \text{A} \quad \arg(\mathbf{I}_{bc}) = -154.3 \cdot \deg$$
$$\mathbf{I}_{ca} = (-5.92 + 2.00 j) \, \text{A} \quad |\mathbf{I}_{ca}| = 6.25 \, \text{A} \quad \arg(\mathbf{I}_{ca}) = 161.3 \cdot \deg.$$

The numerical results demonstrate that the phase shifts between the phase currents are *not* 120°, as in the case of a balanced load, nor are the current magnitudes equal.

The currents in the conductors connecting the generator to the loads are the line currents. The line currents can be calculated using KCL. The node point equation for **a** is:

$$\mathbf{I}_a - \mathbf{I}_{ab} + \mathbf{I}_{ca} = 0.$$

From this equation the current in line **a** is:

$$\mathbf{I}_a := \mathbf{I}_{ab} - \mathbf{I}_{ca}.$$

Using the node point equations for nodes **b** and **c** permits the calculation of the line currents **b** and **c**:

$$I_b := I_{bc} - I_{ab} \quad I_c := I_{ca} - I_{bc}.$$

The numerical values are:

$$I_a = (14.96 - 2.00j) \text{ A} \quad |I_a| = 15.09 \text{ A} \quad \arg(I_a) = -7.6 \cdot \deg$$
$$I_b = (-16.08 - 3.39j) \text{ A} \quad |I_b| = 16.43 \text{ A} \quad \arg(I_b) = -168.1 \cdot \deg$$
$$I_c = (1.12 + 5.39j) \text{ A} \quad |I_c| = 5.50 \text{ A} \quad \arg(I_c) = 78.3 \cdot \deg.$$

In the case of an unbalanced load, the line current magnitudes are different in each phase and the phase shifts are not 120°.

The symmetrical components technique (see Section 4.8) is another approach to analyzing unbalanced three-phase systems.

4.5. SUMMARY

The current, voltage, and power relationships for a wye-connected source are summarized in Table 4.1 for the cases of a balanced wye-connected load (see Fig. 4.16) and delta-connected load (see Fig. 4.17).

Wye and delta connections have their advantages and disadvantages. A wye-configured source provides a load with ready access to either the line voltage or the line-to-neutral voltage. The wye connection also provides a grounding point for safety

TABLE 4.1. Wye-Connected Source with Balanced Load

Per phase Parameters	Wye-Connected Load	Delta-Connected Load
Source voltage	$V_{an} = V_p e^{j\delta} = V_p \angle \delta$	
Line voltage	$V_{ab} = \sqrt{3} V_{an} e^{j30°} = \sqrt{3} V_{an} \angle 30° = \sqrt{3} V_p \angle \delta + 30° = V_L \angle \delta + 30°$	
Load impedance	$Z_Y = Z_Y \angle \theta = Z_Y \angle \delta - \phi$	$Z_\Delta = Z_\Delta \angle \theta = 3Z_Y$
Load voltage	V_{an}	V_{ab}
Load current	$I_{an} = I_Y \angle \phi = \dfrac{V_{an}}{Z_Y}$	$I_{ab} = I_\Delta \angle \phi + 30° = \dfrac{V_{ab}}{Z_\Delta}$
	$= \dfrac{V_p \angle \delta}{Z_Y \angle \theta}$	$= \dfrac{I_L}{\sqrt{3}} \angle \phi + 30°$
Line current	$I_a = I_L \angle \phi = I_{an}$	$I_a = I_L \angle \phi = \sqrt{3} I_{ab} \angle -30°$
Load power (per phase)	$S_Y = V_{an} I_a^* = V_p I_L \angle \delta - \phi$	$S_\Delta = V_{ab} I_{ab}^* = V_L I_\Delta \angle \delta - \phi$
	$= \dfrac{V_L I_L}{\sqrt{3}} \angle \theta$	$= \dfrac{V_L I_L}{\sqrt{3}} \angle \theta$
Load power (total)	$S_T = 3S_Y = \sqrt{3} V_L I_L \angle \theta$	$S_T = 3S_\Delta = \sqrt{3} V_L I_L \angle \theta$

SUMMARY 169

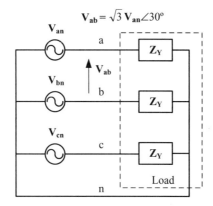

Figure 4.16. Balanced wye–wye connection.

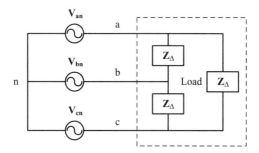

Figure 4.17. Balanced wye–delta connection.

and system protection purposes. In contrast, a load arranged in a delta remains in balance better when serving unbalanced loads and it can trap the third harmonic. An overall delta load configuration permits the addition or removal of a particular load from a specific phase more easily than a wye connection since the neutral point may not be accessible for a Y-connected load.

EXAMPLE 4.8: Analysis of motor operation in delta and wye configurations

A three-phase 60 Hz motor rated at 75 hp and 460 V operates at rated load with an efficiency of 0.9 and a power factor of 0.7 lagging. Assuming the motor is connected (1) in a wye and (2) in a delta, calculate the input power, the complex and reactive powers, and the motor phase and line currents for both cases. Regardless of the connection type, the motor voltage is the line-to-line voltage.

The wye- and delta-connected motor circuits are given in Figure 4.18 and Figure 4.19, respectively. The motor data are assigned to appropriate variables:

$$P_m := 75 \text{hp} \quad \eta := 0.9 \quad f_m : 60 \text{ Hz}$$
$$V_m := 460 \cdot e^{j 0 \text{ deg}} \text{ V} \quad pf_m := 0.7 \text{ (lagging)}.$$

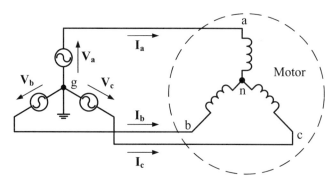

Figure 4.18. Wye-connected three-phase motor.

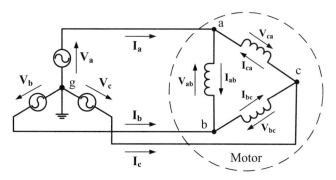

Figure 4.19. Delta-connected three-phase motor.

First, we compute the motor input electric power, complex power, and reactive power, which are independent of the connection type. Every motor has both electrical and mechanical losses in converting the electrical to mechanical energy. The manufacturers provide the motor efficiency specification, which is approximately 80–90%. Therefore, the input electric power to the motor is:

$$P_{m\_in} := \frac{P_m}{\eta} \quad P_{m\_in} = 62.14 \text{ kW}.$$

The real power is measured in watts, but the complex or apparent power is measured in volt-amperes (VA). The reactive power is measured in VAR (volt-amperes reactive). Unfortunately, Mathcad does not recognize the VAR unit. Consequently, the VA unit is used in Mathcad as the measure of reactive power. The three-phase complex and reactive powers are:

$$S_{m\_in} := \frac{P_{m\_in}}{pf_m} \cdot e^{j \cdot acos(pf_m)} \quad S_{m\_in} = (62.14 + 63.40j) \text{ kV} \cdot \text{A}$$

$$Q_{m\_in} := \text{Im}(S_{m\_in}) \quad Q_{m\_in} = 63.4 \text{ kV} \cdot \text{A}.$$

SUMMARY

The reactive power is positive when the power factor is lagging; consequently, the exponential is positive.

Delta Connection Currents

The delta-connected motor has two types of currents: phase (load) and line currents. The line-to-line voltage drives the load current through the phases. Each phase carries one-third of the total power. The current through phase **ab** is:

$$I_{ab\_delta} := \overline{\left(\frac{\frac{S_{m\_in}}{3}}{V_m}\right)} \quad I_{ab\_delta} = (45.0 - 45.9j) \text{ A}$$

$$|I_{ab\_delta}| = 64.3 \text{ A} \quad \arg(I_{ab\_delta}) = -45.6 \text{ deg.}$$

For a balanced load, the other two delta currents are shifted by 120° and 240°:

$$I_{bc\_delta} := I_{ab\_delta} \cdot e^{-j120 \text{ deg}} \quad I_{ca\_delta} := I_{ab\_delta} \cdot e^{-j240 \text{ deg}},$$

$$|I_{bc\_delta}| = 64.3 \text{ A} \quad \arg(I_{bc\_delta}) = -165.6 \text{ deg}$$

$$|I_{ca\_delta}| = 64.3 \text{ A} \quad \arg(I_{ca\_delta}) = 74.4 \text{ deg.}$$

Referring to Figure 4.19, the line current in phase **a** for a balanced delta connection is:

$$I_{a\_delta} := I_{ab\_delta} - I_{ca\_delta} \quad I_{a\_delta} = (27.8 - 107.9j) \text{ A}$$

$$|I_{a\_delta}| = 111.4 \text{ A} \quad \arg(I_{a\_delta}) = -75.6 \text{ deg.}$$

Wye Connection Current

The 460 V line-to-line voltage is the reference; the line-to-neutral voltage is obtained by division of the line-to-line voltage with $\sqrt{3}$ and shifting by −30° using Equation (4.26):

$$V_{m\_ln} := \frac{V_m}{\sqrt{3}} \cdot e^{-j30 \text{ deg}} \quad |V_{m\_ln}| = 265.6 \text{ V} \quad \arg(V_{m\_ln}) = -30.0 \text{ deg.}$$

The line-to-neutral voltages drive the current through the phase windings. The motor has only one type of current, that is, the phase and line currents are identical. Each phase winding carries one-third of the power. The line (or phase) current is:

$$I_{m\_wye} := \overline{\left(\frac{\frac{S_{m\_in}}{3}}{V_{m\_ln}}\right)} \quad I_{m\_wye} = (27.8 - 107.9j) \text{ A}$$

$$|I_{m\_wye}| = 111.4 \text{ A} \quad \arg(I_{m\_wye}) = -75.6 \text{ deg.}$$

The motor represents a balanced load; consequently, the voltage difference between the motor and the supply neutral points is zero. The zero voltage difference permits the elimination of the neutral conductor.

Comparison of the line current for both the wye and delta connections reveals that the line current is independent of the connection of the motor, but the winding or phase current is less in a delta connection than a wye. This is an advantage because the smaller current reduces the winding losses or permits the use of smaller wire for the delta-connected motor windings.

EXAMPLE 4.9: Two three-phase loads connected in parallel

A wye-connected grounded source of 480 V supplies two three-phase loads connected in parallel. One load is wye connected and the other is delta connected. The circuit diagram of the system is provided in Figure 4.20, where the load impedances are as follows:

Wye Load:

$$Z_a := (100 + j \cdot 20) \cdot \Omega \quad Z_b := -j \cdot 75 \cdot \Omega \quad Z_c := j \cdot 80 \cdot \Omega,$$

Delta Load:

$$Z_{ab} := (150 + j \cdot 70) \cdot \Omega \quad Z_{bc} := (150 + j \cdot 70) \cdot \Omega \quad Z_{ca} := 100 \cdot \Omega.$$

We intend to calculate the wye load currents, the delta phase currents, the delta line currents, the supply currents, the supply complex power, and the power factor.

The first step of the calculation is the determination of the source line-to-neutral and line-to-line voltages. The line-to-neutral voltage will be used to calculate the currents of the wye load, and the line-to-line voltage is necessary for the computation of the current in the delta load.

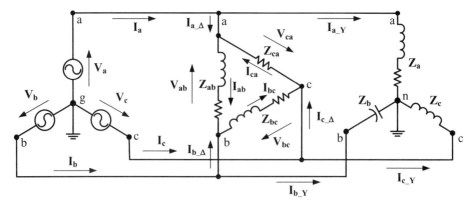

Figure 4.20. Generator supplies two loads connected in parallel.

SUMMARY

The line-to-neutral phasor voltages are:

$$V_a := \frac{480\ V}{\sqrt{3}} \qquad V_b := V_a \cdot e^{-j\,120\ deg} \qquad V_c := V_a \cdot e^{-j\,240\ deg}.$$

The corresponding line-to-line voltages are:

$$V_{ab} := V_a - V_b \qquad V_{bc} := V_b - V_c \qquad V_{ca} := V_c - V_a,$$
$$|V_{ab}| = 480\ V \qquad |V_{bc}| = 480\ V \qquad |V_{ca}| = 480\ V$$
$$\arg(V_{ab}) = 30\ deg \quad \arg(V_{bc}) = -90\ deg \quad \arg(V_{ca}) = 150\ deg.$$

Wye Load Currents

The wye load currents can be calculated by dividing each line-to-neutral voltage with the wye impedance for that phase:

$$I_{a\_Y} := \frac{V_a}{Z_a} \qquad I_{b\_Y} := \frac{V_b}{Z_b} \qquad I_{c\_Y} := \frac{V_c}{Z_c},$$
$$|I_{a\_Y}| = 2.72\ A \qquad |I_{b\_Y}| = 3.70\ A \qquad |I_{c\_Y}| = 3.46\ A$$
$$\arg(I_{a\_Y}) = -11.3\ deg \quad \arg(I_{b\_Y}) = -30.0\ deg \quad \arg(I_{c\_Y}) = 30.0\ deg.$$

Delta Load Currents

The delta load currents can be calculated by dividing each line-to-line voltage with the corresponding delta impedance:

$$I_{ab} := \frac{V_{ab}}{Z_{ab}} \qquad I_{bc} := \frac{V_{bc}}{Z_{bc}} \qquad I_{ca} := \frac{V_{ca}}{Z_{ca}},$$
$$|I_{ab}| = 2.90\ A \qquad |I_{bc}| = 2.90\ A \qquad |I_{ca}| = 4.80\ A$$
$$\arg(I_{ab}) = 5.0\ deg \quad \arg(I_{bc}) = -115.0\ deg \quad \arg(I_{ca}) = 150.0\ deg.$$

Referring to Figure 4.20 and using KCL, the line currents into the delta load are:

$$I_{a\_\Delta} := I_{ab} - I_{ca} \qquad I_{b\_\Delta} := I_{bc} - I_{ab} \qquad I_{c\_\Delta} := I_{ca} - I_{bc},$$
$$|I_{a\_\Delta}| = 7.37\ A \qquad |I_{b\_\Delta}| = 5.02\ A \qquad |I_{c\_\Delta}| = 5.82\ A$$
$$\arg(I_{a\_\Delta}) = -17.0\ deg \quad \arg(I_{b\_\Delta}) = -145.0\ deg \quad \arg(I_{c\_\Delta}) = 120.2\ deg.$$

Supply Currents

Each supply current is the sum of the delta load line current and the wye load current:

$$I_a := I_{a\_\Delta} + I_{a\_Y} \qquad I_b := I_{b\_\Delta} + I_{b\_Y} \qquad I_c := I_{c\_\Delta} + I_{c\_Y},$$
$$|I_a| = 10.1\ A \qquad |I_b| = 4.8\ A \qquad |I_c| = 6.8\ A$$
$$\arg(I_a) = -15.4\ deg \quad \arg(I_b) = -101.0\ deg \quad \arg(I_c) = 89.4\ deg.$$

Supply Complex Power and Power Factor

The supply complex power is calculated by multiplying each phase voltage by the supply current conjugate. The supply power factor is the cosine of the complex power angle:

$$S_a := V_a \cdot \overline{I_a} = (2.69 + 0.74j) \text{ kV} \cdot \text{A} \quad pf_a := \cos(\arg(S_a)) = 0.964$$
$$S_b := V_b \cdot \overline{I_b} = (1.26 - 0.44j) \text{ kV} \cdot \text{A} \quad pf_b := \cos(\arg(S_b)) = 0.945$$
$$S_a := V_c \cdot \overline{I_c} = (1.61 + 0.95j) \text{ kV} \cdot \text{A} \quad pf_c := \cos(\arg(S_c)) = 0.861.$$

The lead/lag relation for the power factor is extracted from the complex power of the specific phase. In particular, the positive reactive powers of phases **a** and **c** mean that pf_a and pf_c are lagging, whereas pf_b is leading because of the negative reactive power of phase **b**. The total three-phase supply power is the sum of the phase powers:

$$S_T := S_a + S_b + S_c \quad S_T = (5.57 + 1.26j) \text{ kV} \cdot \text{A}.$$

Overall, the supply is providing both real and reactive power to the system.

4.6. THREE-PHASE POWER MEASUREMENT

Power measurements generally require *wattmeters*. An analog wattmeter has two inputs: current and voltage. The current input is connected in series with and the voltage input in parallel with the supply or load. The wattmeter provides a measurement of the average power (P). In addition, a voltmeter and an ammeter are used to determine the rms values of the voltage and current (V_{rms} and I_{rms}).

Digital power meters sample the voltage and current waveforms, and a microprocessor calculates the rms values of the voltage and current. Moreover, the instantaneous voltage and current readings are multiplied to calculate the instantaneous power. The average value of this product gives the real power (P).

The measured data permit the calculation of the power factor (pf) and the apparent power ($|S|$). The magnitude of the complex power is:

$$|S| = |\mathbf{VI}^*| = V_{rms} I_{rms}. \tag{4.34}$$

The power factor is:

$$pf = \frac{P}{|S|}. \tag{4.35}$$

The three- and four-wire power measurements described in the succeeding sections are valid for either balanced or unbalanced three-phase systems.

THREE-PHASE POWER MEASUREMENT

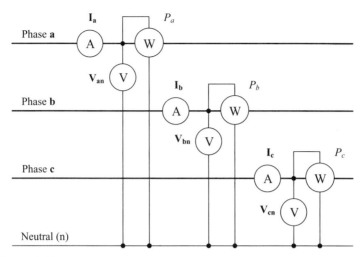

Figure 4.21. Measurement connections for a four-wire, three-phase system.

4.6.1. Four-Wire System

For a four-wire system, the loads can be connected between the phase conductors, or between the phase conductors and the neutral. This may result in unbalanced loading, which requires the measurement of each phase separately, that is, three wattmeters are required. Figure 4.21 depicts the typical measuring system using analog meters. The measurement system provides a current, a voltage, and a power reading for each of the three phases. The sum of the measured three powers gives the total three-phase power:

$$P_T = P_a + P_b + P_c. \tag{4.36}$$

4.6.2. Three-Wire System

In a three-wire system, the loads are connected between the phase conductors. The sum of the three line currents is zero, which permits the measurement of the three-phase power with just two wattmeters. Figure 4.22 sketches the connection diagram for a three-wire system. The current of phase **a** and the voltage between phases **a** and **b** feed one of the wattmeters. The current of phase **c** and the voltage between phases **c** and **b** supply the other wattmeter. The wattmeter multiplies the voltage and current values, and determines the average power value.

The readings of the wattmeters can be expressed in complex notation. The power is the real value of the voltage times the current conjugate. Accordingly, the reading of the wattmeter that is connected to current $\mathbf{I_a}$ and voltage $\mathbf{V_{ab}}$ is:

$$P_{ab} = \mathrm{Re}(\mathbf{V_{ab}I_a^*}) = \mathrm{Re}[(\mathbf{V_a} - \mathbf{V_b})\mathbf{I_a^*}] = \mathrm{Re}(\mathbf{V_a I_a^*} - \mathbf{V_b I_a^*}). \tag{4.37}$$

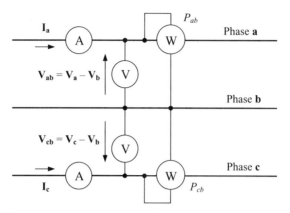

Figure 4.22. Measurement connections for a three-wire, three-phase system.

Similarly, the reading of the wattmeter that is connected to current $\mathbf{I_c}$ and voltage $\mathbf{V_{cb}}$ is:

$$P_{cb} = \operatorname{Re}(\mathbf{V_{cb}}\mathbf{I_c^*}) = \operatorname{Re}[(\mathbf{V_c} - \mathbf{V_b})\mathbf{I_c^*}] = \operatorname{Re}(\mathbf{V_c}\mathbf{I_c^*} - \mathbf{V_b}\mathbf{I_c^*}). \tag{4.38}$$

We shall prove that the sum of the two wattmeter readings is the total three-phase real power:

$$P_T = P_{ab} + P_{cb}. \tag{4.39}$$

The substitution of the two power values results in:

$$P_T = \operatorname{Re}(\mathbf{V_a}\mathbf{I_a^*} - \mathbf{V_b}\mathbf{I_a^*}) + \operatorname{Re}(\mathbf{V_c}\mathbf{I_c^*} - \mathbf{V_b}\mathbf{I_c^*}). \tag{4.40}$$

Simplification of the expression yields:

$$P_T = \operatorname{Re}[\mathbf{V_a}\mathbf{I_a^*} + \mathbf{V_c}\mathbf{I_c^*} - \mathbf{V_b}(\mathbf{I_a^*} + \mathbf{I_c^*})]. \tag{4.41}$$

Recall that the sum of the three line currents is zero such that:

$$\mathbf{I_b} = -(\mathbf{I_a} + \mathbf{I_c}). \tag{4.42}$$

Substituting the current formula into the power equation gives:

$$P_T = \operatorname{Re}(\mathbf{V_a}\mathbf{I_a^*} + \mathbf{V_c}\mathbf{I_c^*} + \mathbf{V_b}\mathbf{I_b^*}). \tag{4.43}$$

The real value of the components can be separated individually, which results in:

$$\begin{aligned} P_T &= \operatorname{Re}(\mathbf{V_a}\mathbf{I_a^*}) + \operatorname{Re}(\mathbf{V_b}\mathbf{I_b^*}) + \operatorname{Re}(\mathbf{V_c}\mathbf{I_c^*}) \\ &= P_a + P_b + P_c. \end{aligned} \tag{4.44}$$

Consequently, the sum of the two wattmeter readings is the three-phase real power.

A similar derivation discloses that the difference of the two wattmeters is related to the reactive power. The equation for the total three-phase reactive power is:

$$Q_T = \sqrt{3}(P_{cb} - P_{ab}). \tag{4.45}$$

4.7. PER-UNIT SYSTEM

Most power engineering calculations use the per-unit system, where every value (ohms, amperes, volts, watts, etc.) are divided by a base value and expressed in per-unit values. The base values are not phasors, they represent the magnitudes only. This simplifies the analyses of the power system. Further simplification is that the per-unit system eliminates the transformers from the network.

Use of the per-unit system first requires the selection of base values. The base power S_{base} and base voltage V_{base} for one point in the system are selected. The corresponding base current and impedance are calculated using Ohm's law and the power relations. If V_{base} is the line-to-line voltage and S_{base} is the three-phase complex power, then the base phase current in a three-phase circuit is:

$$I_{base} = \frac{S_{base}}{\sqrt{3}V_{base}}. \tag{4.46}$$

The base impedance is:

$$Z_{base} = \frac{V_{base}}{\sqrt{3}I_{base}} = \frac{V_{base}^2}{S_{base}}. \tag{4.47}$$

Per-unit values of the power, current, voltage, and impedance are calculated by dividing each value by the corresponding base value. The per unit values are:

$$S_{pu} = \frac{|S|}{S_{base}}, \tag{4.48}$$

$$I_{pu} = \frac{|I|}{I_{base}}, \tag{4.49}$$

$$V_{pu} = \frac{|V|}{V_{base}}, \tag{4.50}$$

$$Z_{pu} = \frac{|Z|}{Z_{base}}. \tag{4.51}$$

The impedance of a generator or a transformer is frequently given in percent or per unit. The bases for these quantities are the rated power and voltage. The per-unit impedance values are frequently converted to ohms. For the conversion of unit impedance to ohms, we substitute Equation (4.47) into Equation (4.51). The result is:

$$Z_{pu} = \frac{|Z|}{Z_{base}} = \frac{S_{base}}{V_{base}^2} |Z|. \tag{4.52}$$

Solving this equation for **Z**, we obtain:

$$|Z| = Z_{base} Z_{pu} = \frac{V_{base}^2}{S_{base}} Z_{pu}. \tag{4.53}$$

The use of the per-unit system for circuit calculations is beyond the scope of this book. However, the conversion of a per-unit impedance value to the ohm value is important because the generator and transformer impedances are given in percent (%) or per unit. The simplest use of the per-unit system requires the conversion of the per-unit impedance values to ohms and solution of the circuit using the methods described in this chapter. A more efficient method is to convert all quantities to per-unit values and solve the circuit using the per-unit values.

EXAMPLE 4.10: This numerical example illustrates the use of the per-unit method. A three-phase generator supplies a load through a transmission line (see Figure 4.23). The generator terminal and excitation voltages together with the voltage regulation must be calculated for a given load. The active power, the load voltage, and (lagging) power factor define the load.

After defining a mega prefix (which is not available as a built-in multiplier), the generator data are: $M := 10^6$,

$$S_g := 50 \, M \cdot V \cdot A \quad V_g := 22 \, kV \quad x_g := 125\%.$$

The transmission line impedance is: $Z_{line} := (0.42 + j \cdot 0.915) \cdot \Omega$.
  The three-phase load data are:

$$P_{load} := 40 \, M \cdot W \quad pf_{load} := 0.8 \quad V_{load} := 22 \, kV.$$

The per-unit method is applied below using two different approaches.

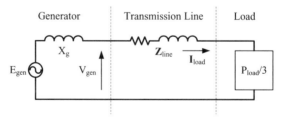

Figure 4.23. Generator supplies a load through a short transmission line.

Approach 1

The generator impedance is converted to ohms and the circuit is calculated in the traditional manner. The generator impedance in ohms per phase is calculated using Equation (4.53):

$$X_g := x_g \cdot \frac{V_g^2}{S_g} \quad X_g = 12.1\,\Omega.$$

The voltage reference given is the line-to-line voltage at the load. The load current is obtained from the power per phase and the line-to-neutral voltage:

$$I_{load} := \frac{\frac{P_{load}}{3}}{\frac{V_{load}}{\sqrt{3}} \cdot pf_{load}} \cdot e^{-j \cdot acos(pf_{load})}$$

$$|I_{load}| = 1.31\text{ kA} \quad arg(I_{load}) = -36.9\text{ deg},$$

where the exponent in the formula is negative due to the lagging power factor. The generator excitation voltage is the sum of the load line-to-neutral voltage and the voltage drop on the line and generator impedances. The calculation results in:

$$E_{gen} := \frac{V_{load}}{\sqrt{3}} + I_{load} \cdot (Z_{line} + jX_g) \quad E_{gen} = 23.4 + 13.3j\text{ kV}.$$

The line-to-line value of the excitation voltage is:

$$\sqrt{3} \cdot |E_{gen}| = 46.63\text{ kV} \quad arg(E_{gen}) = 29.7\text{ deg}.$$

The generator terminal voltage is:

$$V_{gen} := \frac{V_{load}}{\sqrt{3}} + I_{load} \cdot Z_{line} \quad V_{gen} = (13.86 + 0.63j) \cdot \text{kV}$$

$$|V_{gen}| = 13.88 \cdot \text{kV} \qquad arg(V_{gen}) = 2.6 \cdot \text{deg}.$$

The generator line-to-line voltage is:

$$V_{gen\_ll} := \sqrt{3} \cdot |V_{gen}| \quad V_{gen\_ll} := 24.04\text{ kV}.$$

Note that this generator voltage is higher than the rated value, which is an unacceptable condition. Inspection of Figure 4.23 reveals that the no-load voltage is equal to the generator voltage ($V_{no\text{-}load} = V_{gen\_ll}$). Therefore, the voltage regulation is:

$$\text{Reg} := \frac{|V_{\text{gen\_ll}}| - |V_{\text{load}}|}{|V_{\text{load}}|} \quad \text{Reg} = 9.3\%.$$

Note that Mathcad automatically converts the fractional voltage regulation to a percentage value; therefore, the usual 100% multiplier is not needed in this formula.

Approach 2
All parameters are converted to per unit. For this conversion, base power and voltage values are selected. Traditionally, the base voltage is the rated voltage of the system, and the base power can be either the generator-rated complex power or a value of 100 MVA, which is traditionally used by power companies. In this example, we follow utility tradition and select:

$$S_{\text{base}} := 100 \, \text{M} \cdot \text{V} \cdot \text{A} \quad V_{\text{base}} := 22 \, \text{kV}.$$

Using the selected base values, the base current and impedance are calculated with Equation (4.46) and Equation (4.47):

$$I_{\text{base}} := \frac{\dfrac{S_{\text{base}}}{3}}{\dfrac{V_{\text{base}}}{\sqrt{3}}} \quad I_{\text{base}} = 2.62 \, \text{kA},$$

$$Z_{\text{base}} := \frac{V_{\text{base}}}{\sqrt{3} \cdot I_{\text{base}}} \quad Z_{\text{base}} = 4.84 \, \Omega.$$

The substitution of the base current into the base impedance equation yields an alternate expression:

$$Z_{\text{base}} := \frac{V_{\text{base}}^2}{S_{\text{base}}} \quad Z_{\text{base}} = 4.84 \, \Omega.$$

The generator impedance is recalculated using the new base values. First, the generator impedance in ohms is calculated using Equation (4.53):

$$X_g := x_g \cdot \frac{V_g^2}{S_g} \quad X_g = 12.1 \, \Omega.$$

The new per-unit value of the generator impedance is the ratio of the generator impedance in ohms and the new base impedance, Equation (4.51):

$$x_{ge} := \frac{X_g}{Z_{\text{base}}} \quad x_{ge} = 2.5,$$

PER-UNIT SYSTEM

The substitution of the base impedance relation (4.53) and the previously mentioned generator impedance expression into this equation results in a practical formula for the direct recalculation of a per-unit impedance to a new base:

$$x_{ge} := \frac{x_g \cdot \dfrac{V_g^2}{S_g}}{\dfrac{V_{base}^2}{S_{base}}} \qquad x_{ge} = 2.5.$$

The equation is rearranged to obtain a simple form:

$$x_{ge} := x_g \cdot \left(\frac{V_g}{V_{base}}\right)^2 \cdot \frac{S_{base}}{S_g} \qquad x_{ge} = 2.5.$$

Note: To eliminate errors when using the computer to perform the earlier derivation, it is useful to recalculate the generator per-unit impedance after each step. The correct numerical value indicates the validity of the equation.

The per-unit value of the line impedance is calculated by dividing the given line impedance with the base impedance:

$$z_{line} := \frac{Z_{line}}{Z_{base}} \qquad z_{line} = 0.087 + 0.189j.$$

The per unit values of the load voltage and power are:

$$v_{load} := \frac{V_{load}}{V_{base}} \qquad v_{load} = 1.0$$

$$p_{load} := \frac{P_{load}}{S_{base}} \qquad p_{load} = 0.4.$$

The load current is calculated from the per unit values:

$$i_{load} := \frac{p_{load}}{v_{load} \cdot pf_{load}} \cdot e^{-j \cdot acos(pf_{load})} \qquad |i_{load}| = 0.50 \quad \arg(i_{load}) = -36.9 \text{ deg}.$$

The generator excitation and terminal voltages are:

$$e_g := v_{load} + i_{load} \cdot (jx_{ge} + z_{line}) \qquad e_g = 1.84 + 1.05j$$
$$|e_g| = 2.12 \qquad \qquad \arg(e_g) = 29.7 \cdot \text{deg}$$
$$|e_g| \cdot V_{base} = 46.63 \cdot kV,$$

$$v_g := v_{load} + i_{load} \cdot z_{line} \qquad\qquad v_g = 1.09 + 0.05j$$
$$|v_g| = 1.093 \quad \text{or} \quad |v_g| = 109.3\% \quad \arg(v_g) = 2.6 \cdot \text{deg}$$
$$V_{generator} := V_{base} \cdot |v_g| \qquad\qquad V_{generator} = 24.0 \text{ kV}.$$

The voltage regulation is:

$$\text{Regulation} := |v_g| - 1 \quad \text{Regulation} = 9.3\%.$$

The two methods result in the same numerical values. The power industry uses the per-unit method (Method 2) for most calculations, because it reduces the computation time and permits an easy comparison of the results.

4.8. SYMMETRICAL COMPONENTS

The method of symmetrical components is used to analyze an unsymmetrical, unbalanced electrical power network. Charles Fortescue proposed the technique in 1918, and it has been used to evaluate asymmetrical system conditions, such as single-phase to ground faults. The resulting currents and voltages are used to design system protection. Advancements in computer-based analysis of electric power networks have reduced the use of the symmetrical components-based approach. Even so, the symmetrical components method is very useful to make quick analyses of asymmetrical system conditions.

The basic assumption is that an unbalanced three-phase voltage or current can be replaced by three sets of balanced voltages or currents referred to as:

- *Zero Sequence Components* (V_{a0}, V_{b0}, V_{c0}) are three phasors with equal amplitude and phase angle such that $\mathbf{V}_0 = \mathbf{V}_{a0} = \mathbf{V}_{b0} = \mathbf{V}_{c0}$.
- *Positive Sequence Components* (V_{a1}, V_{b1}, V_{c1}) are three phasors with equal amplitude and $\pm 120°$ phase angle between them. This is a symmetrical three-phase system with phase sequence of *abc* such that $\mathbf{V}_1 = \mathbf{V}_{a1} = \mathbf{V}_{b1}e^{j120°} = \mathbf{V}_{c1}e^{j240°}$.
- *Negative Sequence Components* (V_{a2}, V_{b2}, V_{c2}) are three phasors with equal amplitude and $\pm 120°$ phase angle among them. This is a symmetrical three-phase system with phase sequence of *acb* such that $\mathbf{V}_2 = \mathbf{V}_{a2} = \mathbf{V}_{b2}e^{j240°} = \mathbf{V}_{c2}e^{j120°}$.

4.8.1. Calculation of Phase Voltages from Sequential Components

The unbalanced phase voltages ($\underline{\mathbf{V}}_{ph}$) and balanced sequential voltages ($\underline{\mathbf{V}}_{sq}$) are described by the following phasor vectors:

$$\underline{\mathbf{V}}_{ph} = \begin{bmatrix} V_a \\ V_b \\ V_c \end{bmatrix} \quad \underline{\mathbf{V}}_{sq} = \begin{bmatrix} V_0 \\ V_1 \\ V_2 \end{bmatrix}, \tag{4.54}$$

where V_a, V_b, and V_c are the unbalanced line-to-neutral voltages, and V_0, V_1, and V_2 are the zero, positive, and negative sequence voltages.

SYMMETRICAL COMPONENTS

The three unbalanced phase voltages (V_a, V_b, V_c) or currents (I_a, I_b, I_c) can be calculated from the balanced sequential voltages by three transformation relations:

$$\begin{aligned} V_a &= V_{a0} + V_{a1} + V_{a2} = V_0 + V_1 + V_2, \\ V_b &= V_{b0} + V_{b1} + V_{b2} = V_0 + a^2 V_1 + a V_2, \\ V_c &= V_{c0} + V_{c1} + V_{c2} = V_0 + a V_1 + a^2 V_2, \end{aligned} \quad (4.55)$$

where $a = e^{j120°} = 1/2(-1 + j\sqrt{3})$ while noting that $a^{-1} = a^{-2}$. The matrix form of the equations is:

$$\begin{bmatrix} V_a \\ V_b \\ V_c \end{bmatrix} = \begin{bmatrix} 1 & 1 & 1 \\ 1 & a^2 & a \\ 1 & a & a^2 \end{bmatrix} \begin{bmatrix} V_0 \\ V_1 \\ V_2 \end{bmatrix}. \quad (4.56)$$

Introducing the transformation matrix $\underline{\underline{A}}$, the previous formula can be presented in compact form as:

$$\underline{V}_{ph} = \underline{\underline{A}} \underline{V}_{sq}, \quad \text{where} \quad \underline{\underline{A}} = \begin{bmatrix} 1 & 1 & 1 \\ 1 & a^2 & a \\ 1 & a & a^2 \end{bmatrix}. \quad (4.57)$$

The replacement of the voltage phasors by the current phasors gives the symmetrical components of an unbalanced current phasor system:

$$\underline{I}_{ph} = \underline{\underline{A}} \underline{I}_{sq}, \quad (4.58)$$

where $\underline{I}_{ph}$ are the unbalanced phase currents, and $\underline{I}_{sq}$ are the balanced sequential currents.

4.8.2. Calculation of Sequential Components from Phase Voltages

The inverse of matrix $\underline{\underline{A}}$ is used to compute the balanced sequential components of the unbalanced phase voltages or currents. The matrix equation is:

$$\underline{V}_{sq} = \underline{\underline{A}}^{-1} \underline{V}_{ph}, \quad \text{where} \quad \underline{\underline{A}}^{-1} = \frac{1}{3} \begin{bmatrix} 1 & 1 & 1 \\ 1 & a & a^2 \\ 1 & a^2 & a \end{bmatrix}. \quad (4.59)$$

The resulting expressions for the sequential components from the phase voltages or currents are:

$$\begin{aligned} V_0 &= (V_a + V_b + V_c)/3, \\ V_1 &= (V_a + a V_b + a^2 V_c)/3, \\ V_2 &= (V_a + a^2 V_b + a V_c)/3. \end{aligned} \quad (4.60)$$

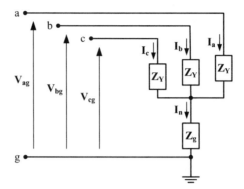

Figure 4.24. Impedance load.

4.8.3. Sequential Components of Impedance Loads

The three phase loads are represented here by their load impedance, which are either wye or delta connected. The wye-connected loads may be grounded through an impedance. Typically, most delta-connected loads are ungrounded. For the analysis, the ungrounded symmetrical delta loads are transferred to an ungrounded wye using the delta–wye transformation presented earlier in this chapter. The equivalent wye-connected impedance of the delta-connected load is $\mathbf{Z_Y} = \mathbf{Z_\Delta}/3$.

Figure 4.24 depicts a wye-connected load grounded through an impedance $\mathbf{Z_g}$. In the case of an ungrounded delta, after the wye–delta transformation, the equivalent grounding impedance is infinite since there is no ground return path.

The neutral current in the circuit is the sum of the load currents:

$$\mathbf{I_n} = \mathbf{I_a} + \mathbf{I_b} + \mathbf{I_c}. \tag{4.61}$$

The voltage equations are:

$$\begin{aligned}
\mathbf{V_{ag}} &= \mathbf{Z_Y}\mathbf{I_a} + \mathbf{Z_g}\mathbf{I_n} = \mathbf{Z_Y}\mathbf{I_a} + \mathbf{Z_g}(\mathbf{I_a} + \mathbf{I_b} + \mathbf{I_c}) \\
&= (\mathbf{Z_Y} + \mathbf{Z_g})\mathbf{I_a} + \mathbf{Z_g}\mathbf{I_b} + \mathbf{Z_g}\mathbf{I_c}, \\
\mathbf{V_{bg}} &= \mathbf{Z_Y}\mathbf{I_b} + \mathbf{Z_g}\mathbf{I_n} = \mathbf{Z_g}\mathbf{I_a} + (\mathbf{Z_Y} + \mathbf{Z_g})\mathbf{I_b} + \mathbf{Z_g}\mathbf{I_c}, \\
\mathbf{V_{cg}} &= \mathbf{Z_Y}\mathbf{I_c} + \mathbf{Z_g}\mathbf{I_n} = \mathbf{Z_g}\mathbf{I_a} + \mathbf{Z_g}\mathbf{I_b} + (\mathbf{Z_Y} + \mathbf{Z_g})\mathbf{I_c}.
\end{aligned} \tag{4.62}$$

The matrix form of these relations is:

$$\underline{\mathbf{V}}_{ph} = \underline{\underline{\mathbf{Z}}}_{ph}\underline{\mathbf{I}}_{ph},$$

$$\begin{bmatrix} \mathbf{V_{ag}} \\ \mathbf{V_{bg}} \\ \mathbf{V_{cg}} \end{bmatrix} = \begin{bmatrix} \mathbf{Z_Y} + \mathbf{Z_g} & \mathbf{Z_g} & \mathbf{Z_g} \\ \mathbf{Z_g} & \mathbf{Z_Y} + \mathbf{Z_g} & \mathbf{Z_g} \\ \mathbf{Z_g} & \mathbf{Z_g} & \mathbf{Z_Y} + \mathbf{Z_g} \end{bmatrix} \begin{bmatrix} \mathbf{I_a} \\ \mathbf{I_b} \\ \mathbf{I_c} \end{bmatrix}, \tag{4.63}$$

SYMMETRICAL COMPONENTS

where $\underline{V}_{ph}$ is the vector of the line-to-ground voltages; $\underline{I}_{ph}$ is the vector of the line currents; and $\underline{\underline{Z}}_{ph}$ is the *impedance matrix*.

The symmetrical components or sequence components of the voltage and current vectors are:

$$\underline{V}_{ph} = \underline{\underline{A}}\,\underline{V}_{sq} \quad \underline{I}_{ph} = \underline{\underline{A}}\,\underline{I}_{sq}. \tag{4.64}$$

These two relations are substituted in the matrix formulation of Equation (4.63), which yields:

$$\underline{\underline{A}}\,\underline{V}_{sq} = \underline{\underline{Z}}_{ph}\,\underline{\underline{A}}\,\underline{I}_{sq}. \tag{4.65}$$

Multiplying the previous expression by $\underline{\underline{A}}^{-1}$ gives:

$$\underline{V}_{sq} = \left(\underline{\underline{A}}^{-1}\underline{\underline{Z}}_{ph}\underline{\underline{A}}\right)\underline{I}_{sq} = \underline{\underline{Z}}_{sq}\,\underline{I}_{sq}, \tag{4.66}$$

where $\underline{\underline{Z}}_{sq} = \underline{\underline{A}}^{-1}\underline{\underline{Z}}_{ph}\underline{\underline{A}}$ is the sequence matrix of the impedance load.

After substituting the impedance matrix into the previous equation and performing the matrix multiplication, a simplified expression results:

$$\begin{bmatrix} V_0 \\ V_1 \\ V_2 \end{bmatrix} = \begin{bmatrix} Z_Y + 3Z_g & 0 & 0 \\ 0 & Z_Y & 0 \\ 0 & 0 & Z_Y \end{bmatrix} \begin{bmatrix} I_0 \\ I_1 \\ I_2 \end{bmatrix}. \tag{4.67}$$

This is alternatively written as:

$$\begin{aligned} V_0 &= (Z_Y + 3Z_g)I_0, \\ V_1 &= Z_Y I_1, \\ V_2 &= Z_Y I_2. \end{aligned} \tag{4.68}$$

Accordingly, in the case of a balanced load, these three equations define three independent networks, which is the advantage of symmetrical components:

1. a zero sequence network with an impedance of $Z_0 = Z_Y + 3Z_g$,
2. a positive sequence network with an impedance of $Z_1 = Z_Y$, and
3. a negative sequence network with an impedance of $Z_2 = Z_Y$.

This means that the zero sequence current can be calculated by dividing the zero sequence voltage by the zero sequence impedance ($I_0 = V_0/Z_0$), and the positive sequence current is computed from the ratio of the positive sequence voltage and the positive sequence impedance ($I_1 = V_1/Z_1$), and the negative sequence current is determined by dividing the negative sequence voltage by the negative sequence impedance ($\mathbf{I}_2 = \mathbf{V}_2/\mathbf{Z}_2$).

EXAMPLE 4.11: Symmetrical components analysis

A grounded wye-connected balanced load is supplied by a grounded wye-connected generator with an unbalanced voltage. Calculate:

1. the load impedances and currents directly using the loop equations,
2. the symmetrical components of the supply voltage,
3. the symmetrical components of the load current, and
4. the phase components of the load current.

The circuit diagram is presented in Figure 4.25. The system data are first defined. The unbalanced supply voltages, in both magnitude and phase angle, are:

$$V_a := 7.2 \text{ kV} \quad V_b := 6.8 \text{ kV} \cdot e^{-j 100 \text{ deg}} \quad V_c := 7.5 \text{ kV} \cdot e^{-j 260 \text{ deg}}.$$

The load parameters are:

$$V_L := 7.2 \cdot \sqrt{3} \text{kV} = 12.47 \cdot \text{kV} \quad P_L := 1800 \text{ kW} \quad pf_L := 0.8 \text{ (lagging)}.$$

Direct Solution Using KVL

The load complex power and impedance for each phase are found directly from:

$$M := 10^6 \quad S_L := \frac{P_L}{pf_L} \cdot e^{j \cdot a\cos(pf_L)} = (1.80 + 1.35j) \text{ M} \cdot \text{V} \cdot \text{A}$$

$$Z_L := \frac{V_L^2}{\overline{S_L}} = (55.30 + 41.47j) \, \Omega.$$

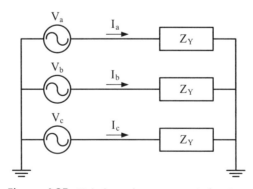

Figure 4.25. Unbalanced wye-connected system.

SYMMETRICAL COMPONENTS

Using the loop voltage equations (KVL), the load currents in the case of asymmetrical supply voltages are:

$$I_a := \frac{V_a}{Z_L} = (83.3 - 62.5j)\, A \quad |I_a| = 104.2\, A \quad \arg(I_a) = -36.9 \cdot \deg$$

$$I_b := \frac{V_b}{Z_L} = (-71.8 - 67.3j)\, A \quad |I_b| = 98.4\, A \quad \arg(I_b) = -136.9 \cdot \deg$$

$$I_c := \frac{V_c}{Z_L} = (49.0 + 96.8j)\, A \quad |I_c| = 108.5\, A \quad \arg(I_c) = 63.1 \cdot \deg.$$

Solution with Symmetrical Components

First, we calculate the symmetrical components of the unbalanced supply voltages. By definition, the unbalanced supply voltages are:

$$V_{ph} := \begin{pmatrix} V_a \\ V_b \\ V_c \end{pmatrix} = \begin{pmatrix} 7.20 \\ -1.18 - 6.70j \\ -1.30 + 7.39j \end{pmatrix}\, kV.$$

The transformation matrix is established as:

$$a := e^{j120\,\deg} \quad A := \begin{pmatrix} 1 & 1 & 1 \\ 1 & a^2 & a \\ 1 & a & a^2 \end{pmatrix}.$$

Using Equation (4.59), the supply voltage sequential or symmetrical components are:

$$V_{sq} := A^{-1} \cdot V_{ph} \quad V_{sq} = \begin{pmatrix} 1.57 + 0.23j \\ 6.88 - 0.08j \\ -1.25 - 0.15j \end{pmatrix} \cdot kV.$$

For reference, the magnitudes and phase angles of the individual sequential components are:

$$k := 0..2 \quad V_k := V_{sq_k}$$
$$|V_0| = 1.59 \cdot kV \quad \arg(V_0) = 8.3 \cdot \deg$$
$$|V_1| = 6.88 \cdot kV \quad \arg(V_1) = -0.7 \cdot \deg$$
$$|V_2| = 1.26 \cdot kV \quad \arg(V_2) = -173.2 \cdot \deg.$$

Each of the sequential network voltages supplies the symmetrical load impedance. The symmetrical components of the load current are calculated using Ohm's law:

$$I_{sq_k} := \frac{V_k}{Z_L} \qquad I_{sq} = \begin{pmatrix} 20.2 - 11.0j \\ 78.9 - 60.6j \\ -15.8 + 9.1j \end{pmatrix} A$$

$k := 0..2 \qquad I_k := I_{sq_k}$

$|(I_0)| = 23.0$ A $\qquad \arg(I_0) = -28.6 \cdot \deg$
$|(I_1)| = 99.5$ A $\qquad \arg(I_1) = -37.5 \cdot \deg$
$|(I_2)| = 18.2$ A $\qquad \arg(I_2) = 150.0 \cdot \deg$.

Referring to Equation (4.58), the phase components of the load currents are:

$$I_{ph} := A \cdot I_{sq} = \begin{pmatrix} 83.3 - 62.5j \\ -71.8 - 67.3j \\ 49.0 + 96.8j \end{pmatrix} A.$$

The current values obtained by direct calculation and by the symmetrical component method are identical:

$I_{pha} := I_{ph_0} \qquad |I_{pha}| = 104.2$ A $\qquad \arg(I_{pha}) = -36.9 \cdot \deg$
$I_{phb} := I_{ph_1} \qquad |I_{phb}| = 98.4$ A $\qquad \arg(I_{phb}) = -136.9 \cdot \deg$
$I_{phc} := I_{ph_2} \qquad |I_{phc}| = 108.5$ A $\qquad \arg(I_{phc}) = 63.1 \cdot \deg$.

Although in this analysis the direct solution was more efficient, the purpose of this example was to illustrate the use of symmetrical components. The major application of symmetrical components is the calculation of the short-circuit current in the case of asymmetrical faults; however, such a calculation is beyond the scope of this textbook.

4.9. APPLICATION EXAMPLES

Two example MATLAB calculations are presented here. The first shows the effect of a capacitor connected in series with the line, and the second demonstrates the voltage drop (regulation) calculation on a feeder with multiple loads. The first example expresses the generator reactance on a per-unit (p.u.) basis. The MATLAB examples are followed by a PSpice network analysis.

EXAMPLE 4.12: Investigation of the effect of capacitive line compensation

A wye-connected generator or network supplies a three-phase delta-connected motor through a short transmission line. A capacitor is connected in series with the

APPLICATION EXAMPLES

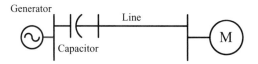

Figure 4.26. Motor supplied through a compensated line.

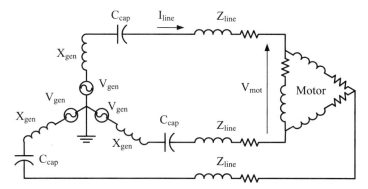

Figure 4.27. Equivalent circuit for compensated motor.

transmission line at the generator side to improve voltage regulation, as shown in the single-line diagram of Figure 4.26. The system frequency is $f = 60$ Hz. The motor line-to-line voltage, power, and power factor are $V_{mot} = 208$ V, $P_{mot} = 25$ hp, and $pf_{mot} = 0.7$ lagging, respectively. The 0.8-mile-long transmission line has a per-unit length impedance of $0.2 + j0.6$ Ω/mile. The wye-connected generator parameters are $S_g = 100$ kVA, $V_{g\_rated} = 208$ V, and $x_g = 10\%$.

The major steps of the analysis are:

(a) draw the equivalent circuit diagram;
(b) calculate the motor current in the delta and the transmission line current;
(c) compute and plot the voltage regulation as a function of capacitance, C_{cap}; and
(d) determine the capacitance value that is needed to reduce the voltage drop to 5%.

An equivalent circuit for the system is drawn in Figure 4.27. The balanced circuit allows us to perform the following calculations by analyzing a single phase. In reality, the series connected capacitor is not used for low voltages, but rather for long (>100 miles) high-voltage (345–500 kV) interconnection between neighboring networks.

MATLAB Program

First, the circuit values given earlier are initialized in the MATLAB program:

```
%   LineCompensation.m
clear all
```

```
% System frequency (Hz and rad/sec)
f = 60;   w = 2*pi*f;

% Motor line-to-line voltage, power and lagging power factor
Vmot = 208;                % volts
Pmot = 25;                 % hp (1 hp = 745.7 W)
Pmot = Pmot*745.7          % watts
PFmot = 0.7;

% Transmission line impedance
Zline = (0.2+0.6j) * 0.8;  % ohms

% Wye-connected generator parameters
Sg = 100e3;                % VA
Vg_rated = 208;            % volts
Xg = 0.1;                  % in p.u.

% Calculate generator reactance from per unit value
Xgen = Xg*Vg_rated^2/Sg;   % ohms
```

The generator reactance computation uses Equation (4.53), where the base values are the generator ratings, that is,

$$X_{gen} = \frac{V_{g\_rated}^2}{S_g} x_g.$$

The phasor current through a single branch of the motor delta is first computed:

$$\mathbf{I}_{mot} = \frac{P_{mot}/3}{pf_{mot} V_{mot}} e^{-j\arccos(pf_{mot})}, \quad (4.69)$$

where the three-phase motor power (P_{mot}) is divided by 3 to find the single-phase equivalent value. To calculate the line current ($\mathbf{I}_{line}$) and the line-to-ground voltage ($\mathbf{V}_{motln}$) of the motor, we make prudent use of the equations summarized in Table 4.1; specifically:

$$\mathbf{I}_{line} = \mathbf{I}_{mot}\sqrt{3}e^{-j30°},$$
$$\mathbf{V}_{motln} = \frac{V_{mot}\angle 0°}{\sqrt{3}e^{j30°}}. \quad (4.70)$$

where the motor voltage is taken as the reference angle of zero for the system.

```
% Calculate the motor (delta) current
Imot = (Pmot/3)/(PFmot*Vmot)*exp(-j*(acos(PFmot)))

% Calculate the line current
Iline = Imot*sqrt(3)*exp(-j*30/180*pi)
```

APPLICATION EXAMPLES

```
% Calculate motor line-to-ground voltage
Vmotln = Vmot/(sqrt(3)*exp(j*30/180*pi))
```

We vary the compensating capacitance and compute the regulation values. We vary the capacitance C_{cap} from 0.5 mF to 2.5 mF in steps of 0.1 mF to find the optimum value:

```
% Initializing for 5% regulation search
found = 0;           %   index to where Ccap value is found
% Vary the capacitor values and create an array
Ccap = 0.0005 : 0.0001 : 0.0025;    % in farad
for k=1: size(Ccap,2)
```

The combined series impedance of the generator and capacitor reactances and the line impedance is computed:

$$Z_{system} = Z_{gen} + Z_{cap} + Z_{line} = jX_{gen} + \frac{1}{j\omega C_{cap}} + Z_{line}.$$

Using KVL, the voltage across the source in the phase of interest is determined:

$$V_{gen} = I_{line} Z_{system} + V_{motln}.$$

Finally, the relative voltage drop across the system (i.e., from the generator to the motor) is calculated from:

$$\text{Regulation} = \frac{|V_{gen}| - |V_{motln}|}{|V_{motln}|} \times 100\%$$

```
% compute impedance from generator to motor
Zsystem = j*Xgen + 1/(j*w*Ccap(k)) + Zline;    % ohms
% compute generator voltage
Vgen = Iline*Zsystem + Vmotln;                 % volts
% compute voltage regulation in percent
Regulation(k) = (abs(Vgen)-abs(Vmotln))/abs(Vmotln)*100;
% find the capacitor corresponding to 5% regulation
```

We search for the capacitance required for a 5% voltage drop and compute the system efficiency (ε) at that operating point using:

$$\varepsilon = \frac{P_{mot}}{P_{gen}}.$$

The power used by the motor is given; the active power supplied by the generator can be determined from the real part of the complex generator power:

$$S_{gen} = 3V_{gen}I_{line}^*$$

```
    if Regulation(k) < 5 & found == 0
        Ccap_5percent_reg = Ccap(k)        % farads
        % generator three-phase complex power
        Sgen = 3*Vgen*conj(Iline)          % volt-amps
        % compute generator real power
        Pgen = real(Sgen)                  % watts
        % determine system efficiency
        efficiency = Pmot/Pgen*100         % percent
        found = k;
    end
end
```

Finally, we plot the regulation versus compensating capacitance and write the capacitance needed for 5% voltage regulation as the plot title:

```
plot(Ccap*1000,Regulation,'LineWidth',2.5);
set(gca,'fontname','Times','fontsize',12);
xlabel('Compensating Capacitor (mF)')
ylabel('Regulation (percent)')
title(['Capacitance at 5% Voltage Drop is ', ...
        num2str(Ccap_5percent_reg*1000),' mF']);
```

MATLAB Results

MATLAB computes the following numerical results:

```
>> LineCompensation
Pmot =      1.8643e+004
Imot =      29.8758 -30.4794i
Iline =     18.4178 -71.5923i
Vmotln =    1.0400e+002 -6.0044e+001i
Ccap_5percent_reg =    0.0010
Sgen =      2.1266e+004 -1.5889e+004i
Pgen =      2.1266e+004
efficiency =    87.6653
```

The MATLAB plot of the regulation versus the compensating capacitance is given in Figure 4.28, along with the minimum capacitance value for a 5% voltage drop. The graph shows that at first, the series capacitor reduces the voltage drop by compensating for the circuit inductance. The voltage drop reaches a minimum when the capacitor and circuit (line and generator) reactances are equal ($C \approx 1.6$ mF). When the capacitor impedance becomes larger than the line impedance, the line is overcompensated and the voltage drop increases. The traditional 5% regulation at this load requires a compensating capacitor of ≈ 1.0 mF.

APPLICATION EXAMPLES

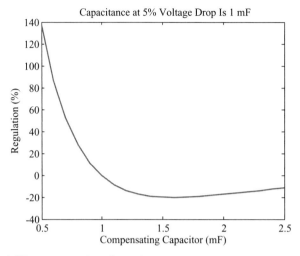

Figure 4.28. MATLAB plot of regulation versus compensation capacitance.

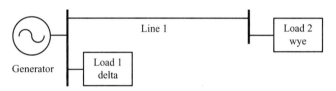

Figure 4.29. Three-phase generator supplying two loads.

EXAMPLE 4.13: Three-phase feeder with two loads

A three-phase source ($f = 50$ Hz) supplies a delta-connected and a wye-connected load through a feeder as depicted in the one-line diagram of Figure 4.29. The source is a grounded, wye-connected generator.

Load 1 is delta-connected and is an asymmetrical load:

Phase **ab** series connected inductance and resistance: $L_{ab} = 3.5$ H, $R_{ab} = 2200$ Ω;
Phase **bc** parallel capacitance and inductance: $C_{bc} = 0.5$ μF, $L_{bc} = 6$ H;
Phase **ca** series capacitance and resistance: $C_{ca} = 1.0$ μF, $R_{ca} = 1000$ Ω.

Load 2 is a grounded, wye-connected, symmetrical load:

$$V_{load2} = 13.8 \text{ kV}, P_{load2} = 30 \text{ kW}, pf_{load2} = 0.75 \text{ lagging}.$$

Line 1 (feeder to Load 2): impedance of $0.45 + j0.65$ Ω/mile, length = 6 miles

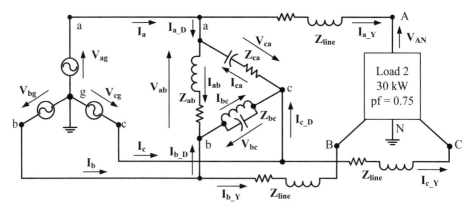

Figure 4.30. Equivalent circuit for three-phase generator supplying two loads.

The objective of the calculation is the determination of the required supply voltage, current, and power. The major steps of the calculation are:

1. draw the three-phase connection diagram and calculate the impedances of the delta load;
2. calculate the Load 2 currents (phases **a, b, c**) and the voltages (phases **a, b, c**) at the generator terminal;
3. compute the generator line-to-line voltages and delta load currents; and
4. determine the delta-connected Load 1 line currents, generator currents, and complex power in each phase.

The equivalent circuit diagram is given in Figure 4.30.

MATLAB Program
The circuit parameters are set first:

```
% PhaseCircuit.m
clear all

% System frequency (Hz and rad/sec)
f = 50;      omega = 2*pi*f;

% Load #1: Delta connected, asymmetrical load
% Phase ab is series inductor and resistor
Lab = 3.5;   Rab = 2200;     % (H) and (ohm)
% Phase bc is parallel capacitor and inductor
Cbc = 0.5e-6;   Lbc = 6;     % (F) and (H)
% Phase ca is series capacitor and resistor
Cca = 1.0e-6;   Rca = 1000;  % (F) and (ohm)
% Load #2: Grounded, wye-connected, symmetrical load
% load voltage (V), power (W), and power factor (inductive)
```

APPLICATION EXAMPLES

```
Vload2 = 13.8e3;   Pload2 = 30e3;   pf_load2 = 0.75;
% Feeder Data (Line 1 Data)
Zline = (0.45 + j*0.65) * 6;    % ohms
% Generator is grounded, wye-connected
```

Referring to Figure 4.30, the impedance on each leg of the delta load is computed:

$$\mathbf{Z}_{ab} = R_{ab} + j\omega L_{ab},$$

$$\mathbf{Z}_{bc} = \frac{1}{1/Z_{\text{L}bc} + 1/Z_{\text{C}bc}} = \frac{1}{1/(j\omega L_{bc}) + j\omega C_{bc}},$$

$$\mathbf{Z}_{ca} = R_{ca} + \frac{1}{j\omega C_{ca}},$$

```
% Calculate the impedances of delta load
Zab = Rab + j*omega*Lab
XLbc = omega*Lbc;
XCbc = -1/(omega*Cbc);
Zbc = 1/(1/(j*XCbc) + 1/(j*XLbc))
Zca = Rca + 1/(j*omega*Cca)
```

Next, the wye load (Load 2) currents and voltage are calculated. The current in phase A is calculated from the single-phase complex power, which is:

$$\mathbf{S}_a = \frac{P_{\text{load2}}/3}{pf_{\text{load2}}} e^{j\arccos(pf_{\text{load2}})}. \tag{4.71}$$

The line-to-neutral voltage of phase A of the wye load is set as the reference voltage phase angle of zero. Unlike the previous MATLAB example that had a balanced load, this problem requires calculation of the individual currents and voltages of the three phases because Load 1 is not balanced. Load 2 is balanced; therefore, the other phase currents lag by 120° and 240°. From Chapter 3, we recall that:

$$\mathbf{I}_{a\_Y} = \left(\frac{\mathbf{S}_a}{\mathbf{V}_{AN}}\right)^*,$$

where $\mathbf{V}_{AN} = V_{\text{load2}}/\sqrt{3} \angle 0°$.

```
% Calculate Load 2 currents
% Compute complex power of Phase A (V*A)
Sa = Pload2*exp(j*acos(pf_load2))/(3.0*pf_load2);
% Compute phase A line-to-neutral voltage (V)
VAN = Vload2/sqrt(3.0);
% Compute phase A current to Load 2
Ia_Y = conj(Sa/VAN)
Ib_Y = Ia_Y*exp(-j*120*pi/180)
Ic_Y = Ia_Y*exp(-j*240*pi/180)
```

The generator terminal voltages (with respect to ground) are computed using KVL from the source to the wye-connected load neutral, which is grounded as seen in Figure 4.30:

$$\mathbf{V}_{ag} = \mathbf{V}_{AN} + \mathbf{I}_{a\_Y}\mathbf{Z}_{Line}.$$

Because the wye load is balanced, the other line-to-neutral load voltages ($\mathbf{V}_{BN}$ and $\mathbf{V}_{CN}$) lag $\mathbf{V}_{AN}$ by 120° and 240°, respectively. The generator line-to-line voltages are also calculated from Kirchhoff's voltage law.

$$\mathbf{V}_{ab} = \mathbf{V}_{ag} - \mathbf{V}_{bg},$$
$$\mathbf{V}_{bc} = \mathbf{V}_{bg} - \mathbf{V}_{cg},$$
$$\mathbf{V}_{ca} = \mathbf{V}_{cg} - \mathbf{V}_{ag}.$$

```
% Calculate voltages at the generator terminal
Vag = VAN + Ia_Y*Zline
Vbg = VAN*exp(-j*120*pi/180) + Ib_Y*Zline
Vcg = VAN*exp(-j*240*pi/180) + Ic_Y*Zline

% Calculate the generator line-to-line voltages
Vab = Vag - Vbg
Vbc = Vbg - Vcg
Vca = Vcg - Vag
```

Since the line-to-line voltages are across the individual delta load impedances, the current through each delta leg is easily determined with Ohm's law:

$$\mathbf{I}_{ab} = \frac{\mathbf{V}_{ab}}{\mathbf{Z}_{ab}} \quad \mathbf{I}_{bc} = \frac{\mathbf{V}_{bc}}{\mathbf{Z}_{bc}} \quad \mathbf{I}_{ca} = \frac{\mathbf{V}_{ca}}{\mathbf{Z}_{ca}}.$$

The line currents into the delta load are obtained from applying KCL at the three delta nodes:

$$\mathbf{I}_{a\_D} = \mathbf{I}_{ab} - \mathbf{I}_{ca},$$
$$\mathbf{I}_{b\_D} = \mathbf{I}_{bc} - \mathbf{I}_{ab},$$
$$\mathbf{I}_{c\_D} = \mathbf{I}_{ca} - \mathbf{I}_{bc}.$$

```
% Calculate the currents in the delta load
Iab = Vab/Zab
Ibc = Vbc/Zbc
Ica = Vca/Zca

% Calculate the delta load line currents
Ia_D = Iab - Ica
Ib_D = Ibc - Iab
Ic_D = Ica - Ibc
```

… APPLICATION EXAMPLES

From KCL, the total current from each source is the sum of the currents to the two loads:

$$\mathbf{I}_a = \mathbf{I}_{a\_D} + \mathbf{I}_{a\_Y},$$
$$\mathbf{I}_b = \mathbf{I}_{b\_D} + \mathbf{I}_{b\_Y},$$
$$\mathbf{I}_c = \mathbf{I}_{c\_D} + \mathbf{I}_{c\_Y}.$$

Finally, the source-generated complex power is determined for each phase:

$$S_{g\_a} = V_{ag} I_a^*,$$
$$S_{g\_b} = V_{bg} I_b^*,$$
$$S_{g\_c} = V_{cg} I_c^*,$$

```
% Calculate the generator currents
Ia = Ia_D + Ia_Y
Ib = Ib_D + Ib_Y
Ic = Ic_D + Ic_Y

% Calculate the generator complex powers
Sg_a = Vag * conj(Ia)
Sg_b = Vbg * conj(Ib)
Sg_c = Vcg * conj(Ic)
```

MATLAB Results
The MATLAB results from **PhaseCircuit.m** are:

```
>> PhaseCircuit
Zab =    2.2000e+003 +1.0996e+003i
Zbc =              0 +2.6778e+003i
Zca =    1.0000e+003 -3.1831e+003i

Ia_Y =     1.2551 - 1.1069i
Ib_Y =    -1.5862 - 0.5335i
Ic_Y =     0.3311 + 1.6404i

Vag =    7.9751e+003 +1.9063e+000i
Vbg =   -3.9859e+003 -6.9076e+003i
Vcg =   -3.9892e+003 +6.9057e+003i

Vab =    1.1961e+004 +6.9095e+003i
Vbc =    3.3018e+000 -1.3813e+004i
Vca =   -1.1964e+004 +6.9038e+003i

Iab =     5.6062 + 0.3387i
Ibc =    -5.1584 - 0.0012i
Ica =    -3.0488 - 2.8009i
```

```
Ia_D =     8.6550 + 3.1396i
Ib_D =   -10.7646 - 0.3400i
Ic_D =     2.1096 - 2.7997i

Ia   =     9.9101 + 2.0327i
Ib   =   -12.3507 - 0.8735i
Ic   =     2.4406 - 1.1593i

Sg_a =    7.9038e+004 -1.6193e+004i
Sg_b =    5.5263e+004 +8.1833e+004i
Sg_c =   -1.7742e+004 +1.2230e+004i
```

EXAMPLE 4.14: PSpice example

A 230 kV high-voltage, 60 Hz network supplies a 69 kV local network through four substations. Figure 4.31 presents the one-line diagram of the 69 kV system. This system supplies three distribution loads, which have constant impedance.

The power system has different types of loads such as constant power, constant current, or a combination of these two. Most frequently, the loads are defined by their constant apparent power, power factor, and line voltage. The equivalent load impedance can be calculated from these data.

Using load flow studies, the utility determines the magnitude and phase angle of voltages at the 69 kV buses: 1, 2, 3, and 4 of the supplying 230 kV substations. The utility also provides the short-circuit current at the 69 kV bus.

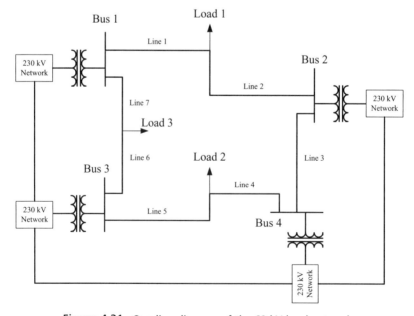

Figure 4.31. One-line diagram of the 69 kV local network.

APPLICATION EXAMPLES

Calculation of Equivalent Circuit Parameters

For the analysis of the 69 kV system, the 230 kV network is represented by its Thévenin equivalent circuit at each substation. The Thévenin equivalent is a voltage source and reactance connected in series. The source voltage is the utility-provided line-to-ground voltage (magnitude and phase angle) at the 69 kV buses. The equivalent Thévenin impedance of the 230 kV network at the 69 kV bus is calculated from the utility-provided short-circuit currents. Furthermore, we assume that the voltage of the 230 kV system is constant and unaffected by the 69 kV loads.

The voltages at the 69 kV buses are:

$$\omega := 2 \cdot \pi \cdot 60 \text{ Hz} \quad M := 10^6 \quad n := 1..4.$$

Substation 1: $V\text{sub}_1 := 67.5 \text{ kV} \quad \delta_1 := 0 \text{ deg}.$

Substation 2: $V\text{sub}_2 := 62 \text{ kV} \quad \delta_2 := 10 \text{ deg}.$

Substation 3: $V\text{sub}_3 := 69 \text{ kV} \quad \delta_3 := 18 \text{ deg}.$

Substation 4: $V\text{sub}_4 := 65 \text{ kV} \quad \delta_4 := 22 \text{ deg}.$

The short-circuit currents are:

Substation 1: $I_{\text{short}_1} := 12 \text{ kA}$ Substation 2: $I_{\text{short}_2} := 7.5 \text{ kA}.$

Substation 3: $I_{\text{short}_3} := 5 \text{ kA}$ Substation 4: $I_{\text{short}_4} := 7.5 \text{ kA}.$

The loads are (all lagging *pf*): k := 1 .. 3.

Load 1: $P_1 := 67 \text{ MW} \quad pf_1 := 0.83.$

Load 2: $P_2 := 83 \text{ MW} \quad pf_2 := 0.76.$

Load 3: $P_3 := 67 \text{ MW} \quad pf_3 := 0.90.$

The transmission line data are: m := 1 .. 7.
The line impedance per mile is:

$$Z_{\text{line}} := (0.27 + j \cdot 0.78) \frac{\Omega}{\text{mi}}.$$

The line lengths are:

$d_{\text{line}_1} := 15 \text{ mi} \quad d_{\text{line}_2} := 16 \text{ mi} \quad d_{\text{line}_3} := 37 \text{ mi} \quad d_{\text{line}_4} := 9 \text{ mi}$

$d_{\text{line}_5} := 35 \text{ mi} \quad d_{\text{line}_6} := 18 \text{ mi} \quad d_{\text{line}_7} := 18 \text{ mi}.$

Load Impedance Calculation

The load impedance is a resistance and inductance connected in parallel. The impedance values are calculated using the rated nominal voltage of 69 kV.

The resistance in the load impedance is calculated from the given power values:

$$RL_k := \frac{(69\text{ kV})^2}{P_k}$$

$RL_1 = 71.06\ \Omega$ $RL_2 = 57.36\ \Omega$ $RL_3 = 71.06\ \Omega$.

The inductance calculation requires the determination of the reactive power using the power factor and active power values. The equations are:

$$\phi_k := \text{acos}(pf_k) \quad Q_k := \frac{P_k}{\tan(\phi_k)}$$

$$X_k := \frac{(69\text{ kV})^2}{Q_k} \quad LL_k := \frac{X_k}{\omega}.$$

The obtained load inductance values are:

$LL_1 = 126.67\text{ mH}$ $LL_2 = 130.12\text{ mH}$ $LL_3 = 91.29\text{ mH}$.

Calculation of the Thévenin Equivalents

The equivalent Thévenin reactance is the ratio of the nominal line-to-neutral voltage and the short-circuit current. The phase angle of the short-circuit current is assumed as −90°, that is, the active component of the short-circuit current is neglected, as described in Section 3.5.3. In actual circumstances, the resistive component may not be negligible:

$$n := 1..4 \quad X_n := \frac{69\text{ kV}}{\sqrt{3}\cdot I_{short_n}} \quad Ls_n := \frac{X_n}{\omega}.$$

The inductance values are:

$Ls_1 = 8.81\text{ mH}$ $Ls_2 = 14.09\text{ mH}$ $Ls_3 = 21.13\text{ mH}$ $Ls_4 = 14.09\text{ mH}$.

The Thévenin equivalent voltage of the network is the line-to-neutral voltage of the voltage measured at the 69 kV bus of each substation. PSpice uses the peak value of the line-to-neutral voltages. This requires the multiplication of the given voltage values by $\sqrt{2/3}$:

$$Vs_n := \frac{Vsub_n}{\sqrt{3}} \cdot \sqrt{2}.$$

APPLICATION EXAMPLES

The source voltage values are:

$$Vs_1 = 55.11 \text{ kV} \quad Vs_2 = 50.62 \text{ kV} \quad Vs_3 = 56.34 \text{ kV} \quad Vs_4 = 53.07 \text{ kV}$$
$$\delta_1 = 0 \text{ deg} \quad \delta_2 = 10 \text{ deg} \quad \delta_3 = 18 \text{ deg} \quad \delta_4 = 22 \text{ deg}.$$

The line impedance is the product of the specific impedance times the line length:

$$Z_m := Z_{line} \cdot d_{line_m} \quad Rtl_m := Re(Z_m) \quad Ltl_m := \frac{Im(Z_m)}{\omega}.$$

The numerical values are:

$$Rtl^T = (0 \quad 4.05 \quad 4.32 \quad 9.99 \quad 2.43 \quad 9.45 \quad 4.86 \quad 4.86) \, \Omega$$
$$Ltl^T = (0 \quad 31.035 \quad 33.104 \quad 76.554 \quad 18.621 \quad 72.415 \quad 37.242 \quad 37.242) \, \text{mH}.$$

PSpice Analysis of the Circuit

The voltages and currents in this complicated network can be calculated using PSpice. The student version is suitable for this analysis. Figure 4.32 provides the equivalent circuit of the system, which was created by the PSpice *Schematics* editor. For additional information on PSpice, see Appendix D.

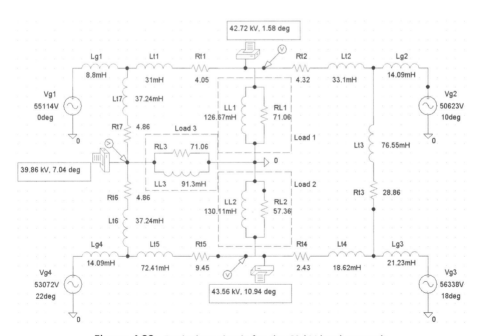

Figure 4.32. Equivalent circuit for the 69 kV local network.

The system is analyzed using the AC sweep analysis. You have to double-click on "Analysis set up" and select "AC sweep." After this, double-click on AC sweep. This opens the "AC sweep and Noise analysis" window, where you should select Total pts = 1, Start Freq: 60 Hz, End Freq: 60 Hz. After this we run PSpice, which first checks the circuit and generates the "netlist" of components. This list gives the numerical values of the components and the instructions for the printing.

The netlist and the results can be obtained after running the simulation from "Schematic 69 kV.dat" file by double-clicking on "View/Output files." The obtained netlist and load voltages for the circuit of Figure 4.32 are:

```
* From [PSPICE NETLIST] section of pspiceev.ini:
.lib "nom.lib"
.INC "69kV system.net"
**** INCLUDING "69kV system.net" ****
* Schematics Netlist *
L_Lt1           $N_0001 $N_0002   31mH
R_Rt1           $N_0002 $N_0003   4.05
L_Lt5           $N_0004 $N_0005   72.41mH
R_Rt5           $N_0005 $N_0006   9.45
L_Lt6           $N_0004 $N_0007   37.24mH
L_Lt7           $N_0008 $N_0001   37.24mH
L_Lg4           $N_0009 $N_0004   14.09mH
R_Rt2           $N_0003 $N_0010   4.32
L_Lt3           $N_0011 $N_0012   76.55mH
L_Lt2           $N_0010 $N_0012   33.1mH
R_Rt4           $N_0006 $N_0013   2.43
V_V2g           $N_0014 0 DC 0V AC 50623V 10deg
V_Vg4           $N_0009 0 DC 0V AC 53072V 22deg
V_Vg1           $N_0015 0 DC 0V AC 55114V 0
R_RL3           0 $N_0016   71.06
L_LL3           0 $N_0016   91.3mH
R_RL1           0 $N_0003   71.06
L_LL1           0 $N_0003   126.67mH
L_LL2           $N_0006 0  130.11mH
L_Lg1           $N_0015 $N_0001   8.8mH
L_Lg2           $N_0012 $N_0014   14.09mH
R_Rt6           $N_0007 $N_0016   4.86
R_Rt7           $N_0016 $N_0008   4.86
R_RL2           $N_0006 0  57.36
R_Rt3           $N_0017 $N_0011   28.86
L_Lt4           $N_0013 $N_0017   18.62mH
V_Vg3           $N_0018 0 DC 0V AC 56338V 18deg
L_Lg3           $N_0017 $N_0018   21.23mH
.PRINT          AC
+ VM([$N_0003])
+ VP([$N_0003])
```

```
.PRINT          AC
+ VM([$N_0006])
+ VP([$N_0006])

.PRINT          AC
+ VM([$N_0016])
+ VP([$N_0016])
```

The results of the calculation are the load voltages. The magnitude (peak value) and phase angle of the load voltages are:

```
Load 1          FREQ     VM($N_0004)  VP($N_0004)
                6.000E+01  3.986E+04   7.035E+00
Load 2          FREQ     VM($N_0005)  VP($N_0005)
                6.000E+01  4.272E+04   1.580E+00
Load 3          FREQ     VM($N_0003)  VP($N_0003)
                6.000E+01  4.356E+04   1.094E+01
```

The obtained voltage values are listed in the text boxes in Figure 4.32.

PSpice outputs the frequency (60 Hz), the magnitude of the peak line-to-neutral voltage, and the phase angle in degrees. Multiplying the peak voltage values by $\sqrt{3/2}$ yields the rms line-to-line voltages. The load voltage values are:

Load 1   48.81 kV,
Load 2   52.32 kV,
Load 3   53.35 kV.

A comparison of the supply voltages ($\approx$69 kV) and load voltages shows that the voltage drop is more than 20%, which indicates that this 69 kV network is overloaded.

In general, PSpice is suitable for analyzing medium-sized networks containing 10–20 voltage sources and lines. The analysis of a large utility network requires the use of a dedicated load flow program that can handle a network with more than a thousand buses.

4.10. EXERCISES

1. Describe the ideal three-phase wye-connected generator. Give the equations for the line-to-line and line-to-neutral voltages.
2. What is an unbalanced four-wire wye-connected system? Draw a sketch.
3. How is the neutral current in a four-wire system calculated?
4. What is the complex power in a four-wire system? Present the equations.
5. Describe a wye-connected balanced load and show an equivalent circuit. Present the equations for calculating the current.
6. Describe a delta-connected balanced load and show an equivalent circuit. Present the equation for calculating the currents.

7. What is an unbalanced three-wire wye-connected system? Draw a sketch.
8. Calculate the voltage differences between the neutral points in a three-wire wye-connected system.
9. Describe the three-phase delta-connected generator.
10. Calculate the currents in a delta-connected unbalanced load. Discuss the phase differences.
11. Discuss the concept of the per-unit system.

4.11. PROBLEMS

Problem 4.1

Use complex exponentials to analytically verify the trigonometric relation of Equation (4.3).

Problem 4.2

A three-phase 5 hp, 208 V motor has an efficiency of 92% and a power factor of 0.87 (lagging) when it runs at full power. Calculate the motor phase currents and line currents if the motor is (a) wye connected and (b) delta connected.

Problem 4.3

Three impedances are connected in a delta and supplied by a balanced wye-connected generator with a line voltage of 4200 V. Calculate the line currents. The 60 Hz system data are:

Branch **ab** is a resistance $R_{ab} = 45\ \Omega$ and an inductance $L_{ab} = 3.9$ H connected in parallel.

Branch **bc** is a resistance $R_{bc} = 15\ \Omega$ and a capacitance $C_{bc} = 8\ \mu F$ connected in series.

Branch **ca** is a resistance $R_{ca} = 30\ \Omega$ and an inductance $L_{bc} = 0.2$ H connected in series.

Problem 4.4

Repeat the previous problem at 50 Hz and a line voltage of 380 V.

Problem 4.5

A three-phase substation bus supplies one wye- and one delta-connected load. The constant impedance loads are connected in parallel. A three-phase feeder with an impedance per phase of $\mathbf{Z} = 0.6 + j1.8\ \Omega$ connects the loads to a source, as seen in the one-line diagram of Figure 4.33. The loads draw the following three-phase power:

PROBLEMS

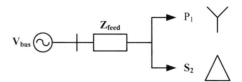

Figure 4.33. One-line diagram for Problem 4.5.

| $P_1 = 35$ kW | $pf_1 = 0.90$ (leading) | wye connected |
| $S_2 = 30$ kVA | $pf_2 = 0.85$ (lagging) | delta connected |

The line-to-line load voltage is 480 V. Determine: (a) the impedance of each load per phase, (b) the total line current flowing through the feeder, (c) the line-to-line voltage at the substation bus, and (d) the total three-phase real and reactive powers supplied by the bus.

Problem 4.6

A delta-connected load and a wye-connected capacitor bank are supplied through a feeder. The delta-connected load consists of three identical resistive–inductive impedances of $Z_\Delta = 50\ \Omega \angle 40°$ per phase. The source is wye connected and consists of a three-phase voltage source of 240 V (line to line). The feeder has a conductor impedance of $Z = 2 + j2\ \Omega$ per phase.
Determine:

(a) The current and line-to-line voltage at the load terminals without the capacitor bank.
(b) The resulting line-to-line voltage at the load terminals if the wye-connected capacitor bank is connected in parallel with the load. The capacitor bank has a reactance of $X = -35\ \Omega$ per phase.
(c) Discuss the effects of the capacitor bank on the voltage at the load terminals.

Problem 4.7

A three-phase generator supplies two three-phase loads. The 2.4 kV generator is rated at 180 kVA. The per-unit impedance of Load 1 is $Z_{\text{Load1}} = 80\% + j75\%$ and of Load 2 is $Z_{\text{Load2}} = 70\% + j25\%$ on 200 kVA and 2.5 kV bases. (a) Draw the equivalent circuit. (b) Determine the ohmic value of the impedances. (c) Calculate the source current in amperes and in percent if the generator operating voltage is 3.8 kV (line to line) and a reactance of $X = 125\%$ is connected in series with the generator voltage.

Problem 4.8

A three-phase generator is rated 12.5 kV and 800 kVA. The generator operates with a line voltage of $V_{ab} = 12.0$ kV. The generator is loaded with three impedances connected

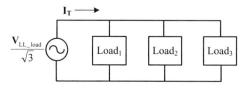

Figure 4.34. Single-phase circuit diagram for Problem 4.9.

in a delta. The impedances are $Z_{ab} = 200 + j300$ Ω, $Z_{bc} = -j250$ Ω, and $Z_{ca} = j150$ Ω. (a) Calculate the generator and load current in each phase. (b) Is the generator overloaded?

Problem 4.9

A balanced three-phase source supplies power to three wye-connected loads, as depicted in the single-phase diagram of Figure 4.34. The loads are:

Load 1:	S_1 = 38 kVA	pf_1 = 0.7 lagging
Load 2:	S_2 = 30 kVA	pf_2 = 0.75 leading
Load 3:	Unknown	

The line voltage at the loads and the line current at the source are 240 V and 15 A, respectively. If the combined power factor at the loads is **unity**, determine: (a) the real, apparent, and reactive powers of the unknown load and (b) the *pf* of the unknown load. (c) Vary the resistive component of Load 3 from 0 to 10 ohms and determine the resistance that corresponds to a power factor of 0.96.

Problem 4.10

Verify that the reactive power of a three-wire three-phase system can be calculated from the two wattmeter measurements using Equation (4.45).

5

TRANSMISSION LINES AND CABLES

Transmission lines are used to transfer large amounts of power over long distances. Extra-high-voltage (EHV) alternating current (ac) lines are utilized for long-distance energy transport, like the interconnection between Arizona and California or the remote James Bay hydropower plant with Montreal. For point-to-point power transport, long-distance bulk energy transfer, high-voltage direct current (HVDC) lines are highly effective because there is no capacitive charging current and no inductive voltage drop, but only resistive voltage drop. Typical example HVDC lines are the Pacific Intertie, which interconnects Los Angeles with Oregon, and the interconnection between France and England using underwater cables. High-voltage ac lines often connect power plants to a city. Typical examples are the high-voltage transmission system between the Palo Verde Nuclear Generating Station and Phoenix, Arizona, and the lines between the Hoover Dam and Las Vegas, Nevada. Within a city, subtransmission lines connect the distribution substations together, and distribution lines, originating from the substations, supply the energy to the residential and commercial customers. In congested cities, underground cables sometimes substitute for overhead transmission lines.

In this chapter, we describe the construction and components of the transmission lines. We address the most important environmental effects of the transmission lines, that is, the electric and magnetic fields. The chapter derives equations for calculation

---

Electrical Energy Conversion and Transport: An Interactive Computer-Based Approach, Second Edition.
George G. Karady and Keith E. Holbert.
© 2013 Institute of Electrical and Electronics Engineers, Inc. Published 2013 by John Wiley & Sons, Inc.

of transmission line resistance, inductance, and capacitance, introduces equivalent circuit for the line, and presents analysis methods to evaluate line performance.

The transmission lines are categorized according to their voltages:

- EHV lines
  - Voltage: 345, 500, 765, and 1000 kV
  - Interconnection between systems
- High-voltage lines
  - Voltage: 115 and 230 kV
  - Interconnection between substations, power plants
- Subtransmission lines
  - Voltage: 46 and 69 kV
  - Interconnection between substations and large industrial customers
- Distribution lines
  - Voltage: from 2.4 to 46 kV, with 15 kV being the most commonly used
  - Supplies residential and commercial customers
- High-voltage direct current (dc) lines
  - Voltage: from ±120 to ±800 kV
  - Interconnection between regions (e.g., from Oregon to California)

5.1. CONSTRUCTION

Figure 5.1 depicts a typical EHV transmission line. The major components of the line are:

- three-phase conductors, which carry the electric current;
- insulators, which support and electrically isolate the conductors;
- tower, which holds the insulators and conductors;
- foundation and grounding; and
- optional shield conductors, which protect against lightning.

Figure 5.2 pictures a transmission line corridor with two different high-voltage transmission lines. The line on the right-hand side uses a steel lattice tower. The line on the left-hand side is erected with a more aesthetically pleasing steel tube tower. The steel tube tower is constructed of a tapered steel tube equipped with banded arms. The arms hold the insulators and the conductors. The left-hand line is built with two-conductor bundles per phase, to reduce the corona effect and generation of radio and television (TV) noise. Grounded *shield conductors* at the top of the towers protect each line against lightning.

The siting of new transmission lines is becoming increasingly difficult because of the environmental effects of the line. Drawn in Figure 5.1 is the tower of an EHV

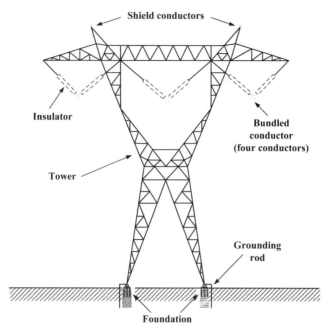

Figure 5.1. Typical EHV transmission line.

line, which is more than 100 ft tall and the line requires a 200- to 300-ft-wide corridor or right-of-way. Figure 5.2 discloses that the lines are not an aesthetically pleasing sight. Most people believe that new lines are necessary as the population is growing. However, the public wants the lines to be located far from their house—"not in my backyard."

An additional problem is the public's aversion to electric and magnetic fields generated by the lines. Although no one has proven any adverse health effects from these fields, they are considered undesirable in populated areas. These considerations have led to the development of transmission line corridors with several lines remote from populated areas.

Sub-transmission lines interconnect substations and locally distribute energy within a city or rural area. These lines are supported with a steel tube or wood tower. The steel tower has a concrete foundation. Frequently, the wood towers are placed in the ground without a foundation. A single conductor is usually used in each phase because the lower subtransmission voltage reduces the corona generation, although multiple conductors per phase may be employed to increase the current-carrying capability of the line. The conductors are supported by post insulators without a cross arm, or may be supported by suspension insulators attached to a cross arm. Lightning protection might be achieved using one grounded shield conductor placed on top of the tower. The shield conductor is grounded at each tower. A plate or vertical tube electrode (ground rod) is utilized for grounding. Figure 5.3 displays a typical double-circuit 69 kV subtransmission line with a wooden pole. This line carries two three-phase circuits and is located

Figure 5.2. Transmission line corridor with two high-voltage lines.

along major city streets and roadways. Note the 12.47 kV distribution line under the 69 kV line and the communication cables under the 12.47 kV circuit.

Distribution lines supply electricity to the residential and commercial buildings. In urban areas, most new construction uses underground cables to distribute the energy. In rural areas, distribution lines mostly employ a wood tower with cross arm(s). The wood is treated with creosote to protect against rotting. Some utilities install concrete towers. A simple concrete block foundation or no foundation is used. Small porcelain or plastic post insulators hold the conductors. The insulator shaft is grounded to eliminate leakage current from causing the wood tower to ignite and burn. A simple steel or brass rod is employed for grounding. A shield conductor is seldom used. Figure 5.4 exhibits typical 15 kV distribution line construction.

Figure 5.5 annotates a distribution line with a 240 V voltage cable connection and a pole-mounted transformer that supplies houses in a neighborhood. The distribution line has four conductors: one of them (at the top of the pole in Fig. 5.5) is the grounded neutral conductor. A 240/120 V service cable is attached under the distribution line. The line supplies each transformer through a fuse and a disconnect switch. The

CONSTRUCTION   211

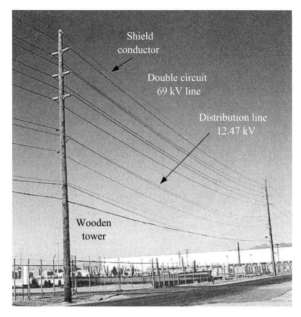

Figure 5.3. Double-circuit subtransmission line.

Figure 5.4. Distribution line construction.

single-phase pole-mounted transformers are installed under the cable connection. The low-voltage 240/120 V line supplies nearby homes. For better utilization of the tower, telephone and/or cable TV lines are often attached to the pole under the transformer.

Figure 5.1, Figure 5.2, Figure 5.3, Figure 5.4, and Figure 5.5 illustrate that the conductors are suspended on the poles. The distance between the poles is called the

Figure 5.5. Distribution line with pole-mounted transformer and service cable connection.

span, and it varies between 100 and 1500 ft, depending on the voltage level. A conductor, which hangs between two points, sags as seen in Figure 5.6. The *sag* varies with the conductor temperature, wind, and icing conditions. In the summer, the sag is significantly larger than in the winter. The sag determines the distance between the conductors and the ground. The National Electrical Safety Code[1] specifies the minimum distance, depending on the line voltage and the use of the land below the line. As an example, below a 22 kV line, the minimum distance must be more than 14.5 ft if the land is used for pedestrian traffic.

High wind and icing can damage the conductors. In particular, high wind can cause lateral motion of the conductors, which could produce flashover since the conductors may swing into close proximity with one another. Similarly, the sudden fall of ice from the conductor generates vertical line motion (called galloping), which can also cause flashover and mechanical damage. In order to limit line damage, the line is divided into sections, as portrayed in Figure 5.6. A *dead end* or *tension tower* terminates the two ends of each section. Between the tension towers are multiple supporting (or suspension) towers that carry the conductors. The insulators are in a vertical position at the

---

[1] Table 232-1, National Electrical Safety Code, ANSI C2-1997, p. 232B1.

CONSTRUCTION

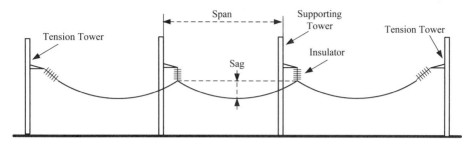

Figure 5.6. Transmission line supporting and tension towers.

Figure 5.7. Transmission line vibration damper (courtesy of AFL-Fujikura Ltd).

supporting towers, and the insulators are aligned with the conductor at the tension towers. The tension tower carries tangential, transverse, and vertical loads. The *supporting tower* carries only transverse and vertical loads. The advantage of this construction is that any storm damage may be limited to one section.

The tension in the conductor depends on the conductor weight, the ambient temperature, wind, ice, and the sag at the time of installation. During operation, the maximum tension occurs at wintertime. The sag calculation must consider the ambient temperature, conductor tensile strength, wind, and ice loading on the line. The mechanical performance analysis of a transmission line is not discussed in this book. The reader is advised to study the literature.[2]

A further problem is wind-produced vibration and oscillation of the conductors, which causes periodic bending of the conductor. This phenomenon, after a few years' time, produces fatigue breakage of the subconductors at towers where the conductor is clamped to the insulator. The oscillation is reduced by installing two vibration dampers in each span close to the insulators. Figure 5.7 exemplifies a typical damper. The damper is tuned to the frequency of the wind-generated vibration. The periodic movement of the weights attenuates the conductor vibration and eliminates conductor failures.

In urban areas, cables are placed in concrete ducts, which protect the cables, and simplify maintenance and replacement of faulty cable sections. Figure 5.8 shows a typical section of an urban distribution cable system installation. The diagram shows the concrete duct banks with low-voltage cables. Manholes, placed in the street or easement (e.g., front or back yard), divide the cable system into sections. In the case

---

[2] Karady, G.G., Transmission system, in *The Electric Power Engineering Handbook*, Grigsby, L.L., ed., CRC Press, Boca Raton, FL, 2001.

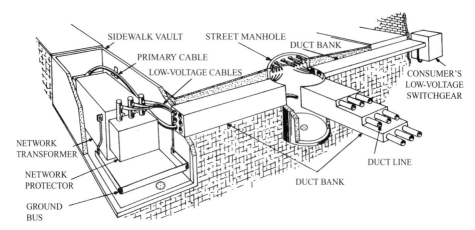

Figure 5.8. Cutaway view of an underground secondary grid network. Source: *Electric Utility Systems and Practices*, John Wiley & Sons, New York, 1978.

Figure 5.9. Typical concrete conduit with cables in a manhole.

of a cable fault, the faulty cable is cut, pulled out, and replaced. The figure points out that the transformer and related switchgear are placed in a sidewalk vault. Figure 5.9 shows a concrete duct bank and 13.8 kV cables down in a manhole.

A typical connection diagram of a residential cable distribution system is depicted in Figure 5.10. The cable loop supplies the residential transformers. A fuse protects each transformer. Each end of the looped cable system is connected to the bus of a distribution substation. However, the loop is normally opened by a disconnect switch. In the case of supply failure (e.g., Supply 1), the fuse cutout is opened at the Supply 1 side, and the disconnect switch is closed at the middle of the feeder. Supply 2 now

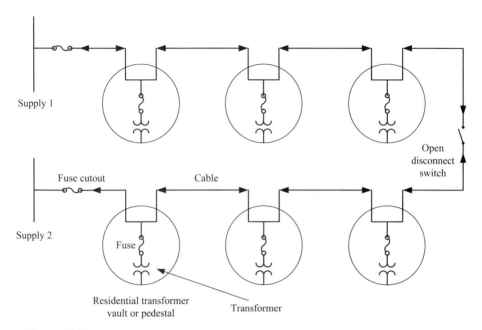

Figure 5.10. Connection diagram of a residential open-loop cable distribution system.

energizes both halves of the cable loop. The advantage of this open-loop system is the elimination of undesirable loop currents, reduction of fault current, and the minimization of customer outages.

High-voltage 230–69 kV cables are used in large municipalities, such as New York, to supply urban substations located in the middle of the city. These cables carry several hundred megawatts of power to supply skyscrapers and other urban loads.

The material presented in this section gives a general overview of the transmission line types and construction. However, utilities are using different types of construction and several variations of the described lines exist.

5.2. COMPONENTS OF THE TRANSMISSION LINES

5.2.1. Towers and Foundations

The most frequently used tower types are:

- lattice tower, used for 220 kV and above;
- guyed lattice tower, 345 kV and above;
- tapered steel tube with cross arm, 230 kV and below;
- concrete tower, for distribution and subtransmission; and
- wood tower, for distribution to 220 kV.

Most of the metal towers are stabilized by a concrete foundation and grounded. Buried copper rods or plates are used for grounding. The grounding resistance determines the lightning strike-caused overvoltage. *Overvoltage* occurs when the voltage is more than 10–20% greater than the rated voltage. Good grounding will limit overvoltage.

5.2.2. Conductors

Transmission lines use stranded aluminum conductors. The use of a stranded conductor replaces the single solid conductor with several small diameter conductors. The stranding improves conductor flexibility.

Typical phase conductors are:

- aluminum conductor steel reinforced (ACSR),
- all-aluminum conductor (AAC),
- all-aluminum alloy conductor (AAAC),
- aluminum conductor steel supported (ACSS),
- aluminum conductor composite core (ACCC),
- aluminum conductor composite reinforced (ACCR), and
- gap-type aluminum conductor steel reinforced (GTACSR).

Typical shield conductors are:

- aluminum-clad steel (Alumoweld), and
- extra-high-strength steel.

Conductors are manufactured with specific lengths. Therefore, during installation, the conductors must be spliced to provide continuity. Figure 5.11 depicts the conductor splicing sleeve before compression. This sleeve is compressed, which assures a connection that is able to carry full current and mechanical loads.

The most frequently used conductor is ACSR, which has a stranded steel center core and one to four outside layers of aluminum strands. The cross section of this conductor is drawn in Figure 5.12. The steel provides mechanical strength while the aluminum offers a low resistance path for current flow. However, the ACSR operating temperature is limited to 90°C because both the Al and steel are subjected to mechanical stress. Overheating the conductor anneals the aluminum, consequently rendering the Al unable to carry the mechanical load.

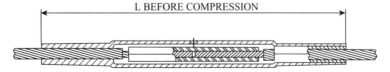

Figure 5.11. Splicing of a conductor (courtesy of AFL-Fujikura Ltd).

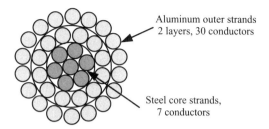

Figure 5.12. Aluminum conductor steel-reinforced (ACSR) conductor.

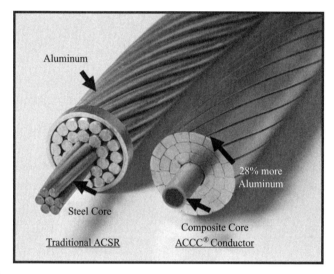

Figure 5.13. ACSR and ACCC conductors (reprinted with permission from CTC Cable Corporation, Irvine, CA).

In the last decade, new high-temperature conductors have emerged. These conductors carry two to three times higher current, operate at higher temperatures (150–210°C) and experience less sag. The major application of these conductors is the upgrading of existing transmission lines by replacing existing conductors with the high-temperature conductors, which increase the line capacity two or three times.

The ACSS conductor has a high strength steel core surrounded by annealed aluminum strand wires. The current is carried by the annealed aluminum and the steel provides all the mechanical strength. Because of the annealed wires, the conductor can operate at elevated temperature.

ACCC employs a lightweight carbon–glass fiber composite core surrounded by one or more layers of trapezoidal-shaped hard drawn and annealed aluminum wires. The aluminum carries the current while the fiber core provides the mechanical strength, even at elevated temperature. Figure 5.13 shows a comparison of a traditional ACSR conductor and an ACCC conductor.

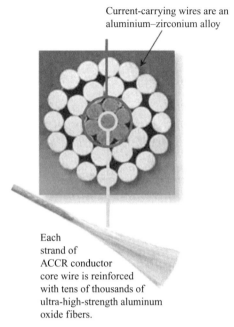

Figure 5.14. Aluminum conductor composite reinforced (ACCR). Photo courtesy of 3M.

The ACCR conductor core consists of aluminum oxide fibers embedded in high purity aluminum, which results in high strength, lightweight, and low resistance. As viewed in Figure 5.14, the core is surrounded by stranded aluminum–zirconium alloy wires with low resistance and high mechanical strength at elevated temperature.

GTACSR uses a steel core surrounded by a tube made out of trapezoidal aluminum segments. Stranded aluminum wires similar to those in the ACSR conductor then surround the tube. Figure 5.15 shows the grease-filled gap between the steel core and the aluminum tube, which reduces friction between the core and surrounding aluminum conductors. The steel core carries the entire mechanical load, whereas the surrounding aluminum carries the current. This results in low sag and permits high temperature operation.

Bundled conductors are utilized for lines above 220 kV to reduce corona and increase current-carrying capacity. The bundled conductor contains two, three, or four conductors in each phase. The distance (d) between the conductors is 12–18 inches, which is maintained by aluminum bars (spacers) placed at 20- to 50-ft distances along a span. Typical bundled conductor arrangements are exhibited in Figure 5.16. Figure 5.17 shows a spacer and damper for three bundled conductors.

5.2.3. Insulators

Ball-and-socket-type insulators composed of porcelain or toughened glass are employed for most high-voltage lines. These are also referred to as "cap-and-pin" insulators.

COMPONENTS OF THE TRANSMISSION LINES

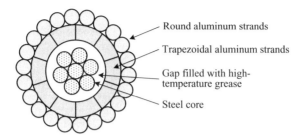

Figure 5.15. Gap-type aluminum conductor steel reinforced (GTACSR) conductor.

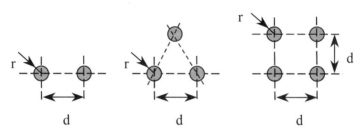

Figure 5.16. Typical bundled conductor arrangements.

Figure 5.17. Spacer and damper for three bundled conductors (courtesy of AFL-Fujikura Ltd).

Figure 5.18 shows the cross section of a ball-and-socket-type insulator. The porcelain skirt provides insulation between the iron cap and steel pin. The upper part of the porcelain is smooth, to promote rain washing and cleaning of the surface. The bottom part is corrugated, which prevents wetting and provides a longer protected leakage path. Portland cement is used to attach the cap and pin.

The ball-and-socket insulators are attached to each other by inserting the ball in the socket and securing the connection by a locking key. Several insulators are connected together to form an *insulator string*. Figure 5.19 shows the ball-and-socket insulators connected together. Strings of insulators are frequently utilized for transmission lines. Insulator strings are arranged vertically on support towers and near

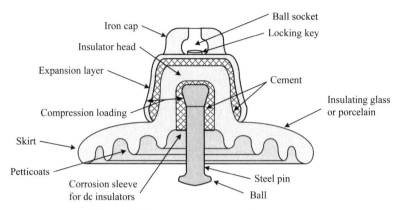

Figure 5.18. Cap-and-pin porcelain insulator.

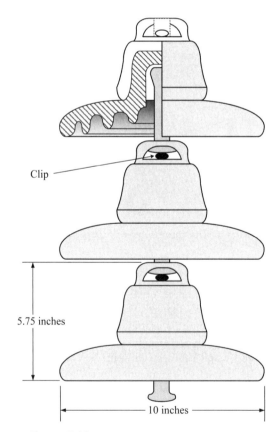

Figure 5.19. Ball-and-socket insulator string.

horizontally on dead-end (tension) towers. The typical number of insulators used by utilities in the United States and Canada in lightly polluted areas are:

Line Voltage (kV)	Number of Insulators per String
69	4–6
115	7–9
138	8–10
230	12
345	18
500	24
765	30–35

Porcelain post insulators are often used to support subtransmission lines. These insulators replace the cross arm of the towers, as presented in Figure 5.3. The post insulator consists of a porcelain column, with weather skirts or corrugation on the outside surface to increase leakage distance. For indoor use, the outer surface is corrugated. For outdoor use, a deeper weather shed is used. The end fitting seals the inner part of the tube to prevent water penetration.

Growing industrial pollution increased the number of flashovers on porcelain insulator strings. *Flashover* occurs when an arc bridges the insulator, thereby providing a conduction path between the line and ground. This motivated the development of pollution-resistant nonceramic (composite) insulators.

Composite insulators are built with mechanical load-bearing fiberglass rods, which are covered by rubber weather sheds to assure high electrical strength. Figure 5.20 shows the cross section of a composite insulator.

End fittings connect the insulator to a tower or conductor. The end fitting is a heavy metal tube with an oval eye, a socket, a ball, a tongue, or a clevis ending. This

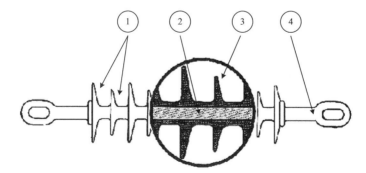

Figure 5.20. Composite insulator. (1) Sheds of alternating diameters prevent bridging by ice, snow, and cascading rain. (2) Fiberglass-reinforced resin rod. (3) Injection-molded EPDM rubber weather sheds and rod covering. (4) Forged steel end fitting, galvanized and joined to rod by swaging process. Source: Sediver.

galvanized forged steel-end fitting tube is swaged and compressed to the fiberglass rod. In this rod, the fiberglass is bound with epoxy or polyester resin, which assures high mechanical strength.

All high-voltage composite insulators use rubber weather sheds installed on fiberglass rods. The interface between the weather shed, fiberglass rod, and the end fittings is carefully sealed to prevent water penetration. The most serious insulator failure is caused by water penetration to the interface. The most frequently used materials are silicon rubber, EPDM (ethylene propylene diene monomer) rubber, and mixtures of the two. The direct molding of the sheds to the fiberglass rod produces high-quality insulators and is the best approach, although other methods are used successfully. The rubber contains fillers and additive agents to prevent discharge-caused tracking and erosion.

The advantages of these nonceramic insulators are:

- lightweight, which lowers construction and transportation costs;
- more vandalism resistant;
- higher strength-to-weight ratio, which allows longer design spans;
- better contamination performance; and
- improved transmission line aesthetics, resulting in better public acceptance of a new line.

The disadvantages are the ultraviolet radiation and surface-discharge-caused aging of the weather shed. This limits the life of present-day composite insulators to 15–20 years.

Post-type composite insulators are frequently employed on medium- and low-voltage lines. The typical line post insulator arrangement is drawn in Figure 5.21. This insulator has a fiberglass core covered by a silicon or EPDM rubber weather shed. The diagram shows the hardware that holds the two phase conductors.

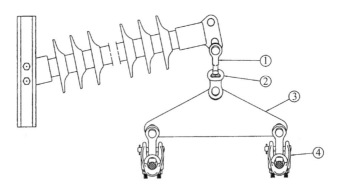

Figure 5.21. Line post-type composite insulator with yoke holding two conductors. (1) Clevis ball, (2) socket for the clevis, (3) yoke plate, and (4) suspension clamp. Source: Sediver.

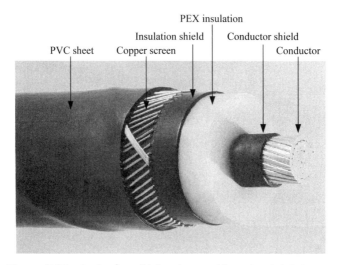

Figure 5.22. Single-phase high-voltage cable with solid dielectric.

5.3. CABLES

In many cities and residential areas, transmission lines are opposed because of environmental and aesthetical concerns, in addition to lack of space for large towers. In these areas, underground cables transport and distribute the electric energy. The underground cables are insulated conductors that are buried in the ground or placed in underground concrete cable ducts. Most cables are placed in concrete ducts, although older installations use directly buried cables. The latter has been found to be undesirable because cable life is reduced significantly.

Figure 5.22 shows a typical, single-phase high-voltage cable with a solid dielectric. The cable uses a stranded aluminum conductor, which is surrounded by a semiconducting shield layer that produces a smooth surface and reduces the stranding-caused electrical field concentration. The cable is insulated by a cross-linked polyethylene (PEX) dielectric. This is covered by a semiconducting layer that terminates the electric field. A grounded metal wire forms the outside shield. This shield is suitable to carry the short-circuit current and serve as a neutral conductor for low- or medium-voltage cables. Finally, a water-resistant external layer protects the cable. This layer is composed of PVC (polyvinyl chloride), high-density PEX, or rubber. Older high-voltage cables use paper insulation, saturated with high-pressure oil.

Figure 5.23 presents a view of a three-phase solid dielectric distribution cable used in residential areas. The cable has three stranded aluminum conductors. A semiconducting shield layer surrounds each of them. Solid PEX dielectric provides insulation for each conductor. A grounded semiconducting layer covers the conductor insulation to assure a cylindrical field distribution. Fillers are placed between the phases to assure a cylindrical cross section. A grounded copper wire surrounding the insulated phase

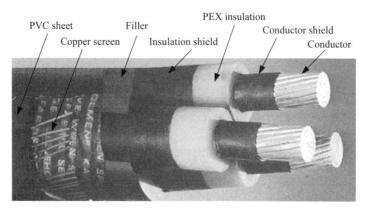

Figure 5.23. Three-phase distribution cable with solid dielectric.

conductors forms the outside shield. Finally, a water-resistant PVC sheet protects the cable from the environment.

5.4. TRANSMISSION LINE ELECTRICAL PARAMETERS

When the transmission lines are energized and carry current, the current produces a magnetic field and the voltage generates charge, which creates an electric field around the line. The results of the generated fields are the line inductance and capacitance. In addition, the conductors have temperature- and frequency-dependent resistance.

Resistance

The calculation of the stranded conductor ac resistance is complicated and the results are inaccurate. The practical method is the use of conductor tables. Table 5.1 lists the technical data of ACSR conductors. As seen in the first column of the table, the conductors are named after birds. The table gives the geometry of each conductor, including the cross sections, number of layers, size of the wires, and so on. These data are followed by the weight and mechanical strength of the conductors. The dc resistance is given at 25°C, while the ac resistance is listed for several different temperatures. These data are employed to determine the transmission line resistance. From the next to the last column, the geometric mean radius (GMR) is used for the line inductance calculation.

For the calculation of the line electrical parameters, the conductor diameter, the GMR, and the conductor ac resistance at the appropriate ambient temperature are needed. The temperature dependence of the resistance has a minor effect on the voltage drop calculation, but significantly affects the line loss.

The "Cardinal" conductors are frequently utilized on high-voltage lines. As an example of using Table 5.1, the line parameters for the Cardinal conductor are extracted.

The diameter of this conductor is $d = 1.196$ inches, the GMR is 0.0404 ft, and the conductor 60 Hz ac resistance at 25°C is 0.0998 Ω/mile and at 75°C is 0.1191 Ω/mile. Considering the desert climate in Arizona, the resistance at 25°C corresponds to winter conditions, if the heating effect of the current is included. The 75°C value represents summer conditions, including the current and direct sunshine-caused radiation heating.

5.5. MAGNETIC FIELD GENERATED BY TRANSMISSION LINES

The current in a conductor produces a magnetic field inside and around the conductor. The magnetic flux lines produced by current flowing in an isolated, long, straight, round conductor are concentric circles. Figure 5.24 depicts the magnetic field distribution inside and outside of a conductor. The right-hand rule, as illustrated in Figure 5.25, gives the field direction. If the thumb of the right hand points in the direction of the current flow, then the fingers of the right hand curl in the magnetic field direction, for example, clockwise, in Figure 5.24.

Both ac and dc lines generate magnetic fields. The intensity of the field is proportional with the current. The dc current-generated magnetic field is constant. The dc line-generated magnetic field has not been seen as a health concern because it is constant and is much less than the Earth's magnetic field, which is about half of a gauss (~0.5 G). However, ac-generated magnetic fields, which vary in time according to the system frequency, have received concern, although their magnitude is also much less than the Earth's magnetic field.

The calculation of the line reactance requires computation of both the internal and external magnetic fields. The evaluation of the environmental effect of the transmission line involves the calculation of the magnetic field at about 1 m above the ground. The people working and moving around the line are exposed to the exterior magnetic field.

The relation between the magnetic field and the current that generates the field is described by Ampere's law or by Maxwell's first equation. Ampere's law states that the integral of the magnetic field intensity vector along a closed path is equal to the sum of the currents within the closed path. The integral form of Ampere's law is given by:

$$\sum I_i = \oint \vec{H} \cdot \overrightarrow{d\ell}, \tag{5.1}$$

where

$\vec{H}$ is the magnetic field intensity vector measured in amp-turn per meter;
I_i is the current(s) that generates the field;
$\overrightarrow{d\ell}$ is the elementary path length vector; and
$\vec{H} \cdot \overrightarrow{d\ell}$ is a scalar (dot) product.

In the case of current in a long, straight, round solid conductor, the magnetic flux lines are circles, as shown in Figure 5.24 and Figure 5.26. The magnetic field is constant

TABLE 5.1. Technical Data for ACSR (Aluminum Conductor Steel-Reinforced) Conductors

	Cross Section			Stranding		Diameter		
Code	Al (kcmil)	Al (mm²)	Total (mm²)	Aluminum	Steel	Conductor (inches)	Core (inches)	Layers
–	2776	1407	1521	84 × 0.1818	19 × 0.1091	2.000	0.546	4
Joree	2515	1274	1344	76 × 0.1819	19 × 0.0849	1.880	0.425	4
Thrasher	2312	1171	1235	76 × 0.1744	19 × 0.0814	1.802	0.407	4
Kiwi	2167	1098	1146	72 × 0.1735	7 × 0.1157	1.735	0.347	4
Bluebird	2156	1092	1181	84 × 0.1602	19 × 0.0961	1.762	0.480	4
Chukar	1781	902	976	84 × 0.1456	19 × 0.0874	1.602	0.437	4
Falcon	1590	806	908	54 × 0.1716	19 × 0.1030	1.545	0.515	3
Lapwing	1590	806	862	45 × 0.1880	7 × 0.1253	1.504	0.376	3
Parrot	1510	765	862	54 × 0.1672	19 × 0.1003	1.505	0.502	3
Nuthatch	1510	765	818	45 × 0.1832	7 × 0.1221	1.465	0.366	3
Plover	1431	725	817	54 × 0.1628	19 × 0.0977	1.465	0.489	3
Bobolink	1431	725	775	45 × 0.1783	7 × 0.1189	1.427	0.357	3
Martin	1351	685	772	54 × 0.1582	19 × 0.0949	1.424	0.475	3
Dipper	1351	685	732	45 × 0.1733	7 × 0.1155	1.386	0.347	3
Pheasant	1272	645	726	54 × 0.1535	19 × 0.0921	1.382	0.461	3
Bittern	1272	644	689	45 × 0.1681	7 × 0.1121	1.345	0.336	3
Grackle	1192	604	681	54 × 0.1486	19 × 0.0892	1.338	0.446	3
Bunting	1193	604	646	45 × 0.1628	7 × 0.1085	1.302	0.326	3
Finch	1114	564	636	54 × 0.1436	19 × 0.0862	1.293	0.431	3
Bluejay	1113	564	603	45 × 0.1573	7 × 0.1049	1.258	0.315	3
Curlew	1033	523	591	54 × 0.1383	7 × 0.1383	1.245	0.415	3
Ortolan	1033	523	560	45 × 0.1515	7 × 0.1010	1.212	0.303	3
Cardinal	954	483	546	54 × 0.1329	7 × 0.1329	1.196	0.399	3
Rail	954	483	517	45 × 0.1456	7 × 0.0971	1.165	0.291	3
Drake	795	403	469	26 × 0.1749	7 × 0.1360	1.108	0.408	2

Source: Electric Power Research Institute (EPRI), *Transmission Line Reference Book 345 kV and Above*, 2nd ed., ed. J.J. LaForest, Palo Alto, CA, 1982, p. 110.

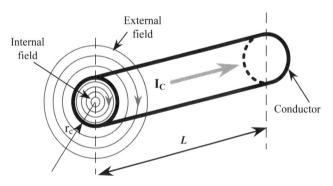

Figure 5.24. Magnetic field generated by a conductor.

MAGNETIC FIELD GENERATED BY TRANSMISSION LINES

Weight (pounds per 1000 ft)	Strength (kips)	Resistance (ohms/mile)					GMR (ft)	Ampacity (A)
		DC	AC at 60 Hz					
		25°C	25°C	50°C	75°C	100°C		
3219	81.6	0.0338	0.0395	0.0421	0.0452	0.0482	0.0667	
2749	61.7	0.0365	0.0418	0.0450	0.0482	0.0516	0.0621	3390
2526	57.3	0.0397	0.0446	0.0482	0.0518	0.0554	0.0595	3218
2303	49.8	0.0424	0.0473	0.0511	0.0550	0.0589	0.0570	3080
2511	60.3	0.0426	0.0466	0.0505	0.0544	0.0584	0.0588	3106
2074	51.0	0.0516	0.0549	0.0598	0.0646	0.0695	0.0534	2751
2044	54.5	0.0578	0.0602	0.0657	0.0712	0.0767	0.0521	2545
1792	42.2	0.0590	0.0622	0.0678	0.0734	0.0790	0.0497	2543
1942	51.7	0.0608	0.0631	0.0689	0.0748	0.0806	0.0508	2460
1702	40.1	0.0622	0.0652	0.0711	0.0770	0.0830	0.0485	2459
1840	49.1	0.0642	0.0663	0.0725	0.0787	0.0849	0.0494	2375
1613	38.3	0.0656	0.0685	0.0747	0.0810	0.0873	0.0472	2375
1737	46.3	0.0680	0.0700	0.0765	0.0831	0.0897	0.0480	2288
1522	36.2	0.0695	0.0722	0.0788	0.0855	0.0922	0.0459	2289
1635	43.6	0.0722	0.0741	0.0811	0.0881	0.0951	0.0466	2200
1434	34.1	0.0738	0.0764	0.0835	0.0906	0.0977	0.0445	2200
1533	41.9	0.0770	0.0788	0.0863	0.0938	0.1013	0.0451	2108
1344	32.0	0.0787	0.0811	0.0887	0.0963	0.1039	0.0431	2110
1431	39.1	0.0825	0.0842	0.0922	0.1002	0.1082	0.0436	2015
1255	29.8	0.0843	0.0866	0.0947	0.1029	0.1111	0.0416	2017
1331	36.6	0.0909	0.0924	0.1013	0.1101	0.1190	0.0420	1924
1165	27.7	0.0909	0.0930	0.1018	0.1106	0.1195	0.0401	1921
1229	33.8	0.0984	0.0998	0.1094	0.1191	0.1287	0.0404	1825
1075	25.9	0.0984	0.1004	0.1099	0.1195	0.1291	0.0385	1824
1094	31.5	0.1180	0.1190	0.1306	0.1422	0.1538	0.0375	1662

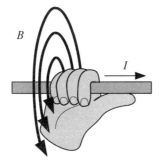

Figure 5.25. Right-hand rule for magnetic field direction from a current-carrying conductor.

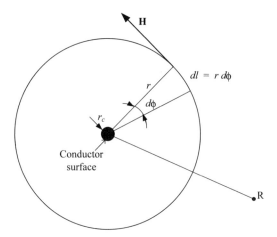

Figure 5.26. Application of Ampere's law for a small cylindrical conductor-generated magnetic field (current flowing out of the page).

along a flux line, which simplifies the integration. This case is diagrammed in Figure 5.26, which illustrates the definitions of the quantities used here.

In this case, the integral is simplified by:

$$I = \int_0^{2\pi} Hr d\phi = Hr \int_0^{2\pi} d\phi = 2\pi r H. \qquad (5.2)$$

The *magnetic field intensity* can be calculated from the previous equation as:

$$H = \frac{I}{2\pi r}. \qquad (5.3)$$

The field analysis requires the calculation of the *magnetic flux density*, B, which in air is the product of the free space permeability (μ_0) and the magnetic field intensity. The flux density is:

$$B = \mu_0 H = \mu_0 \frac{I}{2\pi r}. \qquad (5.4)$$

The flux density is measured in either tesla (T) or gauss (10^4 G = 1 T). From Appendix C, the *free space permeability* is:

$$\mu_0 = 4\pi 10^{-7} \frac{H}{m}. \qquad (5.5)$$

MAGNETIC FIELD GENERATED BY TRANSMISSION LINES

The *magnetic flux* is the integral of the flux density on a surface F:

$$\Phi = \int_F \vec{B} \times \overrightarrow{dF}. \tag{5.6}$$

In this equation, $\vec{B} \times \overrightarrow{dF}$ is a vector (cross) product.

As an example, the flux between the conductor surface at r_c and point R in Figure 5.26 is:

$$\Phi = \int_F \vec{B} \times \overrightarrow{dF} = \mu_0 \int_{r_c}^{R} \frac{I}{2\pi r} dr = I \frac{\mu_0}{2\pi} \ln\left(\frac{R}{r_c}\right). \tag{5.7}$$

This equation assumes the length of the conductor is unity. The magnetic flux is measured in weber (Wb) or tesla-square meters. A weber is a volt-second (Wb = V · s) and a tesla is a volt-second per meter squared (T = V · s/m²).

5.5.1. Magnetic Field Energy Content

The magnetic field is generated by a current. The maintenance of the magnetic field requires electric energy input. The magnetic field stores the energy. The instantaneous electric energy input is:

$$dW = e\, i\, dt = \frac{d\Phi}{dt} i\, dt = i\, d\Phi = i\, dB\, F, \tag{5.8}$$

where

B is the flux density;
F is the selected area;
e is the induced voltage; and
i is the current that generates the field.

Using Ampere's law, i is replaced by the magnetic field intensity H using $i = H\ell$, and B is replaced by the product of magnetic field intensity and the free-space permeability. After substituting these expressions into the above equation, we obtain:

$$dW = i\, dB\, F = (H\ell)(\mu_0\, dH)F = \mu_0 H\, dH\, (\ell F) = \mu_0 H\, dH\, V_{ol}, \tag{5.9}$$

where ℓ is the selected path length. In this equation $\ell F = V_{ol}$ is the volume of the selected area, where the energy stored by the magnetic field is calculated.

The integration of the previous expression gives the magnetic field energy in the selected volume:

$$W = \mu_0 \int_{V_{vol}} H\, dH\, V_{ol} = \mu_0 \frac{H^2}{2} V_{ol}. \tag{5.10}$$

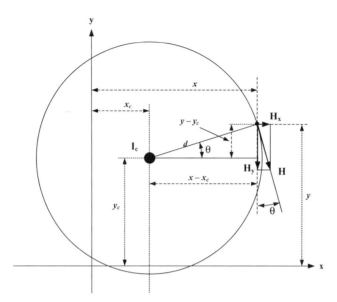

Figure 5.27. Magnetic field vector (current into the page).

5.5.2. Single Conductor Generated Magnetic Field

Figure 5.27 shows the magnetic field generated around a conductor and the magnetic field vector at an arbitrarily selected point. The conductor coordinates are x_c and y_c; the selected point coordinates are x and y. The current flows into the plane.

The magnetic field intensity (**H**) outside the conductor is calculated with Ampere's formula of Equation (5.3):

$$\mathbf{H}(x, y) = \frac{\mathbf{I_c}}{2\pi d} = \frac{\mathbf{I_c}}{2\pi\sqrt{(x-x_c)^2 + (y-y_c)^2}}, \qquad (5.11)$$

where

$\mathbf{I_c}$ is the conductor current;
x_c, y_c are the conductor center-point coordinates; and
x, y are the coordinates of an arbitrarily selected point.

From the larger triangle of Figure 5.27, the cosine and sine of θ are:

$$\cos\theta = \frac{x - x_c}{\sqrt{(x-x_c)^2 + (y-y_c)^2}} \qquad \sin\theta = \frac{y - y_c}{\sqrt{(x-x_c)^2 + (y-y_c)^2}}.$$

MAGNETIC FIELD GENERATED BY TRANSMISSION LINES

From the smaller vector triangle, the x- and y-components of the magnetic field are:

$$\mathbf{H}_x(x, y) = \frac{\mathbf{I}_c}{2\pi d}\sin(\theta) = \mathbf{I}_c \frac{y - y_c}{2\pi\left[(x - x_c)^2 + (y - y_c)^2\right]}, \quad (5.12)$$

$$\mathbf{H}_y(x, y) = \frac{\mathbf{I}_c}{2\pi d}\cos(\theta) = \mathbf{I}_c \frac{x - x_c}{2\pi\left[(x - x_c)^2 + (y - y_c)^2\right]}. \quad (5.13)$$

The total magnetic field intensity is the vector sum of the x and y components:

$$\mathbf{H}(x, y) = \mathbf{H}_x(x, y)\hat{\mathbf{i}} + \mathbf{H}_y(x, y)\hat{\mathbf{j}}, \quad (5.14)$$

where $\hat{\mathbf{i}}$ and $\hat{\mathbf{j}}$ are the unit vectors in the x- and y-directions, respectively. The corresponding magnitude of the total field is:

$$|\mathbf{H}(x, y)| = \sqrt{|\mathbf{H}_x(x, y)|^2 + |\mathbf{H}_y(x, y)|^2}. \quad (5.15)$$

The magnetic flux density (**B**) is:

$$\mathbf{B}(x, y) = \mu_0 \mathbf{H}(x, y). \quad (5.16)$$

EXAMPLE 5.1: MATLAB plot of magnetic field surrounding a conductor

It is useful to plot the magnitude of the magnetic field surrounding a conductor in order to gain a better understanding. First, we create a MATLAB function to calculate the magnetic field at an arbitrary distance from a single conductor using Equation (5.12) and Equation (5.13). This function, entitled "hmagfield.m" will also be utilized in further examples in this chapter.

```
function [Hx,Hy] = hmagfield(Icond, xc, yc, x, y)
% Icond   : conductor current (phasor)
% xc, yc  : conductor center coordinates
% x, y    : location for calculation of magnetic field
% compute distance between conductor and calculation point
d2 = (x-xc)^2 + (y-yc)^2;    % distance squared

% calculate magnetic field in both x and y directions
Hx = Icond*(y-yc)/(2*pi*d2);
Hy = Icond*(x-xc)/(2*pi*d2);

return
```

Suppose a conductor is located at a height of 70 ft from ground level and is carrying 1000 A. The coordinates are set such that $y = 0$ is at ground level and $x = 0$ is directly below the conductor. The earlier function is called using x-values along a range of from

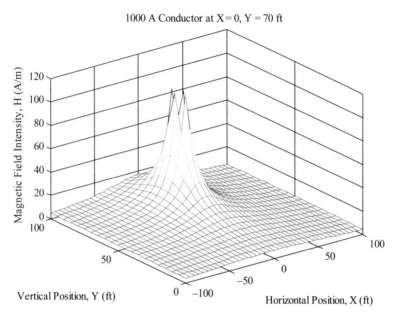

Figure 5.28. Magnetic field intensity around a 1000 A conductor. (The magnetic field strength goes to infinity at the conductor coordinates.)

−100 to +100 ft, and y-values from ground level (y = 0) to a vertical position of 100 ft. Equation (5.15) is used to compute the magnitude of the magnetic field. We then utilize MATLAB to create a plot of the intensity of the magnetic field in the vicinity of the conductor. The resulting graph is given in Figure 5.28.

```
%
%     MagneticFieldPlot.m
%
conv = 3.281;   % conversion from feet to meters
% set the conductor position (ft) and current (A)
xcond = 0;    ycond = 70;
Icond = 1000 + 0j;

% create vectors for x and y positions (ft)
x = (-100 : 5 : 100);
y = (0 : 5 : 100);

% compute magnetic field (A/m) at the desired locations
for i = 1:size(x,2)
    for j = 1:size(y,2)
        [Hx, Hy] = hmagfield(Icond,xcond,ycond,x(i),y(j));
        Hmag(j,i) = sqrt(abs(Hx)^2 + abs(Hy)^2) * conv;
    end
end
```

```
% plot the magnetic field magnitude as a function of position
mesh(x,y,Hmag)
set(gca,'fontname','Times','fontsize',11);
xlabel('Horizontal Position, X (ft)')
ylabel('Vertical Position, Y (ft)')
zlabel('Magnetic Field Intensity, H (A/m)')
title([num2str(abs(Icond)),' Amp Conductor at X=',...
       num2str(xcond),', Y=',num2str(ycond),' ft'])
```

The reader is advised to expect MATLAB to issue a warning when performing the previous calculations since a computation is performed at the center of the conductor where the distance is zero, and hence a division by zero occurs. If necessary, this can be avoided by judiciously creating the x- and y-vectors to avoid including the location of the conductor (i.e., omitting $x = 0$, $y = 70$), for example, by substituting the following statements in the above coding:

```
% create vectors for x and y positions (ft)
% while avoiding the conductor location
x = (-98 : 4 : 98);
y = (0 : 4 : 100);
```

Further, this entire computation could be accomplished in MATLAB with a single function call to draw the mesh plot on a 60 × 60-point grid; specifically:

```
>>ezmesh('1000*3.281/(2*pi*sqrt(x^2+(y-70)^2))',...
         [-100,100,0,100])
```

5.5.3. Complex Spatial Vector Mathematics

To understand the Mathcad and MATLAB computations of magnetic (and later electric) fields around multiple conductors presented in the next two sections, it is useful to pause and examine the addition of complex spatial vectors such as the magnetic and electric fields created by ac lines. Consider two complex spatial vectors:

$$\vec{V}_1 = V_{1x}\hat{i} + V_{1y}\hat{j} = (a+jb)\hat{i} + (c+jd)\hat{j}, \tag{5.17}$$

$$\vec{V}_2 = V_{2x}\hat{i} + V_{2y}\hat{j} = (e+jf)\hat{i} + (g+jh)\hat{j}. \tag{5.18}$$

To add these two complex vectors together requires that the x- and y-directional complex values be added separately. In particular, the sum of the two previously mentioned vectors is:

$$\begin{aligned}
\vec{V}_T = \vec{V}_1 + \vec{V}_2 &= \left(V_{1x}\hat{i} + V_{1y}\hat{j}\right) + \left(V_{2x}\hat{i} + V_{2y}\hat{j}\right) \\
&= [(a+e) + j(b+f)]\hat{i} + [(c+g) + j(d+h)]\hat{j} \\
&= (p+jq)\hat{i} + (s+jt)\hat{j}.
\end{aligned} \tag{5.19}$$

The visualization of the earlier complex spatial vector requires four-dimensional space since there are two pairs of real and imaginary values distributed over two spatial (x- and y-) coordinates. Because of the difficultly in visualizing the earlier complex vectors, it is more convenient to consider only the magnitudes of the x- and y-directional vectors. The x- and y-direction magnitudes (and phase angles) of the total vector are:

$$\vec{v}_T = \sqrt{p^2+q^2}\angle\arctan\left(\frac{q}{p}\right)\hat{i} + \sqrt{s^2+t^2}\angle\arctan\left(\frac{t}{s}\right)\hat{j}, \tag{5.20}$$

The overall magnitude of the total vector is:

$$\begin{aligned}|\vec{v}_T| &= \sqrt{(p^2+q^2)+(s^2+t^2)} = \sqrt{|v_{Tx}|^2+|v_{Ty}|^2} \\ &= \sqrt{(a+e)^2+(b+f)^2+(c+g)^2+(d+h)^2}.\end{aligned} \tag{5.21}$$

In similar fashion, the magnitudes of the original complex vectors are:

$$\begin{aligned}|\vec{v}_1| &= \sqrt{(a^2+b^2)+(c^2+d^2)}, \\ |\vec{v}_2| &= \sqrt{(e^2+f^2)+(g^2+h^2)}.\end{aligned} \tag{5.22}$$

From Equation (5.21) and Equation (5.22), we note the following inequality:

$$|\vec{v}_T| \neq \sqrt{|\vec{v}_1|^2 + |\vec{v}_2|^2}. \tag{5.23}$$

Consequently, the fields cannot be calculated by simply summing the vector magnitudes, but rather both directional components must be separately added, as shown in Equation (5.19).

5.5.4. Three-Phase Transmission Line-Generated Magnetic Field

Each phase conductor of a three-phase line generates its own magnetic field. The total field intensity (H_T) is the sum of the phase-generated magnetic field vectors (H_A, H_B, H_C):

$$\vec{H}_T = \vec{H}_A + \vec{H}_B + \vec{H}_C. \tag{5.24}$$

The calculation method is presented with a numerical example in Mathcad.

EXAMPLE 5.2: Figure 5.29 provides the conductor arrangement of a 500 kV transmission line. The line current is 1500 A, which corresponds to a load of around 1300 MW. The coordinates of the line conductors are:

$$x_A := -35 \text{ ft} \quad x_B := 0 \text{ ft} \quad x_C := 35 \text{ ft}$$
$$y_A := 70 \text{ ft} \quad y_B := y_A \quad y_C := y_A$$

The complex line currents are:

$$I_A := 1500 \text{ A} \quad I_B := I_A \cdot e^{-j \cdot 120 \text{ deg}} \quad I_C := I_A \cdot e^{-j \cdot 240 \text{ deg}}.$$

MAGNETIC FIELD GENERATED BY TRANSMISSION LINES

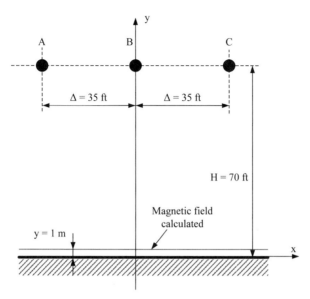

Figure 5.29. Transmission line conductor arrangement.

The magnetic field is measured in milligauss (mG), which we must explicitly define in Mathcad as:

$$mG := 10^{-3} \cdot G.$$

The free space permeability is:

$$\mu_o := 4 \cdot \pi \cdot 10^{-7} \cdot \frac{H}{m}.$$

The magnetic field is normally calculated 1 m above ground level under the transmission line. The calculation will be tested as the equations are entered by calculating the field directly under conductor C; the coordinates are: $x := 35$ ft $y := 1$ m.

The total magnetic field vector ($\mathbf{H_T}$) is computed by first determining the x- and y-directional magnetic field components ($\mathbf{H_x}$ and $\mathbf{H_y}$) from each of the phase conductors. The phase current-generated, x-directional field components are calculated from Equation (5.12). The individual results for the three phase conductors are:

$$H_{Ax}(x, y) := I_A \cdot \frac{y - y_A}{2 \cdot \pi \cdot \left[(x - x_A)^2 + (y - y_A)^2\right]} \qquad H_{Ax}(x, y) = -5.59 \cdot \frac{A}{m}$$

$$H_{Bx}(x, y) := I_B \frac{y - y_B}{2 \cdot \pi \cdot \left[(x - x_B)^2 + (y - y_B)^2\right]} \qquad H_{Bx}(x, y) = (4.60 + 7.97j) \cdot \frac{A}{m}$$

$$H_{Cx}(x, y) := I_C \cdot \frac{y - y_C}{2 \cdot \pi \cdot \left[(x - x_C)^2 + (y - y_C)^2\right]} \qquad H_{Cx}(x, y) = (5.87 - 10.17j) \frac{A}{m}$$

The total x-directional component of the magnetic field is found by summing the three component values:

$$H_x(x, y) := H_{Ax}(x, y) + H_{Bx}(x, y) + H_{Cx}(x, y) \quad H_x(x, y) = (4.88 - 2.19j)\frac{A}{m}.$$

From a practical point of view, the magnitude or absolute value of the field is used to determine the field intensity. The calculated value is:

$$|H_x(x, y)| = 5.355 \frac{A}{m}.$$

The phase current-generated, y-directional field components are computed using Equation (5.13). The results are:

$$H_{Ay}(x, y) := I_A \cdot \frac{x - x_A}{2 \cdot \pi \cdot \left[(x - x_A)^2 + (y - y_A)^2\right]} \quad H_{Ay}(x, y) = 5.86 \frac{A}{m}$$

$$H_{By}(x, y) := I_B \cdot \frac{x - x_B}{2 \cdot \pi \cdot \left[(x - x_B)^2 + (y - y_B)^2\right]} \quad H_{By}(x, y) = (-2.41 - 4.18j)\frac{A}{m}.$$

$$H_{Cy}(x, y) := I_C \cdot \frac{x - x_C}{2 \cdot \pi \cdot \left[(x - x_C)^2 + (y - y_C)^2\right]} \quad H_{Cy}(x, y) = 0.00 \frac{A}{m}$$

The total magnetic field in the y-direction is:

$$H_y(x, y) := H_{Ay}(x, y) + H_{By}(x, y) + H_{Cy}(x, y) \quad H_y(x, y) = (3.45 - 4.18j)\frac{A}{m},$$

and its magnitude is:

$$|H_y(x, y)| = 5.421 \frac{A}{m}.$$

The previous calculation yields the magnetic field x- and y-components (H_x and H_y), which are complex vectors, and which are separated by 90° in space (i.e., orthogonal).

The total magnetic field vector (H_T) is calculated by separately performing a vector addition of the real and imaginary components together (see Eq. 5.19) to obtain the real and imaginary components of the total magnetic field vector. However, if the magnetic flux density or magnetic field magnitudes are the only desired quantities, then the magnitude of the magnetic field can be computed directly from the magnitudes of the x- and y-components using Equation (5.15). The magnitude of the magnetic field at the selected point is:

$$H_{mag}(x, y) := \sqrt{(|H_x(x, y)|)^2 + (|H_y(x, y)|)^2} \quad H_{mag}(x, y) = 7.619 \cdot \frac{A}{m}.$$

The magnetic flux density (B) is used to estimate the magnetic field environmental effect. The flux density is:

$$B_{field}(x, y) := \mu_o \cdot H_{mag}(x, y) \quad B_{field}(x, y) = 95.7 \cdot mG$$
$$\mu := 10^{-6} \quad\quad\quad\quad\quad\quad\quad B_{field}(x, y) = 9.57 \cdot \mu \cdot T$$

The *resultant magnetic field*, as defined by the Institute of Electrical and Electronics Engineers (IEEE) Standard 644-1994 (R2008), is:

$$B_{res}(x, y) := \frac{B_{field}(x, y)}{\sqrt{2}} \quad B_{res}(x, y) = 6.77 \cdot \mu \cdot T.$$

The distribution of the flux density under the transmission line at a 1 m height is plotted in Figure 5.30 for a range of lateral distances from the line.

The maximum flux density occurs directly below the center (phase B) conductor:

$$B_{max} := B_{res}(0 \text{ ft}, 1 \text{ m}) = 77.7 \cdot mG \quad B_{max} = 77.7 \cdot mG.$$

At a 1 m height, the flux densities under conductors A and C are:

$$B_{res}(x_A, 1 \text{ m}) = 67.7 \cdot mG \quad B_{res}(x_C, 1 \text{ m}) = 67.7 \cdot mG,$$

and at the 100- and 250-ft edge of the right-of-way the fluxes are:

$$B_{res}(100 \text{ ft}, 1 \text{ m}) = 30.5 \cdot mG \quad B_{res}(250 \text{ ft}, 1 \text{ m}) = 6.4 \cdot mG.$$

Figure 5.30 reveals that the magnetic field decreases rapidly with distance. In the present example, the flux density is less than 10 mG if the distance is more than 250 ft from the line center.

Several studies have investigated the health effects from exposure to magnetic fields. A few studies linked long-term exposure to an increased incidence of cancer and childhood leukemia.[3] Other researchers, including the U.S. Environmental Protection Agency,[4] report no adverse effects from the magnetic fields. The IEEE Standard (C95.6-2002) for Safety Levels with Respect to Human Exposure to Electromagnetic Fields, 0–3 kHz defines the magnetic maximum permissible exposure (MPE) levels. Table 5.2 gives two values: one for the general public and the other for a controlled environment, which refers to personnel maintaining the power line. According to IEEE Standard C95.6, the magnetic field under the transmission line on the right-of-way should be less than 0.904 mT (9.04 G). This is a very high value—electric utilities typically limit the magnetic field under the transmission line to less than a few hundred milligauss. Industry standards such as this provide acceptable practices but do not carry the force of law.

---

[3] London, S.J., et al., "Exposure to residential electric and magnetic fields and risk of childhood leukemia," *American Journal of Epidemiology, 131*(9), 923–937, 1991.

[4] U.S. Environmental Protection Agency, "Evaluation of the Potential Carcinogenicity of Electromagnetic Fields," EPA/600/6-90/005B, October 1990.

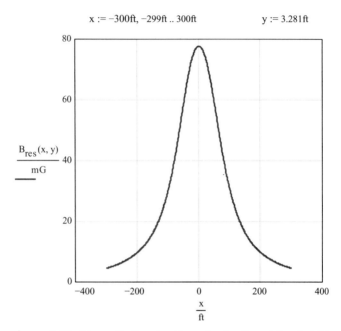

Figure 5.30. Magnetic flux density under the line at 1 m height.

TABLE 5.2. Magnetic Maximum Permissible Exposure (MPE) Levels: Exposure of Head and Torso

Frequency range (Hz)	General Public		Controlled Environment	
	B-rms (mT)	H-rms (A/m)	B-rms (mT)	H-rms (A/m)
<0.153	118	9.39×10^4	353	2.81×10^5
0.153–20	$18.1/f$	$1.44 \times 10^4/f$	$54.3/f$	$4.32 \times 10^4/f$
20–759	0.904	719	2.71	2.16×10^3
759–3000	$687/f$	$5.47 \times 10^5/f$	$2060/f$	$1.64 \times 10^6/f$

f is the frequency in Hz.
Source: IEEE Std C95.6-2002.

EXAMPLE 5.3: MATLAB program for magnetic field from three phase conductors

Here we use the MATLAB function developed in Example 5.1 to compute and plot the combined magnetic field from three phase conductors in the geometry of Figure 5.29. First, the line data are established using:

```
%
%      MagneticFieldThreePhase.m
%
conv = 3.281;    % conversion from feet to meters
```

TRANSMISSION LINE INDUCTANCE

```
xa = -35;    ya = 70;    % Conductor A position (ft)
xb = 0;      yb = ya;    % Conductor B position (ft)
xc = -xa;    yc = ya;    % Conductor C position (ft)
Ia = 1500 + 0j;          % Phase A current (A)
Ib = Ia * exp(-120j*pi/180);  % Phase B current (A)
Ic = Ia * exp(-240j*pi/180);  % Phase C current (A)
```

Next, the *hmagfield* function is repeatedly called for each (*x,y*) position to compute the *x*- and *y*-components of the magnetic field generated by each of the phase conductors. The total magnetic field magnitude is then determined using Equation (5.24) and Equation (5.15).

```
% create vectors for x and y positions (ft)
x = (-97.5 : 5 : 97.5);
y = (2.5 : 5 : 97.5);
% compute magnetic field at the desired locations
for i = 1:length(x)
    for j = 1:length(y)
        % magnetic field (A/ft) due to each phase conductor
        [Hax, Hay] = hmagfield(Ia, xa, ya, x(i), y(j));
        [Hbx, Hby] = hmagfield(Ib, xb, yb, x(i), y(j));
        [Hcx, Hcy] = hmagfield(Ic, xc, yc, x(i), y(j));

        % overall magnetic field (A/m) in x & y directions
        Hx = (Hax + Hbx + Hcx) * conv;
        Hy = (Hay + Hby + Hcy) * conv;

        % determine magnetic field magnitude (A/m)
        Hmag(j,i) = sqrt(abs(Hx)^2 + abs(Hy)^2);
    end
end
```

The results are plotted using the MATLAB code given in the following, and are shown in Figure 5.31:

```
mesh(x,y,Hmag(1:length(y),1:length(x)))
set(gca, 'fontname','Times', 'fontsize',11);
xlabel('Horizontal Position, X (ft)')
ylabel('Vertical Position, Y (ft)')
zlabel('Magnetic Field, H (A/m)')
title([num2str(abs(Ia)),' A Three-Phase Line at height ',...
    'of ',num2str(ya),' ft, and spacing of ',num2str(xc),' ft'])
```

5.6. TRANSMISSION LINE INDUCTANCE

AC current in the conductor produces a magnetic field inside and outside the conductor, as illustrated in Figure 5.24. The magnetic flux lines are concentric circles. Determination of the line inductance requires the calculation of the magnetic flux both inside and outside the conductor.

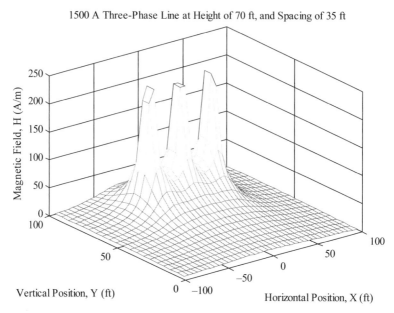

Figure 5.31. Magnetic field intensity around three 1500 A phase conductors.

5.6.1. External Magnetic Flux

Consider a conductor of radius r_c carrying current $\mathbf{I_c}$, as presented in Figure 5.32. For the calculation of the external flux, we select a magnetic flux line at a radius of x. The magnetic field intensity is calculated using Ampere's law. The magnetic field intensity from Equation (5.3) is:

$$\mathbf{H} = \frac{\mathbf{I_c}}{2\pi x}. \tag{5.25}$$

The magnetic flux density from Equation (5.4) is:

$$\mathbf{B} = \mu_0 \mathbf{H} = \mu_0 \frac{\mathbf{I_c}}{2\pi x}. \tag{5.26}$$

For the calculation of the magnetic flux density, we select a small tubular section. The thickness of the section is dx and the length is L. Referring to Equation (5.7), the external field flux in a tubular element of length L and thickness dx is:

$$d\Phi = \mu_0 \frac{\mathbf{I_c} L}{2\pi x} dx. \tag{5.27}$$

TRANSMISSION LINE INDUCTANCE

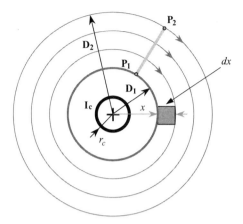

Figure 5.32. Magnetic flux calculation (current direction into page).

The flux passing through the plane placed between P_1 and P_2 is calculated by integrating the flux density between D_1 and D_2. The *magnetic flux* is:

$$\Phi_{12} = \int_{D_1}^{D_2} \mu_0 \frac{I_c L}{2\pi x} dx = \mu_0 \frac{I_c L}{2\pi} \ln\left(\frac{D_2}{D_1}\right). \tag{5.28}$$

This equation will be used to calculate the phase conductor-generated flux and inductance.

5.6.2. Internal Magnetic Flux

For the calculation of the magnetic field inside the conductor, a small tubular section is selected, as drawn in Figure 5.33. If the current (I_c) is uniformly distributed in the conductor, the current (I_x) within a circular section is proportional to the ratio of the cross-sectional areas of that section and the entire conductor:

$$I_x = \frac{\pi x^2}{\pi r_c^2} I_c = \frac{x^2}{r_c^2} I_c. \tag{5.29}$$

Ampere's law is applied to the circular section, marked by the hatched lines. The magnetic field intensity is found by substituting Equation (5.19) into Equation (5.3):

$$H_x = \frac{I_x}{2\pi x} = \frac{x}{2\pi r_c^2} I_c. \tag{5.30}$$

The magnetic flux density is:

$$B_x = \mu_0 H_x = \mu_0 \frac{x}{2\pi r_c^2} I_c. \tag{5.31}$$

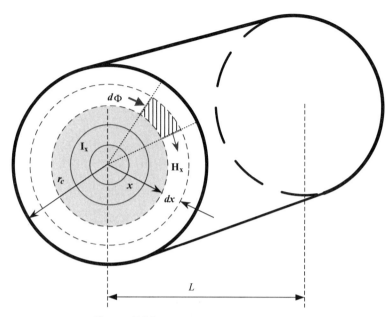

Figure 5.33. Internal flux in a conductor.

The volume of the tubular element of thickness dx and length L is:

$$dV = L2\pi x\, dx. \tag{5.32}$$

According to Equation (5.10), the magnetic energy stored in a tubular element of thickness dx and length L is:

$$dW_x = \frac{1}{2}\mu_0 \mathbf{H}_x^2 dV = \frac{\mu_0}{2}\left(\frac{\mathbf{I}_c x}{2\pi r_c^2}\right)^2 L2\pi x\, dx. \tag{5.33}$$

The magnetic energy in the conductor is found by integrating the previous equation between 0 and r_c. The systematic integration of the expression yields:

$$W_x = \frac{1}{2}\int_V \mu_0 \mathbf{H}_x^2 dV = \frac{1}{2}\int_0^{r_c}\mu_0\left(\frac{\mathbf{I}_c x}{2\pi r_c^2}\right)^2 L2\pi x\, dx = \mu_0 \frac{\mathbf{I}_c^2 L}{4\pi r_c^4}\int_0^{r_c} x^3\, dx, \tag{5.34}$$

$$W_x = \mu_0 \frac{\mathbf{I}_c^2 L}{4\pi r_c^4}\int_0^{r_c} x^3\, dx = \mu_0 \frac{\mathbf{I}_c^2 L}{4\pi r_c^4}\frac{r_c^4}{4} = \mu_0 \frac{\mathbf{I}_c^2 L}{2\pi}\frac{1}{8}. \tag{5.35}$$

The magnetic energy can be expressed either with the internal inductance of the conductor (L_c) or with the internal flux. The magnetic flux and inductance relation is:

$$\Phi_{\text{int}} = \mathbf{I}_c L_c. \tag{5.36}$$

TRANSMISSION LINE INDUCTANCE

The general equation for the magnetic energy in an inductor is:

$$W = \frac{1}{2} I_c^2 L_c. \tag{5.37}$$

The combination of these two equations gives the energy flux relation:

$$W_x = \frac{1}{2} I_c \Phi_{int}. \tag{5.38}$$

The conductor magnetic flux is calculated from substituting Equation (5.35) into the previous relation:

$$\Phi_{int} = \mu_0 \frac{I_c L}{2\pi} \frac{1}{4}. \tag{5.39}$$

The interesting conclusion is that the internal flux in a conductor carrying a current is independent of the radius.

5.6.3. Total Conductor Magnetic Flux

The total conductor-generated magnetic flux is the sum of the internal and external fluxes. Equation (5.28) gives the conductor-generated flux passing through the plane placed between two points P_1 and P_2, as sketched in Figure 5.34. The distances between the center of the conductor, and P_1 and P_2 are D_1 and D_2, respectively.

For the determination of the flux between the conductor and an external point, P_1 is placed on the conductor surface, as seen in Figure 5.34. In this case, D_1 is equal to

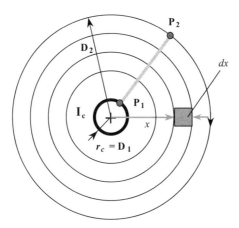

Figure 5.34. Flux between the conductor and an external point.

the conductor radius, r_c. The substitution of $D_1 = r_c$ in Equation (5.28) and replacement of D_2 with D gives a general equation for the external magnetic flux:

$$\Phi_{ext} = \mu_0 \frac{I_c L}{2\pi} \ln \frac{D}{r_c}, \quad (5.40)$$

where

D is the distance from the center of the conductor;
r_c is the conductor radius;
I_c is the conductor current; and
L is the conductor length.

The total conductor-generated flux, which is the sum of the internal and external fluxes, is

$$\Phi_c = \mu_0 \frac{I_c L}{2\pi} \ln\left(\frac{D}{r_c}\right) + \mu_0 \frac{I_c L}{2\pi} \frac{1}{4} = I_c \frac{\mu_0 L}{2\pi}\left[\ln\left(\frac{D}{r_c}\right) + \frac{1}{4}\right]. \quad (5.41)$$

Although this formula is suitable for calculations, the equation is modified by substituting $1/4 = \ln(1/e^{-1/4})$ into the previous expression. The two natural logarithms are combined, which results in a simplified formula. A further simplification is the assumption of an L unit length, which gives the flux in weber per unit length. In the United States, weber per mile is the preferred unit, whereas in Europe, weber per kilometer is used:

$$\frac{\Phi_c}{L} = I_c \frac{\mu_0}{2\pi}\left[\ln\left(\frac{D}{r_c}\right) + \frac{1}{4}\right] = I_c \frac{\mu_0}{2\pi} \ln\left(\frac{D}{e^{-0.25}r_c}\right) = I_c \frac{\mu_0}{2\pi} \ln\left(\frac{D}{GMR}\right), \quad (5.42)$$

where GMR is the *geometric mean radius*.

The GMR for a stranded conductor is GMR = $r_c e^{-0.25}$, but for a stranded conductor, the GMR values are given in conductor tables. As an example, the GMR for the Cardinal conductor is 0.0404 ft from Table 5.1.

Equation (5.42) is used to calculate the flux generated by a conductor. That equation can be interpreted that the actual conductor is replaced by an ideal conductor (tube) with a radius of GMR. The magnetic field inside this ideal conductor tube is zero.

5.6.4. Three-Phase Line Inductance

In the case of a three-phase line, the current in each phase conductor produces a magnetic field. The field lines are concentric circles around the conductors. Conductor A is exposed to three fluxes:

1. Φ_{AA}, the flux generated by the phase **A** current;
2. Φ_{AB}, the flux generated by the phase **B** current; and
3. Φ_{AC}, the flux generated by the phase **C** current.

TRANSMISSION LINE INDUCTANCE

The inductance of a three-phase line for a balanced system is usually referred to as *positive sequence inductance*. It is assumed that the line is transposed and the line currents are balanced.

EXAMPLE 5.4: The approach to compute the line inductance is presented using a numerical example. A three-phase line, built with Cardinal conductors and arranged as illustrated in Figure 5.29, has the following geometrical data:

Phase spacing: $D_{AB} := 25$ ft $D_{BC} := D_{AB}$ $D_{AC} := 2 \cdot D_{AB}$.
Conductor height and GMR: $H_{line} := 70$ ft $GMR_c := 0.0404$ ft.

The phasor line currents are:

$$I_A := 1500 \cdot e^{-j \cdot 0 \text{ deg}} \text{ A} \quad I_B := I_A \cdot e^{-j \cdot 120 \text{ deg}} \quad I_C := I_A \cdot e^{-j \cdot 240 \text{ deg}}.$$

The coordinates of a distant point F (see Fig. 5.35) are:

$$D_{AF} := 1000 \text{ ft} \quad D_{BF} := D_{AF} \quad D_{CF} := D_{AF}.$$

The length of the line is: $L := 100$ mi.

The flux linking conductor A is determined by calculation of the flux paths through the plane between conductor A and an arbitrarily selected distant point F. The current in conductor A produces a flux of Φ_{AA} between A and F, as annotated in Figure 5.35. The flux can be calculated with Equation (5.42):

$$\Phi_{AA} := I_A \cdot \frac{\mu_o}{2 \cdot \pi} \cdot \ln\left(\frac{D_{AF}}{GMR_c}\right) \cdot L. \qquad (5.43)$$

The numerical value is: $\Phi_{AA} = 488.4$ Wb.

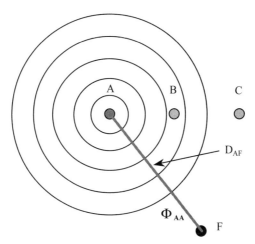

Figure 5.35. Phase A-produced flux linkage between A and F.

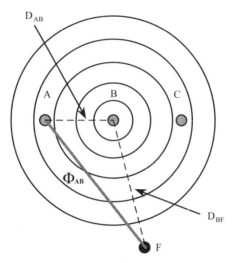

Figure 5.36. Phase B-produced flux linkage between A and F.

The current in conductor B creates a flux of Φ_{AB} between A and F, as exhibited in Figure 5.36. The flux can be calculated with Equation (5.28):

$$\Phi_{AB} := I_B \cdot \frac{\mu_o}{2 \cdot \pi} \cdot \ln\left(\frac{D_{BF}}{D_{AB}}\right) \cdot L. \tag{5.44}$$

The numerical values are: $\Phi_{AB} = (-80.9 - 140.2j)$ Wb $|\Phi_{AB}| = 161.9$ Wb.

Similarly, the current in conductor C generates a flux of Φ_{AC} between A and F, as shown in Figure 5.37. The flux is computed using:

$$\Phi_{AC} := I_C \cdot \frac{\mu_o}{2 \cdot \pi} \cdot \ln\left(\frac{D_{CF}}{D_{AC}}\right) \cdot L. \tag{5.45}$$

The numerical values are: $\Phi_{AC} = (-64.2 + 111.2j)$ Wb $|\Phi_{AC}| = 128.4$ Wb.

The total flux linkage with conductor A is the sum of the fluxes generated by the phase currents, that is, the addition of Equation (5.43), Equation (5.44), and Equation (5.45):

$$\Phi_A = I_A \frac{\mu_0}{2\pi} \ln\left(\frac{D_{AF}}{GMR}\right) L + I_B \frac{\mu_0}{2\pi} \ln\left(\frac{D_{BF}}{D_{AB}}\right) L + I_C \frac{\mu_0}{2\pi} \ln\left(\frac{D_{CF}}{D_{AC}}\right) L. \tag{5.46}$$

Similar equations describe the flux linkage with conductors B and C.

For balanced conditions, the sum of the three currents is zero, which permits the elimination of I_C from the earlier expression. After substitution of $I_C = -I_A - I_B$, simplification results in:

$$\Phi_A = I_A \frac{\mu_0}{2\pi} \ln\left(\frac{D_{AC}}{GMR} \frac{D_{AF}}{D_{CF}}\right) L + I_B \frac{\mu_0}{2\pi} \ln\left(\frac{D_{AC}}{D_{AB}} \frac{D_{BF}}{D_{CF}}\right) L. \tag{5.47}$$

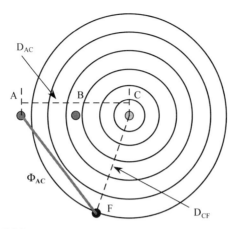

Figure 5.37. Phase C-produced flux linkage between A and F.

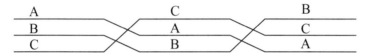

Figure 5.38. Transposed three-phase line.

This equation gives the flux linking phase A by calculation of the flux paths through the plane between conductor A and an arbitrarily selected distant point F. Moving point F to infinity results in $D_{AF} = D_{BF} = D_{CF}$. Substitution of these values yields:

$$\Phi_A := I_A \cdot \frac{\mu_o}{2 \cdot \pi} \cdot \ln\left(\frac{D_{AC}}{GMR_c}\right) \cdot L + I_B \cdot \frac{\mu_o}{2 \cdot \pi} \cdot \ln\left(\frac{D_{AC}}{D_{AB}}\right) \cdot L. \tag{5.48}$$

The numerical value is: $|\Phi_A| = 344.5$ Wb.

Similar equations can be derived for phases B and C. Equation (5.48) demonstrates that if the line is built with unequal conductor spacing, the voltage drop along the line will be unbalanced even if the currents are balanced. In other words, if the line is supplied by balanced three-phase voltage, the phase voltages at the opposite end of the line will be different. Utilities commonly transpose the conductors to avoid these unbalanced voltages. Figure 5.38 explains the concept of *line transposition*. In this example, each phase occupies each position for one-third of the line length.

The transposition equalizes the line and permits the calculation of the flux using an average spacing, called the *geometric mean distance* (GMD). The GMD is calculated by:

$$GMD := \sqrt[3]{D_{AB} \cdot D_{AC} \cdot D_{BC}}. \tag{5.49}$$

The corresponding numerical value is: GMD = 13.44 m.

If line transposition is employed, then this implies that D_{AB}, D_{BC}, and D_{AC} may be replaced by the GMD. The substitution of Equation (5.49) into Equation (5.48) eliminates the second term on the right-hand side of Equation (5.48). The simplified flux equation is:

$$\Phi_A := I_A \cdot \frac{\mu_o}{2\cdot\pi} \cdot \ln\left(\frac{\text{GMD}}{\text{GMR}_c}\right) \cdot L. \tag{5.50}$$

The corresponding numerical value is: $\Phi_A = 337.7$ Wb.

The inductance of phase A is calculated using the flux-inductance relation described by Equation (5.36). This equation gives the inductance:

$$L_{A\_ind} := \frac{\Phi_A}{I_A}. \tag{5.51}$$

The corresponding numerical value is: $L_{A\_ind} = 0.225$ H.

Substituting Equation (5.50) into Equation (5.51) provides the unit length inductance as:

$$L_{A\_inductance} := \frac{\mu_o}{2\cdot\pi} \cdot \ln\left(\frac{\text{GMD}}{\text{GMR}_c}\right). \tag{5.52}$$

The corresponding numerical value is:

$$L_{A\_inductance} = 2.25 \frac{\text{mH}}{\text{mi}}.$$

The per phase reactance of the line per unit length is:

$$X_A := \omega \frac{\mu_o}{2\cdot\pi} \cdot \ln\left(\frac{\text{GMD}}{\text{GMR}_c}\right). \tag{5.53}$$

For a system frequency of $\omega = 2\pi 60$ Hz, the corresponding numerical value is:

$$X_A = 0.849 \frac{\Omega}{\text{mi}}.$$

Many high-voltage transmission lines are built with bundled conductors. The application of bundled conductors reduces the line reactance. An equivalent conductor replaces the bundled conductors. The GMR of this conductor depends on the number of conductors in the bundle and the distance between the conductors within the bundle. The equivalent GMRs of the bundles are:

$$\text{Two-conductor bundle: } \text{GMR}_{eq} = \sqrt{d\text{GMR}_C}, \tag{5.54}$$

$$\text{Three-conductor bundle: } \text{GMR}_{eq} = \sqrt[3]{d^2\text{GMR}_C}, \tag{5.55}$$

$$\text{Four-conductor bundle: } \text{GMR}_{eq} = 1.09\sqrt[4]{d^3\text{GMR}_C}, \tag{5.56}$$

where GMR_C is the conductor GMR obtained from the conductor tables and d is the distance between the conductors in the bundle. For bundled conductors, the GMR_{eq} value is used in place of the GMR_C in Equation (5.52) and Equation (5.53) to determine the line inductance and reactance, respectively.

5.7. TRANSMISSION LINE CAPACITANCE

The conductors of an energized transmission line carry charges. These charges produce an electric field around the conductors. The capacitance of a long, straight, solid round conductor is calculated in this section.

5.7.1. Electric Field Generation

The electric field is generated by electric charge. Typically, a charged spherical-shaped particle generates radial electric field vectors, as drawn in Figure 5.39. The direction of the field vectors depends on the polarity of the charge, as seen in the diagram. Specifically, for a positive charge, the electric field radiates outward, whereas for a negative charge, the field is directed toward the charge. Equipotential surfaces are concentric spheres.

The *electric flux density*, $\vec{D}$, on a surface enclosing the charge is determined by Gauss' law:

$$q = \int_F \vec{D} \times \overline{dF}, \qquad (5.57)$$

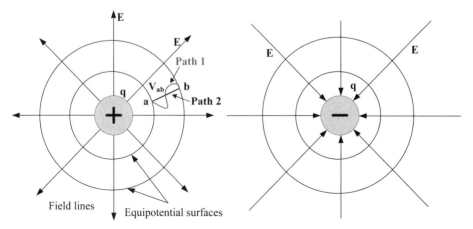

Figure 5.39. Electric field generated by spherical-shaped charges (particles).

where

q is the charge, measured in coulomb (C) or ampere-second;
$\vec{D}$ is the flux density vector measured in C/m²; and
F is the area surface surrounding the charge, measured in square meters.

The *electric field intensity* generated by a charge can be calculated in air by dividing the flux density by the free space permittivity:

$$\vec{E} = \frac{\vec{D}}{\varepsilon_0}, \quad (5.58)$$

where

$\vec{E}$ is the electric field intensity vector in volts per meter or volts per centimeter; and
ε_0 is the free space permittivity in farad per meter.

The *free-space permittivity* is a universal constant. Its value, which can be used for air, is:

$$\varepsilon_0 = \frac{10^{-9}}{36\pi} \frac{F}{m}. \quad (5.59)$$

The integral of the electric field strength between two points is the voltage or potential difference. The value of this integral is independent from the route. As an example using Figure 5.39, the integration along a field line (Path 2) or any other line (Path 1) gives the same result:

$$V_{ab} = \int_a^b \vec{E} \cdot \vec{d\ell}. \quad (5.60)$$

The energizing of a transmission line generates charges on the conductors. These charges produce electric fields around the line. The potential of earth ground is considered zero. The integration of the electric field between a point in the vicinity of the line and the ground gives the *space potential* of the selected point. The potential difference between the conductor and ground is called the *line-to-ground voltage*, and the voltage difference between two conductors is termed the *line-to-line voltage* in a three-phase system.

5.7.2. Electrical Field around a Conductor

Figure 5.40 presents an energized conductor and the radial electric field lines emanating from the conductor. The field inside the conductor is zero. The constant electric field (potential) lines are concentric circles (not shown). For the calculation of the electric

TRANSMISSION LINE CAPACITANCE

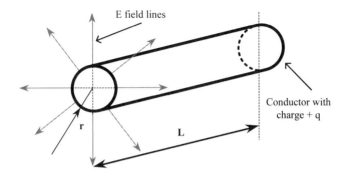

Figure 5.40. Electric field around a conductor.

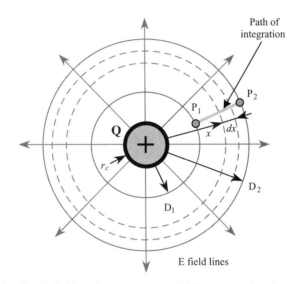

Figure 5.41. Electric field and constant potential lines around a charged conductor.

flux density, **D**, we select a constant potential line at a radius of x, as illustrated in Figure 5.41.

According to Gauss' law, the electric flux density per unit length is:

$$|\vec{D}| = \frac{Q}{2\pi x}, \qquad (5.61)$$

where Q is the electric charge per unit length.

The electric field intensity is the ratio of the flux density and the permittivity:

$$|\vec{E}| = \frac{|\vec{D}|}{\varepsilon_o} = \frac{Q}{2\pi\varepsilon_o}\frac{1}{x}. \qquad (5.62)$$

The permittivity of air can be taken as that for free space. The potential difference between points P_1 and P_2 is the integral of the electric field along the path. The potential difference is:

$$V_{12} = \int_{P_1}^{P_2} |\vec{E}| dx = \frac{Q}{2\pi\varepsilon_o} \int_{D_1}^{D_2} \frac{1}{x} dx = \frac{Q}{2\pi\varepsilon_o} \ln\left(\frac{D_2}{D_1}\right). \quad (5.63)$$

This equation is suitable to calculate the potential difference in the vicinity of a transmission line conductor.

EXAMPLE 5.5: MATLAB visualization of the electric field intensity

Computing the electric field in the vicinity of a single conductor becomes a rather straightforward procedure, given the previous derivation. Recall that the line-to-ground voltage, V_{ln}, for a symmetrical three-phase system is obtained by dividing the line-to-line voltage by $\sqrt{3}$. As the electric field lines extend radially outward from the conductor, the potential V_{12} along a straight path from the conductor to the ground is known to be V_{ln}. The distance (D_2) from the conductor to the ground, however, changes depending on the angle θ denoted in Figure 5.42. To determine the electric field, first substitute Equation (5.63) into Equation (5.62) to eliminate charge from the expression, and set x equal to d to correspond to the distance from the conductor to an arbitrary point (x_o, y_o), as illustrated in Figure 5.42:

$$|\vec{E}| = \frac{V_{ln}}{d \ln\left(\frac{D_2}{D_1}\right)}, \quad (5.64)$$

where D_1 is the conductor radius (r_c). The distance from the conductor center to some arbitrary point (x_o, y_o) is:

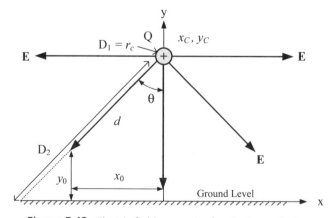

Figure 5.42. Electric field geometry for single conductor.

TRANSMISSION LINE CAPACITANCE

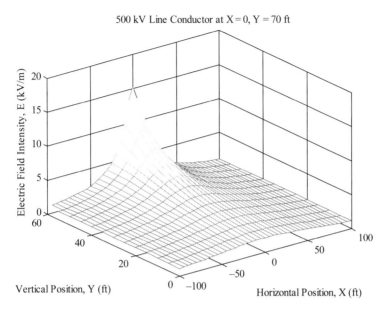

Figure 5.43. Electric field strength below a single 500 kV conductor at a height of 70 ft.

$$d = \sqrt{(x_C - x_o)^2 + (y_C - y_o)^2},$$

and the straight-line distance from the conductor through that arbitrary point to the earth ground is:

$$D_2 = \frac{y_C}{\cos(\theta)} = \frac{y_C}{\left(\dfrac{y_C - y_o}{d}\right)}.$$

Using these expressions, a MATLAB program (given later) is developed to compute and plot the electric field intensity near a single 500 kV conductor that is at a height of 70 ft. Executing the program produces three graphs visualizing the electric field. Figure 5.43 provides a surface plot of the electric field strength below the conductor. The equipotential lines surrounding the conductor are shown in the contour plot of Figure 5.44. MATLAB has the ability to combine the surface and contour plots into a single graph, which is given in Figure 5.45.

```
%
%    EFieldSingleConductor.m
%
conv = 3.281;    % conversion from meters to feet

eps0 = 1e-9/(36*pi);     % permittivity of air (F/m)

% set the conductor position (ft)
xcond = 0;      ycond = 70;
```

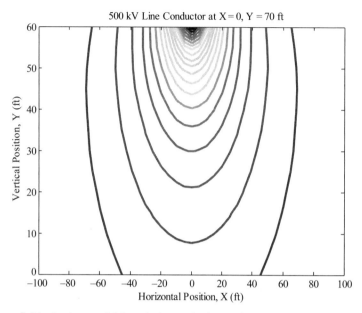

Figure 5.44. Equipotential lines below a single 500 kV conductor at a height of 70 ft.

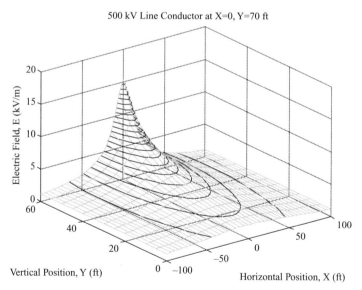

Figure 5.45. Three-dimensional contour of equipotential lines superimposed on a surface plot of the electric field strength from a single 500 kV conductor.

TRANSMISSION LINE CAPACITANCE

```
rc = 0.147;          % conductor radius (ft)
Vline = 500e3;       % line voltage (V)
Vln = Vline/sqrt(3); % line-to-neutral voltage
% create vectors for x and y positions (ft)
x = (-100 : 5 : 100);
y = (0 : 5 : 60);
% compute electric field (V/m) at the desired locations
for i = 1:size(x,2)
    for j = 1:size(y,2)
        % compute distance from conductor to selected point
        % (m)
        d = sqrt((x(i)-xcond)^2 + (y(j)-ycond)^2) / conv;
        % compute the angle normal to ground
        theta = atan2(x(i)-xcond,y(j)-ycond);
        % compute distance from conductor to ground along
        % path (ft)
        dist = ycond / cos(theta);
        % note that cos(theta) also equals (ycond-y(j))/d
        Efield(j,i) = Vln / (d * log(abs(dist/rc)));
    end
end
% plot the electric field magnitude as a function of position
mesh(x,y,Efield/1000)
set(gca, 'fontname','Times', 'fontsize',11);
xlabel('Horizontal Position, X (ft)')
ylabel('Vertical Position, Y (ft)')
zlabel('Electric Field Intensity, E (kV/m)')
title([num2str(Vline/1000),' kV Line Conductor at X=',...
        num2str(xcond),', Y=',num2str(ycond),' ft'])

figure;
contour(x,y,Efield/1000,20,'LineWidth',2.5)
set(gca, 'fontname','Times', 'fontsize',11);
xlabel('Horizontal Position, X (ft)')
ylabel('Vertical Position, Y (ft)')
title([num2str(Vline/1000),' kV Line Conductor at X=',...
        num2str(xcond),', Y=',num2str(ycond),' ft'])

figure
contour3(x,y,Efield/1000,20)
surface(x,y,Efield/1000,'EdgeColor',[0.8 0.8 0.8],...
    'FaceColor','none');
set(gca, 'fontname','Times', 'fontsize',11);
xlabel('Horizontal Position, X (ft)')
ylabel('Vertical Position, Y (ft)')
zlabel('Electric Field, E (kV/m)')
title([num2str(Vline/1000),' kV Line Conductor at X=',...
        num2str(xcond),', Y=',num2str(ycond),' ft'])
```

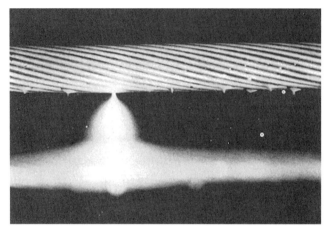

Figure 5.46. Corona discharge on a wet conductor. Source: EPRI *Transmission Line Reference Book*, Red Book.

5.7.3. Three-Phase Transmission Line Generated Electric Field

The energized conductors of a three-phase transmission line generate an electric field. The electric field has undesirable environmental effects.

(a) The high electric field at the conductor surface produces corona discharges that generate radio and TV disturbances.
(b) The electric field at the ground level may produce small shocks and disturb people walking under the line.

5.7.3.1. Corona Discharge. When the electric field intensity is above 20–25 kV/cm at the surface of a conductor, corona discharge is generated. The *corona discharge* is the electric breakdown of the air near the conductor. Figure 5.46 shows a typical corona discharge. In this case, a water droplet on the conductor originates the discharge. Generally, water droplets, dirt, and metal protrusions on the conductor surface increase the local electric field and initiate corona. The high humidity decreases the breakdown strength of the air surrounding the conductors, which further increases the intensity of the discharge. The strong corona discharge on a conductor is visible at night. The corona discharge intensity increases with rain and wet weather.

Other sources of discharge include polluted insulators in foggy weather, during which dry-band arcing may occur. Corona discharge also occurs during fair weather conditions. This occurs mostly where the hardware holding the conductors has sharp edges and surface protrusions, which increase the electric field and initiate discharge.

The electric effect of the corona is high-frequency current pulses, which generate radio and TV disturbances. These effects cause public protest and complaints. The power companies eliminate or reduce the TV and radio disturbances by using bundled conductors, as introduced in Figure 5.16.

5.7.3.2. Electric Field at the Ground Level.
The electric field at ground level produces minor disturbances. As an example, a large ungrounded truck may accumulate charges when parked under or close to a transmission line. A grounded person (i.e., standing on the ground) receives minor electric shocks when (s)he touches the truck. This is not dangerous if the field-produced leakage or discharge current is less than 5 mA. But even small current can cause an accident if the worker drops equipment because of the minor shock. In fact, in Japan, engineers test the level of annoyance by having people walk under the line while carrying an umbrella. Soon afterwards, when the person touches the metallic shaft of the umbrella, he receives a mild shock. If the shock annoys the participant, then the engineer deems the electric field unacceptable.

Pedestrians under a transmission line may experience a tingling sensation on their skin and hair repulsion. Typically, these annoying effects occur when the field at ground level is around 1–3 kV/m. The perception level depends on the person, the weather conditions, and so on.

IEEE Standard (C95.6-2002) for Safety Levels with Respect to Human Exposure to Electromagnetic Fields, 0–3 kHz defines the environmental electric field MPE for whole-body exposure. Table 5.3 gives two values: one for the general public, and another for a controlled environment, which applies to personnel maintaining the power line (i.e., an occupational exposure). According to IEEE Standard C95.6, the electric field under the transmission line on the right-of-way should be less than 5 kV/m for power transmission frequencies of interest. However, a note in the standard permits 10 kV/m for members of the general public walking within the right-of-way. The right-of-way for a 500 kV line is around 125 ft, and for a 220 kV line is about 75–90 ft.

5.7.3.3. Electric Charge Calculation.
The electric field and the line charge are generated by the time-varying supply voltage. This implies that the charge variation is also sinusoidal. The calculation of the transmission line-generated electric field requires the calculation of the root-mean-square (rms) value of charge accumulated on the conductors. Accordingly, we consider only the magnitude of the electric field and not its time-dependent behavior. Energizing the conductors generates charges both on the conductors and in the ground under the conductors. The ground effect is simulated by placing (mirror) image conductors, as portrayed in Figure 5.47. The *image conductors* carry equivalent negative charges and are placed equidistance below ground level.

TABLE 5.3. Environmental Electric Field MPEs, Whole-Body Exposure

General Public		Controlled Environment	
Frequency Range (Hz)	E-rms (V/m)	Frequency Range (Hz)	E-rms (V/m)
1–368	5000	1–272	20,000
368–3000	$1.84 \times 10^6/f$	272–3000	$5.44 \times 10^6/f$
3000	614	3000	1813

f is the frequency in Hz.
Source: IEEE Std. C95.6-2002.

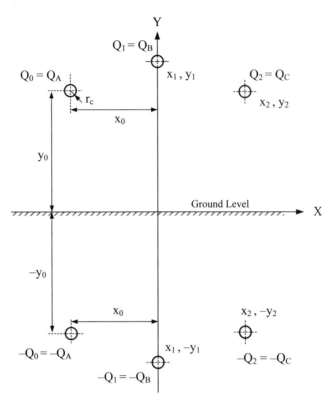

Figure 5.47. Transmission line arrangement for electric field calculation.

5.7.3.4. Concept of Charge Calculation. An electrical charge generates an electric field. In this electric field, the voltage difference between two selected points (1 and 2) can be calculated by Equation (5.63), which is repeated here:

$$V_{1,2} = \frac{Q}{2\pi\varepsilon_o} \ln\left(\frac{D_2}{D_1}\right). \tag{5.65}$$

Figure 5.47 shows that in the case of a three-phase line, six charges (i.e., Q_A, Q_B, Q_C, and their negatives images) produce the electric field.

The concept of the charge calculation is initially presented using a simplified arrangement, which contains only two conductors **k** and **m**, as drawn in Figure 5.48. This diagram annotates the charges and the distances, which are needed for the analysis. The voltage difference between conductor **k** and its image conductor is equal to twice the line-to-neutral voltage:

$$V_{k,-k} = 2V_{\ln,k}. \tag{5.66}$$

TRANSMISSION LINE CAPACITANCE

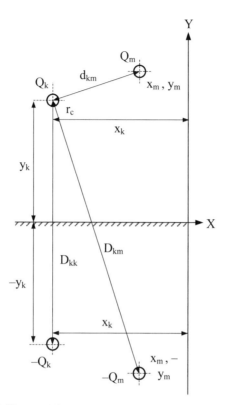

Figure 5.48. Simplified line arrangement with two conductors.

Simultaneously, the voltage difference between conductor **k** and its image conductor **−k** can be calculated by the summation of the voltage differences produced by the charges on all conductors. In this case, the voltage difference between Q_k and its image conductor $(-Q_k)$ is generated by charges Q_k, $-Q_k$, Q_m, and $-Q_m$:

$$V_{k,-k} = \frac{Q_k}{2\pi\varepsilon_0}\ln\left(\frac{D_{kk}}{r_c}\right) + \frac{-Q_k}{2\pi\varepsilon_0}\ln\left(\frac{r_c}{D_{kk}}\right) + \frac{Q_m}{2\pi\varepsilon_0}\ln\left(\frac{D_{km}}{d_{km}}\right) + \frac{-Q_m}{2\pi\varepsilon_0}\ln\left(\frac{d_{km}}{D_{km}}\right). \quad (5.67)$$

The first two terms are identical as well as the last two terms since $-\ln(x) = \ln(1/x)$. This permits the simplification of the equation by combining the first two and the last two terms together, such that:

$$V_{k,-k} = 2\left[\frac{Q_k}{2\pi\varepsilon_0}\ln\left(\frac{D_{kk}}{r_c}\right) + \frac{Q_m}{2\pi\varepsilon_0}\ln\left(\frac{D_{km}}{d_{km}}\right)\right]. \quad (5.68)$$

The combination of Equation (5.66) and Equation (5.68) produces an expression for the charge calculation. After rearrangement the final equation is:

$$V_{\text{ln},k} = \left[\frac{1}{2\pi\varepsilon_0}\ln\left(\frac{D_{kk}}{r_c}\right)\right]Q_k + \left[\frac{1}{2\pi\varepsilon_0}\ln\left(\frac{D_{km}}{d_{km}}\right)\right]Q_m. \quad (5.69)$$

In this equation, the coefficients of the charges are the *potential coefficients*. A further simplification is:

$$V_{\text{ln},k} = p_{kk}Q_k + p_{km}Q_m, \quad (5.70)$$

where:

$$p_{km} = \frac{1}{2\pi\varepsilon_0}\ln\left(\frac{D_{km}}{d_{km}}\right), \quad (5.71)$$

with $d_{km} = r_c$ when $k = m$. The voltage difference between conductor **m** and its image conductor **−m** can be calculated with a similar equation:

$$V_{\text{ln},m} = p_{mk}Q_k + p_{mm}Q_m. \quad (5.72)$$

The two voltage equations form a set of linear equations. The charges on each conductor can be computed from these coupled equations:

$$\begin{aligned}V_{\text{ln},k} &= p_{kk}Q_k + p_{km}Q_m, \\ V_{\text{ln},m} &= p_{mk}Q_k + p_{mm}Q_m.\end{aligned} \quad (5.73)$$

These equations can be represented in a matrix form:

$$\mathbf{V_{\text{ln}}} = \mathbf{PQ}, \quad (5.74)$$

where

$\mathbf{V_{\text{ln}}}$ is the line-to-ground voltage vector;
$\mathbf{Q}$ is the conductor charge vector; and
$\mathbf{P}$ is the potential coefficient matrix of:

$$P = \begin{bmatrix} p_{kk} & p_{km} \\ p_{mk} & p_{mm} \end{bmatrix}.$$

EXAMPLE 5.6: Mathcad conductor charge calculation

The electric charge calculation for a three-phase line is presented using a numerical example. The transmission line rated voltage is 500 kV (line-to-line). The coordinates of the conductors and image conductors are:

$$\begin{aligned}x_0 &:= -35 \text{ ft} & x_1 &:= 0 \text{ ft} & x_2 &:= 35 \text{ ft} \\ y_0 &:= 70 \text{ ft} & y_1 &:= 70 \text{ ft} & y_2 &:= 70 \text{ ft}\end{aligned}.$$

TRANSMISSION LINE CAPACITANCE

The conductor radius and line-to-ground voltage are:

$$r_c := 1.76 \text{ in} \quad V\_{ln} := \frac{500 \text{ kV}}{\sqrt{3}}.$$

The dielectric constant (permittivity) of free space (air) is:

$$\varepsilon_o := \frac{10^{-9}}{36 \cdot \pi} \cdot \frac{F}{m}.$$

The line-to-ground voltages of phases A, B, and C are formed into a vector representation:

$$k := 0..2 \quad V_{ln_k} := V\_{ln} \cdot e^{-j \cdot k \cdot 120 \text{ deg}} \quad V_{ln} = \begin{pmatrix} 288.7 \\ -144.3 - 250.0j \\ -144.3 + 250.0j \end{pmatrix} \text{kV}.$$

The charge calculation requires the distances between the phase conductors (positive charges), which are (refer to Fig. 5.48):

$$d_{k,i} := \sqrt{(x_k - x_i)^2 + (y_k - y_i)^2} \quad \begin{array}{l} i := 0..2 \\ d_{k,k} := r_c \end{array} \quad d = \begin{pmatrix} 0.147 & 35 & 70 \\ 35 & 0.147 & 35 \\ 70 & 35 & 0.147 \end{pmatrix} \text{ft},$$

where the index variable i is used in Mathcad instead of m, to avoid conflicting with the unit of meters. The distances between the phase conductors and image conductors (between positive and negative charges) are:

$$D_{k,i} := \sqrt{(x_k - x_i)^2 + (y_k + y_i)^2} \quad D = \begin{pmatrix} 140.0 & 144.3 & 156.5 \\ 144.3 & 140.0 & 144.3 \\ 156.5 & 144.3 & 140.0 \end{pmatrix} \text{ft}.$$

The potential coefficients are:

$$p_{k,i} := \frac{1}{2 \cdot \pi \cdot \varepsilon_o} \cdot \ln\left(\frac{D_{k,i}}{d_{k,i}}\right) \quad p = \begin{pmatrix} 76.74 & 15.84 & 9.00 \\ 15.84 & 76.74 & 15.84 \\ 9.00 & 15.84 & 76.74 \end{pmatrix} \frac{\text{mi}}{\mu F}.$$

The voltage–charge relation is described by the previously derived matrix expression of Equation (5.74). The solution of the matrix equation gives the charges on the conductor surface:

$$Q := p_{-1} \cdot V_{ln} \quad Q = \begin{pmatrix} 4.48 + 0.38j \\ -2.41 - 4.18j \\ -1.91 + 4.08j \end{pmatrix} \cdot \frac{mC}{mi}.$$

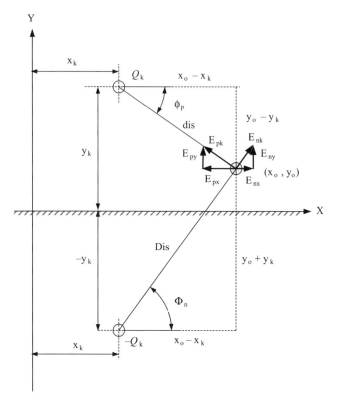

Figure 5.49. Electric field generated by conductor k and its image conductor.

5.7.3.5. Electric Field Calculation. Before computing the electric field for a three-phase line, the concept of the calculation is presented using only one conductor (**k**) and its image conductor (−**k**). Figure 5.49 shows the conductor arrangement. The field is calculated at a selected point having coordinates x_o and y_o. The diagram exhibits the two electrical field vectors E_{pk} and E_{nk} generated by Q_k and $-Q_k$, respectively, where the p and n signify positive and negative charges, respectively. From Figure 5.49 it can be seen that the vectors are divided into x- and y-components. The total field is the vector sum of these components.

For the calculation of three-phase quantities, a variable $k = 0,1,2$ is introduced into the equations. The $k = 0$ denotes the field generated by phase A, $k = 1$ signifies the field produced by phase B, and $k = 2$ represents the field created by phase C. The basic equation for the electric field is given by Equation (5.62), which is repeated here:

$$|\vec{E}_k| = \frac{Q_k}{2\pi\varepsilon_0 D_{ko}}, \tag{5.75}$$

TRANSMISSION LINE CAPACITANCE

where

Q_k is the charge per unit length on conductor **k**; and
D_{ko} is the distance between conductor **k** and a selected point (**o**).

The electric field near a three-phase line is generated by the six charges seen in Figure 5.47. Each charge generates a field vector. The total field is the sum of the field vectors generated by the six charges, that is,

$$\vec{E}_T = \sum_k \vec{E}_k + \vec{E}_{-k}. \tag{5.76}$$

The following example calculates the three-phase generated fields.

EXAMPLE 5.7: Mathcad electric field calculation

This calculation uses the rms values of the voltages, charges, and field vectors. The time variation is not considered. The conductor **k** generated electric field is calculated at a selected point (**o**), whose coordinates are:

$$x_o := 40 \text{ ft} \quad y_o := 3 \text{ ft.}$$

Field Generated by the Positive Charges on the Conductors

Referring to Figure 5.49, the distance between conductor **k**, with positive charge, and the selected point is:

$$\text{dis}(x_o, y_o, k) := \sqrt{(x_o - x_k)^2 + (y_o - y_k)^2} \quad \text{dis}(x_o, y_o, k) = \begin{pmatrix} 100.6 \\ 78.0 \\ 67.2 \end{pmatrix} \cdot \text{ft.}$$

The magnitude of the positive charge generated electric field vector is:

$$E_p(x_o, y_o, k) := \frac{Q_k}{2 \cdot \pi \cdot \varepsilon_o \cdot \text{dis}(x_o, y_o, k)} \quad E_p(x_o, y_o, k) = \begin{pmatrix} 1.64 + 0.14j \\ -1.13 - 1.97j \\ -1.04 + 2.23j \end{pmatrix} \frac{\text{kV}}{\text{m}}.$$

The field vector's space angle, ϕ_p, is:

$$\phi_p(x_o, y_o, k) := \operatorname{asin}\left(\frac{y_o - y_k}{\text{dis}(x_o, y_o, k)}\right) \quad \phi_p(x_o, y_o, k) = \begin{pmatrix} -41.8 \\ -59.2 \\ -85.7 \end{pmatrix} \cdot \text{deg.}$$

The x component of the positive field vector is:

$$E_{px}(x_o, y_o, k) := E_p(x_o, y_o, k) \cdot \cos(\phi_p(x_o, y_o, k))$$

$$E_{px}(x_o, y_o, k) = \begin{pmatrix} 1.22 + 0.10j \\ -0.58 - 1.01j \\ -0.08 + 0.17j \end{pmatrix} \frac{\text{kV}}{\text{m}}.$$

The y component of the positive field vector is:

$$E_{py}(x_o, y_o, k) := E_p(x_o, y_o, k) \cdot \sin(\phi_p(x_o, y_o, k))$$

$$E_{py}(x_o, y_o, k) = \begin{pmatrix} -1.09 - 0.09j \\ 0.97 + 1.69j \\ 1.04 - 2.22j \end{pmatrix} \frac{kV}{m}.$$

Field Generated by the Negative (Image) Charges on the Conductors

The negative charges generated fields are calculated by similar equations. The distance between image conductor $-k$, with negative charge, and the selected point is:

$$Dis(x_o, y_o, k) := \sqrt{(x_o - x_k)^2 + (y_o + y_k)^2} \quad Dis(x_o, y_o, k) = \begin{pmatrix} 104.7 \\ 83.2 \\ 73.2 \end{pmatrix} \cdot ft.$$

The magnitude of the negative charge generated electric field vector is:

$$E_n(x_o, y_o, k) := \frac{-Q_k}{2 \cdot \pi \cdot \varepsilon_o \cdot Dis(x_o, y_o, k)} \quad E_n(x_o, y_o, k) = \begin{pmatrix} -1.57 - 0.13j \\ 1.06 + 1.84j \\ 0.96 - 2.04j \end{pmatrix} \frac{kV}{m}.$$

From Figure 5.49, the field vector's space angle, Φ_n, is:

$$\Phi_n(x_o, y_o, k) := \operatorname{asin}\left(\frac{y_o + y_k}{Dis(x_o, y_o, k)}\right) \quad \Phi_n(x_o, y_o, k) = \begin{pmatrix} 44.2 \\ 61.3 \\ 86.1 \end{pmatrix} \cdot deg.$$

The x- and y-components of the negative field vector are:

$$E_{nx}(x_o, y_o, k) := E_n(x_o, y_o, k) \cdot \cos(\Phi_n(x_o, y_o, k))$$

$$E_{nx}(x_o, y_o, k) = \begin{pmatrix} -1.13 - 0.10j \\ 0.51 + 0.89j \\ 0.07 - 0.14j \end{pmatrix} \frac{kV}{m}$$

$$E_{ny}(x_o, y_o, k) := E_n(x_o, y_o, k) \cdot \sin(\Phi_n(x_o, y_o, k))$$

$$E_{ny}(x_o, y_o, k) = \begin{pmatrix} -1.10 - 0.09j \\ 0.93 + 1.62j \\ 0.95 - 2.04j \end{pmatrix} \frac{kV}{m}$$

TRANSMISSION LINE CAPACITANCE

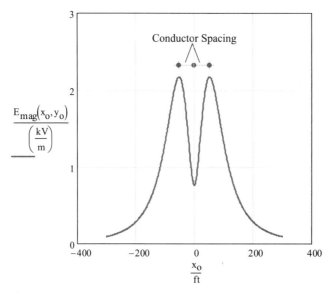

Figure 5.50. Electric field distribution at 3 ft from the ground.

The x- and y-components of the total electric field are found from the sum of all the x- and y-components, respectively, of the conductors and their images:

$$E_x(x_o, y_o) := \sum_k \left(E_{px}(x_o, y_o, k) + E_{nx}(x_o, y_o, k)\right) \quad E_x(x_o, y_o) = (10.9 - 88.1j)\frac{V}{m}$$

$$E_y(x_o, y_o) := \sum_k \left(E_{py}(x_o, y_o, k) + E_{ny}(x_o, y_o, k)\right) \quad E_y(x_o, y_o) = (1.7 - 1.1j)\frac{kV}{m}.$$

The x- and y-components of the electric field are complex vectors, which are orthogonal. The total field magnitude can be directly computed from:

$$E_{mag}(x_o, y_o) := \sqrt{(|E_x(x_o, y_o)|)^2 + (|E_y(x_o, y_o)|)^2} \quad E_{mag}(x_o, y_o) = 2.06 \frac{kV}{m}$$

The total field is plotted at a 3-ft height, that is, $y_o = 3$ ft, as shown in Figure 5.50. The dots denote the approximate position of the phase conductors. The figure reveals that the maximum field occurs under the conductors A and C. The field decreases rapidly by increasing the distance from the conductors. In this case, it can be seen that the field is negligible beyond 300 ft from the conductors.

EXAMPLE 5.8: MATLAB program for electric field calculation

The electric and magnetic fields have significant environmental effects, ranging from the annoyance of pedestrians to the fear of producing leukemia. Using the equations derived earlier, a MATLAB program is developed to calculate the electric fields produced by a three-phase line.

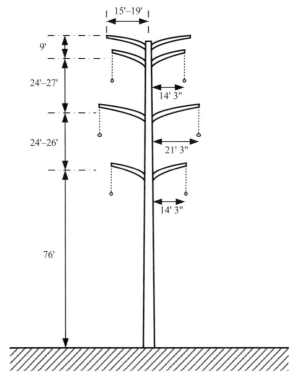

Figure 5.51. Tower arrangement of a double circuit 345 kV line. Data source: EPRI *Transmission Line Reference Book*, Red Book.

The application of MATLAB is demonstrated by calculating the electric fields generated by a 345 kV double-circuit transmission line, which has a tower arrangement as depicted in Figure 5.51. The coordinates of the phase conductors, which each have a diameter of 1.762 in., are:

Circuit 1 (right side)	x (ft)	y (ft)
Phase A1	14.3	70
Phase B1	21.3	94
Phase C1	14.3	118
Circuit 2 (left side)		
Phase A2	−14.3	70
Phase B2	−21.3	94
Phase C2	−14.3	118

TRANSMISSION LINE CAPACITANCE

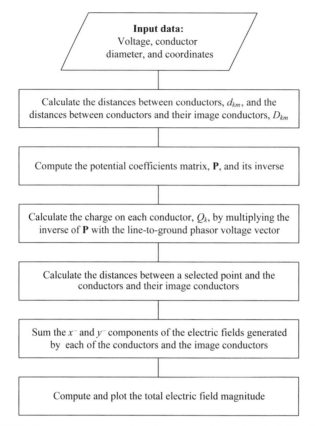

Figure 5.52. Computer program flowchart to perform electric field calculation.

A flowchart for the MATLAB program is given in Figure 5.52. The MATLAB program begins with establishing the physical data:

```
%
%     EFieldDoubleCircuitLine.m
%
clear all
conv = 3.281;       % conversion from meters to feet
eps0 = 1e-9/(36*pi);    % permittivity of air (F/m)
% input data
nc = 6;                 % number of conductors
rc = 1.762 / 12;        % conductor radius (ft)
Vline = 345e3;          % line voltage (V)
Vln = Vline/sqrt(3);    % line-to-neutral voltage

% create vectors for the conductor positions (ft)
xc = [14.3, 21.3, 14.3, -14.3, -21.3, -14.3];
yc = [70, 94, 118, 70, 94, 118];
```

Next, the distances from each conductor to the other five conductors and the six image conductors are computed. From those distances, the elements of the potential coefficients matrix (**P**) are formed using:

$$p_{km} = \frac{1}{2\pi\varepsilon_0} \ln\left(\frac{D_{km}}{d_{km}}\right), \tag{5.77}$$

where D_{km} is the distance from conductor **k** to image conductor **m**, and d_{km} is the distance from conductor **k** to conductor **m**. These distances are computed from:

$$D_{km} = \sqrt{(x_k - x_m)^2 + (y_k + y_m)^2}, \tag{5.78}$$

$$d_{km} = \begin{cases} r_c & k = m \\ \sqrt{(x_k - x_m)^2 + (y_k - y_m)^2} & k \neq m \end{cases}, \tag{5.79}$$

where r_c is the conductor radius. The voltage vector is formed from the line-to-ground phasor voltages:

$$\mathbf{V} = \begin{bmatrix} V_{ln}e^{-j0°} & V_{ln}e^{-j120°} & V_{ln}e^{-j240°} & \cdots \end{bmatrix}. \tag{5.80}$$

The charge vector is calculated from the product of the inverted potential coefficient matrix and the previous voltage vector, that is, $\mathbf{Q} = \mathbf{P}^{-1}\mathbf{V}$.

The MATLAB realization of this is:

```
% create the potential coefficients matrix
for k=1 : nc
    for m = k:nc
        % determine distances between conductors k and m
        if k == m            % for k=m, the distance is
            dkm = rc;        % the conductor radius
        else
            dkm = sqrt((xc(k)-xc(m))^2 + (yc(k)-yc(m))^2);
        end
        % find distance between conductor k and image m
        Dkm = sqrt((xc(k)-xc(m))^2 + (yc(k)+yc(m))^2);
        % compute potential coefficient
        P(k,m) = log(Dkm/dkm) / (2*pi*eps0);
        P(m,k) = P(k,m);    % symmetric matrix
    end
end
% form the line-to-ground voltage vector
vlg = [Vln; Vln*exp(-j*pi*2/3); Vln*exp(-j*pi*4/3); ...
       Vln; Vln*exp(-j*pi*2/3); Vln*exp(-j*pi*4/3)];
% calculate charges on each conductor
q = inv(P) * vlg;
```

TRANSMISSION LINE CAPACITANCE

With the charges on each conductor now known, the electric fields created by the positive charges on conductor **k** and the negative charges on its image conductor are calculated using Equation (5.62). Specifically, the electric field at point **o** due to conductor **k** is:

$$\vec{E}_{ko} = \frac{Q_k}{2\pi\varepsilon_0 d_{ko}}. \tag{5.81}$$

This electric field is a complex vector quantity that can be broken into its *x*- and *y*-components using the geometrical relations portrayed in Figure 5.49:

$$\begin{aligned}\vec{E}_{ko} &= E_{kx}\hat{i} + E_{ky}\hat{j} \\ &= \frac{Q_k}{2\pi\varepsilon_0 d_{ko}}\left[\cos(\phi_p)\hat{i} + \sin(\phi_p)\hat{j}\right],\end{aligned} \tag{5.82}$$

where $\hat{i}$ and $\hat{j}$ are orthogonal unit vectors. The total electric field is found by summing the *x*- and *y*-components from all the conductors and their image conductors:

$$\begin{aligned}\mathbf{E}_x &= \sum_{k=1}^{N} \mathbf{E}_{kx} + \mathbf{E}_{-kx} \\ &= \sum_{k=1}^{N} \frac{Q_k}{2\pi\varepsilon_0 d_{ko}}\cos(\phi_{kp}) + \frac{-Q_k}{2\pi\varepsilon_0 D_{ko}}\cos(\Phi_{kp}) \\ &= \sum_{k=1}^{N} \frac{Q_k}{2\pi\varepsilon_0}\left[\frac{x_o - x_k}{d_{ko}^2} - \frac{x_o - x_k}{D_{ko}^2}\right],\end{aligned} \tag{5.83}$$

$$\begin{aligned}\mathbf{E}_y &= \sum_{k=1}^{N} \mathbf{E}_{ky} + \mathbf{E}_{-ky} \\ &= \sum_{k=1}^{N} \frac{Q_k}{2\pi\varepsilon_0 d_{ko}}\sin(\phi_{kp}) + \frac{-Q_k}{2\pi\varepsilon_0 D_{ko}}\sin(\Phi_{kp}) \\ &= \sum_{k=1}^{N} \frac{Q_k}{2\pi\varepsilon_0}\left[\frac{y_o - y_k}{d_{ko}^2} - \frac{y_o + y_k}{D_{ko}^2}\right].\end{aligned} \tag{5.84}$$

The electric field magnitude is computed from *x*- and *y*-field components:

$$|\vec{E}| = \sqrt{|\mathbf{E}_x|^2 + |\mathbf{E}_y|^2}. \tag{5.85}$$

This is accomplished in MATLAB using:

```
% create vectors for x and y positions (ft)
x = (-100 : 5 : 100);
y = (0 : 5 : 60);
```

```
% compute electric field (V/m) at the desired locations
for i = 1:size(x,2)
    for j = 1:size(y,2)
        Ex = 0;  Ey = 0;
        for k = 1:nc
            % distance from conductor to selected point (ft)
            dist = sqrt((x(i)-xc(k))^2 + (y(j)-yc(k))^2);
            % distance from image conductor to location (ft)
            Dist = sqrt((x(i)-xc(k))^2 + (y(j)+yc(k))^2);
            % calculate x-component electric field
            Ex = Ex + q(k)/(2*pi*eps0) * ...
                (x(i)-xc(k)) * (1/dist^2 - 1/Dist^2);
            % calculate y-component electric field
            Ey = Ey + q(k)/(2*pi*eps0) * ...
                ((y(j)-yc(k))/dist^2 - (y(j)+yc(k))/Dist^2);
        end
        % compute total electric field magnitude (V/m)
        Efield(j,i) = sqrt(abs(Ex)^2+abs(Ey)^2) * conv;
    end
end
```

A contour-enhanced surface plot of the results is created in Figure 5.53 using the following MATLAB coding:

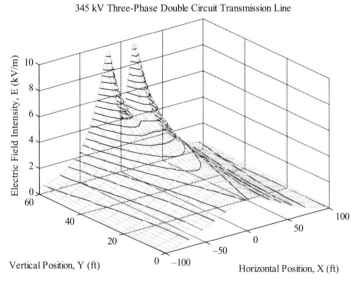

Figure 5.53. Three-dimensional contour of potential lines superimposed on a surface plot of the electric field strength from a three-phase 345 kV double-circuit transmission line.

TRANSMISSION LINE CAPACITANCE

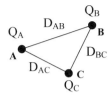

Figure 5.54. Simplified three-phase line.

```
% plot electric field magnitude as a function of position
contour3(x,y,Efield/1000,20)
surface(x,y,Efield/1000,'EdgeColor',[0.8 0.8 0.8],...
    'FaceColor','none');
set(gca, 'fontname','Times', 'fontsize',11);
xlabel('Horizontal Position, X (ft)')
ylabel('Vertical Position, Y (ft)')
zlabel('Electric Field Intensity, E (kV/m)')
title([num2str(Vline/1000),...
        'kV Three-Phase Double Circuit Transmission Line'])
```

5.7.4. Three-Phase Line Capacitance

The three-phase line capacitance for a balanced system is usually referred to as *positive sequence capacitance*, which neglects the ground effect and assumes that the line is transposed.

Figure 5.54 shows a simplified three-phase transmission line without ground. The charges (Q) on the conductor surface produce voltage differences between the conductors. These voltage differences are equal to the line-to-line voltages.

The charge on conductor A produces a voltage difference between conductor A and B. This voltage difference is calculated by integrating the Q_A generated electric field between A and B, which is achieved by the substitution of $D_1 = r_c$ and $D_2 = D_{AB}$ into Equation (5.63). The voltage difference produced by Q_A between A and B is:

$$V_{AB\_A} = \frac{Q_A}{2\pi\varepsilon_0} \ln\left(\frac{D_{AB}}{r_c}\right). \tag{5.86}$$

The charge on conductor B also creates a voltage potential between A and B. This voltage is computed by integrating the Q_B-generated electric field between B and A, which is achieved by the substitution of $D_1 = D_{AB}$ and $D_2 = r_c$ in Equation (5.63) to yield:

$$V_{AB\_B} = \frac{Q_B}{2\pi\varepsilon_0} \ln\left(\frac{r_c}{D_{AB}}\right). \tag{5.87}$$

Similarly, the charge on conductor C generates a voltage between A and B. This voltage difference is determined by integrating the Q_C-produced electric field between A and

B, which is achieved by the substitution of $D_1 = D_{AC}$ and $D_2 = D_{BC}$ in Equation (5.63) to give:

$$V_{AB\_C} = \frac{Q_C}{2\pi\varepsilon_0} \ln\left(\frac{D_{BC}}{D_{AC}}\right). \tag{5.88}$$

The total voltage between A and B is the sum of the voltages generated by the three phase conductors. The result of the summation is:

$$V_{AB} = V_{AB\_A} + V_{AB\_B} + V_{AB\_C}$$
$$= \frac{Q_A}{2\pi\varepsilon_0} \ln\left(\frac{D_{AB}}{r_c}\right) + \frac{Q_B}{2\pi\varepsilon_0} \ln\left(\frac{r_c}{D_{AB}}\right) + \frac{Q_C}{2\pi\varepsilon_0} \ln\left(\frac{D_{BC}}{D_{AC}}\right). \tag{5.89}$$

The voltage difference between A and C can be derived using an approach analogous to the above derivation. The voltage difference between A and C is:

$$V_{AC} = \frac{Q_A}{2\pi\varepsilon_0} \ln\left(\frac{D_{AC}}{r_c}\right) + \frac{Q_B}{2\pi\varepsilon_0} \ln\left(\frac{D_{BC}}{D_{AB}}\right) + \frac{Q_C}{2\pi\varepsilon_0} \ln\left(\frac{r_c}{D_{AC}}\right). \tag{5.90}$$

If the transmission line is transposed, the distances between the conductors are replaced by an average distance called the GMD. The GMD is calculated by Equation (5.49), which is repeated here:

$$\text{GMD} = \sqrt[3]{D_{AB} D_{BC} D_{AC}}. \tag{5.91}$$

The substitution of the GMD yields:

$$V_{AB} = \frac{Q_A}{2\pi\varepsilon_0} \ln\left(\frac{\text{GMD}}{r_c}\right) + \frac{Q_B}{2\pi\varepsilon_0} \ln\left(\frac{r_c}{\text{GMD}}\right) = \frac{(Q_A - Q_B)}{2\pi\varepsilon_0} \ln\left(\frac{\text{GMD}}{r_c}\right), \tag{5.92}$$

$$V_{AC} = \frac{Q_A}{2\pi\varepsilon_0} \ln\left(\frac{\text{GMD}}{r_c}\right) + \frac{Q_C}{2\pi\varepsilon_0} \ln\left(\frac{r_c}{\text{GMD}}\right) = \frac{(Q_A - Q_C)}{2\pi\varepsilon_0} \ln\left(\frac{\text{GMD}}{r_c}\right). \tag{5.93}$$

The sum of the two voltages is:

$$V_{AB} + V_{AC} = \frac{(2Q_A - Q_B - Q_C)}{2\pi\varepsilon_0} \ln\left(\frac{\text{GMD}}{r_c}\right) = \frac{3Q_A}{2\pi\varepsilon_0} \ln\left(\frac{\text{GMD}}{r_c}\right), \tag{5.94}$$

because $Q_A = -Q_B - Q_C$ for balanced sources.

In a balanced system the relation between the line-to-line and line-to-neutral voltages is:

$$V_{AB} = \sqrt{3} V_{AN} e^{j30°} = \sqrt{3} V_{AN} (\cos 30° + j \sin 30°), \tag{5.95}$$

$$V_{AC} = -V_{CA} = \sqrt{3} V_{AN} e^{-j30°} = \sqrt{3} V_{AN} (\cos 30° - j \sin 30°). \tag{5.96}$$

The sum of these two voltages is:

$$V_{AB} + V_{AC} = 2\sqrt{3}V_{AN}\cos(30°) = 2\sqrt{3}V_{AN}\frac{\sqrt{3}}{2} = 3V_{AN}. \quad (5.97)$$

Combining Equation (5.94) and Equation (5.97) results in the line-to-neutral voltage:

$$V_{AN} = \frac{V_{AB} + V_{AC}}{3} = \frac{Q_A}{2\pi\varepsilon_0}\ln\left(\frac{GMD}{r_c}\right). \quad (5.98)$$

The *capacitance-to-neutral* per unit length is:

$$C_{AN} = \frac{Q_A}{V_{AN}} = \frac{2\pi\varepsilon_0}{\ln\left(\dfrac{GMD}{r_c}\right)}. \quad (5.99)$$

Many high-voltage transmission lines are built with bundled conductors. The application of bundled conductors increases the line capacitance. An equivalent conductor replaces the bundled conductors. The radius of this equivalent conductor depends on the number of conductors in the bundle and the distance between the conductors within the bundle. The *equivalent radii* of the bundles are:

$$\text{Two-conductor bundles: } r_{eq} = \sqrt{dr_c}, \quad (5.100)$$

$$\text{Three-conductor bundles: } r_{eq} = \sqrt[3]{d^2 r_c}, \quad (5.101)$$

$$\text{Four-conductor bundles: } r_{eq} = 1.09\sqrt[4]{d^3 r_c}, \quad (5.102)$$

where r_c is the conductor radius and d is the distance between the conductors in the bundle. In the case of bundled conductors, the conductor radius r_c in Equation (5.99) is replaced by the appropriate r_{eq}.

5.8. TRANSMISSION LINE NETWORKS

The line inductance and capacitance obtained from the magnetic and electric field analyses, respectively, are now utilized to model transmission lines of various lengths.

5.8.1. Equivalent Circuit for a Balanced System

The equivalent transmission line circuits for a balanced system are presented in this section. These circuits are usually referred to as the *positive sequence transmission line equivalent circuits*. The *transmission line parameters* are the resistance and reactance of the conductors, and the capacitance to neutral. These parameters are calculated per unit line length and per phase, and assume that the line carries balanced current and is

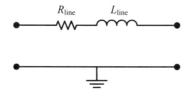

Figure 5.55. Equivalent network of a short line.

transposed. This permits the representation of the line by a single-phase circuit, which corresponds to phase A. The currents and voltages in phases B and C are shifted by 120° and 240°, respectively. The single-phase equivalent circuit carries *one-third of the line power* and is supplied by *the line-to-neutral voltage*.

The parameters are distributed evenly along the line. The type of transmission line model depends on length:

Short	0–50 miles
Medium	50–150 miles
Long	150 miles or more

The previously mentioned values are typical; however, whether the capacitance is negligible depends on the voltage. The line voltage affects the capacitive current; hence, the selected line model must be appropriate in terms of accurately representing the capacitive current as compared to the load current. For example, if the capacitive current is negligible compared to the load current, then the short line model is adequate. Likewise, for high voltage, if the capacitive current is comparable to the load current, the long line model should be used even if the line is less than 150 miles.

The capacitance of a short line is negligible. This simplifies the equivalent circuit, as given in Figure 5.55. This circuit is utilized when the line length is less than 50 miles. The resistance may be obtained directly from the conductor data of Table 5.1. The line inductance per unit length is computed from Equation (5.52):

$$L_{line} = \frac{\mu_0}{2\pi} \ln\left(\frac{\text{GMD}}{\text{GMR}_{eq}}\right), \quad (5.103)$$

where the GMD is the cube root of the product of the distances between the (bundled) conductors (see Eq. 5.49). For the bundling of up to $n = 3$ conductors, the equivalent GMR is:

$$\text{GMR}_{eq} = \left(d^{n-1}\text{GMR}_C\right)^{1/n} \quad \text{for } n = 1, 2, 3, \quad (5.104)$$

with d being the distance between conductors in a bundle, and the conductor GMR_C being available in Table 5.1.

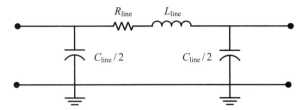

Figure 5.56. Equivalent network of a medium-length line.

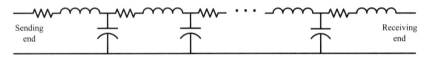

Figure 5.57. Equivalent network for a long transmission line.

A medium-length line can be represented by a nominal Π (pi) circuit. Figure 5.56 shows the single-phase equivalent circuit of a medium length line. This circuit is employed when the line length is 50–150 miles. The use of the medium-length equivalent circuit will be demonstrated through numerical examples in Section 5.10. In this circuit, the line resistance and inductance are concentrated in the middle section, and half of the line capacitance is placed at the beginning and the other half at the end of the line. The capacitance per unit length is calculated with Equation (5.99):

$$C_{line} = \frac{2\pi\varepsilon_0}{\ln\left(\dfrac{\text{GMD}}{r_{eq}}\right)}, \qquad (5.105)$$

where the equivalent conductor radius for up to $n = 3$ bundled conductors is:

$$r_{eq} = \left(d^{n-1} r_c\right)^{1/n} \quad \text{for } n = 1, 2, 3, \qquad (5.106)$$

with the individual conductor radius (r_c) being found in Table 5.1.

A long line is divided into short segments, as illustrated in Figure 5.57. Each segment consists of a reactance and a resistance connected in series, and a capacitance connected in parallel to the ground. An equivalent Π (pi) model can be developed by solving the system differential equations, as will be presented in the following subsection. This is necessary for lines longer than 150–200 miles.

EXAMPLE 5.9: Determine the transmission line resistance at 50°C, inductance, and capacitance for a three-phase 345 kV transmission line that utilizes two bundled Rail conductors arranged horizontally with 25.5-ft lateral phase separation and an 18-inch gap between conductors within a bundle.

Solution: Table 5.1 directly provides the basic data—resistance at 50°C, GMR, and conductor diameter—for a single Rail conductor:

$$R_{rail} := 0.1099 \frac{\Omega}{mi} \quad GMR_{rail} := 0.0385 \text{ ft} \quad d_{rail} := 1.165 \text{ in.}$$

The line was specified as having a separation between bundles, distance between conductors within the bundle, and number of conductors in a bundle of:

$$D_{345} := 25.5 \text{ ft} \quad d := 18 \text{ in} \quad n := 2.$$

Since the conductors within a bundle are electrically in parallel, the line resistance of each phase is:

$$R_{345} := \frac{R_{rail}}{n} = 0.0550 \cdot \frac{\Omega}{mi}$$

The GMD is found using Equation (5.49):

$$GMD_{345} := \sqrt[3]{D_{345} \cdot D_{345} \cdot 2 \cdot D_{345}} = 32.13 \cdot \text{ft}.$$

The equivalent GMR for the bundle is:

$$GMR_{bun} := \sqrt{GMR_{rail} \cdot d} = 2.884 \text{ in.}$$

The inductance per length is therefore:

$$L_{345} := \frac{\mu_0}{2 \cdot \pi} \cdot \ln\left(\frac{GMD_{345}}{GMR_{bun}}\right) = 1.576 \cdot \frac{mH}{mi}.$$

The corresponding inductive reactance at 60 Hz is:

$$X_{L345} := \omega \cdot L_{345} = 0.594 \cdot \frac{\Omega}{mi}.$$

After computing the individual conductor radius, the radius of the bundled conductors is computed:

$$r_{rail} := \frac{d_{rail}}{2} = 0.583 \cdot \text{in} \quad r_{bun} := \sqrt{r_{rail} \cdot d} = 3.238 \text{ in.}$$

Finally, the capacitance per unit length is:

$$C_{345} := \frac{2 \cdot \pi \cdot \varepsilon_0}{\ln\left(\frac{GMD_{345}}{r_{bun}}\right)} = 18.73 \cdot \frac{nF}{mi}.$$

TRANSMISSION LINE NETWORKS

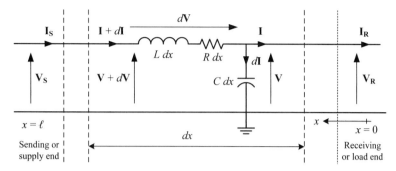

Figure 5.58. Equivalent network of a dx long section of a long transmission line.

The corresponding capacitive reactance is:

$$X_{C345} := \frac{-1}{\omega \cdot C_{345}} = -141.6 \cdot k\Omega \cdot mi.$$

Notice how the inductive reactance is multiplied by the line length, whereas the capacitive reactance must be divided by the line length.

5.8.2. Long Transmission Lines

The transmission line resistance, inductance, and capacitance are distributed along the line. A model of the transmission line is created by dividing the line into small, dx long sections. Each section has resistance and reactance connected in series and capacitance connected in parallel. Figure 5.58 shows a dx long section of a long transmission line with distributed parameters. The resistance (R) is measured in ohm per mile, the inductance (L) is given as henry per mile, and the capacitance (C) as farad per mile. In the steady-state condition, the impedance of the unit length line section is $\mathbf{z}_{RL} = R + j\omega L$. The capacitive reactance of the unit length section is $\mathbf{z}_C = 1/(j\omega C)$.

Referring to Figure 5.58, the voltage loop (Kirchhoff's voltage law, KVL) equation for the dx length equivalent circuit is:

$$(\mathbf{V} + d\mathbf{V}) - \mathbf{I}\mathbf{z}_{RL}dx - \mathbf{V} = 0. \quad (5.107)$$

The rearrangement of this formula yields a first-order differential equation for the voltage:

$$\frac{d\mathbf{V}}{dx} = \mathbf{z}_{RL}\mathbf{I}. \quad (5.108)$$

The node point (Kirchhoff's current law, KCL) equation using the equivalent network is:

$$(\mathbf{I} + d\mathbf{I}) - \frac{\mathbf{V}dx}{\mathbf{z}_C} - \mathbf{I} = 0, \quad (5.109)$$

since $1/(j\omega C dx) = \mathbf{z}_C/dx$. The rearrangement of this expression yields a first-order differential equation for the current

$$\frac{d\mathbf{I}}{dx} = \frac{\mathbf{V}}{\mathbf{z}_C}. \tag{5.110}$$

The node point equation, Equation (5.110), is substituted into the derivative of the voltage equation, Equation (5.108), resulting in a second-order differential equation describing the voltage variation along the line

$$\frac{d}{dx}\left(\frac{d\mathbf{V}}{dx}\right) = \frac{d}{dx}\mathbf{z}_{RL}\mathbf{I}$$

$$\frac{d^2\mathbf{V}}{dx^2} = \frac{\mathbf{z}_{RL}}{\mathbf{z}_C}\mathbf{V} \tag{5.111}$$

The solution of the second-order homogeneous differential equation is:

$$\mathbf{V}(x) = \mathbf{A}e^{\gamma x} + \mathbf{B}e^{-\gamma x}, \tag{5.112}$$

where $\mathbf{A}$ and $\mathbf{B}$ are constants determined by the boundary conditions; and $\gamma = \sqrt{\mathbf{z}_{RL}/\mathbf{z}_C}$ is the *propagation constant*.

The substitution of the voltage solution into the original voltage differential equation, Equation (5.108), yields an expression for the current as a function of line position:

$$\mathbf{I}(x) = \frac{1}{\mathbf{z}_{RL}}\frac{d\mathbf{V}}{dx} = \frac{1}{\mathbf{z}_{RL}}\frac{d}{dx}\left(\mathbf{A}e^{\gamma x} + \mathbf{B}e^{-\gamma x}\right)$$

$$= \frac{\gamma}{\mathbf{z}_{RL}}\left(\mathbf{A}e^{\gamma x} - \mathbf{B}e^{-\gamma x}\right). \tag{5.113}$$

The equation is simplified by introducing the *surge impedance*:

$$\mathbf{Z}_{sur} = \frac{\mathbf{z}_{RL}}{\gamma} = \frac{\mathbf{z}_{RL}}{\sqrt{\mathbf{z}_{RL}/\mathbf{z}_C}} = \sqrt{\mathbf{z}_{RL}\mathbf{z}_C}. \tag{5.114}$$

The simplified current equation is:

$$\mathbf{I}(x) = \frac{1}{\mathbf{Z}_{sur}}\left(\mathbf{A}e^{\gamma x} - \mathbf{B}e^{-\gamma x}\right). \tag{5.115}$$

The boundary conditions are the voltage and current at the receiving (load) end:

$$\mathbf{I}(0) = \mathbf{I}_R \quad \mathbf{V}(0) = \mathbf{V}_R. \tag{5.116}$$

Substitution of the boundary values in the voltage and current equations yields:

$$V(0) = Ae^{\gamma 0} + Be^{-\gamma 0} = A + B = V_R, \quad (5.117)$$

$$I(0) = \frac{1}{Z_{sur}}\left(Ae^{\gamma 0} - Be^{-\gamma 0}\right) = \frac{1}{Z_{sur}}(A - B) = I_R. \quad (5.118)$$

Solving the two simultaneous equations above provides formulas for **A** and **B**:

$$\begin{aligned} A &= \frac{1}{2}(V_R + I_R Z_{sur}), \\ B &= \frac{1}{2}(V_R - I_R Z_{sur}). \end{aligned} \quad (5.119)$$

The next step is to substitute **A** and **B** into the voltage relation:

$$\begin{aligned} V(x) &= \frac{1}{2}(V_R + I_R Z_{sur})e^{\gamma x} + \frac{1}{2}(V_R - I_R Z_{sur})e^{-\gamma x} \\ &= V_R\left(\frac{e^{\gamma x} + e^{-\gamma x}}{2}\right) + I_R Z_{sur}\left(\frac{e^{\gamma x} - e^{-\gamma x}}{2}\right) \\ &= V_R \cosh(\gamma x) + I_R Z_{sur} \sinh(\gamma x). \end{aligned} \quad (5.120)$$

Similarly, the substitution **A** and **B** into the current equation gives:

$$\begin{aligned} I(x) &= \frac{1}{Z_{sur}}\left[\frac{1}{2}(V_R + I_R Z_{sur})e^{\gamma x} - \frac{1}{2}(V_R - I_R Z_{sur})e^{-\gamma x}\right] \\ &= \frac{V_R}{Z_{sur}}\left(\frac{e^{\gamma x} - e^{-\gamma x}}{2}\right) + \frac{I_R Z_{sur}}{Z_{sur}}\left(\frac{e^{\gamma x} + e^{-\gamma x}}{2}\right) \\ &= \frac{V_R}{Z_{sur}} \sinh(\gamma x) + I_R \cosh(\gamma x). \end{aligned} \quad (5.121)$$

In summary, the voltage and current variations along a long transmission line are described by:

$$\begin{aligned} V(x) &= V_R \cosh(\gamma x) + I_R Z_{sur} \sinh(\gamma x), \\ I(x) &= \frac{V_R}{Z_{sur}} \sinh(\gamma x) + I_R \cosh(\gamma x), \\ Z_{sur} &= \sqrt{Z_{RL} Z_C} \quad \gamma = \sqrt{Z_{RL}/Z_C}. \end{aligned} \quad (5.122)$$

Figure 5.59 shows the voltage distribution along a 300-mile-long 500 kV transmission line. The receiving or load-end voltage at $x = 0$ is 500 kV. The supply or sending-end voltage is 545 kV. But the maximum voltage for this selected load is around 552 kV, occurring at 215 miles distance from the load or 85 miles distance from the supply.

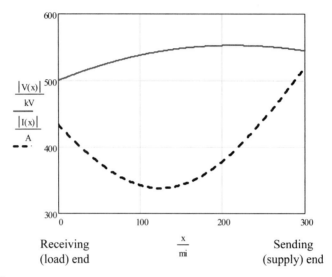

Figure 5.59. Voltage and current distributions along a 300-mile-long 500 kV transmission line.

Figure 5.59 also shows a graph of the current distribution along this 500 kV, 300-mile-long line. The receiving-end or load current is 430 A; the supply or sending-end current is around 520 A. But the current minimum (338 A) occurs at ~125 miles from the load. This example clearly demonstrates that the voltage and current distributions along a long line are not linear.

5.8.2.1. Equivalent Circuit for a Long Transmission Line. Substitution of the line length into the previously mentioned expressions produces a relation between the supply voltage and current and the load voltage and current:

$$\mathbf{V_S} = \mathbf{V}(\ell) = \mathbf{V_R} \cosh(\gamma\ell) + \mathbf{I_R} \mathbf{Z}_{sur} \sinh(\gamma\ell),$$
$$\mathbf{I_S} = \mathbf{I}(\ell) = \frac{\mathbf{V_R}}{\mathbf{Z}_{sur}} \sinh(\gamma\ell) + \mathbf{I_R} \cosh(\gamma\ell). \quad (5.123)$$

The obtained equations for the long transmission line can be represented by the *equivalent pi* circuit drawn in Figure 5.60. The series and parallel impedances of the equivalent Π circuit are:

$$\mathbf{Z}_s = \mathbf{Z}_{sur} \sinh(\gamma\ell),$$
$$\mathbf{Z}_p = \frac{\mathbf{Z}_s}{\cosh(\gamma\ell) - 1}. \quad (5.124)$$

The reader is encouraged to verify the validity of the equivalent pi circuit. This network can be used to calculate the transmission line operation parameters when the line length is more than 150 miles.

TRANSMISSION LINE NETWORKS

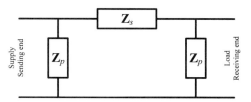

Figure 5.60. Equivalent pi circuit for a long transmission line.

EXAMPLE 5.10: Calculate the transmission line parameters for a 270-mile-long line having a resistance of 0.03 Ω/mile, an inductance of 1.57 mH/mile, and a capacitance of 19 nF/mile in a 60 Hz network.

Solution: The system frequency and line length are first defined:

$$f := 60 \text{ Hz} \quad \omega := 2 \cdot \pi \cdot f = 377.0 \frac{1}{s} \quad \text{Len}_{\text{line}} := 270 \text{ mi}.$$

The basic line data were given as:

$$R_{\text{line}} := 0.03 \frac{\Omega}{\text{mi}} \quad L_{\text{line}} := 1.57 \frac{\text{mH}}{\text{mi}} \quad C_{\text{line}} := 19 \frac{\text{nF}}{\text{mi}}.$$

The unit length line impedance and capacitance are:

$$z_{\text{line}} := R_{\text{line}} + j \cdot \omega \cdot L_{\text{line}} = (0.03 + 0.592j) \cdot \frac{\Omega}{\text{mi}}$$

$$z_{\text{cap}} := \frac{1}{j \cdot \omega \cdot C_{\text{line}}} = -139.6j \cdot k\Omega \cdot \text{mi}.$$

The long line propagation constant and surge impedance are found from Equation (5.122):

$$\gamma_{\text{line}} := \sqrt{\frac{z_{\text{line}}}{z_{\text{cap}}}} = (0.052 + 2.06j) \cdot \frac{10^{-3}}{\text{mi}}$$

$$Z_{\text{surge}} := \sqrt{z_{\text{line}} \cdot z_{\text{cap}}} = (287.5 - 7.3j) \; \Omega.$$

Using Equation (5.124), the equivalent pi circuit parameters are determined:

$$Z_{\text{ser}} := Z_{\text{surge}} \cdot \sinh(\gamma_{\text{line}} \cdot \text{Len}_{\text{line}}) = (7.3 + 151.7j) \; \Omega$$

$$Z_{\text{par}} := \frac{Z_{\text{ser}}}{\cosh(\gamma_{\text{line}} \cdot \text{Len}_{\text{line}}) - 1} = (1.4 - 1007.4j) \; \Omega.$$

For comparison, we compute the line parameters had this long line been treated as a medium-length line. The equivalent medium line parameters are:

$$Z_{MedRL} := z_{line} \cdot Len_{line} = (8.1+159.8j)\,\Omega$$

$$Z_{MedC} := \dfrac{1}{j \cdot \omega \cdot \dfrac{C_{line} \cdot Len_{line}}{2}} = -1034.1j\Omega.$$

The medium line parameters are in good agreement with the long line values, thereby providing a validation of the long line parameters computed.

5.9. CONCEPT OF TRANSMISSION LINE PROTECTION

Figure 5.61 depicts the typical architecture of the electric system. The diagram shows that the circuit breaker (CB), current transformer, and disconnect switches are connected to an outgoing transmission line. The CB is the major element of line protection. In case of a fault, the modern CB is able to interrupt the fault current and switch off the line within two to five cycles. The current and voltage transformers actuate the protection relay, which today is a digital relay. Some protection requires only current signals (e.g., overcurrent protection), while others need both current and voltage signals (e.g., distance protection). The relay senses the fault and transmits an open signal to the CB.

5.9.1. Transmission Line Faults

A transmission line fault is a short circuit between a conductor and ground or between conductors. The fault initiates a large current. In the distribution system, the fault current is less than 10 kA, but at high voltage and close to generators, the fault current can be 40–50 kA. The faults can kill or injure operators, start fires, damage equipment, and produce long outages

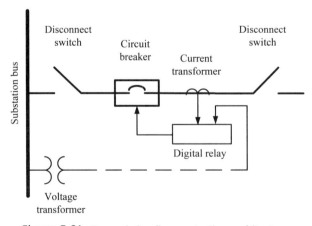

Figure 5.61. Transmission line protection architecture.

CONCEPT OF TRANSMISSION LINE PROTECTION

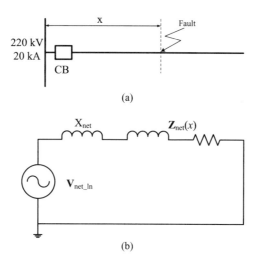

Figure 5.62. Short-circuit current dependence on fault location. (a) Single-line diagram; (b) equivalent circuit.

The short-circuit current depends on the location of the fault because the transmission line impedance limits the short-circuit current. This is demonstrated with a numerical example.

EXAMPLE 5.11: A 220 kV substation with 20 kA short circuit current supplies a 150-mile-long line. The line impedance is $0.06 + j\, 0.65$ Ω/mile. Figure 5.62 gives the single-line diagram and equivalent circuit.

The line-to-neutral voltage and Thévenin equivalent of the supply network are:

$$V_{net\_ln} := \frac{220 \cdot kV}{\sqrt{3}} = 127.0 \cdot kV \quad I_{short} := 20\, kA$$

$$X_{net} := \frac{V_{net\_ln}}{I_{short}} = 6.351\, \Omega \quad Z_{net} := j \cdot X_{net} = 6.351 j\Omega$$

The line impedance is proportional with the distance, where $x = 0$ is located at the supply end:

$$Z_{lin}(x) := (0.06 + j \cdot 0.65) \cdot \frac{\Omega}{mi} \cdot x.$$

The short-circuit current as a function of the distance from the supply is:

$$I_{short}(x) := \frac{V_{net\_ln}}{Z_{net} + Z_{lin}(x)}$$

$$|I_{short}(1\, mi)| = 18.1 \cdot kA \quad |I_{short}(150\, mi)| = 1.22 \cdot kA$$

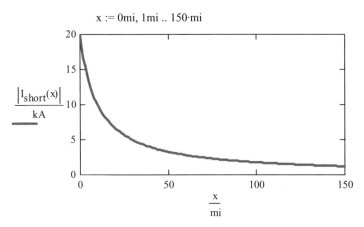

Figure 5.63. Short-circuit current transmission line length function.

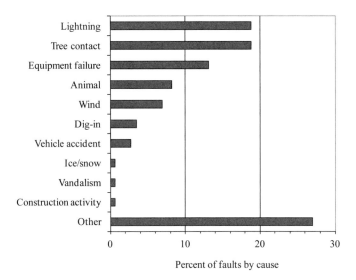

Figure 5.64. Fault causes measured in the EPRI fault study (data from J. J. Burke, D. J. Lawrence, IEEE Trans. Power Apparatus and Systems, vol. 103, Jan. 1984, pp. 1–6.; EPRI 1209–1, 1983).

Figure 5.63 plots the variation of the short-circuit current with the distance from the supply. According to this graph and the earlier numerical results, the short-circuit current 1 mile from the substation is 18.1 kA, and a short at the end of the line ($x = 150$ mi) produces only a 1.22 kA short-circuit current. The protection must recognize the fault independently from the location and initiate CB operation.

Figure 5.64 shows the results of an Electric Power Research Institute (EPRI) study listing the cause of transmission line faults. As shown in Figure 5.64, the most frequent transmission line fault is a phase-to-ground short circuit caused by a lightning strike or

CONCEPT OF TRANSMISSION LINE PROTECTION

tree contact. The lightning or tree contact produces a flashover, which initiates large current. The protection triggers the CB, which interrupts the current at the next zero crossing and thereby switches off the line. In most cases, the flashover and arcing do not damage the insulator or other parts of the line; consequently, the line can be reclosed immediately. Reclosing means to close the CB after current interruption. The CB may be equipped with a special switch called a recloser, which attempts three reclosings. The nomenclature is 0–15–30 reclosing, which means:

- the first reclose, indicated by the "0," is made after no intentional delay for current interruption (this is an immediate reclose),
- the second attempt is made following a 15-second delay time, and
- the final attempt is made after a 30-second delay time.
- If the fault is still present, the CB opens and locks open.

Statistics show that the reclosing is successful in most cases and restores the electric supply within a few seconds.

5.9.2. Protection Methods

The protection of transmission lines and electric networks is an advanced subject, requiring further studies, and it is beyond the scope of this book. However, the basic short circuit protection methods and the concept of selective transmission line protection are now described.

5.9.3. Fuse Protection

Most of the medium-voltage overhead distribution lines on side streets are protected by a fuse as was seen in Figure 5.5. The fuse cutout is mounted on the distribution tower and connected in series with the distribution feeder. Figure 5.65 shows a picture of a modern fuse cutout with a composite insulator holding the explosive-type fuse.

Typical nonlinear fuse characteristics are graphed in Figure 5.66. The plot provides the characteristics of a 10 A fuse with a minimum melting time curve and maximum clearing time curve. The fuse is damaged when the current exceeds the value on the minimum melting time curve. The fuse clears the fault when the current exceeds the maximum clearing time curve.

As presented in Figure 5.66, this 10 A fuse starts to melt just above 20 A, with a melting time over 100 seconds. When the current is 30 A, the fuse is damaged after about 5 seconds and clears the 30 A current after 10 seconds. If the short-circuit current is 100 A, the clearing time is reduced to 0.15 second. This implies that the clearing time is significantly shorter when the fault occurs near to the supply.

5.9.4. Overcurrent Protection

The main feeders and the higher voltage radial lines are protected by overcurrent relays. Radial lines have voltage sources from one end. Today, most companies use digital

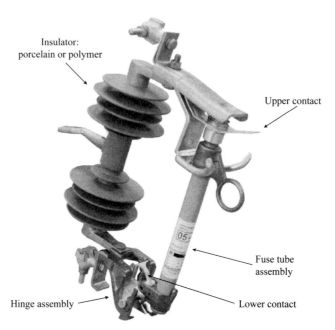

Figure 5.65. Fuse cutout with composite insulator and melting fuse (courtesy of Maclean Power Systems).

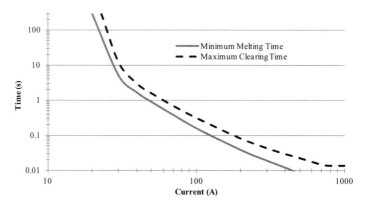

Figure 5.66. Ten-ampere fuse characteristics.

relays. The setup for the overcurrent protection is depicted in Figure 5.67. The current transformer (CT) reduces the line current to a measurable value. The typical CT primary current is selected to be larger than the maximum load current on the line. The secondary current at the rated primary current is 1–5 A. The current signal supplies the digital relay, which digitizes and evaluates the input current. If the current is larger than the relay setpoint value, the relay triggers the CB to open the line. Most of the overcurrent

CONCEPT OF TRANSMISSION LINE PROTECTION

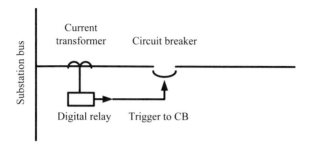

Figure 5.67. Concept of overcurrent protection.

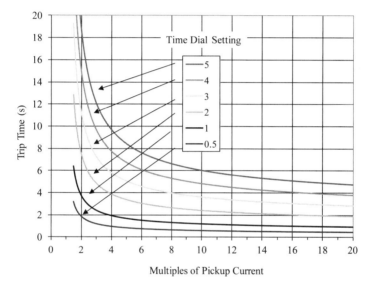

Figure 5.68. Overcurrent relay operation time for moderately inverse characteristics; curve parameters are from IEEE Standard C37.112-1996.

protections use nonlinear characteristics exemplified in Figure 5.68. The minimum operation time is from 1.5 to 2 seconds and the time is reduced when the current increases.

As an example, we select time dial setting 5. The operation time will be around 9 seconds when the multiple of pickup (MOP) is 5. When the MOP is 10 times, the operation time is reduced to about 6.5 seconds. The MOP depends on the maximum expected load current and the current transformer ratio. This method will eliminate the short circuits with time delay depending on the fault location.

For nonradial lines, overcurrent protection performance can be improved by adding a directional element, which assures that the protection does not operate if the fault is behind the relay.

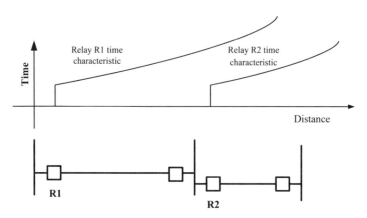

Figure 5.69. Concept of transmission overcurrent protection.

Figure 5.69 illustrates the concept of using overcurrent protection of transmission lines. The staggering of the time delays assures that each line segment is protected, although the delay in the first segment is more than in the last one.

5.9.5. Distance Protection

The high- and extra-high-voltage lines are protected by comparison relaying with distance protection as backup. Most 500 and 765 kV lines use some type of phase comparison or directional comparison relays as the primary protection. For phase comparison, the phases of the currents at the two line ends are compared and the breakers at the two ends of the line are tripped if the phases at the two ends differ significantly. For directional comparison, the powers at the two ends of the line are compared; this is often referred to as pilot relaying. Communication channels between the two ends are required to provide for comparison and transfer trip signals. The typically used communication channels are telephone lines, microwave transmission, or a high-frequency carrier signal on the line itself. Secondary (backup) protection is usually provided by impedance (distance) relays.

The distance protection relay measures the current and voltage and calculates the apparent impedance or admittance. In normal operation, the distance relay measures the combined line and load impedance. When a fault occurs, the relay measures the fault impedance, which depends on the distance between the fault and the relay. The advantages of this protection are that it is independent from the amplitude of the current, and it can discriminate between fault and normal operation as well as a fault in a specific area or other location.

A distance relay typically has three time zones, as seen in Figure 5.70. The diagram shows that Zone 1 in relay R1 is adjusted to 80–90% of the line length, while Zone 2 is adjusted to terminate in the middle or 60% of the second line. About 80% of the second line is protected by Zone 1 of R2. This arrangement assures fast protection for

APPLICATION EXAMPLES

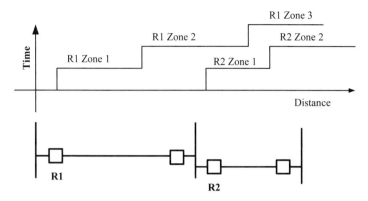

Figure 5.70. Differential relay time–distance characteristic.

a fault occurring on the major part of the line, and the second and third zones provide backup protection.

5.10. APPLICATION EXAMPLES

5.10.1. Mathcad Examples

A coupled set of Mathcad examples is presented in this section. The first computation involves the calculation of transmission line parameters, which are utilized in the transmission line operation analyses that follow.

5.10.1.1. Calculation of Transmission Line Parameters. A 500 kV transmission line is built with lattice towers supporting a two-conductor Bluebird bundle in each phase. Figure 5.71 presents the typical tower arrangement. The line data are:

The rated line voltage is: $V_{line} := 500$ kV.
The line length is: $L_{Line} := 103$ mi.
The operating frequency is: $f := 60$ Hz $\omega := 2 \cdot \pi \cdot f$.
The distance between conductors in the bundle is: $d := 18$ in.
The number of conductors in each bundle is: $n := 2$.
The distance between the phases is: $D := 32$ ft.
The Bluebird conductor data are obtained from Table 5.1.

The conductor radius is:

$$r_c := \frac{1.762 \text{ in}}{2}.$$

The GMR is: $GMR_c := 0.0588$ ft.

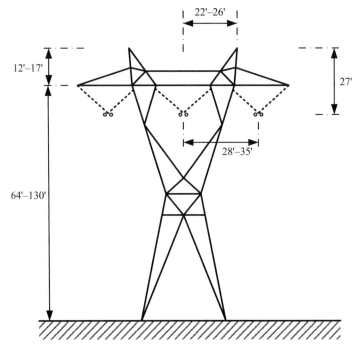

Figure 5.71. Cholla-Saguaro transmission line tower.

The ac resistance per mile at 75°C is:

$$R_{75} := 0.0544 \frac{\Omega}{mi}.$$

The universal constants are:

$$\varepsilon_o := \frac{10^{-9}}{36 \cdot \pi} \cdot \frac{F}{m} \quad \mu_o := 4 \cdot \pi \cdot 10^{-7} \cdot \frac{H}{m}.$$

The GMR of the two-conductor bundle is found using Equation (5.54):

$$GMR := \sqrt{d \cdot GMR_c} \quad GMR = 3.56 \text{ in.}$$

The equivalent radius of the two-conductor bundle is determined from Equation (5.100):

$$r_{equ} := \sqrt{d \cdot r_c} \quad r_{equ} = 3.98 \text{ in.}$$

Using Equation (5.49), the GMD between the phase conductors is:

$$GMD := \sqrt[3]{D \cdot D \cdot 2 \cdot D} \quad GMD = 40.3 \text{ ft.}$$

APPLICATION EXAMPLES

The line reactance per unit length is calculated by using Equation (5.53). The reactance is:

$$X_L := \omega \cdot \frac{\mu_o}{2 \cdot \pi} \ln\left(\frac{GMD}{GMR}\right) \quad X_L = 0.596 \frac{\Omega}{mi}.$$

The total reactance of the line is:

$$X_{Line} := X_L \cdot L_{Line} \quad X_{Line} = 61.4 \, \Omega.$$

The line operating temperature is determined by the ambient temperature, sun radiation, and the load current produced heating. In Arizona, it is prudent to use the 75°C (167°F) temperature for the calculation of the conductor ac resistance.

The two conductors in the bundle are connected in parallel. The line resistance of this two-conductor bundle is:

$$R_{Line} := \frac{R_{75}}{n} \cdot L_{Line} \quad R_{Line} = 2.80 \, \Omega.$$

The line resistance and reactance are connected in series. The overall impedance of this line is:

$$Z_{Line} := R_{Line} + j \cdot X_{Line} \quad Z_{Line} = (2.8 + 61.4j) \, \Omega.$$

The capacitance-to-ground per unit length is calculated by Equation (5.99). The capacitance to ground is:

$$C_{Line} := \frac{2 \cdot \pi \cdot \varepsilon_o}{\ln\left(\frac{GMD}{r_{equ}}\right)} \quad C_{Line} = 18.6 \frac{nF}{mi}.$$

The line performance is studied using a single-phase equivalent circuit. In this circuit, the line capacitance is divided into two parts. Half of the total line capacitance is placed at the beginning and the other half at the end of the line, as diagrammed in Figure 5.72.

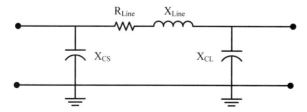

Figure 5.72. Diagram of single-phase equivalent line.

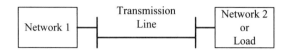

Figure 5.73. Circuit for transmission line operation analysis.

The capacitive reactance at the supply side is:

$$X_{CS} := \frac{-1}{\omega \cdot \frac{C_{Line}}{2} \cdot L_{Line}} \qquad X_{CS} = -2.77 \text{ k}\Omega.$$

The capacitive reactance at the load side is the same:

$$X_{CL} := X_{CS} = -2.77 \text{ k}\Omega.$$

The capacitance term may be expressed in terms of its impedance:

$$Z_{CS} := j \cdot X_{CS} = -2.77 \text{jk}\Omega \qquad Z_{CL} := j \cdot X_{CL} = -2.77 \text{jk}\Omega.$$

5.10.1.2. Transmission Line Operation Analysis. The transmission line operation is analyzed using the circuit exhibited in Figure 5.73. In this circuit, Network 1 supplies a load or another network (Network 2) through the line. The Π (pi) circuit given in Figure 5.72 represents the medium-length transmission line. The two networks are modeled by their Thévenin equivalent circuits, which contain a voltage source and reactance connected in series.

The open-circuit voltage of a large network is nearly equal to the rated line-to-neutral voltage. The network reactance is calculated from the short-circuit current, which can be obtained from the utility operating the network. For this numerical example, we assume that Networks 1 and 2 have the following voltage and short-circuit current values:

$$V_{net1} := 500 \text{ kV} \qquad I_{net1\_short} := -j \cdot 10 \text{ kA}$$
$$V_{net2} := V_{net1} \qquad I_{net2\_short} := -j \cdot 15 \text{ kA},$$

where the angular reference is the open-circuit line-to-line voltage for each network. It has been assumed earlier that the short-circuit current is an inductive current, which is a good approximation. Using these values, the network impedances are calculated with Ohm's law while converting the line voltage to a line-to-neutral value. The obtained impedance values are:

$$V_{net1\_ln} := \frac{V_{net1}}{\sqrt{3}} = 288.7 \cdot \text{kV} \qquad V_{net2\_ln} := \frac{V_{net2}}{\sqrt{3}} = 288.7 \cdot \text{kV}$$

$$Z_{net1} := \frac{V_{net1\_ln}}{I_{net1\_short}} \qquad Z_{net1} = 28.9 \text{j}\Omega$$

$$Z_{net2} := \frac{V_{net2\_ln}}{I_{net2\_short}} \qquad Z_{net2} = 19.2 \text{j}\Omega$$

APPLICATION EXAMPLES

In the succeeding sections, we will perform four different analyses:

1. *Short-Circuit Current Calculation.* First, the line current is determined when a short circuit occurs on the line. This short-circuit current is used for setting protection equipment and selecting CBs.
2. *Open-Circuit Voltage Computation.* Next, the open-circuit voltage generated by the Network 1 supply is determined for use in finding both the Thévenin equivalent and the voltage regulation.
3. *Network 1 Supplies a Load.* When the line is connected to a load, the objective of the analysis is to determine the voltage regulation and transmission efficiency. The voltage regulation should be less than 5%. The transmission efficiency is the ratio of the real power absorbed by the load and the real power supplied by the source.
4. *Transmission Line Interconnects Two Networks.* When the line is connected to a second network (Network 2), the goal of the analysis is to determine the power that can be transferred from one network to the other.

5.10.1.3. Short-Circuit Current Calculation. The selection of CBs and protection coordination requires calculation of the short-circuit current. Figure 5.74 presents the equivalent circuit for the short-circuit current calculation. The short circuit occurs at the receiving end of the line. The short circuit eliminates the line end capacitance.

Figure 5.74 shows that the line impedance and the capacitance at the sending end of the line are connected in parallel. Further, the Network 1 reactance is connected in series with this impedance. The overall equivalent impedance is:

$$Z_{short} := Z_{net1} + \frac{1}{\dfrac{1}{Z_{CS}} + \dfrac{1}{Z_{Line}}} \qquad Z_{short} = (2.93 + 91.63j)\ \Omega.$$

The Network 1 current is:

$$I_{S\_short} := \frac{V_{net1\_In}}{Z_{short}} \qquad |I_{S\_short}| = 3.15\ \text{kA} \qquad \arg(I_{S\_short}) = -88.2\ \text{deg}.$$

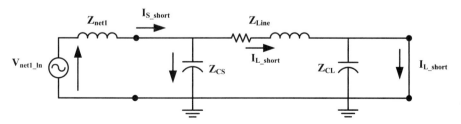

Figure 5.74. Equivalent circuit for short-circuit calculation.

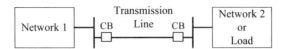

Figure 5.75. Circuit breaker (CB) installation at transmission line ends.

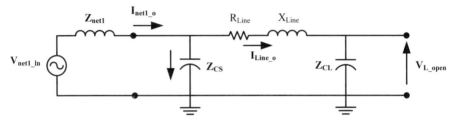

Figure 5.76. Equivalent circuit for open-circuit voltage calculation.

Using current division, the line current due to the short circuit is:

$$I_{L\_short} := I_{S\_short} \cdot \frac{Z_{CS}}{Z_{CS} + Z_{Line}}$$

$$|I_{L\_short}| = 3.22 \cdot kA \quad \arg(I_{L\_short}) = -88.2 \cdot \deg.$$

This short-circuit current level can produce significant ohmic heating, which may anneal the line conductor if the current is not switched off within a few cycles. The high current produces mechanical forces between the conductors, which can also endanger the system if the line is not adequately designed. This example demonstrates that the fast interruption of the short circuit current is necessary to maintain the safety of the electrical system. The short-circuit current is interrupted by CBs connected in series with the line. Typically, CBs are installed at both ends of the line, as shown in Figure 5.75. A short circuit on the line triggers the operation of both CBs. The CBs normally interrupt the (shorted) circuit within three to five cycles.

5.10.1.4. Open-Circuit Voltage Calculation. In preparation for ensuring that adequate voltage regulation is maintained with the variable load, the actual open-circuit voltage for a constant supply voltage is determined. Figure 5.76 presents the equivalent circuit for the open-circuit voltage calculation. The line impedance and the capacitance at the line end are connected in series. The impedance of this branch is:

$$Z_{Line\_CL} := Z_{Line} + Z_{CL} = (2.8 - 2703.7j) \, \Omega.$$

This branch is connected in parallel with the capacitance at the beginning of the line. The network reactance is connected in series with this impedance. The result of the summation of the system impedance is:

APPLICATION EXAMPLES

$$Z_{Line\_CL\_CS} := \cfrac{1}{\cfrac{1}{Z_{Line\_CL}} + \cfrac{1}{Z_{CS}}} + Z_{net1} = (0.72 - 1338.17j)\ \Omega.$$

The Network 1 current is the voltage divided by the calculated impedance:

$$I_{net1\_o} := \frac{V_{net1\_In}}{Z_{Line\_CL\_CS}} \quad |I_{net1\_o}| = 215.7\ \text{A} \quad \arg(I_{net1\_o}) = 89.97\ \text{deg}.$$

Figure 5.76 reveals that the network current is divided between the line and the sending-end capacitance. The line current is calculated by the current division equation:

$$I_{Line\_o} := I_{net1\_o} \cdot \frac{Z_{CS}}{Z_{Line\_CL} + Z_{CS}}$$

$$|I_{Line\_o}| = 109.1\ \text{A} \quad \arg(I_{Line\_o}) = 89.94 \cdot \text{deg}.$$

The open-circuit voltage is the voltage drop across the line receiving-end capacitance:

$$V_{L\_open} := I_{Line\_o} \cdot Z_{CL} \quad |V_{L\_open}| = 301.6\ \text{kV} \quad \arg(V_{L\_open}) = -0.06\ \text{deg}.$$

This voltage is also used later to determine the Thévenin equivalent network.

5.10.1.5. Network 1 Supplies a Load through the Transmission Line. Figure 5.77 shows the equivalent circuit of the system. The load power is variable, but the power factor is constant. The strategy of this calculation is to determine the currents and load voltage as a function of the constant supply voltage.

Nodal analysis is applied to this network, with the application of KCL at the node above the sending-end capacitance, yielding:

$$I_{net1} = I_{CS} + I_{Line}$$

$$\frac{V_{net1\_In} - V_{S\_n}}{Z_{net1}} = \frac{V_{S\_n}}{Z_{CS}} + \frac{V_{S\_n} - V_{L\_n}}{Z_{Line}}.$$

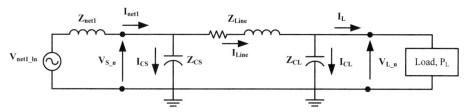

Figure 5.77. Equivalent circuit for a power system (a network supplies a load through a line).

Employing KCL at the node above the receiving-end capacitance provides a second independent relation:

$$I_{Line} = I_{CL} + I_L$$

$$\frac{V_{S\_n} - V_{L\_n}}{Z_{Line}} = \frac{V_{L\_n}}{Z_{CL}} + \frac{P_L}{3 \cdot V_{L\_n} \cdot pf_L} e^{-j \cdot a\cos(pf_L)}.$$

The division by 3 in the latter term is due to the fact that the single-phase equivalent circuit carries one-third of the three-phase power.

A numerical solution for the two unknown node voltages can be obtained using the Mathcad *Find* equation solver:

$$\text{Guess values:} \quad V_{L\_n} := \frac{500 \text{ kV}}{\sqrt{3}} + j \cdot 2 \text{ kV} \quad V_{S\_n} := V_{L\_n}$$

Given

$$\frac{V_{net1\_In} - V_{S\_n}}{Z_{net1}} = \frac{V_{S\_n}}{Z_{CS}} + \frac{V_{S\_n} - V_{L\_n}}{Z_{Line}}$$

$$\frac{V_{S\_n} - V_{L\_n}}{Z_{Line}} = \frac{V_{L\_n}}{Z_{CL}} + \frac{P_L}{3 \cdot V_{L\_n} \cdot pf_L} e^{-j \cdot a\cos(pf_L)}$$

$$\begin{pmatrix} V_{S\_n}(P_L) \\ V_{L\_n}(P_L) \end{pmatrix} := \text{Find}(V_{S\_n}, V_{L\_n}).$$

The load power is variable between 0 and 500 MW with a power factor of $pf_L := 0.8$ (lagging). The nodal voltages at a test load value of: $P_L := 400$ MW are:

$$|V_{S\_n}(P_L)| = 287.3 \text{ kV} \quad \arg(V_{S\_n}(PL)) = -3.2 \text{ deg}$$
$$|V_{L\_n}(P_L)| = 279.3 \text{ kV} \quad \arg(V_{L\_n}(P_L)) = -10.2 \text{ deg}$$

The voltage regulation at this test value is:

$$\text{Reg}(P_L) := \frac{|V_{L\_n}(0 \text{ MW})| - |V_{L\_n}(P_L)|}{|V_{L\_n}(P_L)|} \quad \text{Reg}(P_L) = 8.0 \cdot \%.$$

The voltage regulation should be less than 5%. The load power and voltage at which the 5% regulation is reached is determined from:

$$\text{Guess value} \quad P_L := 200 \text{ MW}$$
$$P_{L\_5\%} := \text{root}(\text{Reg}(P_L) - 5\%, P_L) = 200.3 \cdot \text{MW}$$
$$V_{L\_5\%} := V_{L\_n}(P_{L\_5\%}) \quad |V_{L\_5\%}| = 287.2 \cdot \text{kV}.$$

The voltage regulation is plotted against the load in Figure 5.78 to estimate the load value at 5% regulation. The graph demonstrates that the voltage regulation is a nearly

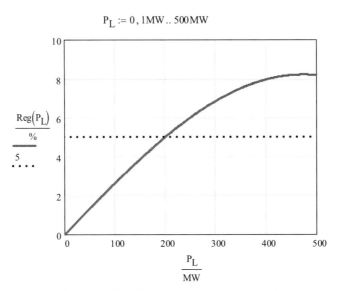

Figure 5.78. Voltage regulation versus load.

linear function of the load until approximately 200 MW. This calculation is a good approximation of power system behavior within this linear region. Above 200 MW, the regulation exceeds 5%, the trend is nonlinear, and the solution becomes inaccurate, because in actuality, the load depends on the load voltage.

Next, the system operating (transmission) efficiency, which is the ratio of the output (load) to input (supply) powers, is evaluated. To compute the transmission efficiency, it is necessary to determine the supply power. The current from the supply can be calculated using a portion of the nodal analysis performed earlier; in particular, we find the input power at the 5% regulation condition:

$$I_{net1}(P_L) := \frac{V_{S\_n}(P_L) - V_{L\_n}(P_L)}{Z_{Line}}$$

$$I_{net1\_5\%} := I_{net1}(P_{L\_5\%}) \quad |I_{net1\_5\%}| = 259.0 \text{ A}.$$

The corresponding real power supplied to the network from all three phases is:

$$P_{net1}(P_L) := 3 \cdot \text{Re}\left(\overline{I_{net1}(P_L)} \cdot V_{net1\_ln}\right)$$

$$P_{net1\_5\%} := P_{net1}(P_{L\_5\%}) = 219.8 \text{ MW}$$

This leads to a transmission efficiency at the 5% regulation of:

$$\text{effic}(P_L) := \frac{P_L}{P_{net1}(PL)} \quad \text{effic}(P_{L\_5\%}) = 91.1 \cdot \%.$$

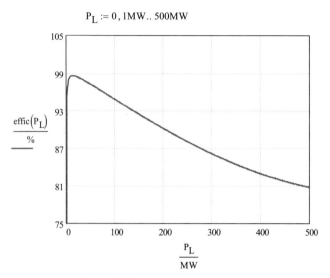

Figure 5.79. Efficiency versus load.

The efficiency is plotted against the load in Figure 5.79 to determine the trend and estimate the maximum efficiency. The graph reveals that the system (line) efficiency in the 0–500 MW range peaks at low load and decreases with increasing load. The maximum efficiency is calculated by the Mathcad *Maximize* function. A guess value for the load is $P_L := 2.5$ MV:

Given

$P_L > 0$

$P_{L\_max} := \text{Maximize}(\text{effic}, P_L) \quad P_{L\_max} = 13.8 \cdot \text{MW}$

$\text{effic}_{Max} := \text{effic}(P_{L\_max}) \quad \text{effic}_{Max} = 99.6 \cdot \%$

The high efficiency indicates that the transmission line is an economical means for long-distance energy transfer.

5.10.1.6. Transmission Line Interconnects Two Networks. Figure 5.80 shows the equivalent circuit of the system, where a 500 kV transmission line interconnects Network 1 and Network 2. Both networks are capable of generating or absorbing real and reactive power. In other words, both networks can work as a generator or as a load.

Power engineers use various models to approximate the network behavior. The simplest model is the Thévenin equivalent of the network. In the case of a large network, a relatively small load has little effect on the voltage; consequently, the open-circuit voltage can be approximated by the network line-to-neutral voltage (either the rated or the actual operating value). The Thévenin impedance is approximated by a reactance, which is the ratio of the rated line-to-neutral voltage and the short-circuit current.

APPLICATION EXAMPLES

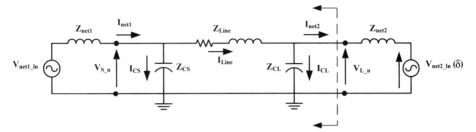

Figure 5.80. Equivalent circuit of a power system where a transmission line interconnects two networks.

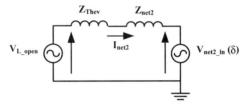

Figure 5.81. Thévenin equivalent of the system of Figure 5.80.

We assume that the voltage of Network 1 is equal to the rated voltage and its phase angle is zero. The Network 2 voltage is 0.95 times the rated voltage but its phase angle varies between $-180°$ and $+180°$.

The medium-length transmission line is represented by a Π circuit and each network by a voltage source and a reactance. The parameter values were given in the example found in Section 5.10.1.1.

The purpose of this analysis is the determination of the power transfer between the two networks as the phase angle (δ) of Network 2 varies. Particularly important is the determination of the maximum transferable power. The maximum transferable power is called the *steady-state stability limit*.

The phase angle of Network 2 is variable, but for the verification of the equations, we initially select a phase angle of $\delta := -50$ deg.

The network line-to-neutral voltages are:

$$V_{net1\_ln} := \frac{500 \text{ kV}}{\sqrt{3}} = 288.7 \text{ kV} \quad V_{net2\_ln}(\delta) := V_{net1\_ln} \cdot 0.95 \cdot e^{j\delta}$$

$$V_{net2\_ln}(\delta) = 274.2 \text{ kV} \quad \arg(V_{net2\_ln}(\delta)) = -50.0 \text{ deg}$$

The first step of the analysis is the replacement of Network 1 and the transmission line (to the left of the dashed line in Fig. 5.80) by its Thévenin equivalent circuit. The Thévenin equivalent contains a voltage source and impedance connected in series, as seen in Figure 5.81. The voltage of the source is the open-circuit voltage of the system when Network 2 is removed. The impedance is the input impedance of the system if the voltage sources are short circuited.

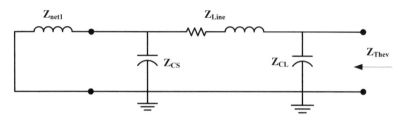

Figure 5.82. Equivalent circuit for Thévenin impedance calculation.

CALCULATION OF THE THÉVENIN IMPEDANCE. The equivalent circuit for the Thévenin impedance calculation is show in Figure 5.82. A short circuit has replaced the Network 1 voltage source.

The reactance of Network 1 and the sending-end line capacitance are connected in parallel. That equivalent parallel impedance is connected in series with the line impedance. The result of combining these impedances is:

$$Z_{net\_CS\_Line} = \frac{1}{\frac{1}{Z_{net1}} + \frac{1}{Z_{CS}}} + Z_{Line} \quad Z_{net\_CS\_Line} = (2.8 + 90.5j)\,\Omega.$$

The equivalent circuit shows that the load capacitance is connected in parallel with the impedance obtained earlier. Combining these impedances gives the Thévenin impedance, which is:

$$Z_{Thev} := \frac{1}{\frac{1}{Z_{net\_CS\_Line}} + \frac{1}{Z_{CL}}} \quad Z_{Thev} = (2.99 + 93.61j)\,\Omega.$$

With the Thévenin impedance and open-circuit voltage found, the equivalent circuit of Figure 5.81 can be used to calculate the current flowing from Network 1 to Network 2. The current is the voltage difference divided by the sum of the Thévenin and Network 2 impedances:

$$I_{net2}(\delta) := \frac{V_{L\_open} - V_{net2\_ln}(\delta)}{Z_{Thev} + Z_{net2}}$$

$$|I_{net2}(\delta)| = 2.16\text{ kA} \quad \arg(I_{net2}(\delta)) = -29.3 \text{ deg}.$$

POWER TRANSFER. The direction of real and reactive power transfer can be determined using the consequences of the passive sign convention. This analysis permits the determination of which network operates as a supplier (generator) and which operates as a load (absorber) of power. Figure 5.81 shows that the chosen current direction for I_{net2} is into the positive voltage terminal of the Network 2 source. Therefore, if the

APPLICATION EXAMPLES 301

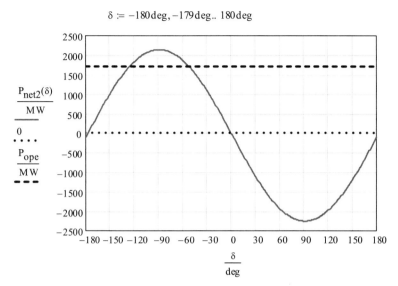

Figure 5.83. Network 2 real power versus power angle.

resultant Network 2 power is positive valued, then the network is absorbing power and, conversely, if the power is negative, Network 2 supplies power. The complex power of Network 2 is:

$$S_{net2}(\delta) := 3 \cdot V_{net2\_ln}(\delta) \cdot \overline{I_{net2}(\delta)} \quad S_{net2}(\delta) = (1666 - 628j) \text{ M.V.A.}$$

The real power of Network 2 is:

$$P_{net2}(\delta) := \text{Re}(S_{net2}(\delta)) \quad P_{net2}(\delta) = 1666.1 \text{ MW}.$$

To analyze the direction of power flow, we plot the Network 2 real power versus the power angle in Figure 5.83. The graph shows that Network 2 power is almost always positive when the power angle is negative. This means that the power flows from Network 1 to Network 2, as alluded to in Figure 5.81. The consequence is that *Network 1 is the supplier* and *Network 2 is the absorber (load)* when $\delta < 0$. Figure 5.83 shows that the Network 2 real power is negative when the power angle is positive. This means that *Network 2 is the supplier* and *Network 1 is the load*, when the power angle is positive.

In a similar manner, the reactive power of Network 2 is:

$$Q_{net2}(\delta) := \text{Im}(S_{net2}(\delta)) \quad Q_{net2}(\delta) = -628.4 \text{ M.V.A.}$$

The Network 2 reactive power versus power angle is plotted in Figure 5.84. The graph shows that Network 2 absorbs reactive power for power angles from about −25° to

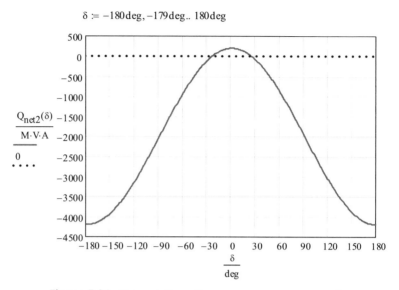

Figure 5.84. Network 2 reactive power versus power angle.

+25°, and for other angles, Network 2 produces reactive power. In practice, reactive power flow is controlled by voltage.

Referring to Figure 5.83, the maximum real power transfer occurs around −90°. The exact value is calculated by the Mathcad *Maximize* function. The guess value is $\delta := -90$ deg. The actual angle for maximum power transfer and the maximum power transfer are:

$$\delta_{max} := \text{Maximize}(P_{net2}, \delta) \quad \delta_{max} = -88.5 \cdot \text{deg}$$
$$P_{max} := P_{net2}(\delta_{max}) \quad P_{max} = 2145 \cdot \text{MW}$$

The safe operation of the system requires that the transferred power is lower than the maximum power. Note that for the analysis of a real system, transient stability, voltage regulation, and the thermal load conditions must be analyzed to determine the safe operating limits—an analysis that is beyond the scope of this book.

5.10.2. PSpice: Transient Short-Circuit Current in Transmission Lines

The most frequent transmission line fault is a lightning-caused insulator flashover and the follow-on short circuit. The short circuit produces a large current that heats the conductors and produces mechanical stresses. The calculation of the short-circuit current is important because the resultant values are used for protection coordination. As discussed in Chapter 1, transmission lines are protected against short circuits. Generally, CBs are installed at both ends of the line, as seen in Figure 5.75. In the case of a

APPLICATION EXAMPLES

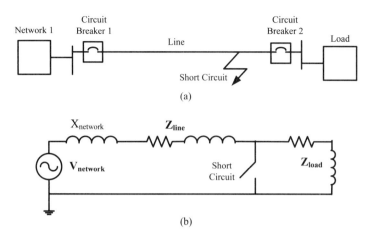

Figure 5.85. Transient short circuit current calculation. (a) Connection diagram; (b) equivalent circuit.

short circuit, the protection detects abnormal current or voltage, and activates both CBs. The short-circuit current is normally interrupted after three to five cycles.

The short circuit causes transient current. Typically, the current is the highest in the first cycle and gradually decreases to a steady-state value. The calculation of the transient short-circuit current is presented in this example. The transient current calculation requires the solution of the network differential equations. This can be achieved by using the transient analysis option of the PSpice circuit simulation program. This calculation is demonstrated using a numerical example.

Figure 5.85a depicts a network that supplies a transmission line. The equivalent circuit of this system is presented in Figure 5.85b. The network line-to-line voltage is 500 kV, which yields a supply line-to-neutral voltage magnitude of:

$$V_{network} = \frac{500 \text{ kV}}{\sqrt{3}} = 288.68 \text{ kV} \cdot \text{rms} = 408 \text{ kV(peak)}.$$

The network inductance value, taken from the example of Section 5.10.1.2, is:

$$L_{network} = \frac{X_{network}}{\omega} = \frac{28.868 \text{ }\Omega}{2\pi 60 \text{ Hz}} = 0.07657 \text{ H}.$$

A line impedance of $5.603 + j100.676$ Ω is utilized to compute the line inductance:

$$L_{line} = \frac{X_{line}}{\omega} = \frac{100.676 \text{ }\Omega}{2\pi 60 \text{ Hz}} = 0.26705 \text{ H}.$$

The inductive load is taken to be $266 + j200$ Ω, which yields a load inductance of 0.5305 H. These circuit component values are used to establish the PSpice circuit of

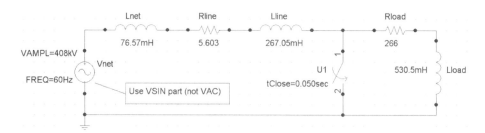

Figure 5.86. PSpice circuit diagram for simulation of a transmission line short circuit.

Figure 5.86. A normally open switch that can be closed at a user-defined time instant represents the short circuit. To perform a transient ac analysis the VSIN part must be utilized instead of the normal VAC source. The switch (U1) is set to close to 50 ms into the simulation.

Figure 5.87 shows a graph of the currents through the line and switch for the 400 ms simulation. Prior to the switch activation, the line current oscillates between about ±1 kA, and the switch current is zero. After the switch is closed, line and switch currents are identical for all practical purposes. The peak current is found using the *Cursor Max* function in Probe, and then is annotated on the plot of Figure 5.87 using the *Mark Label* function in Probe. The maximum current of about 5.13 kA occurs at 58.3 ms into the simulation (i.e., 8.3 ms after the short circuit occurs). About 8–10 cycles are needed to reach the steady-state short circuit current value of about 3 kA peak to peak. The maximum value of the short circuit current approaches nearly twice the steady-state short circuit current.

In the earlier analysis, the source impedance is assumed to be constant. Actually, the inductance value is changing from 0.15 to 1.1 per unit (p.u.).

5.10.3. PSpice: Transmission Line Energization

The sudden energizing of a transmission line can produce serious overvoltage at the end of the line. This phenomenon is investigated using PSpice. The typical scenario occurs when an out-of-service line is returned to service. Figure 5.88 depicts a deenergized transmission line, which is supplied by a network. The CBs at both ends of the line are open. To return the line into service the CB at one end is first closed. This action may cause serious overvoltage at the open end of the line.

Typically, a transmission line can be represented, with reasonable accuracy, by an equivalent Π network. A longer line is divided into 5- to 10-mile long sections and a Π circuit models each section. The resultant Π circuits are connected in series. A more accurate method is the representation of a long line with a distributed circuit. In this example, a 10-mile-long line is selected and modeled by a single Π circuit. The electric network, supplying the line, is represented by its Thévenin equivalent circuit: a voltage source and impedance connected in series. Figure 5.89 shows the equivalent circuit.

APPLICATION EXAMPLES

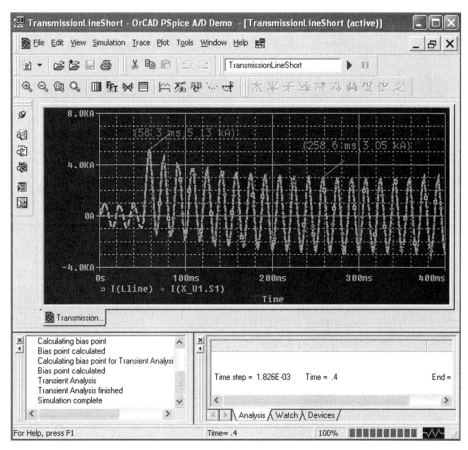

Figure 5.87. Results of PSpice simulation.

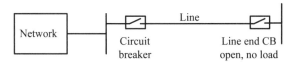

Figure 5.88. Transmission line energization.

The system circuit parameters are given as follows. The network Thévenin equivalent data are:

$$V_{source} = 408.25 \text{ kV} \quad R_{source} = 0.01 \text{ }\Omega \quad L_{source} = 98.483 \text{ mH}.$$

The transmission line data are:

$$L_{line} = 15.783 \text{ mH} \quad R_{line} = 0.254 \text{ }\Omega \quad C_{line1} = C_{line2} = 93.292 \text{ nF}.$$

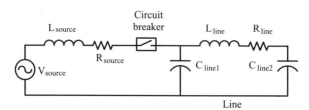

Figure 5.89. Equivalent circuit for line energization.

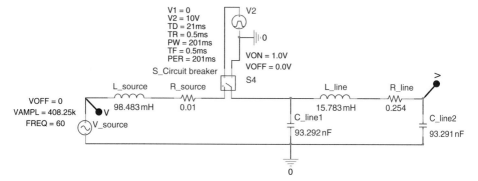

Figure 5.90. PSpice model for calculating line energization-generated overvoltage.

Figure 5.90 presents the PSpice model, which is based on the equivalent circuit of Figure 5.89.

The CB is controlled by a pulse generator, which turns on the supply-side breaker after a delay of TD = 21 ms, which corresponds to the approximate occurrence of the peak value of the supply voltage. The rise time (TR) and fall time (TF) of the control pulse are arbitrarily set to 0.5 ms. The pulse width (PW) and period (PER) are selected to be long enough to permit the evaluation of the damping of the transients. The chosen values are PW = 201 ms and PER = 201 ms. The source is a sine wave (VSIN), where the bias dc voltage is assumed to be zero (VOFF = 0 V). The amplitude of the sine wave is V_{source}. The frequency is 60 Hz and the delay angle is zero.

Figure 5.91 shows the simulation results, which include the supply voltage and the voltage at the line end. It can be seen that the line energization produces a high-frequency transient line voltage, which is superimposed with the 60 Hz source voltage. The maximum amplitude of the transient voltage is about twice the peak value of the supply voltage, or 850 kV. This high voltage may damage insulation. The peak value of the overvoltage depends on the time of energization. The overvoltage is maximum when the line is energized at the instant of the peak of the supply voltage, and minimum when it is energized at the zero crossing of the supply voltage.

The reader is encouraged to vary the delay time TD and study the effect of switching time. The attenuation of the high-frequency transient voltage is slow. The reader can also determine the time constant of the attenuation and the frequency of the transient voltage.

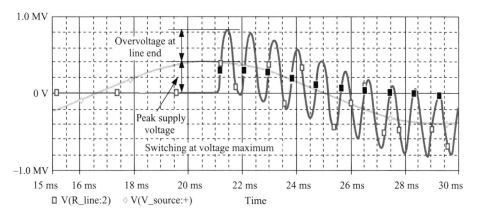

Figure 5.91. Line energization-generated overvoltage.

5.11. EXERCISES

Transmission Line Construction

1. Identify the components of a transmission line. Draw a sketch of a high-voltage line.
2. Explain the use of suspension and dead-end towers. How are these structures arranged to improve transmission line survivability?
3. Draw a double-circuit high-voltage transmission line.
4. Discuss the grounding methods used for transmission lines.
5. Describe the high-voltage conductor structure.
6. Describe the ACSR conductor.
7. Describe the high-temperature conductors and their application.
8. Describe the bundled conductor.
9. What is the structure of composite insulators? Draw a sketch.
10. Describe the standard porcelain suspension insulator. Draw a sketch.
11. Describe a post-type porcelain insulator. Draw a sketch.
12. Describe a post-type composite insulator. Draw a sketch.
13. Describe the conductor holder for single and bundled conductors.

Transmission Line Parameters (Resistance, Reactance, Capacitance, Magnetic and Electric Field)

14. Why is the stranded conductor ac resistance larger than the dc resistance?
15. What data are needed to determine the conductor resistance, and how is the resistance determined?
16. What is the effect of temperature on the conductor resistance?
17. What are the GMD, the GMR of a single conductor, and the GMR of a bundled conductor?

18. Draw an equivalent circuit used for transmission line reactance calculation.
19. Describe the procedure used to calculate the reactance of a line.
20. Explain the concept of magnetic field calculation. What are the basic equations?
21. Explain the environmental effect of magnetic fields.
22. Describe the concept of line capacitance calculation.
23. Explain the method used for calculation of electric charge on the conductors of a three-phase line.
24. Explain the concept of electric field calculation. Give the basic equations.
25. Explain the environmental effects of the electric field.
26. Draw the equivalent circuit for both short- and medium-length transmission lines.
27. What is the concept of a long transmission line equivalent circuit?
28. Discuss the long transmission line equations.
29. Explain the voltage and current distribution along a long transmission line.
30. What is the no-load voltage of a long line? Is it less or greater than the supply voltage?
31. What is the definition of the voltage drop or regulation?
32. What is the definition of the line loss?
33. Describe the fuse protection method.
34. What is the overvoltage protection?
35. Describe the distance protection.

Corona Discharge and Lightning Protection

36. Describe the concept of corona generation and list the effects of corona generation.
37. What is the effect of rain on corona generation?
38. What is the radio and TV interference?
39. Describe the methods used for transmission line lightning protection.

5.12. PROBLEMS

Problem 5.1

Calculate the inductance and reactance of a 60 Hz distribution line. The distance between phases A and B is 24 inches, and between phases B and C is 12 inches The GMR of the conductor utilized is 0.4 ft.

Problem 5.2

Calculate the capacitance of a horizontally arranged distribution line. The distance between phases A and B is 24 inches, and between phases B and C is 12 inches (consequently, the A-to-C distance is 24 + 12 = 36 inches). The diameter of the conductor is 1.15 inches. Discuss whether the capacitance can be considered negligible.

PROBLEMS   309

Problem 5.3

Compute the transmission line resistance at 25°C, inductance, and capacitance per mile for a three-phase 220 kV transmission line that utilizes nonbundled Drake conductors arranged vertically with 23-ft phase separation.

Problem 5.4

Calculate the transmission line resistance at 75°C, inductance, and capacitance per kilometer, for a three-phase 500 kV transmission line that utilizes three bundled Bluebird conductors arranged triangularly with a 9.5-m lateral phase separation and a 30-cm gap between conductors within a bundle.

Problem 5.5

A 300 km transmission line operates at 60 Hz. The resistance, inductance, and capacitance of the line per kilometer are 0.015 Ω, 1.0 mH, and 10 nF, respectively. Determine: (a) the equivalent long line parameters and (b) the equivalent pi circuit parameters.

Problem 5.6

A short three-phase distribution line supplies a balanced load. Calculate the necessary supply voltage and regulation if the load voltage is 13.8 kV line to line. The load is 3 MW with a power factor of 0.82 (lagging). The line length is 12 miles. The line resistance is 0.12 Ω/miles and the line inductance is 0.35 Ω/mile.

Problem 5.7

A short three-phase distribution line supplies a balanced load. Calculate the load voltage and regulation if the supply voltage is 13.8 kV (line to line). The load is 3 MW, with a power factor of 0.82 (lagging). The line length is 12 miles. The line resistance is 0.12 Ω/mile, and the line inductance is 0.35 Ω/mile.

Problem 5.8

A fuel cell, located 15 miles outside a town, supplies the town's three-phase distribution network. The fuel cell voltage is maintained at 12.47 kV. The local network voltage is also kept at this level. Calculate the required phase angle difference between the fuel cell and local network voltage if the fuel cell supplies 15 MW to the network. The line impedance is 0.23 + j0.45 Ω/mile.

Problem 5.9

Figure 5.92 depicts a transmission line tower and the conductor arrangement. The line is built with three bundled conductors per phase, horizontally arranged, with a separation of D = 9 m between centers of adjacent bundles. The average conductor height is 20 m. The equilateral spacing between conductors in each bundle is d = 45 cm. The line is constructed with Bluebird conductors. The line length is 120 km. The supply voltage of the transmission line is 345 kV at 60 Hz.

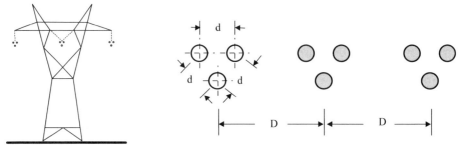

Figure 5.92. Transmission line tower and conductor arrangement.

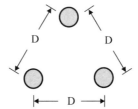

Figure 5.93. Conductor spacing for Problem 5.10.

Determine: (a) the resistance and inductive reactance of the transmission line in ohms per kilometer and (b) the capacitance of the line per phase in microfarads per kilometer and capacitive admittance in microsiemens per kilometer. (c) Draw the equivalent circuit. (d) If the transmission line is connected to a balanced three-phase wye-connected load with impedance of $Z_Y = 150 + j250$ Ω per phase, calculate the load current and load voltage.

Problem 5.10

A three-phase, 60 Hz transmission line is built with an ACSR conductor. The conductor diameter is 0.680 inch, GMR = 0.0230 ft, and ac resistance of 0.342 Ω/mile at 50°C. The conductors are equilaterally spaced (triangle) with D = 1.7 m between conductor centers, as illustrated in Figure 5.93. The line is 50 km long and supplies a 5 MW load. The load voltage is 13.2 kV line to line.

(a) Find the total impedance of the line. (b) Determine the sending-end voltage, current, real power, and reactive power for each of the following load conditions: (1) $pf_1 = 0.85$ lagging, (2) $pf_2 = 0.91$ leading, and (3) $pf_3 = 1$. (c) Plot the sending voltage (obtained with pf_1) versus the length of the transmission line, assuming the length varies from 1 to 50 km.

Problem 5.11

A three-phase, 60 Hz transmission line with a length of 180 miles consists of three Cardinal conductors per bundle with equilateral spacing of 35 cm between conductors.

PROBLEMS

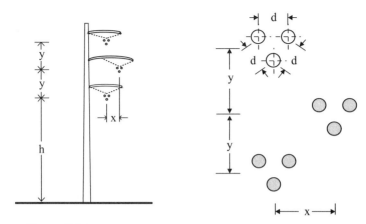

Figure 5.94. Three-phase transmission line and bundle spacing.

The bundles are arranged vertically with spacing between centers, as sketched in Figure 5.94 ($x = 5$ m, $y = 6.5$ m). A 500 kV network supplies the line. The supply short circuit current is 22 kA.

(a) Determine the total resistance, inductive reactance, and capacitive admittance of the transmission line. (b) Draw the equivalent circuit. (c) Obtain the Thévenin equivalent circuit as viewed from the load terminals. (d) Plot the short-circuit current versus distance, assuming that a short circuit may occur at any point on the transmission line. Explain the tendency of the curve. (e) If the short circuit occurs at the end of the line, determine the magnitude of the current.

Problem 5.12

A 230 kV transmission line interconnects two 60 Hz power sources. The following data are available:

Source 1: short-circuit duty = 12,000 MVA
Source 2: short-circuit duty = 19,000 MVA

The *supply short-circuit duty* is three times the product of the line-to-neutral voltage and the short-circuit current. The transmission line length is 125 mi. There are three Falcon conductors per bundle with a distance of 14 inches between the conductors. The distance between phases is 22 ft using a horizontal arrangement, as depicted in Figure 5.92. (a) Find the transmission line and power supply parameters. (b) Draw the equivalent circuit. (c) If the absolute values of the Source 1 and Source 2 voltages are equal to the rated voltage, calculate and plot the active power transferred versus the power angle relation, and determine the maximum power transferred. The power angle here is defined as the angle between the voltages of Sources 1 and 2.

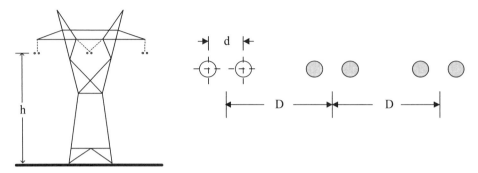

Figure 5.95. Transmission line tower and conductor arrangement.

Problem 5.13

A three-phase, 50 Hz transmission line is built with two Bluejay conductors per bundle, as drawn in Figure 5.95. The distance between conductors in the bundle is 15 in. The separation distance between the center points of adjacent phases is 30 ft. The height of the line is 60 ft. The line has a length of 90 miles and delivers a maximum of power of 1000 MVA with a power factor of 0.85 lagging to the load. The voltage at the load terminals is 345 kV.

(a) Determine the total resistance, inductance, and capacitance of the transmission line. (b) Draw the equivalent circuit. (c) Find the required supply voltage and complex power. (d) Plot the supply voltage and the voltage regulation versus the load, and determine the load value that causes a voltage regulation of 10%.

Problem 5.14

Calculate and plot the total magnetic field intensity under a 500 kV transmission line. The balanced three-phase 500 kV transmission line phase conductors are arranged vertically at the same horizontal coordinate $x_0 = 15$ ft. The phase conductors carrying 1000 A are located at heights of 80, 110, and 140 ft from ground level. Determine the maximum H value at 1 m above ground level.

6

ELECTROMECHANICAL ENERGY CONVERSION

Electromechanical energy conversion is employed in several devices designed by electrical engineers. In many cases, the energy conversion requires the assistance of a magnetic or electric field. Devices using electromechanical energy conversion are categorized as:

- *Transducers* that convert another energy form into a signal; typical examples are microphones, loudspeakers, sensors, and electrical pickups;
- *Actuators* that utilize an electrical signal to move or control another device; these include relays, solenoids, and electromagnets; and
- Continuous energy conversion equipment such as motors and generators.

This chapter addresses the principles of electromechanical energy conversion and presents the major equations used for analyzing the operation of different devices. The practical application of the principles is demonstrated through the analysis of actuator operation. Transducers and actuators can also be constructed of piezoelectric (PE) materials that generate charge when they are strained. Continuous energy conversion equipment like motors and generators are detailed later in this book. First, magnetic

---

Electrical Energy Conversion and Transport: An Interactive Computer-Based Approach, Second Edition.
George G. Karady and Keith E. Holbert.
© 2013 Institute of Electrical and Electronics Engineers, Inc. Published 2013 by John Wiley & Sons, Inc.

circuit analysis is detailed in order to provide a working knowledge of the coupling of electromagnetic fields.

6.1. MAGNETIC CIRCUITS

Electric power equipment, such as transformers, generators, and motors, as well as appliances, have magnetic components and use magnetic coupling for energy transfer. This necessitates the study of magnetic circuits. The typical magnetic circuit contains a laminated iron core with one or more windings. The magnetic property of the iron core is described by the B-H curve, which gives the flux density (B) and magnetic field intensity (H) relation, as graphed in Figure 6.1. Of the four materials graphed in the figure, the quality sheet steel is the best because its permeability is the largest of the three materials. The quality sheet steel is a very low carbon steel with silicon and manganese content of 3% or less. For instance, silicon steel is the material of exclusive use in power transformers.

The iron core is magnetized by a coil that is supplied by an alternating current (ac) current. The ac magnetization increases the flux in the positive cycle and reduces the flux in the negative cycle, which rotates the dipoles in the iron at every cycle. Figure 6.2 shows the process by sketching the B-H behavior for ferromagnetic material where the path from a to b is the initial magnetizing curve. If the magnetic field is relaxed, the magnetic flux density (B) does not return to zero, but rather follows the path from b to c, where a remnant flux density exists—a *hysteresis* effect. The applied magnetic field must be reversed in order to drive B to zero (i.e., point d). Simultaneously, the flux change induces eddy current in the iron core itself. The eddy current generation

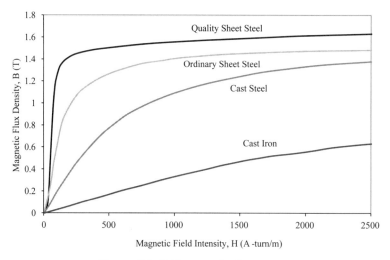

Figure 6.1. B-H magnetization curves.

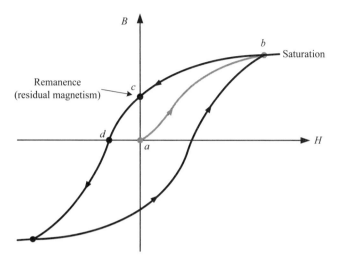

Figure 6.2. Hysteresis effect in ferromagnetic material.

prompted the lamination of the iron core to block the eddy current. Both phenomena generate losses in the iron core, specifically:

1. hysteresis loss, which is caused by the reorientation, rotation of the dipoles in the iron in every cycle; and
2. eddy current losses, which are due to the currents generated by the ac flux in the iron core. The lamination of the core significantly reduces the eddy current losses.

These losses cause heating of the core. The typical loss curve of silicon steel is presented in Figure 6.3.

6.1.1. Magnetic Circuit Theory

Before performing a complete analysis of a magnetic circuit, this section presents the governing relationships of a typical magnetic circuit, as sketched in Figure 6.4. The magnetizing current (I_m) supplied to the coil winding generates a magnetic field. Using the right-hand rule, the magnetic flux (Φ) is in the clockwise direction in the torus. The magnetic field intensity is related to the current using Ampere's circuital law. If the magnetic field intensity (H) is constant in the gap, the line integral can be replaced by the product of the field intensity and length:

$$I_m N_m = H\ell, \qquad (6.1)$$

where N_m is the number of turns in the coil winding and ℓ is the magnetic path length. The *magnetic path length* is the average distance traveled by the magnetic flux. If the

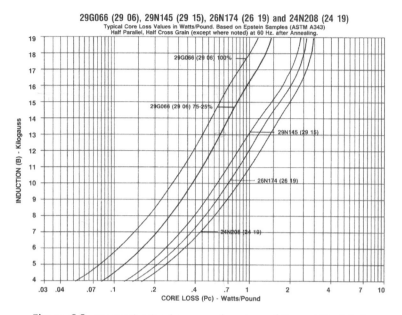

Figure 6.3. Magnetization loss curve (courtesy of Tempel Steel Co.).

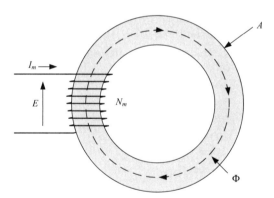

Figure 6.4. Simple toroidal magnetic circuit.

permeability (μ) of the core material is known, the magnetic flux density (B) can be determined in the linear region of the B-H characteristics using:

$$B = \mu H. \qquad (6.2)$$

Permeability is defined as a measure of the ability of a material to carry magnetic flux. The *relative permeability* (μ_r) is defined based on the permeability (μ) of a material in comparison to the *permeability of the free space* (μ_0):

MAGNETIC CIRCUITS

$$\mu_r = \frac{\mu}{\mu_0}, \qquad (6.3)$$

where $\mu_0 = 4\pi \times 10^{-7}$ H/m. The magnetic flux is the product of B and the cross-sectional area (A) of the torus core material:

$$\Phi = BA. \qquad (6.4)$$

Faraday's law of induction provides a means to relate the induced voltage (E) in the coil to the magnetic flux:

$$E = N_m \frac{d\Phi}{dt}. \qquad (6.5)$$

We substitute Equation (6.1), Equation (6.2), and Equation (6.4) into the previous expression (in reverse order) to find:

$$E = N_m \frac{d(BA)}{dt} = AN_m \frac{d(\mu H)}{dt} = \mu AN_m \frac{d}{dt}\left(\frac{I_m N_m}{\ell}\right) = \frac{\mu AN_m^2}{\ell} \frac{dI_m}{dt}. \qquad (6.6)$$

The induced voltage is also related to the inductance (L) created by the coiled winding:

$$E = L \frac{dI_m}{dt}. \qquad (6.7)$$

Equating the results of this expression and Equation (6.6) yields a formula for the inductance of the coil:

$$L = \frac{\mu AN_m^2}{\ell}. \qquad (6.8)$$

Such analysis will be applied to a more involved magnetic circuit in the next section, and we will find identical results. In addition, we will prove that the energy stored in the coil at any time instant is:

$$\text{Energy} = \frac{LI_m^2}{2}. \qquad (6.9)$$

6.1.2. Magnetic Circuit Analysis

Figure 6.5 depicts a magnetic circuit with an iron core equipped with an excitation winding. AC current $\mathbf{I_m}$ supplies the winding, which generates a counterclockwise ac magnetic flux Φ_m in the iron core. The magnetic path is the route traversed by Φ_m in

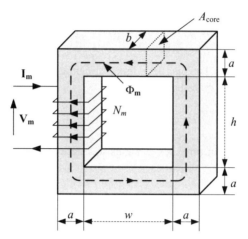

Figure 6.5. Iron core magnetic circuit.

the iron core, as annotated in Figure 6.5. The corresponding magnetic path length, which is the average distance traveled by the magnetic flux, is:

$$\ell = 4a + 2w + 2h. \tag{6.10}$$

The core cross section is:

$$A_{core} = ab. \tag{6.11}$$

The ac current-generated magnetic field is evenly distributed within the core cross section. The magnetic field intensity is calculated using Ampere's law of Equation (6.1):

$$H = \frac{I_m N_m}{\ell}. \tag{6.12}$$

The magnetic flux density can be determined from a B-H curve or from the material permeability. Figure 6.6 plots the magnetization B-H and relative permeability versus magnetic field intensity curves for a representative iron core. In this graph, the magnetic field strength (intensity) is in oersted (Oe) units, where 1 Oe = 79.577 A(turn)/m. Instead of determining the magnetic flux density from the B-H curves, an alternative method is to calculate the flux density from the product of the magnetic field intensity and permeability:

$$B = \mu_0 \mu_r H. \tag{6.13}$$

MAGNETIC CIRCUITS

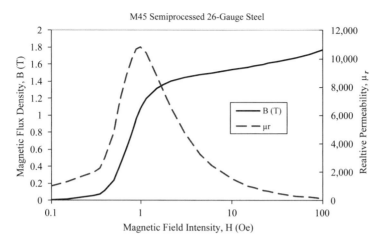

Figure 6.6. Magnetization B-H and relative permeability curves of high-quality transformer steel sheets.

Substitution of Equation (6.12) into Equation (6.13) results in a more compact expression for the flux density:

$$B = \mu_0 \mu_r \frac{I_m N_m}{\ell}. \tag{6.14}$$

6.1.2.1. Magnetic Flux. The magnetic flux is the product of the core cross-sectional area and the flux density if the flux density is uniformly distributed in the iron core:

$$\Phi_m = BA_{\text{core}}. \tag{6.15}$$

The substitution of Equation (6.14) into Equation (6.15) yields:

$$\Phi_m = \mu_0 \mu_r \frac{I_m N_m}{\ell} A_{\text{core}}. \tag{6.16}$$

The magnetizing current is an ac that varies sinusoidally. The time function of the current is:

$$I_{\text{mag}}(t) = \sqrt{2} I_m \cos(\omega t). \tag{6.17}$$

The corresponding magnetic flux time function is:

$$\Phi_m(t) = \mu_0 \mu_r \frac{A_{\text{core}} N_m}{\ell} I_{\text{mag}}(t). \tag{6.18}$$

Substituting Equation (6.17) into the previous expression yields:

$$\Phi_m(t) = \mu_0 \mu_r \frac{A_{core} N_m}{\ell} \sqrt{2} I_m \cos(\omega t). \qquad (6.19)$$

The maximum value of the flux occurs when the cosine is unity is:

$$\Phi_{max} = \mu_0 \mu_r \frac{A_{core} N_m}{\ell} \sqrt{2} I_m \qquad (6.20)$$

Combining Equation (6.19) and Equation (6.20), the magnetic flux time function can be expressed as:

$$\Phi_m(t) = \Phi_{max} \cos(\omega t). \qquad (6.21)$$

6.1.2.2. Induced Voltage. According to Faraday's law ($d\Phi = edt$), when a flux passes through a coil, it induces a voltage in each turn of that coil and the induced voltage is proportional with the rate of flux change. A total induced voltage is obtained from all the induced voltages of the coil turns:

$$E_{ind}(t) = N_m \frac{d}{dt} \Phi_{mag}(t). \qquad (6.22)$$

An alternative expression for the total induced voltage can be derived. First, substitute Equation (6.21) into Equation (6.22):

$$E_{ind}(t) = N_m \frac{d}{dt} \Phi_{max} \cos(\omega t). \qquad (6.23)$$

Then, differentiating the previous equation yields:

$$E_{ind}(t) = -N_m \Phi_{max} \omega \sin(\omega t). \qquad (6.24)$$

The root-mean-square (rms) value of this sinusoidal induced voltage is:

$$E_{rms} = \frac{N_m \Phi_{max} \omega}{\sqrt{2}}. \qquad (6.25)$$

The substitution of $\omega = 2\pi f$ into Equation (6.25) and the reduction of the expression provide a simplified relation for the induced voltage:

$$E_{rms} = \sqrt{2} \pi f N_m \Phi_{max}. \qquad (6.26)$$

MAGNETIC CIRCUITS

If the same core (Fig. 6.5) is equipped with a second coil (not shown), the same flux goes through its winding. For a secondary winding with N_2 turns, then from Equation (6.25), the rms voltage induced in the secondary coil is:

$$E_{rms,2} = \frac{N_2 \Phi_{max} \omega}{\sqrt{2}}. \tag{6.27}$$

6.1.2.3. Coil Inductance.

The inductance of a coil is frequently necessary for electrical circuit calculations. The inductance can be calculated from the induced voltage equation, which is:

$$E_{ind}(t) = L_{ind} \frac{d}{dt} I_{mag}(t). \tag{6.28}$$

Substituting Equation (6.22) into the earlier formula results in the following expression:

$$N_m \frac{d}{dt} \Phi_{mag}(t) = L_{ind} \frac{d}{dt} I_{mag}(t). \tag{6.29}$$

The previous relation is integrated and solved for the inductance. The result is:

$$L_{ind} = \frac{N_m \Phi_{mag}(t)}{I_{mag}(t)}. \tag{6.30}$$

The substitution of the current and flux time functions from Equation (6.17) and Equation (6.21), respectively, into the previous expression gives:

$$L_{ind} = \frac{N_m \Phi_{max} \cos(\omega t)}{\sqrt{2} I_m \cos(\omega t)} = \frac{N_m \Phi_{max}}{\sqrt{2} I_m}. \tag{6.31}$$

Recall that $I_m \sqrt{2}$ is the peak value of the current. The simplification of the previous formula gives another practical relation for the inductance:

$$L_{ind} = \frac{N_m \Phi_{max}}{I_{max}}. \tag{6.32}$$

According to Equation (6.32), the inductance of a coil can be calculated from the product of the number of turns and the flux, divided by the current that generates the flux. However, the inductance is a geometrical quantity that depends only on the dimensions of the coil and the iron core material. Actually, the inductance is independent from the current. This can be demonstrated by the substitution of Equation (6.20) into Equation (6.31), which yields:

$$L_{ind} = \frac{N_m \mu_0 \mu_r \frac{A_{core} N_m}{\ell} \sqrt{2} I_m}{\sqrt{2} I_m}. \tag{6.33}$$

The simplification of the previous expression results in a new formula for the inductance:

$$L_{ind} = \mu_0 \mu_r \frac{A_{core} N_m^2}{\ell}. \tag{6.34}$$

EXAMPLE 6.1: Magnetic circuit analysis

The magnetic circuit of Figure 6.5 is analyzed for a specific situation. The physical dimensions of the iron core are:

$$w := 3 \text{ in} \quad h := w \quad a := 1 \text{ in} \quad b := 1.5 \text{ a}.$$

The circuit is supplied by an rms magnetizing current I_m through a winding having N_m turns:

$$I_m := 2 \text{ A} \quad N_m := 20 \quad f := 60 \text{ Hz} \quad \omega := 2 \cdot \pi \cdot f.$$

The magnetic path length (L_m) is:

$$L_m := 2 \cdot (w + a) + 2 \cdot (h + a) = 40.64 \text{ cm}.$$

The core cross-section (annotated in Fig. 6.5) is:

$$A_{core} := a \cdot b = 9.677 \text{ cm}^2.$$

The magnetic field intensity is computed from Ampere's law:

$$H_m := \frac{I_m \cdot N_m}{L_m} = 1.237 \text{ Oe}.$$

The magnetic flux density can be extracted from Figure 6.6. The obtained value is $B_m := 1.22 \text{T}$ for $H_m = 1.24$ Oe. Alternatively, the relative permeability for steel from Figure 6.6 is approximately $\mu_r := 10^4$ for $H_m = 1.24$ Oe, from which the magnetic flux density is determined to be:

$$B_m := \mu_0 \cdot \mu_r \cdot H_m = 1.237 \text{ T}.$$

MAGNETIC CIRCUITS

where the permeability of free space and air is:

$$\mu_o := 4 \cdot \pi \cdot 10^{-7} \cdot \frac{H}{m}.$$

The small deviation between the two numerical values for B_m is due to the inaccuracy of reading values from the graph of Figure 6.6.

From the product of the magnetic flux density and core cross-sectional area, the magnetic flux is readily computed as:

$$\Phi_m := B_m \cdot A_{core} = 0.001 \text{ Wb}.$$

Small flux values such as this can be observed in Mathcad with more detail by using units of milliweber by defining a general milli prefix: m := 10^{-3}, such that the flux in milliweber is $\Phi_m = 1.197 \cdot m \cdot Wb$. For this situation, the maximum flux is:

$$\Phi_{max} := \sqrt{2} \cdot \left(\mu_o \cdot \mu_r \cdot \frac{A_{core} \cdot N_m}{L_m} \right) \cdot I_m = 1.693 \text{ m} \cdot \text{Wb}.$$

Finally, the inductance is found from several formula, including:

$$L_{ind} := \mu_o \cdot \mu_r \cdot \frac{A_{core} \cdot N_m^2}{L_m} = 11.97 \text{ mH}.$$

6.1.3. Magnetic Energy

The instantaneous power transferred from the electrical supply to the magnetic coil is:

$$p(t) = I_{mag}(t) E_{ind}(t). \tag{6.35}$$

Substituting Equation (6.28) into the previous expression yields:

$$p(t) = I_{mag}(t) L_{ind} \frac{d}{dt} I_{mag}(t). \tag{6.36}$$

The integration of this expression results in the energy function:

$$\begin{aligned}
\text{Energy}(t) &= \int_{-\infty}^{t} L_{ind} I_{mag}(t) \frac{dI_{mag}(t)}{dt} dt \\
&= L_{ind} \int_{I_{mag}(-\infty)}^{I_{mag}(t)} I_{mag} dI_{mag} \\
&= \frac{1}{2} L_{ind} I_{mag}^2 \bigg|_{I(-\infty)}^{I(t)}.
\end{aligned} \tag{6.37}$$

Since the inductor must have initially been unenergized, $I(-\infty) = 0$, then:

$$\text{Energy}(t) = \frac{1}{2} L_{\text{ind}} I_{\text{mag}}^2(t). \tag{6.38}$$

Interestingly, the energy stored in an inductor at some particular time instant, t, is simply a function of the instantaneous value of the current. This implies that in the positive (ac) cycle, the source supplies energy to the inductance, and in the negative cycle, the inductor returns energy to the supply, if the inevitable losses are neglected. This energy fluctuation is the source of the reactive power.

6.1.4. Magnetization Curve

A magnetization curve is a plot of the magnetic flux density as a function of the applied magnetic field intensity. The magnetization curve is typically nonlinear, and is also referred to as a B–H curve. The magnetization (B–H) curves are divided into three regions:

1. the linear region, where the permeability of the material is a constant;
2. the transition or knee region, where the material permeability reaches saturation; and
3. the saturation region.

Figure 6.7 graphs the three regions for silicon steel. Most equipment, such as transformers and motors, operate in the linear region. Transformers are made out of silicon steel,

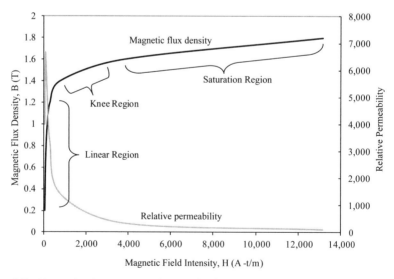

Figure 6.7. Magnetization curve regions and relative permeability of a high-silicon transformer steel.

ically is an alloy with a silicon content of 1–3%. High-frequency transformers and coils use ferrite cores. The ferrite is a ceramic material.

The derivative of the B-H curve is the permeability, which is the slope of the B-H curve. The permeability is the highest in the linear region and reduces significantly in the saturation area. This is demonstrated in Figure 6.7.

The quality of the iron is better if the saturation starts at a higher B value. Also the higher the relative permeability, the steeper the slope in the linear region, which is advantageous. This is demonstrated in a numerical example.

EXAMPLE 6.2: Compare the inductance from using ordinary sheet steel versus cast steel. The data needed to calculate the inductance are:

$$L_{iron} := 15 \text{ cm} \quad A_{iron} := 4 \text{ cm}^2 \quad N_m := 150 \quad I_{exc} := 0.6 \text{ A}.$$

The magnetic field strength is calculated from Ampere's law:

$$H_{iron} := \frac{I_{exc} \cdot N_m}{L_{iron}} \quad H_{iron} = 600 \frac{A}{m}.$$

Using the B-H curves of Figure 6.1, the flux density for cast and silicon steel are:

$$B_{C\_steel} := 0.8 \text{ T} \quad B_{Si\_steel} := 1.2 \text{ T}.$$

The flux in the core is:

$$\Phi_{C\_Steel} := B_{C\_steel} \cdot A_{iron} = 3.2 \times 10^{-4} \text{ Wb}$$
$$\Phi_{Si\_steel} := B_{Si\_steel} \cdot A_{iron} = 4.8 \times 10^{-4} \text{ Wb}.$$

The inductance of each coil is:

$$L_{ind\_C\_Steel} := \frac{\Phi_{C\_Steel} \cdot N_m}{I_{exc}} = 80 \cdot \text{mH}$$
$$L_{ind\_Si\_Steel} := \frac{\Phi_{Si\_steel} \cdot N_m}{I_{exc}} = 120 \cdot \text{mH}.$$

This example reveals that the inductance of the coil made out of silicon steel is significantly higher than the one made out of carbon steel.

EXAMPLE 6.3: Magnetic circuit with air gap

The application of the equations derived for the analysis of a magnetic circuit is demonstrated here in a numerical example. Figure 6.8 presents a magnetic circuit with an

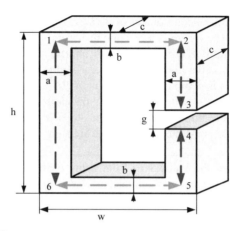

Figure 6.8. Magnetic circuit with air gap and variable cross section.

air gap and variable cross section. The iron core is equipped with an excitation winding, which is not shown in the diagram.

The dimensions of the iron core are:

$$a := 3 \text{ cm} \quad b := 2 \text{ cm} \quad c := 4 \text{ cm}$$
$$g := 0.2 \text{ cm} \quad w := 10 \text{ cm} \quad h := 11 \text{ cm}.$$

The magnetic circuit operational (rms) data are: $B_{gap} := 0.8$ T $I_m := 8$ A.

The permeability of air (free space) is defined as:

$$\mu_o := 4 \cdot \pi \cdot 10^{-7} \frac{H}{m}.$$

In this example, we will calculate:

1. the magnetic flux (Φ) and the flux densities (B) in the different sections of the iron core;
2. the magnetic field intensities (H) in the different sections of the core;
3. the number of turns needed to maintain the flux if the current is $I_m = 8$ A; and
4. the inductance when the iron core is both neglected and included.

(1) Magnetic Flux and the Flux Densities Calculation

To compute the magnetic flux and the flux densities in the iron core, the magnetic circuit is divided into three sections: vertical section (paths 1–6, 2–3, and 4–5), horizontal section (paths 1–2 and 5–6), and the gap. The cross section of the iron core is different in the vertical and horizontal sections. The gap and vertical sections have identical cross sections. The magnetic flux (Φ) is identical in all sections.

MAGNETIC CIRCUITS

The length and cross-sectional area of the vertical section are:

$$L_{vert} := (h-b)+(h-b-g) \quad L_{vert} = 17.8 \cdot cm$$
$$A_{vert} := c \cdot a \quad A_{vert} = 12 \cdot cm^2.$$

The length and cross-sectional area of horizontal section are:

$$L_{horz} := 2 \cdot (w-a) \quad L_{horz} = 14 \cdot cm$$
$$A_{horz} := c \cdot b \quad A_{horz} = 8 \cdot cm^2.$$

The cross-sectional area of the gap is: $A_g := c \cdot a = 12 \text{ cm}^2$.

The flux for the magnetic circuit is calculated using Equation (6.15) with the known flux density in the gap:

$$\Phi_{gap} := B_{gap} \cdot A_g = 9.6 \times 10^{-4} \text{ Wb}.$$

Similarly, the flux densities in the vertical and horizontal sections are:

$$B_{vert} := \frac{\Phi_{gap}}{A_{vert}} \quad B_{vert} := 0.8 \text{ T}$$

$$B_{horz} := \frac{\Phi_{gap}}{A_{horz}} \quad B_{horz} = 1.2 \text{ T}.$$

(2) Magnetic Field Intensity Calculation

Here, we determine the magnetic field intensities (H) in the different sections of the core. Using Equation (6.13), the field strength in the air gap is:

$$H_{gap} := \frac{B_{gap}}{\mu_o} = 6.366 \times 10^5 \frac{A}{m}.$$

The field strengths (H) in the vertical and horizontal sections are determined directly from reading the B-H curves of Figure 6.9 at B_{vert} and B_{horz}:

$$H_{vert} := 140 \frac{A}{m} \quad H_{horz} := 400 \frac{A}{m}.$$

Note that the obtained quantities are rms values.

(3) Computation of the Number of Turns

To calculate the number of turns needed to maintain the flux, an expanded formulation of Equation (6.12) for Ampere's law ($\oint H \, dl = I$) is utilized. The product of the current

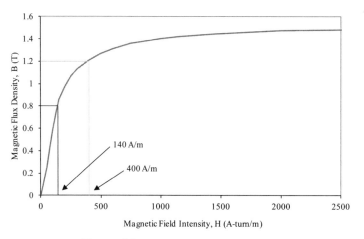

Figure 6.9. Magnetization B-H curves.

and the number of turns is equal with the sum of the magnetic field strengths and section length products:

$$N_m := \frac{H_{gap} \cdot g + H_{vert} \cdot L_{vert} + H_{horz} \cdot L_{horz}}{I_m} \qquad N_m = 169.3.$$

The relationship between B and H is linear below the knee point. At this point, the iron saturates, as seen in Figure 6.7. The permeability of the iron in the linear region is a few thousand, but in the saturation region, it is significantly less. The ampere-turns needed to maintain the flux in the iron is significantly less than the ampere-turns needed for the gap. Therefore, if the iron core is neglected, then the number of turns required is:

$$N_{no\_iron} := \frac{H_{gap} \cdot g}{I_m} \qquad N_{no\_iron} = 159.2.$$

Neglecting the magnetic core slightly reduces the number of turns.

(4) Inductance Calculation

We calculate the inductance for two situations: (1) when the iron core is neglected and (2) when the iron core is included. Using Equation (6.32), the inductance when an iron core is present is:

$$L_{with\_iron} := \frac{\Phi_{gap} \cdot N_m}{I_m} \qquad L_{with\_iron} = 20.3 \cdot mH.$$

Neglecting the magnetic core yields an inductance of:

$$L_{without\_iron} := \frac{\Phi_{gap} \cdot N_{no\_iron}}{I_m} \qquad L_{without\_iron} = 19.1 \cdot mH.$$

The percentage deviation between the two inductance values is:

$$\frac{L_{with\_iron} - L_{without\_iron}}{L_{without\_iron}} = 6.4\%.$$

The relatively small deviation indicates that the iron core may be neglected in many cases.

6.1.5. Magnetization Curve Modeling

The magnetization curves, which are obtained by measurement, represent average values. These curves can be digitized and equations can be developed for the $B(H)$ or $H(B)$ functions using curve-fitting methods. Such fitted equations significantly simplify the analysis of magnetic circuits with a saturable iron core.

EXAMPLE 6.4: To demonstrate this technique, the corresponding B and H values are read from the sheet steel B-H curve of Figure 6.6. The $B(H)$ values obtained from the steel B-H curve are:

$$H_o := \begin{pmatrix} 0.3 \\ 0.4 \\ 0.5 \\ 0.6 \\ 0.8 \\ 1 \\ 1.5 \\ 2 \\ 3 \\ 5 \\ 10 \\ 20 \\ 40 \\ 50 \\ 80 \\ 100 \end{pmatrix} \qquad B_o := \begin{pmatrix} 0.06 \\ 0.12 \\ 0.24 \\ 0.42 \\ 0.8 \\ 1.06 \\ 1.32 \\ 1.4 \\ 1.44 \\ 1.48 \\ 1.52 \\ 1.59 \\ 1.66 \\ 1.68 \\ 1.72 \\ 1.76 \end{pmatrix}.$$

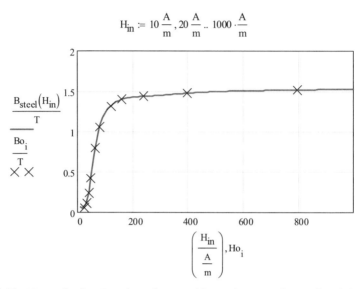

Figure 6.10. Magnetization B and H values read from Figure 6.1 for steel and the resultant spline fit of the data.

The B and H values obtained from the magnetization curve are in oersted and kilogauss units, respectively. The obtained H and B values are converted to amperes per meter and tesla, respectively:

$$\text{Ho} := H_o \cdot \text{Oe} \quad \text{Bo} := B_o \cdot T.$$

The B-H curve is then plotted in Figure 6.10 to validate the procedure.

Next, the Mathcad cubic polynomial interpolation function is called to fit the extracted data:

$$v := \text{cspline}(\text{Ho}, \text{Bo}).$$

After that, the Mathcad interpolation function may be used to create user function(s) that describe the $B(H)$ and/or $H(B)$ curves, specifically:

$$B_{\text{steel}}(H_{\text{steel}}) := \text{interp}(v, \text{Ho}, \text{Bo}, H_{\text{steel}})$$
$$H_{\text{iron}}(B_{\text{iron}}) := \text{interp}(v, \text{Bo}, \text{Ho}, B_{\text{iron}}).$$

These functions approximate the original B-H curve with high accuracy (as seen in Fig. 6.10) and allow the calculation of the B or H values without lengthy reading from a diagram. For example, the functions may be used to find the flux density (B) for a magnetic field intensity (H) of 300 A/m:

$$H_{\text{steel}} := 300 \cdot \frac{A}{m} \quad B_{\text{steel}}(H_{\text{steel}}) = 1.45 \text{ T}.$$

MAGNETIC CIRCUITS

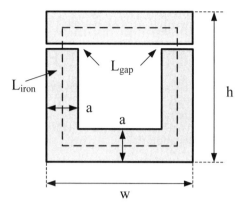

Figure 6.11. Small transformer with air gap.

Likewise, the magnetic field intensity (*H*) for a specific flux density (*B*) can be determined using:

$$B_{iron} := 1.32 \cdot T \quad H_{iron}(B_{iron}) = 119 \frac{A}{m}.$$

MATLAB also has a similar interpolation function, called *spline*, to approximate the B-H curve.

EXAMPLE 6.5: Demonstration of the use of *B*(*H*) equations

Consider the case when the iron core of a small transformer has a small gap due to manufacturing inaccuracy. The no-load or magnetizing current of the transformer is measured. The designer worries about iron saturation. The objective of this exercise is the calculation of the magnetic flux density in the iron. Figure 6.11 depicts the arrangement of the transformer.

The transformer geometrical data are:

$$a := 2 \text{ cm} \quad h := 7 \text{ cm} \quad w := 8 \text{ cm}$$
$$L_{gap} := 1 \text{ mm} \quad A_{gap} := 4 \text{ cm}^2 \quad A_{iron} := A_{gap}.$$

Referring to Figure 6.11 and neglecting the small gap thickness (L_{gap}), the magnetic path length is:

$$L_{iron} := 2 \cdot (h - a + w - a) \quad L_{iron} = 22 \cdot \text{cm}.$$

The coil current and number of primary coil turns are given as:

$$I_c := 5 \text{ A} \quad N_p := 150 \quad \mu_o := 4 \cdot \pi \cdot 10^{-7} \frac{H}{m}.$$

To apply Ampere's circuital law ($\oint H \, d\ell = I$), we separate the contributions from the gap and iron core sections to obtain the following equation:

$$I_c \cdot N_p = H_{gap} \cdot 2 \cdot L_{gap} + H_{iron}(B_{iron}) \cdot L_{iron}. \tag{6.39}$$

The iron core and gap fluxes (Φ) are identical, which permits the calculation of the flux density in each region using Equation (6.15):

$$B_{iron} = \frac{\Phi}{A_{iron}} \quad B_{gap} = \frac{\Phi}{A_{gap}}.$$

The field strength in the gap region is:

$$H_{gap} = \frac{B_{gap}}{\mu_o} \quad H_{gap} = \frac{\Phi}{\mu_o \cdot A_{gap}}.$$

These expressions are substituted into Equation (6.39). The result is:

$$I_c \cdot N_p = \frac{B_{gap}}{\mu_o} \cdot 2 \cdot L_{gap} + H_{iron}(B_{iron}) \cdot L_{iron}$$

$$I_c \cdot N_p = \frac{\Phi}{\mu_o \cdot A_{gap}} \cdot 2 \cdot L_{gap} + H_{iron}\left(\frac{\Phi}{A_{iron}}\right) \cdot L_{iron}.$$

The flux is the unknown in this equation. The flux can be calculated using the Mathcad *Find* equation solver since the spline interpolation function for the magnetization curve is available from the previous example. A guess value for the flux is:

$$\Phi := A_{gap} \cdot 1.3 \, T$$

Given

$$I_c \cdot N_p = \frac{\Phi}{\mu_o \cdot A_{gap}} \cdot 2 \cdot L_{gap} + H_{iron}\left(\frac{\Phi}{A_{iron}}\right) \cdot L_{iron}$$

$$\Phi_{iron} := Find(\Phi) \quad \Phi_{iron} = 1.86 \times 10^{-4} \, Wb.$$

Using Equation (6.15), the magnetic flux density is:

$$B_{iron} := \frac{\Phi_{iron}}{A_{iron}} \quad B_{iron} = 0.464 \, T.$$

MAGNETIC CIRCUITS

The flux density is less than 1 T; this indicates that the iron is not saturated (see Fig. 6.6). The magnetic field strength in the gap is much larger than that in the iron core:

$$H_{iron}(B_{iron}) = 49.6 \, \frac{A}{m} \quad H_{gap} := \frac{B_{iron}}{\mu_o} = 3.70 \times 10^5 \, \frac{A}{m}.$$

EXAMPLE 6.6: Magnetic circuit with parallel paths

The silicon steel iron core of a small transformer has air gaps due to manufacturing errors, as illustrated in Figure 6.12. These gaps produce an uneven flux distribution in the E-type iron core when the primary winding of the transformer is supplied by the rated voltage. The uneven distribution can produce saturation and excessive magnetizing current.

The objective of this study is the determination of the magnetizing current and flux density in each transformer leg. Figure 6.12 views the iron core of the small transformer in which the primary and secondary coils (not shown) are placed on the middle limb.

The transformer geometrical and constant data are:

$$a := 2 \, cm \quad b := 4 \, cm \quad L_{gap1} := 0.2 \, mm \quad \mu_o := 4 \cdot \pi \cdot 10^{-7} \, \frac{H}{m}$$

$$h := 8 \, cm \quad w := 15 \, cm \quad L_{gap2} := 0.1 \, mm \quad L_{gap3} := 1 \, mm.$$

The transformer circuit data are:

$$N_p := 250 \quad N_s := 500 \quad V_p := 120 \, V \quad V_s := 240 \, V.$$

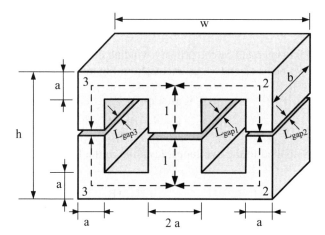

Figure 6.12. Iron core of a small transformer with air gaps and parallel magnetic paths.

Referring to Figure 6.12, the length of the magnetic paths and the iron core cross sections are:

$$L_1 := h - a - L_{gap1} = 5.98 \cdot cm \qquad\qquad A_1 := 2a \cdot b = 16.00 \cdot cm^2$$

$$L_2 := 2\left(\frac{w}{2} - \frac{a}{2}\right) + (h-a) - L_{gap2} = 19.0 \cdot cm \qquad A_2 := a \cdot b = 8.00 \cdot cm^2$$

$$L_3 := 2\left(\frac{w}{2} - \frac{a}{2}\right) + (h-a) - L_{gap3} = 18.90 \cdot cm \qquad A_3 := A_2.$$

The maximum flux created by the primary winding in the center leg (path 1) is calculated from the induced voltage using Equation (6.25) for a frequency of 50 Hz:

$$\omega := 2 \cdot \pi \cdot 50 \text{ Hz} \qquad \Phi_1 := \frac{\sqrt{2} \cdot V_p}{N_p \cdot \omega} = 2.16 \times 10^{-3} \text{ Wb}.$$

The flux density in path 1 obtained using Equation (6.15) is:

$$B_1 := \frac{\Phi_1}{A_1} \qquad B_1 = 1.35 \text{ T}.$$

The magnetic field intensity in gap 1 is found from Equation (6.13):

$$H_{gap\_1} := \frac{B_1}{\mu_o} \qquad H_{gap\_1} = 1.07 \times 10^6 \frac{A}{m}.$$

Finding the field strength in the iron core of path 1 is made simpler using the B-H curve spline interpolation for silicon steel, which is given in Figure 6.13:

$$H_1 := H_{iron}(B_1) \qquad H_1 = 135 \frac{A}{m}.$$

The magnetic field generated by the primary winding coil, which is located in path 1, is divided into two paths (2 and 3), as annotated in Figure 6.12. The flux equation is:

$$\Phi_1 = \Phi_2 + \Phi_3$$
$$B_1 \cdot A_1 = B_2 \cdot A_2 + B_3 \cdot A_3.$$

Ampere's law is applied to the two loops, resulting in the following two equations:

$$I_p \cdot N_p = H_1 \cdot L_1 + H_{gap\_1} \cdot L_{gap1} + H_{iron}(B_2) \cdot L_2 + \frac{B_2}{\mu_o} \cdot L_{gap2}$$

$$I_p \cdot N_p = H_1 \cdot L_1 + H_{gap\_1} \cdot L_{gap1} + H_{iron}(B_3) \cdot L_3 + \frac{B_3}{\mu_o} \cdot L_{gap3}.$$

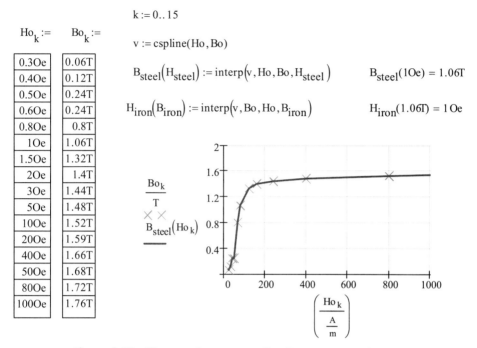

Figure 6.13. Silicon steel B-H curve spline fit equations and graph.

The unknown magnetizing current I_p and the flux densities, B_1 and B_2, are calculated using the Mathcad *Find* equation solver, which is explained in more detail in Appendix A. The guess values are:

$$I_p := 7 \text{ A} \quad B_3 := 0.5 \text{ T} \quad B_2 := 0.4 \text{ T}.$$

The equation solver is employed with appropriate constraints on the primary current and the field densities via:

Given $\quad I_p > 0 \text{ A} \quad B_2 > 0.0 \text{ T} \quad B_3 > 0.0 \text{ T}$

$$B_1 \cdot A_1 = B_2 \cdot A_2 + B_3 \cdot A_3$$

$$I_p \cdot N_p = H_1 \cdot L_1 + H_{gap\_1} \cdot L_{gap1} + H_{iron}(B_2) \cdot L_2 + \frac{B_2}{\mu_o} \cdot L_{gap2}$$

$$I_p \cdot N_p = H_1 \cdot L_1 + H_{gap\_1} \cdot L_{gap1} + H_{iron}(B_3) \cdot L_3 + \frac{B_3}{\mu_o} \cdot L_{gap3}$$

$$\begin{pmatrix} I_p \\ B_2 \\ B_3 \end{pmatrix} := \text{Find}(I_p, B_2, B_3)$$

$I_p = 4.22 \text{ A} \quad B_2 = 1.67 \text{ T} \quad B_3 = 1.03 \text{ T}.$

The results indicate that due to the uneven gap distribution, the magnetic field density in path 2 increases significantly and it is in the saturation region (see Fig. 6.6). Simultaneously, the flux density decreases from 1.35 to 1.01 T in path 3.

This example demonstrates that the B-H curve equations permit easy evaluation of a magnetic circuit with an iron core and parallel paths.

6.2. MAGNETIC AND ELECTRIC FIELD GENERATED FORCES

A charge placed in an electric field is subjected to a force, which may move the charge. Similarly, a force is generated if a charge moves in a magnetic field. These two experimentally observed phenomena can be described by the Lorentz force equations.

6.2.1. Electric Field-Generated Force

The force vector due to an electric field is the product of the charge and the electric field strength. The electric force equation is:

$$\vec{F}_e = q\vec{E}, \qquad (6.40)$$

where

q is the charge of the particle in coulombs;
$\vec{E}$ is the electric field vector in volts per meter; and
$\vec{F}_e$ is the electric field-generated force vector in newtons (N).

Figure 6.14 depicts the electric field-generated force on a positive and a negative charge. If the charge is positive, the electric field moves the charge along the field lines. Conversely, a negative charge moves against the field lines. Two oppositely charged electrodes generate the electric field in the diagram. This electric field can be realized by two parallel direct current (dc) conductors of opposite polarity.

A practical application of the electric field-generated force, called an electrostatic force, is dust removal. Air containing dust particles is blown between two charged electrodes. The force, generated by the high electric field, drives the dust particles to

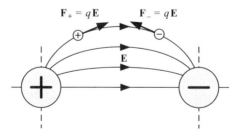

Figure 6.14. Electric field-generated force on a positive charge and a negative charge.

MAGNETIC AND ELECTRIC FIELD GENERATED FORCES

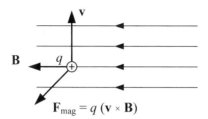

Figure 6.15. Magnetic field-generated force on a positively charged particle.

the electrodes and clears the air. This is the basis for electrostatic precipitators found in some coal-fired power plants.

6.2.2. Magnetic Field-Generated Force

A force is generated if a charge with a speed of v moves in a magnetic field. Figure 6.15 demonstrates the magnetic field-generated force direction. The positively charged particle moves with a velocity of v perpendicular to the homogeneous magnetic field. The magnetic flux density is B. The interaction between the magnetic field and the moving particle generates a force, which is normal to the plane determined by the B field vector and the velocity vector. In other words, the force is perpendicular to both the magnetic field vector and the motion of the particle.

The force vector can be calculated as the cross product of the velocity and magnetic flux density vectors multiplied by the charge. The Lorentz magnetic force equation is:

$$\vec{F}_{mag} = q(\vec{v} \times \vec{B}), \tag{6.41}$$

where

- $\vec{v}$ is the velocity of the positively charged particle in meters per second;
- $\vec{B}$ is the magnetic flux density in tesla; and
- $\vec{F}_{mag}$ is the magnetic field-generated force in newtons.

The direction of the force can be determined by the right-hand rule, which is illustrated in Figure 6.16. The force is perpendicular from the palm of the right hand if the thumb aims toward the motion direction of the positively charged particle and the index finger points with the direction of the magnetic field. This orthogonality is indicated by the cross product (×) in Equation (6.41).

A practical application of this rule is when a current-carrying conductor is placed in a magnetic field, as shown in Figure 6.17. The generated force is perpendicular to both the conductor and the magnetic field. The current can be expressed as the product of the number of charges per unit length and their speed. This modifies the magnetic force equation to:

$$\vec{F}_{mag} = I\vec{\ell} \times \vec{B}, \tag{6.42}$$

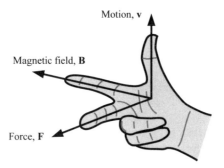

Figure 6.16. Right-hand rule for the force (F) direction on a positively charged particle of velocity **v** moving in a magnetic field (**B**).

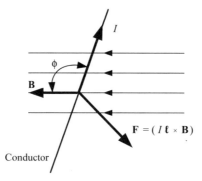

Figure 6.17. Magnetic field-produced force on a current-carrying conductor.

where

ℓ is the conductor length in meters; and
I is the current in amperes.

The magnitude of the magnetic force is:

$$F_{mag} = B\ell I \sin(\phi), \qquad (6.43)$$

where ϕ is the smaller angle between the conductor and the magnetic field. If the conductor moves in the field, the cutting of the magnetic field lines induces a voltage in the conductor. The induced voltage is:

$$V = B\ell v, \qquad (6.44)$$

where v is the conductor speed perpendicular to the field. This is an alternative formulation of Maxwell's second equation. In practically all cases of engineering importance, B, ℓ, and v are mutually perpendicular by design.

MAGNETIC AND ELECTRIC FIELD GENERATED FORCES

EXAMPLE 6.7: Calculation of electrical motor torque

Equation (6.43) can be used to calculate the torque in an electric motor. Figure 6.18 presents a simplified electric motor. Two poles generate the magnetic field. The N–S stator poles produce a more or less homogeneous magnetic field in the air gap, because a cylindrical rotor, made out of high-permeability silicon iron, is placed between the poles. The high permeability of the rotor assures that the magnetic field is normal to the rotor cylindrical surface. The cylindrical rotor has a single coil placed in two slots: **a** and **b**. DC current flows in this coil.

The current and magnetic field directions are perpendicular. The interaction of the magnetic field and the current generates a force in both halves of the conductor coil. The direction of the force is annotated in Figure 6.18. The latter diagram depicts the upper (+) conductor and displays the current, magnetic field, and force vectors in three dimensions. The angle between the current and magnetic field is 90°. In the upper (+) conductor, the current flows into the paper; in the lower (−) conductor, the current flows out from the page. According to the right-hand rule, two tangential forces are generated. Figure 6.18a reveals that the two forces produce a counterclockwise torque in the rotor. The high permeabilities of the stator and rotor assure that the magnetic flux lines are perpendicular to both surfaces, that is, the flux is normal to the rotor and stator surfaces in the air gap where the current-carrying conductors travel.

The calculation of the force and torque is demonstrated with a numerical example. The coil length and the distance between the conductors are:

$$L_{mot} := 1.2 \text{ m} \quad D_{mot} := 30 \text{ cm}.$$

In this example, the single conductor is replaced by a coil that has multiple turns of a wire such that the total length of the coiled conductor is the product of the number of turns and the length of a single turn. The numbers of turns in the coil and the coil current are:

$$N_o := 20 \quad I_c := 700 \text{ A}.$$

The magnetic flux density is $B := 1.2$ T.

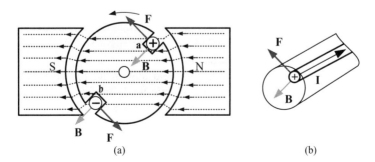

Figure 6.18. Force generation in a simplified motor.

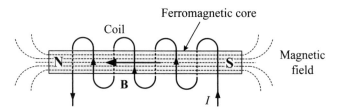

Figure 6.19. A coil-generated magnetic field.

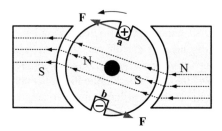

Figure 6.20. Rotor coil-generated magnetic field.

Using Equation (6.43) and the fact that $\phi = 90°$, the force generated by the coil is:

$$F_{mag} := I_c \cdot N_o \cdot L_{mot} \cdot B \quad F_{mag} = 20.16 \text{ kN}.$$

From mechanics, the torque generated by the two conductors is the product of the force and the rotor diameter:

$$T_c := F_{mag} \cdot D_{mot} \quad T_c = 6048 \text{ Nm}.$$

The force generation can be explained by analyzing a coil (*solenoid*) generated magnetic field. Figure 6.19 depicts the magnetic field produced by a coil surrounding a ferromagnetic core. It can be seen that the coil creates a south (S) and a north (N) pole.

In the case of the simplified motor, the coil on the rotor generates a magnetic field, as illustrated in Figure 6.20. The result is the formation of an N pole and an S pole on the rotor. When the stator is excited, the S pole on the stator attracts the N pole on the rotor, which results in a force that turns the rotor.

An alternative explanation is magnetic field alignment. Figure 6.21 shows two ferromagnetic material rods placed in a magnetic field. The magnetic field generates forces, which try to align the rods. The alignment shortens the flux lines and simultaneously reduces the system energy.

Figure 6.22 overlays the rotor coil and the stator-generated magnetic fields. If both the stator and the rotor create a magnetic field, forces are produced that turn the rotor to align the magnetic fields of the rotor and stator. The diagram also indicates that the

MAGNETIC AND ELECTRIC FIELD GENERATED FORCES

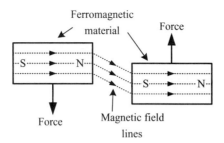

Figure 6.21. Magnetic field alignment.

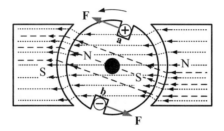

Figure 6.22. Stator- and rotor-generated fields.

rotation stops when the two fields are aligned. In a real dc motor, a switch (the commutator) reverses the coil current direction to maintain the rotation.

A typical example of this type of operation is the *stepper motor*, which is designed to turn the rotor by a precise angle. Common applications of stepper motors include machine tools, and X–Y or pen-driven plotter control. Often the stepper motors are digitally controlled. The stepper motor has a variable reactance or permanent magnet rotor and dc current excited stator with poles. The number of poles on the stator and rotor differs, and frequently the motor is built with three- or four-phase stator windings.

The stepper (or step) motor is controlled by a pulse train of short duration dc pulses. Each pulse moves the rotor by one step-angle forward or backward. A pulse train with specific number of pulses is switched to the motor, thereby turning the rotor with an angle equal to the product of the step-angle and the number of pulses. Typical step angles are 15°, 7.5°, 5°, and 2.5° per pulse. The movement direction is adjusted by the polarity of the pulse train.

The operational concept is explained using a simple stepper motor with three dc current excited stator poles and a dual-pole permanent magnet or ferromagnetic steel rotor. The step angle of this motor, drawn in Figure 6.23, is 60°. Supplying Coil 1 with positive current makes Pole 1 a north (N) pole. This pole attracts the ferromagnetic or permanent magnet south (S) pole of the rotor, which aligns with Pole 1, as seen in Figure 6.23a. If Coil 1 is deenergized and Coil 2 is supplied with negative current, Pole

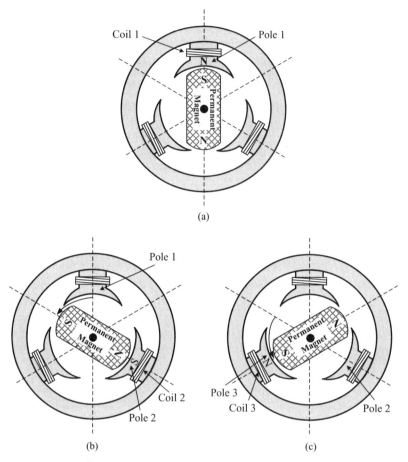

Figure 6.23. Elementary stepper motor concept.

2 becomes an S pole. The newly generated force spins the rotor counterclockwise 60°, as demonstrated in Figure 6.23b. The subsequent energizing of Coil 3, with positive current, makes Pole 3 an N pole. This force turns the rotor counterclockwise an additional 60° as depicted in Figure 6.23c. The rotation direction can be changed by the polarity of the current: positive polarity produces a north pole, and negative polarity a south pole.

Figure 6.24 pictures the stator of a step motor that has 12 poles. Each stator pole can be energized by an external source through a printed circuit terminal. The rotor has four permanent magnets, forming four—N, S, N, S—magnetic poles. The proper energization of the four stator poles (e.g., pole 1 is S, pole 4 is N, pole 7 is S, pole 10 is N) generates forces because the N and S poles on the stator and rotor attract each other. These forces align the stator- and rotor-generated magnetic fields. At this instant, the adjacent stator poles (2, 5, 8, and 11) are energized, which maintains the rotating force.

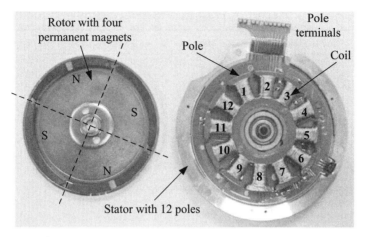

Figure 6.24. Twelve-pole step motor with four permanent magnets.

Figure 6.25. Electromechanical system.

The sequential switching of the appropriate stator poles results in constant rotation speed. The sequential switching requires monitoring the rotor position.

6.3. ELECTROMECHANICAL SYSTEM

In an electromechanical system, both an electrical and a mechanical system are coupled through an electric or magnetic field, as illustrated in Figure 6.25. An energy balance for the overall system can be stated as

$$dW_f = dW_e + dW_m, \qquad (6.45)$$

where W_f is the energy stored in the field, and W_e and W_m are the electrical and mechanical energy inputs, respectively. The differential electric energy input is the power, $P_e(t)$, delivered during a dt period:

$$dW_e = P_e(t)dt = i(t)v(t)dt. \qquad (6.46)$$

Besides field energy storage, some equipment configurations may include potential or kinetic energy storage within spring and moving components, respectively. Using a motor as an example, the input is electric energy and the output is mechanical energy, while the stator and rotor windings hold magnetic energy and the rotor mass stores

TABLE 6.1. Capacitor and Inductor Relations

Relation	Capacitor	Inductor
I–V	$i(t) = \dfrac{d(Cv(t))}{dt}$	$v(t) = L\dfrac{di(t)}{dt}$
Power	$p(t) = Cv(t)\dfrac{dv(t)}{dt}$	$p(t) = Li(t)\dfrac{di(t)}{dt}$
Energy	$w(t) = \tfrac{1}{2}Cv(t)^2$	$w(t) = \tfrac{1}{2}Li(t)^2$

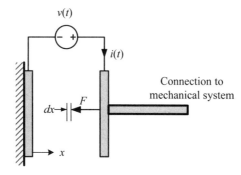

Figure 6.26. Capacitor-based electromechanical system.

kinetic energy. If the rotor speed remains constant, then there is no change in stored kinetic energy. In a motor, there are frictional losses and resistive heating. If the electrical losses (e.g., due to ohmic heating) and the mechanical losses (e.g., resulting from friction) are neglected, then this electromechanical system becomes a *lossless system*.

Two general electromechanical systems are analyzed in the following subsections: (1) a capacitor with electric field-based energy storage and (2) an inductor with magnetic field energy storage. In anticipation of these analyses, general relationships for both the inductor and capacitor are provided in Table 6.1.

6.3.1. Electric Field

The energy balance is first applied to an electromechanical system with an electric field stored in a capacitor. Here, one rigid plate and another movable plate form the capacitor. Often in circuit analyses, the capacitance is assumed constant. However, if the capacitor forms part of an electromechanical device such as that depicted in Figure 6.26, then the capacitance varies according to the separation distance (x) between the parallel plates:

$$C = \frac{\varepsilon A}{x}, \qquad (6.47)$$

where ε is the permittivity of the dielectric (air) and A is the plate surface area.

The current through the variable capacitance may be written as:

$$i(t) = \frac{dq}{dt} = \frac{d}{dt}(Cv) = C\frac{\partial v}{\partial t} + v\frac{\partial C}{\partial t}. \quad (6.48)$$

Utilizing Equation (6.45), the resulting basic energy balance for this system is:

$$dW_f = dW_e + dW_m,$$
$$d\left(\frac{1}{2}Cv^2\right) = ivdt + Fdx, \quad (6.49)$$

where F is an external force applied to the movable plate. As the movable plate travels a distance dx, the separation between plates decreases, thereby increasing the capacitance and correspondingly increasing the stored energy in the electric field.

Substituting Equation (6.48) into Equation (6.49), and expanding the first term using the derivative chain rule yields:

$$Cvdv + \frac{1}{2}v^2 dC = vCdv + v^2 dC + Fdx,$$

which upon simplification reduces to:

$$F = -\frac{1}{2}v^2 \frac{dC}{dx}. \quad (6.50)$$

Substituting the derivative of Equation (6.47) into the above expression gives:

$$F = -\frac{1}{2}v^2 \frac{d}{dx}\left(\frac{\varepsilon A}{x}\right) = \frac{1}{2}v^2 \frac{\varepsilon A}{x^2}. \quad (6.51)$$

This result indicates that an external (input) force must be applied to close the gap between the plates. Noteworthy is that the force is independent of the voltage polarity.

6.3.2. Magnetic Field

Figure 6.27 shows a magnetic actuator that draws a bar composed of magnetic material inward to the closed position. The energy transfer from the electric system to the inductor representing the magnetic circuit is:

$$dW_e = ivdt = i\left(L\frac{di}{dt}\right)dt = Lidi. \quad (6.52)$$

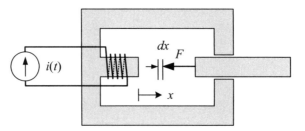

Figure 6.27. Magnet-based electromechanical system.

In the case of a magnetic field, the stored energy is:

$$W_f = \frac{1}{2}Li^2. \qquad (6.53)$$

Taking the differential of the magnetic field energy gives:

$$dW_f = d\left(\frac{1}{2}Li^2\right) = Lidi + \frac{1}{2}i^2 dL. \qquad (6.54)$$

Substituting the relations for dW_e and dW_f into the basic energy balance of Equation (6.45) yields:

$$\begin{aligned} dW_f &= dW_e + dW_m, \\ Lidi + \frac{1}{2}i^2 dL &= Lidi - Fdx. \end{aligned} \qquad (6.55)$$

The force term is negative since the mechanical system is not supplying a force (as in the earlier capacitor-based electric field device) but rather an output force is being applied to the mechanical component. Simplifying the earlier equation provides:

$$F = -\frac{i^2}{2}\frac{dL}{dx}. \qquad (6.56)$$

Using (a) the flux linkage definition of $\lambda = iL = N\Phi$, (b) that the magnetic flux is $\Phi = BA = \mu_0 HA$, and (c) Ampere's circuital law of $Hx = Ni$, the inductance as a function of the air gap length is:

$$L(x) = \frac{N\Phi}{i} = \frac{N\mu_0 HA}{Hx/N} = \frac{\mu_0 N^2 A}{x}. \qquad (6.57)$$

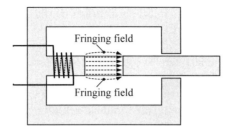

Figure 6.28. Fringing fields outside the gap cross-sectional area.

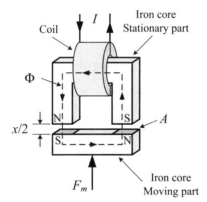

Figure 6.29. Magnetic switch.

This assumes that the stray magnetic field is neglected since *fringing fields*, such as that depicted in Figure 6.28, would increase the effective cross-sectional area. Therefore, the force can be rewritten as:

$$F = -\frac{i^2}{2}\frac{dL}{dx} = -\frac{(Hx/N)^2}{2}\left(\frac{-\mu_0 N^2 A}{x^2}\right) = \frac{\mu_0 H^2 A}{2} = \frac{B^2 A}{2\mu_0}. \quad (6.58)$$

In addition to determining the force exerted by the magnetic field, the methodology presented here can also be utilized to determine the external force required to move the actuator from the closed to the open position.

6.4. CALCULATION OF ELECTROMAGNETIC FORCES

An electromagnet converts electric energy to mechanical energy. The concept is demonstrated in the singly excited system of Figure 6.29. The diagram shows an iron core, made out of high permeability material, and a coil. The iron core has a stationary part and a moving part. When a current (I) supplies the coil, a magnetic flux (Φ) is generated. The flux forms poles in the iron. When the flux exits from the iron, an N pole is generated; when the flux enters the iron, an S pole is produced. The opposing poles attract each other. This results in a force (F_m) that pulls the moving part of the iron core toward the stationary part. The generated force reduces the total gap (x), which minimizes the system energy content.

Designating each of the two gaps as having an $x/2$ separation is judicious since the total gap is simply x. Consequently, the formulae describing this situation are identical to those presented for the magnetic field-based electromechanical system of Section 6.3.2. An alternative approach, however, is presented in the succeeding paragraphs and leads to the same final expression previously given in Equation (6.58). In particular, if the current to the coil is constant, then $di = 0$ and, according to Equation (6.52), $dW_e = Lidi = 0$. This means that in this situation, the basic energy balance of Equation (6.45) may be reduced to:

$$dW_f = dW_m. \quad (6.59)$$

A magnetic field has an energy content that was derived in Section 6.1.3 and given in Table 6.1. The magnetic field energy content is described by the following expressions:

$$W_{mag} = \frac{1}{2}I^2 L = \frac{1}{2}\Phi NI, \quad (6.60)$$

where I is the coil current; L is the coil inductance; Φ is the coil-generated magnetic flux; and N is the number of turns in the coil.

The previous formula can be simplified if we assume that the iron permeability is very high and the flux density is below the saturation level. This implies that the iron core is neglected and that Φ, B, and H are calculated using the air gap data only, that is, $\Phi = BA$, $H = NI/x$, and $B = \mu_0 H$. Figure 6.29 indicates that the two air gaps are connected in series since the same flux passes through each gap. This means that the total air gap length is x, but the gap area is simply A. Substituting these relations into Equation (6.60) gives the alternate expression for the magnetic field energy content:

$$W_{mag} = \frac{1}{2}\Phi NI = \frac{1}{2}(BA)(Hx) = \frac{1}{2}\mu_0 H^2 Ax = \frac{\mu_0 (NI)^2 A}{2x}, \quad (6.61)$$

where B is the flux density in the gap; H is the magnetic field strength in both gaps; A is the cross-sectional area of each air gap; and x is the total length of the air gap.

The magnetic field generates a force, which tries to close the gap. This force represents a mechanical energy (W_m), which is described by:

$$dW_m = -F_m dx. \quad (6.62)$$

CALCULATION OF ELECTROMAGNETIC FORCES 349

Consequently, the force tries to generate a movement. The change in magnetic energy produces an equivalent change in the mechanical energy:

$$dW_{mag} = dW_m = -F_m dx. \quad (6.63)$$

Rearrangement of this expression yields the force generated by the magnetic field:

$$F_m = -\frac{dW_{mag}}{dx}. \quad (6.64)$$

The magnetic field-generated force is equal to the derivative of the magnetic field energy. Substituting Equation (6.61) into the previous relation gives:

$$F_m = -\frac{dW_{mag}}{dx} = -\frac{d}{dx}\frac{\mu_0(NI)^2 A}{2x} = \frac{\mu_0(NI)^2 A}{2x^2}, \quad (6.65)$$

which is identical to Equation (6.58) since:

$$F_m = \frac{\mu_0(NI)^2 A}{2x^2} = \frac{\mu_0(Hx)^2 A}{2x^2} = \frac{B^2 A}{2\mu_0}. \quad (6.66)$$

Similar to the capacitive device, this force is independent of the current direction.

In the derivation of the earlier expressions for F_m, we assumed that only the total gap length (x) changes when the coil is energized. However, in some electromechanical systems, the gap area changes and the gap length remains constant. Figure 6.34 shows an example when the area changes, which prohibits the use of the earlier simplified force relation.

In reality, such electromechanical systems are not lossless; the input electric energy (time-integrated power) produces magnetic energy while incurring an energy loss. The energy balance then becomes:

$$dW_{mag} - dW_{loss} = dW_e + dW_m. \quad (6.67)$$

Neglecting the mechanical and magnetic losses, the electrical loss due to the coil resistance (R_c) is $W_{loss} = I^2 R_c$. Focusing on the differential loss term (dW_{loss}), an interesting result appears; specifically:

$$dW_{loss} = d(I^2 R_c) = R_c 2I dI = 0. \quad (6.68)$$

Hence, for the case of constant current, neglecting the ohmic heating is appropriate.

EXAMPLE 6.8: Electrostatic vs. electromagnetic devices

For structures of identical field areas (A) over equal displacements (x), we compare the electric supply requirements to achieve the same force. A dc voltage, V_e, and a dc current, I_m, power the electric and magnetic field devices, respectively. Comparing Equation (6.51) and Equation (6.65) for the electric and magnetic field device-produced forces, respectively, provides:

$$F_e \lessgtr F_m,$$
$$\frac{\varepsilon_0 A V_e^2}{2x^2} \lessgtr \frac{\mu_0 A (NI_m)^2}{2x^2},$$
$$\varepsilon_0 V_e^2 \lessgtr \mu_0 (NI_m)^2.$$

Recalling that μ_0 is five orders of magnitude larger than ε_0 implies that the applied voltage (V_e) for a capacitor-based actuator, such as that depicted in Figure 6.26, would need to be noticeably larger than the input current (I_m) for a magnetic switch, like that of Figure 6.29, to produce the same force. Moreover, the voltage is limited by the electric field breakdown, which is a much more severe restriction than that placed on the current due to possible overheating of the coil winding. The breakdown electric field in air is around 3 kV/mm. Hence, the force generated by an electrostatic device is significantly less than that achievable with an electromagnetic apparatus, which explains the reason that the latter enjoys more widespread adoption.

EXAMPLE 6.9: Magnetic force calculation

The calculation of the magnetic force is demonstrated using a numerical example for the case of constant gap area. Figure 6.30 exhibits an electromagnet that can be used as an actuator. The dimensions and operational data of this magnetic switch are:

$$h := 3 \text{ cm} \quad w := 4 \text{ cm} \quad d := 1 \text{ cm} \quad c := 1 \text{ cm}$$
$$N_o := 800 \quad I_o := 4 \text{ A} \quad x := 6 \text{ mm}.$$

The magnetic field and the magnetic flux density generated by the coil current are calculated using Ampere's circuital law, if the effect of *the iron core is neglected*. It is

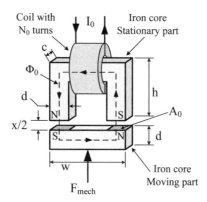

Figure 6.30. Electromagnet-based actuator.

CALCULATION OF ELECTROMAGNETIC FORCES   351

assumed that the permeability of the iron core is very high. This implies that the following equations derived are valid only if the flux density is below the saturation level. The magnetic field strength and flux density as functions of the applied current and total gap length are:

$$H_o(x, I_o) := \frac{I_o \cdot N_o}{x} \qquad H_o(x, I_o) = 5.33 \times 10^5 \frac{A}{m}$$

$$B_o(x, I_o) := \mu_0 \cdot H_0(x, I_o) \qquad B_o(x, I_o) = 0.67 \text{ T}.$$

The magnetic flux is calculated by multiplying the flux density with the cross-sectional area of the iron core ($\Phi = B A$). The result is:

$$A_o := d \cdot c = 1 \text{ cm}^2$$

$$\Phi_o(x, I_o) := B_o(x, I_o) \cdot A_o \qquad \Phi_o(x, I_o) = 6.70 \times 10^{-5} \cdot \text{Wb}.$$

Using Equation (6.60), the magnetic field energy is:

$$W_{mag}(x, I_o) := \frac{1}{2} \cdot \Phi_o(x, I_o) \cdot N_o \cdot I_o \qquad W_{mag}(x, I_o) = 0.107 \text{ J}.$$

From Equation (6.64), the magnetic field-generated force is the derivative of the magnetic energy:

$$F_{mech}(x, I_o) := -\frac{d}{dx} W_{mag}(x, I_o) \qquad F_{mech}(x, I_o) = 17.87 \text{ N}.$$

Mathcad is capable of internally combining and differentiating the earlier expressions to reach the previously mentioned numerical value for F_{mech}. A hand calculation would follow the process exemplified in Equation (6.65). As a check of the numerical result for F_{mech}, two forms of Equation (6.66) are employed:

$$F_{ma} := \frac{\mu_0 \cdot (N_o \cdot I_o)^2 \cdot A_o}{2 \cdot x^2} = 17.87 \text{ N} \qquad F_{ma} = 1.822 \text{ kgf}$$

$$F_{mb}(x, I_o) := \frac{B_o(x, I_o)^2 \cdot A_o}{2 \cdot \mu_0} \qquad F_{mb}(x, I_o) = 17.87 \text{ N}.$$

From a practical point of view, the variation of the force with the distance is important. Figure 6.31 and Figure 6.32 plot the force and flux density variations as a function of the gap distance x. Figure 6.31 reveals that the force increases as the gap length is reduced. According to Equation (6.66) derived earlier, the force becomes infinity at a zero gap length. However, Figure 6.32 shows that the flux density also increases with the reduction of the gap length. This produces saturation when the flux density is larger than about 1.5 T. The saturation limits the force. The derivation previously presented

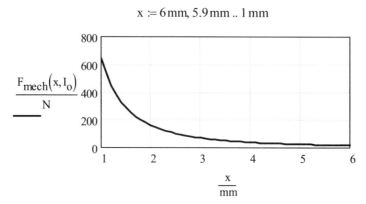

Figure 6.31. Magnetic field-generated force versus gap distance.

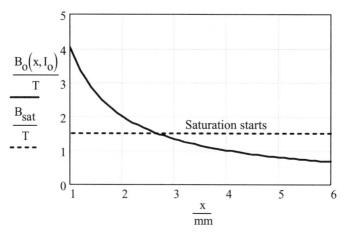

Figure 6.32. Magnetic field-generated flux density versus gap distance. Saturation occurs for $B_0 > 1.5$ T.

neglected the effect of the iron core. This is a good approximation if the core is not saturating. However, the equations derived for the force are not valid if the iron is saturated, which will occur as the gap closes. The saturation limits B, which in turn restricts the force.

The previous analyses have neglected the effects of magnetic hysteresis, including the energy losses.

6.5. APPLICATIONS

This section presents several example applications of electromechanical devices, including actuators, transducers, and sensors.

APPLICATIONS 353

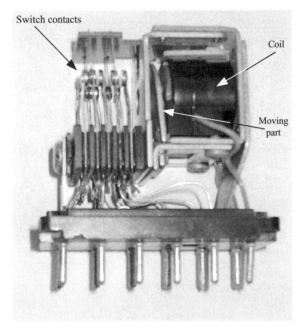

Figure 6.33. Magnetic switch.

6.5.1. Actuators

Industry frequently uses magnetic actuators. A typical example is the magnetic switch pictured in Figure 6.33. Energizing the coil with dc or ac current attracts the moving part and closes the gap. Regardless of the current direction, the magnetic field is created and it endeavors to bring the N and S poles together. The moving part correspondingly closes or opens the switch contacts through mechanical linkage.

The automotive industry uses several magnetic actuators in every vehicle. Typical applications are the magnetic actuators used to open and close car door locks or operate windshield wipers. The actuator concept is depicted in Figure 6.34. This actuator consists of a solenoid coil, a cylindrical plunger, and magnetic yokes that close the magnetic circuit. The plunger and the yokes are made of ferromagnetic material. Energizing the coil pulls the plunger up to close the upper air gap (x).

EXAMPLE 6.10: Magnetic actuator

This example demonstrates the operation of a typical magnetic actuator. The lower gap of Figure 6.34 is constant regardless of the plunger position. For simplification of the calculations, we assume that the lower gap is zero. Thus, this actuator effectively has only one gap, with variable area. The dimensions needed for the force calculation are:

$$d := 3 \, cm \quad g := 4 \, mm \quad r := 1.5 \, cm.$$

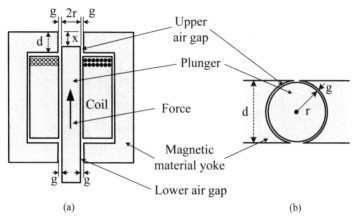

Figure 6.34. Magnetic actuator. (a) Cutaway side view; (b) top view.

The current and the number of turns in the cylindrical solenoid coil are:

$$I_c := 5 \text{ A} \quad N_c := 800.$$

The area of the gap between the yokes and plunger is variable. The gap area increases as the plunger moves upward. The relation for the variable area is:

$$A_{gap}(x) := 2\pi \cdot r \cdot (d-x) \quad A_{gap}(1 \text{ cm}) = 18.85 \text{ cm}^2.$$

The magnetic flux strength is calculated using Ampere's circuital law:

$$H_{gap} := \frac{I_c \cdot N_c}{g} \quad H_{gap} = 1000 \frac{\text{kA}}{\text{m}}.$$

The magnetic flux density is:

$$B_{gap} := \mu_0 \cdot H_{gap} \quad B_{gap} = 1.26 \text{ T}.$$

The flux density is not variable because the gap thickness (g) is constant; only the gap area varies as the plunger moves. The magnetic flux is:

$$\Phi_{gap}(x) := B_{gap} \cdot A_{gap}(x) \quad \Phi_{gap}(1 \text{ cm}) = 2.37 \times 10^{-3} \text{ Wb}.$$

Using Equation (6.60), the magnetic energy in the gap is:

$$W_{gap}(x) := \frac{1}{2} \cdot N_c \cdot \Phi_{gap}(x) \cdot I_c \quad W_{gap}(1 \text{ cm}) = 4.74 \text{ J}.$$

APPLICATIONS

The substitution of the magnetic flux equation into the previous expression yields the following formula:

$$W_{gap}(x) := \frac{1}{2} \cdot N_c \cdot (B_{gap} \cdot A_{gap}(x)) \cdot I_c \quad W_{gap}(1\,\text{cm}) = 4.74\,\text{J}.$$

Finally, the relation describing the variable area is substituted in the previous expression. This results in:

$$W_{gap}(x) := \frac{1}{2} \cdot N_c \cdot B_{gap} \cdot [2\pi \cdot r \cdot (d-x)] \cdot I_c \quad W_{gap}(1\,\text{cm}) = 4.74\,\text{J}.$$

This formula reveals that the magnetic field energy is a linear function of the plunger movement. This relation also reveals that the magnetic field energy is concentrated in the air gap itself rather than in the magnetic material.

Like the derivation of Equation (6.64), assuming constant current means that the force is:

$$F_g(x) := -\frac{d}{dx} W_{gap}(x) \quad F_g(1\,\text{cm}) = 236.9\,\text{N}.$$

Taking the derivative of the energy with respect to x provides the following force relation:

$$\text{Force} := \frac{1}{2} \cdot N_c \cdot B_{gap} \cdot (2\pi \cdot r) \cdot I_c \quad \text{Force} = 236.9\,\text{N}.$$

The derivation of the energy equation results in a constant force that is independent from the plunger movement and position. This is expected because the gap and the flux density are constant as the plunger moves. The force equation can be further simplified to:

$$\text{Force} := \mu_0 \cdot I_c^2 \cdot N_c^2 \cdot \frac{\pi \cdot r}{g} \quad \text{Force} = 236.9\,\text{N}$$

$$\text{Force} = 24.15 \cdot \text{kgf}.$$

It must be pointed out that use of the simplified force expression, Equation (6.58), results in erroneous values, because that gap was variable and that area was constant. If the force is computed using Equation (6.58), the result would be:

$$F_{gap}(x) := \frac{B_{gap}^2 \cdot A_{gap}(x)}{2 \cdot \mu_0} \quad F_{gap}(1\,\text{cm}) = 1184\,\text{N}.$$

This formula (incorrectly) yields a much larger force.

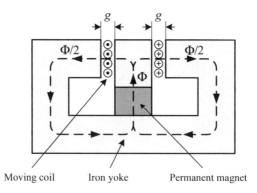

Figure 6.35. Permanent magnet-excited moving-coil transducer.

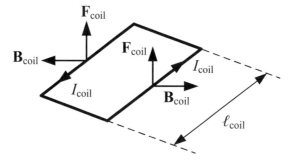

Figure 6.36. Magnetic force generation on the transducer coil of Figure 6.35.

6.5.2. Transducers

Industry uses transducers that have a movable coil and fixed iron core. In most cases, the iron core is excited by a permanent magnet. The most frequently used transducers are the loudspeakers and the drive moving the arm of a computer fixed disk. Before the popularity of digital instruments, some voltmeters and ammeters used moving coil transducers.

Figure 6.35 illustrates the concept of a permanent magnet-excited transducer. The diagram shows the permanent magnet inserted in the middle leg of the iron core. The moving coil is placed in the gap. It can be seen that the magnetic field is perpendicular to the coil conductors. The force is generated by the interaction of the cross magnetic field and the coil current, as depicted in Figure 6.36. The drawing demonstrates that force depends on the current direction. If the coil has a small mass, the coil movement follows the current variation. Typically, the coil in a loudspeaker follows the current up to 20 kHz. The force equation, if the magnetic field is homogeneous in the gap, is:

$$F_{coil} = 2\ell_{coil} I_{coil} N_{coil} B_{gap}, \tag{6.69}$$

where

B_{gap} is the magnetic flux density in the gap;
ℓ_{coil} is the length of the coil, as annotated in Figure 6.36;
I_{coil} is the coil current; and
N_{coil} is the number of turns.

6.5.2.1. Permanent Magnet-Generated Field. Permanent magnets are frequently used to generate constant magnetic field instead of a dc current-excited coil. The advantage is the maintenance-free operation, which is particularly advantageous if the magnet is placed on a rotor or moving object. The disadvantage is that the permanent magnet can provide a magnetic field for only relatively small equipment because the size of the magnet is limited by the available manufacturing technology and the high cost of the magnetic material. In addition, the excitation coil current produces some demagnetization, which can result in instability. The detailed discussion of permanent magnet design is beyond the objectives of this book, but we present the basic concept in the succeeding paragraphs.

Permanent magnets are described by their B-H curve. Figure 6.37 graphs the B-H curves of the most frequently used magnetic materials. The materials can be divided into four groups:

1. Alnico 5, 8 magnets: iron alloys with **al**uminum, **ni**ckel and **co**balt;
2. Ferrite (ceramic) magnets: iron oxide, barium, or strontium carbonate;
3. Rare earth magnets: neodymium–iron–boron; and
4. Samarium cobalt magnet.

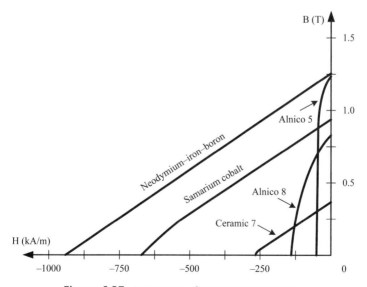

Figure 6.37. B-H curves of permanent magnets.

For the analysis of a magnetic system with a permanent magnet, the curves can be approximated by equations using curve-fitting techniques. Simpler yet, the neodymium–iron–boron and samarium cobalt magnet material B-H curves are straight lines that can be described by linear equations (i.e., $B = b + (b/a)H$), as given later. The magnetic flux density is constrained to nonnegative values using the Mathcad "if" statement:

(a) Neodymium–iron–boron (Nd–Fe–B):

$$a_{NdFeB} := 940 \frac{kA}{m} \quad b_{NdFeB} := 1.25 \text{ T}$$

$$B_{NdFeB}(H_{pmag}) := b_{NdFeB} + \frac{b_{NdFeB}}{a_{NdFeB}} \cdot H_{pmag}$$

$$B_{NdFeB}(H_{pmag}) := \text{if}(B_{NdFeB}(H_{pmag}) < 0, 0, B_{NdFeB}(H_{pmag})).$$

(b) Samarium cobalt (SmCo):

$$a_{SmCo} := 675 \frac{kA}{m} \quad b_{SmCo} := 0.94 \text{ T}$$

$$B_{SmCo}(H_{pmag}) := b_{SmCo} + \frac{b_{SmCo}}{a_{SmCo}} \cdot H_{pmag}$$

$$B_{SmCo}(H_{pmag}) := \text{if}(B_{SmCo}(H_{pmag}) < 0, 0, B_{SmCo}(H_{pmag})).$$

Figure 6.37 discloses that the SmCo equation is actually only an approximation, since the B-H curve for SmCo is not entirely linear.

Typically, the permanent magnet is inserted in an iron core, as shown in Figure 6.38. If the permeability of the iron core is high, Ampere's law can calculate the magnetic field strength in the gap:

$$H_{gap} \cdot gap + H_{pmag} \cdot L_{pmag} = 0 \quad H_{gap} = \frac{-H_{pmag} \cdot L_{pmag}}{gap}.$$

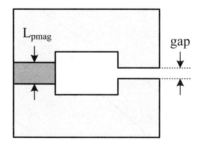

Figure 6.38. Permanent magnet-excited iron core with gap.

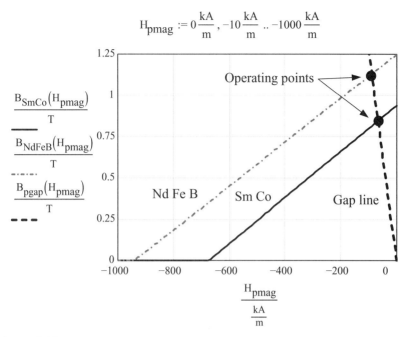

Figure 6.39. Intersection of the gap line and the B-H curves of SmCo and Nd-Fe-B.

The magnetic flux density in the gap is:

$$B_{gap} = \mu_0 \cdot H_{gap} = -\mu_0 \cdot \frac{H_{pmag} \cdot L_{pmag}}{gap}.$$

The steady-state operating point occurs when the flux density in the air gap is equal to the flux density in the permanent magnet. The previous expression describes a straight line. The equation can be rewritten as:

$$B_{pgap}(H_{pmag}) := -\mu_0 \cdot \frac{H_{pmag} \cdot L_{pmag}}{gap}.$$

The intersection of the permanent magnet B-H curve with this straight line gives the operating point. In Figure 6.39, both the neodymium–iron–boron (Nd–Fe–B) and the samarium cobalt (SmCo) B-H curves are plotted together with the gap B-H line. This defines two possible operating points if the coil current-produced demagnetization is negligible.

The B and H at the operating point can be determined from Figure 6.39 or they can be calculated by solving the equations. This latter approach is demonstrated for

SmCo. At the intersection point, the magnetic flux density in the SmCo magnet and in the air gap is the same. This can be expressed by the following equation:

$$-\mu_0 \cdot \frac{H_{pmag} \cdot L_{pmag}}{gap} = b_{SmCo} + \frac{b_{SmCo}}{a_{SmCo}} \cdot H_{pmag}.$$

The algebraic rearrangement of the previous formula yields:

$$H_{pmag} \cdot \left(\mu_0 \cdot \frac{L_{pmag}}{gap} + \frac{b_{SmCo}}{a_{SmCo}}\right) = -b_{SmCo}.$$

The magnetic field strength is calculated from this equation. The result is:

$$H_{pmag} := \frac{-b_{SmCo}}{\left(\mu_0 \cdot \frac{L_{pmag}}{gap} + \frac{b_{SmCo}}{a_{SmCo}}\right)} \qquad H_{pmag} = -67.3 \frac{kA}{m}.$$

The flux density that corresponds to this H value is:

$$B_{pgap}(H_{pmag}) = 0.846 \text{ T}.$$

If we place a moving coil in the gap, the force versus current function can be calculated with Equation (6.69). The coil data are:

$$L_{coil} := 4 \text{ cm} \qquad N_{coil} := 20.$$

The force versus current equation is:

$$F_{coil}(I_{coil}) := 2B_{pgap}(H_{pmag}) \cdot L_{coil} \cdot N_{coil} \cdot I_{coil} \qquad F_{coil}(1 \text{ A}) = 1.35 \text{ N}.$$

The current could be a high-frequency signal that is used in a microphone or loudspeaker, or a dc signal that controls the position of the coil.

6.5.2.2. Permanent Magnet-Based Applications. Several different energy conversion techniques exist to accomplish sound wave-to-electrical signal transformation and vice versa. An example approach that employs a permanent magnet is the loudspeaker illustrated in Figure 6.40. The permanent magnet generates a constant magnetic field. The *voice coil* is used to construct an electromagnet. The speaker converts a varying current into a lateral movement of the voice coil. The voice coil is mechanically attached to the speaker cone such that movement of the coil–cone unit displaces air in front of the speaker, thereby creating pressure waves (i.e., sound). By alternating the current direction through the voice coil, the polar orientation of the electromagnet is correspondingly switched between repulsive and attractive with

APPLICATIONS

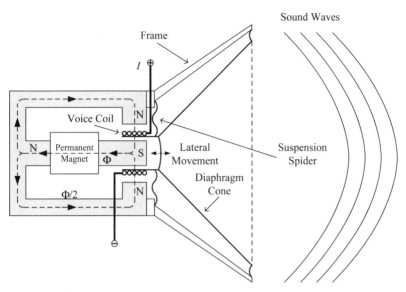

Figure 6.40. Loudspeaker employing permanent magnet.

respect to the field generated by the permanent magnet. The cross magnetic field and current interaction generates the force. This process was discussed in conjunction with Figure 6.35. The rate and displacement of the voice coil, which determine the frequency and amplitude of the sound waves, are a function of the frequency and magnitude of the coil current. A dynamic microphone can be constructed in a very similar fashion to the loudspeaker.

An alternative design of a microphone using a permanent magnet is the moving-armature microphone, which is exemplified in Figure 6.41. The diaphragm is mechanically connected to an armature using a driver rod. Sound waves create lateral movement in the diaphragm-to-armature assembly. A wire coil is wound around the armature. Compression and rarefaction waves deflect the coil-wound armature to the right and left, respectively. Such vibrations vary the magnetic flux (Φ) in the armature and thereby through the coil. When the diaphragm is at rest, the coil is midway between the N–S poles created by the permanent magnet, and although a magnetic flux is established across the air gap between the poles, no flux is created in the armature. However, when the armature is deflected, some of the flux moves through the armature. Specifically, when a compression wave deflects the armature to the right, some fraction of the magnetic flux flows from the north pole at the upper right across the reduced gap and down through the armature. Conversely, a rarefaction wave causes the armature to be deflected to the left, resulting in some of the flux from the N pole of the permanent magnet traveling up through the armature, across the reduced gap, and then into the south pole located at the upper left. Thus, the diaphragm vibrations cause a flux change in the armature which, in turn, induces a voltage in the coil. This voltage signal has an identical waveform to that of the sound waves impacting the diaphragm. Although the

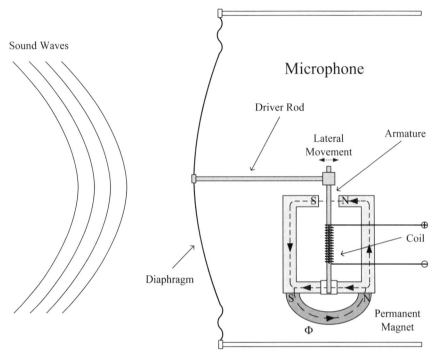

Figure 6.41. Moving-arm microphone.

moving-arm microphone has similar characteristics to the dynamic microphone in terms of sensitivity, impedance, and frequency response, the former is more shock and vibration resistant.

Another interesting application of the permanent magnet concepts presented here is the drive that controls the computer hard drive arm. Figure 6.42 shows such a fixed disk drive arm. The moving coil is shown on the right-hand side of the figure. The magnetic field is generated by permanent magnets and a vertical column constructed of ferromagnetic material assures the cross field. The coil current controls the position of the arm. A feedback loop in the control circuit ensures that the sensors at the end of the arm find the required data on the disk.

6.5.3. Permanent Magnet Motors and Generators

Another application is the permanent magnet motors and generators, which utilize permanent magnets to produce continuously rotating machines driving typically smaller, a few kilowatt-rated machines. Permanent magnets produce dc excitation flux that can be used to excite synchronous machines as well as dc motors and generators. The permanent magnet excitation simplifies the construction, reduces size, and, in some cases, lowers the cost.

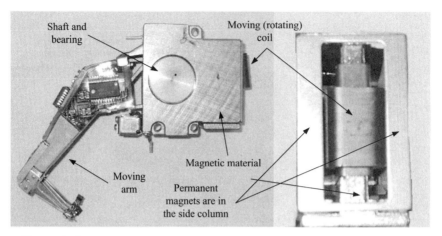

Figure 6.42. Hard drive arm with permanent magnet-assisted moving-coil drive.

A *small synchronous generator* can be built by replacing the rotor poles with permanent magnets, which provide magnetic pole flux without dc excitation. The permanent magnets can be grain-oriented ferrite magnets or higher quality and more expensive neodymium and samarium–cobalt or alnico magnets. The permanent magnet-excited generator operates the same way as the synchronous generators that will be described in Chapter 8. The major operational problem is the demagnetization of the permanent magnets in the event of a short circuit. Using permanent magnets reduces the generator size and significantly simplifies the construction by eliminating the need to supply the rotor with dc current.

Permanent magnet excitation can also be used for small synchronous motors. The startup of these motors requires squirrel cage short-circuited winding, which starts the motor as an induction motor. The permanent magnets are placed inside the rotor cage.

Permanent magnet dc motors can be built by replacing the stator dc-excited poles with permanent magnets, which provide constant excitation flux. This motor with constant excitation functions as a shunt motor. Figure 6.43 depicts a possible arrangement for the permanent magnet shunt motor. Strong permanent magnets are incorporated both in the main poles and in the commutating poles. The rotor winding and the commutator remain the same as other dc motors that will be analyzed in Chapter 10. This machine is frequently used in control circuits because of its linear speed versus torque characteristics. Figure 6.44 exhibits the components of a dc permanent magnet motor used to propel radio-controlled model airplanes.

6.5.3.1. Brushless dc Permanent Magnet Motor.

The weakest component in a dc permanent magnet motor is the commutator, which interrupts the current, when it transfers current from one coil to the other. The commutator is replaced by an electronic switching circuit in the brushless dc permanent magnet motors. Figure 6.24

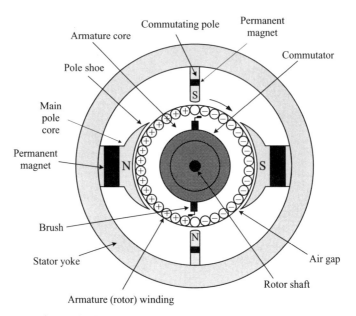

Figure 6.43. DC shunt motor with permanent magnets.

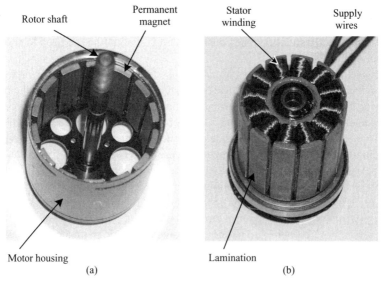

Figure 6.44. Components of brushless dc permanent magnet motor for which the rotor revolves around the stator. (a) Fourteen-pole rotor. (b) Twelve-pole stator.

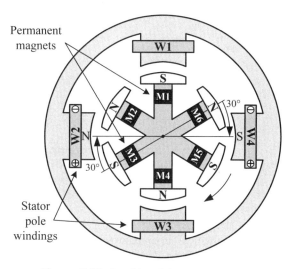

Figure 6.45. Brushless DC motor concept.

presented a brushless dc motor with four (4) permanent magnet rotor poles and twelve (12) stator poles.

Figure 6.45 portrays the operation concept of the brushless dc motor. The rotor of this motor has six poles equipped with permanent magnets M1–M6. The stator has only four poles excited by coils W1–W4. The rotor and energized stator poles are deliberately not aligned in the diagram to permit the force generation. Energizing stator series-connected coils W2 and W4 produces an N pole and an S pole, respectively, toward the rotor. Supplying these two windings generates forces between the stator and rotor poles. In particular, the S pole of M3 is attracted to the N pole of W2 which also repels the N pole of M2; similar interactions simultaneously occur between the poles of M6, W4, and M5. Since series-connected windings W1 and W3 are not energized at this time, the forces turn the motor clockwise 30°. Continued sequential switching of the dc voltage to the coils rotates the motor. This motor is supplied by square-shaped dc pulses. The motor speed is regulated by the frequency of the supply square wave voltages. An electronic circuit produces the square waveforms. Because of the high cost of powerful electronic excitation circuits, these types of motors are mostly used for low power drives. It is important to notice that this type of motor must be started and stopped by gradually increasing or reducing the frequency of the supply voltage, that is, the motor speed must be ramped up or down.

6.5.4. Microelectromechanical Systems

Some transducers are incorporated as part of a sensor, and many of these sensors are electromechanical devices that do not utilize magnetic fields. Most instrumentation today renders the measurand into an electrical signal for transmission and processing. As such, at the heart of these sensors are energy conversion devices. Sensing

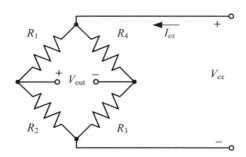

Figure 6.46. Four-arm Wheatstone bridge network.

technologies include, but are not restricted to, those based on PE and piezoresistive materials that convert mechanical stress and strain into electrical outputs. More recently, these devices have been manufactured in miniaturized form as microelectromechanical systems (MEMS) sensors using semiconductor fabrication techniques.

Piezoresistance is a material property for which a change in electrical resistance occurs when stress (e.g., movement, pressure, or vibration) is applied to the substance. Most materials exhibit this effect, but it is highly exploitable in semiconductor materials such as silicon and germanium. Piezoresistive sensors convert stress into resistance change, which is subsequently transformed to a voltage output (V_{out}), typically using a Wheatstone bridge circuit as drawn in Figure 6.46. Network analysis shows that the normal bridge output due to an excitation voltage (V_{ex}) is:

$$V_{out} = V_{ex} \frac{R_2 R_4 - R_1 R_3}{(R_1 + R_2)(R_3 + R_4)}. \tag{6.70}$$

Piezoresistive MEMS accelerometers are utilized in automobile airbag deployment systems. In the accelerometer of Figure 6.47, two piezoresistors are located on opposite sides of the hinge and they traverse the air gap between the fixed core and movable inertial mass. Acceleration-induced force on the movable central mass produces tensile and compressive stress, respectively, in the two piezoresistors. Another application of piezoresistive sensing technology is the MEMS pressure transducer depicted in Figure 6.48. Pressure applied to the top face of the pressure sensor causes a resistance change in the four embedded silicon piezoresistors.

Piezoelectricity is the property possessed by some materials to become electrically charged when subjected to a mechanical stress. Small, lightweight transducers based on the PE properties of certain oxides are in wide use for accelerometers, pressure transducers, and acoustic emission sensors. Acoustic emission sensors are used for process monitoring in applications such as electrical machinery condition monitoring and incipient failure detection. Commonly produced PE ceramics for transducer applications include lead titanate (PT) and lead zirconate titanate (PZT). Sensor manufacturers stack multiple PE discs in a transducer, as seen in Figure 6.49, to achieve larger output signals. The discs are mechanically coupled in series which produces more charge, thereby increasing sensitivity. The discs are electrically connected in parallel

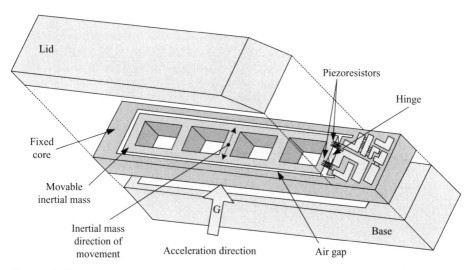

Figure 6.47. Exploded view of an MEMS accelerometer; the "G" arrow denotes the axis sensitive to acceleration.

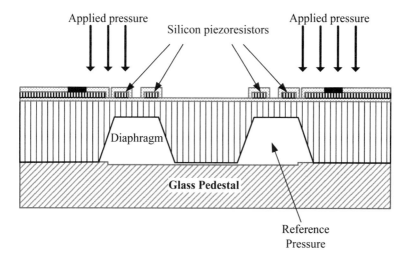

Figure 6.48. Side view of MEMS pressure transducer structure.

such that the overall transducer capacitance is larger, and thus the sensor impedance is lower.

An expanding application area for PE materials is micropower supplies. The PE material can be incorporated into a MEMS structure to produce power for an external circuit or for utilization in the device itself (i.e., a self-powered autonomous system). A common implementation goal is scavenging energy from movement and vibration

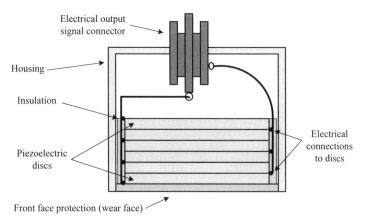

Figure 6.49. Acoustic emission sensor with stacked PE discs.

in the immediate environment. Example applications for so-called *energy harvesting* include medical implants, embedded sensors, and mobile electronics. MEMS devices generate power in the microwatts (μW), whereas larger devices (e.g., within footwear and backpacks) are designed to produce milliwatt (mW) levels. A typical motion-driven design employs a mass–spring system that resonates with the device frame vibration. Besides PE-based construction, microgenerators utilizing electromagnetic and electrostatic methods have been utilized to convert the movement to electricity.

6.6. SUMMARY

This chapter has described energy transfer and conversion techniques using magnetic fields. The theory presented here is fundamental to major equipment such as transformers, motors, and generators. The next chapter, which addresses transformers, is successively followed by chapters on synchronous machines, induction machines, and dc machines. As an example application, present electric and hybrid vehicle manufacturers utilize all three types of machines, including brushless dc motors, ac induction machines, and permanent magnet synchronous machines.

6.7. EXERCISES

1. Describe Ampere's law.
2. Define the magnetic flux, flux density, and magnetic field strength. Give the equations that describe the relations among these quantities.
3. Define the inductance of a coil and present equations for the inductance calculation.

4. Discuss the air gap effect on a magnetic circuit.
5. Describe the B-H curve. What is saturation?
6. What is the hysteresis loss? How does hysteresis loss relate to the B-H curves?
7. Explain the concept of electrical field-generated mechanical force. Present the equations. Is there any practical use of this force?
8. Explain the Lorentz magnetic force equation.
9. Calculate the torque in a simplified electric motor. A dc source produces a uniform radial magnetic field. The rotor winding is supplied by a constant dc current. Draw a sketch and present the equations.
10. What is the magnetic field alignment generated force? Give an example.
11. Explain the concept of a magnetic actuator or switch. Identify the components and draw a sketch.
12. Present the equations for the magnetic field energy.
13. Derive an equation for the magnetic field-generated force using the energy equations.
14. What is a transducer and what is it used for?
15. Explain the force generation in a transducer.
16. Derive a permanent magnet-generated flux in a simple magnetic circuit containing a gap and an iron core. Neglect the effect of the iron core.
17. Draw a sketch showing a magnetic transducer, excited by a permanent magnet.

6.8. PROBLEMS

Problem 6.1

In Example 6.5, the magnetic path length was calculated by assuming that the gap thickness was negligible. Recompute the magnetic fluxes, field strengths, and flux densities if the gap thickness is not neglected. What is the percentage error in the results if the gap thickness is neglected?

Problem 6.2

A circular cross-section, toroidal core, made from quality sheet steel (see Fig. 6.1) is used as an inductor. Its outer and inner diameters are 25 and 15 cm, respectively. An excitation winding with 280 turns is placed around the core, as illustrated in Figure 6.50. A magnetic flux density (B) of 1.25 T is required along the mean diameter of the torus. Determine: (a) the magnetic field strength (H) from the B–H curve, and the relative permeability μ_r, (b) the magnetic flux Φ and excitation current through the coil needed to produce the magnetic flux, and (c) the inductance of the coil. (d) Assuming that an air gap of 8 mm is cut across the torus, calculate the excitation current through the coil needed to produce the same magnetic flux.

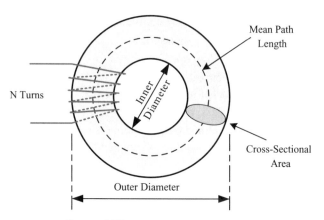

Figure 6.50. Toroidal magnetic circuit.

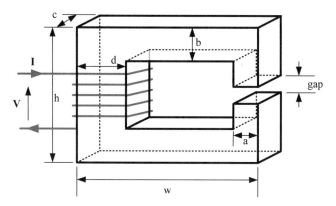

Figure 6.51. Magnetic circuit.

Problem 6.3

The inductor drawn in Figure 6.51 has a coil with 300 turns around the cast iron core. The coil is supplied by a magnetizing current of 4.5 A. The dimensions of the system are:

a = 4.5 cm	b = 4.5 cm	c = 5.5 cm	d = 5.5 cm
w = 23 cm	h = 28 cm	gap = 2.5 mm	

Neglecting the iron core, determine: (a) the total magnetic flux Φ in webers, (b) the H and B across the gap, and (c) the coil inductance L. If the actual permeability of the iron core is considered, calculate: (d) the current through the coil needed to maintain the same B value from part (b), and (e) the coil inductance L_b.

Problem 6.4

A dc bus carries 15 kA during short circuit. The spacing between the positive and negative buses is 25 cm. The distance between the supporting insulators on each bus is 3 m. (a) Draw a sketch showing the conductor arrangement and the direction of the forces. (b) Calculate the magnetic field strength and flux density at each conductor site. (c) Calculate the force between the conductors.

Problem 6.5

A three-phase high-voltage bus carries 6 kA during short circuit. Post insulators support the bus of each phase, and the distance between the support points is 5 m. The spacing between the adjacent, horizontally arranged phases with an A–B–C orientation is 0.5 m. The current amplitudes are the same in each phase, but the phase shift between A and B is −120°, and between A and C is −240°. (a) Draw a sketch showing the physical arrangement and the direction of the forces. (b) Calculate the force between each phase pairing. (c) Which is the most stressed phase?

Problem 6.6

Audio transformer linearity is assured by an air gap inserted in the iron core, as depicted in Figure 6.52. The center leg width (a) and the depth (b) of the core are identically 5 cm. The gap dimension (g) is 1 mm. The 60-Hz primary voltage is 15 V, and the number of coil turns is 40. Calculate the force developed between the two parts of the iron core.

Problem 6.7

The magnetic actuator illustrated in Figure 6.53 is used to operate car door locks. When the 800-turn solenoid is energized by a 2 A dc current, the plunger moves upward and closes the door lock. The variable gap (g_1) ranges from 0.5 to 2.5 cm, and the other gap (g_2) is fixed at 1 mm. The cross-sectional area (A_1) perpendicular to g_1 is 25 cm^2,

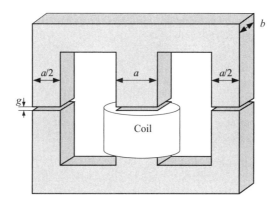

Figure 6.52. Audio transformer with gap.

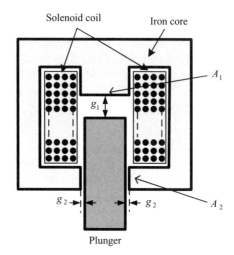

Figure 6.53. Magnetic actuator.

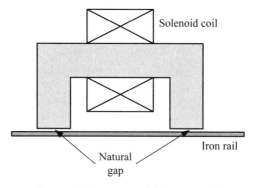

Figure 6.54. Magnet lifting iron rail.

and each A_2 area is $A_1/2$. Neglecting the iron core, calculate the flux density and the force variation as the gap is reduced.

Problem 6.8

A large magnet is used to lift a 200-kg_f iron rail, as shown in Figure 6.54. A solenoid coil produces the magnetic field. At each of the two magnet-to-rail contact locations, the average gap is 3 mm and the contact area is 50 cm². The 200-turn coil has a resistance of 4 Ω. Calculate the coil voltage needed to lift the rail. Neglect the iron core.

Problem 6.9

Figure 6.55 presents the cross-section of a loudspeaker. The iron core is excited by a 4-A dc current through a 300-turn coil, which produces the cross-magnetic field in the

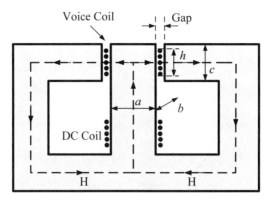

Figure 6.55. Loudspeaker cross section.

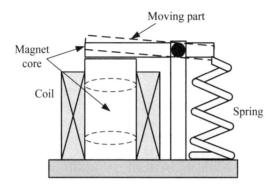

Figure 6.56. Magnetically operated relay.

gaps. The 30-turn loudspeaker voice coil has a height (h) of 4 cm and is placed in the gap. The magnet core dimensions are: $a = b = 2$ cm, where b is the core thickness, $c = 5$ cm, and a gap size of 4 mm. Determine: (a) the flux density in the gap and (b) the force on the coil versus the coil current and plot the results.

Problem 6.10

Figure 6.56 shows a magnetic relay, which is used to switch motors on and off. The relay is operated by a 300-turn solenoid with a cylindrical iron core, which has a diameter of 1.5 cm. The hinged moving part of the relay is normally kept open by a spring. The excitation of the solenoid attracts the hinged moving part of the relay, which closes the relay contacts (not shown). The average gap between the cylindrical iron core and the moving part ranges from 1 to 4 mm. A force of 0.5 kg_f is required to initiate the closing when the gap is at its maximum (4 mm). (a) Calculate the current required to operate the relay. (b) Neglecting the effect of the spring, plot the force versus gap. What is the maximum force? (c) Plot the magnetic flux density versus gap. Does saturation occur?

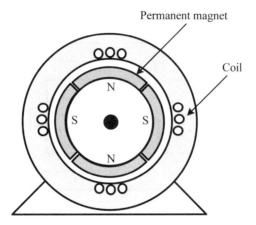

Figure 6.57. Permanent magnet generator.

Problem 6.11

The dc excitation coil for the loudspeaker shown in Figure 6.55 is replaced by a samarium–cobalt permanent magnet. The samarium–cobalt has a linear B-H curve: at $H = 0$, $B = 0.87$ T, and at $B = 0$, $H = -640$ kA/m. (a) Calculate the length of the permanent magnet that generates the same magnetic flux density (0.377 T) as the dc coil in Problem 6.9. Neglect the iron core, consider only the gaps. (b) Calculate the voice coil current that produces a force of 0.5 N on the voice coil.

Problem 6.12

The magnet shown in Figure 6.54 lifts iron rails. Due to the nonuniform cross section of the rail, the natural gap between the N–S poles and the rail is 2 and 5 mm, respectively, at the left and right sides of the magnet. Calculate the forces between the poles and the rail if the coil current is 20 A. Use the magnet data given in Problem 6.8.

Problem 6.13

Four samarium–cobalt permanent magnets, placed on the rotor, are used to produce a magnetic field in the 5 mm stator-to-rotor gap of the permanent magnet generator shown in Figure 6.57. The samarium–cobalt has a linear B-H curve: at $H = 0$, $B = 0.87$ T, and at $B = 0$, $H = -640$ kA/m. The permanent magnet thickness is 4 cm. The generator diameter is 10 cm, its length is 20 cm, and its speed is 3600 rpm. (a) Calculate the magnetic flux density in the gap while neglecting the iron core. (b) Calculate the voltage induced in a coil placed on the stator if the number of turns is 200.

7

TRANSFORMERS

At the beginning of the 20th century, the distance for energy transmission was limited by the transmission line voltage drop. The invention of the transformer revolutionized electric energy transmission. A *transformer* can increase or decrease the voltage, which correspondingly results in the decrease or increase of the current.

The generator voltage is limited to 20–26 kV. This voltage is suitable to transport 5–10 MW of power for a distance of 5–10 miles (8–16 km). The development of the transformer permitted the increase of the transmission voltage to a few hundred kilovolts while simultaneously reducing the current, consequently allowing the transport of large amounts of energy over a few hundred miles.

An important safety feature is that the transformer insulates the primary and secondary electric circuits and permits the separate grounding of the circuits, which prevents high-voltage-caused accidents.

This chapter describes the construction of the transformer, develops the transformer equivalent circuit, and presents operation analyses for both single- and three-phase transformers.

Electrical Energy Conversion and Transport: An Interactive Computer-Based Approach, Second Edition. George G. Karady and Keith E. Holbert.
© 2013 Institute of Electrical and Electronics Engineers, Inc. Published 2013 by John Wiley & Sons, Inc.

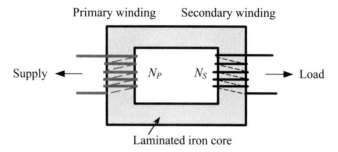

Figure 7.1. Basic components of a single-phase transformer.

7.1. CONSTRUCTION

Figure 7.1 depicts the components of a single-phase transformer. The transformer has a laminated iron core, and primary and secondary windings. The *primary winding* is supplied by alternating current (ac) voltage, which generates an ac magnetic flux in the iron core. This magnetic flux generates ac voltage in the secondary winding. The loading of the *secondary winding* produces current in both the secondary and primary windings. The magnetic field transports the electric energy from the primary to the secondary. The ratios of voltages and currents in the primary and secondary sides depend on the *turns ratio* (T), which is the ratio of the number of turns in the primary (N_P) and secondary (N_S) coils:

$$T = \frac{N_P}{N_S}. \quad (7.1)$$

In a real transformer, the primary and secondary windings are placed on the same leg of the core, as illustrated in Figure 7.2. In this diagram, the secondary winding surrounds the primary winding. This arrangement improves magnetic coupling and reduces the leakage flux of the windings. Historically, the primary winding terminals are marked H_1 and H_2, and the secondary terminals are denoted as X_1 and X_2. The practical application of this nomenclature is the determination of when an ac voltage connected to H_1 and H_2 is in phase with the voltage appearing between X_1 and X_2.

Each transformer coil can be wound either clockwise or counterclockwise. The winding directions determine the voltage polarity. Figure 7.3 demonstrates the effect of winding directions on polarity. In Figure 7.3a, both windings are wound in the same direction; consequently, the secondary and primary voltages are in phase as shown in the graph. Figure 7.3b presents a change of polarity; in this case, the secondary winding is wound in the opposite direction of the primary, which results in the primary and secondary voltages being reversed (180° out of phase). The transformer polarity is important when transformers are connected in parallel. The parallel connection requires that the respective secondary and primary voltages must be equal. Two transformers can be connected in parallel if, in addition to the turns ratio, the primary and secondary winding polarities are the same in both transformers. When two windings are connected

CONSTRUCTION

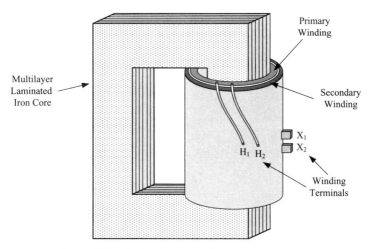

Figure 7.2. Single-phase transformer arrangement.

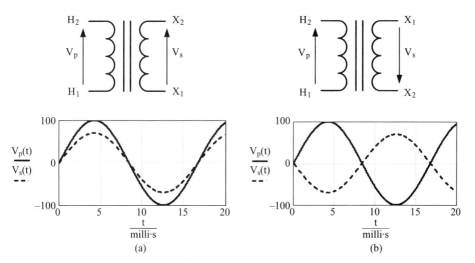

Figure 7.3. Polarity for transformer.

in series, the voltages are added if the winding polarities are the same, and the voltages are subtracted if the polarities oppose each other.

The laminated iron *core* is built with thin silicon iron sheets. The silicon steel sheets are stacked together to form the core. The sheet material consists of around 3% silicon. A thin varnish or film layer placed on one side insulates the sheets. This lamination is needed to reduce the eddy currents in the core. The magnetic flux that generates voltage in the windings produces undesired circulating current in the core. The lamination blocks and breaks up the current path.

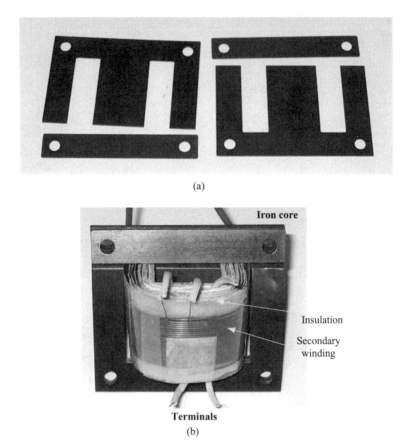

Figure 7.4. Small transformer construction. (a) Lamination; (b) iron core winding.

Figure 7.4 pictures a small transformer, the E- and I-shaped lamination and the winding. The E and I sheets are stacked and insulated screws press the laminated iron core together. The winding construction is seen in Figure 7.5. The winding holder is made out of plastic or, for cheaper units, paper is used. The layered winding is built with enamel-insulated wires. Thin paper or plastic insulation is placed between the layers. Similar but thicker insulation containing several layers separates the primary and secondary windings.

A larger dry-type three- phase distribution transformer is shown in Figure 7.6. The transformer windings are vacuum dried and dipped in epoxy resin insulation.

Large transformers are oil cooled and insulated. The assembled transformer is dried and placed in a steel tank. The tank is filled, under vacuum, with heated transformer oil. The ends of the windings are connected to bushings. The oil circulates through outside radiators for cooling. For increased transformer capacity, fans may blow air through the radiators and the insulating oil may be pumped through the radiators. Figure 7.7 exemplifies an oil-insulated and cooled transformer. This transformer is equipped

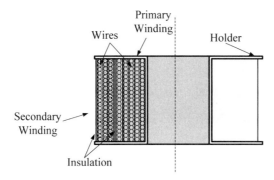

Figure 7.5. Winding construction.

Figure 7.6. Dry-type three-phase distribution transformer (courtesy of Siemens, Erlangen, Germany).

with cooling radiators, which are cooled by forced ventilation. Cooling fans are installed under the radiators. The oil temperature and pressure are monitored to assess transformer performance.

Bushings connect the windings to the electrical system. The transformer bushing is a hollow porcelain insulator. The insulator has a corrugated surface to increase leakage distance and flashover voltage in bad weather. The insulator is filled with transformer oil, which provides insulation. An aluminum or copper bar is threaded

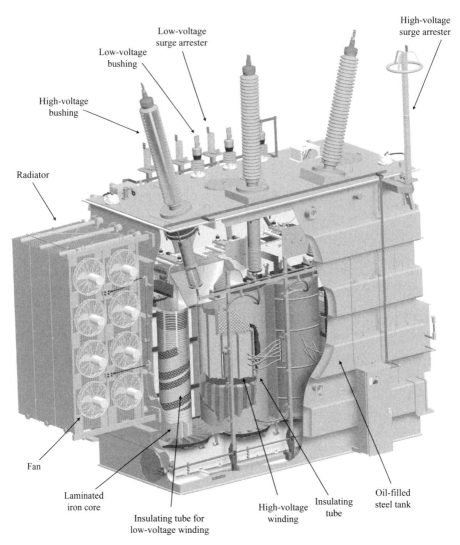

Figure 7.7. Oil-insulated and cooled transformer (© SPX Transformer Solutions, Inc., all rights reserved).

through the porcelain to connect the ends of the transformer winding to the outside bus. Figure 7.8 is a drawing of a high-voltage capacitive-type transformer bushing in which the insulation is enforced by a built-in capacitor that is located in the mounting flange and grounded sleeve region. The capacitor is constructed with conductive layers of aluminum foil with high dielectric paper wound around the conductor and into the bushing core to produce a capacitor that distributes the electric field uniformly at the flange area. Figure 7.9 pictures a large oil-cooled transformer with its low-voltage and high-voltage bushings.

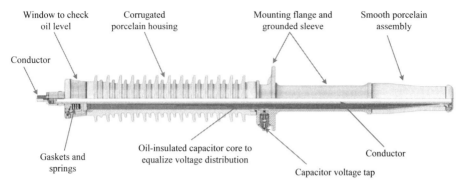

Figure 7.8. Porcelain high-voltage transformer capacitive bushing (courtesy of Hubbell Power Systems, Inc.).

Figure 7.9. Large oil-cooled high-voltage transformer.

7.2. SINGLE-PHASE TRANSFORMERS

The single-phase transformer has a laminated iron core with two windings. In a core-type transformer, the windings are placed on different legs of the core, as indicated in Figure 7.10a. This transformer is made of two L-shaped, or a U-shaped and an I-shaped magnetic sheets. The shell-type transformer is made with E- and I-shaped core stacks and both windings are on the middle leg of the core, as diagrammed in Figure 7.10b. The shell-type transformer has less leakage inductance than the core type.

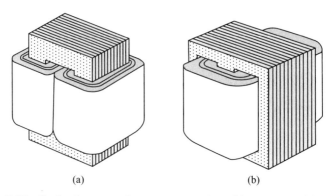

Figure 7.10. Single-phase transformer constructions. (a) Core type; (b) shell type.

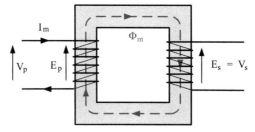

Figure 7.11. Current, voltages, and flux in an unloaded ideal transformer.

7.2.1. Ideal Transformer

The winding of an *ideal transformer* has no resistance, and the iron core has no losses. The transformer is supplied by an ac voltage (V_p), which drives magnetizing current through the primary winding if the secondary is not loaded. The *magnetizing current* (I_m) generates an ac flux in the iron core. This flux induces voltage in both the secondary and primary windings. The supply voltage is equal with the voltage induced in the primary winding (E_p). Figure 7.11 exhibits the magnetizing current, induced voltages, and the flux. The direction of the main flux Φ_m is determined using the right-hand rule. Specifically, the fingers of the right hand are pointed and curled in the direction of current travel through the winding, thus pointing the right thumb in the direction of the flux. In Figure 7.11, the flux is in the clockwise direction.

The induced root-mean-square (rms) voltage can be calculated using Equation (6.25). The primary and secondary induced voltages are:

$$E_p = \frac{N_p \Phi_m \omega}{\sqrt{2}} \quad \text{and} \quad E_s = \frac{N_s \Phi_m \omega}{\sqrt{2}}, \tag{7.2}$$

where

N_p is the number of turns in the primary winding;
N_s is the number of turns in the secondary winding;
Φ_m is the main flux;
f is the frequency, 60 Hz in the United States, $\omega = 2\pi f$;
E_p is the induced rms voltage in the primary winding; and
E_s is the induced rms voltage in the secondary winding.

The *turns ratio* (T) of the transformer is defined as the ratio of the turns in the primary and secondary windings. The division of the equations for the absolute value of the induced voltages produces an expression for the turns ratio:

$$T = \frac{N_p}{N_s} = \frac{E_p}{E_s}. \tag{7.3}$$

When the transformer is loaded at the secondary, the induced secondary voltage drives a current through the load. The secondary current generates a flux Φ_s, which opposes the main flux, as seen in Figure 7.12. The main flux cannot be reduced because the primary induced voltage E_p is equal with the constant supply voltage V_p. To keep the main flux constant, a primary current (I_p) starts to flow. This current generates a primary flux Φ_p, which compensates for the secondary current generated flux Φ_s. The described phenomenon is demonstrated in Figure 7.12, which displays the secondary and primary currents, and the generated fluxes.

An ideal transformer has no losses; consequently, the input and output complex powers are equal ($S_p = S_s$) if the magnetizing current is neglected (i.e., $I_m = 0$). The complex power equation permits the determination of the relation between the primary currents and induced voltages. The primary and secondary complex powers are:

$$\mathbf{E}_p \mathbf{I}_p^* = \mathbf{S}_p = \mathbf{S}_s = \mathbf{E}_s \mathbf{I}_s^*. \tag{7.4}$$

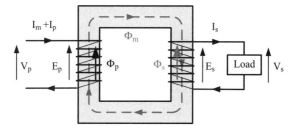

Figure 7.12. Currents and fluxes in a loaded ideal transformer.

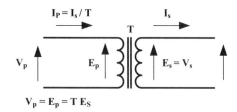

Figure 7.13. Equivalent circuit of an ideal transformer.

Considering only the absolute values or magnitudes of the currents and voltages, the rearrangement of this equation results in:

$$T = \frac{E_p}{E_s} = \frac{I_s}{I_p}. \tag{7.5}$$

Further rearrangement of the equation gives the voltage and current equations of an ideal transformer:

$$E_p = TE_s \quad \text{and} \quad I_p = \frac{I_s}{T}. \tag{7.6}$$

This equation reveals that when the transformer increases the voltage, the current decreases and vice versa. Figure 7.13 provides the equivalent circuit of an ideal transformer if the magnetizing current is neglected. It is important to note that the previous equations depend upon the current direction and voltage polarity annotated in the circuit diagram of Figure 7.13. Specifically, the primary and secondary voltage polarities are the same; however, the primary and secondary current directions are opposite. In particular, the primary current direction is into the positive terminal of the transformer winding (and obeys the passive sign convention), whereas the secondary current is exiting the transformer winding at the positive terminal (in opposition to the passive sign convention).

7.2.1.1. Transferring Impedances through a Transformer.

A practical technique associated with the ideal transformer is the transfer of impedances (and sources) from the transformer secondary side to the primary side and vice versa. Consider the network presented in Figure 7.14a. The goal is to move the secondary impedance (Z_{load}) to the primary side while simultaneously removing the transformer from the network. Essentially, a Thévenin equivalent for the ideal transformer and the load impedance is found in this analysis. To create a Thévenin equivalent of the entire right-hand side (rhs) of the network, the current–voltage relationship on the primary side must be maintained. This means that the Thévenin equivalent impedance Z_P will be such that:

$$Z_P = \frac{V_P}{I_P}. \tag{7.7}$$

SINGLE-PHASE TRANSFORMERS

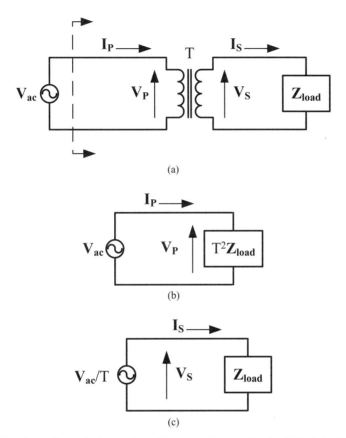

Figure 7.14. Thévenin equivalents of an ideal transformer circuit. (a) Original circuit with ideal transformer; (b) equivalent circuit when secondary impedance is transferred to primary side and ideal transformer eliminated; (c) equivalent circuit when primary source is transferred to secondary side and ideal transformer eliminated.

The relationships between the primary and secondary currents and voltages are known from the turns ratio, such that:

$$\mathbf{Z_P} = \frac{\mathbf{V_P}}{\mathbf{I_P}} = \frac{T\mathbf{V_S}}{\left(\frac{\mathbf{I_S}}{T}\right)} = T^2 \frac{\mathbf{V_S}}{\mathbf{I_S}}. \tag{7.8}$$

Finally, the secondary (load) impedance is defined as the ratio of the secondary voltage and current. Using this fact yields an equivalent impedance that can be placed on the primary side to eliminate the transformer and secondary side, as disclosed in Figure 7.14b:

$$\mathbf{Z_P} = T^2 \mathbf{Z}_{\text{load}}. \tag{7.9}$$

A similar procedure can be carried out to move the primary source to the secondary side to eliminate the transformer. The transferred primary voltage would have a phasor voltage of $\mathbf{V}_{ac}/T$ when moved to the secondary side, as indicated in Figure 7.14c.

The previously mentioned technique can be generalized to transferring any (parallel or series) impedance from the primary side to the secondary side, and vice versa, using the square of the turns ratio. The entire primary or secondary impedances need not be moved. Such an impedance transfer is not accompanied by the transformer elimination when any of the circuit elements (sources and/or impedances) are not transferred. To transfer a secondary impedance, $\mathbf{Z}_S$, to the primary, we generalize the above equation to:

$$\mathbf{Z}_P = T^2 \mathbf{Z}_S. \tag{7.10}$$

Likewise, to transfer a primary side impedance to the secondary, the same formula is employed. If the transferred impedance was originally in series with the transformer winding, then the transferred value must be placed in series on the other side; similarly, if the original impedance is in parallel with the winding, the transferred impedance must be positioned in parallel.

7.2.1.2. Reduction of Power System Losses with Transformer Use.
The transformer initiated the rapid development of electrical power networks because of its ability to change the voltage level. Typically, an electric load is supplied by a transmission line. The permitted length and the load of the line depend on the voltage level. The voltage at the customer level must be higher than 95% of the rated voltage, and the supply voltage shall not be higher than 105% of the rated voltage. This generates a constraint that the percentage voltage drop or system regulation shall be less than 10%. Simultaneously, the loss on the line should be minimized. The increased voltage level reduces the current, which correspondingly reduces both the transmission line voltage drop and losses.

The effect of voltage level is illustrated with a numerical example. Figure 7.15a depicts a transmission system where a generator supplies a load through a line with a transformer at each end. The single-phase equivalent circuit of the system is given in Figure 7.15b. The system can be simplified by assuming ideal transformers and generator. Further, the generator can be transferred to the secondary of the supply transformer (Tr 1) and the load is transferred to the primary side of the transformer (Tr 2) adjacent to the load (see Fig. 7.15c).

From Chapter 3, the voltage regulation is defined as:

$$\text{Voltage Regulation} = \frac{|\mathbf{V}_{\text{no-load}}| - |\mathbf{V}_{\text{load}}|}{|\mathbf{V}_{\text{load}}|} \times 100\%. \tag{7.11}$$

Referring to Figure 7.15c, the no-load voltage is equal to the transferred generator voltage; therefore:

SINGLE-PHASE TRANSFORMERS

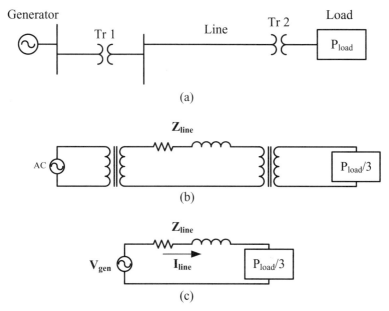

Figure 7.15. Simple transmission line between generator and load with transformers at each end of the line (a) System one-line diagram; (b) single-phase equivalent; (c) single-phase equivalent after transferring source voltage and load impedance.

$$\text{Voltage Drop} = \frac{|V_{gen}| - |V_{load}|}{|V_{load}|} \times 100\%. \tag{7.12}$$

To calculate the generator voltage, we first compute the load current using:

$$\mathbf{I}_{line} = \frac{P_{load}/3}{pf_{load} \dfrac{V_{load}}{\sqrt{3}}} e^{-j\arccos(pf_{load})}.$$

The generator line-to-neutral voltage is the sum of the load line-to-neutral voltage and the voltage drop on the line impedance:

$$\mathbf{V}_{gen} = \mathbf{V}_{load\_ln} + \mathbf{I}_{line}\mathbf{Z}_{line}.$$

The total three-phase line losses are:

$$\text{Loss} = 3|\mathbf{I}_{line}|^2 R_{line}.$$

The following Mathcad calculations use the current and voltage as a function of the variables: line length, load voltage, and load power. The objective of this example is to investigate the effect of these parameters. This is achieved by solving the equations under predetermined constraints, such as the calculation for a given condition is terminated if the regulation is 10% or more.

EXAMPLE 7.1: Transformer impact

The system operation is analyzed using a numerical example. The three-phase load is variable:

$$P_{load} := 0 \text{ W}, 100 \text{ kW} .. 20 \text{ MW} \quad pf_{load} := 0.8 \text{ lagging}.$$

The typical generator voltage is 22 kV. The transformer can increase this voltage to one of the standard transmission voltages. For this example, we select line voltages of 22, 120, 220, and 500 kV. In order to simplify the calculation, we assume that the load voltage is the rated voltage of the system. The calculation will determine the required generator voltage at the high-voltage side of the sending-end transformer.

The line impedance changes for lines at different voltage levels, but for simplicity, we assume an average impedance independent of the voltage. The assumed resistance and reactance for an unbundled line are:

$$R_{line} = 0.1 \frac{\Omega}{mi} \quad X_{line} := 0.8 \frac{\Omega}{mi}.$$

The transmission line impedance per-unit (p.u.) length is:

$$Z_{line} := R_{line} + j \cdot X_{line}.$$

The line length is variable:

$$L_{line} := 0 \text{ mi}, 1 \text{ mi} .. 150 \text{ mi}$$

For the verification of the equations, we use the following set of conditions:

$$V_{sup} := 120 \text{ kV} \quad L_{line} := 120 \text{ mi} \quad P_{load} := 20 \text{ MW} \quad V_{load} := 122 \text{ kV}.$$

Note that the 22 kV case is actually the instance of no transformers present, that is, a direct connection of the generator and load through the transmission line exists.

The single-phase line current for a lagging power factor of 0.8, three-phase load is:

$$I_{line}(P_{load}, V_{load}) := \frac{P_{load}}{\sqrt{3} \cdot pf_{load} \cdot V_{load}} \cdot e^{-j \cdot acos(pf_{load})}$$

$$|I_{line}(P_{load}, V_{load})| = 118.3 \text{ A} \quad arg(I_{line}(P_{load}, V_{load})) = -36.9 \cdot deg.$$

SINGLE-PHASE TRANSFORMERS

The required generation voltage is:

$$V_{gen\_ln}(P_{load}, L_{line}, V_{load}) := \frac{V_{load}}{\sqrt{3}} + Z_{line} \cdot L_{line} \cdot I_{line}(P_{load}, V_{load})$$

$$|V_{gen\_ln}(P_{load}, L_{line}, V_{load})| = 78.8 \cdot kV$$

$$\arg(V_{gen\_ln}(P_{load}, L_{line}, V_{load})) = 6.0 \cdot \deg.$$

To determine the load voltage for a given load power and line length, we utilize the Mathcad *root* function to solve the earlier equation for a specified supply voltage:

$$V_{load\_1}(P_{load}, V_{sup}, L_{line}) := \operatorname{root}\left(V_{gen\_ln}(P_{load}, L_{line}, V_{load}) - \frac{V_{sup}}{\sqrt{3}}, V_{load}\right).$$

The open-circuit voltage and the load voltage at 20 MW are:

$$V_{load\_1}(0\ MW, V_{sup}, L_{line}) = 120.0 \cdot kV$$

$$|V_{load\_1}(P_{load}, V_{sup}, L_{line})| = 109.0 \cdot kV.$$

The voltage regulation is:

$$\operatorname{Reg}(P_{load}, L_{line}, V_{sup}) := \frac{|V_{load\_1}(0\ MW, V_{sup}, L_{line})| - |V_{load\_1}(P_{load}, V_{sup}, L_{line})|}{|V_{load\_1}(P_{load}, V_{sup}, L_{line})|}$$

$$\operatorname{Reg}(P_{load}, L_{line}, V_{sup}) = 10.0 \cdot \%.$$

The regulation should be less than 10%; consequently, we solve for the maximum load that corresponds to a 5% regulation using:

$$P_{load\_5\%}(L_{line}, V_{sup}) := \operatorname{root}(\operatorname{Reg}(P_{load}, L_{line}, V_{sup}) - 5\%, P_{load})$$

$$P_{load\_5\%}(L_{line}, V_{sup}) = 8.62 \cdot MW.$$

Another metric is the transmission efficiency, which can be computed from the line losses. The current through the transmission line is:

$$I_{Line}(P_{load}, L_{line}, V_{sup}) := \frac{P_{load}}{\sqrt{3} \cdot pf_{load} \cdot V_{load\_1}(P_{load}, V_{sup}, L_{line})} \cdot e^{-j \cdot acos(pf_{load})}$$

$$|I_{Line}(P_{load}, L_{line}, V_{sup})| = 132.4\ A.$$

The total resistive power loss from all three phases is:

$$\operatorname{Loss}(P_{load}, L_{line}, V_{sup}) := 3 \cdot (|I_{Line}(P_{load}, L_{line}, V_{sup})|)^2 \cdot L_{line} \cdot R_{line}.$$

At the load power for achieving a 5% regulation, these losses are:

$$\text{Loss}\left(P_{\text{load\_5\%}}\left(L_{\text{line}}, V_{\text{sup}}\right), L_{\text{line}}, V_{\text{sup}}\right) = 106.6 \cdot \text{kW}.$$

These quantities are now used to calculate the transmission system efficiency:

$$\text{effi}\left(P_{\text{load}}, L_{\text{line}}, V_{\text{sup}}\right) := \frac{P_{\text{load}}}{P_{\text{load}} + \text{Loss}\left(P_{\text{load}}, L_{\text{line}}, V_{\text{sup}}\right)}$$

$$\text{effi}\left(P_{\text{load\_5\%}}\left(L_{\text{line}}, V_{\text{sup}}\right), L_{\text{line}}, V_{\text{sup}}\right) = 98.8 \cdot \%.$$

With the interrelationships defined, the performance of the four different cases can be compared. First, the loads that produce 5% regulation are calculated using the Mathcad *root* finder for each of the four line voltages for a 20-mile line length:

Case 1: 22 kV

$$V_{\text{sup}} := 22 \text{ kV} \quad P_{\text{load\_5\%}}\left(L_{\text{line}}, V_{\text{sup}}\right) = 9.79 \cdot \text{MW}$$

$$\left|V_{\text{load\_1}}\left(P_{\text{load\_5\%}}\left(L_{\text{line}}, V_{\text{sup}}\right), V_{\text{sup}}, L_{\text{line}}\right)\right| = 21.0 \cdot \text{kV}$$

Case 2: 120 kV

$$V_{\text{sup}} := 120 \text{ kV} \quad P_{\text{load\_5\%}}\left(L_{\text{line}}, V_{\text{sup}}\right) = 51.7 \cdot \text{MW}$$

$$\left|V_{\text{load\_1}}\left(P_{\text{load\_5\%}}\left(L_{\text{line}}, V_{\text{sup}}\right), V_{\text{sup}}, L_{\text{line}}\right)\right| = 114.3 \cdot \text{kV}$$

Case 3: 220 kV

$$V_{\text{sup}} := 220 \text{ kV} \quad P_{\text{load\_5\%}}\left(L_{\text{line}}, V_{\text{sup}}\right) = 173.8 \cdot \text{MW}$$

$$\left|V_{\text{load\_1}}\left(P_{\text{load\_5\%}}\left(L_{\text{line}}, V_{\text{sup}}\right), V_{\text{sup}}, L_{\text{line}}\right)\right| = 209.5 \cdot \text{kV}$$

Case 4: 500 kV

$$V_{\text{sup}} := 500 \text{ kV} \quad P_{\text{load\_5\%}}\left(L_{\text{line}}, V_{\text{sup}}\right) = 897.7 \cdot \text{MW}$$

$$\left|V_{\text{load\_1}}\left(P_{\text{load\_5\%}}\left(L_{\text{line}}, V_{\text{sup}}\right), V_{\text{sup}}, L_{\text{line}}\right)\right| = 476.2 \cdot \text{kV}$$

The results clearly demonstrate that an increased transmission line voltage significantly increases the maximum load value.

The effect of line length is demonstrated in Figure 7.16 for the latter three line voltages with a plot of the regulation versus line length when the load is 10 MW. The 10 MW load produces a 5% voltage drop on a 22 kV line if the length is:

$$\text{Guess} \quad L_{\text{line}} := 1 \text{ mi}$$

$$\text{root}\left(\text{Reg}(10 \text{ MW}, L_{\text{line}}, 22 \text{ kV}) - 5\%, L_{\text{line}}\right) = 3.5 \cdot \text{mi}.$$

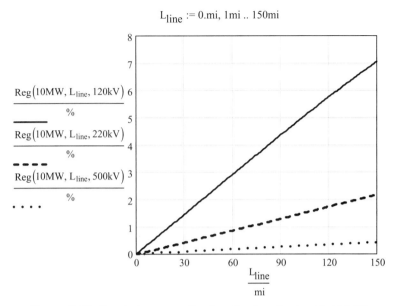

Figure 7.16. Regulation versus line length when the load is 10 MW.

This calculation and the plot of Figure 7.16 clearly establish the advantages of the increasing voltage. The 120 kV line can transport 10 MW 100 miles, while the 500 kV line to several hundred miles.

Figure 7.17 plots the power loss in the transmission line as a function of line length. The graphs demonstrate that increased line voltage decreases the line power loss for a given load power, hence allowing the system to transport power more efficiently. The curves of Figure 7.16 also show the limits imposed from meeting the voltage regulation requirements. The power system can be optimized for a particular line length considering the cost and loss, by changing the voltage, conductor type, and so on.

7.2.2. Real Transformer

The purpose of this section is the development of an equivalent circuit for the *real transformer*. The primary and secondary windings of the real transformer have resistance and leakage reactance, and the magnetizing current and losses in the iron core may not be negligible. The equivalent circuit should contain an ideal transformer and a network of reactance and resistances.

Like the ideal transformer, the unloaded real transformer is supplied by a primary voltage that drives a magnetizing current through the primary winding. The magnetizing current generates the main flux that induces voltage in both the primary and secondary windings. Excitation for the main flux can be represented by a current through an equivalent magnetizing reactance X_m, which is connected in parallel at the primary side of an ideal transformer. The *magnetizing inductance* can be calculated by Equation

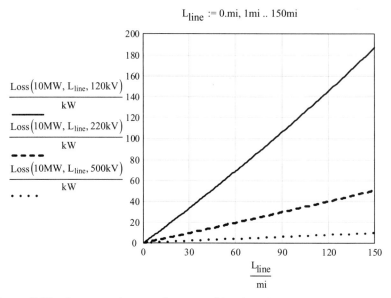

Figure 7.17. Line power loss as a function of line length when the load is 10 MW.

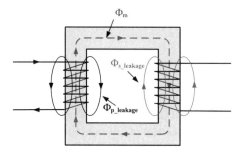

Figure 7.18. Leakage flux in the real transformer.

(6.34), using the iron core dimensions and the number of turns in the primary. The magnetizing reactance is:

$$X_m = \omega L = \omega \mu_0 \mu_r \frac{A_{core} N_p^2}{\ell_p}. \tag{7.13}$$

The real iron core has eddy-current and hysteresis losses, which can be represented by an equivalent core resistance R_c that is connected in parallel with the magnetizing reactance.

The real iron core permeability is not infinitely large, which produces leakage flux, as drawn in Figure 7.18. The diagram illustrates that parts of the leakage flux do not link with the primary or secondary windings. The primary and secondary leakage fluxes are each represented by an equivalent reactance. These reactances (X_p and X_s) are extracted from the real transformer and connected in series at the primary and the secondary sides, respectively, of the ideal transformer model.

The primary and secondary windings have resistances. These resistances are in series with the equivalent leakage reactance of the transformer model. The primary winding resistance (R_p) is connected in series at the primary; the secondary winding resistance (R_s) is connected in series at the secondary side.

The primary and secondary components are connected by an ideal transformer, which is characterized with the turns ratio T. Figure 7.19 provides the equivalent circuit of a real transformer.

The circuit can be modified by transferring the secondary leakage reactance and winding resistance to the primary side, as diagrammed in Figure 7.20a. The equations for the impedance transfer are:

$$R_{st} = R_s T^2 \quad \text{and} \quad X_{st} = X_s T^2 \quad \text{or} \quad \mathbf{Z}_{st} = \mathbf{Z}_s T^2, \tag{7.14}$$

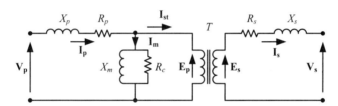

Figure 7.19. Equivalent circuit of a real transformer.

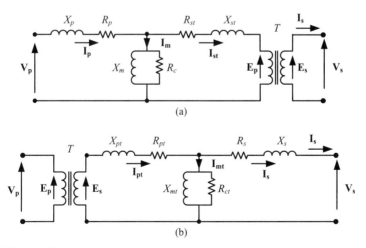

Figure 7.20. Modified real transformer equivalent circuits. (a) Transformer parameters transferred to primary side; (b) transformer parameters transferred to secondary side.

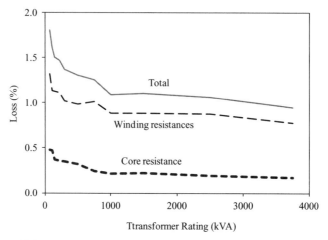

Figure 7.21. Relative losses from a 480/277 V pad-mounted transformer.

where the subscript s denotes the secondary side quantities, and st designates the value of those quantities upon transfer to the primary side. Alternatively, the primary circuit parameters, including the magnetizing resistance and reactance, can be transferred to the secondary, as illustrated in Figure 7.20b. In this case, the resulting modified equivalent circuit parameters for the real transformer are:

$$R_{pt} = \frac{R_p}{T^2} \quad \text{and} \quad X_{pt} = \frac{X_p}{T^2} \quad \text{or} \quad \mathbf{Z}_{pt} = \frac{\mathbf{Z}_p}{T^2}, \tag{7.15}$$

$$R_{ct} = \frac{R_c}{T^2} \quad \text{and} \quad X_{mt} = \frac{X_m}{T^2}. \tag{7.16}$$

A real transformer experiences power losses due to the primary and secondary windings as well as the core resistance. Figure 7.21 plots the losses for a typical three-phase transformer as a function of its rating. The graph shows that the smaller transformers incur slightly higher losses, but overall the power loss is approximately 1–2%.

EXAMPLE 7.2: Single-phase transformer operation analysis

The analysis of single-phase transformer operation is presented using a numerical example and the interactive derivation of the equations to enhance understanding.

System Characteristics
A network with constant voltage supplies a small transformer. A motor that operates with variable load but constant power factor loads the transformer. The equivalent circuit of the described system is displayed in Figure 7.22.

SINGLE-PHASE TRANSFORMERS

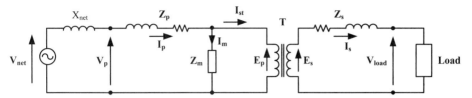

Figure 7.22. Network supplies a load through transformer.

The goal of this analysis is the calculation of the input power and power factor, voltage regulation, and efficiency versus variable load. An additional question is: what are the load and efficiency when the voltage drop is 5%?

The transformer rating data are:

$$S_{tr} := 45 \text{ kV} \cdot \text{A} \quad V_{p\_rated} := 440 \text{ V} \quad V_{s\_rated} := 220 \text{ V}.$$

The primary side impedance is: $Z_p := (0.03 + j \cdot 0.14) \, \Omega$.

The parallel-connected resistance and reactance of the magnetizing impedance are combined together. The obtained magnetizing impedance at the primary side is: $Z_m := (300 + j \cdot 120) \, \Omega$.

The secondary side impedance is: $Z_s := (0.01 + j \cdot 0.04) \, \Omega$.

The variable load data are:

$$P_{load} := 0 \text{ kW}, 1 \text{ kW} \ldots 1.4 \cdot S_{tr} \quad pf_{load} := 0.8 \text{ lagging} \quad V_{load} := 220 \text{ V}.$$

The network supply rating is: $V_{net\_rated} := 440 \text{ V} \quad I_{net\_short} := 3000 \text{ A}$.

The rated current of the transformer is the maximum current that the transformer can carry without overheating. The current ratings, obtained from the transformer rating data, can be used to judge the soundness of the later calculations. The rated currents are:

$$I_{p\_rated} := \frac{S_{tr}}{V_{p\_rated}} \quad I_{p\_rated} = 102.3 \text{ A}$$

$$I_{s\_rated} := \frac{S_{tr}}{V_{s\_rated}} \quad I_{s\_rated} = 204.5 \text{ A}.$$

System Analysis

Finding both the regulation and the efficiency requires the calculation of the network and load currents. The first step of this analysis is the calculation of the network impedance and load current as a function of the load. The equations will be validated using a load value of $P_{load} := 40 \text{ kW}$.

The supply network reactance is found from the network rating. It is assumed that the network impedance is purely reactive:

$$X_{net} := \frac{V_{net\_rated}}{I_{net\_short}} \quad X_{net} = 0.147 \, \Omega.$$

The complex power of the load for a lagging power factor is:

$$S_{load}(P_{load}) := \frac{P_{load}}{pf_{load}} e^{j \cdot acos(pf_{load})}.$$

The numerical value at 40 kW is $S_{load}(P_{load}) = (40.0 + 30.3j) \cdot kV \cdot A$.
The load (or secondary) current is:

$$I_s(P_{load}) := \frac{\overline{S_{load}(P_{load})}}{V_{load}}.$$

The numerical current value at 40 kW is:

$$|I_s(P_{load}, V_{load})| = 227.3 \, A \quad \arg(I_s(P_{load}, V_{load})) = -36.9 \cdot \deg.$$

Applying Kirchhoff's voltage law (KVL) to the equivalent circuit of Figure 7.22 reveals that secondary voltage (E_s) at the transformer is the sum of the voltage drop on the secondary impedance and the load voltage:

$$E_s(P_{load}, V_{load}) := V_{load} + I_s(P_{load}, V_{load}) \cdot Z_s.$$

The numerical voltage value is:

$$|E_s(P_{load}, V_{load})| = 227.3 \, V \quad \arg(E_s(P_{load}, V_{load})) = 1.5 \cdot \deg.$$

The transformer turns ratio is computed from the transformer rating using Equation (7.3):

$$T_R := \frac{V_{p\_rated}}{V_{s\_rated}} \quad T_R = 2.$$

The primary induced voltage (E_p) and secondary transfer current (I_{st}) are calculated using the turns ratio and Equation (7.6):

$$E_p(P_{load}, V_{load}) := T_R \cdot E_s(P_{load}, V_{load})$$

$$I_{st}(P_{load}, V_{load}) := \frac{I_s(P_{load}, V_{load})}{T_R}.$$

The numerical values are:

$$|E_p(P_{load}, V_{load})| = 454.7 \text{ V} \quad \arg(E_p(P_{load}, V_{load})) = 1.5 \cdot \deg$$
$$|I_{st}(P_{load}, V_{load})| = 113.6 \text{ A} \quad \arg(I_{st}(P_{load}, V_{load})) = -36.9 \cdot \deg.$$

The magnetizing current is determined using Ohm's law:

$$I_m(P_{load}, V_{load}) := \frac{E_p(P_{load}, V_{load})}{Z_m}.$$

The numerical value is $|I_m(P_{load}, V_{load})| = 1.41$ A $\quad \arg(I_m(P_{load}, V_{load})) = -20.3 \cdot \deg$.

Applying Kirchhoff's current law (KCL), the primary current is the sum of the magnetizing current and secondary transfer current:

$$I_p(P_{load}, V_{load}) := I_m(P_{load}, V_{load}) + I_{st}(P_{load}, V_{load}).$$

The numerical value is $|I_p(P_{load}, V_{load})| = 115.0$ A $\quad \arg(I_p(P_{load}, V_{load})) = -36.7 \cdot \deg$.

Applying KVL, the network supply voltage is the sum of the primary induced voltage and the voltage drop across the network and primary impedances:

$$V_{net}(P_{load}, V_{load}) := E_p(P_{load}, V_{load}) + I_p(P_{load}, V_{load}) \cdot (j \cdot X_{net} + Z_p).$$

The numerical value is:

$$|V_{net}(P_{load}, V_{load})| = 478.4 \text{ V} \quad \arg(V_{net}(P_{load}, V_{load})) = 4.3 \cdot \deg.$$

With the interrelationships established, the load voltage for a given supply voltage and load power can be determined by solving the earlier set of relations using the Mathcad *root* function:

$$V_{load.o}(P_{load}) := \text{root}(V_{net}(P_{load}, V_{load}) - 440 \text{ V}, V_{load}).$$

The voltages at both no-load and loaded conditions can be determined from the earlier:

$$|V_{load.o}(0 \cdot MW)| = 219.9 \text{ V} \quad \arg(V_{load.o}(0 \cdot MW)) = -0.0 \cdot \deg$$
$$|V_{load.o}(40 \cdot kW)| = 198.6 \text{ V} \quad \arg(V_{load.o}(40 \cdot kW)) = -5.2 \cdot \deg.$$

Therefore, the voltage regulation is computed from:

$$\text{reg}(P_{load}) := \frac{|V_{load.o}(0 \cdot kW) - V_{load.o}(P_{load})|}{|V_{load.o}(P_{load})|} \quad \text{reg}(40 \text{ kW}) = 14.3 \cdot \%.$$

The previous regulation is excessive, operationally speaking. The load corresponding to the 5% regulation is:

$$P_{load\_5\%} := \text{root}(\text{reg}(P_{load}) - 5\%, P_{load}) = 15.9 \cdot \text{kW}.$$

The load voltage at this load is:

$$V_{load\_5\%} := V_{load.o}(P_{load\_5\%})$$
$$|V_{load\_5\%}| = 212.2 \text{ V} \quad \arg(V_{load\_5\%}) = -2.0 \cdot \text{deg}.$$

Next, the transmission efficiency, which is the ratio of delivered (load) power to input (supply) power, is determined. Based on the voltage regulation results, the following formulae are tested using a load power of $P_{load} := 10$ kW. With a method in place to determine the load voltage for a given load, it is now necessary to work backward from the load to the supply using the earlier process.

The secondary current due to the load and its transfer to the primary side are:

$$I_s(P_{load}) := \overline{\frac{S_{load}(P_{load})}{V_{load.o}(P_{load})}} \quad I_{st}(P_{load}) := \frac{I_s(P_{load})}{T_R}.$$

The voltages at the transformer secondary and primary terminals are:

$$E_s(P_{load}) := V_{load.o}(P_{load}) + I_s(P_{load}) \cdot Z_s$$
$$E_p(P_{load}) := T_R \cdot E_s(P_{load}).$$

The magnetizing and supply currents are thus:

$$I_m(P_{load}) := \frac{E_p(P_{load})}{Z_m} \quad I_p(P_{load}) := I_m(P_{load}) + I_{st}(P_{load}).$$

For operation at the rated network voltage, the complex power input is:

$$S_{net}(P_{load}) := 440 \text{ V} \cdot \overline{I_p(P_{load})}.$$

The required real power input at the voltage at the 5% regulation is:

$$P_{net}(P_{load}) := \text{Re}(S_{net}(P_{load})) \quad P_{net}(P_{load\_5\%}) = 16.6 \cdot \text{kW}.$$

The power factor is the cosine of the complex power angle:

$$pf_{net}(P_{load}) := \cos(\arg(S_{net}(P_{load}))).$$

The numerical value at the 5% regulation is $pf_{net}(P_{load\_5\%}) = 0.784$.

SINGLE-PHASE TRANSFORMERS

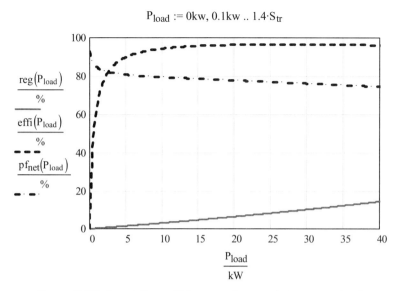

Figure 7.23. Regulation, efficiency, and power factor versus load.

Alternatively, the power factor may be computed from the ratio of the real power and absolute value of the complex power:

$$\text{pf2}_{net}(P_{load}) := \frac{P_{net}(P_{load})}{|S_{net}(P_{load})|} \quad \text{pf2}_{net}(P_{load\_5\%}) = 0.784.$$

The system efficiency is the ratio of the output (load) power to the input (supply) power:

$$\text{effi}(P_{load}) := \frac{P_{load}}{P_{net}(P_{load})}.$$

The efficiency at the 5% regulation operating point is effi($P_{load\_5\%}$) = 95.8%.

Figure 7.23 graphs the regulation, efficiency, and power factor variation with the load. The regulation is a linear function of the load. The efficiency increases rapidly with the load and becomes more or less constant as the load increases.

7.2.3. Determination of Equivalent Transformer Circuit Parameters

The equivalent transformer circuit parameters can be determined from open- and short-circuit measurements. In each case, the supply voltage, input current, and power are measured. The equivalent circuit values for the transformer are calculated from the measured data. The measurements can be performed on either the primary or the secondary side depending on the available power supply. The equivalent circuit parameters obtained are placed on the same side where the measurements were performed. In this

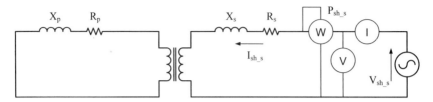

Figure 7.24. Equivalent circuit for short-circuit test, measurement at the secondary side.

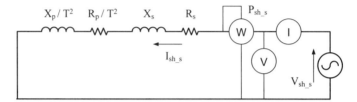

Figure 7.25. Simplified equivalent circuit for short-circuit test.

example, for the open-circuit test, the primary is supplied, and for the short-circuit test, the secondary is supplied.

The method of determining transformer parameters by open- and short-circuit measurement is presented using an interactive derivation method and a numerical example. The single-phase transformer rating data are:

$$M := 10^6 \quad S_{tr} := 20 \, M \cdot V \cdot A \quad V_p := 12.7 \, kV \quad V_s := 69.28 \, kV.$$

7.2.3.1. Short-Circuit Test Measurement.
Figure 7.24 displays the equivalent circuit for the short circuit test, when the measurements are performed at the secondary side. The transformer primary is short circuited and a reduced voltage supplies the secondary side. The reduced supply voltage is selected such that it drives a current close to the rated current through the transformer. The current, voltage, and input real power are measured at the secondary side. The measurement results are:

$$V_{sh\_s} := 6 \, kV \quad I_{sh\_s} := 290 \, A \quad P_{sh\_s} := 0.8 \, MW.$$

7.2.3.2. Short-Circuit Test Evaluation.
The evaluation of the short-circuit test results leads to the series impedance of the transformer. To begin, the primary impedance and short circuit are transferred to the secondary side using Equation (7.14), which eliminates the transformer. The simplified circuit is given in Figure 7.25. Combining the transferred primary and the secondary resistances and reactances further simplifies the circuit, as seen in Figure 7.26, where:

SINGLE-PHASE TRANSFORMERS

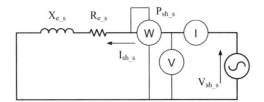

Figure 7.26. Simplified circuit for calculation of series impedance.

$$R_{e\_s} = \frac{R_p}{T^2} + R_s,$$

$$X_{e\_s} = \frac{X_p}{T^2} + X_s.$$

The equivalent circuit in Figure 7.26 shows that the short-circuit impedance consists of a resistance and reactance connected in series. In this circuit, the input power is equal with the loss in the resistance. Consequently, the resistance can be calculated directly from the input power and current measured during the short-circuit test:

$$R_{e\_s} := \frac{P_{sh\_s}}{I_{sh\_s}^2}. \tag{7.17}$$

The corresponding numerical value is $R_{e\_s} = 9.51\ \Omega$.

The absolute value of the impedance is the ratio of the voltage and current:

$$Z_{sh\_s} := \frac{V_{sh\_s}}{I_{sh\_s}}. \tag{7.18}$$

The numerical value is $Z_{sh\_e} = 20.7\ \Omega$.

The absolute value of the impedance, which consists of a resistance and a reactance connected in series, is $|Z| = \sqrt{R^2 + X^2}$. Using this equation, the reactance is:

$$X_{e\_s} := \sqrt{Z_{sh\_s}^2 - R_{e\_s}^2}. \tag{7.19}$$

The corresponding numerical value is $X_{e\_s} = 18.4\ \Omega$.

The short-circuit measurement is performed at the secondary side. Consequently, the calculated resistance and reactance are placed on the secondary side, as illustrated in Figure 7.29.

7.2.3.3. Open-Circuit Test Measurement.
To perform the open-circuit test, the secondary side of the transformer is opened, and a source with a voltage close to

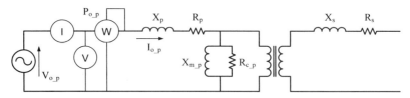

Figure 7.27. Equivalent circuit for open-circuit test, measurement at the primary side.

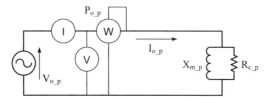

Figure 7.28. Simplified equivalent circuit for open-circuit test.

the rated voltage supplies the primary, as portrayed in Figure 7.27. The current, voltage, and input power are measured at the primary side. The measured values are:

$$V_{o\_p} := 13 \text{ kV} \quad I_{o\_p} := 170 \text{ A} \quad P_{o\_p} := 1.1 \text{ MW}.$$

7.2.3.4. Open-Circuit Test Evaluation. Figure 7.27 introduced the equivalent circuit for the open-circuit test, when the measurements are performed at the primary side. The evaluation of these test results leads to the magnetizing reactance and core loss resistance.

The current in the secondary circuit is zero, which permits the elimination of the ideal transformer from the circuit of Figure 7.27 since the current through the primary winding is correspondingly zero ($I_p = I_s/T$). (Note that although the secondary is open, its voltage is *not* zero.) Further, the primary impedance is negligible compared with the magnetizing impedance, so the primary impedance may be neglected. The simplified equivalent circuit is drawn in Figure 7.28.

Figure 7.28 shows that the magnetizing branch contains a resistance (R_{c_p}) and reactance (X_{m_p}) connected in parallel. In this circuit, the *core resistance* can be calculated directly from the input power and voltage because it is connected directly in parallel with the primary voltage source:

$$R_{c\_p} := \frac{(V_{o\_p})^2}{P_{o\_p}}. \tag{7.20}$$

The numerical value is $R_{c\_p} = 153.6 \, \Omega$. This value is significantly larger than the entire series resistance ($R_{e\_s} = 9.51 \, \Omega$), which justifies that we have neglected the series resistance (R_p) in Figure 7.27.

The absolute value of the magnetizing branch admittance is the ratio of the current and voltage measured in the open-circuit condition. The admittance of the magnetizing branch is:

$$Y_{o\_p} := \frac{I_{o\_p}}{V_{o\_p}}. \tag{7.21}$$

The numerical value is $Y_{o\_p} = 0.013$ S.

The absolute value of an admittance from a resistance and reactance connected in parallel can be expressed by the equation $|Y| = \sqrt{G^2 + B^2}$, where $G = 1/R$ and $B = 1/X$. The substitution of the conductance (G) and the susceptance (B) into the admittance (Y) equation yields:

$$|Y| = \sqrt{\frac{1}{R^2} + \frac{1}{X^2}}. \tag{7.22}$$

The magnetizing reactance is calculated from rearranging this admittance equation. The result is:

$$X_{m\_p} := \frac{1}{\sqrt{Y_{o\_p}^2 - \frac{1}{R_{c\_p}^2}}}. \tag{7.23}$$

The numerical value is $X_{m\_p} = 88.2$ Ω. This value is significantly larger than the primary series reactance, which justifies that we have neglected the series reactance (X_p) in Figure 7.27.

The resultant magnetizing reactance and the core loss resistance are connected to the primary side of the transformer, as shown in Figure 7.29, because the open-circuit measurement is performed on the primary side.

The previous calculations have resulted in the equivalent circuit for the real transformer. The circuit in Figure 7.29 is suitable for analyzing transformer operation. However, the circuit can be further modified to a form similar to the traditional equivalent circuit. One method that has historically been used starts with the transfer of the calculated series impedance from the secondary to the primary side using Equation

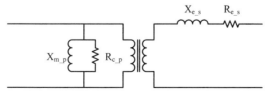

Figure 7.29. Equivalent circuit for a real transformer resulting from the open- and short-circuit tests.

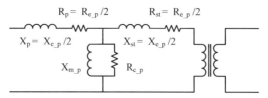

Figure 7.30. Modified equivalent circuit for the open-circuit measurement.

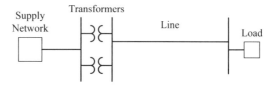

Figure 7.31. Network to analyze operation of transformers connected in parallel.

(7.14), and then dividing the transferred impedance by two. Half of the transferred impedance is located before and the other half is placed after the magnetizing impedance. This is an approximation that is often used when more accurate data are not available. The modified circuit is provided in Figure 7.30.

EXAMPLE 7.3: Transformers in parallel

An electric network supplies a load through two mismatched transformers connected in parallel and a transmission line. Figure 7.31 gives the one-line diagram of the system. The turns ratio and reactance of each transformer are slightly different. The mismatch may produce overloading of the transformers.

The objective of this exercise is the calculation of the required supply voltage and the load current in the transformers, using a numerical example. The system data are first defined in the succeeding paragraphs.

The network rated voltage and short circuit current are:

$$V_{net} := 7.97 \text{ kV} \quad I_{net\_short} := 4 \text{ kA}.$$

The ratings of the transformers are:

$$S_{tr1} := 60 \text{ kV} \cdot \text{A} \quad V_{tr1\_p} := 7.4 \text{ kV} \quad V_{tr1\_s} := 460 \text{ V} \quad x_{tr1} := 5.0\%$$
$$S_{tr2} := 60 \text{ kV} \cdot \text{A} \quad V_{tr2\_p} := 7.8 \text{ kV} \quad V_{tr2\_s} := 460 \text{ V} \quad x_{tr2} := 7.5\%,$$

where the transformer reactances are given on a p.u. basis relative to the transformer rating, that is, $x_{tr1} := 5\%$ equates to 0.05 p.u.

SINGLE-PHASE TRANSFORMERS

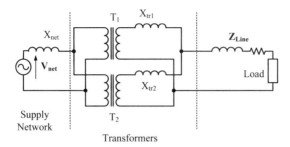

Figure 7.32. Equivalent circuit of the network of Figure 7.31.

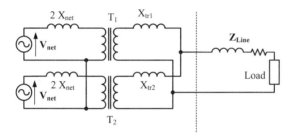

Figure 7.33. Simplification of the equivalent circuit by splitting the supply network.

The transmission line impedance and length are:

$$Z_{line} := (0.2 + j \cdot 0.5) \cdot \frac{\Omega}{mi} \quad L_{line} := 1500 \text{ ft.}$$

The load power, power factor, and voltage are:

$$P_{load} := 80 \text{ kW} \quad pf_{load} := 0.8 \text{ (lagging)} \quad V_{load} := 440 \text{ V.}$$

The equivalent circuit of the network of Figure 7.31 is displayed in Figure 7.32. In this circuit, a Thévenin equivalent represents the network, which is a voltage source (V_{net}) and a reactance (X_{net}) connected in series. The network impedance is the ratio of network rated voltage and short-circuit current. It is assumed that $X_{net} \gg R_{net}$:

$$X_{net} := \frac{V_{net}}{I_{net\_short}} \quad X_{net} = 1.99 \ \Omega.$$

Each transformer's reactance is placed at the secondary side, and the transmission line is represented by an impedance (Z_{Line}).

The equivalent circuit is simplified by splitting the supply network into two identical networks, each of which supplies a transformer. The supply voltages of the new networks are the same as the original source, but the network reactances are doubled. Figure 7.33 shows the supply network separated into two parts.

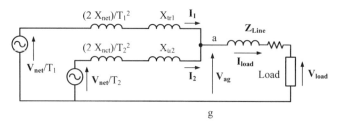

Figure 7.34. Simplified equivalent circuit for parallel transformer operation analyses.

The turns ratio of each transformer is computed from its rating:

$$T_1 := \frac{V_{tr1\_p}}{V_{tr1\_s}} = 16.09 \quad T_2 := \frac{V_{tr2\_p}}{V_{tr2\_s}} = 16.96.$$

The series reactance of each transformer is:

$$X_{tr1} := x_{tr1} \cdot \frac{V_{tr1\_s}^2}{S_{tr1}} \quad X_{tr1} = 0.176 \, \Omega$$

$$X_{tr2} := x_{tr2} \cdot \frac{V_{tr2\_s}^2}{S_{tr2}} \quad X_{tr2} = 0.265 \, \Omega.$$

The series resistance and magnetizing impedance of the transformers are neglected. A further simplification involves the transfer of the network voltage and reactance to the secondary side of the transformers using the appropriate turns ratio and Equation (7.6) and Equation (7.14), respectively. Figure 7.34 shows the simplified circuit. Because the network is split into two parts, as seen in Figure 7.33, this doubles the impedance of the component networks. The impedance of the component networks is transferred to the secondary side of the transformers:

$$X_{net1} := \frac{2X_{net}}{T_1^2} \quad X_{net1} = 0.015 \, \Omega$$

$$X_{net2} := \frac{2X_{net}}{T_2^2} \quad X_{net2} = 0.014 \, \Omega.$$

The phasor load current for an inductive (lagging) load is:

$$I_{load} := \frac{P_{load}}{V_{load} \cdot pf_{load}} \cdot e^{-j \cdot acos(pf_{load})}$$

$$|I_{load}| = 227.3 \, A \quad arg(I_{load}) = -36.9 \cdot deg.$$

SINGLE-PHASE TRANSFORMERS

Using KVL, the voltage between point **a** and the ground (**g**) is:

$$V_{ag} := V_{load} + I_{load} \cdot Z_{line} \cdot L_{line} = (469.7 + 18.1j) \text{ V}$$
$$|V_{ag}| = 470.0 \text{ V} \quad \arg(V_{ag}) = 2.204 \cdot \deg.$$

The impedance of supply branch 1 is: $Z_1 := j \cdot X_{tr1} + j \cdot X_{net1} = 0.192j \ \Omega$.
The impedance of supply branch 2 is: $Z_2 := j \cdot X_{tr2} + j \cdot X_{net2} = 0.278j \ \Omega$.

The required network voltage is calculated from the node point equation. The branch current is the ratio of the voltage difference and the branch impedance. The sum of the two branch currents is the load current. The node point (KCL) equation for point **a** is:

$$\frac{V_{net} - V_{ag}}{T_1}{Z_1} + \frac{\frac{V_{net}}{T_2} - V_{ag}}{Z_2} = I_{load}.$$

This equation is solved using the Mathcad *Find* equation solver. The guess value for the solver includes an imaginary term to inform Mathcad that a complex-valued answer is expected:

$$V_{net} := 7200 \text{ V} + j \cdot 10 \text{ V}$$

Given

$$\frac{\frac{V_{net}}{T_1} - V_{ag}}{Z_1} + \frac{\frac{V_{net}}{T_2} - V_{ag}}{Z_2} = I_{load}$$

$$V_{network} := \text{Find}(V_{net})$$
$$|V_{network}| = 8.00 \cdot \text{kV} \quad \arg(V_{network}) = 4.56 \cdot \deg.$$

The phasor current of branch 1 is:

$$I_1 := \frac{\frac{V_{network}}{T_1} - V_{ag}}{Z_1} = (112.0 - 134.8j) \text{ A}$$
$$|I_1| = 175.2 \text{ A} \quad \arg(I_1) = -50.3 \cdot \deg.$$

Figure 7.34 shows that **I₁** is the load current in Transformer 1 (T₁). This is the current that would flow through the transformer secondary winding. The following calculation reveals that the rated current for Transformer 1 is less than this load current. Transformer 1 will be severely overloaded:

$$I_{tr1\_s\_rating} := \frac{S_{tr1}}{V_{tr1\_s}} = 130.4 \text{ A}.$$

The current in branch 2 is:

$$I_2 := \frac{\frac{V_{network}}{T_2} - V_{ag}}{Z_2} = (69.84 - 1.56j) \text{ A}$$

$$|I_2| = 69.9 \text{ A} \quad \arg(I_2) = -1.28 \cdot \deg.$$

Figure 7.34 points out that I_2 is the load current in Transformer 2 (T_2). The following calculation indicates that the rated current of Transformer 2 is greater than the actual load current:

$$I_{tr2\_s\_rating} := \frac{S_{tr2}}{V_{tr2\_s}} = 130.4 \text{ A}.$$

The results demonstrate that Transformer 1 is overloaded because of the mismatch of transformer turns ratio and impedance.

If the transformers are identical, the load current distributes evenly between the transformers, which results in an individual load current of $|I_{load}|/2 = 113.6$ A, which is less than the rated current of each transformer.

This example demonstrates that even a small mismatch between the turns ratios or impedances can produce unexpected overloading of transformers connected in parallel.

7.3. THREE-PHASE TRANSFORMERS

A very large three-phase network transformer is pictured in Figure 7.35. Most three-phase transformers use a three-legged iron core, as illustrated in Figure 7.36. However, three single-phase transformers can also be interconnected to form a three-phase transformer bank. A three-phase transformer is described by its total apparent power capability (|S|) and voltage ratings (V_{rate}). Regardless of the transformer connection, the rated power is the three-phase rating, and the rated voltage is the voltage between the transformer terminals, which is the line-to-line voltage. The voltage between a terminal and the neutral point, if it exists, is not the rated voltage. The current rating is the terminal current corresponding to the rated apparent power. The rated winding current depends on the connection type (wye or delta). In the case of a wye (Y) connection, the rated winding current is the rated line current and is calculated from the line-to-neutral equivalent of the voltage rating:

$$I_{rate}^Y = \frac{|S|/3}{V_{rate}/\sqrt{3}} = \frac{|S|}{\sqrt{3}V_{rate}}. \tag{7.24}$$

THREE-PHASE TRANSFORMERS

Figure 7.35. Very large three-phase network transformer (courtesy of Siemens, Erlangen, Germany).

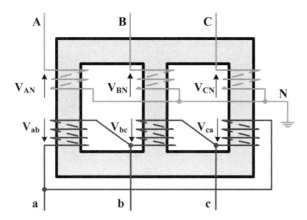

Figure 7.36. A three-phase wye–delta transformer with three-legged iron core.

For a delta (Δ) connection, the rated winding current is:

$$I_{\text{rate}}^{\Delta} = \frac{|\mathbf{S}|/3}{V_{\text{rate}}} = \frac{|\mathbf{S}|}{3V_{\text{rate}}}. \quad (7.25)$$

The corresponding line current for the delta configuration is:

$$I_{\text{line}}^{\Delta} = \sqrt{3} I_{\text{rate}}^{\Delta} = \sqrt{3} \frac{|\mathbf{S}|}{3V_{\text{rate}}} = \frac{|\mathbf{S}|}{\sqrt{3} V_{\text{rate}}}. \quad (7.26)$$

This proves that the line current is independent of the connection type. Note again that V_{rate} is the line-to-line voltage.

The usual connections for three-phase transformers are:

Wye–wye (Y-Y)	Seldom used; imbalance and third harmonics problems
Wye–wye–delta (Y-Y-Δ)	Frequently used to interconnect high-voltage networks (240/345 kV). The delta winding filters the third harmonics, equalizes the unbalanced current, and provides a path for ground current
Wye–delta (Y-Δ)	Frequently used as step down (345/69 kV)
Delta–delta (Δ-Δ)	Used for medium voltage (15 kV); one of the transformers can be removed (open delta)
Delta–wye (Δ-Y)	Step-up transformer in a generation station

In most cases, the neutral point of the Y winding is grounded for safety reasons.

Figure 7.36 depicts a typical three-phase transformer with a three-legged iron core. The iron core has three legs with equal cross sections. Each leg of the core carries primary and secondary windings. In Figure 7.36, the primary (A,B,C) is wye and the secondary (a,b,c) is delta connected. However, transformers can be connected in any of the connections previously listed.

The transformer is usually supplied by a balanced three-phase voltage, which produces a flux in each leg of the iron core. The sum of the fluxes is zero, which eliminates the need for a return path. The flux in the legs induces voltage in the coils. When the transformer is loaded, a current will flow in both the primary and secondary windings. The voltages and currents are *in phase* in the windings placed on the same leg. This implies that in the case of a wye–delta connection, the line-to-ground voltages V_{AN}, V_{BN}, and V_{CN} induce line-to-line voltages (V_{ab}, V_{bc}, and V_{ca}).

The operation of three-phase transformers is demonstrated using numerical examples.

7.3.1. Wye–Wye Connection

Figure 7.37 diagrams a wye–wye connected three-phase transformer. In this transformer, the primary and secondary currents and voltages are in phase. The neutral points are grounded. The transformer turns ratio is:

$$T_{\text{Y-Y}} = \frac{N_p}{N_s} = \frac{\mathbf{V}_{AN}}{\mathbf{V}_{an}} = \frac{\mathbf{I}_a}{\mathbf{I}_A}, \qquad (7.27)$$

where

N_p is the number of turns in the primary winding;
N_s is the number of turns in the secondary winding;
$\mathbf{V}_{AB}$ is the primary line-to-line voltage of phase A and B;

THREE-PHASE TRANSFORMERS

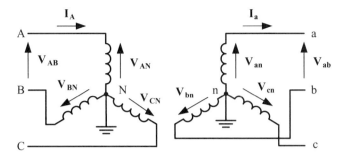

Figure 7.37. Wye-connected three-phase transformer.

V_{ab} is the secondary line-to-line voltage of phase a and b;
V_{AN} is the primary line-to-neutral voltage of phase A;
V_{an} is the secondary line-to-neutral voltage of phase a;
I_A is the primary line current in phase A; and
I_a is the secondary line current in phase a.

The transformer has two meaningful voltages and one current at each side. The ratio of the line-to-line and line-to-neutral voltages is $\sqrt{3}\angle 30°$:

$$\begin{aligned} \mathbf{V}_{AB} &= \sqrt{3}\mathbf{V}_{AN}e^{j30°}, \\ \mathbf{V}_{ab} &= \sqrt{3}\mathbf{V}_{an}e^{j30°}. \end{aligned} \quad (7.28)$$

Because of the earlier interrelations, the wye–wye transformer turns ratio equation can be extended to:

$$T_{Y-Y} = \frac{N_p}{N_s} = \frac{\mathbf{V}_{AB}}{\mathbf{V}_{ab}} = \frac{\mathbf{V}_{AN}}{\mathbf{V}_{an}} = \frac{\mathbf{I}_a}{\mathbf{I}_A}. \quad (7.29)$$

The turns ratio equation could also be expanded to include the voltages and currents from phases B and C:

$$T_{Y-Y} = \frac{N_p}{N_s} = \frac{\mathbf{V}_{AN}}{\mathbf{V}_{an}} = \frac{\mathbf{I}_a}{\mathbf{I}_A} = \frac{\mathbf{V}_{BN}}{\mathbf{V}_{bn}} = \frac{\mathbf{I}_b}{\mathbf{I}_B} = \frac{\mathbf{V}_{CN}}{\mathbf{V}_{cn}} = \frac{\mathbf{I}_c}{\mathbf{I}_C}. \quad (7.30)$$

For balanced cases, a single-phase equivalent circuit is used to simplify the analysis.

EXAMPLE 7.4: Wye–wye transformer

The operation of the Y-Y transformer is demonstrated with a numerical example. A balanced three-phase network supplies a balanced load through a three-phase wye–wye

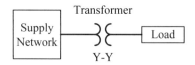

Figure 7.38. Three-phase network supplies a load through a transformer.

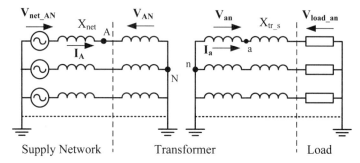

Figure 7.39. Equivalent circuit of the wye–wye transformer system of Figure 7.38.

connected transformer. Figure 7.38 gives the one-line diagram of the system. The objective of the calculation is the determination of the load voltage, given the network voltage, the balanced load power, and the power factor.

A multiplier of $M := 10^6$ and the system data are first defined. The wye transformer ratings are expressed as the three-phase complex power, the primary and secondary line voltages, and the 0.11 per-unit reactance:

$$S_{tr\_3f} := 20 \cdot M \cdot V \cdot A \qquad x_{tr} := 11\%$$
$$V_{p\_ll} := 220 \text{ kV} \qquad V_{s\_ll} := 120 \text{ kV}.$$

The three-phase load parameters are:

$$P_{load} := 15 \cdot M \cdot W \qquad pf_{load} := 0.8 \quad \text{(lagging)}.$$

The three-phase supply network line voltage and short circuit current are:

$$V_{net\_ll} := 225 \text{ kV} \qquad I_{net\_short} := 15 \text{ kA}.$$

The supply network is represented by its Thévenin equivalent circuit, which is a voltage source and a reactance (resistance is neglected). The transformer reactance (X_{tr_s}) is placed on the secondary side of the transformer. The transformer winding resistance, magnetizing reactance, and core loss resistance are neglected. The three-phase equivalent circuit is provided in Figure 7.39.

THREE-PHASE TRANSFORMERS

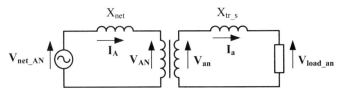

Figure 7.40. Single-phase equivalent circuit for phase A of the three-phase wye–wye transformer system of Figure 7.39.

Figure 7.39 indicates that the grounding of the neutral points provides a return path for the current at both the primary and secondary sides. This permits the representation of each phase with a separate single-phase equivalent circuit. Figure 7.40 presents the equivalent circuit for phase A. Similar circuits describe phases B and C. The supply voltage of each single-phase equivalent circuit is the network *line-to-neutral voltage*.

The analysis starts with the calculation of the equivalent circuit parameters, which are the turns ratio, transformer impedance, network line-to-neutral voltage and impedance, and the single-phase load.

The network line-to-neutral voltage and impedance are:

$$V_{net\_ln} := \frac{V_{net\_ll}}{3} \quad V_{net\_ln} = 129.9 \cdot kV$$

$$X_{net} := \frac{V_{net\_ln}}{I_{net\_short}} \quad X_{net} = 8.66 \, \Omega.$$

In the case of a balanced load, each circuit carries one-third of the total three-phase power. The single-phase load power is:

$$P_{load\_1p} := \frac{P_{load}}{3} \quad P_{load\_1p} = 5.00 \cdot MW.$$

The turns ratio is found from the transformer voltage ratings using Equation (7.27):

$$T_{tr} := \frac{V_{p\_ll}}{V_{s\_ll}} \quad T_{tr} = 1.833.$$

The transformer impedance is:

$$X_{tr\_s} := x_{tr} \cdot \frac{V_{s\_ll}^2}{S_{tr\_3f}} \quad X_{tr\_s} = 79.2 \, \Omega.$$

The calculation of the line voltage for the load is the objective of this example. The load voltage is the variable in this analysis. We will test the calculations using a load voltage of:

$$V_{load} := 120 \text{ kV}.$$

The magnitude of the line-to-neutral value of the load voltage (V_{load_an}) is:

$$V_{load\_ln}(V_{load}) := \frac{V_{load}}{\sqrt{3}} \quad V_{load\_ln}(V_{load}) = 69.3 \text{ kV}.$$

The load current ($\mathbf{I_a}$) is calculated using the single-phase power, line-to-neutral voltage, and lagging power factor:

$$I_{load}(V_{load}) := \frac{P_{load\_1p}}{V_{load\_ln}(V_{load}) \cdot pf_{load}} \cdot e^{-j \cdot a\cos(pf_{load})}$$

$$|I_{load}(V_{load})| = 90.2 \text{ A} \quad \arg(I_{load}(V_{load})) = -36.9 \cdot \deg.$$

The single-phase equivalent circuit demonstrates that the operating voltage of the ideal transformer secondary is the sum of the load voltage and the voltage drop on the transformer reactance. Using KVL, the transformer secondary voltage ($\mathbf{V_{an}}$) is:

$$E_s(V_{load}) := V_{load\_ln}(V_{load}) + j \cdot X_{tr\_s} \cdot I_{load}(V_{load})$$

$$|E_s(V_{load})| = 73.8 \cdot \text{kV} \quad \arg(E_s(V_{load})) = 4.44 \cdot \deg.$$

The transformer secondary current is equal with the load current:

$$I_s(V_{load}) := I_{load}(V_{load})$$

$$|I_s(V_{load})| = 90.2 \text{ A} \quad \arg(I_s(V_{load})) = -36.9 \cdot \deg.$$

The transformer primary voltage (V_{AN}) and current (I_A) are calculated by multiplying the secondary voltage by the turns ratio (T) and dividing the secondary current by T using Equation (7.27). The primary voltage and current are:

$$E_p(V_{load}) := E_s(V_{load}) \cdot T_{tr}$$

$$|E_p(V_{load})| = 135.3 \cdot \text{kV} \quad \arg(E_p(V_{load})) = 4.44 \cdot \deg$$

$$I_p(V_{load}) := \frac{I_s(V_{load})}{T_{tr}}$$

$$|I_p(V_{load})| = 49.2 \text{ A} \quad \arg(I_p(V_{load})) = -36.9 \cdot \deg.$$

The equivalent circuit of Figure 7.40 shows that supply voltage (V_{net_AN}) is the sum of the transformer primary voltage and the voltage drop on the network reactance. The supply voltage as a function of the load voltage is:

$$V_{sup\_ln}(V_{load}) := E_p(V_{load}) + j \cdot X_{net} \cdot I_p(V_{load})$$
$$|V_{sup\_ln}(V_{load})| = 135.6 \cdot kV \quad \arg(V_{sup\_ln}(V_{load})) = 4.6 \cdot \deg.$$

The objective of this calculation is the determination of the load voltage if the supply voltage is equal to the rated network voltage. This can be achieved by solving the equation:

$$V_{net\_ln} = V_{sup\_ln}(V_{load}).$$

This equation is solved using a complex guess value with the Mathcad equation solver:

$$V_{load} := 120 \text{ kV} + j \cdot 10 \text{ kV}$$
$$\text{Given}$$
$$V_{net\_ln} = V_{sup\_ln}(V_{load})$$
$$V_{load\_ll} := \text{Find}(V_{load}).$$

The load line-to-line voltage is:

$$|V_{load\_ll}| = 116.4 \text{ kV} \quad \arg(V_{load\_ll}) = -5.5 \deg.$$

The proper operation of the system requires a voltage regulation less than 5%. The voltage regulation at this load is:

$$\text{Reg} := \frac{V_{net\_ll} - |V_{load\_ll} \cdot T_{tr}|}{|V_{load\_ll} \cdot T_{tr}|} \quad \text{Reg} = 5.4 \cdot \%.$$

The obtained value is acceptable.

The method presented here can be adopted for an unbalanced load, when the supply voltage is calculated for each phase using separate single-phase equivalent circuits. This results in three equations. The Mathcad or MATLAB equation solvers can solve these three equations. Such a calculation provides the load voltage for each phase.

7.3.2. Wye–Delta Connection

Figure 7.41 introduces circuit diagrams for a wye–delta connected three-phase transformer. In this transformer, the primary winding "AN" and the secondary winding "ab" are on the same leg of the iron core, as seen in Figure 7.36. This implies that the primary line-to-ground voltage, V_{AN}, is in phase with the secondary line-to-line voltage, V_{ab}. In

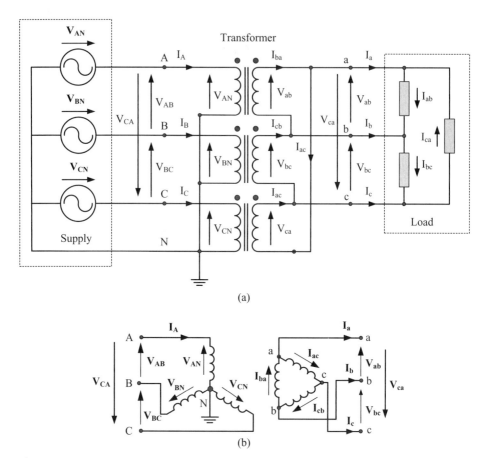

Figure 7.41. Wye–delta connected transformer. (a) Detailed circuit diagram; (b) transformer currents and voltages.

addition, the primary current $\mathbf{I_A}$ is in phase with the secondary current in the delta, $\mathbf{I_{ba}}$. The *actual turns ratio* for the wye–delta transformer is then:

$$T_{Y-\Delta} = \frac{N_p}{N_s} = \frac{\mathbf{V_{AN}}}{\mathbf{V_{ab}}} = \frac{\mathbf{I_{ba}}}{\mathbf{I_A}}. \tag{7.31}$$

Similarly, $\mathbf{V_{BN}}$ and $\mathbf{V_{bc}}$ are on the same leg, in phase, and related through the previous physical turns ratio, and likewise, so are $\mathbf{V_{CN}}$ and $\mathbf{V_{ca}}$ in phase C (see Fig. 7.41a).

At the primary side, the amplitude of the line-to-line voltage is $\sqrt{3}$ times the line-to-neutral voltage and the phase shift between these voltages is 30° for a balanced system:

$$\mathbf{V_{AB}} = \mathbf{V_{AN}} - \mathbf{V_{BN}} = \mathbf{V_{AN}} - \mathbf{V_{AN}}e^{-j120°} = \sqrt{3}\mathbf{V_{AN}}e^{j30°}. \tag{7.32}$$

THREE-PHASE TRANSFORMERS

Further, because the primary line-to-neutral voltage is in phase with the secondary line-to-line voltage, the phase shift between the primary and secondary line-to-line voltages is also 30°.

Figure 7.41a shows that on the secondary side, the line-to-line voltage (e.g., V_{ab}) drives current through the load. This creates a current I_{ab} flowing from node **a** to node **b** in the load and a current (I_{ba}) flowing from node **b** to node **a** in the transformer. In the case of a balanced load, the secondary line current I_a using Figure 7.41 is:

$$I_a = I_{ba} - I_{ac} = I_{ba} - I_{ba}e^{-j240°} = \sqrt{3}I_{ba}e^{-j30°}. \tag{7.33}$$

In this case, the current in the delta secondary winding I_{ba} is in phase with the primary current I_A, which results in a 30° phase shift between the primary and secondary line currents.

This analysis reveals that the wye–delta transformer shifts both the line current and the line-to-line voltage by 30°. In particular, the secondary line currents lag the primary line currents by −30°, and the secondary line voltages lag the primary line voltages by −30°.

Single-Phase Equivalent. In the case of a balanced load and source voltages, a single-phase equivalent circuit, as illustrated in Figure 7.42, can represent the wye–delta transformer. The delta winding at the secondary side can be characterized by an equivalent wye. The line-to-line voltage of this equivalent wye winding is the same as the delta winding. The secondary line-to-neutral voltage is the line-to-line voltage divided by $\sqrt{3}$:

$$V_{ab} = \sqrt{3}V_{an}e^{j30°}. \tag{7.34}$$

The problem is that this equivalent wye circuit does not produce the 30° phase shift. Using a turns ratio that artificially produces the 30° shift can eliminate this problem. The *artificial turns ratio* used in the single-phase equivalent circuit of a wye–delta transformer is defined based on the fact that the transformer line-to-line voltage ratings are specified:

$$T_{Y-\Delta}^{spe} \equiv \frac{V_{AB}}{V_{ab}}. \tag{7.35}$$

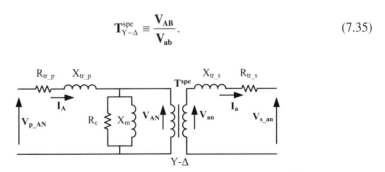

Figure 7.42. Single-phase equivalent circuit of a wye–delta transformer.

This artificial turns ratio is a complex quantity. Progressively substituting Equations (7.32) and Equation (7.34) into the previous formula yields:

$$T^{spe}_{Y-\Delta} = \frac{V_{AB}}{V_{ab}} = \frac{V_{AN}\sqrt{3}e^{j30°}}{V_{ab}} = \frac{V_{AN}}{V_{an}}. \tag{7.36}$$

Comparing the previous formula to Equation (7.31), we recall that V_{AN} and V_{ab} are in phase (and on the same leg); therefore, V_{AB} leads V_{ab} by 30° and V_{AN} leads V_{an} by 30°. By substituting Equation (7.31) into the previous expression, we derive a relation of practical use:

$$T^{spe}_{Y-\Delta} = \frac{V_{AB}}{V_{ab}} = T_{Y-\Delta}\sqrt{3}e^{j30°} = \left|\frac{V_{AB}}{V_{ab}}\right|e^{j30°}. \tag{7.37}$$

Further, the primary and secondary complex powers must be equal such that:

$$\begin{aligned}S_P &= S_S, \\ V_{AN}I_A^* &= V_{an}I_a^*.\end{aligned} \tag{7.38}$$

The second expression in the previous formula can be substituted into Equation (7.36):

$$T^{spe}_{Y-\Delta} \equiv \frac{V_{AN}}{V_{an}} = \frac{I_a^*}{I_A^*}. \tag{7.39}$$

Finally, if desired, Equation (7.33) can be used in the previous equation to produce an additional relation that is only valid for this wye–delta equivalent:

$$T^{spe}_{Y-\Delta} = \frac{I_a^*}{I_A^*} = \frac{\left(I_{ba}\sqrt{3}e^{-j30°}\right)^*}{I_A^*}. \tag{7.40}$$

This artificial turns ratio is a mathematical representation, which allows the three-phase wye–delta transformer to be represented by a single-phase (wye–wye) equivalent. Figure 7.42 provides the complete single-phase equivalent circuit of a wye–delta three-phase transformer for a balanced load.

By utilizing the artificial turns ratio, the method presented for the wye–wye connection (single-phase equivalent) can be used for the calculation of the voltage drop, regulation, efficiency, and so on.

7.3.3. Delta–Wye Connection

Figure 7.43 gives the connection diagram of a delta–wye connected three-phase transformer. In this transformer, the primary winding "AB" and the secondary winding "an" are on the same leg of the iron core. This implies that the primary line-to-line voltage

THREE-PHASE TRANSFORMERS

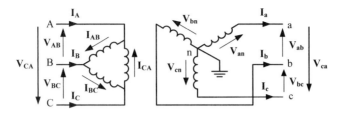

Figure 7.43. Delta–wye connected transformer.

(e.g., V_{AB}) is in phase with the secondary line-to-ground voltage (e.g., V_{an}). In addition, the primary delta current (e.g., I_{AB}) is in phase with the secondary current (e.g., I_a). Hence, the physical turns ratio for the Δ-Y transformer is:

$$T_{\Delta-Y} = \frac{N_p}{N_s} = \frac{V_{AB}}{V_{an}} = \frac{I_a}{I_{AB}}. \quad (7.41)$$

This transformer produces a −30° phase shift. In the case of a balanced load and balanced source, the delta source can be represented by an equivalent wye-connected source. Similar to the earlier wye–delta transformer, a single-phase equivalent circuit can be used in conjunction with an *artificial turns ratio* that produces a −30° phase shift. The single-phase equivalent circuit for this Δ-Y transformer is the same as for the Y-Δ case; hence, we again refer to Figure 7.42 to derive the complex turns ratio, which is defined by Equation (7.35) and repeated here:

$$T_{\Delta-Y}^{spe} = \frac{V_{AB}}{V_{ab}}. \quad (7.42)$$

The substitutions utilized to form Equation (7.36) and Equation (7.39) for the wye–delta transformer are also valid here for the delta–wye artificial turns ratio. Making the substitutions in a slightly different order yields:

$$T_{\Delta-Y}^{spe} = \frac{V_{AB}}{V_{ab}} = \frac{V_{AB}\, e^{-j30°}}{V_{an}\sqrt{3}} = \frac{V_{AN}}{V_{an}} = \frac{I_a^*}{I_A^*}. \quad (7.43)$$

Comparing the previous formula with Equation (7.41), we recall that V_{AB} and V_{an} are in phase (and on the same leg); therefore, V_{AB} leads V_{ab} by −30°, V_{AN} leads V_{an} by −30°, and I_A leads I_a by −30°. Substituting the relation, Equation (7.41), for the physical turns ratio of the Δ-Y transformer into the earlier formula gives:

$$T_{\Delta-Y}^{spe} = \frac{V_{AB}}{V_{ab}} = \frac{T_{\Delta-Y}}{\sqrt{3}} e^{-j30°} = \left|\frac{V_{AB}}{V_{ab}}\right| e^{-j30°}. \quad (7.44)$$

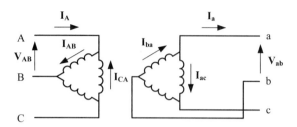

Figure 7.44. Delta–delta connected transformer.

Equation (7.40) is not applicable here, since there is no such current as $\mathbf{I}_{ba}$ in the wye secondary; however, there is a phase current $\mathbf{I}_{AB}$ in the primary delta winding such that from Figure 7.43:

$$\mathbf{I}_A = \mathbf{I}_{AB} - \mathbf{I}_{CA} = \mathbf{I}_{AB} - \mathbf{I}_{AB}e^{-j240°} = \sqrt{3}\mathbf{I}_{AB}e^{-j30°}. \quad (7.45)$$

Substituting this expression into Equation (7.43) yields a relation unique to the Δ-Y transformer:

$$T_{\Delta-Y}^{spe} = \frac{\mathbf{I}_a^*}{\mathbf{I}_A^*} = \frac{\mathbf{I}_a^*}{\left(\sqrt{3}\mathbf{I}_{AB}e^{-j30°}\right)^*}. \quad (7.46)$$

7.3.4. Delta–Delta Connection

Figure 7.44 introduces the delta–delta connected three-phase transformer. The line-to-line voltages and the currents in the delta windings are in phase. The transformer does not produce any phase shift. The delta windings can be replaced by an equivalent wye. In the case of a balanced load, a single-phase equivalent can be used. The turns ratio is the primary line-to-line voltage divided by the secondary line-to-line voltage:

$$T_{\Delta-\Delta} = \frac{N_p}{N_s} = \frac{\mathbf{V}_{AB}}{\mathbf{V}_{ab}} = \frac{\mathbf{I}_a}{\mathbf{I}_A} = \frac{\mathbf{I}_{ba}}{\mathbf{I}_{AB}}. \quad (7.47)$$

Because there is no phase shift, an artificial (complex) turns ratio is unnecessary for the Δ-Δ transformer.

7.3.5. Summary

The three-phase transformer connections with their resulting characteristics are summarized in Table 7.1. The applicable standard provides for the high voltage to lead the low voltage of a Y-Δ or Δ-Y transformer by 30°. In the case of a balanced load, the single-phase equivalent circuit can be used for these transformers. In the equivalent circuit, the turns ratio has to be multiplied by $e^{\pm j30°}$, as is illustrated in Figure 7.42.

THREE-PHASE TRANSFORMERS

TABLE 7.1. Three-Phase Transformer Relations

Transformer Connection	Primary and Secondary Voltage Phase Relation	Phase Shift[a]	Windings Placed on the Same Leg
Wye–wye	V_{AN} is in phase with V_{an}	Zero	A and a
Wye–delta	V_{AN} is in phase with V_{ab}	30°	A and ab
Delta–wye	V_{AB} is in phase with V_{an}	−30°	AB and a
Delta–delta	V_{AB} is in phase with V_{ab}	Zero	AB and ab

[a]Note: Phase shift lead between line-to-line voltages (V_{AB} and V_{ab}) and line-to-neutral voltages (V_{AN} and V_{an}), and line currents (I_A and I_a).

7.3.6. Analysis of Three-Phase Transformer Configurations

As noted earlier, three single-phase transformers can be connected to form a three-phase transformer bank. For the examples presented here, this bank is loaded with a balanced inductive load. For demonstration purposes, the transformers are treated as ideal transformers with the series impedances and exciting current neglected. We calculate the primary and secondary currents and voltages for each of the four possible transformer connections:

1. wye–wye
2. wye–delta
3. delta–wye
4. delta–delta

The rating of the single-phase transformers is identical:

$$S_{tr} := 100 \text{ kV} \cdot \text{A} \quad V_p := 7.967 \text{ kV} \quad V_s := 120 \text{ V}.$$

The three-phase load data are:

$$P_{load} := 240 \text{ kW} \quad pf_{load} := 0.8 \quad \text{(lagging)}.$$

Each single-phase transformer carries one-third of the load:

$$S_{tr\_load} := \frac{P_{load}}{3 \cdot pf_{load}} \cdot e^{j \cdot a\cos(pf_{load})} = (80 + 60j) \text{ kV} \cdot \text{A} \quad |S_{tr\_load}| = 100 \text{ kV} \cdot \text{A}.$$

EXAMPLE 7.5: Wye–wye connection

To analyze the wye–wye connected transformer, we refer to Figure 7.37. On the primary side, the line-to-neutral voltages of the supply are balanced and equal to the transformer rated primary voltage. In this case, the transformer voltage ratings correspond to the nominal line-to-neutral voltage of the system because the three single-phase transformers are wye connected. We set V_{AN} to be the reference with a phase angle of zero:

$$V_{AN} := V_p = 7.967 \cdot kV \quad |V_{AN}| = 7.97 \cdot kV \quad \arg(V_{AN}) = 0.0 \cdot \deg$$
$$V_{BN} := V_{AN} \cdot e^{-j \cdot 120 \, \deg} \quad |V_{BN}| = 7.97 \cdot kV \quad \arg(V_{BN}) = -120.0 \cdot \deg$$
$$V_{CN} := V_{AN} \cdot e^{-j \cdot 240 \, \deg} \quad |V_{CN}| = 7.97 \cdot kV \quad \arg(V_{CN}) = 120.0 \cdot \deg$$

Applying KVL, the V_{AB} line-to-line voltage of the supply is:

$$V_{AB} := V_{AN} - V_{BN} \quad |V_{AB}| = 13.8 \cdot kV \quad \arg(V_{AB}) = 30.0 \cdot \deg$$

For a balanced system, the V_{BC} and V_{CA} voltages have the same rms value, but their phase angles lag V_{AB} by 120° and 240°, respectively:

$$V_{BC} := V_{BN} - V_{CN} \quad |V_{BC}| = 13.8 \cdot kV \quad \arg(V_{BC}) = -90.0 \cdot \deg$$
$$V_{CA} := V_{CN} - V_{AN} \quad |V_{CA}| = 13.8 \cdot kV \quad \arg(V_{CA}) = 150 \cdot \deg$$

The primary line current in phase A is:

$$I_A := \overline{\left(\frac{S_{tr\_load}}{V_{AN}}\right)} \quad |I_A| = 12.55 \, A \quad \arg(I_A) = -36.9 \, \deg$$

The I_B and I_C currents have the same rms value, but their phase angles are $-36.87° - 120° = -156.87°$ and $-36.87° - 240° = -276.87°$ (or equivalently 83.13°), respectively:

$$I_B := \overline{\left(\frac{S_{tr\_load}}{V_{BN}}\right)} \quad |I_B| = 12.55 \, A \quad \arg(I_B) = -156.9 \cdot \deg$$

$$I_C := \overline{\left(\frac{S_{tr\_load}}{V_{CN}}\right)} \quad |I_C| = 12.55 \, A \quad \arg(I_C) = 83.1 \cdot \deg$$

The secondary phase voltages of the transformer bank are equal to the transformer rated secondary voltages, in conjunction with the proper phase angles:

$$V_{an} := V_s = 120 \, V \quad |V_{an}| = 120 \, V \quad \arg(V_{an}) = 0.0 \cdot \deg$$
$$V_{bn} := V_{an} \cdot e^{-j \cdot 120 \, \deg} \quad |V_{bn}| = 120 \, V \quad \arg(V_{bn}) = -120.0 \cdot \deg$$
$$V_{cn} := V_{an} \cdot e^{-j \cdot 240 \, \deg} \quad |V_{cn}| = 120 \, V \quad \arg(V_{cn}) = 120.0 \cdot \deg$$

Applying KVL, the V_{ab} secondary line-to-line voltage is:

$$V_{ab} := V_{an} - V_{bn} \quad |V_{ab}| = 207.8 \, V \quad \arg(V_{ab}) = 30.0 \cdot \deg$$

Alternatively, this voltage can be calculated directly from:

$$V_{ab} := \sqrt{3} \cdot V_{an} \cdot e^{j \cdot 30 \text{ deg}} \quad |V_{ab}| = 207.8 \text{ V} \quad \arg(V_{ab}) = 30.0 \text{ deg}.$$

The V_{bc} and V_{ca} voltages have the same rms value, but their phase angles are $30° - 120° = -90°$ and $30° - 240° = -210°$ (or $150°$), respectively:

$$V_{bc} := V_{bn} - V_{cn} \quad |V_{bc}| = 207.8 \cdot V \quad \arg(V_{bc}) = -90.0 \cdot \text{deg}$$
$$V_{ca} := V_{cn} - V_{an} \quad |V_{ca}| = 207.8 \cdot V \quad \arg(V_{ca}) = 150.0 \cdot \text{deg}.$$

The secondary line currents are:

$$I_a := \overline{\left(\frac{S_{tr\_load}}{V_{an}}\right)} \quad |I_a| = 833.3 \text{ A} \quad \arg(I_a) = -36.9 \text{ deg}$$

$$I_b := \overline{\left(\frac{S_{tr\_load}}{V_{bn}}\right)} \quad |I_b| = 833.3 \text{ A} \quad \arg(I_b) = -156.9 \text{ deg}$$

$$I_c := \overline{\left(\frac{S_{tr\_load}}{V_{cn}}\right)} \quad |I_c| = 833.3 \text{ A} \quad \arg(I_c) = 83.1 \text{ deg}.$$

The I_b and I_c currents have the same rms value as I_a, but their phase angles lag by $120°$ and $240°$, respectively.

The calculation demonstrates that the primary and secondary voltages and currents are in phase. The ideal wye–wye transformer does not produce phase shifts.

In the case of a balanced load, a single-phase equivalent circuit can represent the wye–wye transformer. For each phase, the turns ratio of the ideal transformer in this circuit is found from:

$$T_{Y\_Y} := \frac{V_{AN}}{V_{an}} \quad T_{Y\_Y} = 66.4.$$

Similarly, the turns ratio could have been computed from several different formulations utilizing Equation (7.27):

$$\frac{V_{AB}}{V_{ab}} = 66.4 \quad \frac{I_a}{I_A} = 66.4 \quad \frac{V_p}{V_s} = 66.4$$

EXAMPLE 7.6: Wye–delta connection

The calculation of the primary variables for the wye–delta transformer connection proceeds identically to that of the wye–wye connection in the previous example since the primaries for both cases are wye connections. To provide the reader with alternate approaches, the voltages and currents are computed here in a slightly different, but equivalent, manner. For the primary windings, the line-to-neutral supply voltages are balanced and equal to the transformer rated voltage:

$$V_{AN} := V_p \quad V_{AN} = 7.967 \text{ kV}.$$

The supply line-to-line voltage (e.g., V_{AB}) can be computed by applying Equation (7.32) rather than computing all three of the line-to-neutral voltages and performing KVL, as was carried out in Example 7.5:

$$V_{AB} := V_{AN} \cdot \sqrt{3} \cdot e^{j \cdot 30 \text{ deg}} \quad |V_{AB}| = 13.8 \text{ kV} \quad \arg(V_{AB}) = 30.0 \text{ deg}$$

The V_{BC} and V_{CA} voltages have the same rms value, but their phase angles are $30° - 120° = -90°$ and $30° - 240° = -210°$ (or 150°), respectively.

The primary current in phase A is:

$$I_A := \overline{\left(\frac{S_{tr\_load}}{V_{AN}}\right)} \quad |I_A| = 12.55 \text{ A} \quad \arg(I_A) = -36.9 \text{ deg}.$$

The I_B and I_C currents have the same rms value, but their phase angles are $-36.87° - 120° = -158.87°$ and $-36.87° - 240° = -276.87°$ (or 83.13°), respectively.

In the case of the delta-connected secondary, the secondary line-to-line voltages of the transformer bank are equal to the transformer rated secondary voltages (see Fig. 7.41a):

$$V_{ab} := V_s = 120 \text{ V} \quad |V_{ab}| = 120 \text{ V} \quad \arg(V_{ab}) = 0.0 \cdot \text{deg}$$
$$V_{bc} := V_s \cdot e^{-j \cdot 120 \text{ deg}} \quad |V_{bc}| = 120 \cdot \text{V} \quad \arg(V_{bc}) = -120.0 \cdot \text{deg}.$$
$$V_{ca} := V_s \cdot e^{-j \cdot 240 \text{ deg}} \quad |V_{ca}| = 120 \cdot \text{V} \quad \arg(V_{ca}) = 120.0 \cdot \text{deg}$$

The currents in the delta windings are:

$$I_{ba} := \overline{\left(\frac{S_{tr\_load}}{V_{ab}}\right)} \quad |I_{ba}| = 833.3 \text{ A} \quad \arg(I_{ba}) = -36.9 \cdot \text{deg}$$

$$I_{cb} := \overline{\left(\frac{S_{tr\_load}}{V_{bc}}\right)} \quad |I_{cb}| = 833.3 \text{ A} \quad \arg(I_{cb}) = -156.9 \cdot \text{deg}$$

$$I_{ac} := \overline{\left(\frac{S_{tr\_load}}{V_{ca}}\right)} \quad |I_{ac}| = 833.3 \text{ A} \quad \arg(I_{ac}) = 83.1 \cdot \text{deg}.$$

These phase currents in the delta secondary windings are identical to the line currents flowing through the wye-connected secondary windings of the wye–wye transformer.

THREE-PHASE TRANSFORMERS

The secondary line currents are obtained by applying KCL at the secondary nodes:

$$I_a := I_{ba} - I_{ac} \quad |I_a| = 1.443 \cdot kA \quad \arg(I_a) = -66.9 \cdot \deg$$
$$I_b := I_{cb} - I_{ba} \quad |I_b| = 1.443 \cdot kA \quad \arg(I_b) = 173.1 \cdot \deg$$
$$I_c := I_{ac} - I_{cb} \quad |I_c| = 1.443 \cdot kA \quad \arg(I_c) = 53.1 \cdot \deg.$$

The phase shifts between the primary and secondary line-to-line voltages and line currents are:

$$\phi_V := \arg(V_{AB}) - \arg(V_{ab}) \quad \phi_V = 30.0 \cdot \deg$$
$$\phi_I := \arg(I_A) - \arg(I_a) \quad \phi_I = 30.0 \cdot \deg.$$

The calculation reveals that the primary and secondary line voltages and currents are shifted by 30°. The ideal wye–delta transformer produces a 30° phase shift. In particular, the high-voltage primary line quantities lead the low-voltage secondary by 30°.

The actual physical turns ratio is:

$$\frac{V_p}{V_s} = 66.4 \quad \frac{V_{AN}}{V_{ab}} = 66.4 \quad \frac{I_{ba}}{I_A} = 66.4.$$

In the case of a balanced load, the wye–delta transformer can be represented by a single-phase equivalent circuit. The secondary line-to-neutral voltage in the wye equivalent is:

$$V_{an} := \frac{V_{ab} \cdot e^{-j \cdot 30 \deg}}{\sqrt{3}} \quad |V_{an}| = 69.3 \text{ V} \quad \arg(V_{an}) = -30.0 \cdot \deg.$$

The artificial (complex) turns ratio of the ideal transformer in such a single-phase circuit is defined by Equation (7.35):

$$T_{Y\_\Delta} := \frac{V_{AB}}{V_{ab}} = 99.6 + 57.5j \quad |T_{Y\_\Delta}| = 115.0 \quad \arg(T_{Y\_\Delta}) = 30.0 \cdot \deg.$$

Identical results are obtained using the formulations of Equation (7.36) and Equation (7.40):

$$\frac{V_{AN}}{V_{an}} = 99.6 + 57.5j \quad \frac{V_{AN}}{V_{ab}} \cdot \sqrt{3} \cdot e^{j \cdot 30 \deg} = 99.6 + 57.5j$$

$$\frac{I_a}{I_A} = 99.6 + 57.5j \quad \frac{\left(I_{ba} \cdot \sqrt{3} \cdot e^{-j \cdot 30 \deg}\right)}{I_A} = 99.6 + 57.5j.$$

EXAMPLE 7.7: Delta–wye connection

For a delta-connected primary, as seen in Figure 7.43, the supply line-to-line voltages are equal to the transformer rated primary voltage:

$$V_{AB} := V_p = 7.967 \cdot kV \quad |V_{AB}| = 7.967 \cdot kV \quad \arg(V_{AB}) = 0.0 \cdot \deg$$
$$V_{BC} := V_p \cdot e^{-j \cdot 120 \deg} \quad |V_{BC}| = 7.967 \cdot kV \quad \arg(V_{BC}) = -120.0 \cdot \deg$$
$$V_{CA} := V_p \cdot e^{-j \cdot 240 \deg} \quad |V_{CA}| = 7.967 \cdot kV \quad \arg(V_{CA}) = 120 \cdot \deg$$

The primary currents in the delta windings are:

$$I_{AB} := \overline{\left(\frac{S_{tr\_load}}{V_{AB}}\right)} \quad |I_{AB}| = 12.55 \text{ A} \quad \arg(I_{AB}) = -36.9 \cdot \deg$$

$$I_{BC} := \overline{\left(\frac{S_{tr\_load}}{V_{BC}}\right)} \quad |I_{BC}| = 12.55 \text{ A} \quad \arg(I_{BC}) = -156.9 \cdot \deg$$

$$I_{CA} := \overline{\left(\frac{S_{tr\_load}}{V_{CA}}\right)} \quad |I_{CA}| = 12.55 \text{ A} \quad \arg(I_{CA}) = 83.1 \cdot \deg$$

Note that the current magnitudes in the primary windings of this (delta) transformer are identical to those currents in the primary windings of the wye-connected transformers of the previous two cases.

Employing KCL, the primary line currents (or the supply currents) are (refer to Fig. 7.43):

$$I_A := I_{AB} - I_{CA} \quad |I_A| = 21.7 \cdot A \quad \arg(I_A) = -66.9 \cdot \deg$$
$$I_B := I_{BC} - I_{AB} \quad |I_B| = 21.7 \cdot A \quad \arg(I_B) = 173.1 \cdot \deg$$
$$I_C := I_{CA} - I_{BC} \quad |I_C| = 21.7 \cdot A \quad \arg(I_C) = 53.1 \cdot \deg$$

These line currents for the delta supply are 73% larger than those of the wye-connected supply.

For the wye-connected secondary windings, the phase voltages of the transformer bank are equal with the transformer rated secondary voltages as in Example 7.5:

$$V_{an} := V_s = 120 \text{ V}.$$

Rather than utilize KVL, as was carried out in Example 7.5, the secondary line-to-line voltage V_{ab} can be directly computed from Equation (7.34):

$$V_{ab} := V_{an} \cdot \sqrt{3} \cdot e^{j \cdot 30 \deg} \quad |V_{ab}| = 207.8 \text{ V} \quad \arg(V_{ab}) = 30.0 \cdot \deg.$$

The V_{bc} and V_{ca} voltages have the same rms value, but their phase angles are $30° - 120° = -90°$ and $30° - 240° = -210°$ (or $150°$), respectively.

THREE-PHASE TRANSFORMERS

The secondary current in phase a is:

$$I_a := \overline{\left(\frac{S_{tr\_load}}{V_{an}}\right)} \quad |I_a| = 833.3 \text{ A} \quad \arg(I_a) = -36.9 \text{ deg}.$$

The I_b and I_c currents have the same rms value, but their phase angles are $-36.87° - 120° = -156.87°$ and $-36.87° - 240° = -276.87°$ (or $83.13°$), respectively.

The phase shifts between the primary and secondary line voltages and line currents are:

$$\phi_V := \arg(V_{AB}) - \arg(V_{ab}) \quad \phi_V = -30.0 \cdot \text{deg}$$
$$\phi_I := \arg(I_A) - \arg(I_a) \quad \phi_I = -30.0 \cdot \text{deg}.$$

The calculation discloses that the secondary line-to-line voltages and line currents lag the primary values by $-30°$. Hence, the ideal delta–wye transformer produces a $-30°$ phase shift. Specifically, the primary line voltage and current lead the secondary line quantities by $-30°$.

In the case of a balanced load, a single-phase equivalent circuit can represent the delta–wye transformer. The single-phase turns ratio of the ideal transformer in this circuit is found using Equation (7.43) and Equation (7.46):

$$T_{\Delta\_Y} := \frac{V_{AB}}{V_{ab}} = 33.2 - 19.2j \quad |T_{\Delta\_Y}| = 38.3 \quad \arg(T_{\Delta\_Y}) = -30 \cdot \text{deg}$$

$$\frac{V_{AN}}{V_{an}} = 33.2 - 19.2j \qquad \frac{V_{AB}}{V_{an}} \cdot \frac{e^{-j30\,\text{deg}}}{\sqrt{3}} = 33.2 - 19.2j$$

$$\frac{\overline{I_a}}{\overline{I_A}} = 33.2 - 19.2j \qquad \frac{\overline{I_a}}{\left(\overline{I_{AB}} \cdot \sqrt{3} \cdot e^{-j30\,\text{deg}}\right)} = 33.2 - 19.2j.$$

EXAMPLE 7.8: Delta–delta connection

As for the delta–wye connected transformer, the line-to-line voltage of the primary supply for the delta–delta configuration is equivalent to the transformer rated voltage:

$$V_{AB} := V_p = 7.967 \text{ kV}.$$

Referring to Figure 7.44, the primary currents in the delta windings are:

$$I_{AB} := \overline{\left(\frac{S_{tr\_load}}{V_{AB}}\right)} \quad |I_{AB}| = 12.55 \text{ A} \quad \arg(I_{AB}) = -36.9 \text{ deg}.$$

The other primary phase currents (I_{BC} and I_{CA}) are the same as those found in the previous example and are not repeated here. In particular, I_{BC} and I_{CA} lag I_{AB} by 120° and 240°, respectively.

In Example 7.7, the primary line current (or the supply current) in phase A was found by using KCL; alternatively, I_A may be computed with Equation (7.45):

$$I_A := I_{AB} - I_{CA} \quad |I_A| = 21.7 \text{ A} \quad \arg(I_A) = -66.9 \text{ deg}.$$

The I_B and I_C currents have the same rms value, but their phase angles are $-66.87° - 120° = -186.87°$ (or 173.13°) and $-66.87° - 240° = -306.87°$ (or 53.13°), respectively.

The secondary line-to-line voltages of the transformer bank are equal with the transformer rated secondary voltages, with V_{bc} and V_{ca} lagging V_{ab} by 120° and 240°, respectively:

$$V_{ab} := V_s = 120 \text{ V}.$$

The secondary current in phase a of the delta windings is:

$$I_{ba} := \overline{\left(\frac{S_{tr\_load}}{V_{ab}}\right)} \quad |I_{ba}| = 833.3 \text{ A} \quad \arg(I_{ba}) = -36.9 \text{ deg}.$$

The secondary line current in phase a may be computed using KCL or from Equation (7.33):

$$I_a := I_{ba} \cdot \sqrt{3} \cdot e^{-j \cdot 30 \text{ deg}} \quad |I_a| = 1.44 \text{ kA} \quad \arg(I_a) = -66.9 \text{ deg}.$$

The I_b and I_c currents have the same rms value, but their phase angles are $-66.87° - 120° = -186.87°$ (or 173.13°) and $-66.87° - 240° = -306.87°$ (or 53.13°), respectively.

The phase shifts between the primary and secondary line voltages and currents are:

$$\phi_V := \arg(V_{AB}) - \arg(V_{ab}) \quad \phi_V = 0.0 \cdot \text{deg}$$
$$\phi_I := \arg(I_A) - \arg(I_a) \quad \phi_I = 0.0 \cdot \text{deg}.$$

The calculation shows that the primary and secondary voltages and currents are in phase. The ideal delta–delta transformer does not produce any phase shift.

In the case of a balanced load, a single-phase equivalent circuit can represent the delta–delta transformer. The turns ratio of the ideal single-phase transformer in this circuit is determined from Equation (7.47):

$$T_{\Delta\_\Delta} := \frac{V_{AB}}{V_{ab}} = 66.4 \quad \frac{V_p}{V_s} = 66.4$$

$$\frac{V_{AB}}{V_{ab}} = 66.4 \quad \frac{I_a}{I_A} = 66.4 \quad \frac{I_{ba}}{I_{AB}} = 66.4.$$

7.3.7. Equivalent Circuit Parameters of a Three-Phase Transformer

The analysis of the different transformer connections in a balanced system points out that a three-phase transformer can be represented by a single-phase equivalent circuit such as those drawn in Figure 7.40 and Figure 7.42. A short-circuit and an open-circuit measurement can determine the parameters of the equivalent circuit.

The general method presented in Section 7.2.3 can be used to determine the transformer circuit parameters. However, the open- and short-circuit tests of this transformer will produce three-phase power, line currents, and line-to-line voltages.

Figure 7.45 and Figure 7.46 provide the circuits that are used for the open- and short-circuit tests, respectively, of a three-phase transformer. A balanced three-phase

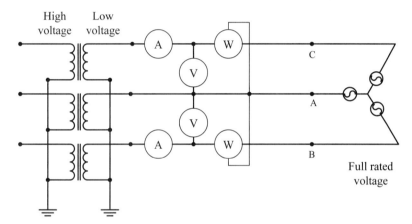

Figure 7.45. Open-circuit test for three-phase transformer parameter measurement.

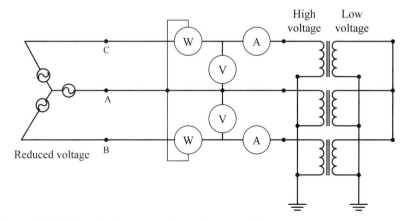

Figure 7.46. Short-circuit test for three-phase transformer parameter measurement.

voltage supplies the transformer. The input power, voltage, and current measurements are needed for the calculation of equivalent circuit parameters.

Consequently, the measurement of the currents and voltages using two voltmeters and two ammeters is sufficient. The readings of the two meters have to be equal. If a small difference is observed, the average value can be used. The voltmeters measure the line-to-line voltages. The calculation requires the line-to-neutral voltage, which is calculated by dividing the line-to-line voltage by $\sqrt{3}$.

The power is measured by the two-wattmeter method. The sum of the two wattmeter readings is the three-phase power; the difference multiplied by $\sqrt{3}$ is the three-phase reactive power (see Section 4.6). For the calculation of the transformer parameters, the measured three-phase power is divided by 3 to obtain the corresponding single-phase power. The actual calculation is presented through a numerical example.

EXAMPLE 7.9: Consider a wye–wye transformer with ratings of:

$$S_{tr} := 100 \text{ kV} \cdot \text{A} \quad V_p := 4600 \text{ V} \quad V_s := 230 \text{ V}.$$

The results of the transformer test measurements are as follows:

Open-Circuit Test: The primary side is opened, and a voltage as annotated in Figure 7.45 supplies the secondary side. The measurement results at the low voltage (secondary) side are:

$$V_{o\_ll} := 230 \text{ V} \quad I_o := 13 \text{ A} \quad P_{o\_3p} := 550 \text{ W}.$$

Short-Circuit Test: The secondary side is shorted and the primary is supplied by reduced voltage, as illustrated in Figure 7.46. The results of the measurements at the high voltage (primary) side are:

$$V_{s\_ll} := 160 \text{ V} \quad I_s := 16 \text{ A} \quad P_{s\_3p} := 1200 \text{ W}.$$

Note that these measurements are on the opposite sides as compared to the example of Section 7.2.3.

Calculation of Transformer Parameters
Open-Circuit Test
The single-phase power and line-to-neutral voltage are calculated from the measured three-phase power and line-to-line voltage. The results are:

$$V_{o\_ln} := \frac{V_{o\_ll}}{\sqrt{3}} \quad V_{o\_ln} = 132.8 \text{ V}$$

$$P_{o\_1p} := \frac{P_{o\_3p}}{3} \quad P_{o\_1p} = 183.3 \text{ W}.$$

THREE-PHASE TRANSFORMERS

The resistance, representing the magnetizing losses or core losses, is calculated from the single-phase power:

$$R_c := \frac{V_{o\_ln}^2}{P_{o\_1p}} \qquad R_c = 96.18 \, \Omega.$$

The admittance of the circuit is:

$$Y_o := \frac{I_o}{V_{o\_ln}} \qquad Y_o = 0.098 \, S.$$

The magnetizing reactance can be calculated using the admittance and the magnetizing (or core loss) resistance:

$$X_m := \frac{1}{\sqrt{Y_o^2 - \frac{1}{R_c^2}}} \qquad X_m = 10.27 \, \Omega.$$

An alternative way to calculate the magnetizing reactance is the determination of the apparent power and reactive power, which are:

$$S_o := V_{o\_ln} \cdot I_o \qquad S_o = 1.73 \cdot kW$$
$$Q_o := \sqrt{S_o^2 - P_{o\_1p}^2} \qquad Q_o = 1.72 \cdot kV \cdot A.$$

The magnetizing reactance can be calculated directly from the reactive power:

$$X_m := \frac{V_{o\_ln}^2}{Q_o} \qquad X_m = 10.27 \, \Omega.$$

The obtained core loss resistance and magnetizing reactance are connected in parallel and placed in the low voltage (secondary) side of the equivalent circuit.

Short-Circuit Test

The single-phase power and line-to-neutral voltage are calculated from the measured three-phase power and line-to-line voltage. The results are:

$$V_{s\_ln} := \frac{V_{s\_ll}}{\sqrt{3}} \qquad V_{s\_ln} = 92.38 \, V$$

$$P_{s\_1p} := \frac{P_{s\_3p}}{3} \qquad P_{s\_1p} = 400 \, W.$$

The combined primary and secondary resistance of the windings is computed using the single-phase power and current. The result is:

$$R_s := \frac{P_{s\_1p}}{I_s^2} \quad R_s = 1.56 \ \Omega.$$

The impedance magnitude of the circuit is:

$$Z_s := \frac{V_{s\_ln}}{I_s} \quad Z_s = 5.77 \ \Omega.$$

The combined primary and secondary leakage reactance of the windings is calculated from the impedance using the combined primary and secondary resistance. The result is:

$$X_s := \sqrt{Z_s^2 - R_s^2} \quad X_s = 5.56 \ \Omega.$$

The obtained combined primary and secondary resistance and the combined primary and secondary leakage reactance are connected in series and placed on the high-voltage (primary) side of the equivalent circuit.

7.3.8. General Program for Computing Transformer Parameters

The approaches for determining the equivalent circuit parameter values for single- and three-phase transformers were presented in Section 7.2.3 and Section 7.3.7, respectively. Here, we develop a general MATLAB program for computing the transformer parameters, regardless of whether the transformer is a single- or three-phase unit. To this end, we create a variable n to facilitate these calculations, where $n = 1$ for single phase and $n = 3$ for three phase. Table 7.2 lists the open-circuit measurements and the

TABLE 7.2. Open-Circuit Transformer Test

Measurements

Variable	Single-Phase Circuit ($n = 1$)	Three-Phase Circuit ($n = 3$)	Parameters
P_o	Single-phase power	Total three-phase power	$R_m = \dfrac{\left(V_o/\sqrt{n}\right)^2}{P_o/n},$
V_o	Line-to-neutral voltage	Line-to-line voltage	$\|Y_m\| = \dfrac{I_o}{V_o/\sqrt{n}},$
I_o	Line current	Line current	$X_m = \dfrac{1}{\sqrt{\|Y_m\|^2 - \dfrac{1}{R_m^2}}}$

THREE-PHASE TRANSFORMERS

TABLE 7.3. Short-Circuit Transformer Test

Measurements			
Variable	Single-Phase Circuit ($n = 1$)	Three-Phase Circuit ($n = 3$)	Parameters
P_s	Single-phase power	Total three-phase power	$R_S = \dfrac{P_s/n}{I_s^2},$
V_s	Line-to-neutral voltage	Line-to-line voltage	$|\mathbf{Z_S}| = \dfrac{V_s/\sqrt{n}}{I_s},$
I_s	Line current	Line current	$X_S = \sqrt{|\mathbf{Z_S}|^2 - R_S^2}$

resultant parameters that can be determined. Similarly, Table 7.3 shows the short-circuit measurements and the equations for those circuit parameters.

A MATLAB m file program can be developed in a straightforward manner using the equations and definitions from the two tables. The program is:

```
%
%    TransformerParameters.m
%
clear all; n=0;

while n ~= 1 & n ~= 3
    % prompt the user to enter the transformer type
    n = input('Enter no. of phases for transformer (1 or 3) > ');
end
% prompt the user to enter the open circuit test results
fprintf('\nEnter the open-circuit test measurement results:\n');
Vo = input('Enter open-circuit voltage measurement (volts) > ');
Io = input('Enter open-circuit current measurement (amps) > ');
Po = input('Enter open-circuit power measurement (watts) > ');

% process open-circuit test measurements
% find magnetizing (core loss) resistance from single-phase
% power
Rm = (Vo/sqrt(n))^2/(Po/n);
fprintf('\nMagnetizing (core loss) resistance = %g ohms', Rm);
% find parallel magnetizing admittance magnitude
Ym = Io/(Vo/sqrt(n));
% extract the magnetizing reactance
Xm = 1/sqrt(Ym^2-1/Rm^2);
fprintf('\nMagnetizing reactance = %g ohms', Xm);

% prompt the user to enter the short-circuit test results
fprintf('Enter the short-circuit test measurement results:\n');
Vs = input('Enter short-circuit voltage measurement (volts) > ');
```

```
Is = input('Enter short-circuit current measurement (amps) > ');
Ps = input('Enter short-circuit power measurement (watts) > ');

% process short-circuit test measurements
% find series winding resistance from single-phase power
Rs = (Ps/n)/Is^2;
fprintf('\nWinding resistance = %g ohms', Rs);
% find series winding impedance magnitude
% line voltage must be changed to line-to-neutral value
Zs = (Vs/sqrt(n))/Is;
% find series transformer winding reactance
Xs = sqrt(Zs^2-Rs^2);
fprintf('\nWinding reactance = %g ohms', Xs);
```

EXAMPLE 7.10: To test the program, we use the measurement results from Section 7.2.3 and Section 7.3.7. The single-phase measurement results were:

Open-circuit test:	V_{oc} = 13 kV	I_{oc} = 170 A	P_{oc} = 1.1 MW
Short-circuit test:	V_{sc} = 6 kV	I_{sc} = 290 A	P_{sc} = 0.8 MW

The input and output dialogue for the single-phase case is:

```
>> transformerparameters
Enter number of phases of the transformer (1 or 3) > 1

Enter the open-circuit test measurement results:
Enter the open-circuit voltage measurement (volts) > 13e3
Enter the open-circuit current measurement (amps) > 170
Enter the open-circuit power measurement (watts) > 1.1e6

Magnetizing (core loss) resistance = 153.636 ohms
Magnetizing reactance = 88.168 ohms

Enter the short-circuit test measurement results:
Enter the short-circuit voltage measurement (volts) > 6e3
Enter the short-circuit current measurement (amps) > 290
Enter the short-circuit power measurement (watts) > 0.8e6

Winding resistance = 9.51249 ohms
Winding reactance = 18.3732 ohms
```

These single-phase results are identical to those of Section 7.2.3.
    The three-phase measurement results were:

Open-circuit test:	V_{oc} = 230 V	I_{oc} = 13 A	P_{oc} = 550 W
Short-circuit test:	V_{sc} = 160 V	I_{sc} = 16 A	P_{sc} = 1200 W

The input and output dialogue for the three-phase case is:

```
>> transformerparameters
Enter number of phases of the transformer (1 or 3) > 3

Enter the open-circuit test measurement results:
Enter the open-circuit voltage measurement (volts) > 230
Enter the open-circuit current measurement (amps) > 13
Enter the open-circuit power measurement (watts) > 550

Magnetizing (core loss) resistance = 96.1818 ohms
Magnetizing reactance = 10.2728 ohms

Enter the short-circuit test measurement results:
Enter the short-circuit voltage measurement (volts) > 160
Enter the short-circuit current measurement (amps) > 16
Enter the short-circuit power measurement (watts) > 1200

Winding resistance = 1.5625 ohms
Winding reactance = 5.55805 ohms
```

These three-phase results are identical to those of Section 7.3.7. The previous program demonstrates that with some forethought, the simple use of the variable n allows the development of equations and programs capable of analyzing both single-phase and three-phase circuits.

7.3.9. Application Examples

Two practical analyses of three-phase transformers are now presented.

EXAMPLE 7.11: Analysis of three-phase transformer operation in a system

In this example, a Y-Y transformer is embedded in a network, which supplies a load. Figure 7.47 portrays a three-phase network that supplies a load through a three-phase

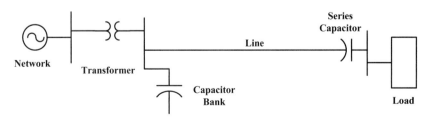

Figure 7.47. One-line diagram of system for MATLAB transformer example.

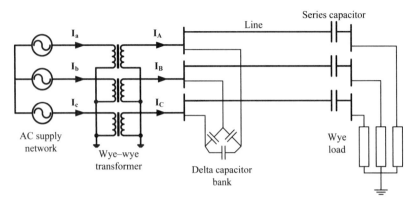

Figure 7.48. Three-phase network supplies a load through three-phase transformer and transmission line.

transformer and transmission line. A shunt capacitor bank, connected in delta, is placed on the transformer secondary terminal to improve the power factor. Another capacitor is connected in series with the transmission line to reduce line impedance and the voltage drop.

The 60 Hz power system under consideration and illustrated in Figure 7.48 consists of:

- a 25-mile transmission line of $0.12 + j0.65$ Ω/mile impedance;
- the supply network with an open-circuit voltage (V_{open}) of 69 kV and a short-circuit current (I_{short}) rating of 4 kA;
- each delta-connected capacitor bank, $C_b = 0.1$ µF;
- the capacitor connected in series with the line, $C_s = 400$ µF; and
- the transformer that is rated at 20 MVA with a rated primary voltage of 69 kV and a rated secondary voltage of 240 kV.

An open-circuit transformer test measurement at the high-voltage (secondary) side yields:

$$V_o = 240 \text{ kV} \quad I_o = 10 \text{ A} \quad P_o = 600 \text{ kW}.$$

The short-circuit test measurement at the low-voltage (primary) side finds:

$$V_s = 7 \text{ kV} \quad I_s = 170 \text{ A} \quad P_s = 1 \text{ MW}.$$

For this problem, it is assumed that V_{net} will vary to maintain 235 kV at the load. The load is variable between 0 and 20 MVA with a power factor of 0.7 lagging. First, the equivalent circuit of this network is drawn in Figure 7.49. Next, we will:

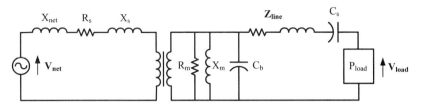

Figure 7.49. Single-phase equivalent of the three-phase network supplies a load through three-phase transformer and transmission line.

- calculate the impedances of the transformer, supply, and capacitors;
- compute all currents and voltages, and voltage regulation versus variable load;
- plot the voltage regulation versus load; and
- determine the load value that causes a system voltage drop of 5%.

The MATLAB program begins with establishing the earlier system data and making a few simple calculations. Specifically, the network reactance (X_{net}) is computed from the open-circuit voltage and short-circuit current for the network; the open-circuit voltage is the line-to-line voltage such that it must be divided by $\sqrt{3}$ to obtain the line-to-neutral voltage (V_{ln}):

$$X_{net} = \frac{|V_{ln}|}{|I_{short}|} = \frac{V_{open}/\sqrt{3}}{I_{short}}, \qquad (7.48)$$

```
%
%       Transformer.m
%
omega = 2*pi*60;    % system frequency (rad/sec)

% transmission line impedance (ohms)
Z_line = (0.12 + j*0.65) * 25;

% primary side network supply
Vnet_open = 69e3;           % volts
Inet_short = 4e3;           % amps
X_net = (Vnet_open/sqrt(3))/Inet_short;
```

To determine the impedance of the delta-connected capacitor bank (Z_{bank}) requires the reactance be divided by 3 to form the desired single-phase wye equivalent ($Z_\Delta = 3Z_Y$):

```
% capacitor bank at transformer
Cbank = 0.1e-6;                 % capacitance (F)
Xbank = -1/(omega*Cbank);       % Delta
Zbank = j*Xbank/3;              % Y connected equivalent
```

The impedance of the series capacitor is computed and added to the line impedance to form an overall impedance for the rhs of the secondary part of the network (not including the variable load):

```
% series capacitor at transmission line end
Cseries = 400e-6;          % capacitance (F)
Zseries = -j/(omega*Cseries);
Z_rhs = Z_line + Zseries;
```

The turns ratio for this Y-Y connected transformer is computed from the ratio of the primary and secondary voltage ratings:

```
% transformer parameters
Str = 20e6;              % rating (V-A)
Vp  = 69e3;              % primary voltage (V)
Vsec = 240e3;            % secondary voltage (V)
Tratio = Vp/Vsec;        % turns ratio
```

The measurements from the transformer open-circuit test allow the computation of the core loss resistance (R_m) and magnetizing admittance ($\mathbf{Y_m}$) from which the magnetizing reactance (X_m) may be extracted and placed on the secondary side of the transformer:

$$R_m = \frac{\left(V_o/\sqrt{3}\right)^2}{P_o/3},$$

$$|\mathbf{Y_m}| = \frac{I_o}{V_o/\sqrt{3}}, \qquad (7.49)$$

$$X_m = \frac{1}{\sqrt{|\mathbf{Y_m}|^2 - \frac{1}{R_m^2}}}.$$

```
% open circuit test
Vo = 240e3;       % voltage (V)
Io = 10;          % current (A)
Po = 600e3;       % power (W)
% find magnetizing (core loss) resistance from single-phase power
Rm = (Vo/sqrt(3))^2/(Po/3);
% find parallel magnetizing admittance magnitude
Ym = Io/(Vo/sqrt(3));
% extract the magnetizing reactance
Xm = 1/sqrt(Ym^2-1/Rm^2);
```

THREE-PHASE TRANSFORMERS

An alternate method to compute the magnetizing reactance is to first determine the quadrature power per phase (Q_o):

$$|\mathbf{S}| = \frac{V_o}{\sqrt{3}} I_o,$$

$$\frac{Q_o}{3} = \sqrt{|\mathbf{S}|^2 - \left(\frac{P_o}{3}\right)^2}, \qquad (7.50)$$

$$X_m = \frac{\left(V_o/\sqrt{3}\right)^2}{Q_o/3}.$$

The results from the transformer short-circuit test lead to the calculation of the transformer winding resistance (R_s) and impedance (magnitude). They are then used to find the winding reactance (X_s), which is placed on the primary side of the transformer:

$$R_S = \frac{P_s/3}{I_s^2},$$

$$|\mathbf{Z_S}| = \frac{V_s/\sqrt{3}}{I_s}, \qquad (7.51)$$

$$X_S = \sqrt{|\mathbf{Z_S}|^2 - R_S^2}.$$

```
% short circuit test
Vs = 7e3;          % line voltage (V)
Is = 170;          % line current (A)
Ps = 1e6;          % three-phase power (W)
% find series winding resistance from single-phase power
Rs = Ps/(3*Is^2)
% find series winding impedance magnitude
% line voltage must be changed to line-to-neutral value
Zs = (Vs/sqrt(3))/Is;
% find series transformer winding reactance
Xs = sqrt(Zs^2-Rs^2)
```

The load voltage and power factor are fixed, and the load power is varied from 0 to 20 MVA in steps of 100 kVA:

```
% Load data
Vload = 235e3;          % volts
pf_load = 0.7;          % lagging
Sload = 6e6 : 0.1e6 : 20e6;     % MVA
for k=1: size(Sload,2);
```

The major portion of this analysis requires working from the load conditions back through the system to determine the network supply voltage and the voltage drop. First, the current through the load, and hence the transmission line, is determined from:

$$\mathbf{I}_{Load} = \frac{|S|/3}{V_{Load}/\sqrt{3}} e^{-j\arccos(pf)}. \tag{7.52}$$

The voltage across the transformer secondary terminals is found by applying KVL:

$$\mathbf{V}_{sec} = \mathbf{I}_{Load}(\mathbf{Z}_{Line} + jX_{Cs}) + \frac{V_{Load}}{\sqrt{3}}.$$

With the transformer secondary voltage now known, the current through each of the parallel circuit elements in the secondary network can be computed from Ohm's law:

$$\mathbf{I}_{bank} = \frac{\mathbf{V}_{sec}}{\mathbf{Z}_{bank}},$$

$$\mathbf{I}_{Rm} = \frac{\mathbf{V}_{sec}}{R_m},$$

$$\mathbf{I}_{Xm} = \frac{\mathbf{V}_{sec}}{jX_m}.$$

Using KCL, the total current through the transformer secondary winding is:

$$\mathbf{I}_{sec} = \mathbf{I}_{Load} + \mathbf{I}_{bank} + \mathbf{I}_{Rm} + \mathbf{I}_{Xm}.$$

The primary supply current is found using the transformer turns ratio (T):

$$\mathbf{I}_{prim} = \frac{\mathbf{I}_{sec}}{T}.$$

KVL is applied to the primary side network to compute the network supply line-to-neutral voltage:

$$\mathbf{V}_{net} = \mathbf{I}_{prim}(jX_{net} + R_s + jX_s) + T\mathbf{V}_{sec}.$$

Finally, the voltage regulation is found from:

$$\text{Voltage Drop} = \frac{|\mathbf{V}_{net}| - T(V_{Load}/\sqrt{3})}{T(V_{Load}/\sqrt{3})},$$

```
    % use load parameters to compute current through load
    % (and line)
    I_load = (Sload(k)/3)/(Vload/...
        sqrt(3))*exp(-j*acos(pf_load));
    % use KVL to compute voltage across transformer secondary
    V_tr_sec = Vload/sqrt(3) + I_load*Z_rhs;
    % Ohm's law to find current through capacitor bank
    I_bank = V_tr_sec/Zbank;
    % Ohm's law to find currents through magnetizing terms
    Ixm = V_tr_sec/(j*Xm);
    Irm = V_tr_sec/Rm;
    % use KCL to find current through transformer secondary
    % winding
    I_tr_sec = I_load + I_bank + Ixm + Irm;
    % use turns ratio to find primary transformer current
    I_supply = I_tr_sec/Tratio;
    % use KVL to find network supply voltage
    Vnet = Tratio*V_tr_sec + I_supply*(Rs+j*Xs+j*X_net);
    % determine system voltage drop
    Vdrop(k) = (abs(Vnet)-abs(Tratio*Vload/sqrt(3)))/...
        abs(Tratio*Vload/sqrt(3))*100;
end
```

The remainder of the MATLAB program searches through the computed regulation values to find the 5% regulation value, and the regulation versus load is plotted. From Figure 7.50, the load value that causes a voltage regulation of 5% is 9.776 MVA.

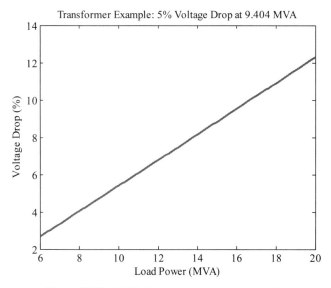

Figure 7.50. MATLAB voltage regulation results.

```
% search for the load at 5% voltage drop using array index
drop = 5;
found = 0;
for k=1: size(Vdrop, 2)
    if (found == 0) && (abs(Vdrop(k)) > drop)
        found = k;
    end
end
% interpolate results to 5% value
Pdrop = Sload(found) - (Sload(found)-Sload(found-1))*...
    (Vdrop(found)-drop)/(Vdrop(found)-Vdrop(found-1));
plot(Sload/1e6,Vdrop,'LineWidth',2.5)
set(gca,'fontname','Times','fontsize',12);
title(['Transformer Example: 5% Voltage Drop at ',...
        num2str(Pdrop/1e6,'%5.3f'),' MVA']);
xlabel('Load Power (MVA)');
ylabel('Voltage Drop (%)');
```

EXAMPLE 7.12: Analysis of three-phase delta–wye transformer operation in a system

Within this example, we illustrate the manner in which the complex turns ratio may be used for the Δ-Y transformer to obtain a more efficient solution. A delta–wye connected transformer supplies two balanced loads through a transmission line. The single-line diagram is given in Figure 7.51. The Δ-Y transformer rating data are:

$$M := 10^6 \qquad S_{tr} := 50 \, M \cdot V \cdot A \qquad x_{tr} := 6\%$$
$$V_p := 220 \, kV \quad \text{delta primary winding}$$
$$V_s := 120 \, kV \quad \text{wye secondary winding.}$$

The transmission line lengths and impedance p.u. length are:

$$L_{12} := 45 \, mi \quad L_{23} := 22 \, mi \quad Z_{line} := (0.107 + j \cdot 0.730) \cdot \frac{\Omega}{mi}.$$

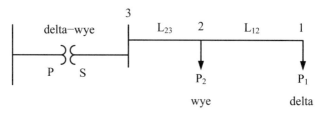

Figure 7.51. One-line diagram of a delta–wye connected transformer supplying two balanced loads through a transmission line.

THREE-PHASE TRANSFORMERS

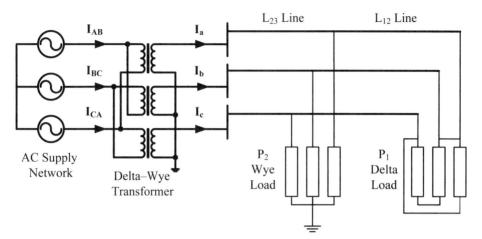

Figure 7.52. Circuit diagram of a delta–wye connected transformer supplying two balanced loads through a transmission line.

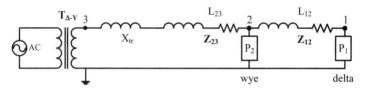

Figure 7.53. Equivalent circuit diagram of a delta–wye connected transformer supplying two balanced loads through a transmission line.

The loads are:

$$P_1 := 10 \text{ MW} \quad pf_1 := 0.75 \text{ (lagging)} \quad V_1 := 115 \text{ kV} \quad \text{delta connected}$$
$$P_2 := 15 \text{ MW} \quad pf_2 := 0.85 \text{ (leading)} \quad \quad \quad \quad \quad \quad \text{wye connected.}$$

The transformer series reactance at the secondary (S) side is:

$$X_{tr\_s} := x_{tr} \cdot \frac{V_s^2}{S_{tr}} \quad X_{tr\_s} = 17.28 \, \Omega.$$

The overall system circuit diagram is provided in Figure 7.52, and the equivalent circuit is given in Figure 7.53.

To solve this problem, we begin the calculation at the point farthest from the transformer, that is, starting at Load 1 where the reference voltage, V_1, is located. The line voltages at this delta load are:

$$V_{ab\_1} := V_1 \quad V_{bc\_1} := V_1 \cdot e^{-j \cdot 120 \text{ deg}} \quad V_{ca\_1} := V_1 \cdot e^{-j \cdot 240 \text{ deg}}.$$

The power in each phase is one-third of the total load power. The currents in the delta branches for a lagging power factor are:

$$I_{ab\_1}(V_{ab\_1}, P_1) := \frac{P_1}{3 \cdot pf_1 \cdot V_{ab\_1}} \cdot e^{-j \cdot acos(pf_1)}$$

$$|I_{ab\_1}(V_{ab\_1}, P_1)| = 38.6 \text{ A}$$

$$I_{bc\_1}(V_{ab\_1}, P_1) := I_{ab\_1}(V_{ab\_1}, P_1) \cdot e^{-j \cdot 120 \text{ deg}}$$

$$I_{ca\_1}(V_{ab\_1}, P_1) := I_{ab\_1}(V_{ab\_1}, P_1) \cdot e^{j \cdot 120 \text{ deg}}.$$

The line current is the difference between the appropriate delta currents:

$$I_{a\_12}(V_{ab\_1}, P_1) := I_{ab\_1}(V_{ab\_1}, P_1) - I_{ca\_1}(V_{ab\_1}, P_1).$$

The corresponding line-to-neutral voltage for Load 1 is:

$$V_{an\_1}(V_{ab\_1}) := \frac{V_{ab\_1}}{\sqrt{3}} \cdot e^{-j \cdot 30 \text{ deg}} \quad |V_{an\_1}(V_{ab\_1})| = 66.4 \text{ kV}.$$

With the currents of Load 1 computed, we turn to the transmission line between Loads 1 and 2. Using KVL, the line-to-neutral voltage at Load 2 is the sum of the line-to-neutral voltage of Load 1 and the voltage drop on Line 12 (L_{12}):

$$V_{an\_2}(V_{ab\_1}, P_1) := V_{an\_1}(V_{ab\_1}) + Z_{line} \cdot L_{12} \cdot I_{a\_12}(V_{ab\_1}, P_1)$$

$$\sqrt{3} \cdot |V_{an\_2}(V_{ab\_1}, P_1)| = 118.0 \cdot \text{kV}.$$

The phase current of wye-connected Load 2 for a leading power factor is:

$$I_{a\_2}(V_{ab\_1}, P_1, P_2) := \frac{P_2}{3 \cdot pf_2 \cdot V_{an\_2}(V_{ab\_1}, P_1)} \cdot e^{j \cdot acos(pf_2)}$$

$$|I_{a\_2}(V_{ab\_1}, P_1, P_2)| = 86.4 \text{ A}.$$

Using KCL, the current of Line 23 is the sum of the Load 1 and Load 2 currents:

$$I_{23}(V_{ab\_1}, P_1, P_2) := I_{a\_2}(V_{ab\_1}, P_1, P_2) + I_{a\_12}(V_{ab\_1}, P_1).$$

If the transformer reactance is placed on the secondary side, the phase **a** voltage at the secondary side of the transformer is:

$$V_{s\_a}(V_{ab\_1}, P_1, P_2) := V_{an\_2}(V_{ab\_1}, P_1) + I_{23}(V_{ab\_1}, P_1, P_2) \cdot (Z_{line} \cdot L_{23} + j \cdot X_{tr\_s}).$$

THREE-PHASE TRANSFORMERS 445

The line-to-neutral voltages of the other phases are:

$$V_{s\_b}(V_{ab\_1}, P_1, P_2) := V_{s\_a}(V_{ab\_1}, P_1, P_2) \cdot e^{-j \cdot 120 \text{ deg}}$$
$$V_{s\_c}(V_{ab\_1}, P_1, P_2) := V_{s\_a}(V_{ab\_1}, P_1, P_2) \cdot e^{-j \cdot 240 \text{ deg}}.$$

The equivalent line-to-line voltage at the transformer secondary is:

$$V_{s\_ab}(V_{ab\_1}, P_1, P_2) := V_{s\_a}(V_{ab\_1}, P_1, P_2) - V_{s\_b}(V_{ab\_1}, P_1, P_2),$$

or by an equivalent alternative method is:

$$V_{s\_ab\_alt}(V_{ab\_1}, P_1, P_2) := \sqrt{3} \cdot V_{s\_a}(V_{ab\_1}, P_1, P_2) \cdot e^{j \cdot 30 \text{ deg}}.$$

At this point, we illustrate two different methods to find the primary line-to-line voltage (V_{AB}). First, V_{AB} is determined using the physical (natural) turns ratio; afterward, we employ the artificial (complex) turns ratio formulation to illustrate its application in finding V_{AB}.

Method 1: Physical Turns Ratio
The physical turns ratio is found from the transformer primary and secondary voltage ratings while recalling that the voltage ratings are line-to-line values, but the voltage for the wye secondary winding is the line-to-neutral equivalent:

$$T_{\text{delta\_wye}} := \frac{V_p}{\frac{V_s}{\sqrt{3}}} = 3.175.$$

The primary voltage in phase "a" is calculated using the natural turns ratio. The secondary line-to-neutral voltage is in phase with the primary line-to-line voltage:

$$V_{p\_AB}(V_{ab\_1}, P_1, P_2) := T_{\text{delta\_wye}} \cdot V_{s\_a}(V_{ab\_1}, P_1, P_2)$$
$$|V_{p\_AB}(V_{ab\_1}, P_1, P_2)| = 212.4 \cdot \text{kV}$$
$$\arg(V_{p\_AB}(V_{ab\_1}, P_1, P_2)) = -27.3 \cdot \text{deg}$$

The phase shift in the transformer is the difference between the phase angles of the primary and secondary line-to-line voltages:

$$\arg(V_{p\_AB}(V_{ab\_1}, P_1, P_2)) - \arg(V_{s\_ab}(V_{ab\_1}, P_1, P_2)) = -30.0 \cdot \text{deg}.$$

The equivalent primary line-to-neutral voltage is then:

$$V_{P\_AN} := V_{p\_AB}(V_{ab\_1}, P_1, P_2) \cdot \frac{e^{-j \cdot 30 \text{ deg}}}{\sqrt{3}}$$
$$|V_{P\_AN}| = 122.6 \cdot \text{kV} \quad \arg(V_{P\_AN}) = -57.3 \cdot \text{deg}.$$

This voltage is needed if an impedance is connected between the transformer and the source. In such a case, the source voltage is calculated by adding any voltage drop through the impedance to this transformer primary line-to-neutral voltage.

Method 2: Complex Turns Ratio
The complex turns ratio is calculated using Equation (7.44):

$$T_{\Delta\_Y} := \frac{V_p}{V_s} \cdot e^{-j30\,deg} = 1.588 - 0.917j.$$

With the secondary line-to-line voltage known, the primary line-to-line voltage is computed directly from the complex turns ratio:

$$V_{AB\_p}(V_{ab\_1}, P_1, P_2) := T_{\Delta\_Y} \cdot V_{s\_ab}(V_{ab\_1}, P_1, P_2)$$
$$|V_{AB\_p}(V_{ab\_1}, P_1, P_2)| = 212.4 \cdot kV$$
$$\arg(V_{AB\_p}(V_{ab\_1}, P_1, P_2)) = -27.3 \cdot deg.$$

The obtained primary line voltage is identical with that calculated using the natural turns ratio.

Referring to Equation (7.43), the complex turns ratio allows the primary line-to-neutral voltage to be computed directly from the secondary line-to-neutral voltage:

$$V_{AN\_p}(V_{ab\_1}, P_1, P_2) := T_{\Delta\_Y} \cdot V_{s\_a}(V_{ab\_1}, P_1, P_2)$$
$$|V_{AN\_p}(V_{ab\_1}, P_1, P_2)| = 122.6 \cdot kV$$
$$\arg(V_{AN\_p}(V_{ab\_1}, P_1, P_2)) = -57.3 \cdot deg.$$

The calculated line-to-neutral voltage is identical to that obtained earlier.

The phase difference between the primary and secondary line-to-line voltages is:

$$\arg(V_{AB\_p}(V_{ab\_1}, P_1, P_2)) - \arg(V_{s\_ab}(V_{ab\_1}, P_1, P_2)) = -30.0 \, deg.$$

The use of complex turn ratio is particularly advantageous when a large network is analyzed.

Regulation or Voltage Drop
As one might surmise, Load 1—the load most distant from the generator—experiences the largest drop in voltage with respect to the supply. For a supply voltage of 220 kV, the corresponding line-to-line voltage at Load 1 is:

$$V_{1\_ll}(P_1, P_2) := \text{root}(V_{AB\_p}(V_{ab\_1}, P_1, P_2) - 220 \, kV, V_{ab\_1})$$
$$|V_{1\_ll}(P_1, P_2)| = 115.1 \cdot kV \quad \arg(V_{1\_ll}(P_1, P_2)) = 28.9 \cdot deg.$$

THREE-PHASE TRANSFORMERS

The voltage drop across the circuit is therefore:

$$\text{Voltage\_drop} := \frac{220 \text{ kV} - \frac{220}{120} \cdot |V_{1\_ll}(P_1, P_2)|}{220 \text{ kV}} \qquad \text{Voltage\_drop} = 4.1 \cdot \%.$$

With two or more loads, proper calculation of the voltage regulation requires determining the no-load and full-load voltages individually at each load.

7.3.10. Concept of Transformer Protection

The protection of transformers is an advanced subject, which is beyond the scope of this book. However, the concept of the transformer protection is described in this section. Transformer protection depends on the transformer rating. Small distribution transformers are protected by a fuse connected in series with the high voltage winding. Figure 1.23 pictures a pole-mounted distribution transformer protected by a fuse cutout. Transformers less than 5–10 MVA are protected by overcurrent protection. Larger units are protected by differential protection, while backup protection is provided by overcurrent protection. In addition, the grounded Y-connected three-phase transformers have dedicated ground fault protection.

7.3.10.1. Overcurrent Protection. The transformer must be protected against overload, which can be produced by a short circuit in the transformer, the transmission line, or a load larger than the transformer rating. The short-circuit current can be 4–10 times the rated current, which requires fast protection operation. A greater-than-rated current overload typically triggers the protection when the current is about twice the transformer rated current. In this case, the protection can operate with significant delay.

Nevertheless, the transformer overload protection should be coordinated with the transmission line protection. A short circuit on the connected transmission line should be cleared by the line protection. The transformer overload protection should be delayed to provide backup protection.

Figure 7.54 illustrates the concept of the overload and ground fault protection of a three-phase transformer. For the overload protection, the transformer phase currents are measured using current transformers (CTs). The obtained current signals supply the overcurrent relay, which has nonlinear characteristics as shown in Figure 5.68. Today, most power companies use digital relays with similar characteristics. The operation time depends on the current value. For simple overload, the time delay is several seconds, but for short-circuit current, the operation time is around 10–20 cycles.

Switching on a large transformer oftentimes produces inrush current because of temporary saturation of the iron core. The peak value of this nonsinusoidal transient current can be 8–30 times the transformer rated current, but it attenuates relatively fast. Figure 7.55 shows a typical transformer inrush current. The magnetic inrush current should not trigger the protection. The inrush current has a high magnitude second harmonic. The relays used for transformer protection may block the trip signal

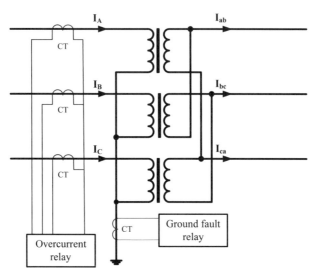

Figure 7.54. Concept of transformer overcurrent and ground fault protection.

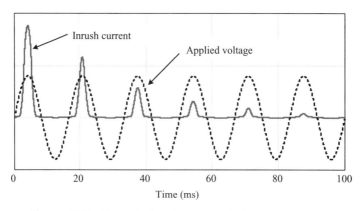

Figure 7.55. Magnetic inrush current of a large transformer.

when second harmonics are detected. The inrush current is important only for larger transformers.

Overvoltage increases the peak magnetic flux density (B) in the transformer, which can penetrate into the iron core saturation region. This will generate a sudden increase in the magnetizing current, which is not sinusoidal and has both positive and negative peaks. Fourier analysis reveals that the saturation generates a large number of harmonics—particularly pronounced is the second harmonic. The harmonics may produce severe disturbances in computers and other sensitive electronic components, which suggests operating below the saturation level.

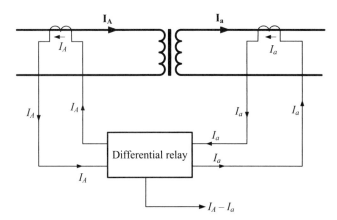

Figure 7.56. Concept of transformer differential protection.

7.3.10.2. Ground Fault Protection.
Ground fault-produced current is driven by the line-to-neutral voltage through the phase conductor and the ground. The current returns to the transformer through the ground conductor in the case of a grounded wye–delta connection. The ground conductor current is sensed by a CT. The CT secondary current supplies a ground fault relay as seen in Figure 7.54.

The ground current is zero in case of balanced load and can be close to the phase current in the case of unbalanced load. The ground fault drives large current through the fault loop. The current can be 5–10 times the rated current. The ground fault relay is an overcurrent relay, which triggers the circuit breaker when the current through the ground is more than the setpoint.

7.3.10.3. Differential Protection.
For large transformers, differential protection guards the transformer in the event of internal fault. Figure 7.56 depicts the concept of differential protection. The primary and secondary currents of the transformer are measured with CTs. The CTs at the primary and the secondary sides reduce the two currents to a preset value. Typically the reduced current value is about 5 A if the transformer carries the rated load. The differential relay subtracts the current coming from the primary-side CT from that originating from the secondary-side CT. In addition, the relay senses the current direction; the current is positive when flowing into the relay and negative if it flows out of the relay.

In the event of an *external fault*, both currents are positive because they flow into the relay. Consequently, the relay finds a current difference that is *zero or close to zero*. In most cases, the turns ratios of the CTs do not match exactly with the transformer turns ratio, which yields a larger than zero current difference.

In the case of an *internal fault*, the current direction changes: one current is positive and the other is negative. Thus, the current difference is the *sum of the two currents*. The relay triggers the circuit breaker when the current difference is larger than a small set value.

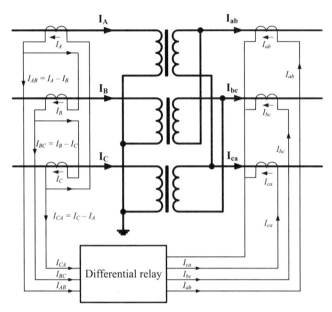

Figure 7.57. Three-phase transformer differential protection.

The differential protection operation is fast—typically, the faulty transformer is switched off within two to three cycles.

Figure 7.57 shows the approach used for differential protection of a three-phase wye–delta connected transformer. The concept is similar to the single-phase case. The differential protection relay subtracts the primary currents from the secondary. If the difference is close to zero, the fault is outside the transformer; if the current difference is large, the fault is inside of the transformer and must be cleared in two to three cycles.

A problem is the 30° difference between the primary and secondary currents, which prevents direct comparison. In order to eliminate the phase shift, the CTs on the *wye side* are connected *in delta*, and the CTs on the *delta side* are connected *in wye*. Figure 7.57 shows that this connection supplies the differential relay with the difference of phase currents at both sides. On the *wye side*, the relay currents are $I_{AB} = I_A - I_B$, $I_{BC} = I_B - I_C$, and $I_{CA} = I_C - I_A$, and on the delta side, the relay currents are $I_{ab} = I_a - I_b$, $I_{bc} = I_b - I_c$, and $I_{ca} = I_c - I_a$. The result is the elimination of the 30° phase shift.

7.4. EXERCISES

1. Why did the invention of transformers accelerate the development and use of electricity?
2. Describe the construction of a small transformer. What are the major components?

3. Draw a sketch describing the construction of windings in a small transformer.
4. Why is lamination needed?
5. Explain the construction of a dry-type transformer.
6. Describe the construction of an oil-insulated and cooled transformer.
7. Describe the construction of a porcelain bushing used for distribution transformers.
8. What is the eddy current loss? Draw a sketch showing a method to reduce the eddy current losses.
9. Why do the magnetizing or iron losses depend on the frequency?
10. What are the copper losses and iron losses in a transformer?
11. Explain the equivalent circuit of a single-phase transformer. Identify the physical representation of each of the circuit parameters.
12. What are the magnetizing reactance and the core loss resistance?
13. What is the leakage inductance? Draw a sketch showing the main flux and the leakage fluxes.
14. Discuss the method used to determine the equivalent transformer circuit parameters by measurement.
15. Which transformer parameters are determined by the open-circuit measurement?
16. Which transformer parameters are determined by the short-circuit measurement?
17. What is the reason that the largest number of distribution transformer failures typically occurs at Christmas time?
18. What is the typical failure mode of a small transformer?
19. What problem occurs if you connect transformers in parallel to supply a large load?
20. Draw the iron core of a three-phase transformer.
21. What connections are used for three-phase transformers?
22. Draw the connection of a wye–delta transformer. What is the phase shift between the primary and secondary line-to-line voltages?
23. Draw the connection of a wye–wye transformer. What is the phase shift between the primary and secondary line-to-line voltages?
24. Draw the connection of a delta–delta transformer. What is the phase shift between the primary and secondary line-to-line voltages?

7.5. PROBLEMS

Problem 7.1

A single-phase transformer supplies a 10 kW motor of a domestic air conditioner that operates with a power factor of 0.75 (lagging). The motor voltage is 240 V. The transformer rating is 15 kVA, with a primary voltage of 7.2 kV, a secondary voltage of 240 V, and a leakage reactance of 10%. Calculate the required transformer supply voltage.

Problem 7.2

A single-phase distribution transformer supplies four houses. The maximum load is 12 kW at a power factor of 0.78 (lagging). The load voltage is 235 V. The transformer rating is 20 kVA, with a primary voltage of 7.2 kV, a secondary voltage of 240 V, and a leakage reactance of 12%. Calculate the transformer supply voltage.

Problem 7.3

A 50 kVA, 7.2 kV/240 V, 7% distribution transformer is supplied by a distribution network. The network line-to-ground voltage is 7.2 kV; its short circuit current is 3 kA inductive. Calculate the short-circuit current if a fault occurs at the transformer secondary side.

Problem 7.4

A three-phase grounded wye–delta connected transformer supplies a load. The load voltage is 480 V (line-to-line); the loads are:

Phase AB is 25 kW with power factor 0.8 lagging;

Phase BC is 10 kVA with power factor 0.75 lagging; and

Phase CA is 47 kW with power factor 0.85 lagging.

The transformer is rated 50 kVA, 12.47 kV/480 V, 9.2%. Calculate: (a) the unbalanced delta currents in each phase (I_{AB}, I_{BC}, I_{CA}), (b) the unbalanced wye currents (I_a, I_b, I_c), (c) the required supply voltages per phase (both line to line and line to neutral), and (d) the ground current at the wye side. *Hint*: place the transformer impedance at the wye side in series in each phase.

Problem 7.5

Calculate the short-circuit current of a three-phase delta–wye connected transformer if the short circuit occurs between phase A and the neutral at the low voltage side, and the transformer is supplied by a 13 kV line-to-line voltage. The transformer rating is 50 kVA, 12.47 kV/480 V, and 9.2%.

Problem 7.6

A 115 V source supplies the primary of a single-phase transformer having a 120/240 V rating. The transformer secondary is connected to a simple resistive load (R). Create a simplified equivalent circuit diagram by eliminating the transformer from the network by transferring the source to the secondary side. What is the voltage of the equivalent source?

Problem 7.7

A single-phase transformer is rated at 7.5 MVA with a primary voltage of 110 kV and a secondary voltage of 15 kV, and the transformer reactance is 15%. (a) Determine the transformer reactance in ohms when it is placed at the high-voltage (primary) side and when it is placed on the low-voltage side. Draw the equivalent circuit for each case. (b) Find the load current and supply voltage at half-load, if the impedance of the transformer is on the low-voltage side and the load voltage is 15 kV with a power factor of unity.

Problem 7.8

A 60-Hz single-phase transformer with capacity of 150 kVA has the following parameters:

$R_P = 0.35\ \Omega$	$R_S = 0.002\ \Omega$	$R_c = 5.2\ k\Omega$
$X_P = 0.5\ \Omega$	$X_S = 0.008\ \Omega$	$X_m = 1.1\ k\Omega$

The primary transformer voltage is 2.8 kV and the secondary transformer voltage is 230 V. The transformer is connected to a variable load (0–300 kW) with a lagging power factor of 0.83 and a load voltage equal to the rated transformer secondary. Determine: (a) the total input impedance of the transformer when the secondary is shorted and (b) the input current, voltage, power, and power factor at full load (150 kW). (c) Plot the voltage regulation versus load and determine the load that produces 5% regulation.

Problem 7.9

The parameters of a single-phase 135 kVA transformer with primary and secondary voltage ratings of 2.6 kV and 230 V, respectively, are:

$R_P = 1.7\ \Omega$	$R_S = 0.017\ \Omega$	$R_c = 4.8\ k\Omega$
$X_P = 1.95\ \Omega$	$X_S = 0.028\ \Omega$	$X_m = 1\ k\Omega$

(a) Determine the voltage regulation of the transformer at a load of 75% of the transformer rating, with a lagging power factor of 0.8, assuming a load voltage equal to the secondary rating. (b) Plot the efficiency-versus-load curve of the transformer for power factors of 0.8 (lagging) and 0.9 (leading) if the load varies between 0% and 120%. (c) What is the maximum efficiency for each case in part (b)?

Problem 7.10

A single-phase 150 kVA transformer with a primary rating of 2.5 kV supplies an inductive load. The load voltage is equal with the secondary rated voltage of 240 V. The transformer parameters are determined by open- and short-circuit tests. The test results are:

Open-Circuit Test (low-voltage side is open; data are measured at the high-voltage side):

$V_{oc} = 2500\ V$	$I_{oc} = 3.7\ A$	$P_{oc} = 2400\ W$

Short-Circuit Test (high-voltage side is shorted; data are measured at the low-voltage side):

$V_{sc} = 28$ V	$I_{sc} = 320$ A	$P_{sc} = 1300$ W

(a) Determine the single-phase equivalent circuit parameters. Draw the equivalent circuit. (b) Calculate the percentage regulation and efficiency as a function of a variable load and lagging power factors of 0.6, 0.8, and 1.0. The load is varied according to $S_{load} = S_{tr} L_f$, where the load factor (L_f) ranges from 0 to 1 (in steps of 0.01) and S_{tr} is the transformer rating. (c) Repeat part (b) if a capacitor of 1000 µF is connected in parallel with the load. (d) Explain the effects of the capacitor on the voltage regulation and on the efficiency.

Problem 7.11

A three-phase 80 kVA transformer is rated to a primary voltage of 2.2 kV and a secondary voltage of 220 V. The transformer parameters are determined by open- and short-circuit tests. The results of the open- and short-circuit tests are:

Short-Circuit Test (low-voltage side short-circuited; data measured at high-voltage side):

$V_{sc} = 48$ V	$I_{sc} = 32$ A	$P_{sc} = 800$ W

Open-Circuit Test (high-voltage side open; data measured at low-voltage side):

$V_{oc} = 220$ V	$I_{oc} = 5.7$ A	$P_{oc} = 220$ W

(a) Determine the parameters of the equivalent circuit. Draw the equivalent circuit of the transformer. (b) Plot the voltage regulation and efficiency versus the load for both lagging and leading power factors of 0.6, 0.8, and 0.9. Obtain the maximum efficiency values. Assume the load voltage is equal to the rated secondary voltage of the transformer. (c) Assuming the primary rated voltage is being supplied to the transformer, determine the load voltage when the transformer carries half-load and the power factor is 0.8, both lagging and leading.

Problem 7.12

Three single-phase transformers are connected to form a three-phase transformer. The voltage ratings of each 7 MVA single-phase transformer are 20 kV (primary) and 2.4 kV (secondary). The single-phase transformers are connected in such a way that the rated line-to-line voltages of the single-phase transformers are 2.4 kV, and the line-to-line voltage of the primary side of the overall three-phase transformer is 34.5 kV. The three-phase transformer supplies a three-phase 10 MW load having a power factor of 0.85 (lagging). The load has a line-to-line voltage equal to the rated line-to-line voltage.

(a) Draw the three-phase circuit diagram showing the transformer connection. What is the transformer connection type? (b) Determine the magnitude and phase angle for the line and phase currents at the primary and secondary sides of the transformer. (c) Calculate the magnitude and phase angle for the line-to-line and phase voltages at the primary and secondary sides of the transformer. *Note*: For the evaluation of (b) and (c), take the phase voltage (phase **ab**) of the load voltage as the reference voltage (0°).

Problem 7.13

A three-phase Y-Y transformer bank is rated at 600 kVA with primary and secondary voltage ratings of 4.5 and 450 kV, respectively. The transformer has an iron loss of 5 kW and a full-load copper loss of 6.8 kW. The transformer carries a connected load that is 82% of the rated transformer capacity, with power factor of 0.87 lagging, and a load voltage equal to the secondary rating. (a) Compute the efficiency of the transformer at the given 82% load. (b) Plot the transformer efficiency versus load and determine the maximum efficiency and the load at which maximum efficiency occurs.

Problem 7.14

A delta–wye connected three-phase transformer is rated at 38 MVA with primary and secondary line voltage ratings of 345 and 125 kV, respectively. The transformer p.u. reactance is 10%, the p.u. no-load current is 0.1, and the iron p.u. power loss is 0.06 per transfomer rating. The transformer carries a load of 21 MW at a power factor of 0.8 lagging. (a) Draw the equivalent circuit. (b) Determine the transformer equivalent circuit parameters at the secondary side. (c) Calculate the magnetizing current. (d) If the transformer is supplied by the rated voltage, compute the load current and voltage.

8

SYNCHRONOUS MACHINES

The *synchronous machine* can be utilized as a motor or as a generator. However, it is most frequently used as a generator. Most electricity is produced by turbine-driven synchronous generators. The generators supply the electrical power system. The generators connected to the network operate at the synchronous speed and produce a voltage at 60 or 50 Hz. Synchronous machines are also installed in airplanes, in which the electrical system frequency is 400 Hz.

8.1. CONSTRUCTION

Two types of synchronous generators are used:

1. *Round Rotor Generator.* These machines are employed in fossil and nuclear power plants. In most cases, steam turbines drive these generators. Typically, these machines rotate at high speed, for example, 1800 and 3600 revolutions per minute (rpm) in nuclear and fossil power plants, respectively. Figure 8.1 pictures a two-pole round rotor generator.

Electrical Energy Conversion and Transport: An Interactive Computer-Based Approach, Second Edition. George G. Karady and Keith E. Holbert.
© 2013 Institute of Electrical and Electronics Engineers, Inc. Published 2013 by John Wiley & Sons, Inc.

CONSTRUCTION

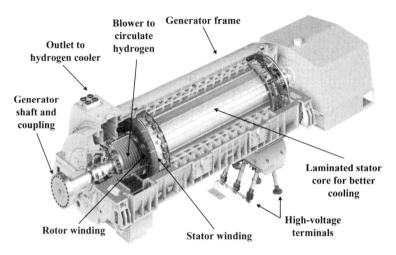

Figure 8.1. Cutaway view of a two-pole round rotor generator (copyright by Siemens AG).

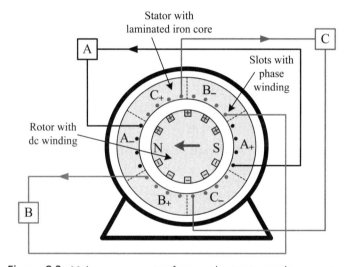

Figure 8.2. Major components of a round rotor two-pole generator.

2. *Salient Pole Generator.* These machines are utilized in hydropower plants. In most cases, hydraulic turbines drive these generators. Generally, these are low-speed machines spinning in the range of 150–900 rpm.

8.1.1. Round Rotor Generator

Figure 8.2 shows an illustration of a two-pole round rotor synchronous generator and its major components, which are the *stator* and *rotor*. The stator is sometimes called the *armature*.

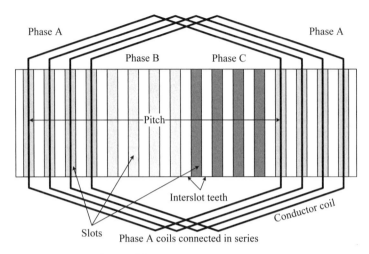

Figure 8.3. Stator coil arrangement.

Stator: The stator is a ring-shaped laminated iron core with slots. Three separate windings are placed in the slots. The specific stator depicted in Figure 8.2 has eight (2 × 4) slots for each phase, with four slots on the positive side and four on the negative side.

Figure 8.3 shows a sketch of a typical stator coil arrangement in the case of 2 × 4 slots per phase. In this diagram, the stator is laid opened and the slots spaced horizontally. The conductors placed in the slots form coils; the distance between the opposing sides of the coils is the pitch, which can be a *full pitch* of 180 electrical degrees or a *fractional pitch* of less than 180 electrical degrees. Many times, when fractional pitch is used, each slot contains more than one coil, which produces a layered arrangement of conductors. One of the advantages of the fractional pitch winding is the better, more sinusoidal induced voltage waveform.

Rotor: The rotor is located within the cavity formed by the stator. The round rotor is a solid iron cylinder with slots. A single field winding is placed in the slots and supplied by dc current. DC current enters the positive terminals (slots). The dc field current generates magnetic flux. Using the right-hand rule, the flux direction is marked on the rotor from the south (S) to the north (N) poles of Figure 8.2.

The cross section of a turbo generator is illustrated in Figure 8.4. The steel housing supports the iron stator core. Cooling ducts are placed between the iron core and the housing. This generator is cooled by air, which is circulated by fan. Larger generators are cooled by hydrogen gas. The hydrogen provides efficient cooling, but may create fire and explosion hazards. The generator housing has to seal in the hydrogen. Large generators can also be liquid cooled. The housing supports the bearings at the two sides of the generator. The bearing is lubricated by high-pressure oil. The stator has a laminated iron core, with three-phase windings. At the left- and right-hand sides, we can see the outgoing stator conductors that connect the generator to the network. Figure 8.4

CONSTRUCTION

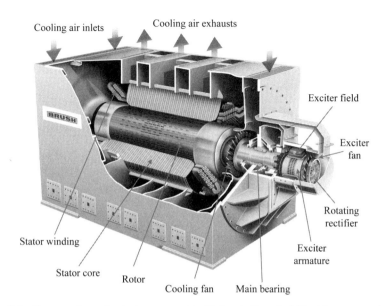

Figure 8.4. Cross-section of a turbo generator (© Brush Electrical Machines, reproduced by permission).

also shows the rotor and the shaft of the generator. The rotor is a large steel cylinder mounted on the shaft, which is supported by the bearings.

Figure 8.5 shows a picture of the details of a stator. We can see the steel housing and the laminated iron core with the slots. Some of the slots are empty, while others are filled with the stator conductors. The stator conductors are square-shaped copper bars insulated with mica tape. The insulating tape is wrapped around the conductors.

The rotor block of a synchronous generator is presented in Figure 8.6. Pictured is the forged steel cylinder with the slots and the shaft. The slots are empty. The dc excitation winding is placed in the slots. This excitation winding is made out of tape-insulated copper bars. Figure 8.7 exhibits the rotor with the dc winding placed in the slots. Steel wedges cover the slots to protect the conductors. At the end of the winding, the conductors are bent to form a loop.

The completely assembled rotor is shown in Figure 8.8. Note that steel retaining rings protect the rotor end turns, and the dc excitation terminals are located at the end of the shaft. This end of the shaft is coupled to the exciter. An electronically controlled rotating transformer excites this particular generator. There are other types of excitation systems.

8.1.2. Salient Pole Generator

Figure 8.9 shows a diagram of the major components of a two-pole salient pole synchronous generator.

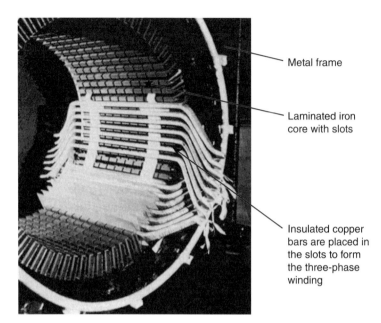

Figure 8.5. Details of a generator stator (Source: G. McPherson and R.D. Laramore, *An Introduction to Electrical Machines and Transformers*, 2nd ed., John Wiley & Sons, New York, 1990).

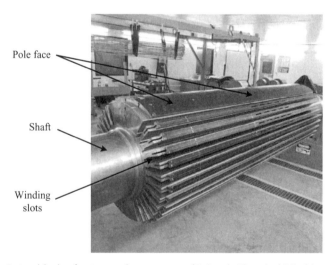

Figure 8.6. Rotor block of a two-pole generator (© Brush Electrical Machines, reproduced by permission).

CONSTRUCTION

Figure 8.7. Generator rotor with conductors placed in the slots (© Brush Electrical Machines, reproduced by permission).

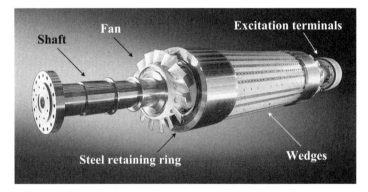

Figure 8.8. Large generator rotor completely assembled (courtesy of Siemens, Erlangen, Germany).

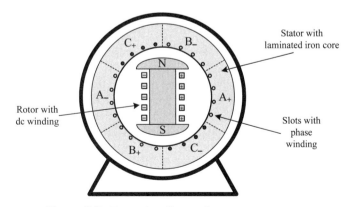

Figure 8.9. Two-pole salient pole generator concept.

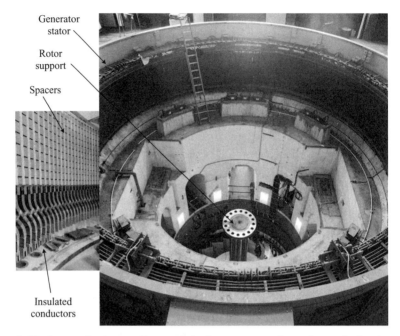

Figure 8.10. Stator of a large salient pole hydro generator; inset shows the insulated conductors and spacers (photo by Hydro-Québec).

Stator: The stator of a salient pole generator is similar to the stator of a round rotor generator. It has a ring-shaped laminated iron core with slots. The three-phase windings are placed in the slots. To illustrate the size of a large salient pole generator, Figure 8.10 shows the stator of a multipole machine. Figure 8.11 displays the corresponding multipole rotor of this large hydrogenerator. Salient pole machines are normally only used when there are more than four poles.

Rotor: DC excitation winding is wound on an iron core to generate the main magnetic flux. Each pole has an iron core and winding. The rotor of a four-pole salient machine is exemplified in Figure 8.12. The poles, the pole winding, and the slip rings can be seen.

Figure 8.13 illustrates a four-pole salient generator for comparison with the two-pole unit of Figure 8.9.

The stator and rotor are swapped in small machines that are utilized in computers and other consumer electronics. The dc field windings and poles are placed in the stator and the alternating current (ac) windings are located on the rotor. These machines may use permanent magnets instead of dc-excited field windings.

8.1.3. Exciter

The operation of a synchronous generator requires supplying the rotor field winding with dc current, which produces the magnetic flux. This process is called *excitation*,

Figure 8.11. Large hydro generator rotor with view of the vertical poles (photo by Hydro-Québec).

Figure 8.12. Rotor of a four-pole salient pole generator (courtesy of Siemens, Erlangen, Germany).

and the dc current is termed excitation current or *field current*. When a turbine spins the rotor, the dc field current-produced flux induces voltage in the stator windings. Excitation of the rotor means the generation of magnetic field by dc field current. Typically, the large generator field current is in the range of 10 kA in steady-state conditions, but up to 15 kA may be required for a short period of time. The dc voltage is around 500 V and the exciter is rated at 5 MW.

The excitation system must be protected against overheating, ground faults, and rotor open circuits. Typically, the rotor is ungrounded; consequently, the first ground fault does not produce a fault current, but following a second ground fault, a part of

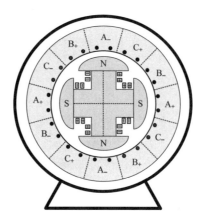

Figure 8.13. Four-pole salient pole generator concept.

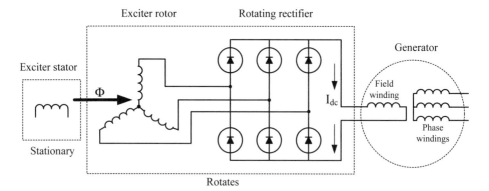

Figure 8.14. Concept of the brushless exciter system.

the rotor winding is short circuited, thereby generating asymmetrical field excitation. This causes vibration that may damage the rotor bearings. A further problem is that the asymmetrical excitation produces unbalanced three-phase stator current, which has a negative sequence component. This produces double-frequency current in the rotor iron and overheating. The loss of excitation or a field open circuit reduces excitation, which causes the generator to operate as an induction generator, supplying power with a leading power factor. This causes loss of synchronization, which requires switching off the generator.

Some generators use rectifiers to produce the dc voltage and excitation current. The voltage and current are supplied to the machine rotor through slip rings and brushes. Figure 8.12 exhibits the slip rings on a salient pole synchronous machine rotor. A *brushless excitation system* has been developed, where the rectifiers are mounted on the generator shaft and the energy is transferred through magnetic coupling.

The brushless excitation systems use an ac alternator and rotating rectifiers to produce the dc needed for the generator field windings on the rotor. Figure 8.14 presents

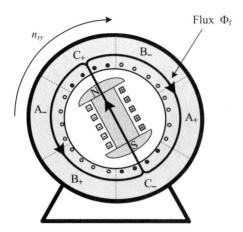

Figure 8.15. Operating concept of a synchronous generator.

the connection diagram of the Westinghouse-developed brushless excitation systems. The excitation system stator has a dc winding, which produces a dc magnetic flux. The rotor attached to the generator shaft has a three-phase winding. The rotation of the three-phase winding in the dc field generates a three-phase ac voltage. This voltage is rectified by a diode bridge, which is mounted on a wheel attached to the generator shaft. The output dc current of the rectifier is connected to the generator field winding through conductors placed inside the generator shaft.

8.2. OPERATING CONCEPT

The operating concept of a generator is explained using Figure 8.15, which depicts a simplified synchronous generator. This figure shows a salient two-pole rotor and a stator with 24 slots, 2×4 for each of the three phases.

8.2.1. Main Rotating Flux

The salient pole rotor is supplied by dc current, which generates a magnetic flux (Φ_f). Figure 8.15 sketches the approximate flux lines and represents the flux with a vector. The flux is directed from the south (S) pole to the north (N) pole internal to the field winding, and then splits in both directions through the stator. After traversing the stator, the flux recombines as it returns to the south pole.

A turbine drives the rotor at synchronous speed. The turning of Φ_f produces a rotating flux, Φ_{rot}. As this flux rotates, the flux linkage with the phase windings varies, as illustrated in Figure 8.16. In Figure 8.16a, the flux vector is perpendicular to phase A, which causes all the flux to go through (or link with) phase A. In Figure 8.16b, the

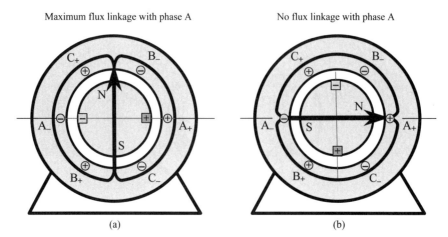

Figure 8.16. Rotation-produced flux linkage variation. (a) Flux is perpendicular to phase A; (b) flux is parallel to phase A.

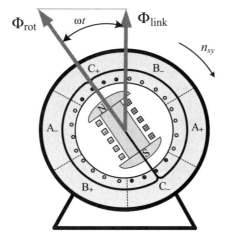

Figure 8.17. Rotating flux linkage to phase A.

flux vector is parallel with the phase A winding, which means that none of the flux links with phase A.

As the rotor turns, sinusoidal voltages are produced in the three-phase windings. If the rotor turns in the clockwise direction, then a *positive phase sequence* (A, B, C) is produced. Conversely, if the rotor spins in the counterclockwise direction, a *negative phase sequence* (A, C, B) is generated.

The flux linkage at an arbitrarily selected instant is annotated in Figure 8.17. In this case, only the vertical component of the flux links with phase A. This vertical component (Φ_{link}) is perpendicular to phase A. The angle between the rotating flux vector (Φ_{rot}) at the vertical direction is ωt. The angular speed of the rotor is ω in radians per second. The angular speed is calculated by:

$$\omega = 2\pi n_{sy}, \tag{8.1}$$

where n_{sy} is the synchronous speed of the generator in revolutions per second.

The linking vertical flux component is:

$$\Phi_{link}(t) = \Phi_{rot}\cos(\omega t), \tag{8.2}$$

where Φ_{rot} is the rotating flux generated by the dc field current. This equation demonstrates that the flux linkage is a cosine function. The rotation-caused change of the flux linkage induces voltage in the phase windings. The frequency of the induced voltage depends upon the rotation speed. The induced voltage is the product of the number of turns in the stator phase winding and the time derivative of the linkage flux. The induced voltage in phase A is:

$$E_s(t) = N_{sta}\frac{d\Phi_{link}(t)}{dt}, \tag{8.3}$$

where N_{sta} is the number of turns in each phase winding of the stator. Substituting Equation (8.2) into Equation (8.3) and taking the derivative yields:

$$\begin{aligned}E_s(t) &= -N_{sta}\Phi_{rot}\omega\sin(\omega t)\\ &= N_{sta}\Phi_{rot}\omega\cos(\omega t + 90°).\end{aligned} \tag{8.4}$$

The root-mean-square (rms) value of this sinusoidal voltage is:

$$E_{sta} = \frac{N_{sta}\Phi_{rot}\omega}{\sqrt{2}}, \tag{8.5}$$

where E_{sta} is the induced voltage generated by the rotor flux.

Equation (8.4) reveals that the phase shift between the flux linkage and the induced voltage is 90°. Therefore, the induced voltage in a phase winding will be maximum when the flux linkage is zero, which occurs as the rotor is parallel with the phase winding as in Figure 8.16b. Vice versa, the induced voltage is zero when the rotor is perpendicular to the particular phase winding (as in Fig. 8.16a) and the flux linkage is maximum.

The phase difference between the voltages generated in the phase windings are 120° because the phase windings are physically placed 120° apart. The phase windings are connected in either a wye or a delta. In the no-load condition, the generator terminal voltage (V_t) is equal to the induced voltage.

The frequency of the induced voltage depends on the generator speed. One rotation generates one sine wave in a two-pole machine. In a four-pole machine, one rotation generates two sine waves. This leads to the conclusion that machine speed is equal to the ratio of frequency (*f*) and half of the number of poles (*p*):

$$n_{sy} = \frac{f}{p/2}. \qquad (8.6)$$

In this formula, the generator speed is measured by revolutions per second. Usually the rotation speed is expressed in rpm, which requires multiplication of the results from Equation (8.6) by 60 s/min. Note that if Mathcad is used, the computer converts the units automatically. The synchronous generators supply the local network, which operates with a frequency of 60 Hz in the United States and 50 Hz in Europe. In the United States, the two-pole generators run with 3600 rpm, the four-pole units operate with 1800 rpm, and a 36-pole hydrogenerator spins at only 200 rpm.

8.2.2. Armature Flux

The loading of the generator produces current in the phase windings. Balanced three-phase load currents produce a rotating (armature) flux with amplitude proportional to the load current. This armature flux also rotates with synchronous speed in the same direction as Φ_{rot}. Figure 8.18 shows the field- and the load-generated fluxes. The generation of this rotating flux will be derived in the next chapter.

The flux linking the phase conductors due to the load-generated rotating flux is defined as $\Phi_{arm}(t)$. Applying the same method used to develop Equation (8.2) results in a relation for the time variation of the armature flux linking to phase A:

$$\Phi_{arm}(t) = \Phi_{ar} \cos(\omega t), \qquad (8.7)$$

where Φ_{ar} is the load current-generated rotating flux when $I_{arm}(t) = \sqrt{2} I_{sta} \cos(\omega t)$ with I_{sta} being the rms load (stator) current.

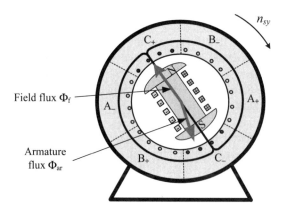

Figure 8.18. Field (Φ_f) and load-generated (Φ_{ar}) rotating fluxes.

This load-generated rotating flux induces three phase voltages in the phase windings. The induced voltage in phase A can be calculated using an expression like Equation (8.3), yielding:

$$E_{ar}(t) = N_{sta}\frac{d\Phi_{arm}(t)}{dt} = -N_{sta}\Phi_{ar}\omega\sin(\omega t). \quad (8.8)$$

The rms value of the sinusoidal voltage is:

$$E_{arm} = \frac{N_{sta}\Phi_{ar}\omega}{\sqrt{2}}, \quad (8.9)$$

where E_{arm} is the induced voltage generated by the armature flux.

The armature flux-generated induced voltage must be subtracted from the main flux-induced voltage (E_{sta}) to obtain the generator terminal voltage. The generator terminal voltage is:

$$\mathbf{V}_t = \mathbf{E}_{sta} - \mathbf{E}_{arm}, \quad (8.10)$$

where $\mathbf{E}_{sta}$ is the induced voltage produced by the rotor flux and $\mathbf{E}_{arm}$ is the induced voltage produced by the load flux.

The load current-generated voltage, calculated with Equation (8.8), is proportional with the load current, since the load current produces the armature flux. This permits the introduction of the armature inductance and calculation of the induced voltage using Equation (8.11).

As in the development of Equation (8.8), it is assumed that the phase A current is a cosine function, that is, $I_{arm}(t) = \sqrt{2}I_{sta}\cos(\omega t)$. The induced armature voltage is:

$$\begin{aligned}E_{ar}(t) &= L_{arm}\frac{dI_{arm}(t)}{dt} = L_{arm}\frac{d}{dt}\sqrt{2}I_{sta}\cos(\omega t)\\ &= -L_{arm}\omega\sqrt{2}I_{sta}\sin(\omega t)\\ &= -X_{arm}\sqrt{2}I_{sta}\sin(\omega t),\end{aligned} \quad (8.11)$$

where

L_{arm} is the *armature inductance*;
X_{arm} is the *armature reactance*; and
I_{sta} is the rms stator load current.

A comparison of Equation (8.11) and Equation (8.8) gives the relation between the load current-generated flux and the armature reactance:

$$X_{arm} = \frac{N_{sta}\Phi_{ar}\omega}{\sqrt{2}I_{sta}}. \quad (8.12)$$

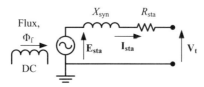

Figure 8.19. Single-phase equivalent circuit of a synchronous generator.

This is a fictitious inductive reactance that represents the effect of flux created by armature current. This effect is often called an *armature reaction*.

The stator winding has leakage inductance that must be added to the armature reactance. The result is the *synchronous reactance*, which describes the load-produced change of the terminal voltage:

$$X_{syn} = X_{arm} + X_{leakage}. \tag{8.13}$$

The voltage drop in this reactance represents the total effect of the load-generated flux. The voltage drop is:

$$\mathbf{E}_{arm\text{-}syn} = \mathbf{I}_{sta}(jX_{syn}). \tag{8.14}$$

If the leakage reactance is neglected, $\mathbf{E}_{arm\text{-}syn}$ is equal to $\mathbf{E}_{arm}$. The terminal voltage of the generator is the voltage drop across the synchronous reactance subtracted from the induced voltage:

$$\mathbf{V}_t = \mathbf{E}_{sta} - \mathbf{E}_{arm\text{-}syn} = \mathbf{E}_{sta} - \mathbf{I}_{sta} jX_{syn}. \tag{8.15}$$

Equation (8.15) suggests that the generator equivalent circuit is a voltage source ($\mathbf{E}_{sta}$) and a reactance (X_{syn}) connected in series. The stator coil has resistance (R_{sta}), which is added to the equivalent circuit. Figure 8.19 presents the single-phase equivalent circuit of a synchronous generator.

EXAMPLE 8.1: Single-phase equivalent generator

The application of the single-phase equivalent circuit for a synchronous generator is demonstrated with a numerical example. A large four-pole, 250 MVA wye-connected synchronous generator carries rated load, with a power factor of 0.8 lagging. The terminal voltage of the generator is the rated voltage (V_{gen}). Calculate the necessary induced stator voltage if the generator data are:

$$f := 60 \text{ Hz} \qquad p := 4 \qquad M := 10^6$$

$$S_{gen} := 250 \, M \cdot V \cdot A \qquad V_{gen} := 24 \text{ kV} \qquad rpm := \frac{1}{min}$$

$$X_{syn} := 125\% \qquad pf_{gen} := 0.8 \, (\text{lagging}).$$

OPERATING CONCEPT

The generator shaft speed is measured in rpm, which is explicitly defined in Mathcad as given earlier. The first step of the calculation is the conversion of the per-unit (p.u.) synchronous reactance to ohms:

$$X_{syn} := x_{syn} \cdot \frac{V_{gen}^2}{S_{gen}} \quad X_{syn} = 2.88 \, \Omega.$$

In this calculation, the single-phase equivalent circuit of Figure 8.19 is used. The power in the single-phase equivalent circuit is one-third of the three-phase power, and the line-to-neutral voltage supplies the circuit since the generator is wye connected. The line-to-line voltage is used in a delta-connected generator.

The complex or apparent power of phase A, if the generator carries the rated load and rated lagging power factor, is:

$$S_{load} := \frac{S_{gen}}{3} \cdot e^{j \cdot acos(pfgen)} \quad S_{load} = (66.7 + 50.0j) \, M \cdot V \cdot A.$$

The line-to-neutral generator terminal voltage is:

$$V_{gen\_ln} := \frac{V_{gen}}{\sqrt{3}} \quad V_{gen\_ln} = 13.86 \, kV.$$

The generator current at full load is:

$$I_{gen} := \overline{\left(\frac{S_{load}}{V_{gen\_ln}}\right)} \quad I_{gen} = (4.811 - 3.608j) \cdot kA$$

$$|I_{gen}| = 6.014 \cdot kA \quad arg(I_{gen}) = -36.9 \cdot deg.$$

Using Equation (8.15), the induced voltage from the equivalent circuit is:

$$E_{sta} := V_{gen\_ln} + I_{gen} \cdot (j \cdot X_{syn}) \quad E_{sta} = (24.25 + 13.86j) \cdot kV$$
$$|E_{sta}| = 27.93 \cdot kA \quad arg(E_{sta}) = 29.7 \cdot deg.$$

The angle between the terminal voltage and the induced voltage is termed the *power angle* and, in this example, was previously computed as 29.7°. The ratio of the induced line-to-neutral voltage to the terminal voltage is:

$$\frac{|E_{sta}|}{V_{gen\_ln}} = 2.016.$$

The calculation indicates that the induced voltage is about twice the line-to-neutral voltage. Applying Equation (8.6), the generator speed is:

$$n_{sy} := \frac{f}{\frac{p}{2}} \quad n_{sy} = 30\frac{1}{\sec} \quad n_{sy} = 1800 \text{ rpm}.$$

8.3. GENERATOR APPLICATION

The synchronous machine can operate as a generator or as a motor. The motor and the generator implementations have the same construction. If a generator is mechanically disconnected from its turbine, it will continue to operate as a *synchronous condenser* to provide reactive power. If the shaft is directly connected to a mechanical load, the synchronous machine will drive the load at synchronous speed. However, most synchronous machines operate as a generator and supply energy to the electric network. In a large network, several hundred synchronous generators operate in parallel. Each generator operates at synchronous speed (e.g., 60 Hz in the United States) by controlling the mechanical power input to the system.

The synchronous generators presented in this chapter are primarily intended for large-scale electricity generation, but this does not imply that they cannot be used on a smaller scale. In fact, an automobile *alternator* is a three-phase synchronous generator with typically 8–12 poles. An alternator employs permanent magnets for the poles on the rotor, and multiphase windings on the stator. The rotation of the permanent magnets generates a rotating flux that induces variable frequency voltage in the stator windings. This voltage is diode rectified to produce dc current. The current magnitude is electronically controlled.

8.3.1. Loading

Increasing the input mechanical power increases the output electrical power of a generator. The output power is frequently termed the *generator load*. In the case of a steam turbine, this requires an increase of the steam flow to the turbine. The power increase drives up the power angle (δ), which is the angle between the terminal voltage and the induced voltage. The speed of the generator remains nearly constant. The operational power angle must be less than 90°. Typically, the power angle is around 30°. The generator loading will be numerically demonstrated in Example 8.2.

8.3.2. Reactive Power Regulation

Synchronous generators are able to produce or consume reactive power. This can be achieved by the regulation of the induced voltage by the dc excitation current (generator excitation). Increasing the rotor dc current increases the magnitude of the induced voltage. The decrease of this current reduces the induced voltage. When the excitation is:

- *Increased*, the generator reactive power also increases;
- *Decreased*, the generator reactive power also decreases.

A synchronous generator has a power factor rating, which is usually around 0.8, and a megawatt rating. When the machine produces reactive power, the field current must be increased. This is limited by the rotor current rating to avoid overheating. This correspondingly restricts the maximum reactive power that the generator can produce. An additional limitation is the current-carrying capacity of the stator, which establishes the generator megavolt ampere rating. This forms a circle on a *P–Q* plot. The above two limits constrain the valid operating region. This phenomenon is illustrated using the reactive capability curve in Figure 8.20. In addition, there is a lower limit established by the underexcitation limit and stator end turn heating (not shown in the figure). A further restriction is imposed by the maximum turbine power output.

8.3.3. Synchronization

The synchronous machine has to be started by a mechanical device (turbine, reciprocating engine, etc.), which increases the machine speed to the synchronous speed. The

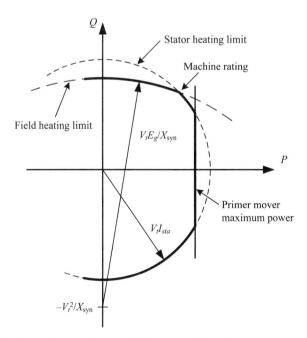

Figure 8.20. Generator reactive capability curve with turbine maximum power limit.

proper interconnection of a rotating machine with the network is called *synchronization*. The steps of the synchronization procedure are:

- Verify that the phase sequences of the two systems are the same.
- Adjust the machine speed with the turbine that drives the generator until the generator voltage frequency is nearly the same as the frequency of the network voltage.
- Adjust the terminal voltage of the generator by changing the dc field (rotor) current until the generator terminal voltage is almost equal to the network voltage. Acceptable limit is 5%.
- Adjust the phase angle of the generator terminal voltage by regulating the input power until it is nearly equal with the phase angle of the network voltage. Acceptable limits are about 15°.

The voltage magnitude and angle between the terminals of the generator circuit breaker are measured. When the voltage magnitude is small, within about 5%, and the phase angle is within 15° and changing slowly, the circuit breaker (CB) is closed. The generator is synchronized and interconnected to the network.

In the past, lamps connected across the open breaker were used to detect the voltage differences. Today, electronic circuits compare the voltages and control the generator. However, some operators still prefer to synchronize the generator manually. A faulty synchronization can severely damage a generator. Further regulation of the mechanical input power and induced voltage adjusts the generator active and reactive powers to the desired values.

8.3.4. Static Stability

The electrical network interconnects the generators and loads. The output powers of the generators are constantly adjusted to supply the continuously varying load. The total power generated must be exactly equal with the load and network losses because the network cannot store energy. In a practical network, the most frequently used operating strategy is that most generators supply predetermined active power. The sum of these generator output powers is close to the required load. One or more generators supply the difference.

The concept of static stability is explained by analyzing the case where a generator supplies a large infinite network with constant voltage. A performance metric is the generator static stability limit, which is the maximum real power that the generator can deliver. To determine this limit, the generator is modeled as a voltage source, $\mathbf{E}_{gen}(\delta)$, and a synchronous reactance, X_{syn}, connected in series. The generator resistance is small and neglected in this ideal case. As illustrated in Figure 8.21, the generator is connected to the electrical network, which is represented as a voltage source, $\mathbf{V}_{net}$.

The generator-induced voltage is:

$$\mathbf{E}_{gen}(\delta) = E_g[\cos(\delta) + j\sin(\delta)]. \tag{8.16}$$

GENERATOR APPLICATION

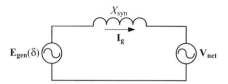

Figure 8.21. Generator supplying a large infinite network.

The generator output current is:

$$\mathbf{I}_g = \frac{\mathbf{E}_{gen}(\delta) - \mathbf{V}_{net}}{jX_{syn}} = \frac{E_g[\cos(\delta) + j\sin(\delta)] - \mathbf{V}_{net}}{jX_{syn}}. \quad (8.17)$$

The complex power supplied to the network by the generator is:

$$\mathbf{S}_g = \mathbf{V}_{net}\mathbf{I}_g^* = \mathbf{V}_{net}\left(\frac{E_g[\cos(\delta) + j\sin(\delta)] - \mathbf{V}_{net}}{jX_{syn}}\right)^*. \quad (8.18)$$

Letting $\mathbf{V}_{net} = V_{net}\angle 0°$ reduces the earlier expression to:

$$\mathbf{S}_g = V_{net}\left(\frac{E_g[\cos(\delta) + j\sin(\delta)] - V_{net}}{jX_{syn}}\right)^* = V_{net}\left(\frac{E_g\cos(\delta) - V_{net} - jE_g\sin(\delta)}{-jX_{syn}}\right). \quad (8.19)$$

The real power delivered to the network is:

$$P_g = \frac{V_{net}E_g\sin(\delta)}{X_{syn}}. \quad (8.20)$$

The maximum real power is produced when $\delta = 90°$, assuming that the resistance is negligible.

The synchronous generator interaction with the network is investigated in this section using two numerical examples:

1. a synchronous generator supplies power to a large network with constant voltage, and
2. three synchronous generators supply a large network.

These examples will demonstrate the major problems related to the interaction between the network and generators in steady-state conditions. However, the understanding of the synchronous generator dynamic operation requires further studies beyond the introductory material in this book.

EXAMPLE 8.2: A synchronous generator supplies a network (small system stability)

The application of the single-phase equivalent circuit of a synchronous generator is illustrated in this example. A small three-phase generator is interconnected to the local network through a distribution line. Figure 8.22 shows the system one-line diagram. This is a typical case of *cogeneration*, when a factory installs a generator, which supplies the factory and sells the surplus energy to the local network.

Figure 8.23 provides the single-phase equivalent circuit of this system. In this circuit, a Thévenin equivalent models the network, which is a source (V_{net}) and a reactance (X_{net}) connected in series. The distribution line is represented by a reactance and resistance connected in series (Z_{line}), and the generator is characterized by a voltage source producing the induced voltage (E_{sta}) and the synchronous reactance (X_{syn}).

The system rating data are:
Generator: $S_{gen} := 150 \text{ kV·A}$ $V_{gen} := 12.47 \text{ kV}$ $x_{gen} := 128\%$.
Network: $V_{net} := 12.47 \text{ kV}$ $I_{net\_short} := 2 \text{ kA}$.
Distribution line:

$$L_{line} := 48 \text{ mi} \quad z_L := (0.5 + j \cdot 0.67) \cdot \frac{\Omega}{\text{mi}}.$$

The network voltage is constant and selected as the reference voltage with a zero phase angle. An increase of the generator input power increases the power angle (δ). The purpose of this study is to calculate the generator power as a function of the power angle.

Figure 8.22. One-line diagram of a simple power system.

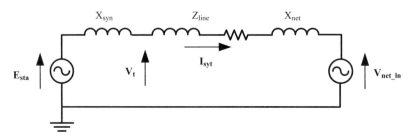

Figure 8.23. Single-phase equivalent circuit of the network in Figure 8.22.

GENERATOR APPLICATION

The first step of the study is to calculate the impedances. The generator synchronous reactance is:

$$X_{syn} := x_{gen} \cdot \frac{V_{gen}^2}{S_{gen}} \quad X_{syn} = 1.327 \text{ k}\Omega.$$

The line impedance is:

$$Z_{line} := z_L \cdot L_{line} \quad Z_{line} = (24.00 + 32.16j) \, \Omega.$$

The network line-to-neutral voltage and reactance are:

$$V_{net\_ln} := \frac{V_{net}}{\sqrt{3}} \quad V_{net\_ln} = 7.20 \cdot \text{kV}$$

$$X_{net} := \frac{V_{net\_ln}}{I_{net\_short}} \quad X_{net} = 3.60 \, \Omega.$$

The generator dc excitation is kept constant, and in this case, the induced stator line-to-neutral voltage is selected to be twice the line-to-neutral voltage of the network ($E_{sta} = 2V_{net\_ln}$). This selection is justified by the fact that the synchronous reactance of most generators is above 100%. The generator-induced voltage is:

$$E_{sta}(\delta) := 2 \cdot (V_{net\_ln} \cdot e^{j\delta}) \quad E_{sta}(60 \text{ deg}) = (7.20 + 12.47j) \, \text{kV},$$

where this solution is tested at a power angle of 60°. The equivalent circuit of Figure 8.23 shows that the current in this system is the voltage difference divided by the total system impedance:

$$I_{syt}(\delta) := \frac{E_{sta}(\delta) - V_{net\_ln}}{j \cdot X_{syn} + Z_{line} + j \cdot X_{net}} \quad I_{syt}(60 \text{ deg}) = (9.148 + 0.161j) \, \text{A}$$

$$|I_{syt}(60 \text{ deg})| = 9.150 \, \text{A} \quad \arg(I_{syt}(60 \text{ deg})) = 1.01 \cdot \text{deg}.$$

The three-phase complex power of the generator and network are:

$$S_g(\delta) := 3 \cdot E_{sta}(\delta) \cdot \overline{I_{syt}(\delta)} \quad S_g(60 \text{ deg}) = (203.6 + 338.8j) \, \text{kV} \cdot \text{A}$$

$$S_{net}(\delta) := 3 \cdot V_{net\_ln} \cdot \overline{I_{syt}(\delta)} \quad S_{net}(60 \text{ deg}) = (197.6 - 3.5j) \, \text{kV} \cdot \text{A}.$$

The three-phase real power of the generator and network are:

$$P_g(\delta) := \text{Re}(S_g(\delta)) \quad P_g(60 \text{ deg}) = 203.6 \cdot \text{kW}$$

$$P_{net}(\delta) := \text{Re}(S_{net}(\delta)) \quad P_{net}(60 \text{ deg}) = 197.6 \cdot \text{kW}.$$

The generator terminal line-to-neutral voltage and corresponding line-to-line voltage are:

$$V_t(\delta) := E_{sta}(\delta) - jX_{syn} \cdot I_{syt}(\delta) \quad V_t(60 \text{ deg}) = (7.413 + 0.331j) \cdot kV$$
$$V_{t\_ll}(\delta) := \sqrt{3} \cdot |V_t(\delta)| \quad V_{t\_ll}(60 \text{ deg}) = 12.85 \cdot kV.$$

The voltage regulation at the generator terminals should be less than 5%:

$$\text{Reg}(\delta) := \frac{V_{t\_ll}(\delta) - V_{net}}{V_{net}} \quad \text{Reg}(60 \text{ deg}) = 3.1\%.$$

The generator power and the power delivered to the network versus the power angle are plotted in Figure 8.24. The graph shows that an increase of both the generator-delivered and the network-absorbed power increases the power angle until a maximum power is reached around a 90° power angle. At this point, the operation becomes unstable. Further increase of the power angle results in decreasing power, which causes the generator to fall out of synchronism.

The maximum generator power is calculated using the Mathcad *Maximize* function. A guess value for the power angle is $\delta := 60$ deg. First, the power angle for maximum power delivery is found:

$$\delta_{max} := \text{Maximize}(P_g, \delta) \quad \delta_{max} = 91.0 \text{ deg}.$$

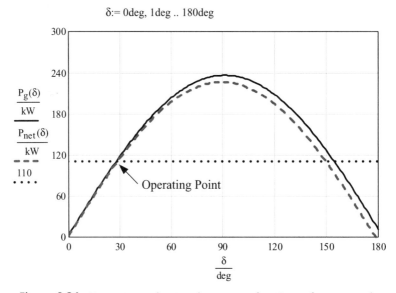

Figure 8.24. Generator and network power as functions of power angle.

This value is confirmed from the graph of Figure 8.24. Next, the maximum power is computed:

$$P_{g\_max} := P_g(\delta_{max}) \quad P_{g\_max} = 236.2 \text{ kW}.$$

This *static stability curve* can be used to evaluate operating conditions. For instance, if the generator rating is larger than the maximum power, the generator cannot be fully utilized. In practical conditions; a power reserve of 10–20% is desired. As an example, in Figure 8.24, the selected load is 110 kW. The operating point is the intersection of the stability curve (generator power) with the 110 kW line. The power angle at a load of 110 kW can be calculated by the Mathcad *root* function:

$$\delta_{110\,kW} := \text{root}(P_g(\delta) - 110 \text{ kW}, \delta) \quad \delta_{110\,kW} = 27.55 \text{ deg}.$$

The voltage drop of a transmission system should be less than 5%. This requires the evaluation of the voltage regulation. The voltage regulation versus power angle function is presented in Figure 8.25. The range of the voltage drop can be determined from:

$$\delta := 90 \text{ deg}$$

Given

$$\text{Maximize}(\text{Reg}, \delta) = 33.9 \cdot \text{deg}$$
$$\text{Reg}_{max} := \text{Reg}(\text{Maximize}(\text{Reg}, \delta)) = 3.7 \cdot \%$$
$$\text{Minimize}(\text{Reg}, \delta) = 213.9 \cdot \text{deg}$$
$$\text{Reg}_{min} := \text{Reg}(\text{Minimize}(\text{Reg}, \delta)) = -9.0 \cdot \%.$$

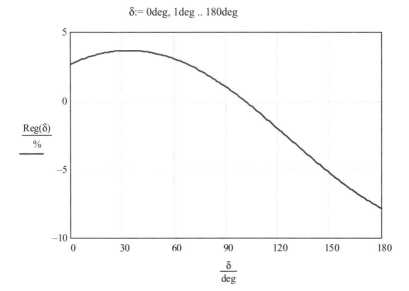

Figure 8.25. Voltage regulation as a function of power angle.

Figure 8.25 shows that the regulation is less than 5% in the 0° to ~150° power angle range. Of course, the plot is only meaningful for δ < 90° since the system is unstable beyond 90°.

EXAMPLE 8.3: Three synchronous generators supply a large network

Three generators supply a large electric network through an interconnected 220 kV system, as seen in Figure 8.26. The large network absorbs the power generated by the units and maintains the voltage constant at its terminal voltage, $V_0 = 220$ kV.

Each generator is connected to the 220 kV system through a transformer. It is assumed that the generators maintain the required voltage (V_n for n = 1,2,3) at the high-voltage side (220 kV). The generator impedance is transferred to the high-voltage (secondary) side and combined with the transformer impedance. This combined value (X_g) is used for the calculation.

The purpose of this extended exercise is to determine the active and reactive power flow and the current through each line, and the terminal voltage angles of the generators as well as the excitation (induced) voltages and their angles.

The system data are given below. The network line voltage (V_0) and corresponding line-to-neutral voltage are:

$$V_{net} := 220 \text{ kV} \quad V_{net\_ln} := \frac{V_{net}}{\sqrt{3}} = 127.0 \cdot \text{kV} \quad M := 10^6.$$

For each generator, the power and voltage ratings are specified along with the generator per unit reactance. The generator ratings are:

Generator 1: $S_{g_1} := 400 \text{ M} \cdot \text{V} \cdot \text{A} \quad V_{gen_1} := 220 \text{ kV} \quad x_1 := 120\%$
Generator 2: $S_{g_2} := 300 \text{ M} \cdot \text{V} \cdot \text{A} \quad V_{gen_2} := 220 \text{ kV} \quad x_2 := 115\%$
Generator 3: $S_{g_3} := 250 \text{ M} \cdot \text{V} \cdot \text{A} \quad V_{gen_3} := 220 \text{ kV} \quad x_3 := 95\%$

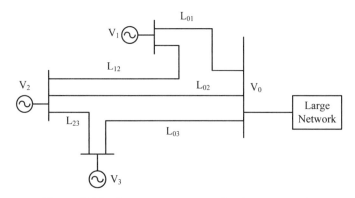

Figure 8.26. Three generators supply a large network.

GENERATOR APPLICATION

The generator single-phase impedances at the network voltage level are:

$$n := 1\ldots 3 \quad X_{g_n} := x_n \cdot \frac{\left(\frac{V_{gen_n}}{\sqrt{3}}\right)^2}{\frac{S_{g_n}}{3}} \quad X_g^T = (145.2 \quad 185.5 \quad 183.9)\,\Omega.$$

The transmission lines have the same impedance per unit length, but different lengths:

$$Z_{line} := 0.06\,\frac{\Omega}{mi} + j \cdot 0.59\,\frac{\Omega}{mi} \quad L_{12} := 35\text{ mi} \quad L_{23} := 40\text{ mi}$$

$$L_{01} := 70\text{ mi} \quad L_{02} := 60\text{ mi} \quad L_{03} := 45\text{ mi}.$$

The line impedances are then:

$$Z_{01} := L_{01} \cdot Z_{line} \quad Z_{02} := L_{02} \cdot Z_{line} \quad Z_{03} := L_{03} \cdot Z_{line}$$

$$Z_{12} := L_{12} \cdot Z_{line} \quad Z_{23} := L_{23} \cdot Z_{line}.$$

The generators are operating at voltages and powers of:

$$V_{rms\_n1} := 225\text{ kV} \quad V_{rms\_n2} := 232\text{ kV} \quad V_{rms\_n3} := 228\text{ kV}$$

$$P_1 := 365\text{ MW} \quad P_2 := 225\text{ MW} \quad P_3 := 178\text{ MW}.$$

Generator Currents Calculation

The calculation starts with an estimation of the guess value for the voltage and reactive power at each of the three nodes where the generators are connected to the 220 kV system. We estimate that the reactive powers are about 100 MVA, and that the bus voltages are near the rated value but have a small imaginary component. The estimation follows a trial–and-error procedure. A change of the guess values can help to assure the convergence of the equation solver algorithm.

The guess values for the single-phase reactive powers and line-to-neutral voltages are:

$$V_{n1} := \frac{V_{rms\_n1} + j \cdot 10\text{ kV}}{\sqrt{3}} \quad Q_1 := 100 \cdot M \cdot V \cdot A$$

$$V_{n2} := \frac{V_{rms\_n2} + j \cdot 10\text{ kV}}{\sqrt{3}} \quad Q_2 := 100 \cdot M \cdot V \cdot A$$

$$V_{n3} := \frac{V_{rms\_n3} + j \cdot 10\text{ kV}}{\sqrt{3}} \quad Q_3 := 100 \cdot M \cdot V \cdot A.$$

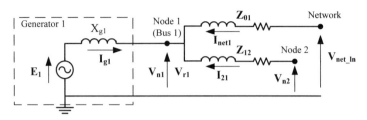

Figure 8.27. Single-phase equivalent circuit at Generator 1.

Node Point Equations

Nodal analysis is performed at each bus that is connected to a generator. As an example and referring to Figure 8.27, Bus 1 (at Generator 1) is interconnected with the network and with Generator 2. The node point (Kirchhoff's current law, KCL) equation for Node 1 is the sum of the Generator 1 current ($\mathbf{I}_{g1}$), the current of Line 12 ($\mathbf{I}_{21}$), and the current of Line 01 ($\mathbf{I}_{net1}$). The node point equations for Nodes 2 and 3 can be derived in a similar way.

The line currents are calculated by dividing the voltage difference between the node points by the line impedance. The generator current is calculated from the complex conjugate of the ratio of the generator three-phase complex power divided by 3 and the line-to-neutral voltage of the generator.

The unknown variables are the reactive powers of the generators (Q_i for $i = 1,2,3$) and the generator voltage phase angle, or complex value of the generator voltage (V_{ni}). The solution is constrained by the fact that the absolute value of each generator voltage must be equal to the given generator operating voltage (V_{rms_ni}).

The node point equations are solved by the Mathcad *Find* equation solver:

Given

$$Q_1 = \mathrm{Re}(Q_1) \quad Q_2 = \mathrm{Re}(Q_2) \quad Q_3 = \mathrm{Re}(Q_3)$$

$$|V_{n1}| = \frac{V_{rms\_n1}}{\sqrt{3}} \quad |V_{n2}| = \frac{V_{rms\_n2}}{\sqrt{3}} \quad |V_{n3}| = \frac{V_{rms\_n3}}{\sqrt{3}}$$

$$\overline{\left(\frac{P_1 + j \cdot Q_1}{3 \cdot V_{n1}}\right)} + \frac{V_{n2} - V_{n1}}{Z_{12}} + \frac{V_{net\_ln} - V_{n1}}{Z_{01}} = 0$$

$$\overline{\left(\frac{P_2 + j \cdot Q_2}{3 \cdot V_{n2}}\right)} + \frac{V_{n1} - V_{n2}}{Z_{12}} + \frac{V_{net\_ln} - V_{n2}}{Z_{02}} + \frac{V_{n3} - V_{n2}}{Z_{23}} = 0$$

$$\overline{\left(\frac{P_3 + j \cdot Q_3}{3 \cdot V_{n3}}\right)} + \frac{V_{n2} - V_{n3}}{Z_{23}} + \frac{V_{net\_ln} - V_{n3}}{Z_{03}} = 0$$

$$\begin{pmatrix} Q_{r1} \\ Q_{r2} \\ Q_{r3} \\ V_{r1} \\ V_{r2} \\ V_{r3} \end{pmatrix} := \mathrm{Find}(Q_1, Q_2, Q_3, V_{n1}, V_{n2}, V_{n3}).$$

GENERATOR APPLICATION

The Mathcad solution is:

Generator 1:

$V_{r1} = (126.9 + 27.7j) \cdot kV \quad \sqrt{3} \cdot |V_{r1}| = 225 \cdot kV$

$\delta_{r1} := \arg(V_{r1}) \quad \delta_{r1} = 12.3 \cdot \deg$

$Q_{r1} = -56.0 \cdot M \cdot V \cdot A$

Generator 2:

$V_{r2} = (132.0 + 22.5j) \cdot kV \quad \sqrt{3} \cdot |V_{r2}| = 232 \cdot kV$

$\delta_{r2} := \arg(V_{r2}) \quad \delta_{r2} = 9.68 \cdot \deg$

$Q_{r2} = 198.2 \cdot M \cdot V \cdot A$

Generator 3:

$V_{r3} = (130.5 + 17.6j) \cdot kV \quad \sqrt{3} \cdot |V_{r3}| = 228 \cdot kV$

$\delta_{r3} := \arg(V_{r3}) \quad \delta_{r3} = 7.69 \cdot \deg$

$Q_{r3} = 30.3 \cdot M \cdot V \cdot A.$

The absolute value of each resultant line-to-neutral bus voltage (which is $\sqrt{3}|V_{ri}|$) is the same as the specified generator operating voltage. This validates the results.

Generator Current and Power

The generator current is calculated by dividing the generator complex power by both a factor of 3 (to determine the single-phase apparent power) and the generator line-to-neutral voltage, and then taking the conjugate of the obtained complex number. The generator single-phase currents are:

$I_{g1}(V_{r1}, Q_{r1}) := \overline{\left(\dfrac{P_1 + j \cdot Q_{r1}}{3 \cdot V_{r1}}\right)} \quad I_{g1}(V_{r1}, Q_{r1}) = (884.5 + 339.9j) \text{ A}$

$I_{g2}(V_{r2}, Q_{r2}) := \overline{\left(\dfrac{P_2 + j \cdot Q_{r2}}{3 \cdot V_{r2}}\right)} \quad I_{g2}(V_{r2}, Q_{r2}) = (634.9 - 392.2j) \text{ A}$

$I_{g3}(V_{r3}, Q_{r3}) := \overline{\left(\dfrac{P_3 + j \cdot Q_{r3}}{3 \cdot V_{r3}}\right)} \quad I_{g3}(V_{r3}, Q_{r3}) = (456.9 - 15.7j) \text{ A}.$

The three-phase complex power of each generator is:

$S_{g1} := 3 \cdot V_{r1} \cdot \overline{I_{g1}(V_{r1}, Q_{r1})} \quad S_{g1} = (365.0 - 56.0j) \cdot M \cdot V \cdot A$

$|S_{g1}| = 369.3 \cdot M \cdot V \cdot A \quad P_1 = 365 \cdot MW$

$S_{g2} := 3 \cdot V_{r2} \cdot \overline{I_{g2}(V_{r2}, Q_{r2})} \quad S_{g2} = (225.0 + 198.2j) \cdot M \cdot V \cdot A$

$|S_{g2}| = 299.9 \cdot M \cdot V \cdot A \quad P_2 = 225 \cdot MW$

$S_{g3} := 3 \cdot V_{r3} \cdot \overline{I_{g3}(V_{r3}, Q_{r3})} \quad S_{g3} = (178.0 + 30.3j) \cdot M \cdot V \cdot A$

$|S_{g3}| = 180.6 \cdot M \cdot V \cdot A \quad P_3 = 178 \cdot MW.$

The obtained real parts of the complex power values are the same as the specified generator powers (P_n). This validates the results.

The operation of the system can be limited by the capacities of the generators. Comparison of the obtained apparent power value with the generator rating permits the assessment of the generator loading. In this example, the generator ratings are 400, 300, and 250 MVA, respectively. A comparison of the ratings with the calculated values ($|S_{gi}|$) demonstrates that none of the generators is overloaded.

However, the operating limits can be tested by increasing the load on the generators and the corresponding increase of system load until the capacity is reached. In fact, while arriving at the system conditions for this example, we increased the power output of Generator 2 in steps until the generator apparent power became $S_{g2} = 300$ MVA. This occurred around $P_2 = 225$ MW. This trial-and-error method can be used to assess the operating limits. Such a problem would normally be studied with a power flow program.

Transmission Line Current and Power Flows

The system operation may be limited by the current carrying capacity of the line. Typically, the conductor tables give an approximate value for the current carrying capacity of the conductors or the utilities specify the current for which the line is designed. As an example, the Arizona Public Service Company designed the 500 kV Cholla-Saguaro line for 2400 A. This line is built with two Bluebird (ACSR, 1.762" 84/19) conductors. In any load condition, the current of this line should be less than 2400 A.

Although in this example the rated line current is not specified, the reader should remember that comparison of the calculated current values with the current carrying capacity of the conductors is needed to assess the line operation security.

The line current in phase A and the three-phase complex power transfer are calculated in this section for each transmission line. For the evaluation of the power transfer through the lines, the reader is reminded that the passive sign convention is obeyed when current enters the positive voltage terminal. In this example, if the current direction is into the bus, then the passive sign convention is met. If the real or reactive power is positive, the appropriate power is being absorbed into that bus from the line. If the real or reactive power is negative, the bus is supplying the respective power.

Line 12:
For Line 12 to obey the passive sign convention at Node 1, current is assumed to flow from Node 2 to Node 1. The current flow from Node 2 to Node 1 is:

$$I_{21} := \frac{V_{r2} - V_{r1}}{Z_{12}} \quad I_{21} = (-221.7 - 270.2j) \text{ A}.$$

The negative current value implies that the actual current direction is from Node 1 to Node 2. The power from Line 12 to Node 1 is:

$$S_{12} := 3 \cdot V_{r1} \cdot \overline{I_{21}} \quad S_{12} = (-106.8 + 84.5j) \text{ M} \cdot \text{V} \cdot \text{A}.$$

The complex power indicates that Bus 1 is supplying real power to the line, and receiving reactive power from the line. Consequently, Bus 1 acts as a supply and Bus 2 serves as a load in this context. In similar fashion, the complex power at Bus 2 is computed, assuming current flow from Line 12 into Node 2. This requires the calculation of the complex power using $\mathbf{I}_{12}$ instead of $\mathbf{I}_{21}$:

$$\mathbf{I}_{12} := -\mathbf{I}_{21} \quad \mathbf{I}_{12} = (221.7 + 270.2\mathrm{j})\,\mathrm{A}$$
$$\mathbf{S}_{21} := 3 \cdot \mathbf{V}_{r2} \cdot \overline{\mathbf{I}_{12}} \quad \mathbf{S}_{21} = (106.1 - 92.1\mathrm{j}) \cdot \mathrm{M} \cdot \mathrm{V} \cdot \mathrm{A}.$$

The application of the passive sign convention leads to the conclusion that Node 1 acts as source of real power and Node 2 as load. Specifically, the real power is supplied by Node 1 ($P_{12} < 0$) and absorbed by Node 2 ($P_{21} > 0$). The reactive power is absorbed by Node 1 ($Q_{12} > 0$) and delivered by Node 2 since ($Q_{21} < 0$). To summarize, real power flows from Node 1 to Node 2, and reactive power flows from Node 2 to 1. The difference between the power sent by Node 1 and received by Node 2 is the line loss.

Line 01:
Since current flow should be toward the large network, we select the current direction from Node 1 to the network (Node 0):

$$\mathbf{I}_{1\mathrm{net}} := \frac{\mathbf{V}_{r1} - \mathbf{V}_{\mathrm{net\_ln}}}{\mathbf{Z}_{01}} \quad \mathbf{I}_{1\mathrm{net}} = (662.8 + 69.7\mathrm{j})\,\mathrm{A}.$$

The positive current implies that the current direction is from Node 1 to the network, and that Node 1 acts as a source and the network is a load:

$$\mathbf{S}_{1\_\mathrm{net}} := 3 \cdot \mathbf{V}_{r1} \cdot \overline{-\mathbf{I}_{1\mathrm{net}}} \quad \mathbf{S}_{1\_\mathrm{net}} = (-258.2 - 28.5\mathrm{j}) \cdot \mathrm{M} \cdot \mathrm{V} \cdot \mathrm{A}$$
$$\mathbf{S}_{\mathrm{net}\_1} := 3 \cdot \mathbf{V}_{\mathrm{net\_ln}} \cdot \overline{\mathbf{I}_{1\mathrm{net}}} \quad \mathbf{S}_{\mathrm{net}\_1} = (252.6 - 26.5\mathrm{j}) \cdot \mathrm{M} \cdot \mathrm{V} \cdot \mathrm{A}.$$

The passive sign convention indicates that real power flows from Node 1 to the network. The reactive power is negative at both nodes, which implies that each node delivers reactive power to the line. The line itself absorbs the reactive power.

Line 02:
We assume the current flow is from Node 2 to the network:

$$\mathbf{I}_{2\mathrm{net}} := \frac{\mathbf{V}_{r2} - \mathbf{V}_{\mathrm{net\_ln}}}{\mathbf{Z}_{02}} \quad \mathbf{I}_{2\mathrm{net}} = (644.0 - 76.4\mathrm{j})\,\mathrm{A}$$
$$\mathbf{S}_{2\_\mathrm{net}} := 3 \cdot \mathbf{V}_{r2} \cdot \overline{-\mathbf{I}_{2\mathrm{net}}} \quad \mathbf{S}_{2\_\mathrm{net}} = (-249.9 - 73.8\mathrm{j}) \cdot \mathrm{M} \cdot \mathrm{V} \cdot \mathrm{A}$$
$$\mathbf{S}_{\mathrm{net}\_2} := 3 \cdot \mathbf{V}_{\mathrm{net\_ln}} \cdot \overline{\mathbf{I}_{2\mathrm{net}}} \quad \mathbf{S}_{\mathrm{net}\_2} = (245.4 + 29.1\mathrm{j}) \cdot \mathrm{M} \cdot \mathrm{V} \cdot \mathrm{A}.$$

Based on the complex powers, Node 2 is the source and the network is the load. In this case, both real and reactive power flow is from Node 2 to the network.

Line 23:

$$I_{32} := \frac{V_{r3} - V_{r2}}{Z_{23}} \quad I_{32} = (-212.6 + 45.6j) \text{ A}$$

The negative real part current suggests that real power flow is from Node 2 to Node 3. This motivates the calculation of the complex power using I_{23} instead of I_{32}:

$$S_{23} := 3 \cdot V_{r2} \cdot \overline{I_{32}} \quad S_{23} = (-81.2 - 32.4j) \cdot M \cdot V \cdot A$$
$$I_{23} := -I_{32} \quad\quad\quad I_{23} = (212.6 - 45.6j) \text{ A}$$
$$S_{32} := 3 \cdot V_{r3} \cdot \overline{I_{23}} \quad S_{32} = (80.8 + 29.1j) \cdot M \cdot V \cdot A.$$

The results signify that both real and reactive power flow from Node 2 to Node 3.

Line 03:
Again, anticipating current flow toward the network:

$$I_{3net} := \frac{V_{r3} - V_{net\_ln}}{Z_{03}} \quad I_{3net} = (669.6 - 61.3j) \text{ A}$$
$$S_{3\_net} := 3 \cdot V_{r3} \cdot \overline{-I_{3net}} \quad S_{3\_net} = (-258.8 - 59.4j) \text{ M} \cdot V \cdot A$$
$$S_{net\_3} := 3 \cdot V_{net\_ln} \cdot \overline{I_{3net}} \quad S_{net\_3} = (255.1 + 23.4j) \text{ M} \cdot V \cdot A.$$

The signs of the power values indicate that both powers flow from Node 3 to network.

The overall power flow in the transmission system is annotated in Figure 8.28. The total three-phase complex power supplied to the large network can be calculated from the sum of the powers to Node 0:

$$S_{net} := S_{net\_1} + S_{net\_2} + S_{net\_3} \quad S_{net} = (753.1 + 25.9j) \text{ M} \cdot V \cdot A.$$

The total real power loss within the transmission lines is:

$$P_{loss} := (P_1 + P_2 + P_3) - \text{Re}(S_{net}) \quad P_{loss} = 14.9 \text{ MW}.$$

The percentage of the real power produced by the generators that is absorbed by the lines is:

$$P_{loss\%} := \frac{P_{loss}}{P_1 + P_2 + P_3} \quad P_{loss\%} = 1.94\%$$

Generator-Induced Voltage Phase Angle
In Example 8.2, we calculated the generator phase angle for the determination of the maximum power that a generator can deliver to a network. We learned that the maximum power occurs when the power angle is near 90°. The generator power angle is the phase

INDUCED VOLTAGE AND ARMATURE REACTANCE CALCULATION

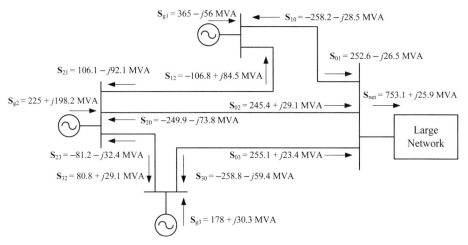

Figure 8.28. System power flow.

angle of the induced voltage. From Equation (8.15) or referring to Figure 8.27, the induced voltage (E_i) is the sum of the voltage drop across the synchronous reactance and the calculated terminal voltage:

Generator 1:
$E_1 := V_{r1} + j \cdot X_{g_1} \cdot I_{g1}(V_{r1}, Q_{r1})$ $E_1 = (77.6 + 156.1j) \cdot kV$
$|E_1| = 174.3 \cdot kV$ $\arg(E_1) = 63.6 \cdot \deg$

Generator 2:
$E_2 := V_{r2} + j \cdot X_{g_2} \cdot I_{g2}(V_{r2}, Q_{r2})$ $E_2 = (204.8 + 140.3j) \cdot kV$
$|E_2| = 248.3 \cdot kV$ $\arg(E_2) = 34.4 \cdot \deg$

Generator 3:
$E_3 := V_{r3} + j \cdot X_{g_3} \cdot I_{g3}(V_{r3}, Q_{r3})$ $E_3 = (133.34 + 101.65j) \cdot kV$
$|E_3| = 167.7 \cdot kV$ $\arg(E_3) = 37.3 \cdot \deg.$

It can be seen that the phase angles are noticeably less than 90°, which suggests that static stability is not a problem at these load values. In this system, the generators are overloaded before the 90° phase angle is reached (since the calculations of this example are based on voltage and current conditions at the generator rated values).

8.4. INDUCED VOLTAGE AND ARMATURE REACTANCE CALCULATION

The generator converts mechanical energy to electrical energy using the magnetic coupling between the stator and rotor. The purpose of this section is to demonstrate a simplified method to calculate generator parameters by analyzing the magnetic circuit.

The *generator parameters* are the induced voltage and the synchronous reactance. As suggested by Equation (8.15), these two quantities form the basis for creating a circuit model representing a synchronous generator in the steady state. If leakage is neglected, the synchronous reactance is simply the armature reactance. We will find that the rms value of the induced voltage (E_{sta}) and the armature reactance (X_{arm}) are functions of the field excitation current (I_f), specifically:

$$E_{sta} = \frac{\sqrt{2}}{\pi} \omega \mu_0 N_{sta} N_{rot} \frac{\ell_{sta} \ell_{rot}}{\ell_{gap}} I_f, \qquad (8.21)$$

$$X_{arm} = \frac{3}{\pi} \omega \mu_0 N_{sta}^2 \frac{\ell_{sta} \ell_{rot}}{\ell_{gap}} I_f, \qquad (8.22)$$

where N_{rot} is the number of turns in the rotor, and ℓ_{sta}, ℓ_{rot}, and ℓ_{gap} are, respectively, the magnetic path lengths in the stator, rotor, and each air gap between the stator and rotor. The effort at deriving these quantities is largely one of determining the magnetic fluxes. This is accomplished in the following two subsections.

The induced voltage-versus-dc excitation current, the load current-generated rotating flux, and the armature reactance are derived using Mathcad in combination with a numerical example. For this analysis, the generator is simplified. Both the stator and rotor windings are replaced by a single loop. The generator three-phase power and line voltage ratings are:

$$M := 10^6 \quad S_{gen} := 10 \cdot M \cdot V \cdot A \quad V_{gen} := 22 \text{ kV}.$$

The generator data used in this example are the rotor diameter (D_{rotor}), the stator length (L_{stator}), the gap thickness (L_{gap}), the number of poles (p), the number of turns in both the rotor and stator (N_{rotor} and N_{stator}), the stator frequency (f_A) and angular speed (ω), the rotor dc excitation current (I_{dc_rotor}), and the stator load current (I_{stator}):

$$D_{rotor} := 75 \text{ cm} \quad L_{stator} := 1.2 \text{ m} \quad L_{gap} := 20 \text{ mm} \quad p := 2$$
$$N_{rotor} := 400 \quad N_{stator} := 50 \quad f_A := 60 \text{ Hz}$$
$$I_{dc\_rotor} := 100 \text{ A} \quad I_{stator} := 300 \text{ A} \quad \omega := 2 \cdot \pi \cdot f_A.$$

8.4.1. Induced Voltage Calculation

The rotor winding is excited by dc current, which produces a magnetic field. When the rotor spins with the synchronous speed, this field induces a 60 Hz voltage in the stator phase windings. Figure 8.29 shows an illustration of the simplified generator and the stator current-generated magnetic field lines. The upward direction of the field through the rotor is determined by the right-hand rule.

INDUCED VOLTAGE AND ARMATURE REACTANCE CALCULATION

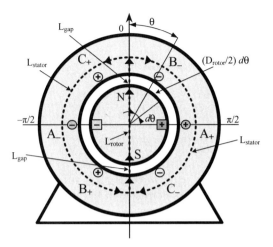

Figure 8.29. Rotor-generated magnetic field in the simplified generator.

The diagram shows that the magnetic circuit is divided into four sections: (1) rotor, (2) top gap, (3) stator, and (4) bottom gap. Application of Ampere's circuital law for this magnetic loop yields:

$$2 \cdot H_{gap} \cdot L_{gap} + H_{rotor} \cdot L_{rotor} + H_{stator} \cdot L_{stator} = I_{dc\_rotor} \cdot N_{rotor}, \quad (8.23)$$

where

- L_{gap} is the thickness of the air gap between the stator and rotor;
- $L_{rotor} := D_{rotor}$ is the magnetic path length in the rotor;
- L_{stator} is the magnetic path length in the stator;
- I_{dc_rotor} is the dc excitation current in the rotor;
- N_{rotor} is the number of turns in the rotor winding; and
- H is the magnetic field strength in the particular section.

Because of its high permeability, the iron of the rotor and stator is neglected, which simplifies the earlier expression to:

$$2 \cdot H_{gap} \cdot L_{gap} = I_{dc\_rotor} \cdot N_{rotor}. \quad (8.24)$$

The goal of this derivation is to determine the induced voltage-versus-dc excitation current relation. This requires expressing the magnetic field strength as a function of the excitation current by rearranging the previous formula:

$$H_{gap}(I_{dc\_rotor}) := \frac{I_{dc\_rotor} \cdot N_{rotor}}{2 \cdot L_{gap}}. \quad (8.25)$$

For the given generator data, this equation yields:

$$H_{gap}(I_{dc\_rotor}) = 1 \times 10^6 \, \frac{A}{m}.$$

The magnetic flux density in the air gap is:

$$B_{gap}(I_{dc\_rotor}) := \mu_0 \cdot H_{gap}(I_{dc\_rotor}), \qquad (8.26)$$

with a corresponding numerical value of $B_{gap}(I_{dc\_rotor}) = 1.257$ T.

The flux density in the gap is constant. In the upper part of the generator, the magnetic field lines extend outward from the rotor. We assume that these lines are positive. In the lower part of the generator, the magnetic field lines are entering the rotor. We assume that these lines are negative. These assumptions generate a square-shaped flux density distribution along the gap, which can be described by:

$$B_{rotor}(I_{dc\_rotor}, \theta) := \text{sign}(\cos(\theta)) \cdot B_{gap}(I_{dc\_rotor}), \qquad (8.27)$$

where $\theta := -90 \text{ deg}, -89 \text{ deg}.. 270 \text{ deg}$.

Figure 8.30 graphs the flux density distribution along the rotor surface. The base component of a square-shaped distribution curve is used for the induced voltage calculation. The *base component* is the 60 Hz fundamental frequency, and is calculated using the Fourier series:

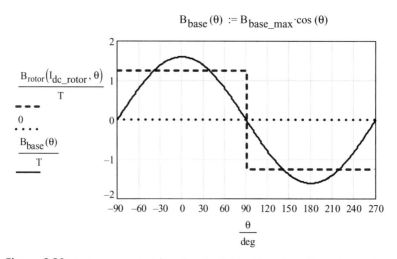

Figure 8.30. Rotor-generated flux density distribution along the rotor surface.

INDUCED VOLTAGE AND ARMATURE REACTANCE CALCULATION

$$B_{\text{base\_max}} := \frac{1}{\pi} \cdot \int_{-\frac{\pi}{2}}^{\frac{3\pi}{2}} B_{\text{rotor}}(I_{\text{dc\_rotor}}, \xi) \cdot \cos(\xi) d\xi = 1.6 \text{ T} \tag{8.28}$$

This equation can be evaluated symbolically using Mathcad. Type the definite integral into the workspace and then insert the symbolic equal sign. This results in:

$$\frac{2}{\pi} \int_{-\frac{\pi}{2}}^{\frac{\pi}{2}} B_{\text{gap}}(I_{\text{dc\_rotor}}) \cdot \cos(\theta) d\theta \rightarrow \frac{4 \cdot B_{\text{gap}}(I_{\text{dc\_rotor}})}{\pi}.$$

Therefore, the base component, which represents the magnitude of a cosinusoidal waveform at the fundamental frequency, is:

$$B_{\text{base\_max}}(I_{\text{dc\_rotor}}) := \frac{4}{\pi} \cdot B_{\text{gap}}(I_{\text{dc\_rotor}}). \tag{8.29}$$

The sinusoidal flux density distribution of the base component, which is graphed in Figure 8.30, is:

$$B_{\text{base}}(\theta) := B_{\text{base\_max}} \cdot \cos(\theta). \tag{8.30}$$

The substitution of Equation (8.25) into Equation (8.26), and placing their results into Equation (8.29), yields the following expression for the maximum flux density:

$$B_{\text{base\_max}}(I_{\text{dc\_rotor}}) := \frac{4}{\pi} \cdot \mu_0 \cdot \frac{I_{\text{dc\_rotor}} \cdot N_{\text{rotor}}}{2 \cdot L_{\text{gap}}}. \tag{8.31}$$

We test this equation with the numerical values and verify its validity:

$$B_{\text{base\_max}}(I_{\text{dc\_rotor}}) = 1.6 \text{ T}.$$

The flux generated by the dc excitation current is the integral of the flux density for the upper surface of the rotor ($\Phi = B \cdot Area$). This flux density integral is:

$$\Phi_{\text{rotor}}(I_{\text{dc\_rotor}}) := \int_{-\frac{\pi}{2}}^{\frac{\pi}{2}} B_{\text{base}}(\theta) \cdot \frac{D_{\text{rotor}}}{2} \cdot L_{\text{stator}} \, d\theta. \tag{8.32}$$

The numerical value of the flux is $\Phi_{\text{rotor}}(I_{\text{dc\_rotor}}) = 1.44$ Wb.

We symbolically determine the flux to obtain a practical expression:

$$\int_{-\frac{\pi}{2}}^{\frac{\pi}{2}} B_{base\_max} \cdot \cos(\theta) \cdot \frac{D_{rotor}}{2} \cdot L_{stator} \, d\theta \rightarrow B_{base\_max} \cdot D_{rotor} \cdot L_{stator}, \quad (8.33)$$

$$\Phi_{rotor}(I_{dc\_rotor}) := B_{base\_max}(I_{dc\_rotor}) \cdot D_{rotor} \cdot L_{stator}.$$

The substitution of Equation (8.31) into Equation (8.33) yields a formula for the rotor-generated flux:

$$\Phi_{rotor}(I_{dc\_rotor}) := \frac{4}{\pi} \cdot \mu_0 \cdot \frac{I_{dc\_rotor} \cdot N_{rotor}}{2 \cdot L_{gap}} \cdot D_{rotor} \cdot L_{stator}. \quad (8.34)$$

The reader can verify the validity of this equation by calculating the numerical value of the flux. The verification is $\Phi_{rotor}(I_{dc\_rotor}) = 1.44$ Wb.

This magnetic flux rotates with synchronous speed. The rotation generates time-varying flux linkage with the stator phase windings. The flux linkage with phase A is given by:

$$\Phi_A(I_{dc\_rotor}, t) := \Phi_{rotor}(I_{dc\_rotor}) \cdot \cos(\omega \cdot t). \quad (8.35)$$

From Faraday's law, the voltage induced in phase A is the derivative of the flux:

$$E_A(I_{dc\_rotor}, t) := N_{stator} \cdot \frac{d}{dt} \Phi_A(I_{dc\_rotor}, t). \quad (8.36)$$

If desired, the rms value of the induced voltage can be found directly from the definition of an rms value (see Chapter 3) using the earlier relation for E_A:

$$T := \frac{1}{60} \sec \quad \sqrt{\frac{1}{T} \cdot \int_{0 \sec}^{T} E_A(I_{dc\_rotor}, t)^2 \, dt} = 19.19 \text{ kV}.$$

The substitution of Equation (8.35) into Equation (8.36) is evaluated symbolically using Mathcad. Enter the expression and evaluate:

$$N_{stator} \cdot \frac{d}{dt}(\Phi_{rotor}(I_{dc\_rotor}) \cdot \cos(\omega \cdot t)) \rightarrow -N_{stator} \cdot \omega \cdot \Phi_{rotor}(I_{dc\_rotor}) \cdot \sin(\omega \cdot t).$$

This means that:

$$E_A(I_{dc\_rotor}, t) := -N_{stator} \cdot \Phi_{rotor}(I_{dc\_rotor}) \cdot \omega \cdot \sin(\omega \cdot t). \quad (8.37)$$

INDUCED VOLTAGE AND ARMATURE REACTANCE CALCULATION

The rms value of the sinusoidal voltage is calculated by dividing the maximum value by $\sqrt{2}$:

$$E_{A\_rms}(I_{dc\_rotor}) := \frac{\omega \cdot N_{stator} \cdot \Phi_{rotor}(I_{dc\_rotor})}{\sqrt{2}}. \tag{8.38}$$

The numerical value is $E_{A\_rms}(I_{dc\_rotor}) = 19.19$ kV.

The substitution of Equation (8.34) into Equation (8.38) yields a closed-form expression for the rms magnitude of the induced voltage. The rms voltage magnitude induced in any of the stator phases is described by:

$$E_{g\_rms}(I_{dc\_rotor}) := \frac{\omega \cdot N_{stator} \cdot \left(\frac{4}{\pi} \cdot \mu_0 \cdot \frac{I_{dc\_rotor} \cdot N_{rotor}}{2 \cdot L_{gap}} \cdot D_{rotor} \cdot L_{stator}\right)}{\sqrt{2}}. \tag{8.39}$$

The numerical verification results in: $E_{g\_rms}(I_{dc\_rotor}) = 19.19$ kV.

Equation (8.39) can be simplified. The final result for the induced generator voltage as a function of dc excitation current in the rotor is:

$$E_{g\_rms}(I_{dc\_rotor}) := \omega \cdot \frac{N_{stator}}{\pi} \cdot \mu_0 \cdot I_{dc\_rotor} \cdot \frac{N_{rotor}}{L_{gap}} \cdot D_{rotor} \cdot L_{stator} \cdot \sqrt{2}. \tag{8.40}$$

It can be seen that the induced voltage is linearly related with the dc excitation current if the iron is neglected. Figure 8.31 shows a plot of the rms stator voltage–dc excitation current function. The graph shows that the induced voltage is directly proportional with the excitation current. However, at higher voltages, the iron starts to

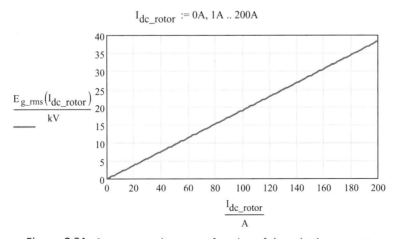

Figure 8.31. Stator rms voltage as a function of dc excitation current.

saturate, which produces saturation of the induced voltage (not shown in the equation or graph).

Simplification of Calculations. The stator has slots distributed along the interior surface. The phase conductors are placed in the slots to form coils as illustrated in Figure 8.2. The distance between the opposite side of the coils is the pitch as described earlier. Generators are built with full pitch (180 electrical degrees) or fractional pitch coils.

The voltage induced in the coils is the derivative of the flux linkage, which is a maximum when the rotor pole is perpendicular to the coil, since all the magnetic flux is linked with this coil. Conversely, the induced voltage is zero when the coil is aligned with the rotor pole. The flux linkage is also proportional with the coil area.

This implies that the voltage induced in a full pitch coil is larger than the voltage induced in a fractional pitch coil. Equation (8.2) calculates the induced voltage in a full pitch coil. The area of the fractional pitch coils is less than the full pitch coils. Consequently, the voltage induced in a fractional pitch coil is multiplied by the *pitch factor*, which is defined as:

$$k_p = \frac{\text{induced voltage of short-pitched coil}}{\text{induced voltage of a full-pitched coil}} = \sin(\alpha/2), \quad (8.41)$$

where α is the fractional pitch coil angle in electrical degrees.

Equation (8.2) assumes that the induced voltage is identical in all the conductors of a phase winding. However, Figure 8.5 reveals that the phase winding conductors are in slots and distributed along the stator interior surface. As an example, Figure 8.32a shows that the rotor pole alignment with conductor 1, when the induced voltage is zero, occurs before the rotor pole alignment with conductors 2 and 3. This implies that the induced ac voltages in conductors 1, 2, and 3 are not in phase. Because of this phase difference, the total voltage induced in the phase conductors is less than the algebraic

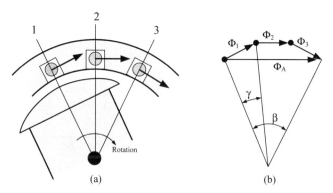

Figure 8.32. Induced voltage in phase A, which consists of three distributed conductors.

INDUCED VOLTAGE AND ARMATURE REACTANCE CALCULATION

sum of the voltages induced in the conductors. The total induced voltage is the vector sum of the voltages induced in the individual conductors.

Multiplication of the induced voltage with the *breadth factor* yields the reduced phase voltage. The breadth factor is:

$$k_b = \frac{\sin(\beta/2)}{n\sin(\gamma/2)}, \qquad (8.42)$$

where, as illustrated in Figure 8.32b, β is the angle encompassing each phase and γ is the angle between adjacent slots; and n is the number of slots per phase. Many times the phases of a generator contain series and parallel-connected coil groups. The phase difference between the induced voltages in these groups causes further reduction of the induced phase voltage. This requires a concomitant decrease in the breadth factor value.

The phase winding is formed by the series connection of the conductor coils in the slots. When fractional pitch coils are used, the induced voltage is reduced further. This suggests the combination of the breadth factor and pitch factor to form a *winding factor*. The winding factor is:

$$k_w = k_b k_p, \qquad (8.43)$$

The actual induced phase voltage is the product of the voltage calculated by Equation (8.5) and the winding factor:

$$E_{sta} = k_w \frac{N_{sta}\Phi_{rot}\omega}{\sqrt{2}}. \qquad (8.44)$$

An additional neglected effect is the increase of magnetic flux density in the teeth between slots holding the conductors, which can produce saturation and affects the induced voltage.

The stator flux generated by phases A, B, and C is calculated by replacing each distributed phase winding by a single coil with a 180-electrical degree pitch and placed in the middle of the specific phase winding. Such an arrangement is depicted in Figure 8.33, which portrays the distributed phase A, B, and C conductors at the bottom and the simplified concentrated conductors. The concentrated conductors generate the square-wave fluxes shown in the upper part of Figure 8.33. The distributed stator coil-generated flux is sketched in Figure 8.34. Each coil generates its own square-shaped flux waveform. For distributed conductors, the superposition of the square-wave flux produces the staircase displayed in Figure 8.34. For calculations, the staircase function may be approximated with a straight line.

The base component of the square wave is calculated using Fourier analysis. This base component is used for the flux calculation. The distributed winding-produced flux approximates the sine wave better than the square-wave flux used for deriving Equation (8.28). For an accurate calculation, the Fourier base component should be computed using the staircase waveform of Figure 8.34. Manufacturers use other methods like

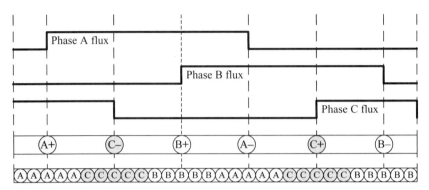

Figure 8.33. Stator current-generated flux calculation.

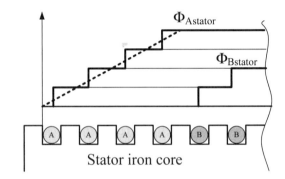

Figure 8.34. Flux distribution with distributed winding.

varying the number of conductors in the different coils to achieve a better sine wave and generate fewer harmonics.

In similar fashion, the distributed windings in the round rotor are arranged in such a way that a sinusoidal wave is generated. Figure 8.35 graphs both the actual flux produced by a round rotor as well as the base waveform equivalent, which is sinusoidal.

8.4.2. Armature Reactance Calculation

The three-phase load current produces a rotating flux, which reduces the generator terminal voltage. The rotating armature flux is proportional with the load current. Customarily, the effect of this flux is represented by an equivalent reactance, called the *armature reactance*. The following derivation is a simplified armature reactance calculation that neglects winding factors and leakage inductance, and uses concentrated stator coils. Nevertheless, the basic method for determining armature reactance is conveyed.

INDUCED VOLTAGE AND ARMATURE REACTANCE CALCULATION

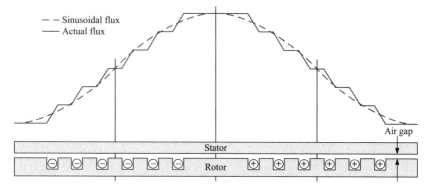

Figure 8.35. Round rotor generator flux.

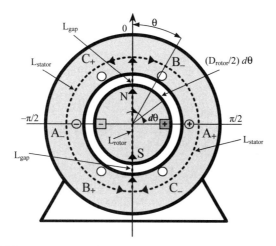

Figure 8.36. Arrangement for calculation of load current-generated flux.

The determination of the stator current-generated flux is similar to the dc rotor current-generated flux calculation. The flux generated by the load current in phase A can be determined using Ampere's circuital law. The flux lines and major dimensions used for this analysis are annotated in Figure 8.36. We continue to use the generator values of the previous section to perform the derivation of the armature reactance.

The generator rated rms current in a single phase is found from one-third of the generator three-phase apparent power and the line-to-neutral voltage:

$$I_{ac} := \frac{\frac{S_{gen}}{3}}{\frac{V_{gen}}{\sqrt{3}}} \quad I_{ac} = 262.4 \text{ A}.$$

This is a sinusoidal current, which for phase A is described by:

$$I_{A\_ac}(t) := \sqrt{2} \cdot I_{ac} \cdot \cos(\omega \cdot t). \qquad (8.45)$$

The ac current in phase A produces an ac flux. Figure 8.36 shows the flux lines when the current is maximum. The magnetic circuit is divided into four sections: (1) rotor, (2) top gap, (3) stator, and (4) bottom gap. Applying Ampere's circuital law to this magnetic loop yields:

$$2 \cdot H_{A\_ac} \cdot L_{gap} + H_{A\_rotor} \cdot L_{rotor} + H_{A\_stator} \cdot L_{stator} = I_{A\_ac}(t) \cdot N_{stator}. \qquad (8.46)$$

Because of its high permeability, the iron is neglected, which means the magnetic field for the rotor and stator are negligible. With this assumption, the magnetic field strength is:

$$H_{A\_ac}(t) := \frac{I_{A\_ac}(t) \cdot N_{stator}}{2 \cdot L_{gap}}. \qquad (8.47)$$

The calculation is verified using:

$$t := \frac{1}{60} \text{ s} \quad H_{A\_ac}(t) = 4.64 \times 10^5 \frac{A}{m}.$$

The magnetic flux density in the gap is:

$$B_{A\_ac}(t) := \mu_0 \cdot H_{A\_ac}(t). \qquad (8.48)$$

which, at the time instant of interest, is $B_{A\_ac}(t) = 0.583$ T.

The low flux density indicates that the load current does not produce iron saturation. The flux density in the air gap is constant, which produces a square-shaped flux density distribution along the gap. The distribution is similar to the flux distribution produced by the rotor dc current. For the analysis of the load current-generated flux, the base component of the square-shaped distribution curve is calculated using a Fourier series. Similar to Equation (8.29) and Equation (8.30), the base component of the flux density is:

$$B_{base\_A}(t, \theta) := \frac{4}{\pi} \cdot B_{A\_ac}(t) \cdot \cos(\theta). \qquad (8.49)$$

The air-gap flux is the integral of the flux density along the rotor surface:

$$\Phi_{A\_ac}(t) := \int_{-\frac{\pi}{2}}^{\frac{\pi}{2}} \frac{4}{\pi} \cdot B_{A\_ac}(t) \cdot \cos(\theta) \cdot L_{stator} \cdot \frac{D_{rotor}}{2} d\theta. \qquad (8.50)$$

The numerical value is $\Phi_{A\_ac}(t) = 0.668$ Wb.

The symbolic evaluation of the integral and substitution of Equation (8.47) and Equation (8.48) in the integral results in a closed-form expression for the flux. A similar calculation has been presented previously. The reader should refer to and apply the derivation used in Equation (8.31) through Equation (8.34):

$$\Phi_{A\_ac}(t) := \frac{4}{\pi} \cdot \mu_0 \cdot \frac{I_{A\_ac}(t) \cdot N_{stator}}{2 \cdot L_{gap}} \cdot L_{stator} \cdot D_{rotor}, \quad (8.51)$$

$$\Phi_{A\_ac}(t) = 0.668 \text{ Wb}.$$

The obtained numerical value validates the derived expression. The substitution of Equation (8.45) into Equation (8.51) produces the time function of the phase A generated ac flux:

$$\Phi_{A\_ac}(t) := \frac{4}{\pi} \cdot \mu_0 \cdot \frac{\sqrt{2} \cdot I_{ac} \cdot N_{stator}}{2 \cdot L_{gap}} \cdot L_{stator} \cdot D_{rotor} \cdot \cos(\omega \cdot t). \quad (8.52)$$

Figure 8.37 shows the location of the ac flux and its magnitude as a function of time (i.e., rotor position). The A phase generates a flux, which is perpendicular to the phase coil and its amplitude changes sinusoidally. This is a pulsating flux, which is represented by a time variable vector.

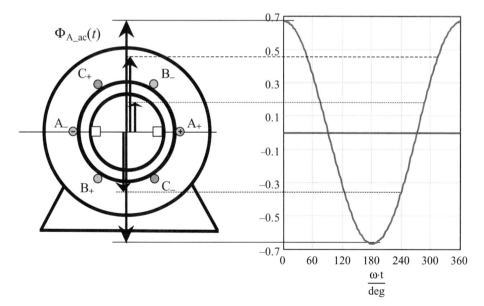

Figure 8.37. Phase A load current-generated ac flux.

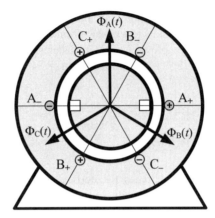

Figure 8.38. AC fluxes generated by the phase currents.

The other two phase currents generate similar sinusoidal ac fluxes. The distinction is the 120° or 240° phase lag of the current. The B and C phase generated ac fluxes are:

$$\Phi_{B\_ac}(t) := \frac{4}{\pi} \cdot \mu_0 \cdot \frac{\sqrt{2} \cdot I_{ac} \cdot N_{stator}}{2 \cdot L_{gap}} \cdot L_{stator} \cdot D_{rotor} \cdot \cos(\omega \cdot t - 120 \text{ deg}), \quad (8.53)$$

$$\Phi_{C\_ac}(t) := \frac{4}{\pi} \cdot \mu_0 \cdot \frac{\sqrt{2} \cdot I_{ac} \cdot N_{stator}}{2 \cdot L_{gap}} \cdot L_{stator} \cdot D_{rotor} \cdot \cos(\omega \cdot t - 240 \text{ deg}). \quad (8.54)$$

Figure 8.37 points out that the flux vector (Φ_{A_ac}) is perpendicular to the phase A winding. Likewise, the flux vectors of phases B and C are perpendicular to their windings. This is demonstrated in Figure 8.38, which exhibits the flux vectors generated by the three phases. From the geometry of the diagram, it can be seen that the flux vectors B and C are shifted by 120° and 240°, respectively, from flux vector A.

Figure 8.38 implies that only the vertical components of the fluxes produced by phases B and C link with phase A. The flux linkages with phase A are described by using Equation (8.52), Equation (8.53), and Equation (8.54), while considering the angle between the A phase and the other phase of interest:

$$\Phi_{AA}(t) := \frac{4}{\pi} \cdot \mu_0 \cdot \frac{\sqrt{2} \cdot I_{ac} \cdot N_{stator}}{2 \cdot L_{gap}} \cdot L_{stator} \cdot D_{rotor} \cdot \cos(\omega \cdot t), \quad (8.55)$$

$$\Phi_{BA}(t) := \frac{4}{\pi} \cdot \mu_0 \cdot \frac{\sqrt{2} \cdot I_{ac} \cdot N_{stator}}{2 \cdot L_{gap}} \cdot L_{stator} \cdot D_{rotor} \cdot \cos(\omega \cdot t - 120 \text{ deg}) \cdot \cos(120 \text{ deg}), \quad (8.56)$$

$$\Phi_{CA}(t) := \frac{4}{\pi} \cdot \mu_0 \cdot \frac{\sqrt{2} \cdot I_{ac} \cdot N_{stator}}{2 \cdot L_{gap}} \cdot L_{stator} \cdot D_{rotor} \cdot \cos(\omega \cdot t - 240 \text{ deg}) \cdot \cos(240 \text{ deg}). \quad (8.57)$$

INDUCED VOLTAGE AND ARMATURE REACTANCE CALCULATION

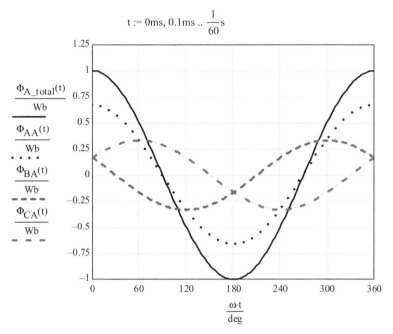

Figure 8.39. Flux linkages with phase A.

The total flux linkage with phase A is the sum of the three phase flux linkages:

$$\Phi_{A\_total}(t) := \Phi_{AA}(t) + \Phi_{BA}(t) + \Phi_{CA}(t). \tag{8.58}$$

Figure 8.39 plots the total flux linkage as well as the A, B, and C phase-generated flux linkage with phase A. It can be seen that all flux linkages are sinusoidal ac functions. Using the trigonometric relation $\cos(\alpha+\beta) = \cos(\alpha)\cos(\beta) - \sin(\alpha)\sin(\beta)$, and Equation (8.55), Equation (8.56), and Equation (8.57), Equation (8.58) can be simplified. The result is:

$$\Phi_{A\_total}(t) := \frac{3}{2} \cdot \left(\frac{4}{\pi} \cdot \mu_0 \cdot \frac{\sqrt{2} \cdot I_{ac} \cdot N_{stator}}{2 \cdot L_{gap}} \cdot L_{stator} \cdot D_{rotor} \cdot \cos(\omega \cdot t) \right). \tag{8.59}$$

The validity of the equation is verified by the obtained numerical value at $t = 0$ ms:

$$\Phi_{A\_total}(0 \text{ ms}) = 1.002 \text{ Wb}.$$

Comparing Equation (8.51) and Equation (8.59) reveals that the ratio of the total flux linkage and the flux linkage of phase A is:

$$\frac{\Phi_{A\_total}(t)}{\Phi_{A\_ac}(t)} = 1.5.$$

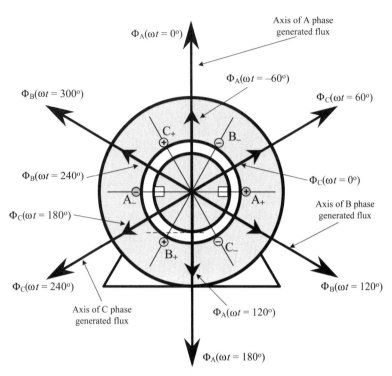

Figure 8.40. Time-varying flux vectors for each phase.

In similar fashion, the total flux linkages of phases B and C are:

$$\Phi_{B\_total}(t) := \frac{3}{2} \cdot \left(\frac{4}{\pi} \cdot \mu_0 \cdot \frac{\sqrt{2} \cdot I_{ac} \cdot N_{stator}}{2 \cdot L_{gap}} \cdot L_{stator} \cdot D_{rotor} \cdot \cos(\omega \cdot t - 120 \text{ deg}) \right), \quad (8.60)$$

$$\Phi_{C\_total}(t) := \frac{3}{2} \cdot \left(\frac{4}{\pi} \cdot \mu_0 \cdot \frac{\sqrt{2} \cdot I_{ac} \cdot N_{stator}}{2 \cdot L_{gap}} \cdot L_{stator} \cdot D_{rotor} \cdot \cos(\omega \cdot t - 240 \text{ deg}) \right). \quad (8.61)$$

Figure 8.40 denotes the direction of the time-varying flux vectors of phases A, B, and C, and the vector magnitudes are shown at $\omega t = 0°, 60°, 120°, 180°, 240°,$ and $300°$. At a given time instant, the vectors are added together to produce the total flux (Φ_{ABC}) generated by the phase currents. It will be proved that the amplitude of the total flux is 1.5 times the amplitude of the maximum of any phase-generated flux. Determination of the total flux vector at any time instant requires the summation of the total flux linkages of the three phases while considering the vector direction of each. Figure 8.41 demonstrates that the sum of the time-varying vectors produces a rotating vector with an amplitude 1.5 times the peak value of the phase vectors.

The production of the rotating flux vector requires the summation of the three flux vectors for the phases, which requires the transformation of Equation (8.59), Equation

INDUCED VOLTAGE AND ARMATURE REACTANCE CALCULATION 503

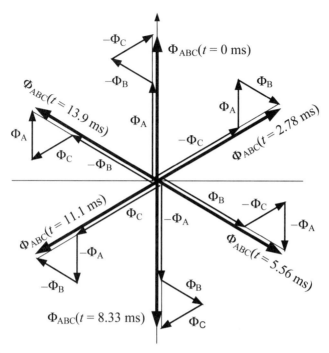

Figure 8.41. Demonstration of the phase currents-generated rotating flux vector for a two-pole, 60 Hz generator.

(8.60), and Equation (8.61) to vectors. The phase angle of Equation (8.59) is zero; consequently, this equation describes a time-varying vector in the zero direction. Equation (8.60) describes the time variation of the B phase-generated flux, whose phase angle should be −120°. The multiplication of the equation by $e^{-j120°}$ produces a time-varying complex vector in the desired direction. Similarly, the multiplication of Equation (8.61) by $e^{-j240°}$ generates a complex, time-variable vector in the −240° direction. A rotating vector is produced by the summation of the flux vectors:

$$\Phi_{ABC}(t) := \Phi_{A\_total}(t) + \Phi_{B\_total}(t) \cdot e^{-j120\,\text{deg}} + \Phi_{C\_total}(t) \cdot e^{-j240\,\text{deg}}. \quad (8.62)$$

The total flux rotates with the synchronous speed and has constant amplitude. This is demonstrated in Figure 8.42, which graphs the total fluxes generated by the phase currents together with the rotating total flux vector amplitude. The conclusion from this analysis is that:

$$\frac{|\Phi_{ABC}(t)|}{\max(\Phi_{A\_total}(t))} = 1.5$$

Figure 8.43 plots the normalized rotating vector magnitude and phase angle time variation. The graph indicates that the magnitude of the flux vector is constant and the phase

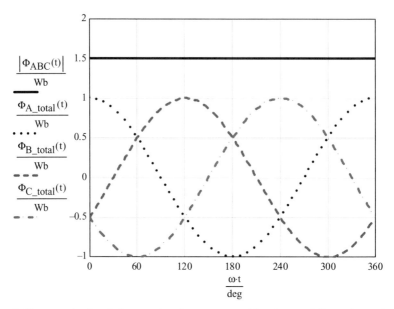

Figure 8.42. Magnitude of the rotating flux generated by the total fluxes from the three phases.

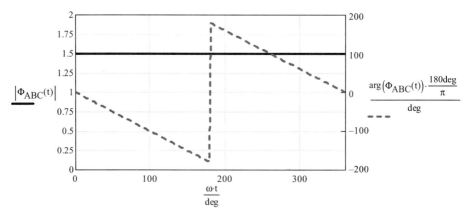

Figure 8.43. Rotating flux vector magnitude and phase angle time variation.

angle varies linearly with the position (or time). Figure 8.44 is a polar plot of the total flux as a function of the angle (i.e., ωt), which produces a circle, demonstrating that the generated $\Phi_{ABC}(t)$ is a rotating vector.

The armature inductance for each phase is determined from the total flux for the given phase. This determination can be performed using either Ampere's circuital law or Faraday's law of induction.

INDUCED VOLTAGE AND ARMATURE REACTANCE CALCULATION

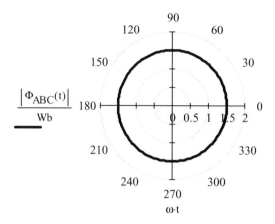

Figure 8.44. Demonstration of rotating flux vector using a polar plot.

8.4.2.1. Faraday's Law of Induction. The phase current-generated flux links with the phase windings and induces voltages in the windings. This voltage is subtracted from the induced voltage to produce the generator terminal voltage. The voltage induced by the phase A load current is:

$$E_{A\_load}(t) := N_{stator} \cdot \frac{d}{dt} \Phi_{A\_total}(t). \tag{8.63}$$

The numerical value is $E_{A\_load}(4ms) = -18.85$ kV.

Equation (8.59) is substituted into Equation (8.63), resulting in:

$$E_{A\_load}(t) := N_{stator} \cdot \frac{d}{dt}\left[\frac{3}{2} \cdot \left(\frac{4}{\pi} \cdot \mu_0 \cdot \frac{\sqrt{2} \cdot I_{ac} \cdot N_{stator}}{2 \cdot L_{gap}} \cdot L_{stator} \cdot D_{rotor} \cdot \cos(\omega \cdot t)\right)\right].$$

The derivative is calculated symbolically to yield:

$$E_{A\_load}(t) := -\frac{3}{2} \cdot \left(\frac{4}{\pi} \cdot \mu_0 \cdot \frac{\sqrt{2} \cdot I_{ac} \cdot N_{stator}^2}{2 \cdot L_{gap}} \cdot L_{stator} \cdot D_{rotor} \cdot \omega \cdot \sin(\omega \cdot t)\right). \tag{8.64}$$

A numerical evaluation checks the validity of the equation: $E_{A\_load}(4ms) = -18.85$ kV.

The ac flux linkage produced induced voltage can be calculated by:

$$E_{A\_load}(t) = L_{arm} \cdot \frac{d}{dt}[I_{A\_ac}(t)],$$

where L_{arm} is the *armature inductance*. Substituting Equation (8.45) gives:

$$E_{A\_load}(t) := L_{arm} \cdot \frac{d}{dt} \cdot \left(\sqrt{2} \cdot I_{ac} \cdot \cos(\omega \cdot t)\right) \tag{8.65}$$

The symbolic evaluation of the previous expression results in:

$$E_{A\_load}(t) := -L_{arm} \cdot \omega \cdot \sqrt{2} \cdot I_{ac} \cdot \sin(\omega \cdot t), \tag{8.66}$$

where $X_{arm} = L_{arm}\omega$ is the *armature reactance*.

Equation (8.66) is equated to Equation (8.64) and the armature reactance is calculated. The result is:

$$X_{arm} := \frac{3}{2} \cdot \left(\frac{4}{\pi} \cdot \mu_0 \cdot \frac{N_{stator}^2}{2 \cdot L_{gap}} \cdot L_{stator} \cdot D_{rotor} \cdot \omega\right). \tag{8.67}$$

8.4.2.2. Ampere's Circuital Law. Alternatively, the armature reactance can be determined using Ampere's circuital law. The rms value of the total rotating flux from Equation (8.59) is:

$$\Phi_{A\_rms} = 1.5 \cdot \frac{4}{\pi} \cdot \mu_0 \cdot \frac{I_{ac} \cdot N_{stator}}{2 \cdot L_{gap}} \cdot L_{stator} \cdot D_{rotor}.$$

The relation between the flux and the inductance is:

$$N_{stator} \cdot \Phi_{A\_rms} = L_{arm} \cdot I_{ac}.$$

The armature reaction produced inductance is:

$$L_{arm} = \frac{N_{stator} \cdot \Phi_{A\_rms}}{I_{ac}} = \frac{3}{\pi} \cdot \mu_0 \cdot \frac{N_{stator}^2}{L_{gap}} \cdot L_{stator} \cdot D_{rotor}.$$

Therefore the armature reactance is:

$$X_{arm} = \frac{3}{\pi} \cdot \mu_0 \cdot \omega \cdot \frac{N_{stator}^2}{L_{gap}} \cdot L_{stator} \cdot D_{rotor}.$$

The numerical value of the armature reactance is $X_{arm} = 50.89\ \Omega$. The synchronous reactance is the sum of the armature and the coil leakage reactances. The coil leakage reactance is typically small, around 10%, while the armature reactance is above 100%. Consequently, the armature reactance can be used in lieu of the synchronous reactance.

This corresponds to a p.u. value of:

$$x_{arm} := X_{arm} \cdot \frac{S_{gen}}{V_{gen}^2} \quad x_{arm} = 105.2\%.$$

This is a reasonable value because the armature reactance is typically around 90–130% based on the rated power and voltage of the generator.

8.5. CONCEPT OF GENERATOR PROTECTION

A large synchronous generator connected to a power system must be protected against short circuits or ground faults in the stator, ground faults in the rotor, and faults in the generator step-up (GSU) transformer. In the event of any of these faults, the generator must be disconnected from the network by the CBs at the high-voltage side of the GSU transformer, and the excitation must be removed. Due to the energy stored in the rotor winding, removing the excitation requires time that prolongs the current flow during the fault and can cause permanent damage to the machine. A stator fault is infrequent but can be devastating when it occurs.

Chapter 7 presented the basic concept of differential protection. In this section, a more in-depth discussion is offered. Figure 8.45 explains the concept of the differential protection in greater detail. As depicted in the diagram, the relay is supplied by two current transformers (CTs), which convert the main currents I_A and I_a into the measured currents I_1 and I_2, respectively. The differential relay develops a *restraint current* in accordance with:

$$I_r = \frac{|I_1| + |I_2|}{2}, \tag{8.68}$$

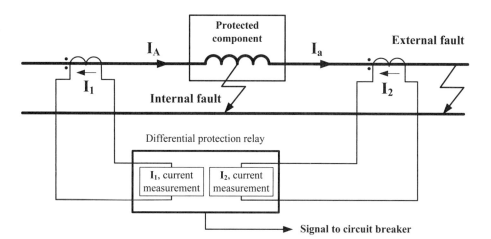

Figure 8.45. Concept of differential protection for a component.

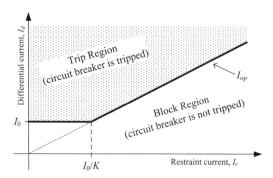

Figure 8.46. Differential relay characteristics.

as well as the *differential current*:

$$I_d = |\mathbf{I}_1 - \mathbf{I}_2|. \tag{8.69}$$

Figure 8.46 depicts typical differential relay characteristics. The combination of I_d and I_r above the line initiates CB tripping and excitation removal; below the curve, no trip signal is generated. The relay characteristic line may be described by:

$$I_{op} = \begin{cases} I_0 & I_r \leq I_0/K \\ KI_r & I_r > I_0/K \end{cases}, \tag{8.70}$$

where I_{op} is the calculated *operating current*; I_0 is a constant threshold value called the minimum pickup, typically equal to 1 A; and K is the slope of the line in Figure 8.46, where typical adjustable values of K are from 10% to 60%. After the relay calculates I_{op} from Equation (8.70) using the measured restraint current, an action is then taken according to:

$$\begin{aligned} I_{op} < I_d & \quad \text{trip signal} \\ I_{op} > I_d & \quad \text{no trip signal} \end{aligned}. \tag{8.71}$$

Digital relays can also implement a more elaborate curve than that presented in Figure 8.46.

In the event of an *external fault*, the main currents entering and leaving the protected equipment (e.g., the generator) are identical. This case produces a zero or near-zero differential current ($I_d \approx 0$). In most cases, the measurement of large short-circuit current produces measurement errors, which yields a larger than zero current difference. For ideal CTs, the differential current is zero, but in the case of CT saturation or mismatch of CT ratings, a small current difference exists. For such a small I_d value, the implementation of the minimum pickup current (I_0) in the relay characteristics prevents inadvertent actuation of the CB. Hence, for an external fault, the relay blocks the CB operation.

CONCEPT OF GENERATOR PROTECTION

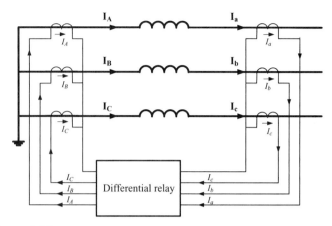

Figure 8.47. A wye-connected generator stator differential protection.

In the case of an *internal fault*, the main currents are both entering the generator. Although the restraint current from an internal fault may be of similar magnitude as that due to an external fault, a large differential current occurs since the direction of $\mathbf{I_a}$ is toward the generator, causing $\mathbf{I_2}$ to be negative valued. The differential current then exceeds the operating current calculated from the relay characteristics. Consequently, the relay triggers the CB and excitation removal.

Figure 8.47 shows the basic connection for three-phase wye-connected generator differential protection. In each phase, two CTs measure the currents at the two ends of the generator stator windings. The CTs reduce the two currents to a preset value. Typically, the reduced current value is about 5 A at rated load. The differential relay records the main current and the differential current in each phase. The differential current is the sum or the difference of the currents entering and leaving the generator.

The differential protection operation is fast—typically, the faulty generator is switched off from the power system within two to three cycles and excitation is removed by opening the field breaker. Fault current continues to flow for a short time due to the stored electromagnetic energy. The digital differential protection used by the power industry employs a more elaborate algorithm.

Figure 8.48 shows a diagram of the circuit for the combined ground fault and overcurrent protection of a generator. Three CTs supply either a three-phase overcurrent protection relay or three individual relays protecting each phase of the generator. The overcurrent protection guards against overload and operates with current dependent time delay. Table 8.1 lists the relations describing the Institute of Electrical and Electronics Engineers (IEEE) standard specified inverse time-dependent delay for overcurrent relays. The current ratio (I_{ratio}) of the fault and setting currents uses the measured fault current from the CT secondary. The setting current is selectable; typically, I_{set} is less than half of the minimum fault current because the relay must operate at this current level. However, the setting current must be larger than the maximum load current. An acceptable value for I_{set} is two to three times the maximum load current. Figure 8.49

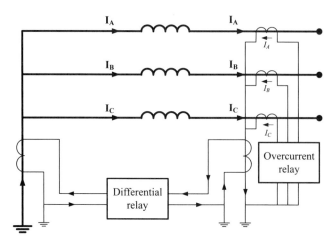

Figure 8.48. Generator ground fault and overcurrent protection.

TABLE 8.1. Overcurrent Protection Relay Equations (International Electrotechnical Commission [IEC] 60255)

Condition	Time Delay, $t(I_r, TD)$
IEEE moderately inverse	$\dfrac{TD}{7}\left(\dfrac{0.0515}{I_{ratio}^{0.02}-1}+0.114\right)$
IEEE very inverse	$\dfrac{TD}{7}\left(\dfrac{19.61}{I_{ratio}^{2}-1}+0.491\right)$
Extremely inverse	$\dfrac{TD}{7}\left(\dfrac{28.2}{I_{ratio}^{2}-1}+0.1217\right)$
US CO8 inverse	$\dfrac{TD}{7}\left(\dfrac{5.95}{I_{ratio}^{2}-1}+0.18\right)$
US CO2 short time inverse	$\dfrac{TD}{7}\left(\dfrac{0.02394}{I_{ratio}^{0.02}-1}+0.01694\right)$

Where TD = time dial setting (1–11); I_{set} = relay setting current; and $I_{ratio} = I_{fault}/I_{set}$ is the ratio of fault and setting currents.

shows examples of the moderately inverse relay characteristics for three different time dial settings. The graph shows that increasing current reduces the operation time delay.

The generator ground fault protection is a zero-sequence differential protection as drawn in Figure 8.48. The zero-sequence current is produced by two CTs:

1. a CT placed in the generator ground conductor at the neutral points and
2. a CT measuring the sum of the phase currents at the generator terminals.

The two CTs supply a differential single-phase relay, which provides an instantaneous trip signal in the event of a generator ground fault.

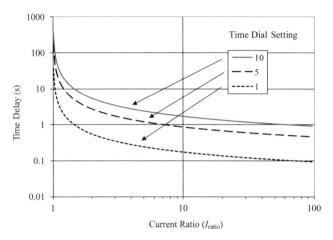

Figure 8.49. IEEE moderately inverse overcurrent relay characteristics when the time dial setting (TD) is 1, 5, and 10 using the same setting current.

The differential, ground differential, and overcurrent protections are the major protection equipment guarding the generator against internal faults. In addition, the generator must be protected against abnormal operating conditions, including:

- overvoltage,
- excitation system failure,
- motoring (loss of turbine drive),
- unbalanced load,
- loss of synchronization,
- frequency deviation (higher or lower frequency),
- transient instability,
- subsynchronous oscillations, and
- mechanical vibration.

Protection against the abnormal operating conditions requires specialized sensors and sophisticated instrumentation.

8.6. APPLICATION EXAMPLES

This section comprises four extended examples, including:

1. Mathcad analysis of a synchronous generator,
2. MATLAB analysis of static stability,
3. MATLAB analysis of generator loading, and
4. PSpice simulation of generator transients.

EXAMPLE 8.4: Mathcad analysis of a synchronous generator

A three-phase generator supplies a large network through a transformer and a transmission line. The system one-line diagram is presented in Figure 8.50.
The generator apparent power and line voltage ratings are:

$$M := 10^6 \quad S_g := 450 \, M \cdot V \cdot A \quad V_g := 28 \, kV,$$

The physical dimensions and specifications of the generator are:

$$N_{rotor} := 22 \quad N_{stator} := 7 \quad p := 2 \quad f_A := 60 \, Hz$$
$$D_{rotor} := 120 \, cm \quad L_{stator} := 10 \, m \quad L_{gap} := 150 \, mm \quad \omega := 2 \cdot \pi \cdot f_A.$$

The transformer rating data are:

$$S_{tr} := 500 \, M \cdot V \cdot A \quad V_p := 27 \, kV \quad V_s := 500 \, kV \quad x_{tr} := 12\%.$$

The transmission line length and p.u. length impedance are:

$$L_{line} := 45 \, mi \quad z_{line} := (0.1 + j \cdot 0.65) \cdot \frac{\Omega}{mi}.$$

The network line-to-line voltage is $V_{net} := 480 \, kV$.
In this example, we will:

1. form and draw an equivalent single-phase circuit, and then simplify the circuit;
2. calculate the generator-induced voltage if the generator carries the rated load at a power factor of 0.8 lagging, and the generator terminal voltage is the rated value;
3. compute and plot the power delivered to the network versus the power angle;
4. determine the maximum power and related power angle, as well as the power angle that is needed to deliver 400 MW to the network; and
5. calculate the dc excitation current in the rotor, which generates the induced voltage determined in part 2.

Accordingly, $pf_g := 0.8 \quad P_{network} := 400 \, MW.$

Figure 8.50. One-line diagram of synchronous generator network.

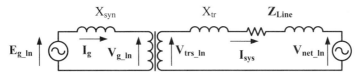

Figure 8.51. Single-phase equivalent circuit of synchronous generator network.

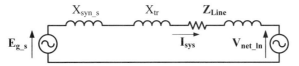

Figure 8.52. Simplified single-phase equivalent circuit of synchronous generator network.

(1) Equivalent Simplified Circuit

The equivalent circuit is provided in Figure 8.51. The induced voltage (E_{g_ln}) and the synchronous reactance (X_{sy}) model the generator. The transformer reactance (X_{tr}) is placed on the secondary side. An impedance (Z_{Line}) characterizes the transmission line and the line capacitance is neglected. The network reactance is neglected; the network is simply represented by a voltage source (V_{net_ln}).

Neglecting leakage, the generator single-phase synchronous reactance is calculated using Equation (8.67) for the armature reactance as well as 10% leakage reactance:

$$X_{syn} := 1.1 \cdot \frac{3}{2} \left(\frac{4}{\pi} \mu_0 \cdot \frac{N_{stator}^2}{2 \cdot L_{gap}} \right) \cdot L_{stator} \cdot D_{rotor} \cdot \omega = 1.951 \, \Omega.$$

The transformer reactance at the secondary side and the turns ratio are:

$$X_{tr} := x_{tr} \cdot \frac{V_s^2}{S_{tr}} := 60 \, \Omega \quad T := \frac{V_p}{V_s} = 0.054.$$

For the simplification of the network, the generator-induced voltage and synchronous reactance are transferred to the secondary side of the transformer. This eliminates the transformer from the circuit, as drawn in Figure 8.52. The generator-induced voltage and synchronous reactance at the secondary side are:

$$E_{g\_s} := \frac{E_{g\_ln}}{T} \quad X_{syn\_s} := \frac{X_{syn}}{T^2} = 669.0 \, \Omega.$$

The line impedance is:

$$Z_{Line} := L_{line} \cdot z_{line} \quad Z_{Line} = (4.5 + 29.25j) \, \Omega.$$

The overall system impedance is the sum of the three impedances connected in series:

$$Z_{system} := j \cdot X_{syn\_s} + j \cdot X_{tr} + Z_{Line} = (4.5 + 758.2j) \, \Omega.$$

The network line-to-neutral voltage is:

$$V_{net\_ln} := \frac{V_{net}}{\sqrt{3}} \quad V_{net\_ln} = 277.1 \, \text{kV}.$$

(2) Generator-Induced Voltage

The generator carries the rated load at a power factor of 0.8 lagging and the generator terminal voltage is the rated value. The rated complex power of the generator A phase is:

$$S_{g\_a} := \frac{S_g}{3} \cdot e^{j \cdot a\cos(pfg)} \quad S_{g\_a} = (120 + 90j) \, \text{M} \cdot \text{V} \cdot \text{A}.$$

The generator line-to-neutral voltage is:

$$V_{g\_ln} := \frac{V_g}{\sqrt{3}} \quad V_{g\_ln} = 16.17 \, \text{kV}.$$

The generator current is:

$$I_g := \overline{\left(\frac{S_{g\_a}}{V_{g\_ln}}\right)} \quad I_g = (7.42 - 5.57j) \cdot \text{kA}$$

$$|I_g| = 9.279 \cdot \text{kA} \quad \arg(I_g) = -36.9 \cdot \text{deg}.$$

Referring to Figure 8.51, the generator-induced line-to-neutral voltage is:

$$E_{g\_ln} := V_{g\_ln} + I_g \cdot (j \cdot X_{syn}) = (27.03 + 14.48j) \, \text{kV}.$$

The generator-induced voltage magnitude and power angle at rated load are:

$$|E_{g\_ln}| = 30.66 \cdot \text{kV} \quad \arg(E_{g\_ln}) = 28.2 \cdot \text{deg}.$$

This voltage is transferred to the secondary side of the transformer, as shown in Figure 8.52:

$$E_{g\_s} := \frac{|E_{g\_ln}|}{T} \quad E_{g\_s} = 567.8 \cdot \text{kV}.$$

(3) Power Delivered to the Network vs. Power Angle

The line current I_{sys} in the simplified equivalent circuit of Figure 8.52 is calculated by dividing the voltage difference by the system impedance. The independent variable is the power angle (δ), which in this case is defined as the phase difference between the induced voltage and the network voltage. To test the results below, a power angle of 60° is temporarily chosen:

$$I_{sys}(\delta) := \frac{|E_{g\_s}| \cdot e^{j\delta} - V_{net\_ln}}{Z_{system}} \quad I_{sys}(60 \text{ deg}) = (648.6 - 5.1j) \text{ A}.$$

The three-phase real power delivered to the network is:

$$P_{net}(\delta) := \operatorname{Re}\left(3 \cdot V_{net\_ln} \cdot \overline{I_{sys}(\delta)}\right) \quad P_{net}(60 \text{ deg}) = 539.2 \text{ MW}.$$

A plot of the network power versus power angle (the *power angle curve*) is shown in Figure 8.53.

(4) Maximum Power and Related Power Angle

The maximum power and related power angle as well as the power angle needed to deliver 400 MW to the network are determined here. The maximum power is found using the Mathcad maximization function with a guess value of $\delta := 80$ deg. First, the power angle for maximum power delivery is found:

$$\delta_{max} := \operatorname{Maximize}(P_{net}, \delta) \quad \delta_{max} = 89.7 \text{ deg}.$$

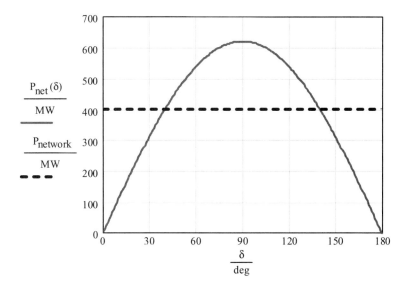

Figure 8.53. Power delivered to the network as a function of power angle.

This value is confirmed from the graph of Figure 8.53. If the line resistance is neglected, the maximum power occurs at exactly 90°.

Next, the maximum power is computed:

$$P_{net\_max} := P_{net}(\delta_{max}) \quad P_{net\_max} = 620.8 \text{ MW}.$$

Finally, the power angle at the specified load is determined to be:

$$\delta_{network} := \text{root}(P_{net}(\delta) - P_{network}, \delta) \quad \delta_{network} = 39.9 \text{ deg}.$$

This angle of 42° matches the value estimated from Figure 8.53.

(5) DC Excitation Current in the Rotor

Here, we calculate the dc excitation current necessary to produce the required generator voltage. Equation (8.40) gives the induced voltage and excitation current relation for a generator:

$$E_{g\_ln} = \omega \cdot \frac{N_{stator}}{\pi} \cdot \mu_0 \cdot I_{dc\_rotor} \cdot \frac{N_{rotor}}{L_{gap}} \cdot D_{rotor} \cdot L_{stator} \cdot \sqrt{2}.$$

The dc current is calculated from this formula using the induced voltage (E_{g_ln}) determined earlier. The result is:

$$I_{dc\_rotor} := \frac{|E_{g\_ln}|}{\omega \cdot \frac{N_{stator}}{\pi} \cdot \mu_0 \cdot \frac{N_{rotor}}{L_{gap}} \cdot D_{rotor} \cdot L_{stator} \cdot \sqrt{2}} = 11.7 \text{ kA}.$$

Using Equation (8.31), the flux density in the gap, as calculated in the following equation, is reasonable:

$$B_{base\_max}(I_{dc\_rotor}) := \frac{4}{\pi} \cdot \mu_0 \cdot \frac{I_{dc\_rotor} \cdot N_{rotor}}{2 \cdot L_{gap}} \quad B_{base\_max}(I_{dc\_rotor}) = 1.369 \text{ T}.$$

EXAMPLE 8.5: MATLAB analysis of static stability

A three-phase synchronous generator supplies a 60 Hz network through a circuit, as sketched in Figure 8.54. The generator is rated at 159 MVA and 23 kV. The stator length is 1.3 m and the rotor diameter is 25 cm with a 3-cm gap between the stator and rotor. The stator and rotor have 60 and 200 turns, respectively, per coil. The rotor is excited with 380 A dc field current. The transformer is rated at 180 MVA with primary and secondary voltages of 24 and 120 kV, respectively. The p.u. reactance of the transformer is 13%. The network line voltage rating is 115 kV with a short circuit current of 5 kA.

APPLICATION EXAMPLES 517

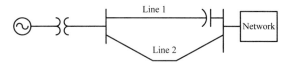

Figure 8.54. One-line diagram of a generator supplying a network through a transformer and two lines.

The transmission lines have a p.u. length impedance of $0.2 + j0.7$ Ω/mile. Line 1 is 70 miles long and Line 2 has a length of 65 miles. Line 1 also has a series capacitor of 70 μF installed.

In this example, we will:

1. calculate the rotor-generated excitation voltage and the synchronous reactance;
2. draw an equivalent circuit and determine its parameters;
3. compute and plot the real power delivered to the network versus power angle;
4. determine the power angle at maximum power and the corresponding maximum power; and
5. calculate the power angle at 80% of the maximum power.

The system parameters are first initialized in MATLAB:

```
%
%    SynchGenerator1.m
%
clear all
omega = 2*pi*60;       % system frequency (rad/sec)
uo = 4*pi*1e-7;        % henries/meter

% System data

% Generator rating data
Sgen = 159e6;          % volt-amps
Vgen = 23e3;           % volts
Lstator = 1.3;         % meter
Drotor = 0.25;         % meter
Lgap = 0.03;           % meter
Irot_dc = 380;         % amps
Nrotor = 200;
Nstator = 60;

% Transformer rating data
Str = 180e6;           % volt-amps
Vprim = 24e3;          % volts
Vsec = 120e3;          % volts
Xtr = 0.13;            % per unit
Tr = Vprim/Vsec;       % turns ratio
```

```
% Network data
% line-to-line voltage
Vnet = 115e3;          % volts
% line-to-neutral voltage
Vnet_ln = Vnet/sqrt(3);   % volts
Ishort_net = 5e3;      % amps

% Transmission line data
L1 = 70;      % miles
L2 = 65;      % miles
Zline = 0.2+0.7j;      % ohms/mile
Cline1 = 70e-6;        % farads
```

(1) Rotor-Generated Excitation Voltage and Synchronous Reactance

Using Equation (8.21) and the generator dimensions, the induced voltage is calculated. The synchronous reactance is computed using Equation (8.22). The flux density should be less than 1.6 T to avoid saturation. The magnetic flux density in the rotor is determined from combining Equation (8.25) and Equation (8.26), that is,

$$B_{gap} = \mu_0 H_{gap} = \mu_0 \frac{N_{rot}}{2\ell_{gap}} I_f$$

```
% Calculate rotor-generated excitation voltage and
% synchronous reactance
% First, compute induced voltage (volts)
Eg = (omega*sqrt(2)*uo*Irot_dc*Nrotor*Nstator*Drotor*...
    Lstator)/(pi*Lgap);
fprintf('\nInduced voltage magnitude = %g volts', Eg);
% Next, compute synchronous reactance (ohms)
Xsync = (omega*3*uo*Nstator^2*Drotor*Lstator)/(pi*Lgap);
fprintf('\nSynchronous reactance = %g ohms', Xsync);
% Finally, compute flux density in the generator (tesla)
Bgen = (uo*Irot_dc*Nrotor)/(2*Lgap);
fprintf('\nFlux density = %g tesla', Bgen);
```

(2) Equivalent Circuit and Parameters

An equivalent single-phase circuit is drawn in Figure 8.55. The impedances of the network, transmission lines, and the transformer are then determined. The transformer reactance is found from its p.u. value:

$$X_{tr\_s} = \frac{V_{sec}^2}{S_{tr}} x_{tr}.$$

The network reactance is obtained from Thévenin equivalency considerations using the line-to-neutral voltage magnitude for the network:

APPLICATION EXAMPLES

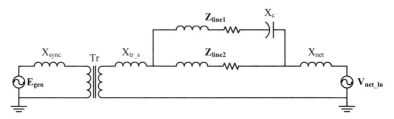

Figure 8.55. Single-phase equivalent circuit for generator supplying a network through a transformer and two lines.

$$X_{net} = \frac{|V_{net\_ln}|}{|I_{short\_net}|}.$$

The combined line impedance is acquired from the parallel combination of Lines 1 and 2, including the capacitor in Line 1:

$$Z_{lines} = \frac{1}{\dfrac{1}{Z_{line1} + jX_c} + \dfrac{1}{Z_{line2}}}, \text{ where } X_c = \frac{-1}{\omega C_{line1}}$$

```
% Calculate the impedances
% transformer impedance
Xtr_s = Xtr*Vsec^2/Str;
fprintf('\n\nTransformer reactance = %g ohms',Xtr_s);
% transmission line impedances
Zline1 = L1*Zline;  % Line 1 (ohms)
Zline2 = L2*Zline;  % Line 2 (ohms)
Xc = -1/(omega*Cline1);
Zlines = 1/(1/(Zline1+j*Xc)+1/Zline2);
fprintf('\nOverall line impedance = %g + j %g ohms',...
     real(Zlines),imag(Zlines));
% network impedance
Xnet = Vnet_ln/Ishort_net;
fprintf('\nNetwork reactance = %g ohms',Xnet);
```

The generator impedance is transferred to the 120 kV (secondary) voltage level using the transformer turns ratio:

$$X_{gen\_s} = \frac{X_{sync}}{T_r^2}.$$

A simplified diagram of the circuit is given in Figure 8.56. The overall system impedance is the sum of the individual impedances:

$$Z_{system} = jX_{gen\_s} + jX_{tr\_s} + Z_{lines} + jX_{net}$$

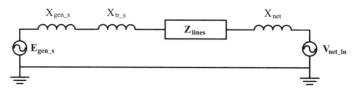

Figure 8.56. Simplified single-phase equivalent circuit for generator supplying a network through a transformer and two lines.

```
% transferred generator impedance
Xgen_s = Xsync/Tr^2;
fprintf('\nTransferred generator reactance = %g ohms',Xgen_s);
% overall system impedance
Zsystem = j*Xgen_s+j*Xtr_s+Zline+j*Xnet;
```

(3) Network Power versus Power Angle

We will assume that the network voltage is the reference with a phase angle of zero. We set $\mathbf{E}_{gen} = |E_g|e^{j\delta}$, where E_g is the magnitude of the excitation voltage calculated in part 1 and δ is the power angle. The power angle is varied from 0° to 180° in steps of 0.01°. The generator excitation voltage is transferred to the secondary side of the transformer:

$$\mathbf{E}_{gen\_s} = \frac{\mathbf{E}_{gen}}{T_r} = \frac{|E_g|e^{j\delta}}{T_r}.$$

The network current is found using Ohm's law:

$$\mathbf{I}_g = \frac{\mathbf{E}_{gen\_s} - \mathbf{V}_{net\_ln}}{\mathbf{Z}_{system}}.$$

The three-phase power, which obeys the passive sign convention, of the network is:

$$P_{net} = \mathrm{Re}(\mathbf{S}_{net}) = \mathrm{Re}(3\mathbf{V}_{net\_ln}\mathbf{I}_g^*)$$

```
% Compute the network current and real power v. power angle
delta = 0 : 0.01 : 180; % degrees
for k=1 : size(delta,2)
    % compute transferred generator phasor voltage
    Egen_s = Eg*exp(j*delta(k)/180*pi)/Tr;
    % compute network current (amps)
    Ig = (Egen_s-Vnet_ln)/Zsystem;
    % determine three-phase real power to network
    Pnet(k) = real(3*Vnet_ln*conj(Ig));
end
```

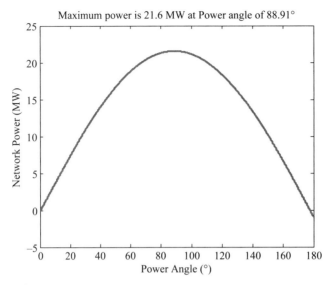

Figure 8.57. Network power as a function of power angle.

```
% plot network power v. power angle
plot(delta,Pnet/1e6,'LineWidth',2.5);
set(gca, 'fontname','Times', 'fontsize',12);
xlabel('Power Angle (°)');
ylabel('Network Power (MW)');
xlim([0 180]);
```

The network real power as a function of the power angle is graphed in Figure 8.57.

(4) Power Angle at Maximum Power

Here we determine the power angle at the maximum power and the corresponding maximum power:

```
% Find the maximum power
[Pmax,J] = max(Pnet);
% Power angle at maximum power
delta_max = delta(J);
fprintf('\n\nMaximum power = %g watts',Pmax);
fprintf('\nPower angle at maximum power = %g°',...
    delta_max);
title(['Maximum power is ',num2str(Pmax/1e6,'%4.1f'),...
    ' MW at Power angle of ',num2str(delta_max),'°']);
```

(5) Power Angle Corresponding to 80% Maximum Power

Finally, we calculate the power angle that corresponds to 80% of the maximum power. Such a power level might be the operational value.

```
% Calculation of the power angle that corresponds to
% 80% of Pmax
Pop = Pmax*0.8;
[error,K] = min(abs(Pnet(1:fix(size(delta,2)/2))-Pop));
delta_op = delta(K);
fprintf('\n\nPower angle at 80%% of maximum power = %g°'
    delta_op);
```

The results from the MATLAB program are as follows:

```
>> synchgenerator1

Induced voltage magnitude = 10535 volts
Synchronous reactance = 17.6432 ohms
Flux density = 1.59174 tesla

Transformer reactance = 10.4 ohms
Overall line impedance = 9.02589 + j 10.017 ohms
Network reactance = 13.2791 ohms
Transferred generator reactance = 441.08 ohms

Maximum power = 2.15657e+007 watts
Power angle at maximum power = 88.91°

Power angle at 80% of maximum power = 52.5°
```

EXAMPLE 8.6: MATLAB analysis of generator loading

Two three-phase generators supply a variable load and a constant load through transformers and lines, as illustrated in the one-line diagram of Figure 8.58. The system data are given as follows:

	Power Rating (MVA)	Voltage Rating (kV)		Reactance (%)
Generator 1	150	22		120
Generator 2	180	22		110
Transformer 1	150	Primary: 23	Secondary: 220	13
Transformer 2	180	Primary: 23	Secondary: 220	11

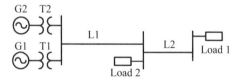

Figure 8.58. One-line diagram of a system with two each of generators, transformers, lines, and loads.

APPLICATION EXAMPLES

The transmission lines have a p.u. length impedance of $0.08 + j0.95$ Ω/mile. Lines 1 and 2 have lengths of 35 and 48 miles, respectively. The load conditions are:

	Power (MW)	Voltage (kV)	Power Factor
Load 1	100	225	0.85 (lagging)
Load 2	60	–	0.75 (lagging)

These system circuit values are input to MATLAB, as shown:

```
%
%    SynchGenerator2.m
%
clear all
% System parameters
% Generator 1
Sg1 = 150e6;      % volt-amps
Vg1 = 22e3;       % volts
xg1 = 1.2;        % p.u.
% Generator 2
Sg2 = 180e6;      % volt-amps
Vg2 = 22e3;       % volts
xg2 = 1.1;        % p.u.
% Transformer 1
Str1 = 150e6;     % volt-amps
Vtr1_p = 23e3;    % primary voltage
Vtr1_s = 220e3;   % secondary voltage
xtr1 = 0.13;      % reactance (p.u.)
% Transformer 2
Str2 = 180e6;     % volt-amps
Vtr2_p = 23e3;    % primary voltage
Vtr2_s = 220e3;   % secondary voltage
xtr2 = 0.11;      % reactance (p.u.)
% Lines
Zline = 0.08+j*0.95;       % ohm/mile
L1 = 35;                   % Line 1 length (miles)
L2 = 48;                   % Line 2 length (miles)
% Load 1
Pload1 = 100e6;   % watts
pfload1 = 0.85;   % inductive
Vload1 = 225e3;   % volts
% Load 2
Pload2 = 60e6;    % watts
pfload2 = 0.75;   % inductive
```

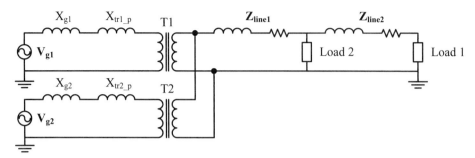

Figure 8.59. Equivalent circuit of a system with two each of generators, transformers, lines, and loads.

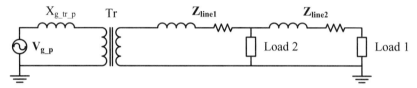

Figure 8.60. Simplified equivalent circuit of a system with two each of generators, transformers, lines, and loads.

In this example, we will:

1. draw the equivalent circuit and calculate impedances, and simplify the circuit by transferring all impedances to the 220 kV level;
2. calculate the load and line currents; and
3. compute the induced voltage, current, and input power of the generators, assuming that the generator-induced voltages are equal.

(1) Equivalent Circuit

We begin by forming the single-phase equivalent circuit. After computing all the impedances, the circuit is simplified by transferring primary-side impedances to the 220 kV level. The equivalent circuit of the system is given in Figure 8.59. The generator and transformer impedances are calculated from the p.u. values:

$$X_{gk} = \frac{V_{gk}^2}{S_{gk}} x_{gk} \quad X_{trk\_p} = \frac{V_{trk\_p}^2}{S_{trk}} x_{trk}.$$

The generator–transformer pairs are connected in parallel and can be replaced by a single unit. The simplified circuit is drawn in Figure 8.60:

$$X_{g\_tr\_p} = \frac{1}{\dfrac{1}{X_{g1} + X_{tr1\_p}} + \dfrac{1}{X_{g2} + X_{tr2\_p}}}.$$

APPLICATION EXAMPLES

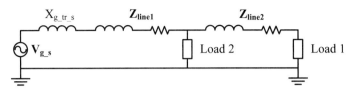

Figure 8.61. Reduced equivalent circuit of a system with two each of generators, transformers, lines, and loads.

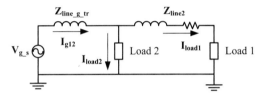

Figure 8.62. Final equivalent circuit of a system with two each of generators, transformers, lines, and loads.

The transfer of the generator–transformer impedance to the secondary side using the turns ratio results in further simplification. The reduced equivalent circuit is given in Figure 8.61:

$$T = \frac{V_{trk\_p}}{V_{trk\_s}} \quad X_{g\_tr\_s} = \frac{X_{g\_tr\_p}}{T^2}.$$

Line 1 and the transformer–generator unit are connected in series; their combined impedance is:

$$Z_{line\_g\_tr} = Z_{line1} + jX_{g\_tr\_s}.$$

The final simplified equivalent circuit is presented in Figure 8.62. The previous values are now computed in MATLAB.

```
% Generator and transformer impedances (ohms)
Xg1 = xg1*Vg1^2/Sg1;
Xg2 = xg2*Vg2^2/Sg2;
% transformer impedance on the primary side
Xtr1_p = xtr1*Vtr1_p^2/Str1;
Xtr2_p = xtr2*Vtr2_p^2/Str2;
% combine the generator-transformer impedances
Xg_tr1 = Xg1 + Xtr1_p;
Xg_tr2 = Xg2 + Xtr2_p;

% combine the two primaries, which are connected in parallel
Xg_tr_p = 1/(1/Xg_tr1 + 1/Xg_tr2);
```

```
% combined impedance is transferred to the secondary side
T = Vtr1_p/Vtr1_s;       % turns ratio
Xg_tr_s = Xg_tr_p/T^2;   % ohms

% Line impedances (ohms)
Zline1 = Zline * L1;
Zline2 = Zline * L2;

% Line 1 and the transformer-generator unit
% are connected in series
Zline_g_tr = Zline1 + j*Xg_tr_s;      % ohms
```

(2) Currents

Whether the loads are wye or delta connected is not specified; however, we recall from Chapter 4 that the line current is independent of the connection type. To calculate the currents throughout the system secondary, the current through Load 1 is first computed from the single-phase equivalent load power and the line-to-neutral voltage, which is selected to have a reference phase angle of zero (i.e., $\mathbf{V}_{load1\_ln} = 225/\sqrt{3} \text{ kV} \angle 0°$):

$$\mathbf{I}_{load1} = \frac{P_{load1}/3}{pf_{load1}\left(V_{load1}/\sqrt{3}\angle 0°\right)} e^{-j\arccos(pf_{load1})}.$$

According to Figure 8.62, the Load 1 current is also the current through Line 2. Using Kirchhoff's voltage law (KVL), the line-to-neutral voltage of Load 2 can be determined:

$$\mathbf{V}_{load2\_ln} = \mathbf{V}_{load1\_ln} + \mathbf{I}_{load1}\mathbf{Z}_{line2}.$$

With the voltage and power at Load 2 known, the current flow into Load 2 is obtained:

$$\mathbf{I}_{load2} = \frac{P_{load2}/3}{pf_{load2}\mathbf{V}_{load2\_ln}} e^{-j\arccos(pf_{load2})},$$

where the negative sign in the exponentials of the formula for $\mathbf{I}_{load1}$ and $\mathbf{I}_{load2}$ are due to the lagging power factor. Finally, application of KCL provides the current through Line 1, which is also the current through the (transferred) generator:

$$\mathbf{I}_{g12} = \mathbf{I}_{load1} + \mathbf{I}_{load2}.$$

```
% Calculate the current of load 1, which is current of line 2
% determine the single-phase power
Pl1_n = Pload1/3;        % watts
% find load 1 line-to-neutral voltage
Vl1_n = Vload1/sqrt(3);  % volts
% load current for an inductive load
Iload1 = (Pl1_n/(pfload1 * Vl1_n)) * exp(-j*acos(pfload1));   % amps
```

```
% use KVL to compute load 2 voltage
Vl2_n = Vl1_n + Iload1 * Zline2;      % volts
% determine single-phase power in load 2
Pl2_n = Pload2/3;          % watts
% calculate the current of load 2 for inductive load
Iload2 = (Pl2_n/(pfload2 * Vl2_n)) * exp(-j*acos(pfload2));
% amps

% use KCL to compute the current through line 1
Ig12 = Iload1 + Iload2;     % amps
```

(3) Generator-Induced Voltage and Power

We calculate the induced voltages of the generators and the generator powers, assuming that the generator-induced voltages are equal. Referring to Figure 8.62, the generator voltage is:

$$\mathbf{V_{g\_s}} = \mathbf{V_{load2\_ln}} + \mathbf{I_{g12}} \mathbf{Z_{line\_g\_tr}}.$$

Having computed the equivalent generator voltage on the secondary side ($\mathbf{V_{g\_s}}$), this line-to-neutral voltage is transferred to the primary side ($\mathbf{V_{g\_p}}$) using the turns ratio. The secondary current ($\mathbf{I_{g12\_s}}$) is also transferred to the primary side ($\mathbf{I_{g12\_p}}$):

$$\mathbf{V_{g\_p}} = \mathbf{V_{g\_s}} T \quad \mathbf{I_{g12\_p}} = \frac{\mathbf{I_{g12\_s}}}{T}.$$

Current division is applied to the transferred generator current to obtain the current from each generator. The current supplied by Generator 1 is:

$$\mathbf{I_{g1\_p}} = \mathbf{I_{g12\_p}} \frac{X_{g\_tr2}}{X_{g\_tr1} + X_{g\_tr2}}.$$

The current from Generator 2 can be determined using a formula similar to the previous one, or by applying KCL (i.e., $\mathbf{I_{g2\_p}} = \mathbf{I_{g12p}} - \mathbf{I_{g1\_p}}$). Finally, the three-phase power produced by each generator is computed:

$$P_{gk} = \operatorname{Re}(\mathbf{S_{gk}}) = \operatorname{Re}(3\mathbf{V_{g\_p}} \mathbf{I_{gk\_p}^*})$$

```
% Calculate the induced voltage of the generators and
% the generator powers assuming that the generator-induced
% voltages are equal to the equivalent generator voltage
% on the secondary side
Vg_s = Vl2_n + Ig12 * Zline_g_tr;     % volts
% the equivalent line-to-line voltage
Vg_s_line = sqrt(3) * abs(Vg_s);       % volts
```

```
% The line-to-neutral voltage is transferred to the primary side
Vg_p = Vg_s * T;           % volts
fprintf('\nInduced line voltage = %g kV at %g degrees.', ...
    sqrt(3)*abs(Vg_p)/1e3, angle(Vg_p)*180/pi);
% The generator currents are
% transfer line 1 current to primary side
Ig12_p = Ig12/T;           % amps
% the current from each generator is found from current
% division
Ig1_p = Ig12_p * Xg_tr2/(Xg_tr1 + Xg_tr2);      % amps
Ig2_p = Ig12_p * Xg_tr1/(Xg_tr1 + Xg_tr2);      % amps
fprintf('\n\nGenerator 1 current = %g amps at %g degrees.', ...
    abs(Ig1_p), angle(Ig1_p)*180/pi);
fprintf('\nGenerator 2 current = %g amps at %g degrees.', ...
    abs(Ig2_p), angle(Ig2_p)*180/pi);
% The generator three-phase input real powers
Pg1 = real(3*conj(Ig1_p)*Vg_p);      % watts
Pg2 = real(3*conj(Ig2_p)*Vg_p);      % watts
fprintf('\n\nGenerator 1 real power = %g MW.', Pg1/1e6);
fprintf('\nGenerator 2 real power = %g MW.', Pg2/1e6);
```

The results from the MATLAB program are:

```
>> synchgenerator2

Induced line voltage = 39.6927 kV at 24.2093 degrees.

Generator 1 current = 2023.24 amps at -37.3127 degrees.
Generator 2 current = 2670.35 amps at -37.3127 degrees.

Generator 1 real power = 66.3244 MW.
Generator 2 real power = 87.5376 MW.
```

EXAMPLE 8.7: PSpice simulation of generator transients

Investigation of transients in a generator circuit requires the analysis of a large nonlinear system, which is beyond the scope of this book. However, the transient phenomena can be demonstrated using simplified circuits. One of the most important transients in a generator is a short circuit and the clearing of the short circuit by a CB.

Figure 8.63 presents the typical circuit arrangement. A generator supplies a transmission line through a transformer. The line is protected by a CB connected in series. A lightning stroke or human interference causes a short circuit on the line. Most often, the short circuit occurs between a phase conductor and the ground (termed a *single line-to-ground fault*), but the simultaneous short circuit of the three phases is rare but

APPLICATION EXAMPLES 529

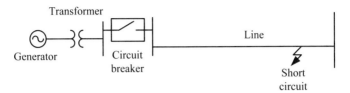

Figure 8.63. Short-circuit clearing by a CB.

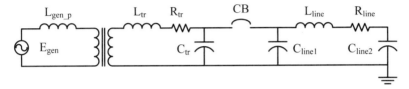

Figure 8.64. Single-phase equivalent circuit for clearing three-phase short circuit.

happens. The *Westinghouse Transmission and Distribution Reference Book* gives the fault distribution statistics for a high voltage system as:

Line-to-ground fault	70%
Line-to-line fault	15%
Three-phase fault	5%
Line-to-line-to-ground fault	10%

Although investigation of a ground fault is beyond the scope of this book, the latter three-phase short circuit can be investigated using the single-phase equivalent circuit drawn in Figure 8.64.

The difficulty in performing an exact analysis of this short circuit is that the generator impedance changes during the short circuit. The impedance is smallest at the beginning of the short circuit and increases with time. Historically, three impedances have been used to represent this time-dependent system. Generally, the *subtransient impedance* is utilized between cycles 0 and 3; the *transient impedance* is employed between 3 and 20 cycles; and the *synchronous impedance* is used beyond 20 cycles. The subtransient p.u. impedance is 8–10%, the transient impedance is around 12–20%, and the synchronous impedance is 80–130%. The resistive parts of these impedances are very small and often neglected. In this example, the generator is represented by its subtransient reactance since the CB operates within three cycles.

It is assumed that a three-phase short circuit occurs on the transmission line. The short circuit generates large current, which is cleared by the CB. The sudden interruption of the short-circuit current generates an overvoltage between the CB terminals and the ground. From a practical point of view, it is the maximum short-circuit current and the peak overvoltage that are needed to assess the system operation.

The single-phase equivalent circuit of Figure 8.64 represents this system. The line is modeled by an equivalent Π circuit; the CB is depicted by a switch; and the transformer is represented by an inductance (L_{tr}) and resistance (R_{tr}) connected in series with an ideal transformer and a capacitor (C_{tr}) connected in parallel. It is assumed that the line is not loaded. During an actual fault interruption, an arc is generated in the CB and current flow is stopped at a current zero crossing. In this simulation, an ideal switch that can interrupt the current at any instant is used.

The transformer has capacitance between the windings, and between the windings and the ground. These capacitances are neglected in steady-state operation, but they may not be neglected for transient studies. In this case, the transformer winding capacitance-to-ground is represented by a capacitance (C_{tr}) connected in parallel with the high-voltage terminals. The capacitance between the windings is neglected.

The PSpice simulation is presented as a numerical example. The system data are given here. The generator ratings are:

$$\omega := 2 \cdot \pi \cdot 60 \text{ Hz} \qquad M := 10^6 \qquad pf_{gen} := 0.8 \text{ lagging}$$
$$S_{gen} := 1559 \text{ M} \cdot \text{V} \cdot \text{A} \qquad V_{gen} := 24 \text{ kV} \qquad X_{subtransient} := 21\%.$$

The transformer ratings are:

$$S_{tr} := 1550 \cdot \text{M} \cdot \text{V} \cdot \text{A} \qquad V_{prim} := 24 \text{ kV} \qquad V_{sec} := 525 \text{ kV}$$
$$x_{tr} := 8\% \qquad C_{tr} := 500 \text{ nF}.$$

The capacitance is on the high-voltage side.
The transmission line data are:

$$\text{Len} := 10 \text{ mi} \qquad Y_{len} := 7.034 \times 10^{-6} \frac{S}{\text{mi}}$$
$$R_{len} := 0.0254 \frac{\Omega}{\text{mi}} \qquad X_{len} := 0.595 \frac{\Omega}{\text{mi}}.$$

For the PSpice simulation of the system operation, the circuit parameters must be calculated. The generator transient reactance and corresponding inductance are:

$$X_{gen} := X_{subtransient} \cdot \frac{V_{gen}^2}{S_{gen}} = 0.078 \text{ } \Omega \qquad L_{gen\_p} := \frac{X_{gen}}{\omega} = 0.206 \text{ mH}.$$

The generator inductance is transferred to the secondary side of the transformer using the turns ratio. The generator inductance at the secondary side is:

$$T := \frac{V_{prim}}{V_{sec}} = 0.046 \qquad L_{gen} := \frac{L_{gen\_p}}{T^2} = 98.483 \text{ mH}.$$

APPLICATION EXAMPLES 531

The generator-induced voltage under transient conditions is calculated by adding the voltage drop on the transient reactance to the rated line-to-neutral voltage. The generator rated line-to-neutral terminal voltage is:

$$V_{gen\_ln} := \frac{V_{gen}}{\sqrt{3}} = 13.856 \text{ kV}.$$

The voltage drop is calculated using the generator rated voltage and rated current. The generator rated current is:

$$I_{gen} := \frac{\frac{S_{gen}}{3}}{V_{gen\_ln}} \cdot e^{-j a \cos(pf_{gen})} = (30.00 - 22.50j) \text{ kA}.$$

The generator-induced voltage is:

$$E_{gen} := V_{gen\_ln} + I_{gen} \cdot j \cdot X_{gen} \quad |E_{gen}| = 15.78 \text{ kV}.$$

This voltage is transferred to the secondary side by dividing the voltage with the turns ratio:

$$E_{gen\_s} := \frac{E_{gen}}{T} \quad |E_{gen\_s}| = 345.08 \text{ kV}.$$

The PSpice voltage source requires the peak value of the generator voltage, which is:

$$V_{gen} := \sqrt{2} \cdot |E_{gen\_s}| = 488.015 \text{ kV}.$$

The transformer reactance and inductance at the secondary side are:

$$X_{tr} := x_{tr} \cdot \frac{V_{sec}^2}{S_{tr}} = 14.226 \text{ }\Omega \quad L_{tr} := \frac{X_{tr}}{\omega} = 37.735 \text{ mH}.$$

The transformer has resistance, which is assumed to be one-tenth of the reactance at 60 Hz: $R_{tr} := 1.45 \text{ }\Omega$.

The transmission line resistance, inductance and capacitance are:

$$R_{line} := R_{len} \cdot Len = 0.254 \text{ }\Omega$$

$$L_{line} := \frac{X_{len}}{\omega} \cdot Len = 15.783 \cdot \text{mH}$$

$$C_{line} := \frac{Y_{len}}{2 \cdot \omega} \cdot Len = 93.291 \cdot \text{nF}.$$

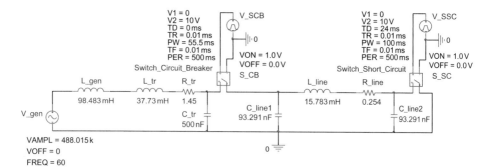

Figure 8.65. PSpice model for generator short-circuit analysis.

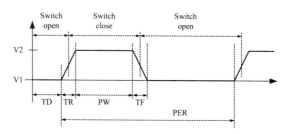

Figure 8.66. PSpice voltage pulse.

The transfer of the generator voltage and reactance to the high-voltage side eliminates the transformer. The PSpice schematic capture interface is used to build the Simulation Program with Integrated Circuit Emphasis (SPICE) circuit model. The calculated circuit data are incorporated in the model. Figure 8.65 shows the PSpice circuit in which the short circuit is assumed to occur at the receiving end of the line.

This example simulates the development of short-circuit current and the overvoltages generated by clearing of a three-phase temporary fault. The scenario is that a three-phase fault occurs on an unloaded transmission line. This produces large transient fault current, which triggers the CB that protects the line. The CB clears the fault after about two cycles. This operation can produce overvoltages on both the line and the generator sides of the breaker as well as between the terminals of the CB. In a real system, the actual CB not only clears the fault, but also recloses after fault clearing with a few cycles of delay. For a temporary fault, the reclosing assures the continuation of the service with only a few cycles of interruption time.

In the PSpice model of Figure 8.65, the short circuit is produced by an ideal voltage-controlled switch (S_SC) at the line end. A pulse generator (V_SSC) controls this switch. The CB is also represented by an ideal voltage-controlled switch (S_CB) connected in series with the line. A pulse generator (V_SCB) controls the CB switch. The general form of a PSpice voltage pulse is seen in Figure 8.66. After an initial delay time (TD), the pulse has rise time (TR), fall time (TF), and duration or width (PW). Figure 8.67 shows the actual control pulses for this simulation. Figure 8.67a shows

APPLICATION EXAMPLES 533

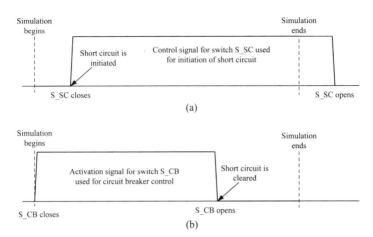

Figure 8.67. Voltage pulse control signals for switches. (a) Short-circuit activation signal; (b) CB activation signal.

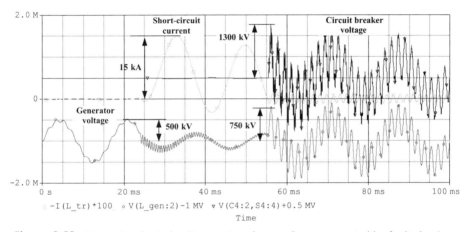

Figure 8.68. Generator short-circuit current and overvoltage generated by fault clearing.

how the pulse initiates the short circuit after a brief delay (TD of V_SSC). The switch representing the CB (S_CB) closes immediately at the beginning of the simulation and opens after a variable delay (PW of V_SCB). Figure 8.67b exhibits the control pulse used for the CB. The period (PER) of both pulses is chosen to be longer than the simulation time to avoid repetition of the switch operation. A Π circuit models the transmission line. The transmission line representation can be improved by connecting several Π circuits in series. The transformer is depicted by an impedance connected in series with the line and a capacitance connected in parallel. The generator is described by a voltage source and an inductance connected in series.

Figure 8.68 and Figure 8.69 plot the results of two different simulations, which encompass the fault occurrence and the breaker opening. The graphs show the generator

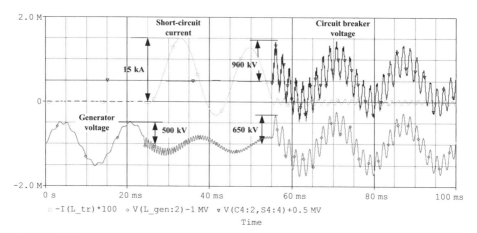

Figure 8.69. Overvoltage generated by current interruption close to current zero crossing.

short-circuit current, the generator terminal voltage to ground, and the voltage between the CB terminals. Both the generator current and voltage shown in these graphs are for the quantities as transferred to the high-voltage side of the GSU transformer. Using the turns ratio of $T = 0.0457$, the actual generator short-circuit current is $15\ \text{kA}/T = 328\ \text{kA}$ and the generator terminal voltage is $500\ \text{kV} \times T = 22.9\ \text{kV}$. In Figure 8.68, the actual generated overvoltage is $750\ \text{kV} \times T = 34.3\ \text{kV}$. Comparing this overvoltage with the generator rated line-to-line voltage of 24 kV will demonstrate that slight current chopping produces significant overvoltage.

The generator short-circuit current has two components: an ac and a dc component. The ac component has constant amplitude because the generator impedance increase during the short circuit is neglected. The initial dc component is equal to the ac current at the instant the short circuit occurs ($t = 0^+$), and thereafter the dc component attenuates due to the circuit resistance. Consequently, the short-circuit current is maximum when the short circuit occurs at the instant the voltage is zero. In both simulations, the short circuit occurs when the generator voltage is very close to zero. The reader is encouraged to modify the pulse generator delay (e.g., change TD = 24 ms to 16.67 ms or other values, per Problem 8.14) and observe the variation of the short-circuit current. The short-circuit current has significant dc component that attenuates over time.

For the first simulation, the switch representing the CB opened after less than two cycles just before current zero crossing. This produces current chopping, which generates significant overvoltage, as seen in Figure 8.68. The maximum voltage between the generator terminals and the ground is around 750 kV and between the CB terminals is around 1300 kV. The voltages have a 60 Hz signal and superimposed higher frequency components, which produce oscillatory voltage waveforms. This type of voltage is commonly called *switching overvoltage*. The voltage across the breakers is referred to as the transient recovery voltage.

The switching overvoltage can be reduced by interrupting the current at its zero crossing, which is demonstrated in the second simulation. In this case, the short-circuit current is interrupted after less than two cycles of delay at current zero crossing. Figure 8.69 reveals that the short-circuit interruption at current zero crossing reduces the overvoltages. The maximum generator terminal voltage at the high-voltage side is around 650 kV (actual generator voltage is 29.7 kV) and the maximum voltage between the CB terminals is around 900 kV. For comparison, the peak value of the 60 Hz generator voltage at the high-voltage side is around 500 kV, which corresponds to an actual voltage of 22.9 kV. The reader is encouraged to modify the PW (pulse width) value of S_CB and observe the effect of current chopping (see Problem 8.15).

In the real system after a short circuit, an actual CB opens and draws an arc between the contacts. This arc is automatically extinguished at current zero crossing. The CB prevents the arc reignition by injecting SF_6 or high-pressure air. The injected gas blows out the arc plasma and deionizes the gap between the contacts. This results in the interruption of the short-circuit current at zero crossing. *Current chopping* occurs when a powerful CB interrupts low short-circuit current. This happens in the case of air-blast CBs, when the injected high-pressure air blows out the arc before zero crossing.

The switching-generated overvoltages can be dangerous because the high switching overvoltages between the terminals and the ground can produce insulator or air-gap flashover, which can cause service interruption. A more severe case is insulation breakdown in the transformer that requires the replacement of the unit.

The overvoltage between the CB terminals can cause breakdown between the separated contacts, which prevents short circuit interruption and may result in CB explosion or damage.

8.7. EXERCISES

1. What are the major components of a synchronous generator?
2. What is the difference between the round rotor and salient pole generators? Draw sketches.
3. Describe the construction of the stator of a large generator.
4. Describe the construction of the rotor of a large generator.
5. How are the large generators cooled?
6. What is a hydrogenerator? Describe its rotor construction.
7. What kind of current flows in the rotor of a generator?
8. What kind of current flows in the stator of a generator?
9. What is the effect of the rotor current?
10. Draw a sketch and explain the generation of the induced voltage in the stator.
11. What is the synchronous speed? Define and present an equation for its calculation.

12. Derive equations for calculating the induced voltage using the synchronous reactance and terminal voltage.
13. Discuss the effect of load on the machine magnetic flux and terminal voltage.
14. What is the synchronous reactance? Derive the equation for the synchronous reactance.
15. Present the single-phase equivalent circuit of a synchronous generator.
16. What is the effect of loading on the synchronous generator operation?
17. Describe the process of synchronization.
18. How is the reactive power regulated in a synchronous generator?
19. How is the active power regulated in a synchronous generator?
20. Describe the concept of electric network operation with several hundred synchronous generators.
21. How are the synchronous generators connected in a large electric network?
22. What is the static stability problem?
23. What is the terminal voltage and induced voltage relation when the generator absorbs or generates reactive power?
24. Outline the calculation method to determine the limit of static stability.
25. What is the power angle? For approximately what power angle value is the generator power maximum?
26. Show the stable and unstable parts of the generator power versus power angle curve.
27. Derive equations for the induced voltage calculation using the machine dimensions.
28. Derive equations for the synchronous reactance calculation using the machine dimensions.
29. Describe the three-phase current generated rotating flux.
30. How is the rotating stator flux represented in the generator equivalent circuit?

8.8. PROBLEMS

Problem 8.1

Determine the synchronous speed of a six-pole generator in the United States.

Problem 8.2

Determine the synchronous speed of an eight-pole generator in Europe.

Problem 8.3

Show whether the power at each of the four buses of Figure 8.28 is balanced.

Problem 8.4

A 150 MVA, 24 kV, 123% three-phase synchronous generator carries the rated load at a power factor 0.83 (lagging). Calculate the necessary excitation voltage and input complex power.

Problem 8.5

A 150 MVA, 24 kV, 123% three-phase synchronous generator operates with an induced voltage that is twice the rated line-to-neutral voltage. Calculate the three-phase steady-state short-circuit current at the generator terminals.

Problem 8.6

A 150 MVA, 24 kV, 123% three-phase synchronous generator supplies a large network. The network voltage is 27 kV. The phase angle between the network voltage and the generator-induced voltage is 60°. Calculate the generator induced voltage if the network receives 300 MW of power.

Problem 8.7

A 150 MVA, 24 kV, 123% three-phase synchronous generator supplies a large network. The network voltage is 27 kV. The generator operates with an induced voltage, which is twice the rated line-to-neutral voltage. Calculate the power angle if the network receives an inductive reactive power of 50 MVA.

Problem 8.8

A 250 MVA, 24 kV, 125% three-phase synchronous generator supplies a large network. The network voltage is 27 kV. The generator operates with an induced voltage, which is twice the rated line-to-neutral voltage. The phase angle between the network voltage and induced voltage is 56°. Calculate the network real and reactive powers.

Problem 8.9

A three-phase wye-connected synchronous generator supplies a network through a transmission line. The network can absorb or deliver power while maintaining its terminal voltage constant. The four-pole, 60 Hz generator data rating are 40 MVA and 26 kV with a 0.85 p.u. reactance. The field current of the generator can be adjusted to regulate the excitation (induced) voltage from 0.75 to 1.5 times the rated voltage. The network voltage rating is 24 kV. The transmission line impedance and length are $0.07 + j0.5$ Ω/mile and 8 miles.

(a) Calculate the generator-induced voltage versus power factor if the network voltage is at the rated value and the bus absorbs the generator rated power. After this, plot the voltage regulation of the system versus the power factor. Use the induced voltage and network voltage for plotting the voltage regulation. The leading power factor varies from 0.5 to 1. What power factor corresponds to 10% regulation?

(b) Plot the maximum power (power angle is 90°) versus induced voltage. Determine the required induced voltage that allows the safe transfer of maximum power from the generator to the network (not to exceed the generator rating). Assume that the network voltage is the same as in part (a). The excitation voltage is the variable.

Problem 8.10

A three-phase 60 Hz synchronous generator rated at 15 MVA and 2.2 kV has a synchronous reactance per phase of 13 Ω and spins at 1800 rpm.

(a) Calculate and plot the induced voltage versus load function if the terminal voltage is the rated value and the power factor is 0.8 lagging. Calculate the induced voltage at the *rated load* and at *open-circuit* conditions.

(b) Calculate and plot the stator short-circuit current for variable load conditions if the short circuit occurs across the machine terminals. Use the induced voltage function calculated in part (a). Using the resultant plot, determine the short-circuit current at both the *open-circuit* condition and at *rated load*.

Problem 8.11

A three-phase synchronous generator is rated at 120 MVA and 20 kV. The 60 Hz generator has two poles with 120 turns per pole. The induced voltage (line-to-ground) is twice the rated voltage (line-to-line). The rotor diameter is 2.7 m with a gap of 2.2 cm between the rotor and stator. The stator length is 13 m with 60 turns per phase in the stator. Determine: (a) the generator speed (in rpm, and radians per second), (b) the dc flux, flux density, and field intensity (H), (c) the dc excitation current through the rotor, and (d) the synchronous reactance. (e) Draw the equivalent circuit with the parameter values.

Problem 8.12

A synchronous generator supplies a network through two transformers and a transmission line, as drawn in Figure 8.70. The generator is rated at 380 MVA and 22 kV with a p.u. reactance of 1.2. The identical transformers are rated at 480 MVA with primary

Figure 8.70. One-line diagram for Problem 8.12.

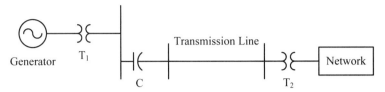

Figure 8.71. One-line diagram for Problem 8.13.

and secondary voltages of 22 and 340 kV, respectively, and a 15% reactance. The transmission line is 45 miles long with an impedance of $0.07 + j0.5$ Ω/mile. The transmission line capacitances are neglected. The network voltage is 21 kV.

(a) Determine the transmission line, generator, and transformer parameters. Draw the equivalent circuit. (b) Compute the generator excitation (induced) voltage at rated load. Assume that the terminal voltage is the same as the rated voltage and the power factor is 0.88 lagging. (c) Calculate the power transmitted to the network if the absolute value of the excitation voltage is the same as the value found in part (b), and the angle between the network and generator excitation voltage varies. Plot the transmitted power versus power angle and determine the power angle at 350 MW.

Problem 8.13

A synchronous generator supplies a network through two transformers and a transmission line, as shown in Figure 8.71. The generator is rated at 750 MVA with a terminal voltage of 22.5 kV and a reactance of 107%. The identical transformers are rated at 800 MVA with primary and secondary voltages of 22.5 and 220 kV, respectively, and a 16% reactance. The transmission line is 120 miles long with line voltage of 220 kV and an impedance of $0.07 + j0.5$ Ω/mile. The total transmission line capacitance is 3 µF. The network line voltage is 22 kV. A variable capacitor from 50 to 500 µF is connected in series with the transmission line.

(a) Determine the impedances of the system. Draw the single-phase equivalent circuit.
(b) Find the induced voltage of the generator at rated load and 0.83 lagging, assuming the terminal voltage of the generator is the same as the rated voltage. The absolute value of the obtained induced voltage is used for part (c).
(c) Plot the power transmitted to the network versus the capacitance in series with the line, at a power angle of 60° (angle between the network voltage and generator excitation voltage). Determine the capacitance value that is needed to transfer 950 MW.

Problem 8.14

Using the generator data from Example 8.7, change the initiation time of the short circuit using the V_SSC pulse generator time delay (TD). Specifically, choose initiation times when the voltage phase angle is approximately 0°, 45°, 90°, and 135°. Calculate and plot the generator short-circuit current and terminal voltage. Estimate and compare both the peak and dc components of the short-circuit currents from these simulations to those of Example 8.7.

Problem 8.15

Using the generator data from Example 8.7, study the effect of CB current interruption timing on the overvoltage caused by the current chopping. In particular, vary the voltage pulse width (PW) of the V_SCB source to the CB from 40 to 60 ms in steps of 5 ms. Use plots of the CB voltage and short-circuit current time variation to estimate the short-circuit current (i.e., current chopped) at the switch opening and the peak CB overvoltage.

9

INDUCTION MACHINES

9.1. INTRODUCTION

The single-phase induction motor is the most often used motor in the world. Most appliances, such as washing machines and refrigerators, use a single-phase induction machine. Figure 9.1 shows a small single-phase induction motor, which is frequently used in household appliances. For industrial applications, the three-phase induction motor is used to drive machines. Figure 9.2 shows a picture of a large three-phase motor used in an industrial drive. The introduction of electronic drives has permitted accurate speed and torque control of the induction machine, which further extended the application of these motors. As an example, the electronically controlled induction motor is replacing the direct current (dc) drives traditionally used by the textile industry. The induction motor is reliable and has simple construction, which advanced its widespread use. This motor is the workhorse of industry. Important to note is that induction machines can also be used as generators—this operational mode is presented near the end of this chapter.

Electrical Energy Conversion and Transport: An Interactive Computer-Based Approach, Second Edition. George G. Karady and Keith E. Holbert.
© 2013 Institute of Electrical and Electronics Engineers, Inc. Published 2013 by John Wiley & Sons, Inc.

Figure 9.1. Single-phase induction motor.

Figure 9.2. Large three-phase induction motor (courtesy of Siemens, Erlangen, Germany).

Figure 9.3. Induction motor components.

9.2. CONSTRUCTION

Figure 9.3 provides a breakout of the parts of an induction motor, which include the stator, rotor, end bells, and fan. The alternating current (ac) induction motor has three main components: motor housing, stator, and rotor.

The motor housing consists of three parts: the cylindrical middle piece that holds the stator iron core and the two bell-shaped end covers holding the ball bearings. This motor housing is made of cast aluminum or cast iron. Long screws hold the three parts together. The legs at the middle section permit the attachment of the motor to a base. A cooling fan is attached to the shaft, as annotated in Figure 9.3. This fan blows air over the ribbed stator frame.

9.2.1. Stator

The stator of a large induction motor with three-phase windings is pictured in Figure 9.4. The construction of the large induction machine windings is similar to the synchronous machine windings. Figure 9.5 shows the stator iron core of a single-phase motor. The iron core has a cylindrical shape and is laminated with slots. In this photograph, the iron core is without windings, but has paper liner insulation placed in some of the slots.

In a three-phase motor, the three phase windings are placed in the slots. A single-phase motor has two windings: the main and the starting windings. Figure 9.6 displays a single-phase stator installed in its housing with the main windings placed in the slots.

Figure 9.4. Stator of a large induction motor (courtesy of Siemens, Erlangen, Germany).

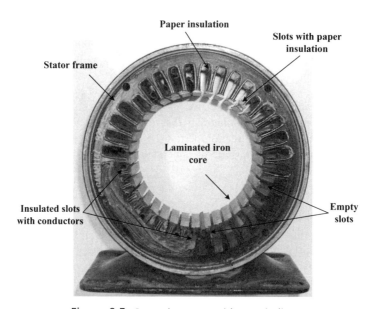

Figure 9.5. Stator iron core without windings.

CONSTRUCTION

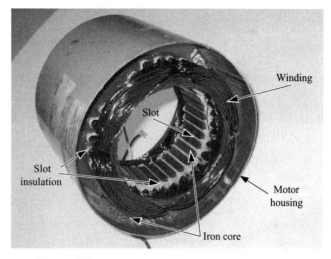

Figure 9.6. Single-phase stator with main windings.

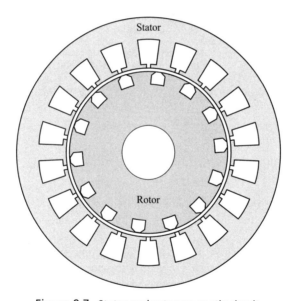

Figure 9.7. Stator and rotor magnetic circuit.

The photograph also points outs the coils of the windings. Typically, thin enamel insulated wires are used.

The elements of the laminated iron core are punched from a silicon iron sheet. The stator and rotor laminated sheets are assembled and held together by insulated bolts. Figure 9.7 illustrates the rotor and stator iron core assembly, which forms the magnetic circuit of the machine.

Figure 9.8. Rotor of a large induction motor (courtesy of Siemens, Erlangen, Germany).

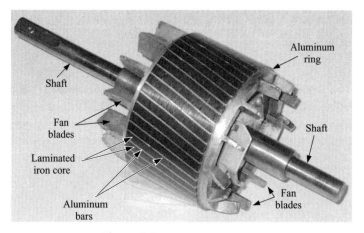

Figure 9.9. Squirrel cage rotor.

9.2.2. Rotor

The rotor of a large induction motor is exemplified in Figure 9.8. Induction motor manufacturers employ two different types of rotor configurations:

Squirrel Cage Rotor. This rotor has a laminated iron core with slots and is mounted on a shaft. Aluminum bars are molded in the slots and the bars are short circuited with two end rings. The bars are slanted on a small rotor to reduce audible noise. Fins are placed on the ring that shorts the bars. These fins work as a fan and improve cooling. An alternate construction uses copper bars placed in the slots and short circuited with two rings. Figure 9.9 exhibits a squirrel cage rotor with slanted aluminum bars and fins on the end rings. Figure 9.10 shows an illustration of the squirrel cage bars and the end rings without the iron core and fins.

Wound Rotor. Most motors use the squirrel cage rotor because of the robust and maintenance-free construction. However, large, older motors use a wound rotor with three-phase windings placed in the rotor slots. The windings are connected in a three-wire wye. The ends of the windings are connected to three slip rings.

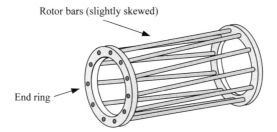

Figure 9.10. Squirrel cage rotor concept.

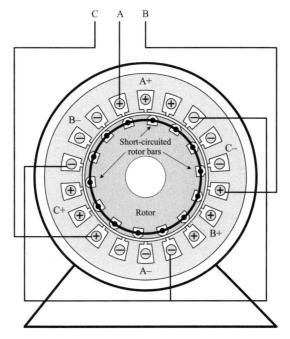

Figure 9.11. Connection diagram of a two-pole induction motor with squirrel cage rotor.

Resistors or power supplies are connected to the slip rings through brushes for reduction of starting current and speed control. However, recent advancements in semiconductor technology have reduced the cost of the electronic motor drives and made the wound rotor construction obsolete.

9.3. THREE-PHASE INDUCTION MOTOR

9.3.1. Operating Principle

Figure 9.11 depicts a three-phase, two-pole, squirrel cage induction motor electrical connection diagram. This two-pole motor has three stator phase windings connected in a three-phase wye. Each phase has 2 × 3 = 6 slots. The phases are shifted by 120°. The squirrel cage rotor has short-circuited bars. The motor is supplied by symmetrical, balanced three-phase voltage at the terminals. The stator three-phase windings can also be connected in a delta configuration.

9.3.1.1. Concept of Motor Operation.

The operation of a three-phase motor is discussed first. The major stages of its operation are as follows:

1. The three-phase stator is supplied by balanced three-phase voltage that drives an ac magnetizing current through each phase winding.
2. The magnetizing current in each phase generates a pulsating ac flux. The flux amplitude varies sinusoidally and the direction of the flux is perpendicular to the phase winding. The three fluxes generated by the phase windings are separated by 120° in space and in time for a two-pole motor.
3. The total flux in the machine is the sum of the three fluxes. The summation of the three ac fluxes results in a rotating flux, which turns with constant speed and has constant amplitude.
4. The rotating flux induces a voltage in the short-circuited bars of the rotor. This voltage drives current through the bars.
5. The interaction of the rotating flux and the rotor current generates a force that drives the motor.

The direction of the motor shaft rotation can be reversed by interchanging any two phases. For example, assuming that the phase sequence of the motor is *abc*, then swapping phases *b* and *c* will reverse the motor direction (i.e., change the sequence from *abc* to *acb*).

9.3.1.2. Discussion of Motor Operation.

The previous chapter discussed the three-phase stator winding-generated magnetic field in detail. Figure 9.12 presents the resulting rotating magnetic field, which generates voltage and current in the rotor bars. The diagram displays the three components of the magnetic field at a phase angle of −51°. Each phase generates a magnetic field vector. The vector sum of the component vectors Φ_a, Φ_b, and Φ_c gives the resulting rotating magnetic field vector Φ_{mag}, whose amplitude is 1.5 times the maximum individual phase vector amplitudes (see previous chapter for detailed derivation of this), and Φ_{mag} rotates clockwise with constant (synchronous) speed.

PHYSICS OF INDUCED VOLTAGE AND FORCE GENERATION. Before analyzing the induction motor operation, we recall Faraday's law, which states that a voltage will be induced in a conductor that moves perpendicular to a magnetic field, as illustrated in Figure 9.13. The induced voltage in this case is:

$$V = B\ell v, \tag{9.1}$$

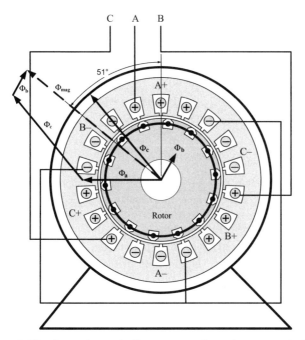

Figure 9.12. Three-phase winding-generated rotating magnetic field.

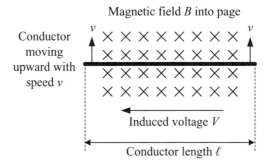

Figure 9.13. Voltage induced in a conductor moving through a magnetic field.

where B is the magnetic flux density, v is the conductor speed, and ℓ is the conductor length.

The other relation necessary for the analysis is the magnetic field–current generated force (the Lorentz force). The magnetic field and current interaction generates an electromagnetic force. In a conductor, the current (I) produces distributed forces along the length of the conductor. The total force on a conductor with a length of ℓ is:

$$F = B\ell I \sin(\phi), \quad (9.2)$$

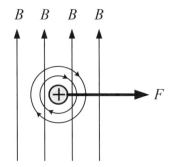

Figure 9.14. Force direction on a current-carrying conductor placed in a magnetic field (B) that is stronger than the conductor-generated circular magnetic field (current into the page).

where ϕ is the angle between B and $I\ell$. In an induction motor, $\phi = 90°$ because B is perpendicular to the conductor. The direction of the force is provided in Figure 9.14 for a magnetic flux density vector that is perpendicular to the current-carrying conductor.

MOTOR INDUCED VOLTAGE. Figure 9.12 shows a sketch of the arrangement of a three-phase motor. The three-phase windings on the stator generate a rotating field. The rotor bars cut the magnetic field lines as the field rotates. The rotating field induces a voltage in the short-circuited rotor bars. The induced voltage is proportional to the speed difference between the rotating field and the spinning rotor. Using Faraday's equation, the induced voltage in a single rotor bar is:

$$V_{bar} = B\ell_{rot}(v_{syn} - v_{mot}), \qquad (9.3)$$

where

B is the rotating magnetic flux density;
ℓ_{rot} is the rotor bar length;
v_{syn} is the speed of the rotating field at the rotor bar radial position; and
v_{mot} is the motor (rotor) speed at the rotor bar radial position.

The speed of flux cutting is the difference between the magnetic field speed and the rotor speed. The two speeds can be calculated by using the radius (r_{rot}) at the rotor bar location and the rotational speed:

$$v_{syn} = 2\pi r_{rot} n_{syn},$$
$$v_{mot} = 2\pi r_{rot} n_{mot}, \qquad (9.4)$$

where n_{syn} and n_{mot} are the synchronous and motor speeds, respectively. The substitution of these formulae into Faraday's induced voltage equation results in:

THREE-PHASE INDUCTION MOTOR

$$V_{bar} = 2\pi r_{rot} B \ell_{rot}(n_{syn} - n_{mot}). \tag{9.5}$$

The voltage and current generation in the rotor bar require a speed difference between the rotating field and the rotor. Consequently, the rotor speed is always less than the magnetic field speed. The relative speed difference is the *slip*, which is calculated using:

$$s = \frac{n_{syn} - n_{mot}}{n_{syn}} = \frac{\omega_{syn} - \omega_{mot}}{\omega_{syn}}, \tag{9.6}$$

where

- s is the slip;
- n_{syn} is the synchronous (magnetic field) speed;
- n_{mot} is the actual motor (rotor) speed;
- $\omega_{syn} = 2\pi n_{syn}$ synchronous angular speed of the field; and
- $\omega_{mot} = 2\pi n_{mot}$ angular speed of the motor.

The synchronous speed depends on the frequency of the supply voltage and the number of poles in the motor. The motor shown in Figure 9.11 has two poles because only one set of three-phase windings is on the stator. However, the stator can be divided into two halves and two sets of three-phase windings can be placed in the slots. This type of motor has four poles. Similarly, we can build large motors with 6, 8, 10, or more poles.

As an example, if a stator has 36 slots and it is built as a two-pole motor, each phase will occupy 2 × 6 slots. Specifically, the number of slots per phase is 36/3 = 12. Since each coil has two sides, the number of slots per phase is 2 × 6 = 12. In a two-pole motor, the distance between phases A and B is 120°. For a four-pole motor, the number of slots per phase would be 36/(2 × 3) = 6. Again, each coil winding has two sides; consequently, in the case of four poles, each phase occupies 2 × 3 slots. The physical distance between phases A and B is 60° for a four-pole motor.

The *synchronous speed* is calculated using:

$$n_{syn} = \frac{f}{p/2}, \tag{9.7}$$

where f is the frequency of the supply network and p is the number of poles. The angular speed of the field is:

$$\omega_{syn} = 2\pi n_{syn}. \tag{9.8}$$

The combination of Equation (9.5) and Equation (9.6) results in:

$$V_{bar} = 2\pi r_{rot} B \ell_{rot} s n_{syn}. \tag{9.9}$$

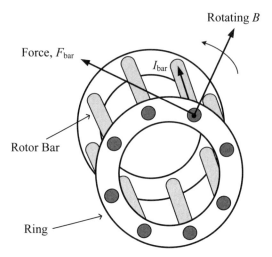

Figure 9.15. Rotating magnetic field-generated driving force.

This induced voltage drives a current through the short-circuited bars of the rotor, as shown in Figure 9.15. From Ohm's law, the current is:

$$I_{bar} = \frac{V_{bar}}{|\mathbf{Z}_{bar}|}, \qquad (9.10)$$

where $\mathbf{Z}_{bar}$ is the impedance of the rotor bar. The substitution of the induced bar voltage in this formula yields:

$$I_{bar} = \frac{2\pi r_{rot} B \ell_{rot} s n_{syn}}{|\mathbf{Z}_{bar}|}. \qquad (9.11)$$

MOTOR FORCE GENERATION. The interaction between the magnetic field **B** and the current $\mathbf{I}_{bar}$ in the rotor bar generates a force, $\vec{F} = \vec{B} \times \vec{I}_{bar} \ell_{rot}$, that drives the motor. This force drives the rotor in the same direction as the magnetic field rotation. More specifically, the rotating magnetic field drags the rotor. Figure 9.15 points out the magnetic flux density, the rotor current, and force directions. The force magnitude is:

$$F_{bar} = B \ell_{rot} I_{bar}. \qquad (9.12)$$

The substitution of the bar current into this expression results in:

$$F_{bar} = \frac{2\pi r_{rot} B^2 \ell_{rot}^2 s n_{syn}}{|\mathbf{Z}_{bar}|}. \qquad (9.13)$$

THREE-PHASE INDUCTION MOTOR

This equation indicates that the force is increasing with the magnetic flux density, which suggests using magnetic material that has a high saturation point. In other words, the better material results in a smaller motor for the same performance.

Actually, the same force is generated on the opposite side of the motor. This results in a driving torque. A synopsis of the force generation process is:

1. the three-phase windings generates a rotating field;
2. the rotating field induces a current in the rotor bars;
3. the current generation requires a speed difference between the rotor and the magnetic field; and
4. the interaction between the field and the current produces the driving force.

9.3.2. Equivalent Circuit

An induction motor has two magnetically coupled circuits: the stator and the rotor. The latter is short circuited. This is similar to a transformer, whose secondary is rotating and short circuited. The motor has balanced three-phase circuits; consequently, a single-phase representation is sufficient.

Both the stator and rotor have windings or conductors, which have resistance and leakage inductance. This suggests that the induction motor can be represented with an equivalent circuit, which has a stator circuit represented by a winding resistance (R_{sta}) and a leakage inductance (L_{sta}) connected in series, and a rotor circuit represented by a winding resistance (R_{rot}) and a leakage inductance (L_{rot}) connected in series. A transformer represents the magnetic coupling between the two circuits. The stator produces a rotating magnetic field that induces voltage in both windings. A magnetizing reactance (X_m) and a core resistance connected in parallel represent the magnetic field generation. The resistance (R_c) represents the eddy current and hysteresis losses in the iron core. Figure 9.16 presents the single-phase equivalent circuit of a three-phase induction motor.

In this circuit, the flux that links with both the stator and the rotor induces a voltage in both circuits. The magnetic flux rotates with constant amplitude and synchronous speed. This flux cuts the stationary conductors of the stator with the synchronous speed and induces a 60 Hz voltage in the stator windings. The root-mean-square (rms) value of the voltage induced in the stator is:

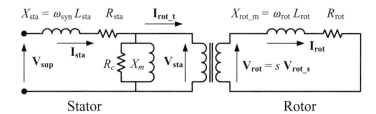

Figure 9.16. Single-phase equivalent circuit of a three-phase induction motor.

$$V_{sta} = \frac{N_{sta}\Phi_{max}\omega_{syn}}{\sqrt{2}}, \qquad (9.14)$$

where

N_{sta} is the number of turns in phase A of the stator; and
Φ_{max} is the peak value of the rotating flux (= Φ_{mag}).

The flux rotates with the synchronous speed and the rotor with the motor speed. Consequently, the flux cuts the rotor conductors with the speed difference between the rotating flux and the rotor. Using Equation (9.6), the speed difference is calculated as:

$$\omega_{rot} = \omega_{syn} - \omega_{mot} = \omega_{syn}s. \qquad (9.15)$$

The voltage induced in the rotor circuit is:

$$V_{rot} = \frac{N_{rot}\Phi_{max}(\omega_{syn} - \omega_{mot})}{\sqrt{2}} = \frac{N_{rot}\Phi_{max}\omega_{syn}s}{\sqrt{2}}. \qquad (9.16)$$

The division of Equation (9.16) with Equation (9.14) results in:

$$V_{rot} = \frac{N_{rot}}{N_{sta}}V_{sta}s = V_{rot\_s}s, \qquad (9.17)$$

where V_{rot_s} is the induced voltage when the motor starts, at which point $s = 1$.

The speed difference determines the frequency of the rotor current, which is:

$$f_{rot} = \frac{\omega_{rot}}{2\pi} = \frac{\omega_{syn} - \omega_{mot}}{2\pi} = \frac{\omega_{syn}s}{2\pi} = sf_{syn}. \qquad (9.18)$$

The slip is from 2% to 5%, which indicates that the rotor current frequency is around $(0.02) \times 60$ Hz = 1.2 Hz to $(0.05) \times 60$ Hz = 3 Hz.

The rotor circuit leakage reactance is:

$$X_{rot\_m} = L_{rot}\omega_{rot} = L_{rot}\omega_{syn}s = X_{rot}s, \qquad (9.19)$$

where

X_{rot_m} is the rotor reactance at the motor frequency; and
X_{rot} is the rotor reactance at the synchronous frequency (f_{syn}).

The rotor voltage and the rotor reactance are marked on Figure 9.16. The relation between the rotor current and the rotor-induced voltage is calculated by the loop voltage equation (Kirchhoff's voltage law, KVL):

$$\mathbf{V}_{rot} = \mathbf{V}_{rot\_s}s = \mathbf{I}_{rot}(R_{rot} + jX_{rot}s). \qquad (9.20)$$

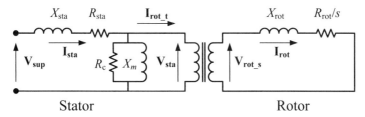

Figure 9.17. Modified equivalent circuit of a three-phase induction motor.

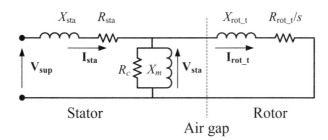

Figure 9.18. Simplified equivalent circuit of a three-phase induction motor.

The division of this equation with the slip yields:

$$\mathbf{V}_{rot\_s} = \mathbf{I}_{rot}\left(\frac{R_{rot}}{s} + jX_{rot}\right). \tag{9.21}$$

The implementation of this expression for the equivalent circuit results in Figure 9.17. In this circuit, the stator and rotor are connected through an ideal transformer with a turns ratio of N_{sta}/N_{rot}.

Further simplification is to transfer the rotor impedances to the stator side of the ideal transformer by multiplying the impedances with the turns ratio squared. The results of the impedance transfer are:

$$\frac{R_{rot\_t}}{s} = \left(\frac{N_{sta}}{N_{rot}}\right)^2 \frac{R_{rot}}{s} \quad \text{and} \quad X_{rot\_t} = \left(\frac{N_{sta}}{N_{rot}}\right)^2 X_{rot}. \tag{9.22}$$

This transfer permits the elimination of the ideal transformer, which yields the simplified equivalent circuit of Figure 9.18.

The last modification of the equivalent circuit is the separation of the rotor resistance into two parts:

$$\frac{R_{rot\_t}}{s} = R_{rot\_t} + \left(\frac{1-s}{s}\right)R_{rot\_t}. \tag{9.23}$$

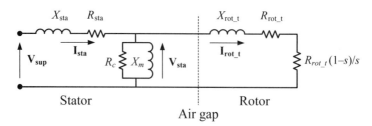

Figure 9.19. Final single-phase equivalent circuit of a three-phase induction motor.

Implementing this partition yields the final single-phase equivalent circuit of Figure 9.19. In this circuit, the resistance $R_{\text{rot\_t}}(1 - s)/s$ represents the outgoing mechanical power generated by the motor.

9.3.3. Motor Performance

This single-phase equivalent circuit of Figure 9.19 represents phase A of a three-phase motor. The leakage reactance (X_{sta}) and the resistance (R_{sta}) of the stator winding, together with the magnetizing reactance (X_m) and a resistance (R_c), represent the stator. The magnetizing resistance is used to calculate iron losses. The leakage reactance and resistance of the rotor bars are transferred to the stator. This equivalent circuit is used to evaluate motor performance.

Before performing a detailed analysis of a motor, several terms are defined in the context of the single-phase equivalent circuit of a three-phase induction motor. Referring to Figure 9.19, the input real electrical power from all three phases to the motor is:

$$P_{\text{sup}} = \text{Re}(S_{\text{sup}}) = \text{Re}\left(3 V_{\text{sup}} I_{\text{sta}}^*\right). \tag{9.24}$$

If the motor is wye connected, the supply voltage (V_{sup}) is the line-to-neutral voltage. Whereas if the motor is delta connected, the supply voltage is the line-to-line voltage.

Figure 9.20 is an illustration of the energy balance in a motor. The power transferred through the air gap by the magnetic coupling is the input power (P_{sup}) minus the stator copper loss and the magnetizing (stator iron) loss. The electrically developed power (P_{dev}) is the difference between the air-gap power (P_{ag}) and rotor copper loss. The electrically developed power can be computed from the power dissipated in the second term of Equation (9.23) that represents the rotor resistance:

$$P_{\text{dev}} = 3|I_{\text{rot\_t}}|^2 \left(R_{\text{rot\_t}} \frac{1-s}{s} \right). \tag{9.25}$$

The subtraction of the mechanical ventilation and friction losses (P_{mloss}) from the developed power gives the mechanical output power:

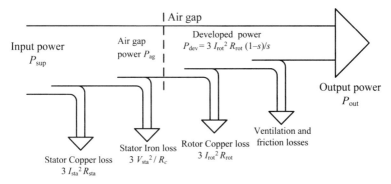

Figure 9.20. Motor energy balance flow diagram.

$$P_{out} = P_{dev} - P_{mloss}. \tag{9.26}$$

The motor output power is often expressed in terms of horsepower (hp), where 1 hp = 745.7 W. The overall efficiency of the motor is the ratio of the output and input real powers:

$$\eta = \frac{P_{out}}{P_{sup}}. \tag{9.27}$$

Often motor performance is also specified in terms of the torque developed. The *torque* is a measure of the rotational force generated by the motor, and is mathematically defined as:

$$T = \frac{P_{out}}{\omega_{mot}}. \tag{9.28}$$

Although the units of torque equate to joules, torque is expressed in units of newton-meter (or pound-foot).

9.3.4. Motor Maximum Output

Expressions for the maximum power and torque outputs from an induction motor can be developed using a Thévenin equivalent circuit model. Figure 9.21 delineates the part of the motor circuit to be represented by a Thévenin equivalent. Determination of the Thévenin equivalent network consists of finding two basic quantities: (1) the open-circuit voltage and (2) the equivalent impedance.

Referring to Figure 9.22, the open-circuit voltage, $\mathbf{V}_{Th}$, is readily determined from applying voltage division to the network:

$$\mathbf{V}_{Th} = \mathbf{V}_{sup} \frac{\mathbf{Z}_m}{\mathbf{Z}_{sta} + \mathbf{Z}_m}. \tag{9.29}$$

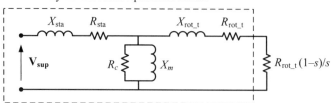

Figure 9.21. Single-phase equivalent circuit of a three-phase induction motor prior to forming Thévenin equivalent.

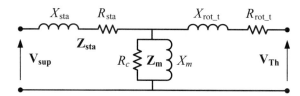

Figure 9.22. Induction motor circuit for open-circuit voltage analysis.

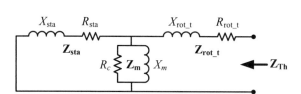

Figure 9.23. Induction motor network for Thévenin equivalent impedance determination.

Similarly, while shorting the supply voltage as depicted in Figure 9.23, the Thévenin equivalent impedance is found by looking back into the original circuit from the terminals where the circuit is separated:

$$\mathbf{Z}_{Th} = \mathbf{Z}_{rot\_t} + \frac{\mathbf{Z}_{sta}\mathbf{Z}_m}{\mathbf{Z}_{sta} + \mathbf{Z}_m}. \quad (9.30)$$

From these two analyses, the Thévenin equivalent circuit is formed as drawn in Figure 9.24. From Ohm's law, the current to the rotor is:

$$\mathbf{I}_{rot\_t} = \frac{\mathbf{V}_{Th}}{\mathbf{Z}_{Th} + R_{rot\_t}(1-s)/s}. \quad (9.31)$$

THREE-PHASE INDUCTION MOTOR

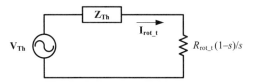

Figure 9.24. Thévenin single-phase equivalent circuit for a three-phase induction motor.

To determine the output power, the previous relation for the rotor current is substituted into Equation (9.25), while noting that in this situation:

$$|\mathbf{I}_{rot\_t}|^2 = \mathbf{I}_{rot\_t}\mathbf{I}_{rot\_t}^* = \left(\frac{\mathbf{V}_{Th}}{\mathbf{Z}_{Th} + R_{rot\_t}(1-s)/s}\right)\left(\frac{\mathbf{V}_{Th}^*}{\mathbf{Z}_{Th}^* + R_{rot\_t}(1-s)/s}\right). \quad (9.32)$$

The developed output power is then:

$$P_{dev} = \frac{3|\mathbf{V}_{Th}|^2 R_{rot\_t}(1-s)/s}{|\mathbf{Z}_{Th}|^2 + (R_{rot\_t}(1-s)/s)^2 + 2R_{Th}R_{rot\_t}(1-s)/s}, \quad (9.33)$$

where R_{Th} is the real part of $\mathbf{Z}_{Th}$. To determine the slip at which the maximum power is developed, the derivative of the previous expression is taken with respect to the slip and set equal to zero (i.e., $dP_{dev}/ds = 0$), yielding a quadratic equation for s:

$$\left(|\mathbf{Z}_{Th}|^2 - R_{rot\_t}^2\right)s^2 + 2R_{rot\_t}^2 s - R_{rot\_t}^2 = 0. \quad (9.34)$$

The corresponding slip for the maximum power output from a motor is:

$$s_{P,max} = \frac{R_{rot\_t}}{R_{rot\_t} + |\mathbf{Z}_{Th}|}. \quad (9.35)$$

Back substituting this relation into Equation (9.33) provides a formula for the maximum power output:

$$P_{max} = \frac{3|\mathbf{V}_{Th}|^2}{2(|\mathbf{Z}_{Th}| + R_{Th})}. \quad (9.36)$$

In a similar manner, a relation for the maximum torque may be obtained. Neglecting the mechanical loss (i.e., $P_{mloss} = 0$), the motor torque of the Thévenin equivalent circuit is:

$$T = \frac{P_{dev}}{2\pi n_{syn}(1-s)} = \frac{3|\mathbf{V}_{Th}|^2 R_{rot\_t} s}{2\pi n_{syn}\left[|\mathbf{Z}_{Th}|^2 s^2 + R_{rot\_t}^2(1-s)^2 + 2R_{Th}R_{rot\_t}s(1-s)\right]}. \quad (9.37)$$

From $dT/ds = 0$, the slip at maximum torque is:

$$s_{T,\max} = \frac{R_{\text{rot\_t}}}{\sqrt{|\mathbf{Z}_{\text{Th}}|^2 + R_{\text{rot\_t}}^2 - 2R_{Th}R_{\text{rot\_t}}}}. \tag{9.38}$$

The corresponding maximum torque is:

$$T_{\max} = \frac{3|\mathbf{V}_{\text{Th}}|^2}{4\pi n_{\text{syn}}\left[R_{Th} - R_{\text{rot\_t}} + \sqrt{|\mathbf{Z}_{\text{Th}}|^2 + R_{\text{rot\_t}}^2 - 2R_{Th}R_{\text{rot\_t}}}\right]}. \tag{9.39}$$

9.3.5. Performance Analyses

The purpose of motor performance analysis is to determine the input and output powers, efficiency, and motor torque as functions of motor speed. These performance analyses are presented here by an interactive derivation of the equations using Mathcad. The reader is asked to use her/his computer and follow the derivation using Mathcad or MATLAB. To avoid typing and dimension errors, the equations are tested with numerical values as the derivation proceeds. The derived formulae can be used as tools for motor analysis. Although the motor speed is the independent variable of interest, we derive the succeeding equations primarily in terms of the slip. We test the equations with a slip of $s_{\text{test}} := 3\%$.

We use the data of a 20 hp, 440 V, wye-connected three-phase motor. The four-pole, 60 Hz motor data are:

$$P_{\text{motor}} := 20 \text{ hp} \qquad V_{\text{mot}} := 440 \text{ V} \qquad p := 4$$
$$R_{\text{sta}} := 0.44 \text{ }\Omega \qquad X_{\text{sta}} := 1.25 \text{ }\Omega \qquad f_{\text{syn}} := 60 \text{ Hz}$$
$$R_{\text{rot\_t}} := 0.40 \text{ }\Omega \qquad X_{\text{rot\_t}} := 1.25 \text{ }\Omega \qquad \omega_{\text{syn}} := 2 \cdot \pi \cdot f_{\text{syn}}$$
$$R_c := 350 \text{ }\Omega \qquad X_m := 27 \text{ }\Omega \qquad P_{\text{mloss}} := 262 \text{ W}.$$

9.3.5.1. Calculation of Motor Impedance as a Function of the Slip. The rotor impedance is the sum of the two resistances and the rotor reactance connected in series, as seen in Figure 9.19.

$$Z_{\text{rot\_t}}(s) := j \cdot X_{\text{rot\_t}} + R_{\text{rot\_t}} + R_{\text{rot\_t}} \cdot \frac{(1-s)}{s}, \tag{9.40}$$

with a numerical value in this case of $z_{\text{rot\_t}}(s_{\text{test}}) = (13.33 + 1.25j) \text{ }\Omega$.

The magnetizing reactance and core resistance are connected in parallel in the magnetizing branch. The magnetizing impedance is independent from the slip:

$$Z_m := \frac{j \cdot X_m \cdot R_c}{j \cdot X_m + R_c}, \tag{9.41}$$

with a numerical result of $z_m = (2.07 + 26.84j) \text{ }\Omega$.

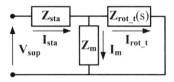

Figure 9.25. Simplified motor equivalent circuit.

The stator impedance is the sum of the stator resistance and reactance, and it is also independent of the slip:

$$Z_{sta} := R_{sta} + j \cdot X_{sta}, \tag{9.42}$$

which is $z_{sta} = (0.44 + 1.25j)\ \Omega$.

The combination of the impedances simplifies the equivalent circuit, which is provided in Figure 9.25.

Combining the parallel connected magnetizing and rotor impedances together and adding to the stator impedance yields the overall motor impedance. The resulting motor impedance depends on the slip:

$$Z_{mot}(s) := Z_{sta} + \frac{Z_m \cdot Z_{rot\_t}(s)}{Z_m + Z_{rot\_t}(s)}, \tag{9.43}$$

and equals $z_{mot}(s_{test}) = (10.22 + 6.82j)\ \Omega$.

9.3.5.2. Calculation of Motor Currents. The motor is wye connected, which implies that the supply voltage is the line-to-neutral voltage:

$$V_{sup} := \frac{V_{mot}}{\sqrt{3}} = 254.0\ \text{V}.$$

The motor input (stator) current is the ratio of the supply (line-to-neutral) voltage and the motor impedance:

$$I_{sta}(s) := \frac{V_{sup}}{Z_{mot}(s)}, \tag{9.44}$$

$$I_{sta}(s_{test}) = (17.20 - 11.48j)\ \text{A}$$

$$|I_{sta}(s_{test})| = 20.7\ \text{A} \quad \arg(I_{sta}(s_{test})) = -33.7 \cdot \deg.$$

The phase angle indicates that the stator current is an inductive and lagging current.

The rotor current is calculated by using the current division rule while referring to Figure 9.25:

$$I_{rot\_t}(s) := I_{sta}(s) \cdot \frac{Z_m}{Z_m + Z_{rot\_t}(s)}, \quad (9.45)$$

$$I_{rot\_t}(s_{test}) = (17.14 - 2.84j) \text{ A}$$

$$|I_{rot\_t}(s_{test})| = 17.4 \text{ A} \quad \arg(I_{rot\_t}(s_{test})) = -9.4 \cdot \deg.$$

The phase angle reveals that the rotor current is also an inductive and lagging current.

The motor starting current can be calculated by substituting $s_{start} := 1$ in the stator current relation of Equation (9.44):

$$I_{start} := I_{sta}(s_{start})$$

$$I_{start} = (30.93 - 93.57j) \text{ A} \quad |I_{start}| = 98.6 \text{ A}.$$

Typically, the ratio of the starting current and rated current is around 6. This large starting current produces a voltage dip if the supply is weak (i.e., supply impedance is high).

The input stator current varies with the motor speed or slip. Figure 9.26 plots the input current–versus-slip function. The typical motor operating range is between slips of 1% and 5%. In this range, the current–slip relation is nearly linear as demonstrated in the graph. The largest starting current occurs when $s = 1$ (100% slip).

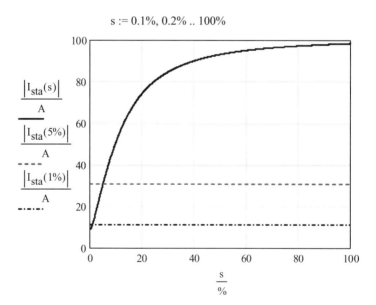

Figure 9.26. Motor input current versus slip.

CALCULATION OF THE MOTOR INPUT POWER. The input complex power is three times the product of the supply voltage and the conjugate of the motor input current:

$$S_{sup}(s) := 3 \cdot V_{sup} \cdot \overline{I_{sta}(s)}, \quad (9.46)$$

$$|S_{sup}(s_{test})| = 15.8 \text{ kV} \cdot \text{A} \quad S_{sup}(s_{test}) = (13.1 + 8.8j) \text{ kV} \cdot \text{A}.$$

The real and reactive powers are:

$$P_{sup}(s) := \text{Re}(S_{sup}(s)), \quad (9.47)$$

$$Q_{sup}(s) := \text{Im}(S_{sup}(s)), \quad (9.48)$$

$$P_{sup}(s_{test}) = 13.1 \text{ kW} \quad Q_{sup}(s_{test}) = 8.8 \text{ kV} \cdot \text{A}.$$

The power factor can be calculated using either the ratio of the real power and apparent power or the cosine of the complex power angle:

$$\text{pf}_{sup}(s) := \cos(\arg(S_{sup}(s))), \quad (9.49)$$

$$\text{Pf}_{sup}(s) := \frac{P_{sup}(s)}{|S_{sup}(s)|}. \quad (9.50)$$

The numerical results for the lagging power factor verify that either approach is valid:

$$\text{pf}_{sup}(s_{test}) = 0.832 \quad \text{Pf}_{sup}(s_{test}) = 0.832.$$

CALCULATION OF THE MOTOR OUTPUT POWER. From Figure 9.20, three times the power dissipated in the load resistance is the developed power:

$$P_{dev}(s) := 3 \cdot (|I_{rot\_t}(s)|)^2 \cdot R_{rot\_t} \cdot \frac{(1-s)}{s}, \quad (9.51)$$

$$P_{dev}(s_{test}) = 11.7 \text{ kW}.$$

This function has a singularity at $s = 0$, when the motor speed reaches the synchronous speed. At that point, the induced voltage becomes zero, which results in zero current and power. This singularity can be disregarded by selecting a nonzero value for the minimum slip studied.

The developed power less the mechanical ventilation and friction losses gives the output mechanical power:

$$P_{mech}(s) := P_{dev}(s) - P_{mloss}, \quad (9.52)$$

$$P_{mech}(s_{test}) = 15.4 \text{ hp}.$$

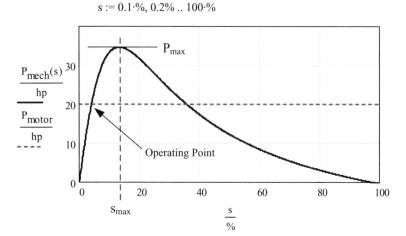

Figure 9.27. Mechanical output power versus slip.

The variation of the mechanical power with the slip is plotted in Figure 9.27. The graph shows that the power peaks at a maximum value of P_{max} at a slip of s_{max}. Before the slip reaches s_{max}, increasing the motor load increases the slip and reduces the speed until the load reaches the maximum power, P_{max}. Further load increase causes a power imbalance; the power generated by the motor is less than the load, which slows and eventually stops the motor. In other words, an increase of the slip past s_{max} causes a reduction of the speed, which reduces the motor power. The stable operating region is from $s = 0$ to s_{max}.

An exact value for the maximum power can be calculated by using the *Maximize* function of Mathcad. The seed value for the maximization is $s := 10\%$:

$$s_{max} := \text{Maximize}(P_{mech}, s) \quad s_{max} = 13.4 \cdot \%$$
$$P_{max} := P_{mech}(s_{max}) \quad P_{max} = 34.6 \cdot \text{hp}.$$

The slip at rated power (P_{motor}) can be calculated by the Mathcad *root* function, which solves the nonlinear equation: $P_{mech}(s) - P_{motor} = 0$. The root function requires a seed value. The selected seed value is $s := 2\%$.

$$s_{rated} := \text{root}(P_{mech}(s) - P_{motor}, s) \quad s_{rated} = 4.1\%.$$

This slip value is annotated on Figure 9.27 as the operating point.

EFFICIENCY OF A THREE-PHASE INDUCTION MOTOR. The motor efficiency is the ratio of the output mechanical power and input electrical power:

$$\eta(s) := \frac{P_{mech}(s)}{P_{sup}(s)}, \tag{9.53}$$

$$\eta(s_{test}) = 87.4\%.$$

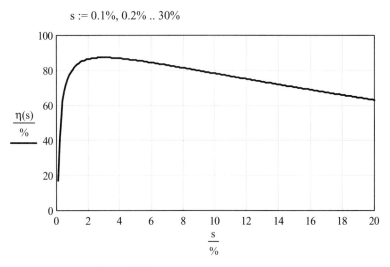

Figure 9.28. Motor efficiency versus slip.

The efficiency-versus-slip function is graphed in Figure 9.28. The function has a maximum value, which can be determined from the plot or calculated using the Mathcad *Maximize* function. The seed value for the maximization is s := 2%:

$$s_{max\_eff} := \text{Maximize}(\eta, s) \quad s_{max\_eff} = 2.9 \cdot \%$$
$$\eta_{max} := \eta(s_{max\_eff}) \quad \eta_{max} = 87.4 \cdot \%.$$

The efficiency at the rated power is close to the maximum:

$$\eta_{rated} := \eta(s_{rated}) \quad \eta_{rated} = 86.7\%.$$

9.3.5.3. Motor Speed. The synchronous speed is calculated using Equation (9.7). The practical unit for the motor speed is rpm, which is revolutions per minute. This unit can be defined in Mathcad:

$$\text{rpm} := \frac{1}{\min} \quad n_{syn} := \frac{f_{syn}}{\frac{p}{2}} \quad n_{syn} = 1800 \text{ rpm}.$$

The motor speed as a function of the slip is calculated from rearranging Equation (9.6):

$$n_{mot}(s) := n_{syn} \cdot (1-s), \tag{9.54}$$
$$n_{mot}(s_{test}) = 1746 \text{ rpm}.$$

The circular speed of the motor is:

$$\omega_{mot}(s) := 2 \cdot \pi \cdot n_{mot}(s). \tag{9.55}$$

The substitution of Equation (9.54) into Equation (9.55) results in:

$$\omega_{mot}(s) := 2 \cdot \pi \cdot n_{syn} \cdot (1-s), \tag{9.56}$$

$$\omega_{mot}(s_{test}) = 182.8 \cdot \frac{rad}{sec}.$$

9.3.5.4. Motor Torque. The motor *torque* is the ratio of the output mechanical power and the motor circular frequency:

$$T_{mot}(s) := \frac{P_{mech}(s)}{\omega_{mot}(s)}. \tag{9.57}$$

This torque function has singularities at $s = 0$ and $s = 1$ due to the numerator and denominator expressions, respectively. Moreover, the torque equation is inaccurate when the slip approaches unity or zero, anyway. This equation has acceptable accuracy roughly between $s = 0.5\%$ and $s = 80-90\%$. In Figure 9.29, the torque is plotted in the practical range, where the equation is accurate.

When the slip approaches zero, the motor speed approaches the synchronous speed, and the induced voltage, rotor current, and developed power decrease to zero. At very low speed, the motor-developed power will be less than the mechanical power

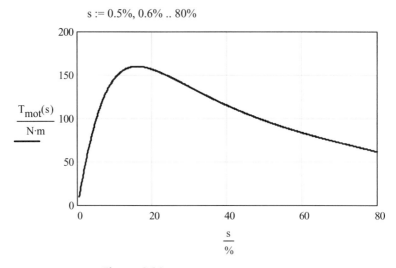

Figure 9.29. Motor torque versus slip.

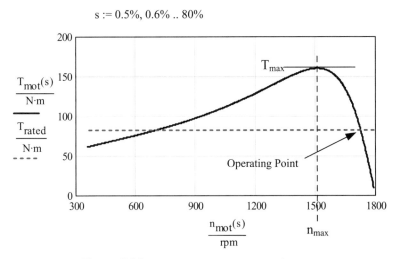

$s := 0.5\%, 0.6\% .. 80\%$

Figure 9.30. Motor torque-versus-speed curve.

loss. The mechanical power loss is a passive power that cannot drive the motor. Consequently, the equation cannot be used at very small slip values, when this torque equation yields negative values.

When the slip approaches unity, the motor slows down. The mechanical loss is not independent from the speed. Actually, the ventilation loss decreases with the speed. This makes the previous torque equation invalid at low speed and at the starting condition.

The torque–speed characteristic is important for practical purposes because it can be used to estimate motor performance. The torque–speed curve is given in Figure 9.30.

The torque has a maximum value (T_{max}) that can be determined from the plot or calculated by the *Maximize* function of Mathcad. In this case, the slip value where the torque is maximum is first calculated. The seed value for the maximization is $s := 20\%$:

$$s_{max} := \text{Maximize}(T_{mot}, s) \quad s_{max} = 16.1\%.$$

Having found the slip, the maximum torque and the corresponding speed are calculated:

$$T_{max} := T_{mot}(s_{max}) \quad T_{max} = 160.4 \cdot N \cdot m$$
$$n_{max} := n_{mot}(s_{max}) \quad n_{max} = 1511 \cdot \text{rpm}.$$

Figure 9.30 shows that the practical motor operating range is between the speed (n_{max}) at maximum torque and the synchronous speed (1800 rpm). The speed and torque at the rated operating power are:

$$n_{rated} := n_{mot}(s_{rated}) \quad n_{rated} = 1726 \cdot rpm$$
$$T_{rated} := T_{mot}(s_{rated}) \quad T_{rated} = 82.5 \cdot N \cdot m$$

The torque at the rated operating point power is annotated in Figure 9.30.

9.3.5.5. Starting Torque. When the motor starts at $s = 1$, the ventilation losses are zero and the friction loss is passive. The negative friction loss does not drive the motor backward. Therefore, the mechanical losses are assumed zero when $s = 1$. This implies that the starting torque is calculated from the developed power instead of the mechanical output power. The singularity of the developed power at $s = 1$ can be eliminated by substituting Equation (9.52), Equation (9.51), and Equation (9.56) into Equation (9.57) and assuming $P_{mloss} = 0$:

$$T_{start}(s) := \frac{3 \cdot (|I_{rot\_t}(s)|)^2 \cdot R_{rot\_t} \cdot \frac{(1-s)}{s}}{2 \cdot \pi \cdot n_{syn} \cdot (1-s)}.$$

The simplification of the previous equation leads to:

$$T_{start}(s) := \frac{3 \cdot (|I_{rot\_t}(s)|)^2 \cdot \frac{R_{rot\_t}}{s}}{2 \cdot \pi \cdot n_{syn}}. \tag{9.58}$$

This equation gives a starting torque at $s = 1$ of $T_{start}(1) = 56.4$ N·m.

EXAMPLE 9.1: Mathcad example of a motor driving a pump

The application of the torque curve is demonstrated by a numerical example. The motor described earlier drives a pump. The pump torque varies with its shaft speed (n_p). The pump torque–speed curve is described by:

$$T_p(n_p) := 30 \, N \cdot m + n_p^2 3 \cdot 10^{-5} \cdot \frac{N \cdot m}{rpm^2}, \tag{9.59}$$

$$T_p(1000 \, rpm) = 60 \cdot N \cdot m.$$

We assume that the motor and pump share a common shaft such that they both turn at the same speed. The motor and pump torque curves are graphed in Figure 9.31. In order to plot the two torques, the pump speed (n_p) in Equation (9.59) has to be replaced by the motor speed (n_{mot}). Because of this, the pump torque is expressed as a function of the motor slip:

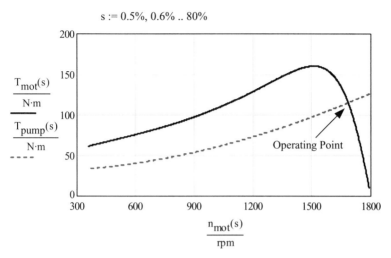

Figure 9.31. Pump and motor torque–speed curves.

$$T_{pump}(s) := 30 \text{ N} \cdot \text{m} + n_{mot}(s)^2 \, 3 \cdot 10^{-5} \cdot \frac{\text{N} \cdot \text{m}}{\text{rpm}^2}. \quad (9.60)$$

Figure 9.31 shows the motor and pump torque–speed characteristics. The operating point is determined by the intersection of the two curves. The graph demonstrates that the motor starting torque and driving torque are larger than the torque required by the pump. The difference between the two torques accelerates the pump until the two torques become equal. The intersection of the torque curves gives the stable operating point. The torque and speed at the operating point can be determined from Figure 9.31 or they can be calculated by solving the equation:

$$T_{mot}(s) - T_{pump}(s) = 0.$$

The seed value for the Mathcad *root* equation solver is $s := 6\%$:

$$s_{p\_op} := \text{root}(T_{mot}(s) - T_{pump}(s), s) \quad s_{p\_op} = 6.4 \cdot \%$$
$$n_{mot}(s_{p\_op}) = 1684 \cdot \text{rpm}$$
$$T_{pump\_op} := T_{pump}(s_{p\_op}) \quad\quad\quad T_{pump\_op} = 115.1 \cdot \text{N} \cdot \text{m}.$$

The motor–pump system settles at one speed. However, if the pump load changes (e.g., the fluid flow is somehow restricted in the pipe), then the system will settle at a different speed based on the altered torque curve.

9.3.6. Determination of Motor Parameters by Measurement

The resistances and reactances in the equivalent circuit (see Figure 9.19) for an induction motor can be determined by a series of measurements. The measurements are:

1. *No-Load Test.* This test determines the magnetizing reactance and core loss resistance.
2. *Blocked-Rotor Test.* This test gives the combined values of the stator and rotor resistance and reactance.
3. *Stator Resistance Measurement.*

An interactive derivation is used for the calculation of the motor parameters. The equations are verified using a numerical example.

The no-load and blocked-rotor test results for a three-phase, four-pole, wye-connected, 60 Hz motor, rated 6 hp, 208 V are:

No-load test:

$$V_{no\_load} := 208 \text{ V} \quad I_{no\_load} := 4.5 \text{ A} \quad P_{no\_load} := 285 \text{ W}.$$

Blocked-rotor test at 15 Hz:

$$V_{blocked} := 38 \text{ V} \quad I_{blocked} := 13 \text{ A} \quad P_{blocked} := 480 \text{ W}.$$

The measured per-phase stator resistance is $R_{sta} := 0.41 \; \Omega$.

9.3.6.1. No-Load Test. For the no-load test, the motor shaft is free and the rated voltage supplies the motor. The supply voltage and the input current and power are measured. In the case of a three-phase motor, the line-to-line voltages, line currents, and the three-phase power using two wattmeters are measured. The three-phase power measurement is described in Chapter 4.

Figure 9.32 presents the single-phase equivalent motor circuit for the no-load test. In the no-load condition, the speed is close to the synchronous speed and the slip is close to zero ($s \sim 0$), which results in very small rotor and stator currents. If $s \sim 0$, then R_{rot} $(1 - s)/s \sim \infty$, which implies that the rotor is an open circuit and it can be eliminated from the equivalent circuit. Because the magnetizing impedance is significantly larger than the stator impedance, the stator impedance can be neglected without significant error. Typically, the no-load current is less than 10% of the rated current. This produces a very small voltage drop on the stator impedance, which also suggests the elimination of the stator impedance from the equivalent circuit of Figure 9.32. The resulting simplified circuit from neglecting the rotor circuit and the stator impedance is illustrated in Figure 9.33.

This circuit reveals that the magnetizing reactance (X_m) and the core loss resistance (R_c) can be calculated from the results of the no-load test. However, Figure 9.33 provides the single-phase equivalent circuit of the motor. The single-phase equivalent circuit represents phase A. Because of the single-phase representation, the measured power has to be divided by 3. In the case of a wye-connected motor, the circuit is

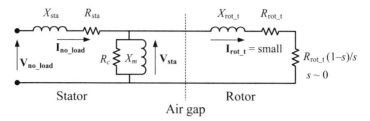

Figure 9.32. Equivalent motor circuit in no-load test condition.

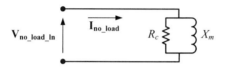

Figure 9.33. Simplified equivalent motor circuit in no-load test condition.

supplied by the line-to-neutral voltage. In the case of a delta-connection, the line-to-line voltage is used.

The line-to-neutral voltage and the power of phase A are:

$$V_{no\_load\_ln} := \frac{V_{no\_load}}{\sqrt{3}} \quad V_{no\_load\_ln} = 120.1 \text{ V}$$

$$P_{no\_load\_A} := \frac{P_{no\_load}}{3} \quad P_{no\_load\_A} = 95 \text{ W}.$$

The core loss resistance is calculated using the input real power and the supply voltage in the no-load condition:

$$R_c := \frac{V_{no\_load\_ln}^2}{P_{no\_load\_A}}. \tag{9.61}$$

The result is $R_c = 151.8 \; \Omega$.

The apparent power (absolute value of the complex power) in the no-load condition is:

$$S_{no\_load\_A} := V_{no\_load\_ln} \cdot I_{no\_load}, \tag{9.62}$$

$$S_{no\_load\_A} = 540.4 \text{ V} \cdot \text{A}.$$

The reactive power is found considering the power triangle:

$$Q_{no\_load\_A} := \sqrt{S_{no\_load\_A}^2 - P_{no\_load\_A}^2}, \tag{9.63}$$

$$Q_{no\_load\_A} = 532.0 \text{ V} \cdot \text{A}.$$

The magnetizing reactance is:

$$X_m := \frac{V_{no\_load\_ln}^2}{Q_{no\_load\_A}}, \quad (9.64)$$

$$X_m = 27.1\ \Omega.$$

Truly in the no-load condition, the motor power balance is:

$$P_{no\text{-}load} = P_{stator\_loss} + P_{core\_loss} + P_{rotate\_loss},$$

where P_{rotate_loss} includes both ventilation and frictional losses. This implies that the core loss resistance calculated earlier using Equation (9.61) includes the core loss, the rotational loss, and even the small no-load stator losses.

9.3.6.2. Blocked-Rotor Test. The rotor is blocked to prevent rotation and the motor is supplied by reduced voltage and reduced frequency. The supply voltage is reduced until the motor current is around the rated value. The supply frequency is typically 15 Hz. Without rotor movement, $n_{mot} = 0$, and hence, the slip is unity, which means that the $R_{rot\_t}(1-s)/s$ term disappears. Figure 9.34 shows the equivalent circuit for the blocked-rotor test.

The reduced voltage and frequency minimize the magnetizing current and core losses. This allows the simplification of the circuit by neglecting the magnetizing impedance. Figure 9.35 presents the simplified circuit, which reveals that the combined

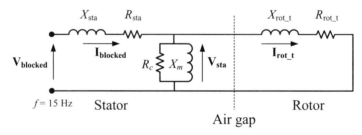

Figure 9.34. Equivalent motor circuit for blocked-rotor test condition.

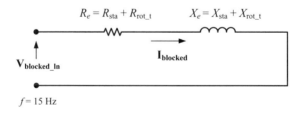

Figure 9.35. Simplified equivalent motor circuit for blocked-rotor test condition.

stator and rotor impedance $(R_e + jX_e)$ can be calculated from the results of the blocked-rotor test. However, this diagram provides the single-phase equivalent circuit of the motor, representing phase A. In the case of a wye-connected motor, the line-to-neutral voltage supplies the circuit. In the case of a delta connection, the line-to-line voltage is used. Because of the single-phase representation, the power must be divided by three. The single-phase values of the voltage and power are:

$$V_{blocked\_ln} := \frac{V_{blocked}}{\sqrt{3}} \quad V_{blocked\_ln} = 21.9 \text{ V}$$

$$P_{blocked\_A} := \frac{P_{blocked}}{3} \quad P_{blocked\_A} = 160 \text{ W}.$$

The equivalent motor resistance (R_e) is calculated using the input real power and current measured during the blocked-rotor test:

$$R_e := \frac{P_{blocked\_A}}{I_{blocked}^2}, \tag{9.65}$$

$$R_e = 0.947 \, \Omega.$$

The stator resistance is measured directly, and the obtained value per phase was given earlier as R_{sta}. The rotor resistance is obtained by subtracting the stator resistance from the results of Equation (9.65):

$$R_{rot\_t} := R_e - R_{sta}, \tag{9.66}$$

$$R_{rot\_t} = 0.537 \, \Omega.$$

The next step is the calculation of the magnitude of the circuit impedance:

$$Z_{blocked} := \frac{V_{blocked\_ln}}{I_{blocked}}, \tag{9.67}$$

$$Z_{blocked} = 1.688 \, \Omega.$$

The equivalent reactance at a frequency of 15 Hz is:

$$X_{e\_15\,Hz} := \sqrt{Z_{blocked}^2 - R_e^2}, \tag{9.68}$$

$$X_{e\_15\,Hz} = 1.397 \, \Omega.$$

The equivalent reactance at 60 Hz is:

$$X_e := X_{e\_15\,Hz} \cdot \frac{60 \text{ Hz}}{15 \text{ Hz}}, \tag{9.69}$$

$$X_e = 5.588 \, \Omega.$$

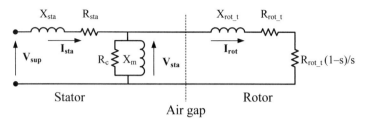

Figure 9.36. Equivalent single-phase circuit of a three-phase induction motor.

The stator and rotor reactance cannot be separated using the results of the previous measurements. A widely accepted assumption is that the two reactances are equal:

$$X_{sta} := \frac{X_e}{2} \quad X_{sta} = 2.794 \, \Omega$$

$$X_{rot\_t} := \frac{X_e}{2} \quad X_{rot\_t} = 2.794 \, \Omega.$$

9.3.6.3. Summary. The results of this analysis are the equivalent circuit parameters that can be used for the motor performance evaluation. The equivalent circuit is given in Figure 9.36. The obtained reactance and resistance values per phase are:

Magnetizing circuit: $R_c = 151.8 \, \Omega$ $X_m = 27.1 \, \Omega$
Stator impedance: $R_{sta} = 0.41 \, \Omega$ $X_{sta} = 2.79 \, \Omega$
Rotor impedance: $R_{rot\_t} = 0.537 \, \Omega$ $X_{rot\_t} = 2.79 \, \Omega$

EXAMPLE 9.2: Induction motor parameters determination with MATLAB

The parameters of a three-phase induction motor with a delta-connected supply are determined in this example. The stator resistance (R_{sta}) is directly measured as 0.6 Ω, and the other test results are as follows:

	Supply Voltage (V)	Supply Current (A)	Input Power (W)
No-load test	208	5	200
Blocked-rotor test	17	10	450

The motor is located in Europe, where the system frequency is 50 Hz. In this example, the motor is connected in a delta such that the supply voltage (V_{sup}) is the line-to-line voltage.

THREE-PHASE INDUCTION MOTOR

The test results are first initialized in MATLAB as given in the following:

```
% InductionMotorParameters.m

% system/synchronous frequency (Hz)
fsyn = 50;

% Measurement results; delta connected supply

% No-load test
Vnoload = 208;          % volts
Inoload = 5;            % amps
Pnoload = 200;          % watts

% Blocked rotor test
Vblock = 17;            % volts
fblock = 15;            % supply frequency
Iblock = 10;            % amps
Pblock = 450;           % watts

% Stator resistance by dc measurement
Rstator = 0.6;          % ohms
```

The no-load test measurements are used to determine the magnetizing reactance (X_m) and core loss resistance (R_c). The core loss resistance for the single-phase equivalent circuit is:

$$R_c = \frac{V_{\text{no-load}}^2}{P_{\text{no-load}}/3}. \tag{9.70}$$

Referring to Figure 9.33, we recall that the magnetizing reactance and core loss resistance are in parallel, hence:

$$\frac{I_{\text{no-load}}}{V_{\text{no-load}}} = |\mathbf{Y}| = \sqrt{G^2 + B^2},$$

$$B = \frac{1}{X_m} = \sqrt{\left(\frac{I_{\text{no-load}}}{V_{\text{no-load}}}\right)^2 - \frac{1}{R_c^2}}. \tag{9.71}$$

Next, the blocked-rotor test results are utilized to compute the combined stator and rotor impedance. The equivalent series resistance and rotor resistance for the single-phase equivalent circuit (see Figure 9.35) are:

$$R_e = \frac{P_{\text{blocked}}/3}{I_{\text{blocked}}^2},$$

$$R_{\text{rot}} = R_e - R_{\text{sta}}. \tag{9.72}$$

The equivalent reactance at the test frequency (f_{test}) and then converted to the synchronous frequency (f_{syn}) is:

$$X_{e\_test} = \sqrt{\left(\frac{V_{\text{blocked}}}{I_{\text{blocked}}}\right)^2 - R_e^2},$$

$$X_{e\_syn} = X_{e\_test} \frac{f_{\text{syn}}}{f_{\text{test}}}.$$

The stator and rotor reactance are then each taken to be half of this equivalent reactance.

The remainder of the MATLAB program to compute the motor parameters is:

```
% Evaluation of no-load test
% calculate core loss resistance (ohms)
Rcore = Vnoload^2/(Pnoload/3);
% compute magnetizing reactance (ohms)
Xmag = 1/sqrt((Inoload/Vnoload)^2-1/Rcore^2);

fprintf('\nCore loss resistance = %5.1f ohms', Rcore);
fprintf('\nMagnetizing reactance = %5.2f ohms', Xmag);

% Evaluation of blocked rotor test
% determine equivalent series resistance (ohms)
Re = (Pblock/3)/Iblock^2;
% extract the rotor resistance (ohms)
Rrotor = Re-Rstator;
% calculate series reactance at the test frequency
Xetest = sqrt((Vblock/Iblock)^2-Re^2);
% convert reactance from test to system frequency
Xesyn = Xetest*fsyn/fblock;
% stator and rotor reactances (ohms) assumed equal
Xstator = Xesyn/2;
Xrotor = Xstator;

fprintf('\n\nStator resistance = %4.1f ohms', Rstator);
fprintf('\nStator reactance = %4.1f ohms', Xstator);

fprintf('\n\nRotor resistance = %4.1f ohms', Rrotor);
fprintf('\nRotor reactance = %4.1f ohms', Xrotor);
```

The MATLAB results are:

```
>> InductionMotorParameters

Core loss resistance = 649.0 ohms
Magnetizing reactance = 41.69 ohms
```

THREE-PHASE INDUCTION MOTOR

```
Stator resistance =   0.6 ohms
Stator reactance  =   1.3 ohms

Rotor resistance  =   0.9 ohms
Rotor reactance   =   1.3 ohms
```

EXAMPLE 9.3: Induction motor operation analysis using MATLAB

A three-phase microturbine-driven generator supplies a three-phase induction motor through a transformer and two parallel transmission lines as depicted in the one-line diagram of Figure 9.37. The 60 Hz generator is rated at 80 kVA with a line voltage of 2200 V and per-unit (p.u.) reactance of 1.2. The transformer is rated at 75 kVA with primary and secondary voltages of 2100 and 460 V, respectively. Open-circuit and short-circuit tests are performed on the transformer with the following measurement results:

Test	Power (W)	Voltage (V)	Current (A)
Open circuit (primary side measurement)	1200	2000	1.3
Short circuit (secondary side measurement)	1800	60	52

The transmission lines have a p.u. length impedance of $0.3 + j0.6$ Ω/mile. The length of Line 1 is 2 miles, and Line 2 is 50% longer than Line 1.

The four-pole motor is rated at 80 hp with a line voltage of 450 V. The per phase motor parameters are:

	Resistance (Ω)	Reactance (Ω)
Stator (series)	0.4	0.9
Rotor (series)	0.3	0.9
Magnetizing (parallel)	600	200

In this example, we will:

1. draw the equivalent circuit;
2. calculate the transformer, line, and generator parameters;

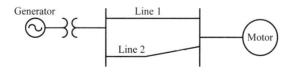

Figure 9.37. One-line diagram of a generator supplying a motor through a transformer and two parallel transmission lines.

3. compute all the currents as a function of motor speed, assuming the generator-induced line-to-ground voltage is twice the rated generator line-to-line voltage;
4. plot the motor torque versus speed characteristics and find the maximum torque;
5. determine the speed and torque when the motor carries the rated load; and
6. graph the motor power versus speed.

The system data are entered into a MATLAB program via the following:

```
% InductionMotor1.m
% Matlab induction motor example 1
% Three-phase generator supplies a three-phase
% induction motor through a transformer and
% two parallel transmission lines
clear all
fsys = 60;         % system frequency (Hz)

% Generator data
Sg = 80e3;         % rating (volt-amps)
Vg = 2200;         % rated line voltage (V)
xg = 1.2;          % reactance (p.u.)

% Transformer data
Str = 75e3;        % rating (volt-amps)
Vprim = 2100;      % primary voltage (V)
Vsec = 460;        % secondary voltage (V)
% Transformer open circuit test measurements at primary side
Pop = 1200;        % power (W)
Vop = 2000;        % line voltage (V)
Iop = 1.3;         % current (A)
% Xformer short circuit test measurements at secondary side
Psh = 1800;        % power (W)
Vsh = 60;          % line voltage (V)
Ish = 52;          % current (A)

% Transmission line data
Zline = 0.3+j*0.6; % ohms/mile
L1 = 2;            % line 1 length (miles)
L2 = L1*1.5;       % line 2 length (miles)
Zline1 = Zline*L1; % line 1 impedance (ohms)
Zline2 = Zline*L2; % line 2 impedance (ohms)

% Motor data
Prated = 80*746;   % rated power (W)
Vrated = 450;      % volts
poles = 4;
Rsta = 0.4;        % stator resistance (ohm)
Xsta = 0.9;        % stator reactance (ohm)
```

THREE-PHASE INDUCTION MOTOR

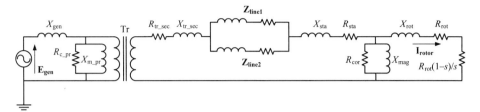

Figure 9.38. Equivalent circuit of a generator supplying a motor through a transformer and two parallel transmission lines.

```
Rrot = 0.3;      % rotor resistance (ohm)
Xrot = Xsta;     % rotor reactance (ohm)
Xmag = 200;      % magnetizing reactance (ohm)
Rcor = 600;      % core loss resistance (ohm)
```

An equivalent circuit diagram of the system is given in Figure 9.38. Referring to the circuit schematic allows us to employ equations from Chapter 7 and Chapter 8, respectively, that compute the transformer and generator parameters. The generator reactance is calculated from the p.u. reactance (x_g) and generator ratings (V_{gen} and S_{gen}):

$$X_{gen} = \frac{V_{gen}^2}{|S_{gen}|} \cdot x_g.$$

The transformer parameters are determined from the open- and short-circuit test results using formulas from Section 7.3.8. Using the line-to-line voltage (V_{op}), current (I_{op}), and three-phase power (P_{op}) of the transformer open-circuit test measurements allows us to compute the core loss resistance (R_{c_pr}) and magnetizing admittance (Y_{m_pr}) from which the magnetizing reactance (X_{m_pr}) may be extracted to form the primary side parameters:

$$R_{c\_pr} = \frac{(V_{op}/\sqrt{3})^2}{P_{op}/3} \quad |Y_{m\_pr}| = \frac{I_{op}}{V_{op}/\sqrt{3}} \quad X_{m\_pr} = \frac{1}{\sqrt{|Y_{m\_pr}|^2 - \frac{1}{R_{c\_pr}^2}}}.$$

According to Figure 9.38 the magnetizing resistance and reactance are in parallel, therefore the primary magnetizing impedance is:

$$Z_{mag\_pr} = \frac{1}{\frac{1}{R_{c\_pr}} + \frac{1}{jX_{m\_pr}}}.$$

The transformer short-circuit test measurements (V_{sh}, I_{sh} and P_{sh}) lead to the calculation of the transformer secondary winding resistance (R_{tr_sec}) and impedance (magnitude). They are then used to find the secondary winding reactance (X_{tr_sec}):

$$R_{tr\_sec} = \frac{P_{sh}/3}{I_{sh}^2} \quad |\mathbf{Z}_{sr}| = \frac{V_{sh}/\sqrt{3}}{I_{sh}} \quad X_{tr\_sec} = \sqrt{|\mathbf{Z}_{sr}|^2 - R_{tr\_sec}^2}$$

```
% Begin analysis; all R, X, Z values in ohms
% Calculation of generator parameter
Xgen = xg*Vg^2/Sg;    % generator reactance per phase (ohm)

% Calculation of transformer parameters
Vop_n = Vop/sqrt(3);   % open circuit line-to-neutral voltage
% Transformer primary magnetizing core per phase values
Rc_pr = Vop_n^2/(Pop/3);      % prim core resistance
Ypr = Iop/Vop_n;              % prim core admittance (S)
Xm_pr = 1/sqrt(Ypr^2 - 1/Rc_pr^2);   % prim core reactance
Zmag_pr = 1/(1/(j*Xm_pr) + 1/Rc_pr); % prim core impedance
% Compute transformer secondary per phase values
Rtr_sec = (Psh/3)/Ish^2;      % sec resistance
Zsr = (Vsh/sqrt(3))/Ish;      % sec impedance
Xtr_sec = sqrt(Zsr^2-Rtr_sec^2);  % sec reactance
```

Next, we calculate the impedances on the secondary side of the transformer. Specifically, we compute the overall motor impedance, the parallel combination of the transmission line impedances, and the overall impedance of the entire secondary. The rotor and magnetizing impedances are:

$$\mathbf{Z}_{rot} = \frac{R_{rot}}{s} + jX_{rot} \quad \mathbf{Z}_{mag} = \frac{1}{\frac{1}{R_{cor}} + \frac{1}{jX_{mag}}}$$

Referring to Figure 9.38, the parallel combination of the rotor and core magnetizing impedances ($\mathbf{Z}_{r\_m}$) are in series with the stator impedance:

$$\mathbf{Z}_{r\_m} = \frac{1}{\frac{1}{\mathbf{Z}_{mag}} + \frac{1}{\mathbf{Z}_{rot}}} \quad \mathbf{Z}_{mot} = R_{sta} + jX_{sta} + \mathbf{Z}_{r\_m}.$$

The two parallel transmission lines are in series with the motor and transformer secondary impedances, and this yields the total impedance on the secondary side of the circuit:

$$\mathbf{Z}_{line\_12} = \frac{1}{\frac{1}{\mathbf{Z}_{line1}} + \frac{1}{\mathbf{Z}_{line2}}} \quad \mathbf{Z}_{sec} = R_{tr\_sec} + jX_{tr\_sec} + \mathbf{Z}_{line\_12} + \mathbf{Z}_{mot}.$$

THREE-PHASE INDUCTION MOTOR

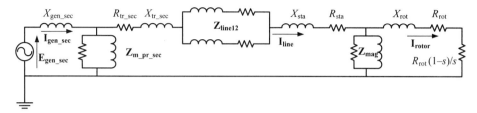

Figure 9.39. Simplified equivalent circuit of a generator supplying a motor through a transformer and two parallel transmission lines.

A slip value of 3% is selected to test the entered formulas:

```
% Select a slip value to test the equations with
s = 0.03;
% Motor impedance
Zrot = Rrot/s + j*Xrot;      % rotor impedance (ohm)
% parallel magnetizing reactance and core loss resistance
Zmag = 1/(1/(j*Xmag)+1/Rcor);   % magnetizing impedance (ohm)
Zr_m = 1/(1/Zmag + 1/Zrot); % rotor and core in parallel
Zmot = Rsta+j*Xsta + Zr_m;   % overall motor impedance (ohm)

% Combine parallel lines 1 and 2 (ohms)
Zline_12 = 1/(1/Zline1 + 1/Zline2);

% Determine overall impedance of secondary (ohm)
Zsec = j*Xtr_sec + Rtr_sec + Zline_12 + Zmot;
```

We now transfer the generator and the transformer magnetizing impedance to the secondary side. This results in the simplified equivalent circuit shown in Figure 9.39. The values of the transferred elements are calculated using:

$$T_{ratio} = \frac{V_{prim}}{V_{sec}} \quad X_{gen\_sec} = \frac{X_{gen}}{T_{ratio}^2} \quad Z_{m\_pr\_sec} = \frac{Z_{mag\_pr}}{T_{ratio}^2}.$$

With the impedance elements transferred to the secondary side, an overall impedance for the entire circuit can be found:

$$Z_{sys} = jX_{gen\_sec} + \frac{1}{\frac{1}{Z_{m\_pr\_sec}} + \frac{1}{Z_{sec}}}.$$

Using the assumption of the generator-induced line-to-ground voltage being twice the rated generator line-to-ground voltage, the induced voltage, which is taken to have a reference angle of zero, is also transferred to the secondary:

$$\mathbf{E}_{gen} = 2V_g \angle 0° \quad \mathbf{E}_{gen\_sec} = \frac{\mathbf{E}_{gen}}{T_{ratio}}.$$

The current from the transferred generator is:

$$I_{gen\_sec} = \frac{E_{gen\_sec}}{Z_{sys}}$$

```
% Transfer transformer primary magnetizing and
% generator impedances to the secondary
Tratio = Vprim/Vsec;            % turns ratio
Xgen_sec = Xgen/Tratio^2;       % generator impedance
Zm_pr_sec = Zmag_pr/Tratio^2;   % magnetizing impedance
% Compute overall system impedance (ohm)
Zsys = j*Xgen_sec + 1/(1/Zm_pr_sec+1/Zsec);
% Assume generator-induced line-to-ground voltage is
% twice the rated generator line-to-line voltage
Egen = 2*Vg;
Egen_sec = Egen/Tratio;         % transferred induced voltage
Igen_sec = Egen_sec/Zsys;       % generator current (A)
```

The line current being supplied to the motor is computed via a current divider from:

$$I_{line} = I_{gen\_sec} \frac{Z_{m\_pr\_sec}}{Z_{m\_pr\_sec} + Z_{sec}}.$$

The current to the rotor is determined using current division between the rotor and motor magnetizing impedance:

$$I_{rotor} = I_{line} \frac{Z_{mag}}{Z_{mag} + Z_{rot}}.$$

```
% Supply or motor line current (amp)
Iline = Igen_sec*Zm_pr_sec/(Zm_pr_sec+Zsec);
% Motor rotor current (amp)
Irotor = Iline*Zmag/(Zmag+Zrot);
```

We now vary the slip from 0.01% to 99.99%, and recompute those quantities which are a function of the slip. By combining Equation (9.6) and Equation (9.7), we create a formula to calculate the motor speed from the slip:

$$n_m = (1-s)n_{sy} = (1-s)\frac{f}{p/2}. \tag{9.73}$$

We obtain the motor torque and mechanical output power as a function of the motor speed using Equation (9.25) and Equation (9.28), respectively:

THREE-PHASE INDUCTION MOTOR

```
% Vary the motor slip; find speed and torque
for k = 1:9999
    s = k/10000;
    Zrot = Rrot/s+j*Xrot;
    Zr_m = 1/(1/Zmag+1/Zrot);
    Zmot = Rsta+j*Xsta+Zr_m;
    Zsec = Zmot+Zline_12+j*Xtr_sec+Rtr_sec;
    Zsys = j*Xgen_sec+1/(1/Zm_pr_sec+1/Zsec);
    Igen_sec = Egen_sec/Zsys;
    Iline = Igen_sec*Zm_pr_sec/(Zm_pr_sec+Zsec);
    Irotor = Iline*Zmag/(Zmag+Zrot);
    % compute total mechanical power output
    Pout(k) = 3*(abs(Irotor))^2*Rrot*(1-s)/s;
    nm(k) = 60*fsys*(1-s)/(poles/2);
    Tmot(k) = Pout(k)/(2*pi*nm(k)/60);
end
```

The motor torque is plotted versus speed, as seen in Figure 9.40:

```
% Plot the motor torque versus speed
plot(nm,Tmot,'LineWidth',1.5)
set(gca,'fontname','Times','fontsize',12);
axis([1200 1800 0 1000])
xlabel('Motor Speed (rpm)');
ylabel('Motor Torque (N·m)');
```

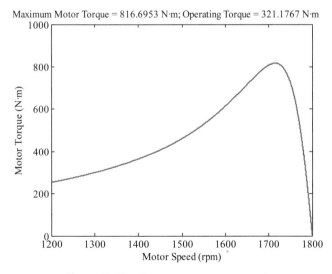

Figure 9.40. Motor torque versus speed.

The MATLAB *max* function is used to find both the maximum torque and the array index of the maximum torque. The array index is then used to determine the corresponding slip and speed at which the maximum torque occurs:

```
% Find the maximum motor torque
[Tmax, imax] = max(Tmot);
smax = imax/10000;
% Determine speed at maximum torque
nmax = 60*fsys*(1-smax)/(poles/2);
fprintf('Operating torque of %g N·m at a speed of %g rpm.',...
    Trated, nrated);
```

The torque at rated power is found by searching through the array of power outputs computed in the *for* loop until P_{rated} is encountered. The maximum and operating torques are printed and placed in the title of the graph of Figure 9.40:

```
% find the torque at rated operating conditions
found = 0;
k = 1;
while (found == 0)
    if Pout(k) > Prated
        found = k;
    end
    k = k + 1;
end
srated = found/10000;
nrated = 60*fsys*(1-srated)/(poles/2);
Trated = Tmot(found);
fprintf('Operating torque of %g N·m at a speed of %g rpm.',...
    Trated, nrated);

title(['Maximum Motor Torque = ',num2str(Tmax),' N·m; ',...
    'Operating Torque = ',num2str(Trated),' N·m']);
figure;
```

Finally, the motor mechanical output power versus speed is plotted in Figure 9.41:

```
% Plot the motor power output versus speed
plot(nm,Pout/1e3,'LineWidth',1.5)
set(gca,'fontname','Times','fontsize',12);
axis([1200 1800 0 150])
xlabel('Motor Speed (rpm)');
ylabel('Motor Mechanical Power Output (kW)');
title(['Rated Motor Power = ',num2str(Prated/1e3),' kW ',...
        'at a speed of ',num2str(nrated),' rpm']);
```

THREE-PHASE INDUCTION MOTOR

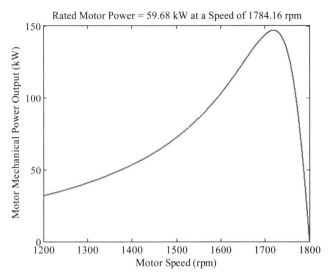

Figure 9.41. Motor mechanical output power versus speed.

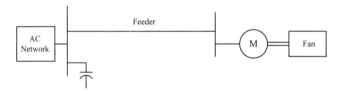

Figure 9.42. One-line diagram of network supplying motor-driven fan through a feeder.

The textual results from the program are:

```
>> InductionMotor1
Maximum torque of 816.695 N·m at a speed of 1714.32 rpm.
Operating torque of 321.177 N·m at a speed of 1784.16 rpm.
```

EXAMPLE 9.4: MATLAB Motor-Driven Fan

A three-phase, wye-connected motor is supplied by a 220 V network through a feeder. The supply power factor is improved by a capacitor connected in parallel with the network. The 60 Hz motor drives a fan. The purpose of this example is to determine the capacitance value necessary to make the power factor equal to unity.

The motor-driven fan is supplied through the feeder by a 220 V ac network as illustrated in the one-line diagram of Figure 9.42. The half-mile long feeder has a p.u. length impedance of $0.25 + j0.55$ Ω/mile. The fan torque–speed characteristic is given by:

$$T_{\text{fan}}(n) = 1.6 \frac{\text{N} \cdot \text{m}}{\text{s}} n + 7 \text{ N} \cdot \text{m}, \tag{9.74}$$

where n is the shaft speed. The six-pole motor is rated at 35 hp with a line voltage of 208 V. The motor stator and rotor impedances are $0.85 + j1.3$ and $0.65 + j1.3$ Ω, respectively, per phase. The motor core loss resistance of 25 Ω/phase and magnetizing reactance of 35 Ω/phase are connected in parallel.

The system data are first entered into a MATLAB M-file:

```
% InductionMotor2.m
% Matlab induction motor example 2
% Motor-driven fan is supplied by a network through a feeder
clear all
fsys = 60;          % system frequency (Hz)

% Network line-to-line voltage
Vnet  =   220;      % volts

% Wye-connected induction motor data
Pmotor = 35;        % hp
Vmotor = 208;       % volts
poles = 6;

Rsta = 0.85;        % stator resistance (ohm)
Xsta = 1.3;         % stator reactance (ohm)
Rrot = 0.65;        % rotor resistance (ohm)
Xrot = Xsta;        % rotor reactance (ohm)
Xmag = 35;          % magnetizing reactance (ohm)
Rcor = 25;          % core loss resistance (ohm)

% Feeder data
Zfeeder = 0.25+j*0.55;       % ohm/mi
Lfeeder = 0.5;               % length (mi)
Zfeed = Zfeeder*Lfeeder;     % feeder impedance (ohm)
```

In this example, we will:

1. draw the equivalent circuit and calculate the system parameters for a slip of 5%;
2. calculate and plot the motor and fan torques versus shaft speed, and determine the maximum torque and the operating point (torque and speed); and
3. compute the capacitance necessary to improve the power factor to unity for a slip of 5%.

The network line-to-neutral voltage is taken to have a reference angle of zero:

$$\mathbf{V}_{net\_ln} = \frac{V_{net}}{\sqrt{3}} \angle 0°.$$

Using the equivalent circuit for the system given in Figure 9.43, we compute the various impedances for 5% slip. These results will also be used later when determining the

THREE-PHASE INDUCTION MOTOR

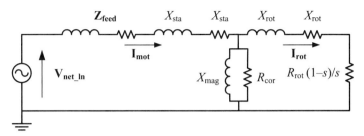

Figure 9.43. Equivalent circuit of network supplying motor-driven fan through a feeder.

required capacitance to achieve a unity power factor. The rotor, magnetizing, and motor impedances are first calculated:

$$Z_{rot} = \frac{R_{rot}}{s} + jX_{rot} \quad Z_{mag} = \frac{R_{cor}jX_{mag}}{R_{cor} + jX_{mag}} \quad Z_{mot} = R_{sta} + jX_{sta} + \frac{1}{\frac{1}{Z_{mag}} + \frac{1}{Z_{rot}}}.$$

From Figure 9.43, the overall system impedance is the series combination of the feeder and motor impedances:

$$Z_{sys} = Z_{feed} + Z_{mot}.$$

The motor current is found using Ohm's law:

$$I_{mot} = \frac{V_{net\_ln}}{Z_{sys}}.$$

The rotor current is determined by applying current division:

$$I_{rot} = I_{mot} \frac{Z_{mag}}{Z_{mag} + Z_{rot}}.$$

With mechanical losses neglected, the mechanical output power is the electrically developed power at the rotor, that is, Equation (9.25):

$$P_{out} = 3|I_{rot}|^2 R_{rot} \frac{1-s}{s}.$$

The input (supply) real power is:

$$P_{in} = \text{Re}(S_{in}) = \text{Re}(3V_{net\_ln}I_{mot}^*).$$

The motor and fan torques are calculated using Equation (9.28) and Equation (9.74), respectively.

```
% Begin analysis
% Slip value for the capacitor calculation is first used
s = 0.05;
Vnet_ln = Vnet/sqrt(3);              % network line-to-neutral
                                     % voltage (V)
nsy = 60*fsys/(poles/2);             % synchronous frequency (Hz)
Zrot = Rrot/s+j*Xrot;                % rotor impedance (ohm)
% parallel magnetizing reactance and core loss resistance
Zmag = Rcor*j*Xmag/(Rcor+j*Xmag);    % magnetizing impedance
                                     % (ohm)
Zmot = Rsta +j*Xsta + 1/(1/Zmag+1/Zrot);   % motor impedance
                                           % (ohm)
Zsys = Zfeed+Zmot;                   % system impedance w/o
                                     % capacitor (ohm)
Imot = Vnet_ln/Zsys;                 % supply current = motor
                                     % current (amp)
Irot = Imot*Zmag/(Zmag+Zrot);        % supply current = motor
                                     % current (amp)
S_in = 3*Vnet_ln*conj(Imot);         % supply complex power (V*A)
P_in = real(S_in);                   % supply real power (W)
Pout = 3*(abs(Irot))^2*(1-s)*Rrot/s; % supply real
                                     % power (W)
nm = nsy*(1-s);                      % motor-fan shaft speed (rpm)
Tmot = Pout/(2*pi*nm/60);            % motor torque (N·m)
Tfan = 7+1.6*nm/60;                  % fan torque (N·m)
```

Now, we vary the slip from 0.0001 to 0.9999 in steps of 0.0001. We calculate the supply and motor currents from which the input power of the supply and motor output power are computed. The motor and fan torques as a function of shaft speed are then determined:

```
% Vary the slip value
for k = 1 : 9999
    s = k/10000;
    Zrot = Rrot/s+j*Xrot;                      % rotor impedance
                                               % (ohm)
    Zmag = Rcor*j*Xmag/(Rcor+j*Xmag);          % magnetizing
                                               % impedance
    Zmot = Rsta+j*Xsta + 1/(1/Zmag+1/Zrot);    % motor impedance
                                               % (ohm)
    Zsys = Zfeed+Zmot;                         % system impedance
                                               % w/o C
```

THREE-PHASE INDUCTION MOTOR

```
    Imot = Vnet_ln/Zsys;              % motor current
                                      % (amp)
    Irot = Imot*Zmag/(Zmag+Zrot);     % rotor current
                                      % (amp)
    Sin = 3*Vnet_ln*conj(Imot);       % supply complex
                                      % power
    Pin = real(Sin);                  % supply real
                                      % power (W)
    Pout = 3*(abs(Irot))^2*(1-s)*Rrot/s;  % motor output
                                      % power (W)
    nm(k) = nsy*(1-s);                % motor-fan speed
                                      % (rpm)
    Tmot(k) = Pout/(2*pi*nm(k)/60);   % motor torque
                                      % (N·m)
    Tfan(k) = 7+1.6*nm(k)/60;         % fan torque (N·m)
end
```

We plot the motor and fan torque–speed curves, as shown in Figure 9.44:

```
% Plot the motor and fan torques versus speed
plot(nm,Tmot,nm,Tfan,'--','LineWidth',1.5);
set(gca,'fontname','Times','fontsize',12);
legend('Motor torque','Fan torque','Location','NorthWest');
xlabel('Shaft speed (rpm)');
ylabel('Torque (N·m)');
```

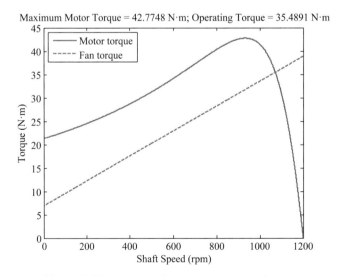

Figure 9.44. Motor and pump torque–speed curves.

We determine the maximum motor torque and the corresponding slip and motor speed:

```
% Find the maximum motor torque
[Tmax, imax] = max(Tmot);
% Slip and speed at maximum torque
smax = imax/10000;
nmax = nsy*(1-smax);
fprintf('Maximum torque of %g N·m at a speed of %g rpm.\n',...
    Tmax,nmax);
```

The operating point occurs where the motor and fan torque curves intersect. After finding that location, the slip and shaft speed are computed:

```
% Find the operating conditions
found = 0;
k = 9999;
while (found == 0)
    if Tfan(k) >= Tmot(k)
        found = k;
    end
    k = k - 1;
end
sop = found/10000;
nop = nsy*(1-sop);
Top = Tmot(found);
fprintf('Operating torque of %g N·m at a speed of %g rpm.\n',...
    Top,nop);
title(['Maximum Motor Torque = ',num2str(Tmax),' N·m; ',...
    'Operating Torque = ',num2str(Top),' N·m']);
```

Finally, a capacitor is connected in parallel with the supply voltage, and the capacitance value necessary to achieve a power factor (*pf*) of unity is calculated. For *pf* = 1, the complex power of the supply is purely real power, and the reactive power supplied by the capacitor must be equal to that absorbed by the circuit prior to the installation of the capacitor, that is,

$$S_{new} = P_{in} \quad Q_{cap} = S_{in} - S_{new} = Q_{in}.$$

The capacitor value can be determined using:

$$X_{cap} = \frac{\left(V_{net}/\sqrt{3}\right)^2}{Q_{cap}/3} = \frac{V_{net}^2}{Q_{cap}} \quad C_{cap} = \frac{1}{\omega X_{cap}}$$

```
% Calculation with parallel capacitor included
pfnew = 1.0;                % supply power factor with capacitor
S_new = P_in/pfnew;         % supply apparent power (V*A)
Q_cap = imag(S_in-S_new);   % supply reactive power (var)
Xcap = Vnet^2/Q_cap;        % capacitor reactance (ohm)
C_cap = 1/(2*pi*fsys*Xcap); % capacitance needed (unit)
fprintf('For unity pf, capacitance of %g µF is required.',...
    C_cap*1e6);
```

The results from MATLAB are

```
>> InductionMotor2
Maximum torque of 42.7748 N·m at a speed of 929.28 rpm.
Operating torque of 35.4891 N·m at a speed of 1068.6 rpm.
For unity pf, capacitance of 113.444 µF is required.
```

9.4. SINGLE-PHASE INDUCTION MOTOR

The single-phase induction machine is the most frequently used motor for refrigerators, washing machines, clocks, drills, compressors, pumps, and so on.

The single-phase motor stator has a laminated iron core with *two windings arranged perpendicularly*. One is the main and the other is the auxiliary winding or *starting winding*, as depicted in Figure 9.45. This means that "single-phase" motors are truly two-phase machines. The motor uses a squirrel cage rotor, which has a laminated iron core with slots. Aluminum bars are molded on the slots and short circuited at both ends with a ring.

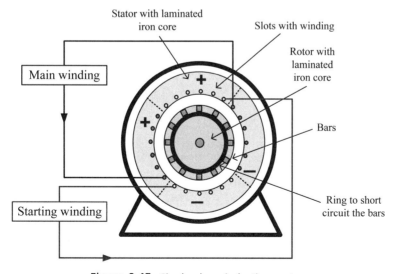

Figure 9.45. Single-phase induction motor.

9.4.1. Operating Principle

The single-phase induction motor operation can be described by two methods:

1. double revolving field theory, and
2. cross-field theory.

Double revolving theory is perhaps the easier of the two explanations to understand. As such, we first present the former theory in detail, and later briefly describe cross-field theory.

9.4.1.1. Double Revolving Field Theory. A single-phase ac current supplies the main winding that produces a pulsating magnetic field. Mathematically, the pulsating field could be divided into two fields, which are rotating in opposite directions. The interaction between the fields and the current induced in the rotor bars generates opposing torque. Under these conditions, with only the main field energized the motor will not start (see the revolving fields in Figure 9.46). However, if an external torque moves the motor in any direction, the motor will begin to rotate. This generates a positive slip related to the forward rotating field:

$$s_{\text{pos}} = (n_{\text{syn}} - n_{\text{mot}})/n_{\text{syn}}. \tag{9.75}$$

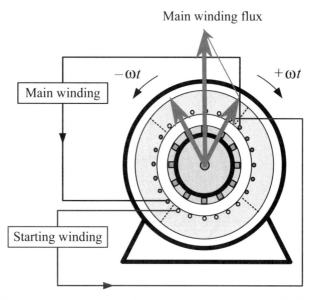

Figure 9.46. Single-phase motor main winding generates two rotating fields, which oppose and counterbalance one another.

SINGLE-PHASE INDUCTION MOTOR

The positive slip is small, 1–5%. Simultaneously, this external torque also generates a negative slip related to the reverse rotating field:

$$s_{neg} = (n_{syn} + n_{mot})/n_{syn}. \tag{9.76}$$

The negative slip is large, 1.95–1.99. Combining Equation (9.75) and Equation (9.76) results in:

$$s_{pos} = 2 - s_{neg}. \tag{9.77}$$

Equation (9.58) showed that the three-phase induction motor starting torque inversely depends on the slip ($T_{start} \propto 1/s$). This implies that a small positive slip (0.01–0.03) generates larger torque than a larger negative slip (1.95–1.99). This torque difference drives the motor, which continues to rotate in a forward direction without any external torque.

Each of the rotating fields induces a voltage in the rotor, which drives current and produces torque. An equivalent circuit can represent each field (i.e., the same circuit used in Figure 9.19 for the three-phase motors). The parameters of the two circuits are the same with the exception of the slip. The two equivalent circuits are connected in series. Figure 9.47 shows the equivalent circuit of a single-phase motor in running condition. The current, power, and torque can be calculated from the combined equivalent circuit, as will be demonstrated in Section 9.4.2. In particular, the input power to the motor is simply:

$$\mathbf{S}_{in} = \mathbf{V}_{sta}\mathbf{I}_{sta}^*. \tag{9.78}$$

However, calculation of the electrically developed power must include the powers dissipated in both the forward and reverse rotating fields:

$$P_{dev} = |\mathbf{I}_{pos}|^2 \frac{R_{rot}}{2} \frac{1-s_{pos}}{s_{pos}} + |\mathbf{I}_{neg}|^2 \frac{R_{rot}}{2} \frac{1-s_{neg}}{s_{neg}}, \tag{9.79}$$

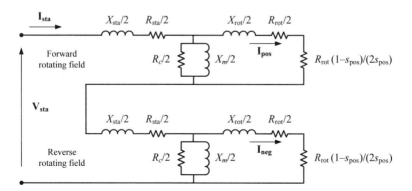

Figure 9.47. Equivalent circuit of a single-phase motor in running condition.

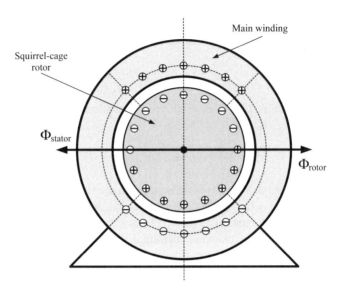

Figure 9.48. Single-phase motor fluxes at standstill (rotor is not moving).

where $\mathbf{I}_{pos}$ and $\mathbf{I}_{neg}$ are the rotor currents flowing through the resistances representing the positive and negative slips, respectively.

9.4.1.2. Cross-Field Theory. The stator of a single-phase motor has a main winding and a starting winding, and the squirrel cage rotor with short-circuited bars, as depicted in Figure 9.45. Figure 9.48 exhibits a motor without the starting winding in standstill. The main winding is supplied by single-phase voltage, which drives current through the winding and generates a pulsating, sinusoidal flux. This flux is represented in Figure 9.48 by the stator flux (Φ_{stator}). The stator flux induces a voltage in the shorted rotor bars. The induced voltage drives a current through the rotor bars. The rotor current produces a rotor flux (Φ_{rotor}), which balances the main winding-generated flux. Figure 9.48 displays the stator and rotor fluxes and the corresponding currents. The interaction between the stator flux and rotor current generates a force that tries to turn the rotor clockwise. Conversely, the interaction between the rotor flux and stator current generates a force in the counterclockwise direction. The two forces balance each other; consequently, the motor will not start.

However, if the motor shaft is moved in either direction, it will begin to rotate in that direction and start to develop power and torque. Figure 9.49 exhibits the conditions when the motor is started and the rotor bars are moving with a speed of v_{mot}.

The revolving of the rotor induces a voltage in the rotor bars. The induced voltage drives currents through the bars. These currents produce pulsating sinusoidal magnetic flux, which is perpendicular to the main winding-generated flux. The direction of the flux can be verified by using the right-hand rule. The generated flux depends on the speed of the motor. There is a phase shift between the two fluxes, as well as a spatial shift of 90°, as shown in Figure 9.49.

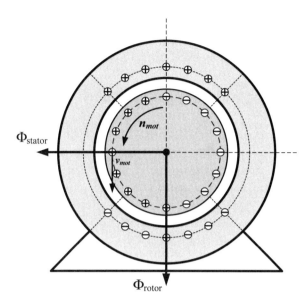

Figure 9.49. Rotation generated flux in a single-phase motor.

It can be proven that the two sinusoidal fluxes generate a rotating flux that drives the motor in the direction that it was originally started. The details of the cross-field theory are too complicated for this book. However, the provided explanation should lead to a better understanding of single-phase motor operation.

9.4.2. Single-Phase Induction Motor Performance Analysis

The single-phase motor performance is investigated using an interactive derivation of the equations. The double revolving field theory analysis is illustrated by a numerical example.

A single-phase one-quarter hp motor is rated to:

$$V_s := 120 \text{ V} \quad P_{rated} := 0.25 \text{ hp} \quad p := 4 \quad f_{syn} := 60 \text{ Hz}.$$

The equivalent circuit parameters are:

$$R_{sta} := 2\,\Omega \quad X_{sta} := 2.5\,\Omega \quad P_{mloss} := 50 \text{ W}$$
$$R_{rot} := 4.1\,\Omega \quad X_{rot} := 2.2\,\Omega$$
$$R_c := 400\,\Omega \quad X_m := 51\,\Omega.$$

The synchronous speed of the motor is calculated with Equation (9.7):

$$n_{syn} := \frac{f_{syn}}{\frac{p}{2}} \quad \text{rpm} := \frac{1}{\text{min}} \quad n_{syn} = 1800 \text{ rpm}.$$

We select the motor speed (n_{mot}) as the independent variable and calculate the positive and negative slips as functions of the motor speed using Equation (9.75) and Equation (9.76). The results are verified by using $n_m := 1760$ rpm:

$$s_{pos}(n_{mot}) := \frac{n_{syn} - n_{mot}}{n_{syn}}, \qquad (9.80)$$

$$s_{neg}(n_{mot}) := \frac{n_{syn} + n_{mot}}{n_{syn}}, \qquad (9.81)$$

where the corresponding numerical values are:

$$s_{pos}(n_m) = 2.22\% \quad s_{neg}(n_m) = 197.8\%.$$

Referring to Figure 9.47, the rotor impedances for positive and negative slips are:

$$Z_{pr}(n_{mot}) := \frac{R_{rot}}{2} + \frac{j \cdot X_{rot}}{2} + \frac{R_{rot}}{2} \cdot \left(\frac{1 - s_{pos}(n_{mot})}{s_{pos}(n_{mot})}\right), \qquad (9.82)$$

$$Z_{nr}(n_{mot}) := \frac{R_{rot}}{2} + \frac{j \cdot X_{rot}}{2} + \frac{R_{rot}}{2} \cdot \left(\frac{1 - s_{neg}(n_{mot})}{s_{neg}(n_{mot})}\right), \qquad (9.83)$$

$$Z_{pr}(n_m) = (92.2 + 1.1j) \, \Omega \quad Z_{nr}(n_m) = (1.04 + 1.10j) \, \Omega.$$

The magnetizing impedance, which is identical for each slip, is:

$$Z_{mag} := \frac{\frac{R_c}{2} \cdot \frac{j \cdot X_m}{2}}{\frac{R_c}{2} + \frac{j \cdot X_m}{2}}, \qquad (9.84)$$

$$Z_{mag} = (3.20 + 25.09j) \, \Omega.$$

The rotor and magnetizing impedances are connected in parallel, as drawn in Figure 9.47. The stator impedance is connected in series with the combined rotor and magnetizing impedances. The motor impedances are therefore:

$$Z_{pmot}(n_{mot}) := \frac{R_{sta}}{2} + \frac{j \cdot X_{sta}}{2} + \frac{1}{\frac{1}{Z_{mag}} + \frac{1}{Z_{pr}(n_{mot})}}, \qquad (9.85)$$

$$Z_{nmot}(n_{mot}) := \frac{R_{sta}}{2} + \frac{j \cdot X_{sta}}{2} + \frac{1}{\frac{1}{Z_{mag}} + \frac{1}{Z_{nr}(n_{mot})}}, \qquad (9.86)$$

$$Z_{pmot}(n_m) = (9.8 + 23.1j) \, \Omega \quad Z_{nmot}(n_m) = (1.95 + 2.33j) \, \Omega.$$

SINGLE-PHASE INDUCTION MOTOR

Applying Ohm's law, the motor current versus motor speed is calculated using:

$$I_{mot}(n_{mot}) := \frac{V_s}{(Z_{nmot}(n_{mot}) + Z_{pmot}(n_{mot}))}, \qquad (9.87)$$

$$I_{mot}(n_m) = (1.79 - 3.89j) \text{ A}.$$

The motor input electrical power is:

$$P_{in\_mot}(n_{mot}) := \text{Re}(V_s \cdot \overline{I_{mot}(n_{mot})}), \qquad (9.88)$$

$$P_{in\_mot}(n_m) = 215.4 \text{ W}.$$

The rotor currents versus speed for the positive and negative slips are calculated by the current division equation:

$$I_{pos}(n_{mot}) := I_{mot}(n_{mot}) \cdot \frac{Z_{mag}}{Z_{mag} + Z_{pr}(n_{mot})}, \qquad (9.89)$$

$$I_{neg}(n_{mot}) := I_{mot}(n_{mot}) \cdot \frac{Z_{mag}}{Z_{mag} + Z_{nr}(n_{mot})}, \qquad (9.90)$$

$$I_{pos}(n_m) = (1.093 + 0.042j) \text{ A} \quad I_{neg}(n_m) = (1.83 - 3.64j) \text{ A}.$$

The electrically developed power is the power dissipated in the load resistances of the two slips:

$$P_{dev}(n_{mot}) := (|I_{pos}(n_{mot})|)^2 \cdot \frac{R_{rot}}{2} \cdot \frac{1 - S_{pos}(n_{mot})}{S_{pos}(n_{mot})} + (|I_{neg}(n_{mot})|)^2 \cdot \frac{R_{rot}}{2} \cdot \frac{1 - S_{neg}(n_{mot})}{S_{neg}(n_{mot})},$$

$$P_{dev}(n_m) = 0.122 \text{ hp}.$$

$$(9.91)$$

This equation has a singularity at synchronous speed (i.e., $n_{mot} = n_{syn}$) since the slips become zero.

The subtraction of the mechanical losses from the developed power gives the mechanical output power:

$$P_{mech}(n_{mot}) := P_{dev}(n_{mot}) - P_{mloss}, \qquad (9.92)$$

$$P_{mech}(n_m) = 0.055 \text{ hp}.$$

Both the mechanical output power and electrically developed power are plotted in Figure 9.50 as a function of motor speed. The graph reveals that both the developed power and the mechanical power have maximum values, which occur at the same speed, and which can be calculated by the Mathcad *Maximize* function. The seed value for this function is $n_m := 1500$ rpm.

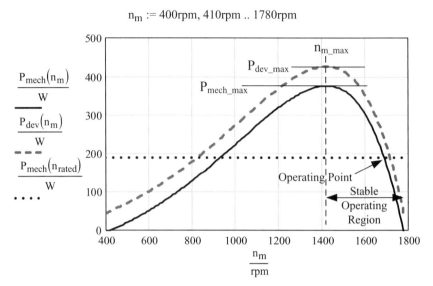

$$n_m := 400\text{rpm}, 410\text{rpm} .. 1780\text{rpm}$$

Figure 9.50. Single-phase motor mechanical output power and electrically developed power versus speed.

$$n_{m\_max} := \text{Maximize}(P_{mech}, n_m) \quad n_{m\_max} = 1420 \text{ rpm}$$
$$P_{mech\_max} := P_{mech}(n_{m\_max}) \quad P_{mech\_max} = 375.9 \text{ W}$$
$$P_{dev\_max} := P_{dev}(n_{m\_max}) \quad P_{dev\_max} = 425.9 \text{ W}.$$

The stable operating range is between the synchronous speed (1800 rpm) and the maximum speed (n_{m_max}) of 1420 rpm. The motor operating speed has to be greater than n_{m_max}. In this region, increasing load reduces the motor speed. The maximum developed power gives guidance on the maximum load that the motor can carry; above this load, the motor will stall. Motor stalling means that the speed is suddenly driven to zero and therefore the motor stops, which consequently produces large current. Most motors are equipped with a thermally operated switch that turns off the current to the windings.

Using the Mathcad *root* function, we compute the motor speed at rated load:

$$n_{rated} := \text{root}(P_{mech}(n_m) - P_{rated}, n_m) \quad n_{rated} = 1689 \text{ rpm}$$
$$P_{mech}(n_{rated}) = 186.4 \text{ W} \quad P_{mech}(n_{rated}) = 0.250 \text{ hp}.$$

The corresponding mechanical power matches the motor specifications and is shown in Figure 9.50.

The motor efficiency is the ratio of output mechanical power to input electrical power:

$$\eta(n_{mot}) := \frac{P_{mech}(n_{mot})}{P_{in\_mot}(n_{mot})}. \tag{9.93}$$

SINGLE-PHASE INDUCTION MOTOR

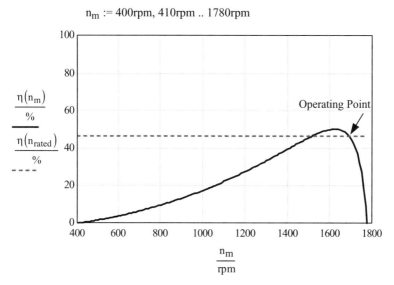

Figure 9.51. Single-phase motor efficiency versus speed.

The efficiency at rated load is $\eta(n_{rated}) = 46.4\%$. In Figure 9.51, the efficiency is plotted as a function of motor speed. Typically, a single-phase motor is less efficient than a three-phase motor.

9.4.2.1. Motor Torque. The angular frequency is calculated using the motor speed:

$$\omega_{mot}(n_{mot}) := 2 \cdot \pi \cdot n_{mot}, \qquad (9.94)$$

$$\omega_{mot}(n_{rated}) = 176.9 \cdot \frac{rad}{sec}.$$

The motor torque is then:

$$T_{mot}(n_{mot}) := \frac{P_{mech}(n_{mot})}{\omega_{mot}(n_{mot})}, \qquad (9.95)$$

$$T_{mot}(n_{rated}) = 1.054 \, N \cdot m.$$

The torque function has singularities at both the synchronous speed and at zero speed. The torque variation with the motor speed is graphed in Figure 9.52.

9.4.2.2. Starting Torque. The single-phase motor starting torque is zero because of the pulsating single-phase magnetic flux. The starting of the motor requires the generation of a rotating magnetic flux similar to the rotating flux in a three-phase motor. Two perpendicular coils that have currents 90° out of phase can generate the necessary rotating magnetic fields that start the motor. Therefore, single-phase motors

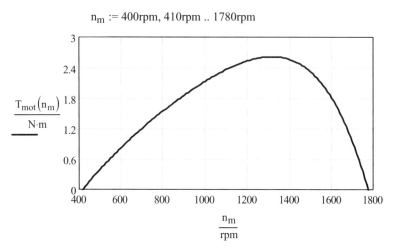

Figure 9.52. Torque–speed characteristic of a single-phase motor.

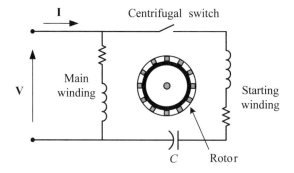

Figure 9.53. Single-phase motor connection.

are built with two perpendicular windings. The phase shift is achieved by connecting a resistance, an inductance, or a capacitance in series with the starting winding.

Figure 9.53 depicts the connection diagram of a motor using a capacitor to generate the starting torque. When the motor reaches the operating speed, a centrifugal switch turns off the starting winding. The centrifugal switch is necessary because most motors use a cheap electrolytic capacitor that can only carry ac current for a short period. A properly selected capacitor produces around 90° phase shift and large starting torque.

A typical torque–speed characteristic of a small motor is presented in Figure 9.54. The capacitor produces high starting torque and rapidly accelerates the motor. As shown in Figure 9.54, when the motor speed becomes 1370 rpm, the centrifugal switch disconnects the starting winding and the capacitor. This reduces the torque. The motor then operates at the main winding torque–speed curve. The acceleration continues until the intersection of the full load torque line and the motor main winding torque–speed characteristic, which is the operating point at full load. During the acceleration, the current is high, which produces considerable heating in both windings.

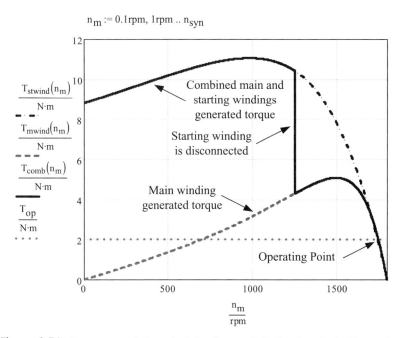

Figure 9.54. Torque–speed characteristic of a small single-phase induction motor.

A less effective but more economical method uses a centrifugal switch and a resistance connected in series with the starting winding, which is built with thin wire in this case. The high resistance in the starting winding circuit produces a current practically in phase with the voltage. The considerable inductance and small resistance in the main winding produces a lagging current. Because of the phase difference, the two windings produce a rotating magnetic field, whose amplitude varies. This is the *split-phase motor*, which has a small starting torque and is used in various domestic appliances.

Using *shaded pole motors* can solve the starting problem of the single-phase induction motor. Figure 9.55 is an illustration of the concept of the shaded pole motor. The motor has two salient poles excited by ac current. Each pole includes a small portion that has a short-circuited winding. This part of the pole is called the *shaded pole*. The main winding produces a pulsating flux that links with the squirrel cage rotor. This flux induces a voltage in the shorted coil. The induced voltage produces a current in the shorted coil. This current generates a flux that opposes the main flux in the shaded pole (the part of the pole that carries the shorted winding). The result is that the flux in the unshaded and shaded parts of the pole will be unequal. Both the amplitude and the phase angle will be different. These two fluxes generate an unbalanced rotating field. The field amplitude changes as it rotates. Nevertheless, this rotating field produces a torque, which starts the motor in the direction of the shaded pole. The starting torque is small but sufficient for fans and other household equipment requiring small starting torque. The motor efficiency is poor but it is inexpensive. Figure 9.56 is a photograph of a shaded pole motor that drives a household fan.

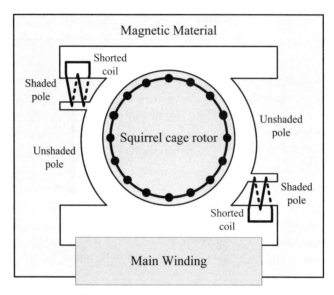

Figure 9.55. Concept of single-phase shaded pole motor.

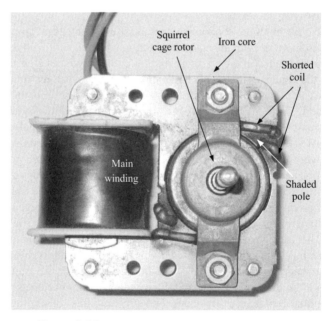

Figure 9.56. Shaded pole motor for household fan.

9.5. INDUCTION GENERATORS

An induction machine can be utilized as a generator when it is driven above the synchronous speed and simultaneously supplied by three-phase voltage. The three-phase voltage source provides magnetizing current or reactive power and maintains the system frequency. As typical examples: a two-pole induction machine must be driven above 3600 rpm and a four-pole machine above 1800 rpm to generate real power. The generator does not need synchronization; rather, the existing network maintains the frequency. The generator-provided power depends on the slip above the synchronous speed. Typically, an induction generator supplies maximum power between slips of 3% and 5% above the synchronous speed. This means that a four-pole generator operates between 1850 and 1890 rpm.

The induction, or asynchronous, generator cannot supply a network alone and is not suitable for black start. A generator with black-start capability can restart a network without other voltage sources. This shortcoming has prevented widespread application of induction generators. However, in recent years, induction generators have been installed on wind turbines and used as microhydrogenerators because the induction machine can operate with variable speed. A gearbox is employed to increase the induction generator speed above the synchronous speed and it provides real power to the network in spite of the wind-produced speed variation. An additional advantage is the simple, strong, and durable construction of the induction machine. An induction generator is very simple since no slip ring, commutator, or bushes are needed.

Operating Principle
The induction generator consists of a three-phase stator connected to a power system and a squirrel cage short-circuited rotor. Inducing three-phase current into the rotor produces a rotating field that induces current in the stator windings. The machine starts to rotate as a motor. In this operation mode, the speed difference induces voltage in the rotor. The prime mover increases the rotor speed.

- Below the synchronous speed, the machine consumes power from the network. When the synchronous speed is reached, no power is exchanged and the rotor current will be practically zero.
- Above the synchronous speed, the rotor conductors are cutting the stator-produced rotating magnetic field lines, which generate voltage and current in the rotor. This is the source of power transfer from the generator to the network. In other words, above the synchronous speed, the speed difference will induce current in the rotor, which generates a magnetic field and drives real power to the network. However, the network maintains the frequency and supplies reactive magnetizing power.

9.5.1. Induction Generator Analysis

The generator operation can be analyzed using the induction machine equivalent circuit of Figure 9.57. The generator operation is analyzed using the typical induction motor

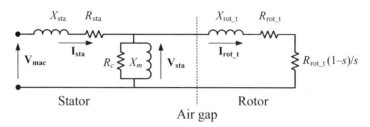

Figure 9.57. Single-phase equivalent circuit of a three-phase induction machine.

data of Section 9.3.5, as the analysis here will be performed independent of whether the induction machine serves as a generator or a motor. The base data for the four-pole, 440 V, 60 Hz three-phase induction machine are:

$$V_{mac} := 440 \text{ V} \quad p := 4 \quad P_{mloss} := 262 \text{ W}$$
$$R_{sta} := 0.44 \text{ }\Omega \quad X_{sta} := 1.25 \text{ }\Omega \quad f_{syn} := 60 \text{ Hz}$$
$$R_{rot\_t} := 0.40 \text{ }\Omega \quad X_{rot\_t} := 1.25 \text{ }\Omega$$
$$R_c := 350 \text{ }\Omega \quad X_m := 27 \text{ }\Omega.$$

The stator and core magnetizing impedances are readily established from:

$$Z_{sta} := R_{sta} + j \cdot X_{sta} = (0.44 + 1.25j) \text{ }\Omega$$
$$Z_m := \frac{R_c \cdot j \cdot X_m}{R_c + j \cdot X_m} = (2.07 + 26.84j) \text{ }\Omega.$$

The rotor impedance is a function of the slip:

$$Z_{rot\_t}(s) := R_{rot\_t} + j \cdot X_{rot\_t} + R_{rot\_t} \cdot \left(\frac{1-s}{s}\right),$$

where a positive slip corresponds to the motor mode while a negative slip value denotes generator operation. Performing nodal analysis yields the following relationship:

$$\frac{V_{mac} - V_{sta}}{Z_{sta}} - \frac{V_{sta}}{Z_m} - \frac{V_{sta}}{Z_{rot\_t}(s)} = 0.$$

Rearranging the equation gives an expression for the stator voltage as a function of the slip:

$$V_{sta}(s) := \frac{V_{mac} \cdot Z_m \cdot Z_{rot\_t}(s)}{Z_m \cdot Z_{sta} + (Z_m + Z_{sta}) \cdot Z_{rot\_t}(s)}.$$

INDUCTION GENERATORS

The stator and rotor currents are:

$$I_{sta}(s) := \frac{V_{mac} - V_{sta}(s)}{Z_{sta}} \quad I_{rot\_t}(s) := \frac{V_{sta}(s)}{Z_{rot\_t}(s)}.$$

Adhering to the passive sign convention, the three-phase complex power and real power for the induction machine are:

$$S_{mac}(s) := 3 \cdot V_{mac} \cdot \overline{I_{sta}(s)} \quad P_{mac}(s) := \operatorname{Re}(S_{mac}(s)).$$

The mechanical input or output power is the developed power less the loss term:

$$P_{mech}(s) := 3 \cdot (|I_{rot\_t}(s)|)^2 \cdot R_{rot\_t} \cdot \left(\frac{1-s}{s}\right) - P_{mloss}.$$

At slips of ±2%, the machine electrical and mechanical powers are:

Generator mode	Motor mode
$P_{mac}(-2\%) = -24.7 \cdot kW$	$P_{mac}(2\%) = 27.5 \cdot kW$
$P_{mech}(-2\%) = -27.9 \cdot kW$	$P_{mech}(2\%) = 24.3 \cdot kW.$

The motor and generator power are calculated and plotted in Figure 9.58. The graph shows that in the motor mode (s > 0), the machine supply power is larger than the mechanical power ($P_{mac} > P_{mech}$), whereas in generator mode (s < 0), the mechanical input power is larger than the generated output electric power ($|P_{mech}| > |P_{mac}|$). In the

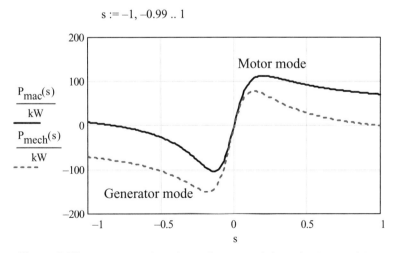

Figure 9.58. Electrical and mechanical powers of the induction machine.

motor mode, the larger input supply power produces starting power at s = 1. In actuality, the induction machine operating range is rather narrow at around s = ±4%.

9.5.2. Doubly Fed Induction Generator

The doubly fed induction motor and generator were originally developed nearly a century ago. Resistances connected to the rotor through slip rings control the machine speed. The doubly fed induction machine attributes include:

1. excellent starting torque for high inertia loads,
2. low starting current compared to the squirrel cage induction motor,
3. resistance changeable speed from over 50% to 100% full speed, but
4. higher maintenance of brushes and slip rings compared to a squirrel cage motor.

However, due to rapid development of electronic motor speed control for the squirrel cage induction motor, as well as the high maintenance cost of the slip rings, the doubly fed induction machine use subsequently declined and the machine became a technology relic.

More recently, the increased use of wind power worldwide and the variable speed of wind turbines led to the reintroduction of the doubly fed induction generator. Particular advantages of the electronically controlled doubly fed machine include its ability to generate or consume reactive power and its capability to transmit power even when wind speed variation alters the turbine speed away from the network synchronous frequency. Electronic control permits the generator to improve power system stability in the event of grid disturbances.

Electronically Controlled Doubly Fed Induction Machine
The doubly fed induction generator is a wound rotor induction machine, which has three-phase windings, described previously in connection with Figure 9.11, on *both* the rotor and the stator. The rotor windings are connected to three slip rings. The rotor of this machine can be loaded by three resistances or can be supplied by current through three brushes connected to the slip rings.

Figure 9.59 exhibits the rotor and the slip rings of a doubly fed machine. The photograph shows that the laminated rotor iron core has slots containing coils forming a three-phase circuit. The ends of the three-phase windings are welded to the slip rings made out of copper rings. The stator and rotor windings are similar, but in most designs, the rotor winding has more turns than the stator. Consequently, the rotor voltage is higher and the rotor current is lower than the stator voltage and current.

The basic connection diagram of a doubly fed induction machine with an electronically controlled rotor is shown in Figure 9.60. The wind turbine drives a gearbox, which increases the induction generator rotor speed. The stator is supplied at the power frequency of the local grid. The three-phase voltage is rectified by a controlled rectifier, which powers an inverter that supplies the induction generator rotor with variable frequency voltage and current in such a way that the generator produces controllable

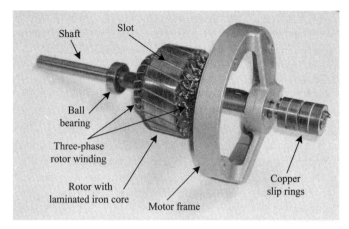

Figure 9.59. Three-phase doubly fed or wound rotor induction machine rotor with slip rings.

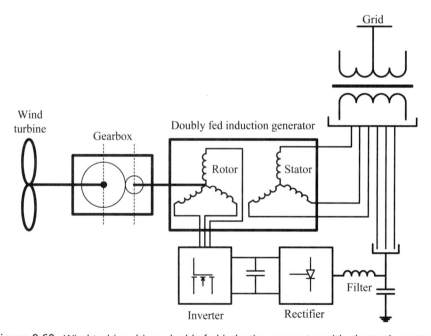

Figure 9.60. Wind turbine-driven doubly fed induction generator with electronic control.

real and reactive power to the grid while the wind turbine speed varies. In addition, this generator can provide transient stability support for the grid in case of severe disturbances.

The rectifier–inverter circuit is equipped with direct axis vector control or with direct torque control. The three-phase inverter uses insulated gate bipolar transistor (IGBT) switches and operates in pulse width modulation (PWM) mode (these topics are presented in Chapter 11). The inverter is typically designed to carry about 25% of the total power, which is around a few megawatts. The PWM drive reduces the filtering requirement and the harmonic pollution of the main grid.

9.6. CONCEPT OF MOTOR PROTECTION

Overheating-produced thermal stress causes the majority of induction motor failures. In most cases, the high temperature destroys the stator winding insulation, which can initiate a ground fault or a short circuit between winding conductors such as an interturn short circuit. All the internal faults produce large short-circuit current typically 8–10 times the rated current, which causes permanent damage of the motor and requires repair. The extremely severe overheating can also produce deformation, and even melting, of the squirrel cage rotor conductors, or it can damage the motor bearing, which may produce misalignment of the rotor and mechanical damage of components.

Typical operating conditions that cause overheating include:

- overloading;
- asymmetrical loading, which can be the result of a loose connection, loss of one phase, distorted supply voltage, or ground fault-caused asymmetrical current;
- repeated startup: typically the peak startup current is six times the rated current, but it lasts only for a few seconds; however, repetitive startup overheats the motor;
- undervoltage operation, which increases the motor current to maintain the mechanical load;
- overvoltage operation;
- motor overload-caused stalling: the severe overload stalls the motor, which increases the current to the locked rotor current level; and
- mechanical jam-caused starting failure, which also increases the current to the locked rotor current level. The locked rotor current level is 6–10 times the load current, which causes burning of the insulation.

Most motors are equipped with thermal temperature sensor protection. In small low-voltage motors, a small bimetallic element is connected in series with the stator winding. Typically, a bimetal switch is placed in each phase of the motor winding at the end. Owing to the differences in the thermal expansion coefficients of two coupled metal strips (e.g., steel and copper), elevated temperature bends the bimetal strip, which in turn interrupts the stator current and switches off the motor temporarily. After a cooling-down period, the bimetal closes the circuit and permits restarting the motor. Figure 9.61 displays a motor stator that employs an automatic reset thermal overload protector on the winding.

Large motors use resistance temperature detectors (RTDs) that are embedded in the stator windings where hot spots are expected. The RTD resistance varies almost

CONCEPT OF MOTOR PROTECTION

Figure 9.61. AC motor stator with on-winding thermal overload protector (courtesy of Bodine Electric Company).

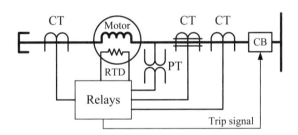

Figure 9.62. Large motor protection.

linearly with temperature. The RTD supplies a thermal relay, which switches off the motor when overheating occurs. In addition, the RTD can be used to detect loss of ventilation, loss of cooling, or high ambient temperature.

Small low-voltage motors are protected with a magnetic or electronic starter/protection relay, which is equipped with an adjustable current sensor. Overcurrent initiates the sensor, which in turn activates the magnetic switch to switch off the motor. A digital electronic starter can regulate the motor voltage (30–80% of the line-to-line voltage) to accelerate or decelerate the motor and eliminate the large starting current. The acceleration and deceleration ramp is typically adjustable between 1 and 20 seconds. In addition, the unit may provide overload, overcurrent, locked rotor, phase loss, and phase sequence protection.

Large motors are protected with combined overcurrent, ground fault, and differential protection similar to the manner in which synchronous generators are protected. Figure 9.62 presents a large motor that is supplied through a circuit breaker and uses one set of current transformers (CTs) for differential protection, three CTs for overload and overcurrent protection, and a potential transformer (PT) for directional protection. The protection relays that may be used for a large motor include:

- stator differential,
- thermal overload,
- RTD biased thermal overload,
- stator RTD sensor,
- bearing RTD sensor,
- mechanical jam,
- instantaneous overcurrent,
- time overcurrent,
- breaker failure,
- current unbalance,
- phase reversal,
- directional power and overcurrent protection,
- under- and overvoltage, and
- under- and overfrequency.

This list proves that depending on the role of the motor, a large number of protection relays can be employed. However, in most cases, the differential, overcurrent, and RTD relays are sufficient for motor protection.

9.7. EXERCISES

1. Describe the construction of the stator of a three-phase induction motor. Draw a sketch.
2. Describe the construction of the rotor of a three-phase induction motor. Draw a sketch.
3. Describe the squirrel cage rotor. Draw a sketch.
4. What is the wound rotor? Where is it used?
5. Describe the construction of the stator of a single-phase induction motor. Draw a sketch.
6. Describe the construction of the rotor of a single-phase induction motor. Draw a sketch.
7. What is the role of the capacitor and centrifugal switch in single-phase motor operation?
8. When does the centrifugal switch operate?
9. How is a small motor cooled?
10. Describe the generation of the rotating magnetic field in a three-phase induction motor.

11. Describe the generation of the torque in a squirrel cage motor.
12. What is the slip? Present typical values and an equation.
13. What is the synchronous speed? Give an equation and define the parameters.
14. Why is the motor speed less than the synchronous speed?
15. Describe the three-phase motor equivalent circuit and identify the parameters.
16. What is the developed power? Explain and provide an equation.
17. Explain the method for calculating the output power.
18. Explain the power balance in an induction motor.
19. Sketch the three-phase motor torque–speed characteristic and identify the operating range.
20. List the losses in an induction motor.
21. What is the blocked-rotor test?
22. What parameters can be calculated from the blocked-rotor test results?
23. What is the no-load test?
24. What parameters can be calculated from the no-load test results?
25. Explain the problems that starting an induction motor can produce.
26. What is the slip when a three-phase motor starts?
27. Present the concept of motor speed control.
28. Explain the operation of a single-phase motor.
29. Draw the equivalent circuit of a single-phase induction motor and explain the parameters.
30. Sketch the single-phase motor torque–speed characteristic and identify the operating range.
31. What is the starting torque of a single-phase motor? Why?
32. Explain the methods used to start a single-phase motor.
33. Show the effect of capacitive starting on the torque–speed characteristic.
34. List typical applications of three-phase and single-phase motors.

9.8. PROBLEMS

Problem 9.1

For an eight-pole, three-phase motor with 48 total slots on the stator, determine (a) the number of slots per phase and (b) the angular distance between phases.

Problem 9.2

A three-phase motor rated 25 hp, 480 V, operates with a power factor of 0.74 lagging and supplies the rated load. The motor efficiency is 96%. Calculate the motor input power, reactive power, and current.

Problem 9.3

A three-phase motor rated 25 hp, 480 V, operates with a power factor of 0.7 lagging and supplies the rated load. The motor efficiency is 96%. Calculate the capacitor needed to improve the motor power factor to 0.95 (lagging).

Problem 9.4

Calculate the starting current of an induction motor if the four-pole, 480 V motor data per phase are:

	Resistance (Ω)	Reactance (Ω)
Rotor (series)	0.125	0.4
Stator (series)	0.1	0.35
Magnetizing (parallel)	65	35

Problem 9.5

Calculate the input current of an induction motor if the motor speed is 1720 rpm. Use the 60 Hz motor from Problem 9.4.

Problem 9.6

An induction motor drives a pump. The motor output power is 23 hp at a speed of 1150 rpm. Calculate the required supply voltage and motor current. Use the motor data from Problem 9.4.

Problem 9.7

A three-phase 60 Hz induction motor rated at 140 hp with line voltage of 240 V is operating at a speed of 3510 rpm and a slip of 2.5% when driving a load. Determine: (a) the synchronous speed, (b) the number of stator poles, (c) the rotor frequency, and (d) the rated current if the motor power factor is 0.8 lagging.

Problem 9.8

An eight-pole, three-phase 60 Hz induction motor with a line voltage of 230 V has the following per-phase parameters:

	Resistance (Ω)	Reactance (Ω)
Rotor (series)	0.2	0.4
Stator (series)	0.3	0.5
Magnetizing (parallel)	120	15

The rotational loss (mechanical loss, or friction and winding loss) is 400 W.

(a) Draw the single-phase equivalent circuit. (b) Calculate the motor speed if the slip is 2.5%. (c) Calculate the input current and input power factor as a function of slip, which varies between 0% and 100%. Verify the values if the slip is 2.5%. (d) Plot the developed torque and the output torque versus speed. Determine the speed when the output torque is 150 N·m. (e) Plot the efficiency versus speed. Determine the maximum efficiency and the corresponding speed.

Problem 9.9

A three-phase, wye-connected 60 Hz induction motor is rated at 50 hp with a line voltage of 240 V. The four-pole motor operates at rated conditions and a 6% slip. The friction and winding losses are 300 W. (a) Calculate the shaft speed in rpm. (b) Determine the air-gap power, the developed torque, and the torque delivered to the load.

Problem 9.10

A three-phase, four-pole 60 Hz induction motor rated at 100 hp with a line-to-line voltage of 2100 V is tested with the no-load test and blocked-rotor test, obtaining the following data:

No-Load Test	Blocked-Rotor Test
60 Hz	17 Hz
2100 V	450 V
5 A	3.5 A
3600 W	120 W

The stator resistance is 1.4 Ω per phase. (a) Calculate the parameters of the single-phase equivalent circuit. (b) Draw the equivalent circuit. (c) Plot the torque, output power, and efficiency versus slip, and determine these parameters if the motor slip is 10%.

Problem 9.11

A three-phase 60 Hz induction motor rated at 6 hp and a line voltage of 240 V operates at full load. The two-pole motor has rotor copper losses of 500 W and mechanical losses of 400 W. Determine: (a) the mechanical power developed, (b) the air-gap power, (c) the shaft speed, and (d) the shaft torque.

Problem 9.12

A three-phase, wye-connected 60 Hz induction motor rated at 40 hp with a line voltage of 440 V is operating at a slip of 5%. The core loss and the combined friction and winding losses at this load are 300 and 150 W, respectively. The four-pole motor parameters, in ohms per phase, are the following: magnetizing reactance, 22 Ω; stator impedance, $0.4 + j1.4$ Ω; and rotor impedance, $0.35 + j1.5$ Ω. The motor is supplied by a network with a voltage of 445 V. The network short-circuit current is 800 A (inductive).

Determine: (a) the line current and power factor and motor terminal voltage (line-to-line, absolute value); (b) the real and reactive input powers; (c) the air-gap power; (d) the mechanical power and torque developed; (e) the shaft mechanical power (hp) and torque; and (f) the efficiency. *Hint*: Convert the core loss to an equivalent resistance R_c and substitute it into the equivalent circuit.

Problem 9.13

A three-phase 60 Hz induction motor is rated at 20 hp with a line-to-line voltage of 240 V. The per-phase parameters of the equivalent circuit of the six-pole motor are given as follows. The stator impedance is $0.25 + j0.38$ Ω and the rotor impedance is $0.2 + j0.3$ Ω. The motor magnetizing reactance is 32 Ω. The combined rotational losses (friction and winding losses plus core loss) are 620 W. A network supplies the motor through a transformer. The network voltage is 13.8 kV and its inductive short-circuit current is 1000 A. The transformer is rated at 20 kW with primary and secondary voltages of 13.8 kV and 250 V, respectively, and a p.u. reactance of 0.12.

(a) Draw the one-line diagram and equivalent circuit. For a motor speed of 1120 rpm, determine: (b) the slip and power factor values, (c) the output torque, (d) the efficiency, and (e) the starting current and torque. (f) If the motor is connected to a variable load, plot the curve of the torque–speed characteristics. Obtain the maximum torque and the related slip, as well as the starting torque from the diagram.

Problem 9.14

A three-phase wye-connected induction motor is rated at 380 hp with a line-to-line voltage of 480 V. The results obtained for the four-pole motor from the no-load and blocked-rotor tests are as follows:

No-Load Test	Blocked-Rotor Test	Stator Resistance Measurement
60 Hz	250 Hz	20 V dc
480 V	48 V	55 A dc
3.8 A	29 A	
850 W	1150 W	

The motor drives a pump. The pump torque–speed characteristic is:

$$T(n) = 60 \text{ N} \cdot \text{m} + 0.2 \text{ N} \cdot \text{m} \cdot \text{s}^2 n^2.$$

Verify the calculation using a motor speed of 1750 rpm.

(a) Determine the parameters of the single-phase equivalent circuit. Draw the equivalent circuit. (b) Calculate the stator and rotor currents versus motor speed. (c) Calculate the motor torque versus motor speed. (d) Determine the operating point, where the motor torque characteristic intersects the pump torque curve. Determine the operating speed.

Problem 9.15

A single-phase, 1/2 hp induction motor is supplied by a 120 V network. The six-pole motor data are: (1) a rotor resistance of 4.2 Ω and reactance of 3.4 Ω after being transferred to the stator side, (2) a stator impedance of 2.3 + j2.9 Ω, and (3) a parallel magnetizing resistance and reactance of 550 and 55 Ω, respectively.

(a) Draw the equivalent circuit and calculate the rated current if the power factor is 0.6 lagging. (b) Calculate and plot the stator and rotor currents versus motor speed. Determine the motor speed at rated current. (c) Calculate and plot the power factor and determine the speed at 0.6 lagging power factor. (d) Calculate the input and output powers versus motor speed, and plot the output power. Determine the speed at half-load. (e) Calculate and plot the output torque versus motor speed. Determine the maximum torque and the corresponding speed.

Problem 9.16

A four-pole, single-phase 3/4 hp motor is supplied by 110 V. The motor operates at a speed of 1750 rpm. The equivalent circuit data are as follows: The stator resistance and reactance are 1.55 and 2.6 Ω, respectively. The magnetizing resistance and reactance are 58 and 62 Ω, respectively. The rotor resistance and reactance are 3.0 and 2.9 Ω, respectively. The friction and winding losses are 12 W. Calculate: (a) the stator and forward and reverse rotor currents, (b) the output and input powers, and (c) the motor efficiency.

10

DC MACHINES

The direct current (dc) machine can be operated as a motor or as a generator. However, it is most often used for a motor. The major advantages of dc machines are the easy speed and torque regulation. However, their application is limited to mills, mines, and trains. As examples, trolleys and underground subway cars may use dc motors. Earlier automobiles were equipped with dc dynamos to charge their batteries. Even today, the starter for an internal combustion engine is a series dc motor. In the past, dc generators were utilized to excite synchronous generators. However, the recent development of power electronics has reduced the use of dc motors and generators. Electronically controlled alternating current (ac) drives are gradually replacing the dc motor drives in factories. Nevertheless, a large number of dc motors are still employed by industry, and several thousand are sold annually.

10.1. CONSTRUCTION

Figure 10.1 shows an illustration of the general arrangement of a dc machine. Unlike an ac motor, the stator of the dc motor has poles, which are excited by dc current to produce magnetic fields. The poles may have two dc excitation windings: one connected

---

Electrical Energy Conversion and Transport: An Interactive Computer-Based Approach, Second Edition.
George G. Karady and Keith E. Holbert.
© 2013 Institute of Electrical and Electronics Engineers, Inc. Published 2013 by John Wiley & Sons, Inc.

CONSTRUCTION

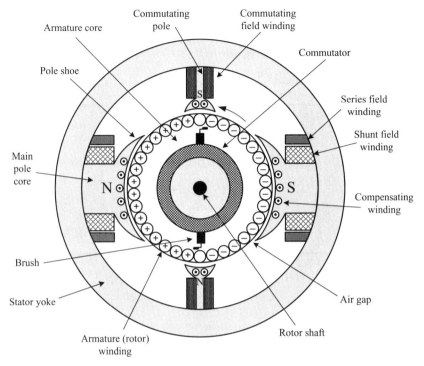

Figure 10.1. General arrangement of a dc motor.

in parallel and another connected in series with the rotor. In the neutral zone, in the middle between the poles, commutating poles are placed to reduce sparking of the commutator. The commutator is a rotating electrical switch that reverses current direction. The commutating poles are also supplied by dc current. Compensating windings are mounted on the main poles. These short-circuited windings damp rotor oscillations. The poles are positioned on an iron core that provides a closed magnetic circuit.

The motor housing supports the iron core, the brushes, and the bearings. The rotor has a ring-shaped laminated iron core with slots. Coils with several turns are placed in the slots. The distance between the two legs of the coil is about 180 electric degrees. The ends of each coil are connected to a commutator segment. The commutator segments connect the coils in series. The commutator consists of insulated copper segments mounted on an insulated tube. Two brushes are pressed to the commutator to permit current flow. The brushes are placed in the neutral zone, where the magnetic field is close to zero, to reduce arcing.

The *commutator* switches the current from one rotor coil to the adjacent coil, which requires the interruption of the coil current. The sudden interruption of an inductive current generates high voltages ($v = L\,di/dt$). The high voltage produces flashover and

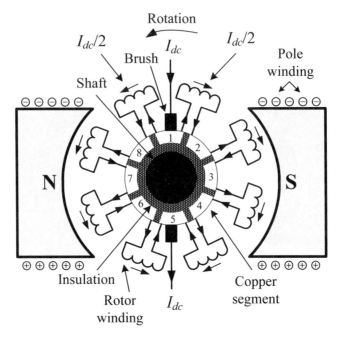

Figure 10.2. Commutator with the rotor coils connections for a two-pole dc motor.

arcing between the commutator segment and the brush. The arcing can be lessened by reducing the magnetic field at the brush site.

Figure 10.2 depicts the commutator and the rotor coil connections for a hypothetical rotor with eight coils. The commutator switches the location of the current entry and exit as the rotor turns. As an example, the motor current (I_{dc}) flows into the rotor at segment 1 and flows out of the rotor from segment 5 in Figure 10.2. Further clockwise rotation switches the entry and exit points to segments 8 and 4, respectively, when the next commutator segment slides to the brush. The brushes divide the rotor into two parallel electrical paths. The current flows in opposite directions in the two halves of the rotor separated by the brushes.

An actual commutator is exhibited in Figure 10.3. In this commutator, the segments are made out of copper, and mica insulation is placed between the segments. The coil endings are welded to the copper segments.

Figure 10.4 shows a picture of the stator of a small dc machine with four main poles. An iron frame supports the iron core. The ring-shaped iron core has poles with an excitation winding. The winding is secured by tape. Large dc motors have commutating interpoles between the main poles. These interpoles reduce the magnetic field in the neutral zone, where the brushes are located. This eliminates arcing of the commutator.

The rotor of a dc motor with the commutator is seen in Figure 10.5. The rotor has a laminated iron core with slots. Coils are placed in the slots. The photograph shows

CONSTRUCTION 619

Figure 10.3. Details of the commutator of a dc motor.

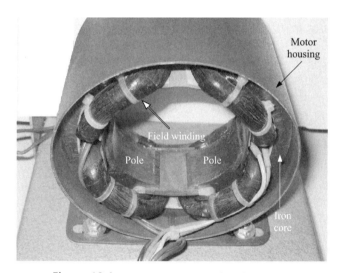

Figure 10.4. DC motor stator with poles visible.

the interconnection between the coil endings, the commutator, and the ball bearing supported shaft.

The rotor, commutator, and brush assembly are presented in Figure 10.6. A fan attached to the rotor shaft at the right-hand side cools this motor. The graphite brush is pushed to the commutator by a spring. The square-shaped brush holder attached to the housing secures the graphite brush. This motor has two brush assemblies.

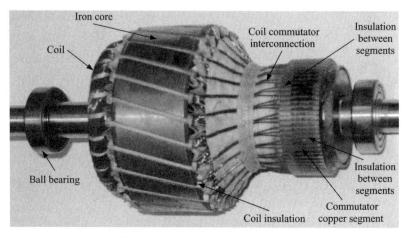

Figure 10.5. Rotor of a dc motor.

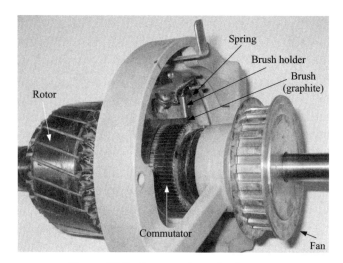

Figure 10.6. Cutaway view of a dc motor.

10.2. OPERATING PRINCIPLE

10.2.1. DC Motor

In a dc motor, the stator poles are supplied by dc excitation current, which produces a dc magnetic field. The rotor is supplied by dc current through the brushes, commutator, and coils. The interaction between the stator magnetic field and rotor current generates a force that drives the motor. This process is explained in detail using a simplified dc machine.

OPERATING PRINCIPLE

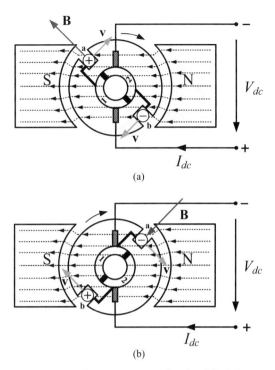

Figure 10.7. Force generation and commutation of a simplified dc motor. (a) Rotor current flow from segment 1 to 2 (from slot **a** to **b**); (b) Rotor current flow from segment 2 to 1 (from slot **b** to **a**).

Figure 10.7 shows a sketch of a simplified machine with two poles and only one rotor coil. The coil has two sides: one is placed in slot **a**, the other in slot **b**. The ends of this coil are connected to a two-segment commutator and supplied by dc current (I_{dc}) through the two brushes. In particular, the coil end located in slot **a** is connected to segment 1, and the coil end at slot **b** is attached to segment 2. The N–S poles have coils, which are also supplied by dc current, and they produce the magnetic field. For simplicity, these pole excitation coils are not shown in Figure 10.7, but do appear in Figure 10.2.

Figure 10.7 demonstrates that the magnetic field lines enter into the rotor from the north pole (N) and exit toward the south pole (S). The stator poles generate a magnetic field that is perpendicular to the current carrying conductors (a single coil in this diagram) in the rotor. The interaction between the magnetic field and the rotor coil current produces a Lorentz force, which is perpendicular to both the magnetic field and conductor. This force drives the motor. The force on an individual conductor is $\vec{F} = I_{dc}\vec{\ell}_{cond} \times \vec{B}$. Fleming's left-hand rule for motors, as illustrated in Figure 10.8, points the second finger in the direction of current flow and the index finger with the magnetic field direction, such that the thumb then specifies the resultant force (and rotational motion) direction.

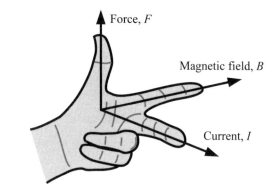

Figure 10.8. Fleming's left-hand rule for motors.

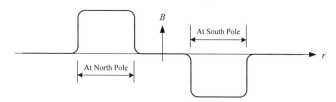

Figure 10.9. Flux density distribution due to stator field winding-generated magnetic field.

The generated force turns the rotor clockwise until the coil reaches the neutral point between the poles. At this point, the magnetic field perpendicular to the rotor current becomes practically zero together with the force. However, inertia drives the motor beyond the neutral zone, where the direction of the magnetic field reverses. To avoid the reversal of the force direction, the commutator changes the current direction, which maintains the clockwise rotation. Before reaching the neutral zone, the current enters in segment 1 and exits from segment 2, as seen in Figure 10.7a. Therefore, current enters the coil end at slot **a** and exits from slot **b** during this stage. After passing the neutral zone, the current enters segment 2 and exits from segment 1, as depicted in Figure 10.7b. This reverses the current direction through the rotor coil, when the coil passes the neutral zone. The result of this current reversal is the maintenance of the rotation.

An actual dc machine has several rotor coils and commutator segments, which assures smooth continuous rotation. The motor rotation direction can be reversed by switching the polarity of the stator field winding voltage. This voltage polarity reversal swaps the north and south poles. Furthermore, the rotation direction can also be changed by reversing the rotor current direction (i.e., the rotor voltage polarity).

The field winding on the stator poles produces a magnetic field that interacts with the rotor current and drives the motor. Figure 10.9 shows a sketch of the approximate distribution of this field in the air gap. It can be seen that the flux density is practically constant in the air gap. The armature (rotor) current also produces a flux, which is more

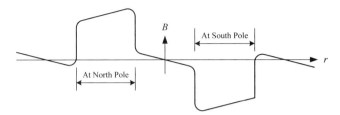

Figure 10.10. Armature current effect on magnetic flux density distribution.

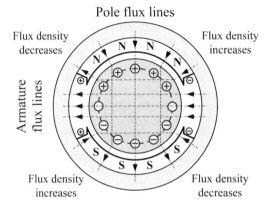

Figure 10.11. Effect of armature current-generated magnetic flux on flux density.

or less perpendicular to the main flux. The approximate overall stator and rotor flux distribution is shown in Figure 10.10. Figure 10.11 demonstrates that the armature-generated flux is added to the pole flux on the right-hand side of each pole and subtracted on the left-hand side. This distorts the magnetic field density in the air gap, as viewed in Figure 10.10. The increase of flux density at the right-hand side may cause saturation, which results in the reduction of the induced voltage. The armature current-caused reduction of induced voltage is called *armature reaction*. In most cases, this effect is small. Consequently, in this book, the armature reaction is neglected.

10.2.2. DC Generator

The simplified machine of Figure 10.12 is used to explain dc generator operation. The N–S poles produce a dc magnetic field and the rotor coil turns in this field. However, in the case of a generator, a turbine or other machine drives the rotor. The conductors in the rotor slots cut the magnetic flux lines, which induce voltage in the rotor coils. The coil has two sides: one is placed in slot **a**, the other in slot **b**. As demonstrated in Figure 10.13, Fleming's right hand rule for generators points the index finger with the magnetic field direction and the thumb toward the rotation direction, such that the second finger then determines the direction of current flow.

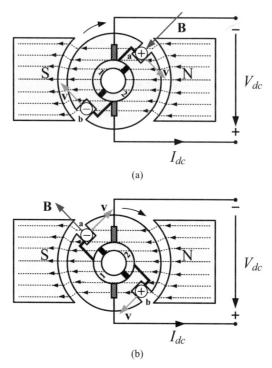

Figure 10.12. DC generator operation. (a) Rotor current flow from segment 1 to 2 (from slot **a** to **b**); (b) Rotor current flow from segment 2 to 1 (from slot **b** to **a**).

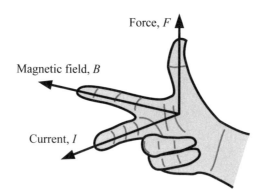

Figure 10.13. Fleming's right-hand rule for generators.

Figure 10.12 illustrates the principle of the induced voltage generation and commutation process. In Figure 10.12a, the conductors in slot **a** are cutting the magnetic field lines entering into the rotor from the north pole, and the conductors in slot **b** are cutting the field lines exiting from the rotor to the south pole.

The cutting of the magnetic field lines generates voltage in the conductors. The voltages generated in the two sides of the coil are added. The induced voltage in the coil is proportional with the magnetic flux derivative, that is, $E = N\, d\Phi/dt$. The flux derivative can be expressed with the motor rotational speed. The induced generator voltage is:

$$E_g = 2N_r B \ell_g v, \qquad (10.1)$$

where

N_r is the number of turns in the rotor coil;
B is the magnetic flux density in the gap;
ℓ_g is the length of the rotor; and
v is the velocity of the rotor.

The induced voltage is connected to the generator terminals through the commutator and brushes. In Figure 10.12a, the induced voltage in **b** is positive and in **a** is negative. The positive terminal is connected to commutator segment 2 and to the conductors in slot **b**. The negative terminal is connected to segment 1 and to the conductors in slot **a**.

However, when the coil passes the neutral zone, conductors in slot **a** are then moving toward the south pole and cut flux lines exiting from the rotor, and conductors in slot **b** cut the flux lines entering the rotor from the north pole. This changes the polarity of the induced voltage in the coil. The voltage induced in **a** is now positive, and in **b** is negative. However, simultaneously, the commutator reverses its terminals, which assures that the output voltage (V_{dc}) polarity is unchanged. In Figure 10.12b, the positive terminal is connected to commutator segment 1 and to the conductors in slot **a**. The negative terminal is connected to segment 2 and to the conductors in slot **b**.

The stator poles are shaped to produce a sinusoidal flux density distribution in the air gap. This assures sinusoidal induced voltage in the rotor winding. The commutator rectifies the generated voltage. An actual dc generator has a large number of slots, which results in constant and ripple-free induced voltage.

10.2.3. Equivalent Circuit

In a generator, the magnetic field produced by the stator poles induces a voltage in the rotor (or armature) coils when the generator is rotated. The dc field current of the poles generates a magnetic flux (Φ_{ag}). This implies that the flux is proportional with the field current (I_f) if the iron core is not saturated:

$$\Phi_{ag} = K_1 I_f, \qquad (10.2)$$

where K_1 is a proportionality constant. The rotor conductors cut the field lines that generate voltage in the coils. The generated voltage can be calculated by Equation (10.1), which is repeated here:

$$E_{ag} = 2N_r B\ell_g v. \tag{10.3}$$

This formula is modified by substituting for the rotor coil speed and the magnetic flux, which are given by:

$$v = \omega \frac{D_g}{2} \tag{10.4}$$

and

$$\Phi_{ag} = B\ell_g D_g, \tag{10.5}$$

where D_g is the rotor diameter and ω is the rotor angular velocity. Equation (10.5) assumes that the magnetic flux densities in the gap and at the center of the rotor are the same. The substitution of Equation (10.4) and Equation (10.5) into Equation (10.3) gives a practical expression for the induced voltage:

$$E_{ag} = 2N_r B\ell_g v = 2N_r B\ell_g \left(\omega \frac{D_g}{2}\right) = N_r (B\ell_g D_g)\omega = N_r \Phi_{ag}\omega. \tag{10.6}$$

For further simplification, we substitute Equation (10.2) into Equation (10.6), and combine the number of rotor turns and the constant K_1 into a new constant K_m, which is called the *machine constant*:

$$E_{ag} = N_r \Phi_{ag}\omega = N_r K_1 I_f \omega = K_m I_f \omega. \tag{10.7}$$

When the generator is loaded, the load current produces a voltage drop on the rotor winding resistance. In addition, there is a more or less constant voltage drop of 1–3 V on the brushes. These two voltage drops reduce the terminal voltage of the generator. The terminal voltage can be calculated by using Kirchhoff's voltage law (KVL):

$$E_{ag} = V_{dc} + I_{ag} R_a + V_{brush}, \tag{10.8}$$

where

V_{dc} is the generator terminal voltage;
I_{ag} is the load current;
R_a is the generator armature (or rotor) winding resistance;
V_{brush} is the voltage drop on the brushes; and
E_{ag} is the field current- (or stator current-) generated induced voltage.

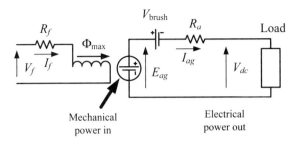

Figure 10.14. Equivalent circuit of a separately excited dc generator.

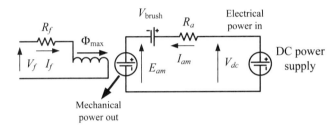

Figure 10.15. Equivalent circuit of a separately excited dc motor.

Equation (10.8) suggests that the dc generator can be represented by a voltage source and a resistance connected in series. It must be noted that the flux is generated by inductances. The exact equivalent circuit includes the rotor and stator inductances as well as the magnetizing inductance. However, for steady-state analysis, the inductances can be neglected since the dc current does not produce a voltage drop on an inductance. This justifies the use of an equivalent circuit that does not include the inductances. Figure 10.14 shows the equivalent circuit of a separately excited dc generator. In this diagram, a separate dc source (V_f) supplies the stator excitation circuit. The generator-induced voltage (E_{ag}) drives current (I_{ag}) through the load. The induced voltage is larger than the terminal or load voltage (V_{dc}), that is, $E_{ag} > V_{dc}$. The generator has to be rotated by an engine or turbine. The input to the generator is mechanical power supplied through the shaft. The output is electrical power, which is delivered to the load.

Figure 10.15 presents the equivalent circuit of a separately excited dc motor. The motor equivalent circuit is similar to the generator circuit, but the current (I_{am}) flows from the dc source into the motor. This requires that the dc terminal voltage is larger than the induced voltage, that is, $V_{dc} > E_{am}$. The input to the motor is electric power, while the motor output is mechanical power.

The motor-induced voltage (E_{am}) is found in the same manner as the generator-induced voltage (E_{ag}). The motor terminal voltage is the sum of the induced voltage, the voltage drop on the armature resistance, and the brush voltage drop. The motor-induced voltage equations are:

$$V_{dc} = E_{am} + I_{am}R_a + V_{brush}, \quad (10.9)$$

$$E_{am} = K_m I_f \omega. \quad (10.10)$$

The motor and generator operation are analyzed in Section 10.3 using numerical examples and an interactive derivation of the equations. For either mode of operation, the rotor angular frequency is related to the shaft speed by:

$$\omega = 2\pi n_m, \quad (10.11)$$

where n_m is the shaft speed, which is generally quantified in revolutions per minute (rpm).

10.2.4. Excitation Methods

There are four different methods for supplying the dc current to the motor or generator poles:

1. separate excitation,
2. shunt connection,
3. series connection, and
4. compound connection.

Figure 10.14 and Figure 10.15 introduce a *separately excited* dc generator and motor, respectively. This connection requires a separate dc power supply for the stator. The advantage of this machine is that in the case of a motor, its speed, or in the case of a generator, its output voltage, can be controlled accurately by the field current (I_f). However, the extra power supply increases the cost of the system.

Figure 10.16 shows the *shunt* connection for a motor. The shunt machine is the most frequently used arrangement. It requires only one power supply. A load change has little effect on the motor speed or the generator voltage.

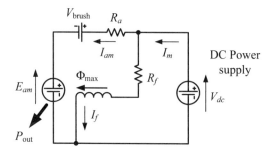

Figure 10.16. Equivalent circuit of a shunt motor connection.

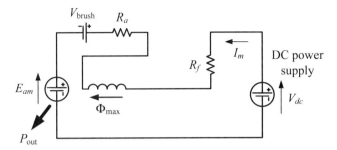

Figure 10.17. Equivalent circuit of a series motor connection.

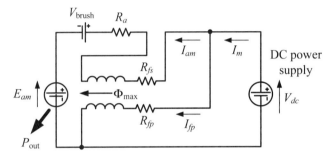

Figure 10.18. Compound motor connection.

Figure 10.17 exhibits the *series* motor connection in which the field current and motor current are the same. This motor has a large starting torque, which is an advantage for trolley and tramway drives. A problem with this motor occurs for a sudden loss of mechanical load, which increases the motor speed rapidly and can lead to mechanical failure and an accident.

Figure 10.18 displays an equivalent circuit for the *compound* motor connection, which has two excitation windings on the stator poles. One field coil is connected in series (R_{fs}) with the rotor, while the other field winding is in parallel (R_{fp}).

10.3. OPERATION ANALYSES

The dc motor and generator operation are presented through numerical examples using interactive derivations. The reader is encouraged to use her or his computer and follow the derivations. This increases the understanding of the subject and accelerates the learning process.

Here, we examine the operation of three machine configurations: (1) the separately excited, (2) the shunt, and (3) the series connections. In each case, circuit analysis is performed on the equivalent circuit while using Equation (10.7). In all three cases, we derive relations for the dc motor operation, and in the case of the shunt connection, we also develop the expressions for its operation as a generator.

Within this section, we shall attempt to write a few general expressions for the motor behavior. Specifically, if the voltage drop on the brushes is neglected (i.e., $V_{brush} \approx 0$), then the motor-induced voltage can be expressed either in terms of the machine constant (K_m) or the application of KVL to the motor circuit:

$$K_m I_f \omega = E_{am} = V_{dc} - I_{am} R_m, \qquad (10.12)$$

where R_m is the total resistance in the path of the armature current (I_{am}) through the motor. The mechanical output power is then:

$$P_{out} = E_{am} I_{am}. \qquad (10.13)$$

The mechanical losses (ventilation and friction) should be subtracted from the output power calculated by Equation (10.13). In this chapter, the mechanical losses are neglected. The reader is encouraged to test the effect of the mechanical losses, which are typically 5–10% of the rated power.

Consequently, the torque developed by the motor is simply:

$$T = \frac{P_{out}}{\omega} = K_m I_{am} I_f. \qquad (10.14)$$

However, as we shall see later, the input electrical power depends on the specific motor excitation connection, and therefore writing a single general expression for input power is unsuccessful. After development later, the general motor relations are summarized in a table at the end of this section.

10.3.1. Separately Excited Machine

First, we analyze the separately excited machine, specifically its motor operation. The separately excited machine data are given in terms of its ratings and a test measurement. The rated values are:

$$P_m := 40 \text{ kW} \quad V_{dc} := 240 \text{ V} \quad R_a := 0.25 \, \Omega \quad R_f := 120 \, \Omega.$$

The dc motor measurement is simpler than the ac motor, and can occur at either a loaded or an unloaded condition. At the particular test conditions, the measured values to be used for the motor constant calculation are:

$$I_{mo} := 8 \text{ A} \quad V_{mo} := V_{dc} \quad \text{rpm} := \frac{1}{\text{min}}$$

$$I_{fo} := 2 \text{ A} \quad n_{mo} := 1000 \text{ rpm}.$$

The brush voltage drop and iron core saturation are neglected.

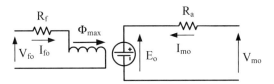

Figure 10.19. Separately excited motor in test condition.

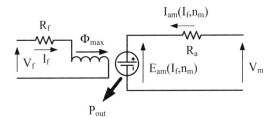

Figure 10.20. Separately excited motor simplified equivalent circuit.

10.3.1.1. Motor Constant. The analysis starts with calculation of the motor constant using the measured currents, speed and voltage. The equivalent circuit of the separately excited motor in test condition is drawn in Figure 10.19. In the test condition, the motor angular speed is:

$$\omega_o := 2 \cdot \pi \cdot n_{mo} = 104.7 \cdot \frac{\text{rad}}{\text{sec}}.$$

The induced voltage is calculated by using the loop voltage equation (KVL) for the rotor circuit:

$$E_o := V_{mo} - I_{mo} \cdot R_a \quad E_o = 238 \text{ V}.$$

The motor constant is calculated from the induced voltage using Equation (10.10):

$$K_m := \frac{E_o}{I_{fo} \cdot \omega_o} \quad K_m = 1.136 \frac{V \cdot s}{A}$$

10.3.1.2. Separately Excited Motor Operation. The equivalent circuit for the operation analyses is provided in Figure 10.20. The motor is supplied with the rated voltage: $V_m := V_{dc}$.

The motor can be controlled by the field current (I_f), which is selected as one of the variables. The other independent variable is the motor speed (n_m). We will investigate the motor current (I_{am}), output power (P_{out}), and torque (T_m) as a function of these two variables. The validity of the equations is tested using $n_m := 1000$ rpm and $I_f := 2A$.

The calculation of the induced voltage versus speed and field current equation requires the motor angular speed, which is given by:

$$\omega_m(n_m) := 2 \cdot \pi \cdot n_m. \tag{10.15}$$

The induced voltage is calculated using Equation (10.10). The induced voltage function is:

$$E_{am}(I_f, n_m) := K_m \cdot I_f \cdot \omega_m(n_m). \tag{10.16}$$

The numerical value at 1000 rpm and a field current of 2 A is $E_{am}(I_f, n_m) = 238$ V.

MOTOR CURRENT. The motor current is calculated from the equivalent circuit using the loop voltage equation (KVL) for the rotor:

$$I_{am}(I_f, n_m) := \frac{V_m - E_{am}(I_f, n_m)}{R_a}. \tag{10.17}$$

The numerical value is $I_{am}(2\text{ A}, 1000\text{ rpm}) = 8$ A.

Figure 10.21 demonstrates that the motor current varies linearly with the speed at different values of the field current. The graph also discloses that the motor current becomes approximately zero at 500, 1000, or 2000 rpm for field currents of 4, 2, and 1 A, respectively. Increasing the field current decreases the speed at which the motor

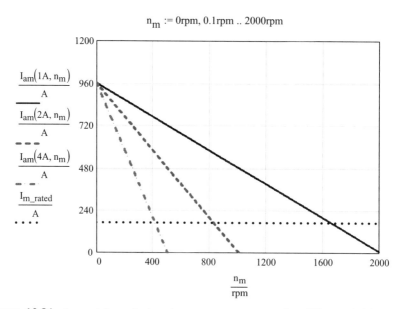

Figure 10.21. Separately excited motor current versus speed at different field currents.

OPERATION ANALYSES

current becomes zero. Above this speed, the motor will theoretically operate as a generator because the direction of the motor current reverses, but generator operation requires an external force to drive the machine.

The motor starting current is calculated by substituting a speed of zero into the current equation (Eq. 10.17):

$$I_{start} := I_{am}(I_f, 0 \text{ rpm}). \qquad (10.18)$$

The numerical value is $I_{start} = 960$ A.

The starting current is independent from the field current, because the induced voltage is zero when the motor starts, that is, $E_{am}(I_f, 0 \text{ rpm}) = 0$ V such that the starting current may also be calculated directly from:

$$I_{start} := \frac{V_m}{R_a} = 960 \text{ A}.$$

The rated current of the motor is the rated motor power divided by the rated voltage:

$$I_{m\_rated} := \frac{P_m}{V_{dc}} \quad I_{m\_rated} = 166.7 \text{ A}.$$

The ratio of the starting current to rated current is:

$$\frac{I_{start}}{I_{m\_rated}} = 5.76.$$

The high starting current can be dangerous because it produces overheating and increases mechanical stresses. This rather high starting current can be reduced by either of the following:

- supplying the motor with reduced voltage during the starting process, and then as the motor gains speed, gradually increasing the voltage, or
- inserting a resistance in series with the motor.

Substituting Equation (10.15) and Equation (10.16) into Equation (10.17) yields an expression for the motor current in terms of the field current and motor speed:

$$I_{am}(I_f, n_m) := \frac{V_m - K_m \cdot I_f \cdot 2 \cdot \pi \cdot n_m}{R_a}. \qquad (10.19)$$

A plot (not shown) of the motor current versus field current at constant speed also exhibits a linear relation if the motor iron core is not saturated.

MOTOR POWER. The input and output powers are the product of the motor current and the appropriate voltage. The input electrical power, including the field excitation, is:

$$P_{in}(I_f, n_m) := V_m \cdot I_{am}(I_f, n_m) + I_f^2 \cdot R_f. \tag{10.20}$$

The numerical value at $I_f = 2$ A and $n_m = 1000$ rpm is $P_{in}(I_f, n_m) = 2.4$ kW.

The output mechanical power is:

$$P_{out}(I_f, n_m) := E_{am}(I_f, n_m) \cdot I_{am}(I_f, n_m). \tag{10.21}$$

The numerical value is $P_{out}(2 \text{ A}, 1000 \text{ rpm}) = 1.90$ kW.

Figure 10.22 shows that the motor output power has a bell-shaped behavior with respect to the speed. Each curve has a maximum mechanical output value. The motor operation is stable to the right side of the maximum, where an increase in load power decreases the motor speed. The graph shows that variation of the field current changes the range of operation. As an example, for a field current of 1 A, the motor operates between 1000 and 2000 rpm.

The stable operating point determination is demonstrated in the following calculation. The motor operates at the rated power of $P_m = 40$ kW. The speed at this load is calculated with the Mathcad *root* function. A guess value of $n_m := 1500$ rpm is used. The motor speed at rated load is a function of the field current:

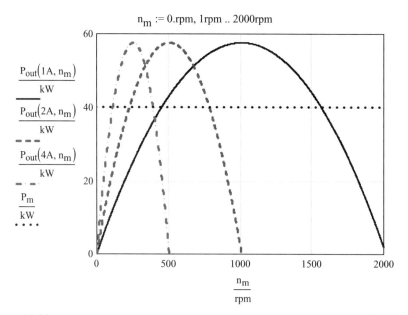

Figure 10.22. Separately excited motor output power versus speed at different field currents.

$$n_{rated}(I_f) := root(P_{out}(I_f, n_m) - P_m, n_m).$$

The speeds at a few different values of the field current are:

$$n_{rated}(1\,A) = 1566 \text{ rpm} \quad n_{rated}(2\,A) = 782.9 \text{ rpm} \quad n_{rated}(4\,A) = 391.5 \text{ rpm}.$$

These results are consistent with the curves in Figure 10.22.

The motor efficiency is the ratio of the output and input powers:

$$\varepsilon(I_f, n_m) := \frac{P_{out}(I_f, n_m)}{P_{in}(I_f, n_m)}. \tag{10.22}$$

The numerical value at the given conditions is $\varepsilon(I_f, n_m) = 79.3\%$.

It is indicated in Figure 10.23 that the efficiency is strongly affected by the field current and the motor speed, although the motor voltage also affects the efficiency.

MOTOR TORQUE. The torque of the motor is calculated by dividing the output power with the angular speed:

$$T_m(I_f, n_m) := \frac{P_{out}(I_f, n_m)}{\omega_m(n_m)}. \tag{10.23}$$

The numerical value is $T_m(2\,A, 1000 \text{ rpm}) = 18.2 \cdot N \cdot m$.

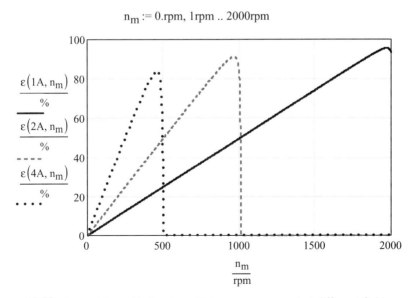

Figure 10.23. Separately excited motor efficiency versus speed at different field currents.

Equation (10.23) has a singularity at zero speed. An alternate formula, which is valid at the starting conditions, can be derived for the torque. First, we substitute Equation (10.16) into Equation (10.21) to find a reduced expression for the output mechanical power:

$$P_{out}(I_f, n_m) := K_m \cdot I_f \cdot \omega_m(n_m) \cdot I_{am}(I_f, n_m). \qquad (10.24)$$

This formula points out that the output power increases as the speed and currents are increased. By substituting Equation (10.24) into Equation (10.23), an equivalent relation for the torque (without a singularity) is found:

$$T_m(I_f, n_m) := K_m \cdot I_f \cdot I_{am}(I_f, n_m). \qquad (10.25)$$

Figure 10.24 denotes that the motor torque varies linearly with the speed for a given field current. The field current affects the motor starting current and the motor operating range. The motor operating point can be adjusted by controlling the field current.

As an example, the following calculation gives the required field current if the motor drives a fan, which requires a torque of 1000 N·m at 500 rpm. A guess value for the field current begins the Mathcad *root* equation solver:

$$I_f := 5 \text{ A} \qquad n_m := 500 \text{ rpm}.$$

The required field current is:

$$\text{root}(T_m(I_f, n_m) - 1000 \text{ N} \cdot \text{m}, I_f) = 2.625 \text{ A}.$$

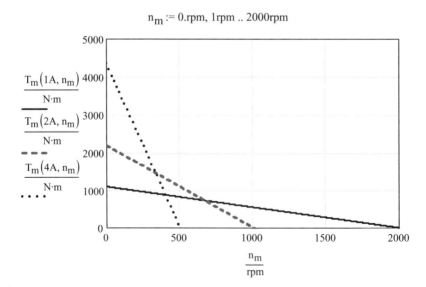

Figure 10.24. Separately excited motor torque versus speed at different field currents.

OPERATION ANALYSES

10.3.2. Shunt Machine

In the case of a shunt motor, the machine field winding (R_f) is connected in parallel with the supply, as drawn in Figure 10.16. For simplicity, we neglect the voltage drop across the brushes and possible saturation of the iron core. We use the same rating data as that for the separately excited motor for the analysis of the shunt motor and generator:

$$P_m := 40 \text{ kW} \quad V_{dc} := 240 \text{ V} \quad R_a := 0.25 \text{ }\Omega \quad R_f := 120 \text{ }\Omega.$$

The objective of this analysis is the derivation of the shunt motor operating characteristics, which are power, efficiency, and torque versus speed.

Motor Constant

The first step of the analysis is the determination of the machine constant using the measured values:

$$I_{mo} := 8 \text{ A} \quad \text{rpm} := \frac{1}{\min} \quad n_{mo} := 1000 \text{ rpm} \quad V_{mo} := V_{dc}.$$

The motor constant calculation was demonstrated earlier as part of the separately excited motor operation analysis. This calculation requires the field excitation current, which is the supply voltage (V_{mo}) divided by the field winding resistance (R_f). From Ohm's law and referring to Figure 10.16, the field current at the steady-state test condition is:

$$I_{fo} := \frac{V_{mo}}{R_f} \quad I_{fo} = 2 \text{ A}.$$

Using Kirchhoff's current law (KCL), the armature current in the test condition is:

$$I_{ao} := I_{mo} - I_{fo} \quad I_{ao} = 6 \text{ A}.$$

The motor angular speed is:

$$\omega_o := 2 \cdot \pi \cdot n_{mo} \quad \omega_o = 104.7 \cdot \frac{\text{rad}}{\text{sec}}.$$

Applying KVL, the induced voltage is:

$$E_{mo} := V_{mo} - I_{ao} \cdot R_a \quad E_{mo} = 238.5 \text{ V}.$$

Using Equation (10.10), the obtained shunt motor constant is:

$$K_m := \frac{E_{mo}}{I_{fo} \cdot \omega_o} \quad K_m = 1.139 \frac{V \cdot s}{A}.$$

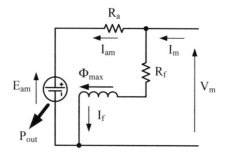

Figure 10.25. Shunt motor simplified equivalent circuit.

10.3.2.1. Shunt Motor Operation. Further calculations utilize the simplified equivalent circuit presented in Figure 10.25, and neglect the brush voltage drop.

The supply voltage is selected as one of the independent variables because it can be used for motor control. The other independent variable is the motor speed. The calculation will be verified by using $n_m := 1000$ rpm and $V_m := V_{dc}$.

The calculation of the induced voltage requires the angular speed, which in general is:

$$\omega_m(n_m) := 2 \cdot \pi \cdot n_m. \tag{10.26}$$

The numerical value at 1000 rpm is:

$$\omega_m(n_m) = 104.7 \cdot \frac{\text{rad}}{\text{sec}}.$$

MOTOR CURRENT. The field current is found from applying Ohm's law at steady-state conditions:

$$I_f := \frac{V_m}{R_f}. \tag{10.27}$$

Using Equation (10.10), the induced voltage is:

$$E_{am}(n_m) := K_m \cdot I_f \cdot \omega_m(n_m). \tag{10.28}$$

The corresponding numerical value at 1000 rpm is $E_{am}(n_m) = 238.5$ V.

The loop voltage equation applied to the rotor circuit in Figure 10.25 permits the calculation of the armature motor current, which is:

$$I_{am}(n_m, V_m) := \frac{V_m - E_{am}(n_m)}{R_a}. \tag{10.29}$$

The numerical value is $I_{am}(1000 \text{ rpm}, 240 \text{ V}) = 6$ A.

OPERATION ANALYSES

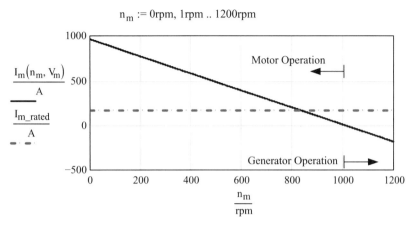

Figure 10.26. Shunt motor current versus motor speed at rated voltage.

Using KCL, the motor input current is the sum of the field current and armature current:

$$I_m(n_m, V_m) := I_f + I_{am}(n_m, V_m). \qquad (10.30)$$

The corresponding numerical value is $I_m(1000 \text{ rpm}, 240 \text{ V}) = 8 \text{ A}$.

The rated motor current is:

$$I_{m\_rated} := \frac{P_m}{V_{dc}} \quad I_{m\_rated} = 166.7 \text{ A}.$$

Using Equation (10.30), the motor current is plotted as a function of speed at rated voltage. Figure 10.26 reveals that the motor current decreases linearly with the speed. The motor draws the maximum current when it starts. The starting current can be calculated by substituting a zero speed in the current equation (Eq. 10.30). The starting current is:

$$I_{start} := I_m(0 \text{ rpm}, V_m). \qquad (10.31)$$

The numerical value of the starting current is $I_{start} = 962 \text{ A}$.

The ratio of the starting current to the rated current is nearly the same as for the separately excited motor. The obtained numerical value is:

$$\frac{I_{start}}{I_{m\_rated}} = 5.77.$$

This rather high starting current can be reduced by supplying the motor with reduced voltage during starting. When the motor gains speed, the voltage is increased gradually to the rated voltage. In most cases, the starting current is limited to three times the rated current. The calculation of the reduced voltage needed is provided:

$$\text{Given}$$
$$I_m(0 \text{ rpm}, V_m) = 3 \cdot I_{m\_rated}$$
$$\text{Find}(V_m) = 124.5 \text{ V}.$$

This reduced voltage is about half the rated value (V_{dc}) of 240 V.

Figure 10.26 indicates that the motor current becomes zero near 1000 rpm. Above this speed, the motor would operate as a generator because the direction of motor current reverses; however, the generator operation requires an external driver.

MOTOR POWER. The input electrical power is calculated using:

$$P_{in}(n_m, V_m) := V_m \cdot I_m(n_m, V_m). \tag{10.32}$$

The output mechanical power is computed from:

$$P_{out}(n_m, V_m) := E_{am}(n_m) \cdot I_{am}(n_m, V_m). \tag{10.33}$$

The corresponding numerical values for the powers at 1000 rpm and 240 V are:

$$P_{in}(n_m, V_m) = 1.92 \cdot \text{kW} \quad P_{out}(n_m, V_m) = 1.43 \cdot \text{kW}.$$

Figure 10.27 plots the shunt motor input and output power variation as a function of the speed. The input power is a maximum when the motor starts, and thereafter the input power declines with increasing speed. The bell-shaped output power curve reaches a maximum value around 500 rpm. For the calculation of the maximum power, the voltage variable is eliminated from the output power relation (Eq. 10.33). This is achieved by the creation of a new power equation, with only one variable and the selection of a constant motor voltage equal to the rated value:

$$P_{ou}(n_m) := P_{out}(n_m, V_{dc}).$$

A guess value for the maximization is $n_m := 500$ rpm:

$$n_{max\_P} := \text{Maximize}(P_{ou}, n_m) \quad n_{max\_P} = 503.1 \text{ rpm}.$$

The maximum power is:

$$P_{out}(n_{max\_P}, V_{dc}) = 77.2 \text{ hp} \quad P_{out}(n_{max\_P}, V_{dc}) = 57.6 \text{ kW}.$$

OPERATION ANALYSES

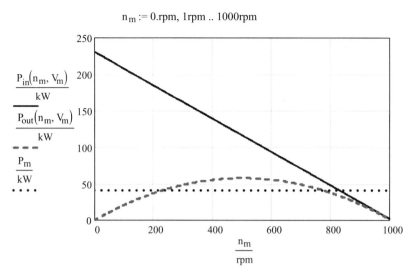

Figure 10.27. Shunt motor input and output powers versus speed at rated voltage.

A common question to be answered is, "What is the motor speed at rated output power?" Figure 10.27 discloses that this problem has two solutions. A guess value for the first solution is $n_m := 200$ rpm. The Mathcad *Find* equation solver solution is:

Given

$$P_{out}(n_m, V_m) = P_m$$

$$n_{Prated} := \text{Find}(n_m) = 225.0 \text{ rpm}.$$

A guess value for the second solution and the solution are:

$$n_m := 700 \text{ rpm}$$

Given

$$P_{out}(n_m, V_m) = P_m$$

$$n_{Prated} := \text{Find}(n_m) = 781.3 \text{ rpm}.$$

Figure 10.27 reveals that the second solution represents a stable operating point since an increased load causes a decrease in the motor speed. The first solution is an unstable operating point.

The motor efficiency is the ratio of the output and input powers:

$$\varepsilon(n_m, V_m) := \frac{P_{out}(n_m, V_m)}{P_{in}(n_m, V_m)}. \tag{10.34}$$

The numerical value at 1000 rpm and 240 V is $\varepsilon(n_m, V_m) = 74.5\%$.

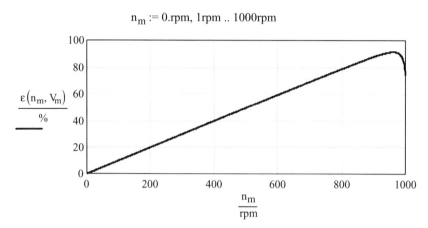

Figure 10.28. Shunt motor efficiency versus speed at rated voltage.

Figure 10.28 shows a graph of the variation of the shunt motor efficiency with the speed. It can be seen that the efficiency is very poor at low speeds and increases with the speed. For the maximum efficiency calculation, the voltage variable is eliminated from the efficiency formula (Eq. 10.34):

$$\text{effic}(n_m) := \varepsilon(n_m, V_{dc}).$$

A guess value for the maximization is $n_m := 970$ rpm.
The speed where maximum efficiency occurs is:

$$n_{max\_\varepsilon} := \text{Maximize}(\text{effic}, n_m) \quad n_{max\_\varepsilon} = 962.4 \text{ rpm}.$$

This speed corresponds to a maximum efficiency of:

$$\text{effic}(n_{max\_\varepsilon}) = 91.3\%.$$

However, this maximum efficiency corresponds to an output power of only $P_{out}(n_{max\_\varepsilon}, V_m) = 9.61$ kW, which is about one-sixth of the maximum power output.

MOTOR TORQUE. The motor torque is the output power divided by the angular speed:

$$T_m(n_m, V_m) := \frac{P_{out}(n_m, V_m)}{\omega_m(n_m)}. \tag{10.35}$$

The numerical value at 1000 rpm and 240 V is $T_m(n_m, V_m) = 13.7$ N·m.

OPERATION ANALYSES

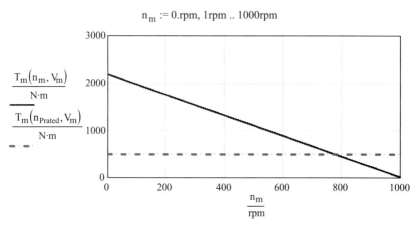

Figure 10.29. Shunt motor torque versus speed at rated voltage.

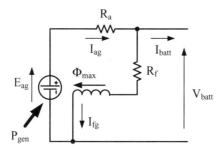

Figure 10.30. Equivalent circuit of a dc shunt generator.

The singularity in the previous formula can be eliminated by substituting an equivalent expression for the output power. Specifically, first express the output power in terms of the current and voltage using Equation (10.33), and then substitute for the induced voltage using Equation (10.28), which yields:

$$T_m(n_m, V_m) := K_m \cdot I_f \cdot I_{am}(n_m, V_m). \qquad (10.36)$$

Figure 10.29 shows that the torque linearly decreases with increasing speed. The curve demonstrates that the starting torque is very high, which is a desirable quality for certain applications. The torque at rated load is:

$$T_m(n_{Prated}, V_m) = 489 \text{ N} \cdot \text{m}.$$

This completes evaluation of the shunt motor performance.

10.3.2.2. Generator Operation. The shunt motor can also be used as a generator. Figure 10.30 shows the dc shunt generator equivalent circuit. The generator operation will be analyzed when the generator charges a battery. The goal here is the calculation of the required generator speed to provide a particular load/battery current, assuming the generator or battery voltage is constant. The equations are tested by using a battery current $I_{batt} := 20$ A and a battery voltage $V_{batt} := 238$ V.

The first step of the analysis is the calculation of the generator field current from Ohm's law at steady-state conditions:

$$I_{fg} := \frac{V_{batt}}{R_f}. \tag{10.37}$$

The numerical value is $I_{fg} = 1.98$ A.

Using KCL, the generator armature current is the sum of the battery current and the field current, as exhibited in Figure 10.30:

$$I_{ag}(I_{batt}) := I_{fg} + I_{batt}. \tag{10.38}$$

The induced voltage is calculated from the rotor circuit loop voltage equation (KVL):

$$E_{ag}(I_{batt}) := V_{batt} + R_a \cdot I_{ag}(I_{batt}). \tag{10.39}$$

The numerical value is $E_{ag}(I_{batt}) = 243$ V.

The required input (mechanical) power to the generator is the product of the induced voltage and the generator-produced current:

$$P_{gen}(I_{batt}) := I_{ag}(I_{batt}) \cdot E_{ag}(I_{batt}). \tag{10.40}$$

Figure 10.31 plots the variation of the input power versus load current curve.

By substituting Equation (10.26) into Equation (10.28), we obtain a modified induced voltage equation, which is used to calculate the generator speed. The result is:

$$n_n(I_{batt}) := \frac{E_{ag}(I_{batt})}{2 \cdot \pi \cdot K_m \cdot I_{fg}}. \tag{10.41}$$

The generator speed at 20 A is $n_n(I_{batt}) = 1029.5$ rpm.

Figure 10.32 reveals that the relation between the generator speed and battery current is linear. The linear relation permits the accurate control of the load by speed variation.

A typical unknown is the speed at a given power, for example, at the rated power of 40 kW. The guess value is $I_{batt} := 150$ A:

$$I_{rated} := \text{root}\left(P_{gen}(I_{batt}) - P_m, I_{batt}\right) \quad I_{rated} = 144 \text{ A}$$
$$n_n(I_{rated}) = 1160 \cdot \text{rpm}.$$

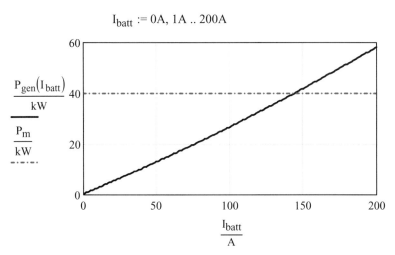

Figure 10.31. Shunt generator input power versus load (battery) current.

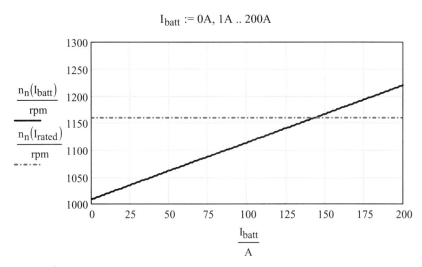

Figure 10.32. Shunt generator speed versus load (battery) current.

10.3.3. Series Motor

The purpose of this analysis is the derivation of the series motor operating characteristics, which are efficiency versus speed, and torque versus speed. These operating analyses are performed using a numerical example. The 20 kW, 240 V series motor has an armature resistance of 0.25 Ω and a field coil resistance of 0.3 Ω. The motor data are:

$$P_m := 20 \text{ kW} \quad V_{dc} := 240 \text{ V} \quad R_a := 0.25 \, \Omega \quad R_f := 0.3 \, \Omega.$$

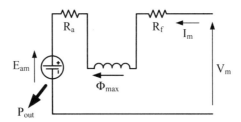

Figure 10.33. Series motor simplified equivalent circuit.

The motor current, speed, and voltage are measured to determine the motor constant. At rated terminal voltage, the motor current is 8 A while running at 500 rpm:

$$V_{mo} := V_{dc} \quad I_{mo} := 8 \text{ A} \quad \text{rpm} := \frac{1}{\min} \quad n_{mo} := 500 \text{ rpm}.$$

Figure 10.33 shows a diagram of the series motor simplified equivalent circuit if the voltage drop across the brushes and the iron core saturation are neglected.

MOTOR CONSTANT. The motor angular speed at the test condition is:

$$\omega_o := 2 \cdot \pi \cdot n_{mo} = 52.4 \cdot \frac{\text{rad}}{\sec}.$$

The induced voltage is calculated by the loop voltage equation using the circuit in Figure 10.33:

$$E_{mo} := V_{mo} - I_{mo} \cdot (R_a + R_f) \quad E_{mo} = 235.6 \text{ V}.$$

The motor constant is calculated using Equation (10.10):

$$K_m := \frac{E_{mo}}{I_{mo} \cdot \omega_o} \quad K_m = 0.562 \frac{V \cdot s}{A}.$$

DERIVATION OF MOTOR EQUATIONS. The series motor can be controlled by regulating the supply voltage. This voltage is selected as the first independent variable; the second variable is the motor speed. The calculations are numerically verified using $n_m := 100$ rpm and $V_m := V_{dc}$.

The first step is the calculation of the angular speed, which is:

$$\omega_m(n_m) := 2 \cdot \pi \cdot n_m \quad \omega_m(n_m) = 10.5 \cdot \frac{\text{rad}}{\sec}.$$

OPERATION ANALYSES

The equivalent circuit of Figure 10.33 indicates that the field current is equal with the motor current, which depends on the speed and supply voltage. Replacing the field current with the motor current in Equation (10.10) results in:

$$E_{am}(n_m) := K_m \cdot I_m(n_m, V_m) \cdot \omega_m(n_m). \tag{10.42}$$

MOTOR CURRENT. The loop voltage equation for the circuit in Figure 10.33 gives the motor voltage–current relation:

$$I_m(n_m, V_m) \cdot (R_a + R_f) = V_m - E_{am}(n_m). \tag{10.43}$$

Substituting Equation (10.42) into Equation (10.43) yields:

$$I_m(n_m, V_m) \cdot (R_a + R_f) = V_m - K_m \cdot I_m(n_m, V_m) \cdot \omega_m(n_m). \tag{10.44}$$

Rearranging the earlier expression provides a formula to calculate the motor current:

$$I_m(n_m, V_m) := \frac{V_m}{R_a + R_f + K_m \cdot \omega_m(n_m)}. \tag{10.45}$$

The numerical value at 100 rpm and 240 V is $I_m(n_m, V_m) = 37.3$ A.
The motor rated current is:

$$I_{m\_rated} := \frac{P_m}{V_{dc}} \quad I_{m\_rated} = 83.3 \text{ A}.$$

Using Equation (10.45), the motor current is graphed as a function of speed at the rated voltage. Figure 10.34 demonstrates that the motor current declines rapidly with

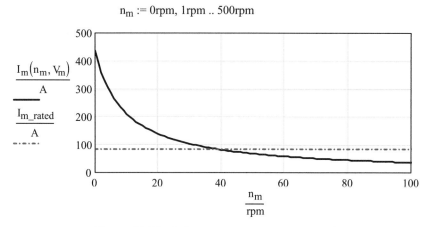

Figure 10.34. Series motor current versus speed.

increasing speed. The characteristics show that this motor operates well at low speed. In this example, the speed at the rated current is only about 40 rpm, as seen in the graph.

The starting current is calculated by substituting a speed of zero into the motor current formula (Eq. 10.45). The result is:

$$I_{start} := I_m(0 \text{ rpm}, V_m) \quad I_{start} = 436 \text{ A}.$$

The ratio of the starting current to rated current is:

$$\frac{I_{start}}{I_{m\_rated}} = 5.24.$$

The numerical value of the induced voltage is calculated by substituting the obtained current in Equation (10.42):

$$E_{am}(n_m) := K_m \cdot I_m(n_m, V_m) \cdot \omega_m(n_m) \quad E_{am}(n_m) = 219.5 \text{ V}.$$

MOTOR POWER. The input electrical and output mechanical powers are the product of the motor current and appropriate voltage:

$$P_{in}(n_m, V_m) := V_m \cdot I_m(n_m, V_m). \tag{10.46}$$

$$P_{out}(n_m, V_m) := E_{am}(n_m) \cdot I_m(n_m, V_m). \tag{10.47}$$

The numerical values at 100 rpm and 240 V are:

$$P_{in}(n_m, V_m) = 8.94 \text{ kW} \quad \text{and} \quad P_{out}(n_m, V_m) = 8.18 \text{ kW}.$$

Figure 10.35 shows a graph of the input and output power variations with speed. It can be seen that the input electrical power is very high during the starting conditions, when the output mechanical power is practically zero. Past the output power peak value, increasing the speed decreases both powers. The maximum (peak) output mechanical power can be determined using the Mathcad *Maximize* function with a guess value of $n_m := 10$ rpm. The new mechanical output power function is defined, consisting of a single variable (i.e., the motor speed):

$$P_{mech}(n_m) := P_{out}(n_m, V_m).$$

The *Maximize* function is employed to find the speed of the peak output power:

Given

$$n_{Pmax} := \text{Maximize}(P_{mech}, n_m) = 9.34 \text{ rpm}.$$

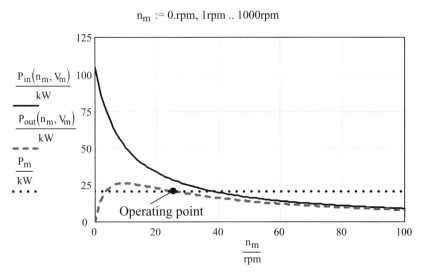

Figure 10.35. Series motor input and output powers versus speed.

The actual maximum output power is:

$$P_{mech}(n_{Pmax}) = 35.1 \text{ hp} \quad P_{mech}(n_{Pmax}) = 26.2 \text{ kW}.$$

Also of interest is the determination of the motor speed at rated output power (P_m). Figure 10.35 shows that numerically there are two possible solutions, but we are only interested in the stable operating point, which is the right-hand side intersection of the P_{out} and P_m curves at ~30 rpm, which is used as the guess value in the *Find* equation solver:

$$n_m := 30 \text{ rpm}$$
$$\text{Given}$$
$$P_{mech}(n_m) = P_m$$
$$n_{Prated} := \text{Find}(n_m) = 27.0 \text{ rpm}.$$

The motor efficiency is the ratio of output and input powers:

$$\varepsilon(n_m, V_m) := \frac{P_{out}(n_m, V_m)}{P_{in}(n_m, V_m)}. \tag{10.48}$$

The numerical value is $\varepsilon(100 \text{ rpm}, 240 \text{ V}) = 91.5 \cdot \%$.

Figure 10.36 shows that the motor efficiency is very low when the motor starts and gradually improves as the motor gains speed.

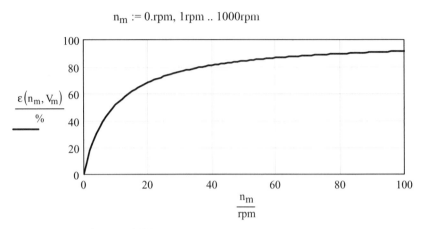

Figure 10.36. Series motor efficiency versus speed.

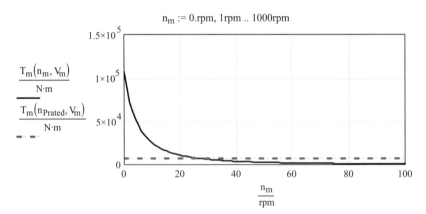

Figure 10.37. Series motor torque versus speed.

MOTOR TORQUE. The torque of the motor is:

$$T_m(n_m, V_m) := \frac{P_{out}(n_m, V_m)}{\omega_m(n_m)}. \tag{10.49}$$

The numerical value is $T_m(100 \text{ rpm}, 240 \text{ V}) = 781 \cdot \text{N} \cdot \text{m}$.

The singularity in the torque formula can be removed by substituting for the output power using Equation (10.42) and Equation (10.47) to yield:

$$T_m(n_m, V_m) := K_m \cdot I_m(n_m, V_m)^2. \tag{10.50}$$

It is demonstrated in Figure 10.37 that the starting torque is very high, but the torque declines rapidly as the motor gains speed. The ratio of the starting and rated torque can be more than 100, which is advantageous for applications requiring large torque like

OPERATION ANALYSES

trains or trolleys. The rapidly decreasing torque curve indicates that the decrease of the torque increases the speed. This is particularly dangerous when the motor suddenly loses its load and the speed increases to an unsafe value. The torque at the rated load is:

$$T_m(n_{Prated}, V_m) = 7.08 \times 10^3 \text{ N} \cdot \text{m}.$$

This concludes the derivation of the series motor equations and the evaluation of the series motor performance.

10.3.4. Summary

At first, the previous derivations for the three motor configurations may appear to lead to significantly different defining formulas; however, a closer examination leads to general equations that are capable of describing the behavior of all three motors when the brush voltage drop and iron core saturation are neglected. The general equations are presented in Table 10.1. Further reductions of some expressions are possible; for example, using the fact that the three currents are the same in the series motor leads to a reduced formula of $T_m = K_m I_f^2$ for the motor torque.

Because of the ease by which dc motors can be controlled, these machines have widespread use, including applications such as robotics. The separately excited dc motor speed can be controlled accurately by regulating the excitation voltage. The torque–speed relationship is practically linear if the motor is not saturated. Similarly, the speed of a shunt-connected dc motor can be regulated by a variable resistance (R_c) inserted in the circuit in series with the field resistance, as drawn in Figure 10.38a. The speed of a series motor can be adjusted by connecting a variable resistance in parallel with the excitation winding, as shown in Figure 10.38b. This reduces the field current

TABLE 10.1. General DC Motor Relations

Quantity	Separately Excited	Shunt Connection	Series Connection
Motor voltage		V_m	
Motor current		I_m	
Armature current	$I_a = I_m$	$I_m = I_a + I_f$	$I_a = I_m = I_f$
Field current	I_f independent		
Motor resistance		$R_m = R_a$	$R_m = R_a + R_f$
Induced voltage		$E_m = K_m I_f \omega_m = V_m - I_a R_m$	
Input electric power	$P_{in} = I_m V_m + I_f^2 R_f$		$P_{in} = I_m V_m$
Output mechanical power		$P_{out} = I_a E_m$	
Efficiency		$\varepsilon = \dfrac{P_{out}}{P_{in}}$	
Torque		$T_m = \dfrac{P_{out}}{\omega_m} = K_m I_a I_f$	

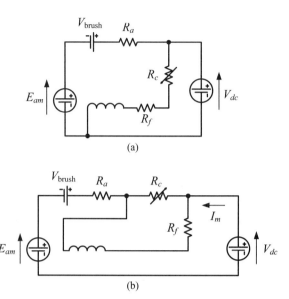

Figure 10.38. Incorporation of variable resistance (R_c) for dc motor control: (a) shunt motor; (b) series motor.

below the armature current, thereby causing a motor speed increase. DC motor speed can also be reduced by lowering the armature voltage. This can be accomplished by electronic control or by simply inserting a variable resistance in series with the armature. Both methods reduce the armature voltage along with the speed. DC motor control is a subject of the next chapter.

10.4. APPLICATION EXAMPLES

This section consists of four extended examples, including:

1. battery supplying DC shunt motor,
2. battery powering car engine starter,
3. series motor driving a pump, and
4. series motor with brush and copper losses.

EXAMPLE 10.1: Battery supplying DC shunt motor

In this example, a battery supplies a dc shunt motor through a feeder. In turn, the motor drives a pump. The pump torque–speed characteristic is:

$$a := 1\,\text{N}\cdot\text{m} \quad b := 0.002\cdot\text{N}\cdot\text{m}\cdot\text{sec}^2 \quad T_{\text{pump}}(n) := a + b\cdot n^2.$$

The feeder resistance is $R_{\text{feeder}} := 0.2\,\Omega$.

APPLICATION EXAMPLES

The fully charged battery voltage and internal resistance are:

$$V_{batt} := 25 \text{ V} \quad R_{batt} := 0.05 \text{ }\Omega.$$

The motor data are:

$$P_{rated} := 1 \text{ hp} \quad V_{rated} := 24 \text{ V} \quad R_a := 0.1 \text{ }\Omega \quad R_f := 10 \text{ }\Omega.$$

The motor is tested and the measurement results are:

$$V_{test} := V_{rated} \quad I_{test} := 90 \text{ A} \quad \text{rpm} := \text{min}^{-1} \quad n_{test} := 850 \text{ rpm}.$$

The aim here is the determination of the operating point, where the motor and pump speeds and torques are equal.

The major steps of the calculation are:

1. draw the equivalent circuit and calculate the motor parameters;
2. calculate the currents and voltages as a function of the speed;
3. compute and plot the motor and pump torques versus speed; and
4. determine the operating point, that is, where the torque–speed curves intersect.

(1) Equivalent Circuit and Motor Constant

The equivalent circuit is diagrammed in Figure 10.39. Using the motor test results and Ohm's law, the field current is:

$$I_{f\_test} := \frac{V_{test}}{R_f} \quad I_{f\_test} = 2.4 \text{ A}.$$

The induced voltage, using the loop voltage equation for the armature, is:

$$E_{test} := V_{test} - R_a \cdot (I_{test} - I_{f\_test}) \quad E_{test} = 15.24 \text{ V}.$$

The motor angular speed is $\omega_{test} := 2 \cdot \pi \cdot n_{test}$.

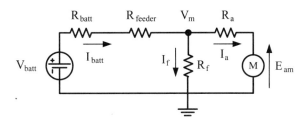

Figure 10.39. Equivalent circuit for battery supplying shunt motor.

The motor constant is calculated from the induced voltage equation (Eq. 10.10):

$$K_{mot} := \frac{E_{test}}{\omega_{test} \cdot I_{f\_test}} \qquad K_{mot} = 0.071 \frac{V \cdot s}{A}.$$

(2) Currents and Voltages as a Function of Motor Speed

Motor Terminal Voltage

The motor terminal voltage is V_m, as seen in Figure 10.39. This is a variable voltage, which is determined by solving the node point equation. The validity of the equations is tested using the rated voltage and an arbitrary speed:

$$V_m := V_{rated} \qquad n_m := 700 \text{ rpm}.$$

The induced voltage is obtained by combining Equation (10.10) and Equation (10.11) and expressing the field current in terms of the motor voltage:

$$E_{am}(n_m) := K_{mot} \cdot 2 \cdot \pi \cdot n_m \cdot \frac{V_m}{R_f}. \qquad (10.51)$$

The node point equation (KCL) for the node labeled V_m in Figure 10.39 is:

$$\frac{V_m - V_{batt}}{R_{batt} + R_{feeder}} + \frac{V_m}{R_f} + \frac{V_m - E_{am}(n_m)}{R_a} = 0. \qquad (10.52)$$

Substituting Equation (10.51) into the node point equation (Eq. 10.52) results in:

$$\frac{V_m - V_{batt}}{R_{batt} + R_{feeder}} + \frac{V_m}{R_f} + \frac{V_m}{R_a} - \frac{K_{mot} \cdot 2 \cdot \pi \cdot n_m \cdot V_m}{R_f \cdot R_a} = 0. \qquad (10.53)$$

This expression can be solved with the Mathcad *Find* equation solver. A guess value for the unknown variable is $V_m := 10$ V:

Given

$$\frac{V_m - V_{batt}}{R_{batt} + R_{feeder}} + \frac{V_m}{R_f} + \frac{V_m}{R_a} - \frac{K_{mot} \cdot 2 \cdot \pi \cdot n_m \cdot V_m}{R_f \cdot R_a} = 0$$

$$V_{mot}(n_m) := \text{Find}(V_m) \qquad V_{mot}(n_m) = 11.3 \text{ V}.$$

Alternatively, the node point equation can be solved directly. The rearrangement of Equation (10.53) results in:

$$\left[\frac{1}{(R_{batt} + R_{feeder})} + \frac{1}{R_f} + \frac{1}{R_a} - \frac{K_{mot} \cdot 2 \cdot \pi \cdot n_m}{R_a \cdot R_f} \right] \cdot V_m = \frac{V_{batt}}{R_{batt} + R_{feeder}}.$$

APPLICATION EXAMPLES

The coefficient of V_m is a conductance that is a function of motor speed:

$$G_m(n_m) := \frac{1}{(R_{batt} + R_{feeder})} + \frac{1}{R_f} + \frac{1}{R_a} - \frac{K_{mot} \cdot 2 \cdot \pi \cdot n_m}{R_a \cdot R_f} \quad G_m(n_m) = 8.87 \text{ S}.$$

The combination of these two equations produces a formula for the motor voltage:

$$V_m(n_m) := \frac{V_{batt}}{G_m(n_m) \cdot (R_{batt} + R_{feeder})} \quad V_m(n_m) = 11.3 \text{ V}.$$

Motor Currents
From Equation (10.51), the induced voltage is:

$$E_{am}(n_m) := K_{mot} \cdot 2 \cdot \pi \cdot n_m \cdot \frac{V_m(n_m)}{R_f} \quad E_{am}(n_m) = 5.90 \text{ V}.$$

Referring to Figure 10.39, the various currents are computed. The motor armature current is:

$$I_a(n_m) := \frac{V_m(n_m) - E_{am}(n_m)}{R_a} \quad I_a(n_m) = 53.8 \text{ A}.$$

The field current is:

$$I_f(n_m) := \frac{V_m(n_m)}{R_f} \quad I_f(n_m) = 1.13 \text{ A}.$$

Using KCL, the battery current is:

$$I_{batt}(n_m) := I_f(n_m) + I_a(n_m) \quad I_{batt}(n_m) = 54.9 \text{ A}.$$

(3) Motor and Pump Torque–Speed Characteristics
The motor output power and torque are:

$$P_{out}(n_m) := E_{am}(n_m) \cdot I_a(n_m) \quad P_{out}(n_m) = 0.425 \cdot \text{hp}$$

$$T_{mot}(n_m) := \frac{P_{out}(n_m)}{2 \cdot \pi \cdot n_m} \quad T_{mot}(n_m) = 4.33 \cdot \text{N} \cdot \text{m}.$$

Alternatively, the motor torque can be computed from the field and armature currents:

$$T_{mot}(n_m) := K_{mot} \cdot I_f(n_m) \cdot I_a(n_m) \quad T_{mot}(n_m) = 4.33 \text{ N} \cdot \text{m}.$$

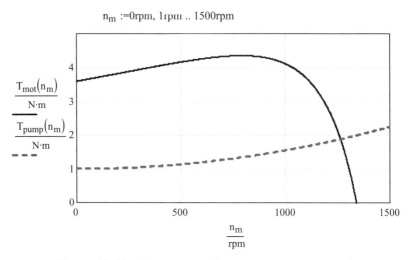

Figure 10.40. Series motor and pump torque versus speed.

The pump torque is:

$$T_{pump}(n) := a + b \cdot n^2 \quad T_{pump}(n_m) = 1.27 \text{ N} \cdot \text{m}.$$

The motor and pump torque–speed curves are plotted in Figure 10.40.

(4) Operating Point, Speed and Torque

The operating point is the intersection of the two torque curves displayed in Figure 10.40. The speed at this point is calculated using the Mathcad *root* function. A guess value for the speed from the graph is n := 1300 rpm.

The operating speed is:

$$n_{op} := \text{root}(T_{mot}(n) - T_{pump}(n), n) \quad n_{op} = 1261 \text{ rpm}.$$

The torque and output power at the operating speed are:

$$T_{op} := T_{mot}(n_{op}) \quad T_{op} = 1.88 \text{ N} \cdot \text{m}$$
$$P_{op} := P_{out}(n_{op}) \quad P_{op} = 0.334 \text{ hp}.$$

The input power to the motor and the motor efficiency are:

$$P_{in} := V_m(n_{op}) \cdot I_{batt}(n_{op}) \quad P_{in} = 310 \text{ W}$$
$$\varepsilon_{mot} := \frac{P_{op}}{P_{in}} \quad \varepsilon_{mot} = 80.3\%.$$

APPLICATION EXAMPLES

At this point, the battery supplies an input power of:

$$P_{batt} := V_{batt} \cdot I_{batt}(n_{op}) \quad P_{batt} = 362 \text{ W}.$$

The overall system efficiency at the operating point is:

$$\varepsilon_{sys} := \frac{P_{op}}{P_{batt}} \quad \varepsilon_{sys} = 68.7\%.$$

EXAMPLE 10.2: Battery powering car engine starter

A nominal 12 V battery supplies the starter of an automobile. The battery is represented by a 12.5 V source in series with a 0.05 Ω resistor. The starter is a dc series motor. A 20 ft starting cable with a resistance of 0.2 Ω is used to connect the battery to the motor. The motor is rated at 50 W and 12 V. The armature and field resistances are 0.1 and 0.2 Ω, respectively. The motor constant is 0.1 V·s/A. The equivalent motor circuit is given in Figure 10.41.

The system data are first initialized in the MATLAB program:

```
% DCseriesMotorStarter.m
% DC series motor used as an automobile starter
clear all

% Battery characteristics
Vbat = 12.5;       % voltage (Vdc)
Rbat = 0.05;       % resistance (ohm)

% Cable resistance (ohm)
Rcable = 0.2;

% Motor rating data
Pmot = 50;         % rated power (W)
Vmot = 12;         % rated voltage (V)
Kmot = 0.1;        % motor constant (V·sec/A)
Ra = 0.1;          % armature resistance (ohm)
Rf = 0.2;          % field resistance (ohm)
```

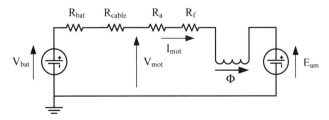

Figure 10.41. Battery supplying series motor (starter) through a cable.

In this MATLAB example, we will:

1. determine the current and torque variation as the motor speeds up;
2. calculate the starting current and starting torque;
3. plot the motor current, output power, and torque versus speed characteristics, when it is supplied by the battery; and
4. compute the speed at rated load.

In contrast to previous MATLAB programs, in this example we take advantage of the vector operations capability of the software. In particular, here we define an array containing motor speeds (n_m) from 0 to 1000 rpm in steps of 1 rpm. After that, we create a corresponding vector for the angular speed using $\omega_m = 2\pi n_m$.

```
% Start Analysis
% Create motor speed (rpm) vector
nm = 0 : 1000;
% Motor angular speed vector (rad/sec)
wm = 2*pi*nm/60;
```

Equation (10.10) provides a formula for the induced voltage in the motor, specifically:

$$E_{am} = K_{mot} I_{mot} \omega_m. \tag{10.54}$$

Referring to Figure 10.41 and using KVL, an expression is developed for the motor current (I_m) in terms of the battery voltage (V_{bat}) and other circuit parameters:

$$V_{bat} - I_{mot}(R_{bat} + R_{cable} + R_a + R_f) - E_{am} = 0. \tag{10.55}$$

Substituting Equation (10.54) into Equation (10.55) and rearranging yields:

$$I_{mot} = \frac{V_{bat}}{R_{bat} + R_{cable} + R_a + R_f + K_{mot}\omega_m}. \tag{10.56}$$

The computed motor current can be compared to the rated value obtained from the rated output power and motor voltage:

$$I_{rate} = \frac{P_{mot}}{V_{mot}}. \tag{10.57}$$

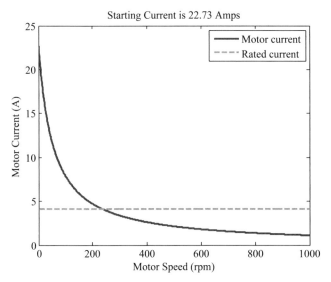

Figure 10.42. Starter current versus speed.

The motor current is plotted as a function of starter speed in Figure 10.42. One can see why automobile batteries are partially specified in terms of their cold cranking power in amperes, since the starting current needed is significant.

```
% Motor current (A) vector
Imot = Vbat ./ (Ra+Rf+Rcable+Rbat+Kmot*wm);
fprintf('Starting current = %g Amps.\n',Imot(1));

% Motor current rating (A)
Irate = Pmot/Vmot;
fprintf('Motor current rating is %g A.\n',Irate);
fprintf('Starting current is %g times rated current.\n',...
    Imot(1)/Irate);

% Plot the motor current vs. speed
plot(nm,Imot,nm,Irate*ones(1,size(nm,2)),'--',...
    'LineWidth',2.5);
set(gca,'fontname','Times','fontsize',12);
legend('Motor current','Rated current');
xlabel('Motor Speed (rpm)');
ylabel('Motor Current (A)');
title(['Starting Current is ',num2str(Imot(1),'%5.2f'),...
    ' Amps']);
```

With the motor current known, the actual induced voltage can be calculated using Equation (10.54). The mechanical output power is then the product of the motor current and induced voltage:

$$P_{out} = I_{mot} E_{am} \tag{10.58}$$

```
% Motor induced voltage (V) vector
Eam = Kmot * Imot .* wm;

% Motor output power (W) vector
Pout = Imot .* Eam;

% Plot the motor output power vs. speed
figure;
plot(nm,Pout,nm,Pmot*ones(1,size(nm,2)),'-',...
    'LineWidth',2.5);
set(gca,'fontname','Times','fontsize',12);
legend('Output power','Rated power');
xlabel('Motor Speed (rpm)');
ylabel('Output Mechanical Power (W)');
```

The motor output power is graphed in Figure 10.43, which also notes the rated power. The speed corresponding to the rated output power is found using the following coding, and is labeled as the graph title. The MATLAB program begins searching from the largest speed (1000 rpm) and tests for the speed at which the output power is greater than or equal to the rated value.

```
nrated = 0;
k = size(nm,2);
```

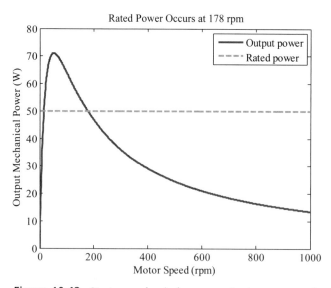

Figure 10.43. Starter mechanical power output versus speed.

APPLICATION EXAMPLES

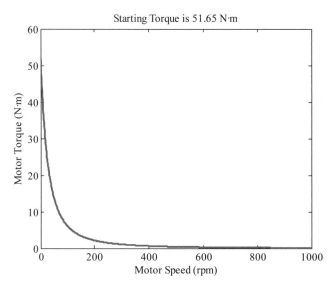

Figure 10.44. Starter torque versus speed.

```
while (nrated == 0)
    if Pout(k) >= Pmot
        nrated = k;
    end
    k = k-1;
end
fprintf('Rated power occurs at %g rpm.\n',nrated);
title(['Rated power occurs at ',num2str(nrated),' rpm']);
```

A general expression without a singularity at zero speed is created for the motor torque output:

$$T_{out} = \frac{P_{out}}{\omega_m} = \frac{I_{mot}E_{am}}{\omega_m} = \frac{I_{mot}K_{mot}I_{mot}\omega_m}{\omega_m} = K_{mot}I_{mot}^2. \quad (10.59)$$

The torque is calculated as a function of motor speed and plotted in Figure 10.44.

```
% Motor torque (N·m) vector
Tout = Kmot * Imot.^2;

fprintf('Starting torque = %g N·m.\n',Tout(1));

% Plot the motor torque vs. speed
figure;
plot(nm,Tout)
xlabel('Motor Speed (rpm)');
ylabel('Motor Torque (N·m)');
title(['Starting Torque is ',num2str(Tout(1)),' N·m']);
```

The printed results from the MATLAB program are:

```
>> DCseriesMotorStarter
Starting current = 22.7273 Amps.
Motor current rating is 4.16667 A.
Starting current is 5.45455 times rated current.
Rated power occurs at 178 rpm.
Starting torque = 51.6529 N·m.
```

EXAMPLE 10.3: Series motor driving a pump

A dc source supplies a series dc motor through a feeder, as illustrated in Figure 10.45. The 1-mile-long feeder has a resistance of 0.5 Ω/mile. The motor drives a pump, whose torque varies linearly with the speed. The pump torque–speed characteristic is described by the following equation:

$$T_{\text{pump}}(n_m) = 250 \text{ N} \cdot \text{m} + (70 \text{ N} \cdot \text{m} \cdot \text{s})n_m. \tag{10.60}$$

The 450 V motor is rated at 45 hp. The motor field and armature resistances are 0.35 and 0.6 Ω, respectively. The measurement results from testing the motor in a no-load condition at 950 rpm are a voltage of 460 V and a motor current of 3.2 A.

The equivalent motor circuit is given in Figure 10.46, and the system parameters are initialized in MATLAB:

```
% DCseriesMotorPump.m
% DC series motor drives a pump
clear all
```

Figure 10.45. One-line diagram of dc source supplying motor-driven pump.

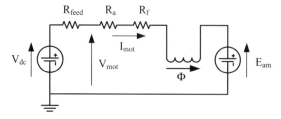

Figure 10.46. Equivalent circuit for dc source supplying motor-driven pump.

APPLICATION EXAMPLES

```
% Pump torque constants
% Torque = G + D * nm
G = 250;         % N·m
D = 70;          % N·m·sec

% Feeder line characteristics
Lfeeder = 1;     % length (mile)
Rfeeder = 0.5;   % resistance (ohm/mile)
Rfeed = Lfeeder*Rfeeder;  % resistance (ohm)

% Motor rating data
Prated = 45;     % power (hp)
Vrated = 450;    % voltage (V)
Ra = 0.6;        % armature resistance (ohm)
Rf = 0.35;       % field resistance (ohm)
```

In this example, we use MATLAB to:

1. evaluate the motor constant from the test results;
2. compute the current and motor terminal voltage versus speed;
3. calculate and plot the motor and pump torque–speed characteristics;
4. determine the operating point (speed and torque) if the feeder is supplied by the motor rated voltage; and
5. calculate the supply voltage and starting torque if the starting current is limited to twice the rated current. Under such a constraint, we determine whether the motor torque is sufficient to start the pump.

The motor constant is determined from the measurements. The induced voltage is calculated by referring to Figure 10.46 and applying KVL at the test condition:

$$E_0 = V_0 - I_0(R_a + R_f). \quad (10.61)$$

The motor constant is then computed from:

$$K_m = \frac{E_0}{I_0 \omega_0} \quad (10.62)$$

```
% Motor test results
Vo = 460;        % test voltage (V)
Io = 3.2;        % test current (A)
no = 950;        % test speed (rpm)

% Evaluate motor test measurements to obtain motor constant
wo = 2*pi*no/60;    % motor angular speed (rads/sec)
Eo = Vo-Io*(Ra+Rf); % induced voltage (V)
Km = Eo/(Io*wo);    % motor constant (V·sec/A)
fprintf('The motor constant (Km) = %g V·sec/A.\n', Km);
```

We begin the analysis by setting the dc source voltage equal to the motor rated voltage. After creating motor speed arrays in revolutions per minute and radians per second, we calculate the motor current. Using Equation (10.10) for the induced voltage and KVL applied to the motor circuit, the motor current is:

$$I_{mot} = \frac{V_{dc}}{R_{feed} + R_a + R_f + K_m \omega_m}. \tag{10.63}$$

Using KVL, the motor terminal voltage is:

$$V_{mot} = V_{dc} - I_{mot} R_{feed} \tag{10.64}$$

```
% Begin analysis
% Set the dc source voltage to the motor rated value (V)
Vdc = Vrated;

% Vary the motor-pump shaft speed (rpm)
nm = 0 : 1000;

% Motor angular speed vector (rad/sec)
wm = 2*pi*nm/60;

% Motor currrent (A) vector
Imot = Vdc ./ (Rfeed+Ra+Rf+Km*wm);

% Motor voltage (V) vector
Vmot = Vdc - Imot*Rfeed;
```

The series motor torque is computed from the previously derived expression of Equation (10.59), and the pump torque is calculated from Equation (10.60). The motor and pump torques are graphed in Figure 10.47.

```
% Motor torque (N·m) vector
Tmot = Km * Imot.^2;

% Pump torque (N·m) vector
Tpump = G+D*nm/60;

plot(nm,Tmot,nm,Tpump,'--','LineWidth',2.5);
set(gca,'fontname','Times','fontsize',12);
legend('Motor','Pump');
xlabel('Motor-Pump Speed (rpm)');
ylabel('Torque (N·m)');
axis([0 500 0 1000]);
```

The operating point occurs where the motor and pump torque curves intersect. The torque values are compared until the intersection is found. The operating point is printed to the screen and written in the plot title of Figure 10.47.

APPLICATION EXAMPLES

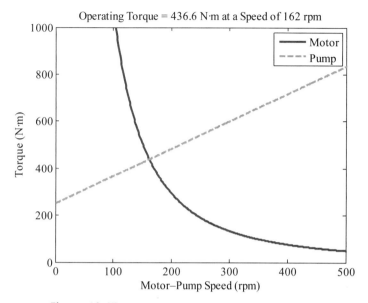

Figure 10.47. Motor and pump torque–speed curves.

```
nop = 0;
k = 1;
while nop == 0
    if Tpump(k) >= Tmot(k)
        nop = k;
    end
    k = k+1;
end
fprintf('Operating torque is %g N·m at a speed of %g rpm.',...
    Tmot(nop), nm(nop));
title(['Operating Torque = ',num2str(Tmot(nop),'%5.1f'),...
    ' N·m at a Speed of ',num2str(nm(nop)),' rpm']);
```

The rated current is calculated using Equation (10.57). The starting current is then limited to twice the rated current. The torque produced from the limited starting current is calculated and compared to the torque required by the pump at a speed of zero:

```
% Motor current rating (A)
Irated = Prated*745.7/Vrated;    % 1 hp = 745.7 W
fprintf('Motor current rating is %g A.\n\n',Irated);

% Limit starting current (A) to twice rated current
Istart = 2*Irated;

% Limited starting motor voltage (V)
Vstart = Istart*(Ra+Rf);
```

```
fprintf('Limited current of %g A at a reduced voltage',...
    'of %g V.\n', Istart, Vstart);
% Limited starting torque (N·m)
Tstart = Km*Istart^2;
fprintf('Limited current produces a starting torque',...
    'of %g N·m.\n', Tstart);
% Minimum pump starting torque
Tpump_start = G;
fprintf('Pump requires a minimum starting torque',...
    'of %g N·m.\n',G);
% Compare limited starting torque to minimum pump starting
% torque
if Tstart >= Tpump_start
    fprintf('Pump can be started with limited',...
        'starting current.\n');
else
    fprintf('Pump cannot be started with limited',...
        'starting current.\n');
end
```

The results from the MATLAB program are:

```
>> DCseriesMotorPump
The motor constant (Km) = 1.43541 V·sec/A.

Operating torque is 436.639 N·m at a speed of 162 rpm.

Motor current rating is 74.57 A.

Limited current of 149.14 A at a reduced voltage of 141.683 V.
Limited current produces a starting torque of 31927.5 N·m.
Pump requires a minimum starting torque of 250 N·m.
Pump can be started with limited starting current.
```

The previous analysis demonstrates that the pump can be successfully started at the lower voltage and current. The procedure could be repeated to find a voltage for which the current rating is not exceeded.

EXAMPLE 10.4: Series motor with brush and copper losses

A series motor is rated 100 hp, 380 V. The field winding and armature resistances are 0.068 and 0.072 Ω, respectively. The voltage drop on the brushes is 3 V total. At rated voltage, the motor operates with a speed of 675 rpm when the motor current is 170 A.

For a copper loss (armature loss) of 1500 W, calculate the motor speed, current, power, torque, and efficiency.

The motor rating data are:

$$V_m := 380 \text{ V} \quad P_m := 100 \text{ hp} \quad R_f := 0.068 \, \Omega \quad R_a := 0.072 \, \Omega.$$

The motor conditions at a single operating point are:

$$V_{mo} := V_m \quad I_{mo} := 170 \text{ A} \quad \text{rpm} := \min^{-1} \quad n_{mo} := 675 \text{ rpm}.$$

Unlike prior examples, the voltage drop on the motor brushes is not neglected:

$$V_{brush} := 3 \text{ V}.$$

The motor constant calculation begins with determining the motor angular speed:

$$\omega_{mo} := 2 \cdot \pi \cdot n_{mo} \quad \omega_{mo} = 70.7 \cdot \frac{\text{rad}}{\text{sec}}.$$

Referring to Figure 10.48, KVL is employed to obtain the induced voltage:

$$E_{mo} := V_{mo} - I_{mo} \cdot (R_f + R_a) - V_{brush} \quad E_{mo} = 353 \text{ V}.$$

Using Equation (10.10), the motor constant is computed:

$$K_m := \frac{E_{mo}}{I_{mo} \cdot \omega_{mo}} \quad K_m = 0.0294 \, \frac{\text{V} \cdot \text{s}}{\text{A}}.$$

In this example, the power loss in the armature is stated as:

$$P_{copper} := 1500 \text{ W}.$$

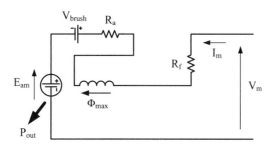

Figure 10.48. Series motor with brush voltage drop.

This known ohmic heating loss in R_a allows the direct determination of the motor current:

$$I_m := \sqrt{\frac{P_{copper}}{R_a}} \quad I_m = 144 \text{ A}.$$

This motor current is less than the rated value of:

$$I_{m\_rated} := \frac{P_m}{V_m} \quad I_{m\_rated} = 196 \text{ A}.$$

The induced voltage at these conditions is:

$$E_{am} := V_m - I_m \cdot (R_f + R_a) - V_{brush} \quad E_{am} = 357 \text{ V}.$$

The motor speed is found by combining Equation (10.10) and Equation (10.11):

$$n_m := \frac{E_{am}}{K_m \cdot 2 \cdot \pi \cdot I_m} \quad n_m = 803 \text{ rpm}.$$

The motor input and output powers are:

$$P_{in} := V_m \cdot I_m \quad P_{in} = 54.8 \text{ kW}$$
$$P_{out} := E_{am} \cdot I_m \quad P_{out} = 69.1 \text{ hp}.$$

The motor efficiency is:

$$\varepsilon := \frac{P_{out}}{P_{in}} \quad \varepsilon = 93.9\%.$$

The torque produced by the motor is:

$$T_m := \frac{P_{out}}{2 \cdot \pi \cdot n_m} \quad T_m = 612 \text{ N} \cdot \text{m}.$$

An alternative formula for computing the torque is found by substituting for the output power:

$$T_{mot} := K_m \cdot I_m^2 \quad T_{mot} = 612 \text{ N} \cdot \text{m}.$$

PROBLEMS

10.5. EXERCISES

1. Describe the construction of the stator of a dc motor. Draw a sketch.
2. Describe the construction of the rotor of a dc motor. Draw a sketch.
3. Describe the commutator.
4. How is a small dc motor cooled?
5. What is the role of the commutator?
6. Describe the generation of the magnetic field in a dc motor.
7. Describe the generation of continuous torque in a dc motor.
8. Explain the interconnection of the rotor and stator in a dc motor.
9. Describe the series motor and draw an equivalent circuit diagram.
10. Describe the shunt motor and sketch an equivalent circuit diagram.
11. Describe the separately excited dc motor and show an equivalent circuit diagram.
12. Describe the compound dc motor and present an equivalent circuit diagram.
13. What is the motor constant and how is it determined?
14. Describe the operation of a shunt dc generator.
15. Show the equations for calculating dc shunt generator currents, power, and torque.
16. Describe the operation of a shunt dc motor.
17. Present the equations for calculating dc shunt motor currents, power, and torque.
18. Describe the operation of a series dc motor.
19. Provide the equations for calculating dc series motor currents, power, and torque.
20. Sketch the torque-versus-speed characteristic of a dc shunt motor.
21. Draw the torque-versus-speed characteristic of a dc series motor.
22. List the losses in a dc motor.
23. What is the starting current of a dc shunt motor?
24. How can the starting current of a dc shunt motor be controlled?
25. What is the starting current of a dc series motor?
26. How can the starting current of a dc series motor be regulated?
27. Explain the speed regulation of a separately excited dc motor.
28. Explain the speed regulation of a dc shunt motor.
29. Explain the speed regulation of a dc series motor.
30. List typical applications for dc motors.

10.6. PROBLEMS

Problem 10.1

Prove that the efficiency of a series motor is simply the ratio of the induced voltage to the motor voltage, that is, $\varepsilon = E_m/V_m$.

Problem 10.2

A dc shunt motor is supplied by 120 Vdc and runs at 800 rpm. The armature resistance is 1.1 Ω and the field winding resistance is 150 Ω. The motor constant is 1.2 V·s/A. Calculate the output power.

Problem 10.3

A dc series motor is supplied by 250 Vdc and runs at 600 rpm. The armature resistance is 1.1 Ω and the field winding resistance is 0.8 Ω. The motor constant is 0.75 V·s/A. Calculate the output power.

Problem 10.4

A dc series motor is supplied by 250 Vdc. The motor current is 6 A when it runs at 600 rpm. The armature resistance 1.1 Ω and the field winding resistance is 0.8 Ω. Calculate the motor constant.

Problem 10.5

A dc shunt generator charges a battery. The battery voltage is 12 V and the required charging current is 50 A. The generator shunt winding resistance is 20 Ω and the armature resistance is 0.05 Ω. The generator speed is 800 rpm. In open-circuit conditions, the generator input current is 5 A. Calculate the required input power.

Problem 10.6

A 12 V car battery supplies a dc shunt motor. The motor shunt coil resistance is 15 Ω and the armature resistance is 0.03 Ω. In open-circuit conditions, the motor speed is 700 rpm and its input current is 5 A. Calculate the motor speed at a load of 0.5 hp.

Problem 10.7

A 12 V car battery supplies a dc series motor. The motor field winding resistance is 0.015 Ω and the armature resistance is 0.03 Ω. In open-circuit conditions, the motor speed is 700 rpm and its input current is 5 A. Calculate the motor speed at a load of 0.5 hp.

Problem 10.8

A dc separately excited motor is supplied by 120 Vdc. The motor load is 20 hp and the motor speed is 850 rpm. The motor data are armature resistance of 0.04 Ω and field winding resistance of 100 Ω. The motor constant is 0.5 V·s/A. Calculate the required field voltage.

Problem 10.9

Using the motor-driven pump data of Example 10.3, calculate the supply voltage and starting torque if the starting current is limited to the rated current. Determine whether the motor torque is sufficient to start the pump.

Problem 10.10

A dc shunt motor rated at 240 V has a field winding resistance of 120 Ω and an armature resistance of 0.12 Ω. The motor is supplied at the rated voltage. Under the no-load condition, the motor draws a line current of 4.75 A. When driving a load, the motor runs at 1350 rpm and draws a current of 29 A. Use the load data for the calculation of the motor constant.

(a) Draw the equivalent circuit of the motor and determine the motor constant. (b) Calculate the no-load speed. (c) Plot the motor speed versus the load current and determine the load current when the speed is 1% below the no-load speed.

Problem 10.11

A 228 Vdc series motor delivers its rated power with a current of 25 A at the rated speed of 1400 rpm. The armature and field resistances are 0.17 and 0.11 Ω, respectively.

(a) Draw the equivalent circuit diagram and calculate the motor constant. (b) Calculate the speed and the torque of the motor when the current is 40 A. (c) Calculate and plot the motor torque and speed versus load current curves and determine the load current and speed when the torque is 20 N·m.

Problem 10.12

A dc shunt motor is rated at 50 hp and 240 V. The voltage drop across the brushes is 3 V, and the armature and field resistances are 0.12 and 125 Ω, respectively. At no load, the motor draws a current of 15 A at a speed of 1800 rpm.

(a) Draw the equivalent circuit. (b) Calculate the motor constant. (c) Calculate and plot the motor torque–speed characteristics. Determine the starting torque and the speed when the torque is 1000 N·m.

Problem 10.13

A dc shunt generator rated at 85 kW produces a voltage of 280 V. The brush voltage drop is 2.5 V, and the armature and field resistances are 0.09 and 115 Ω, respectively. The generator delivers rated current at rated speed and rated voltage.

(a) Calculate the field, armature, and load currents versus load. Test the equations with the rated load. (b) Determine the terminal voltage at no-load and at rated load conditions. (c) Calculate the voltage regulation of the generator. Use the no-load voltage as the base value. (d) Plot the terminal voltage as a function of the load. Determine the load that corresponds to a 5% voltage drop using the no-load voltage as the base.

Problem 10.14

A dc shunt generator operates at 675 rpm with an output of 380 V and 170 A. The armature resistance and the field resistance are 0.072 and 85 Ω, respectively. When the generator supplies power to a load, the armature copper loss is 1500 W. The brush contact voltage is 3 V.

(a) Determine the machine constant K_a and draw the equivalent circuit. (b) Calculate the load current, induced voltage, and speed when the copper loss is the value given earlier and the speed is adjusted such that the terminal voltage is kept constant. (c) Calculate the input and output powers and efficiency.

Problem 10.15

Derive the formula for the generator mode operation of the separately excited machine. Specifically, find the generator-induced voltage and terminal voltage, the input mechanical and output electrical powers, and the efficiency.

11

INTRODUCTION TO POWER ELECTRONICS AND MOTOR CONTROL

The rapid development of semiconductor switches has resulted in the implementation of power electronic circuits for control of motors, lighting, battery chargers, and even the power system. This introductory chapter presents the basic concepts of induction and direct current (dc) motor drive controls and the related electronic circuits. This requires the treatment of rectifiers, and voltage and current commutated inverters. The reduction in the cost of semiconductor devices and increases in their ratings resulted in the electronic control of power systems in the form of flexible alternating current transmission system (FACTS) devices. The most frequently used device is the static volt-ampere reactive (VAR) compensator (SVC) that contains a thyristor-switched capacitor and a thyristor-controlled reactance connected in parallel to regulate reactive power. This chapter presents both FACTS and DC-to-DC converters.

First, the concept of dc and induction motor controls is addressed, along with the definition of the variables that have to be controlled and the relations used for the control. Next, brief descriptions of the semiconductor switches are provided. The operation of the controlled single-phase rectifiers is discussed, and basic equations to analyze their operation are derived. This is followed by a description of pulse width modulation (PWM) and the analyses of PWM converter operation. Later, the application of PWM converters for induction motor control is presented.

Electrical Energy Conversion and Transport: An Interactive Computer-Based Approach, Second Edition.
George G. Karady and Keith E. Holbert.
© 2013 Institute of Electrical and Electronics Engineers, Inc. Published 2013 by John Wiley & Sons, Inc.

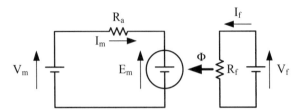

Figure 11.1. Equivalent circuit for a separately excited dc motor.

11.1. CONCEPT OF DC MOTOR CONTROL

DC motor control being simpler than alternating current (ac) motor control, we first explore the former in this chapter. The most frequent objective of dc motor control is speed regulation in such a way that the torque remains constant. In this section, we derive the equations for this case. However, other goals, such as maintaining the speed at variable torque, may be achieved in a similar manner.

The concept of speed control of a separately excited dc motor is illustrated using a numerical example. Figure 11.1 shows the equivalent circuit for a separately excited dc motor. The 250 V, 10 hp motor has a machine constant of 0.9 V·s/A. The armature and field winding resistances are 2 and 125 Ω, respectively. The stator or field voltage (V_f) is a constant 250 V. The motor voltage or armature voltage (V_m), which is supplied to the rotor through the brushes and commutator, is variable.

The dc motor data are initialized using:

$$P_{rated} := 10 \text{ hp} \quad V_{rated} := 250 \text{ V} \quad K_m := 0.9 \frac{V \cdot s}{A}$$
$$V_f := 250 \text{ V} \quad R_f := 125 \text{ } \Omega \quad R_a := 2 \text{ } \Omega.$$

To verify the following calculations, the motor is initially supplied with its rated voltage and an operating speed (n_m) of 1000 rpm is chosen:

$$V_m := V_{rated} \quad \text{rpm} := \text{min}^{-1} \quad n_m := 1000 \text{ rpm}.$$

In Chapter 10, we derived a formula for the induced voltage in terms of the motor constant, specifically, $E_m = K_m \omega_m I_f$. Substituting $\omega_m = 2\pi n_m$ and using Ohm's law at the field winding yields a relation for the induced voltage:

$$E(n_m) := K_m \cdot 2 \cdot \pi \cdot n_m \cdot \frac{V_f}{R_f}. \tag{11.1}$$

For the selected motor speed, the induced voltage is $E(n_m) = 188.5$ V. Except for n_m, the parameters in Equation (11.1) are constants, so a constant C_1 is established for convenience:

$$C_1 := K_m \cdot 2 \cdot \pi \cdot \frac{V_f}{R_f}, \tag{11.2}$$

where C_1 for this motor is $C_1 = 11.31$ Wb. The simplified expression for the induced voltage is:

$$E(n_m) := C_1 \cdot n_m. \tag{11.3}$$

Applying Kirchhoff's voltage law (KVL) to the main circuit loop in Figure 11.1, the motor voltage is equal to the induced voltage plus the armature resistance voltage drop:

$$V_m = R_a \cdot I_m + E(n_m) \tag{11.4}$$

Rearranging the earlier expression, the motor current is:

$$I_m := \frac{V_m - E(n_m)}{R_a}. \tag{11.5}$$

The current at rated motor voltage is $I_m = 30.8$ A. Substituting Equation (11.3) for the induced voltage into Equation (11.5) yields:

$$I_m := \frac{V_m - C_1 \cdot n_m}{R_a}. \tag{11.6}$$

The motor mechanical output power is the product of the induced voltage and motor current:

$$P_m := E(n_m) \cdot I_m, \tag{11.7}$$

which numerically is $P_m = 7.77$ hp. Substituting Equation (11.3) and Equation (11.6) into Equation (11.7) yields:

$$P_m := C_1 \cdot n_m \cdot \frac{V_m}{R_a} - \frac{C_1^2 \cdot n_m^2}{R_a}. \tag{11.8}$$

The motor torque is:

$$T_m := \frac{P_m}{2 \cdot \pi \cdot n_m}, \tag{11.9}$$

with a numerical value of $T_m = 55.4$ N·m. The substitution of the motor power into the torque formula yields:

$$T_m := C_1 \cdot n_m \cdot \frac{V_m}{R_a \cdot (2 \cdot \pi \cdot n_m)} - \frac{C_1^2 \cdot n_m^2}{2 \cdot \pi \cdot n_m \cdot R_a}. \tag{11.10}$$

This expression is simplified by the introduction of two additional constants:

$$C_2 := \frac{C_1}{2 \cdot \pi \cdot R_a} \quad C_2 = 0.900 \cdot A \cdot s$$

$$C_3 := \frac{C_1^2}{2 \cdot \pi \cdot R_a} \quad C_3 = 10.2 \; A \cdot s^2 \cdot V.$$

The simplified torque formula is:

$$T_m := C_2 \cdot V_m - C_3 \cdot n_m. \tag{11.11}$$

The purpose of this analysis is the derivation of a relationship describing the motor speed as a function of torque and motor armature voltage. By rearranging Equation (11.11), the motor speed is calculated from:

$$n_m := \frac{C_2}{C_3} \cdot V_m - \frac{T_m}{C_3}. \tag{11.12}$$

This expression indicates that the motor speed depends on the torque and the motor voltage. Again, another constant may be employed:

$$C_4 := \frac{C_2}{C_3} \quad C_4 = 5.305 \; \frac{\text{rpm}}{V}.$$

The motor speed as a function of motor torque and armature voltage is then:

$$n_{\text{mot}}(V_m, T_m) := C_4 \cdot V_m - \frac{T_m}{C_3}. \tag{11.13}$$

Setting the torque to $T_m := 100 \cdot$N·m, the corresponding motor speed at rated voltage is:

$$n_{\text{mot}}(V_m, T_m) = 736.8 \text{ rpm}.$$

CONCEPT OF DC MOTOR CONTROL

If we substitute the full expressions for the constants into Equation (11.13), we obtain:

$$n_{motor}(V_m, T_m) := \frac{R_f}{2 \cdot \pi \cdot K_m \cdot V_f} \cdot V_m - \frac{R_a \cdot R_f^2}{2 \cdot \pi \cdot K_m^2 \cdot V_f^2} \cdot T_m. \tag{11.14}$$

We assume that the motor cannot be reversed; consequently, the motor speed is always positive. This limitation is expressed in Mathcad using the *if* conditional function, which has the form:

if(condition, true value returned, false value returned).

The motor speed is maintained positive using:

$$n_{motor}(V_m, T_m) := if\left[\left(C_4 \cdot V_m - \frac{T_m}{C_3}\right) < 0, 0, C_4 \cdot V_m - \frac{T_m}{C_3}\right]. \tag{11.15}$$

The equation shows a linear relationship between the speed and armature voltage for a given torque, and this is exhibited in Figure 11.2. The linear association shown demonstrates that the motor speed is simply and easily controlled using the motor rotor (armature) voltage. Although the previous derivation is based upon the separately excited machine, the speed–torque relation for other dc motors can be derived using a similar approach.

Figure 11.3 shows an illustration of the concept of the speed control of a dc motor. The ac voltage is rectified by the thyristor-controlled rectifier that produces variable dc voltage, which is filtered by the capacitor in the dc link. This dc voltage supplies the

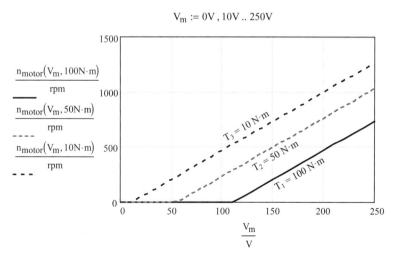

Figure 11.2. Separately excited motor speed versus motor voltage at three different torque values.

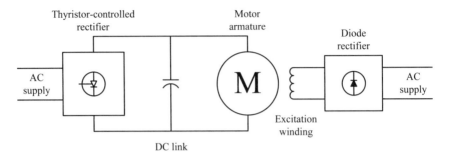

Figure 11.3. Concept of a dc motor drive.

motor armature. The motor speed is proportional with the armature (motor) voltage. The motor excitation (stator field) winding is supplied by constant dc voltage, which is generated by a diode rectifier.

The earlier analysis reveals that dc motor control requires rectification. Both diode- and thyristor-based rectification are presented later in this chapter.

11.2. CONCEPT OF AC INDUCTION MOTOR CONTROL

The most frequent objective of ac induction motor control is the regulation of the motor speed over a wide range. A typical application occurs in industrial process control, where the required response time is relatively long, measured in several seconds. For instance, dampers can regulate the airflow in a duct. Many times, this type of control can produce turbulence and noise. An alternative solution is the control of airflow by regulating the fan speed. A domestic example is an adjustable speed drive for a residential air conditioner. The air conditioner motor speed is controlled instead of switching the unit off and on.

The basic method of induction motor speed control is the variation of the supply voltage frequency (f_{sup}) by a thyristor-controlled rectifier–inverter unit. The speed control range can be separated into two parts:

1. when the supply voltage frequency is below the system frequency ($f_{sup} < f_{sys}$); and
2. when the supply voltage frequency is above the system frequency ($f_{sup} > f_{sys}$).

The concept of speed control of an induction motor is illustrated with a numerical example. A three-phase, wye-connected, squirrel cage induction motor is supplied by a variable frequency three-phase power supply. The motor drives a centrifugal pump, as depicted in Figure 11.4. The pump output is defined by the motor–pump speed, which is regulated by the variation of the supply voltage frequency.

The six-pole ac motor is rated at 15 hp with a line voltage of 460 V. The single-phase equivalent circuit for the ac motor is given in Figure 11.5, and the corresponding circuit values are defined as follows:

CONCEPT OF AC INDUCTION MOTOR CONTROL

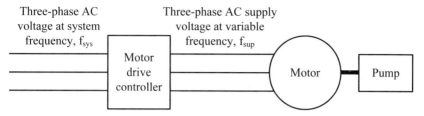

Figure 11.4. Three-phase motor controlled by an adjustable speed drive.

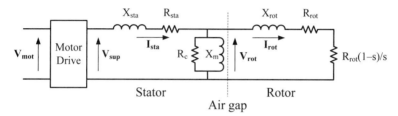

Figure 11.5. Single-phase equivalent circuit for a controlled induction motor.

$$P_{mot} := 15 \text{ hp} \quad V_{mot} := 460 \text{ V} \quad p := 6$$
$$R_{sta} := 0.2 \, \Omega \quad R_{rot} := 0.25 \, \Omega \quad R_c := 317 \, \Omega.$$
$$X_{sta} := 1.2 \, \Omega \quad X_{rot} := 1.29 \, \Omega \quad X_m := 42 \, \Omega$$

The pump torque–speed characteristic is described by:

$$T_{pump}(n_m) := 100 \text{ N} \cdot \text{m} + (0.15 \text{ N} \cdot \text{m} \cdot \text{s}^2) \cdot n_m^2,$$

where n_m is the pump speed. The supply frequency and the motor speed are variable, but for the verification of the equations derived later, we select initial values of:

$$f_{sup} := 40 \text{ Hz} \quad \text{rpm} := \text{min}^{-1} \quad n_m := 775 \text{ rpm}.$$

From Chapter 9, the synchronous speed and slip of the motor as a function of frequency are:

$$n_{syn}(f_{sup}) := \frac{f_{sup}}{\frac{p}{2}}, \tag{11.16}$$

$$s_m(f_{sup}, n_m) := \frac{n_{syn}(f_{sup}) - n_m}{n_{syn}(f_{sup})}. \tag{11.17}$$

The numerical values are:

$$n_{syn}(f_{sup}) = 800 \text{ rpm} \quad s_m(f_{sup}, n_m) = 3.13\%.$$

The reactance of the motor depends on the frequency. The motor data are measured at the system frequency of 60 Hz ($f_{sys} := 60$ Hz); consequently, the reactances must be recalculated for other supply frequencies (f_{sup}). Frequency scaling of the inductive reactances ($X = \omega L$) is accomplished by multiplying the reactance at the system frequency by the ratio of the supply to system frequencies ($X_2 = \omega_2 L = (\omega_2/\omega_1)\omega_1 L = (\omega_2/\omega_1)X_1$):

$$X_{rot}(f_{sup}) := X_{rot} \cdot \frac{f_{sup}}{f_{sys}} \quad X_{sta}(f_{sup}) := X_{sta} \cdot \frac{f_{sup}}{f_{sys}} \quad X_m(f_{sup}) := X_m \cdot \frac{f_{sup}}{f_{sys}}.$$

The motor impedances are calculated by using the equivalent circuit of Figure 11.5. The rotor impedance is:

$$Z_{rot}(f_{sup}, n_m) := j \cdot X_{rot}(f_{sup}) + \frac{R_{rot}}{s_m(f_{sup}, n_m)}$$

$$Z_{rot}(f_{sup}, n_m) = (8 + 0.86j) \, \Omega.$$

The rotor and magnetizing impedances are connected in parallel. The combined impedance is:

$$Z_{m\_rot}(f_{sup}, n_m) := \frac{1}{\frac{1}{R_c} + \frac{1}{j \cdot X_m(f_{sup})} + \frac{1}{Z_{rot}(f_{sup}, n_m)}}$$

$$Z_{m\_rot}(f_{sup}, n_m) = (6.86 + 2.65j) \, \Omega.$$

The total motor impedance is the sum of the stator impedance and the combined rotor and magnetizing impedances:

$$Z_{mot}(f_{sup}, n_m) := j \cdot X_{sta}(f_{sup}) + R_{sta} + Z_{m\_rot}(f_{sup}, n_m)$$

$$Z_{mot}(f_{sup}, n_m) = (7.06 + 3.45j) \, \Omega.$$

The change of ac supply frequency varies the motor magnetic flux ($\Phi \propto I$). Reduction of the supply voltage frequency increases the flux, and an increase of frequency decreases the flux. This implies that motor core saturation can occur if the frequency is reduced. The saturation increases the magnetizing current and disturbs operation. To avoid magnetic saturation, the motor flux is maintained constant. The flux is proportional to the voltage and inversely proportional to the frequency. Consequently, the flux can be kept constant by changing the supply voltage with the frequency. The reduction

CONCEPT OF AC INDUCTION MOTOR CONTROL

of the supply frequency requires the simultaneous reduction of the supply voltage. The variable supply line-to-neutral voltage is:

$$V_{sup}(f_{sup}) := \frac{V_{mot}}{\sqrt{3}} \cdot \frac{f_{sup}}{f_{sys}}.$$

This rule implies that the supply voltage has to be increased when the supply frequency is increased. However, most motors can tolerate only 5–10% overvoltage because the elevated voltage produces overload. The frequency increase reduces the motor flux, which eliminates the danger of saturation. Consequently, if a higher than rated speed is needed, the motor is supplied with higher than 60 Hz frequency, but the voltage is kept constant at the rated voltage level. Simultaneously, if lower than rated speed is required, the motor is supplied with lower than 60 Hz frequency but the voltage is varied proportional with the frequency. The modified current equation to implement this approach utilizes the Mathcad *if* conditional function:

$$V_{sup}(f_{sup}) := \text{if}\left(f_{sup} < f_{sys}, \frac{V_{mot}}{\sqrt{3}} \cdot \frac{f_{sup}}{f_{sys}}, \frac{V_{mot}}{3}\right).$$

From Ohm's law, the motor current is:

$$I_{sta}(f_{sup}, n_m) := \frac{V_{sup}(f_{sup})}{Z_{mot}(f_{sup}, n_m)}$$

$$|I_{sta}(f_{sup}, n_m)| = 22.5 \text{ A} \quad \arg(I_{sta}(f_{sup}, n_m)) = -26.1 \text{ deg}.$$

An induction machine can operate as a motor or as a generator. The generator operation requires driving the motor faster than the synchronous speed. This capability is sometimes used for *regenerative breaking* of electric cars and trains. However, an ordinary motor does not operate in a generator mode. The generator operation can be eliminated from the mathematical solution by assuming that the current is zero above the synchronous speed. The modified current equation utilizes the Mathcad *if* conditional function:

$$I_{sta}(f_{sup}, n_m) := \text{if}(n_m > n_{syn}(f_{sup}), 0 \text{ A}, I_{sta}(f_{sup}, n_m)).$$

Using current division, the rotor current is:

$$I_{rot}(f_{sup}, n_m) := I_{sta}(f_{sup}, n_m) \cdot \frac{Z_{m\_rot}(f_{sup}, n_m)}{Z_{rot}(f_{sup}, n_m)}$$

$$|I_{rot}(f_{sup}, n_m)| = 20.6 \text{ A} \quad \arg(I_{rot}(f_{sup}, n_m)) = -11.1 \text{ deg}.$$

From Chapter 9, the developed output power is:

$$P_{dev}(f_{sup}, n_m) := 3 \cdot (|I_{rot}(f_{sup}, n_m)|)^2 \cdot \frac{1 - s_m(f_{sup}, n_m)}{s_m(f_{sup}, n_m)} \cdot R_{rot}$$

$$P_{dev}(f_{sup}, n_m) = 13.2 \text{ hp}.$$

If the ventilation and friction losses are neglected, the ac motor torque is:

$$T_m(f_{sup}, n_m) := \frac{P_{dev}(f_{sup}, n_m)}{2 \cdot \pi \cdot n_m} \quad T_m(f_{sup}, n_m) = 121.5 \text{ N} \cdot \text{m}.$$

Figure 11.6 shows the motor torque–speed characteristics at different supply frequencies, f_{sup} (i.e., 10, 30, 60, 90, and 120 Hz). It can be seen that both the increase and the reduction of the supply frequency affect both the speed of the motor and the maximum value of the torque curve.

Reduction of the frequency from the rated speed (at 60 Hz) lowers the synchronous speed (which occurs at a torque of zero) and shifts the torque–speed curve to the left (lower speed). Simultaneously, the maximum torque is also reduced. As the frequency is lowered, the corresponding motor voltage is reduced too. The appropriate supply voltage values are given in the graph.

Increasing the frequency from 60 Hz increases the synchronous speed (which occurs at a torque of zero) and shifts the torque–speed curve to the right (higher speed).

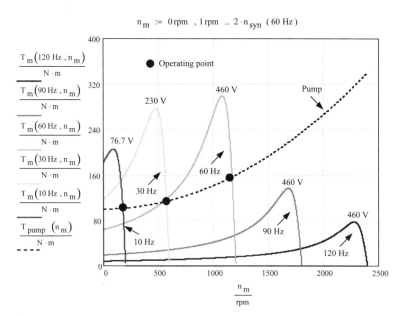

Figure 11.6. Motor and pump torque–speed curves with supply line voltage.

CONCEPT OF AC INDUCTION MOTOR CONTROL

Simultaneously, the maximum torque is reduced. However, the corresponding motor voltage is kept constant.

The intersections of the motor and pump torque–speed curves determine the operating points. Figure 11.6 indicates that at high speed, this motor is unable to drive the pump because the pump torque curve is above the motor torque curves. The motor and pump torque curves do not intersect significantly above the rated speed region. The operating region can be determined by trial and error.

It can be seen that the speed at each operating point is very close to the synchronous speed for that frequency. This, to a first approximation, permits the replacement of the actual operating speed with the synchronous speed of Equation (11.16). The approximate control equations for *reduced frequency* are:

$$n_m(f_{sup}) = \frac{f_{sup}}{\frac{p}{2}} \quad V_{sup}(f_{sup}) = \frac{V_{mot}}{\sqrt{3}} \cdot \frac{f_{sup}}{f_{sys}}. \tag{11.18}$$

If the supply frequency is increased above the system frequency, the supply voltage is not regulated; rather, it is kept constant at the rated value level. Consequently, the ac motor speed control requires the ability to:

- regulate the supply frequency;
- control the supply voltage in proportion to the frequency while the frequency is below the system frequency; and
- maintain the supply voltage at the rated voltage level when the frequency is above the system frequency.

This example reveals two control areas; however, power engineering practice identifies three control ranges for induction motor control. Figure 11.7 shows the approximate variation of torque and motor power versus frequency. The detailed analysis of the torque–speed–frequency relationship is beyond this book's objectives, but we identify the ranges as follows:

- *Constant Torque Range*. In this range, the speed is below the rated speed, and the supply voltage frequency is below the system frequency. In general, the motor can supply more or less constant torque around the rated torque level.
- *Constant Power Range*. In this range, the speed is beyond the rated speed, and the supply voltage frequency is above the system frequency. The torque decreases inversely with the frequency, but the motor power remains more or less constant, as shown in Figure 11.7.
- *High-Speed Range*. If the motor speed is increased beyond two to three times the rated speed, the torque decreases inversely with the square of the supply frequency.

Figure 11.8 exhibits the major components of a typical induction motor drive (controller). The ac supply voltage is rectified and the obtained dc voltage is filtered by the

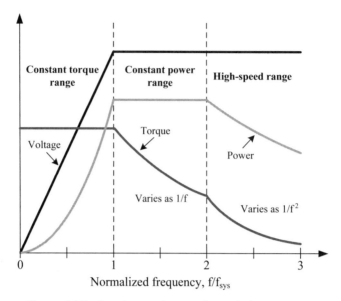

Figure 11.7. Speed control ranges for an induction motor.

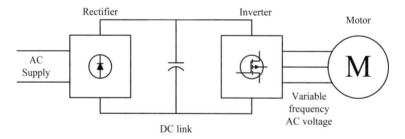

Figure 11.8. Concept of induction motor control characteristics.

capacitor in the dc link. The dc voltage supplies the inverter that generates variable-frequency ac voltage. Both the frequency and the amplitude (root-mean-square [rms] value) of the inverter output are controlled. The inverter-generated three-phase voltage supplies the motor. The frequency of the inverter voltage determines the motor speed. The output voltage of the inverter is regulated to maintain the motor flux constant at lower speed, which requires a constant voltage/frequency ratio.

An alternative solution is to regulate the dc voltage by a controlled rectifier and run the inverter with variable frequency, without voltage control.

These results demonstrate that ac induction motor control requires both a rectifier and an inverter, whereas the dc motor drive requires rectification only.

11.3. SEMICONDUCTOR SWITCHES

Before examining rectifier and inverter operation, an understanding of the devices that make such converters possible is needed. The most important component of a rectifier or inverter is the semiconductor switch. Advances in semiconductor technology frequently produce new switches. Presently, the most commonly used switches are:

- diode,
- thyristor,
- gate turn-off (GTO) thyristor,
- metal–oxide–semiconductor field-effect transistor (MOSFET), and
- insulated gate bipolar transistor (IGBT).

The operating principle of each of these five semiconductor devices is briefly presented in the succeeding sections.

11.3.1. Diode

The *diode* is a two-terminal device (*pn* junction), which is intended to permit current flow only from the anode (p^+) to the cathode (*n*). Figure 11.9 shows the symbol of a

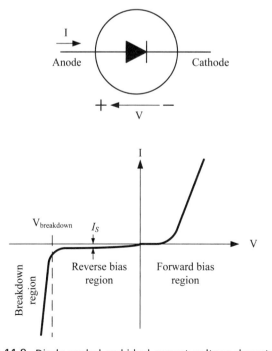

Figure 11.9. Diode symbol and ideal current–voltage characteristic.

Figure 11.10. Typical medium power diode.

diode and a typical current–voltage (I-V) characteristic. For an ideal diode under a forward bias voltage, the current is an exponential function known as Shockley's expression:

$$I = I_S \left(e^{V/nV_T} - 1 \right), \qquad (11.19)$$

where I_S is the reverse bias saturation current, V_T is the thermal voltage, and n is a diode constant ($1 \leq n \leq 2$). The thermal voltage is $V_T = kT/q$ where k is the Boltzmann constant, T is the absolute temperature, and q is the fundamental charge constant.

The *forward current* of a large diode can be several thousand amperes. Conversely, the *reverse current* is only few milliamps. The *forward voltage drop* is less than a volt, when the current is low, but can be a few volts at rated current. In the reverse current direction, the increase of the voltage beyond the rated value produces *breakdown* and destruction of the device.

Figure 11.10 shows a picture of a typical medium power diode. The major part of the loss in a *power diode* is the product of forward voltage drop and forward current. In a large power diode, the power loss can be several hundred watts. This requires that the diode be mounted on an air- or liquid-cooled heat sink. Figure 11.11 displays a power diode mounted on an air-cooled heat sink with fins.

The operation of a power diode is demonstrated in Figure 11.12. The figure shows the operation of a single-phase rectifier supplying a load resistance (R_{load}). The diode conducts during the positive cycle, when it interconnects the ac source (V_{ac}) with the load resistance. The diode current (I_{load}) is equal to the ratio of the ac voltage and resistance, that is, $I_{load} = V_{ac}/R_{load}$. The voltage across the diode is small, less than a volt. In the negative cycle, the diode is not conducting ($I_{load} = 0$). Consequently, the diode voltage will be equal to the ac voltage as displayed on the lower graph.

To a first approximation, the selection of a diode requires the average forward current, the maximum reverse voltage, and the maximum fault current and its duration. For the power diode of Figure 11.12, the maximum reverse voltage is 100 V, which occurs at $\omega t = 270°$, and the average forward current over a full cycle is found by integrating the current:

Figure 11.11. Power diode mounted on an air-cooled heat sink.

$$I_{dc\_ave} := \frac{\omega}{2\cdot\pi} \cdot \int_0^{\frac{2\cdot\pi}{\omega}} I_{load}(t)\,dt = 6.069 \text{ A},$$

where the cycle period is $T = 1/f = 2\pi/\omega$.

11.3.2. Thyristor

The *thyristor* is a three-terminal latching switch, with a four-layer (*npnp*) construction. The thyristor is also known as a *silicon-controlled rectifier* (SCR). The large devices used for high-voltage dc (HVDC) transmission can carry a few thousand amperes of current and withstand voltages up to 10 kV.

Figure 11.13 presents the symbol for a thyristor. The current (I_A) is purposed to flow only from the anode to the cathode, that is, the thyristor conducts in only one direction. A gate current pulse (I_{gate}) can turn the SCR on, if the anode-to-cathode voltage (V_{ACat}) is positive. The device turns off when the current reaches zero or a low holding current value. The thyristor is forward biased when the anode–cathode voltage is positive and reverse biased when this voltage is negative.

Figure 11.14 provides an illustration of the thyristor current–voltage characteristics. The graph shows that a negative bias voltage drives small reverse currents (10–50 mA) through the thyristor. But the increase of negative voltage ultimately causes reverse voltage breakdown, which in most cases destroys the device. In the forward biased direction, when the thyristor is not gated ($i_{g0} = 0$), a small anode current flows

688 INTRODUCTION TO POWER ELECTRONICS AND MOTOR CONTROL

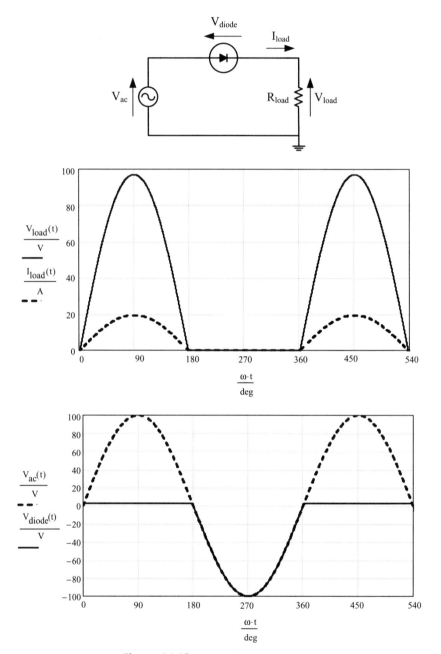

Figure 11.12. Power diode operation.

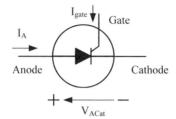

Figure 11.13. Thyristor or silicon-controlled rectifier (SCR) symbol.

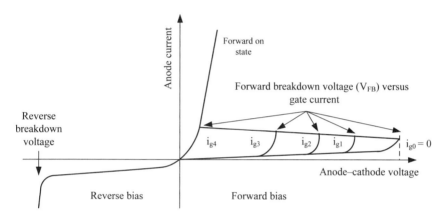

Figure 11.14. Thyristor current–voltage characteristics.

and the device turns on if the anode–cathode voltage is larger than the forward breakdown voltage (V_{FB}). The application of a gate pulse reduces the anode–cathode voltage required to turn on the device. In the graph, $i_{g4} > i_{g3} > i_{g2} > i_{g1} > i_{g0}$. The voltage drop slightly increases with the current when the device turns on but the voltage drop is never larger than a few volts.

The described operation of a thyristor is illustrated in Figure 11.15. An ac source supplies a single-phase thyristor-controlled rectifier loaded by a resistance. The upper plot demonstrates that the thyristor starts to conduct in the positive period when the gate is fired at $\omega t = 30°$ by a short duration voltage pulse (V_{gate}), which is displayed in the lower graph. Typically, the gate pulse is a 10 V positive pulse.

The thyristor turns off (stops conducting), when the current (I_{load}) direction is reversed. When the thyristor is not conducting, the voltage across the device is the ac voltage, as shown in the lower graph of Figure 11.15. The SCR voltage drop during the conduction period is a few volts.

Figure 11.16 exhibits the details of the turn-on and turn-off processes. The turn on is initiated by the gate pulse. The anode–cathode current starts to flow after a few microseconds delay (t_d). The rise time (t_{rise}) of the current depends on the external circuit. But the device manufacturer specifies the maximum |di/dt| values. A faster

690 INTRODUCTION TO POWER ELECTRONICS AND MOTOR CONTROL

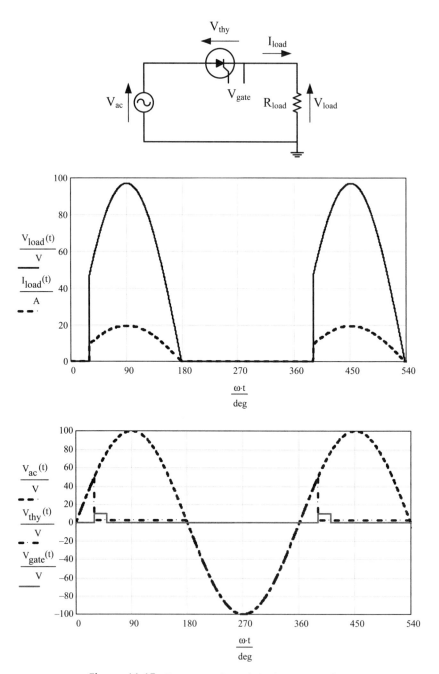

Figure 11.15. Demonstration of thyristor operation.

SEMICONDUCTOR SWITCHES

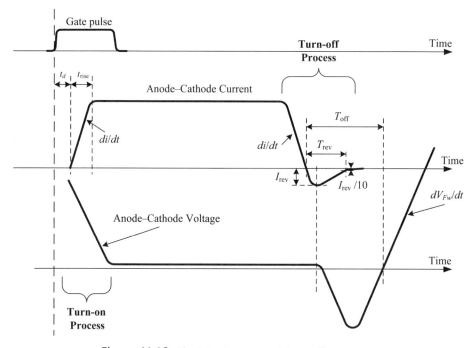

Figure 11.16. Thyristor turn-on and turn-off processes.

current increase or decrease may destroy the device. A typical value for $|di/dt|$ is around 500–1000 A/μs.

An external circuit initiates the turn off of the device. The device turns off when the current is rapidly decreasing. Figure 11.16 shows that the device does not turn off at zero current. For a short period of time (T_{rev}), typically 20–200 μs, a negative current flows. This current neutralizes the charges accumulated in the semiconductor during the conduction period. The integral of the negative current is the storage charge, which is specified by the manufacturer. A typical value is around 500–1000 μC. Figure 11.16 shows that after turn off, the anode–cathode voltage is negative. A positive forward voltage (V_{Fw}) can be applied to the device after the manufacturer-specified turn-off time (T_{off}). The critical parameter is the derivative of the voltage rise (dV_{Fw}/dt). The manufacturer specifies the maximum value of the voltage derivative. A typical value is around 500–1000 V/μs.

In the first approximation, the selection of a thyristor requires the average forward current, the maximum current derivative, and the maximum reverse and forward voltages and their derivatives, together with the maximum fault current and its duration.

Figure 11.17 shows a picture of two flat-pack thyristors mounted on a liquid-cooled heat sink. The two devices are connected in series to permit higher operating voltage. The flat-pack type thyristor is used for high-power applications. The photograph shows three liquid-cooled heat sink elements. This arrangement permits the cooling of both sides of each thyristor.

Figure 11.17. Two flat-pack thyristors mounted between three liquid-cooled heat sinks.

Figure 11.18 exhibits a large light-fired thyristor used for HVDC transmission. The photograph displays the fiber-optic light pipe used for firing the device and the silicon wafer inside the device. This device is rated for 4000 A and 10 kV.

The earlier description illustrates that the thyristor operation requires a square pulse to fire the device. This pulse is generated by an auxiliary circuit, which is described later in this chapter.

11.3.3. Gate Turn-Off Thyristor

The *gate turn-off thyristor* (GTO) is presented in Figure 11.19. The GTO is a thyristor that can be turned on by a positive gate pulse when the anode-to-cathode voltage is positive, and can be turned off by a negative pulse before the polarity of V_{ACat} is reversed. The device is designed to conduct in only one direction from the anode to the cathode. The current–voltage characteristics of the GTO are similar to those of the thyristor.

The operation of the device is demonstrated in Figure 11.20. The graphs show that the GTO conducts only in the positive period, when the current flows from the anode to the cathode. The device is turned on (at $\omega t = 30°$) by a positive pulse and turned off (at $\omega t = 140°$) with a negative pulse before the current reversal in the negative period. The gate signals consist of a positive pulse and a negative pulse. The switching of the GTO requires a powerful auxiliary circuit, which can produce both positive and negative gate signals. The turn on of a GTO requires a gate signal similar to the thyristor

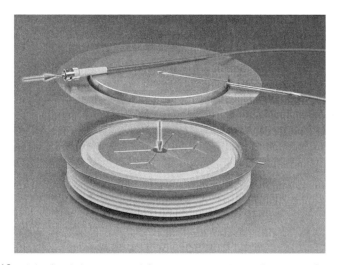

Figure 11.18. Light-fired thyristor used for HVDC transmission (courtesy of Siemens, Erlangen, Germany).

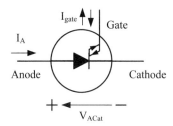

Figure 11.19. Gate turn-off thyristor (GTO) symbol.

(from 100 mA to 1 A). The turn off requires large current. A typical turn-off current is one-third or one-fifth of the rated conduction current.

When the device is off, the ac voltage appears across the GTO, as shown in the lower graph. During the conduction period, the device voltage is only a few volts.

11.3.4. Metal–Oxide–Semiconductor Field-Effect Transistor

The *metal–oxide–semiconductor field-effect transistor* (MOSFET) is a fast-acting switch. Figure 11.21 presents the symbol of the MOSFET, which has three terminals: drain (D), source (S), and gate (G). Current flows from the drain to the source in an n-channel device, and from the source to the drain in a p-channel device. The gate (gate–source) voltage controls the current flow. For protection against reverse voltage-produced breakdown, the device is shunted by a diode in the reverse direction.

694 INTRODUCTION TO POWER ELECTRONICS AND MOTOR CONTROL

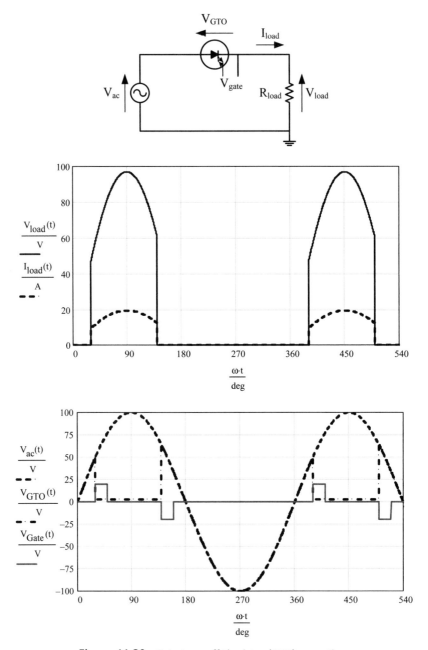

Figure 11.20. Gate turn-off thyristor (GTO) operation.

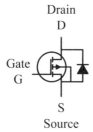

Figure 11.21. Power metal–oxide–semiconductor field-effect transistor (MOSFET) symbol.

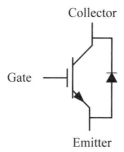

Figure 11.22. Insulated gate bipolar transistor (IGBT) symbol.

In power circuits, the MOSFET is driven by a relatively large gate pulse, which turns on the device and drives it to saturation. For this mode, the voltage drop is a few volts across the device.

Generally, the maximum ratings of power MOSFETS are in the range of a few hundred amperes (200–300 A) and around 1000–1500 V. The operating frequency can be in the megahertz range. A typical application of the power MOSFET is in *pulse-width-modulation* (PWM) circuits.

11.3.5. Insulated Gate Bipolar Transistor

The *insulated gate bipolar transistor* (IGBT) is a three-terminal device, which is used as a high-speed switch. Because of the relatively low reverse breakdown voltage, the device is shunted by a diode in the reverse direction. The IGBT symbol is shown in Figure 11.22. The current–voltage characteristics of a typical 600 A, 600 V IGBT are given in Figure 11.23. The graph indicates that the gate-emitter voltage should be around 20 V, which results in collector-emitter voltage of approximately 2 V at a collector current of 600 A.

The current flows from the collector to the emitter, when the device is forward biased and a positive (gate-emitter) voltage triggers the gate pulse. The removal of the gate pulse interrupts the current.

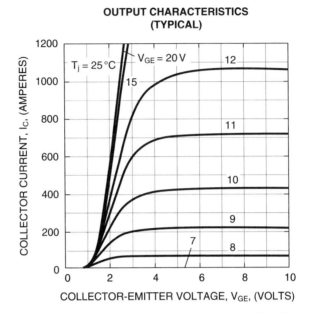

Figure 11.23. Output characteristics of a 600 V, 600 A IGBT module (Source: *Power Semiconductor Data Book*, Powerex, Inc., Youngwood, PA, 1988).

TABLE 11.1. Summary of Important Semiconductor Power Switch Characteristics

Switch	Important Characteristics	Applications
Diode	No gate	Rectifier
Thyristor (SCR)	Gate control	Rectifier and inverter use
GTO	Gate control; on–off control	Rectifier and inverter use
MOSFET and IGBT	On–off control multiple times per cycle (i.e., high speed)	Mostly for inverter, but can use in a rectifier

Generally, the maximum rating of power IGBTs is in the range of a few hundred amperes (400–800 A) and around less than 2000 V. The operating frequency can be in the kilohertz range. A typical application of the power IGBT is in pulse-width-modulation (PWM) circuits.

11.3.6. Summary

A critical characteristic of all these semiconductor switches is that current is allowed to flow in only one direction. A comparison and contrast of other important characteristics and the primary use for these switches are summarized in Table 11.1.

11.4. RECTIFIERS

Rectifiers convert ac voltage and current to dc voltage and current. Typical power applications include:

1. battery chargers,
2. DC motor drives,
3. power supplies (for computers, appliances, uninterruptible power supplies, etc.), and
4. generator excitation systems.

In most cases, the rectifier must control the output dc voltage or must keep it at a constant level even if the load and/or the supply ac voltages change. This is achieved by using controllable switches like thyristors, GTOs, MOSFETs, or IGBTs in the rectifier circuit.

We begin this section with analyses of simple diode rectifiers such as the bridge rectifier. Afterward, both single- and three-phase controlled rectifiers are presented.

11.4.1. Simple Passive Diode Rectifiers

An uncontrolled (passive) rectifier can be formed by simply using diodes. Such diode rectifiers are typically used in the power supplies of popular consumer electronics, and are one of the first circuits studied in an electronics course. These "dc transformers" (a prevalent misnomer) first reduce household voltage (120 Vac) down to around 5–12 Vac, and then use diodes to rectify the voltage to a dc output. These power supplies are sometimes referred to as a power cube transformer because of their cubical shape.

Three diode rectifier circuits are shown in Figure 11.24. A *half-wave rectifier* is constructed using a single diode. The output (V_0) of the half-wave rectifier (Fig. 11.24a) follows the positive half-cycle of the ac voltage, but is zero during the negative portion of the cycle:

$$V_0(t) = \begin{cases} \dfrac{V_{ac}(t)}{T} & 0 < t \leq \dfrac{1}{2f} \\ 0 & \dfrac{1}{2f} < t \leq \dfrac{1}{f} \end{cases}, \qquad (11.20)$$

where f is the ac cycle frequency and T is the turns ratio of the transformer.

Full-wave rectification can be achieved using either the *full-wave rectifier* or the *bridge rectifier*. The full-wave rectifier of Figure 11.24b provides a positive output at all times, since D_1 is activated during the positive ac cycle and D_2 is turned on during the negative cycle. Likewise, the bridge rectifier (Fig. 11.24c) produces the same full-wave output since the combination of D_1 and D_2 allows current flow during the positive

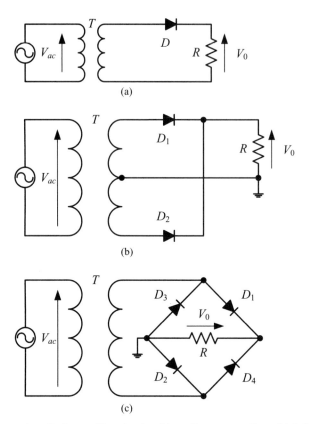

Figure 11.24. Passive diode rectifier circuits. (a) Half-wave rectifier; (b) full-wave rectifier circuit with center-tapped transformer; (c) full-wave bridge rectifier.

half-cycle, and D_3 and D_4 permit current flow in the negative half-cycle. The full-wave rectifier outputs are:

$$V_0(t) = \begin{cases} \dfrac{V_{ac}(t)}{T} & 0 < t \leq \dfrac{1}{2f} \\ \dfrac{-V_{ac}(t)}{T} & \dfrac{1}{2f} < t \leq \dfrac{1}{f} \end{cases}. \qquad (11.21)$$

Clearly, each of the three rectifiers does not produce a constant dc output. The pulsations caused by the ac components of the rectifier output are referred to as a *ripple*. A capacitor placed in parallel with the resistive load of the rectifier circuits will smooth the ripple.

11.4.1.1. Fourier Series Analysis.
Fourier analysis may be employed to quantify the purity of the rectifier output signal. Recall that the Fourier series allow us to represent a periodic signal as the sum of a dc component and an infinite number of sinusoidal components at harmonically related frequencies. A signal is periodic if $x(t + T) = x(t)$, where T is the period. The *Fourier trigonometric series* decomposes a waveform, $x(t)$, into:

$$x(t) = \frac{a_0}{2} + \sum_{n=1}^{\infty} [a_n \cos(n\omega_0 t) + b_n \sin(n\omega_0 t)], \quad (11.22)$$

where $a_0/2$ is the average (dc) value of $x(t)$ and ω_0 is the *fundamental frequency*, which corresponds to the period of $x(t)$ via $\omega_0 = 2\pi f = 2\pi/T$. The frequencies that are integer multiples of the fundamental frequency are the *harmonic frequencies*, that is, $n\omega_0$ for $n > 1$. The Fourier trigonometric series coefficients can be calculated from:

$$a_n = \frac{2}{T} \int_0^T x(t)\cos(n\omega_0 t)\,dt,$$
$$b_n = \frac{2}{T} \int_0^T x(t)\sin(n\omega_0 t)\,dt. \quad (11.23)$$

In practice, only a finite number of terms is incorporated, which means that the Fourier series representation is an approximate method. It can be shown that the rms (effective) value of the function $x(t)$ is equal to:

$$X_{rms} = \sqrt{\left(\frac{a_0}{2}\right)^2 + \frac{1}{2}\sum_{n=1}^{\infty} a_n^2 + b_n^2}, \quad (11.24)$$

since the effective value of the dc component is $|a_0/2|$, and the rms values of the fundamental and harmonic components are $\sqrt{(a_n^2 + b_n^2)/2}$.

Some additional restrictions often make finding certain coefficients unnecessary:

- Case 1. If $x(t)$ is an even function (i.e., $x(-t) = x(t)$), then $b_n = 0$ for all n.
  - Subcase 1a. Further, if $x(t)$ is an even function and $x(t) = x(T/2-t)$, then not only do all $b_n = 0$, but also $a_n = 0$ for all even values of n, including $n = 0$.
- Case 2. If $x(t)$ is an odd function (i.e., $x(-t) = -x(t)$), then $a_n = 0$ for all n.
  - Subcase 2a. Further, if $x(t)$ is an odd function and $x(t) = x(T/2-t)$, then not only do all $a_n = 0$, but also $b_n = 0$ for all even values of n.

The series in the two subcases are known as odd harmonic series, since only the odd harmonics appear. The functions of these subcases are also referred to as half-wave symmetric.

The fast Fourier transform (FFT) is a computationally efficient algorithm for calculating the discrete Fourier transform (DFT), which may be defined as:

$$X(n) = \frac{1}{N}\sum_{k=0}^{N-1} x(k)\exp(-j2\pi nk/N), \quad (11.25)$$

where the signal $x(k)$ represents N discrete samples of $x(t)$, which have been sampled at a time interval of Δt, such that $t = k\Delta t$. This means that $X(n)$ is at a frequency of $n\omega_0 = 2\pi nf = 2\pi n/T = 2\pi n/(N\Delta t)$. The relationship between the earlier Fourier trigonometric series coefficients and the above DFT is:

$$X(0) = \frac{a_0}{2}; \; X(n) = \frac{a_n - jb_n}{2}. \quad (11.26)$$

11.4.1.2. Half-Wave Rectifier Circuit Analysis.
We will analyze the half-wave rectifier circuit of Figure 11.24a, except that the transformer is omitted for simplification of the analysis. The ac voltage input is:

$$V_{ac}(t) = \sqrt{2}V_{rms}\sin(\omega_0 t). \quad (11.27)$$

Ignoring the voltage drop across the diode, the corresponding rectifier output voltage is:

$$V_0(t) = \begin{cases} \sqrt{2}V_{rms}\sin(\omega_0 t) & 0 < t \leq \frac{T}{2} \\ 0 & \frac{T}{2} < t \leq T \end{cases}, \quad (11.28)$$

where T is the period of the source. The Fourier series coefficients for the half-wave rectified sine wave are obtained from standard mathematical tables[1]:

$$V_0(t) = \sqrt{2}V_{rms}\left[\frac{1}{\pi} + \frac{1}{2}\sin(\omega_0 t) - \frac{2}{\pi}\sum_{n=2,4,6,\ldots}\frac{1}{n^2-1}\cos(n\omega_0 t)\right]. \quad (11.29)$$

The mean value of the signal is therefore $\sqrt{2}V_{rms}/\pi = a_0/2$. Table 11.2 lists the Fourier series trigonometric coefficients for a pure sine wave and for its half and full wave rectified waveforms.

---

[1] Beyer, W.H., ed., *CRC Standard Mathematical Tables and Formulae*, 29th ed., CRC Press, Boca Raton, FL, 1991, p. 408.

RECTIFIERS

TABLE 11.2. Fourier Series Trigonometric Coefficients

	Sine Wave	Half-Wave	Full Wave
Waveform, $x(t)$	$V_M \sin(\omega_0 t)$	$V_M \sin(\omega_0 t) \quad 0 < t \leq \dfrac{T}{2}$ $0 \quad \dfrac{T}{2} < t \leq T$	$V_M \lvert \sin(\omega_0 t) \rvert$
Mean, $\dfrac{a_0}{2}$	0	$\dfrac{a_0}{2} = \dfrac{V_M}{\pi}$	$\dfrac{a_0}{2} = \dfrac{2V_M}{\pi}$
a_n	$a_n = 0$	$a_n = \begin{cases} \dfrac{-2V_M}{\pi(n^2-1)} & n = 0, 2, 4, 6, \ldots \\ 0 & n = 1, 3, 5, \ldots \end{cases}$	$a_n = \begin{cases} \dfrac{-4V_M}{\pi(n^2-1)} & n = 0, 2, 4, 6, \ldots \\ 0 & n = 1, 3, 5, \ldots \end{cases}$
b_n	$b_1 = V_M$ $b_n = 0 \quad n > 1$	$b_1 = \dfrac{V_M}{2}$ $b_n = 0 \quad n > 1$	$b_n = 0$

EXAMPLE 11.1: MATLAB calculation of rectified waveform spectrum

In this example, the Fourier coefficients of the ac sinusoidal voltage and its half- and full-wave rectified signals are computed and compared. Using a frequency of 50 Hz and an rms voltage of 1 V, the three waveforms are generated over a single period of 0.02 second. The number of samples for each waveform is set as a power of 2 in anticipation of using an FFT (i.e., $4096 = 2^{12}$). The following code accomplishes this and produces the plot provided in Figure 11.25:

```
% Fourier spectrum of rectified waveforms
clear all; clc;

fsys = 50;        % system ac frequency (Hz)
w0 = 2*pi*fsys;   % angular frequency (rad/sec)
T = 1/fsys;       % signal period (sec)

npts = 4096;      % number of voltage samples
time = linspace(0,T,npts);   % time vector

% create the reference ac voltage signal
Vrms = 1;         % ac rms voltage
Vac = sqrt(2)*Vrms*sin(w0*time);

% create the half-wave rectified signal
Vhalf(1:npts/2) = Vac(1:npts/2);
Vhalf(npts/2+1:npts) = 0;

% create the full-wave rectified signal
Vfull = abs(Vac);
```

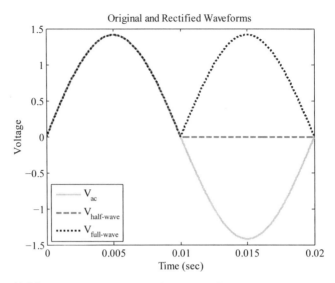

Figure 11.25. AC voltage and its half-wave and full-wave rectified waveforms.

```
% plot the original and rectified waveforms
figure (1)
plot(time,Vac,'g-',time,Vhalf,'m-',time,Vfull,'k:', ...
    'LineWidth',2.5);
set(gca, 'fontname','Times', 'fontsize',12);
xlabel('Time (sec)');
ylabel('Voltage');
title('Original and Rectified Waveforms');
legend('V_{ac}','V_{half-wave}','V_{full-wave}', ...
    'Location','SouthWest');
```

The FFT function in MATLAB is then utilized to determine the DFT of each of the three waveforms. Like many software packages, the MATLAB FFT function does not include the division by N found in Equation (11.25), but which is incorporated in the following code:

```
% perform fast Fourier transform (FFT) analyses
Xac   = fft(Vac)/npts;
Xhalf = fft(Vhalf)/npts;
Xfull = fft(Vfull)/npts;

nfreq = 10;      % number of frequency points
freq  = [0:fsys:nfreq*fsys];   % frequency vector
```

The Fourier series trigonometric coefficients are analytically determined using the relations from Table 11.2, and then compared to the values obtained from the FFT numerical analyses, which are shown in Table 11.3. In order to make a direct

RECTIFIERS

TABLE 11.3. Comparison of Fourier Series Coefficients and FFT Results

	Half Wave Coefficients		Comparison	
	Analytical		Numerical (FFT)	
n	a(n)	b(n)	2X*(n)/N	
0	0.90032		0.90010	
1	0.00000	0.70711	0.00027	0.70702j
2	-0.30011	0.00000	-0.30023	0.00023j
3	0.00000	0.00000	-0.00000	-0.00013j
4	-0.06002	0.00000	-0.06004	0.00009j
5	0.00000	0.00000	-0.00000	-0.00007j
6	-0.02572	0.00000	-0.02573	0.00006j
7	0.00000	0.00000	-0.00000	-0.00005j
8	-0.01429	0.00000	-0.01429	0.00004j
9	0.00000	0.00000	-0.00000	-0.00004j
10	-0.00909	0.00000	-0.00910	0.00003j

	Full Wave Coefficients		Comparison	
	Analytical		Numerical (FFT)	
n	a(n)	b(n)	2X*(n)/N	
0	1.80063		1.80019	
1	0.00000	0.00000	-0.00054	0.00000j
2	-0.60021	0.00000	-0.60045	0.00092j
3	0.00000	0.00000	0.00000	-0.00000j
4	-0.12004	0.00000	-0.12007	0.00037j
5	0.00000	0.00000	0.00000	-0.00000j
6	-0.05145	0.00000	-0.05146	0.00024j
7	0.00000	0.00000	0.00000	-0.00000j
8	-0.02858	0.00000	-0.02859	0.00018j
9	0.00000	0.00000	0.00000	-0.00000j
10	-0.01819	0.00000	-0.01819	0.00014j

	Sine Wave Coefficients		Comparison	
	Analytical		Numerical (FFT)	
n	a(n)	b(n)	2X*(n)/N	
0	0.00000		0.00000	
1	0.00000	1.41421	0.00108	1.41404j
2	0.00000	0.00000	-0.00000	-0.00046j
3	0.00000	0.00000	-0.00000	-0.00026j
4	0.00000	0.00000	-0.00000	-0.00018j
5	0.00000	0.00000	-0.00000	-0.00014j
6	0.00000	0.00000	-0.00000	-0.00012j
7	0.00000	0.00000	-0.00000	-0.00010j
8	0.00000	0.00000	-0.00000	-0.00009j
9	0.00000	0.00000	-0.00000	-0.00008j
10	0.00000	0.00000	-0.00000	-0.00007j

comparison between the trigonometric series coefficients, the FFT results are multiplied by 2 and the complex conjugate is taken in accordance with Equation (11.26). The tabular comparison discloses that the FFT results do not match exactly to those of the analytical. This discrepancy is due to the imperfection of the rectangular band-pass filter represented by the DFT. A discussion of spectral sidelobe leakage is beyond the scope of this book, but the magnitude of the effect can be observed in the FFT results for the pure sine wave for which only the b_1 coefficient should be nonzero:

```
% compute the coefficients for the half wave
a0 = 2/pi *sqrt(2)*Vrms;
for k=2:2:10
    a(k-1) = 0;
    b(k-1) = 0;
    a(k) = -2/pi /(k^2-1) *sqrt(2)*Vrms;
    b(k) = 0;
end
b(1) = 1/2 *sqrt(2)*Vrms;
ListCoeff('Half',a0,a,b,nfreq,Xhalf)
% compute the coefficients for the full wave
a0 = 4/pi *sqrt(2)*Vrms;
for k=2:2:10
    a(k-1) = 0;
    b(k-1) = 0;
    a(k) = -4/pi /(k^2-1) *sqrt(2)*Vrms;
    b(k) = 0;
end
ListCoeff('Full',a0,a,b,nfreq,Xfull)
% compute the coefficients for the pure sine wave
a0 = 0;
a = zeros(1,10);
b = zeros(1,10);
b(1) = sqrt(2)*Vrms;
ListCoeff('Sine',a0,a,b,nfreq,Xac)
```

Finally, the magnitudes of the FFT-obtained Fourier coefficients are plotted in Figure 11.26 using dual x-axes for frequency in Hertz and the harmonic number. The spectrum for the pure ac voltage is a single point at the fundamental harmonic ($n = 1$) at 50 Hz; the strength of this harmonic is seen to be $\sqrt{2}$. The spectral component strengths for the half- and full waves diminish quickly with increasing harmonic number.

```
figure (2)
plot(freq,abs(Xac(1:nfreq+1)),'g-v',...
    freq,abs(Xhalf(1:nfreq+1)),'m-s',...
    freq,abs(Xfull(1:nfreq+1)),'k:o','LineWidth',2.5)
set(gca, 'fontname','Times', 'fontsize',12);
xlabel('Frequency (Hz)');
```

RECTIFIERS

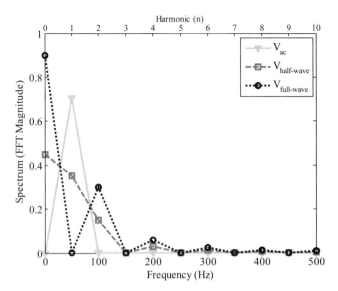

Figure 11.26. Harmonic composition (spectrum) of the sinusoidal voltage and its half- and full-wave rectified signals.

```
ylabel('Spectrum (FFT Magnitude)');
legend('V_{ac}','V_{half-wave}','V_{full-wave}',...
    'Location','NorthEast');
% add a second x-axis for the harmonic number
ax1 = gca;
ax2 = axes('Position',get(ax1,'Position'),...
    'XAxisLocation','top','color','none');
xlimit = [0,nfreq];
set(ax2,'xlim',xlimit,'ytick',get(ax1,'ytick'),...
    'yticklabel','','fontname','Times');
title('Harmonic (n)','fontname','Times');
```

EXAMPLE 11.2: Half-wave rectifier Mathcad analysis

The analysis of the half-wave rectifier is a straightforward process in Mathcad. First, the system frequency and period of the ac voltage are established:

$$f_{sys} := 60 \text{ Hz} \quad \omega := 2 \cdot \pi \cdot f_{sys} \quad T_p := \frac{1}{f_{sys}}.$$

North American ac household voltage (120 V) is taken as the input waveform to the rectifier, and the output of the rectifier is across a 5 Ω resistor:

$$V_{rms} := 120 \text{ V} \quad V_{ac}(t) := \sqrt{2} \cdot V_{rms} \cdot \sin(\omega \cdot t) \quad R_o := 5 \text{ }\Omega.$$

For simplicity in this example, the voltage drop across the diode is assumed negligible. The *if* statement constricts the dc output to follow the ac input only during the positive half-cycle, and the current is computed from Ohm's law:

$$V_{half\_wave}(t) := if\ (V_{ac}(t) > 0,\ V_{ac}(t),\ 0\ V) \quad I_{half\_wave}(t) := \frac{V_{half\_wave}(t)}{R_o}.$$

The dc voltage and current from the rectifier are shown in Figure 11.27a over three cycles (periods). The magnitude of the dc current component is the average of the output current over a one-cycle period:

$$I_{dc} := \frac{1}{T_p} \cdot \int_0^{T_p} I_{half\_wave}(t)\,dt \quad I_{dc} = 10.8\ A.$$

$$t := 0 \cdot s,\ \frac{T_p}{100} \ldots 3 \cdot T_p$$

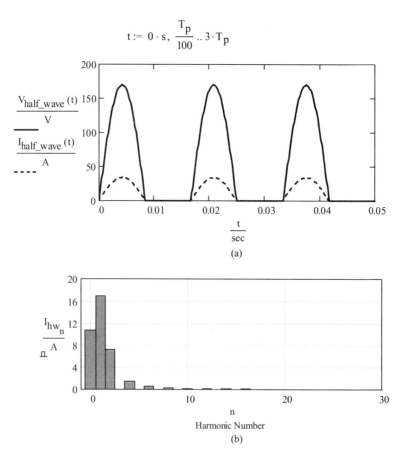

Figure 11.27. Mathcad half-wave rectifier analysis. (a) Rectifier output voltage and current; (b) harmonic content of rectifier current.

As a check of the mean value, $I_{dc} = V_{dc}/R = \left(\sqrt{2}V_{rms}/\pi\right)/R = \sqrt{2}(120\text{ V})/(\pi(5\text{ }\Omega)) = 10.8$ A.

The first 30 Fourier coefficients are calculated using Equation (11.23). The cosine (a_n) coefficients are:

$$n := 0 .. 30$$

$$I_{cos\_a_n} := \frac{2}{T_p} \cdot \int_{0s}^{T_p} I_{half\_wave}(t) \cdot \cos(n \cdot \omega \cdot t) \, dt$$

$$I_{cos\_a}^T = \begin{array}{|c|c|c|c|c|c|c|c|c|} \hline & 0 & 1 & 2 & 3 & 4 & 5 & 6 & 7 & 8 \\ \hline 0 & 21.61 & 0 & -7.2 & 0 & -1.44 & 0 & -0.62 & 0 & \ldots \\ \hline \end{array} \text{ A.}$$

The sine (b_n) coefficients are:

$$I_{sin\_b_n} := \frac{2}{T_p} \cdot \int_0^{T_p} I_{half\_wave}(t) \cdot \sin(n \cdot \omega \cdot t) \, dt$$

$$I_{sin\_b}^T = \begin{array}{|c|c|c|c|c|c|c|c|c|} \hline & 0 & 1 & 2 & 3 & 4 & 5 & 6 & 7 & 8 \\ \hline 0 & 0 & 16.97 & 0 & 0 & 0 & 0 & 0 & 0 & \ldots \\ \hline \end{array} \text{ A.}$$

These trigonometric series coefficients can be used to find the strength of the various frequency components, including the fundamental (or base) and dc components:

$$I_{hw_n} = \sqrt{\left(I_{cos\_a_n}\right)^2 + \left(I_{sin\_b_n}\right)^2} \quad I_{hw_0} := \left|\frac{I_{cos\_a_0}}{2}\right|$$

$$I_{hw}^T = \begin{array}{|c|c|c|c|c|c|c|c|c|} \hline & 0 & 1 & 2 & 3 & 4 & 5 & 6 & 7 & 8 \\ \hline 0 & 10.8 & 16.97 & 7.2 & 0 & 1.44 & 0 & 0.62 & 0 & \ldots \\ \hline \end{array} \text{ A.}$$

The Fourier analysis results are plotted, as shown in Figure 11.27b. Since Ohm's law linearly relates the rectifier output voltage and current, the spectral composition of the voltage will be the same as that of the current.

EXAMPLE 11.3: Half-wave rectifier PSpice analysis

PSpice can also be used to simulate the half-wave rectifier circuit. The circuit schematic is first created, as drawn in Figure 11.28. The VSIN part is used instead of the VAC component so that a transient analysis can be performed, and so that the input signal matches the sine wave used in the previous equations. The 60 Hz circuit is simulated for three periods (50 ms). Figure 11.29a presents the source voltage, the resultant voltage across the resistor, and the current through the diode. The waveform shapes clearly portray from where the name half-wave rectifier originates. PSpice has the capability to compute the FFT of the resultant waveforms. Figure 11.29b presents the

708 INTRODUCTION TO POWER ELECTRONICS AND MOTOR CONTROL

Figure 11.28. PSpice circuit model of half-wave rectifier.

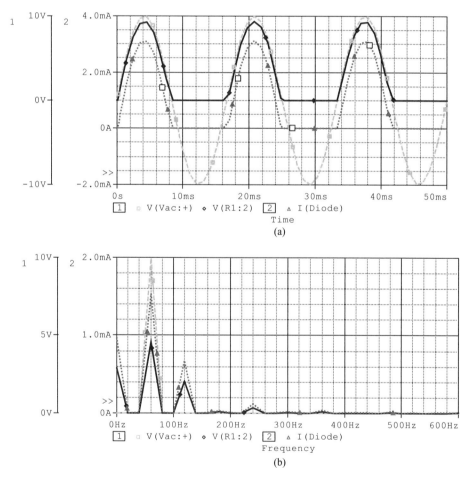

Figure 11.29. PSpice results from half-wave rectifier simulation. (a) AC input voltage, diode current, and rectifier output voltage; (b) FFT analysis of the ac input voltage, diode current, and rectifier output voltage.

FFT analysis results for the three waveforms. Note that the resistor voltage and diode current have the same spectral composition, but different magnitudes, as expected.

11.4.2. Single-Phase Controllable Rectifiers

The single-phase bridge circuit is the most frequently used rectifier. This circuit is built with thyristor switches for line-commutated circuits, and with MOSFETs or IGBTs for PWM circuits.

11.4.2.1. Single-Phase Line-Commutated Rectifiers. Figure 11.30 presents the circuit diagram of a thyristor-controlled, single-phase bridge rectifier circuit. The bridge is supplied by an ac voltage source through a transformer, and the circuit outputs a dc voltage. The transformer provides insulation between the ac and dc circuits, which permits the separate grounding of the ac and dc circuits. Federal regulations require the grounding of the neutral point of an ac circuit. In addition, one point of the dc circuit must be grounded. The transformer allows the free selection of the ground points.

Figure 11.31 shows a sketch of the voltage directions and current paths for the single-phase bridge rectifier. In the positive cycle of the ac supply voltage (see Fig. 11.31a), thyristors Th_1 and Th_2 are turned on by a gate pulse with a *delay angle* of α. In the negative ac supply cycle of Figure 11.31b, thyristors Th_3 and Th_4 are instead turned on with a delay angle of α.

Commutation occurs in the changeover from thyristor pair Th_1–Th_2 to Th_3–Th_4. This is achieved by firing thyristor pair Th_3–Th_4 at the appropriate time. This produces reverse current in thyristor pair Th_1–Th_2, which turns them off. The activation of either thyristor pair connects the ac supply voltage directly to the load. The supply voltage drives current (I_{dc}) through the load while the supply current (I_{ac}) is positive. The change of the supply current direction turns off the thyristors.

Because of the bridge connection, the polarity of the dc load voltage and current is the same in both cycles. Each pair of thyristors can conduct until the other pair is

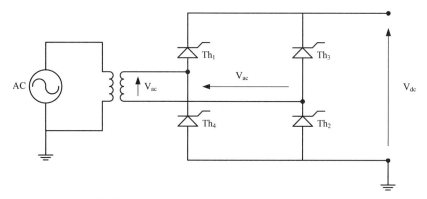

Figure 11.30. Single-phase thyristor-controlled bridge rectifier.

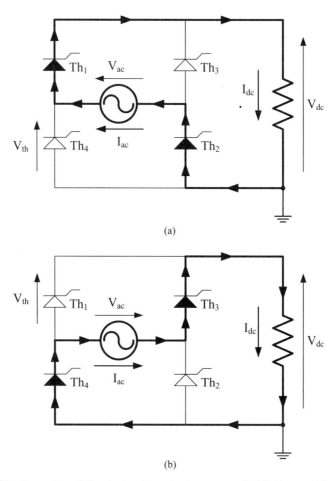

Figure 11.31. Operation of the single-phase thyristor-controlled bridge rectifier. (a) Positive ac supply cycle; (b) negative ac supply cycle.

fired or until the supply current reverses direction. In the case of a resistive load, the turn off is initiated by the change of current direction.

Figure 11.32 shows graphs of both the ac and dc voltages and currents of the single-phase, thyristor-controlled bridge rectifier. The current flow is initiated by the firing of the thyristor pairs. The figure also plots the delayed gate pulse that activates the thyristors. The delay (α) for Th_1–Th_2 is 60°, and for Th_3–Th_4 is 240°.

It can be seen that both the dc voltage and current are discontinuous and pulsating. Figure 11.32b shows that the delayed firing reduces the *average* dc voltage, because the firing delay cuts out the first part of the voltage curve. The instantaneous dc current is calculated by Ohm's law.

RECTIFIERS

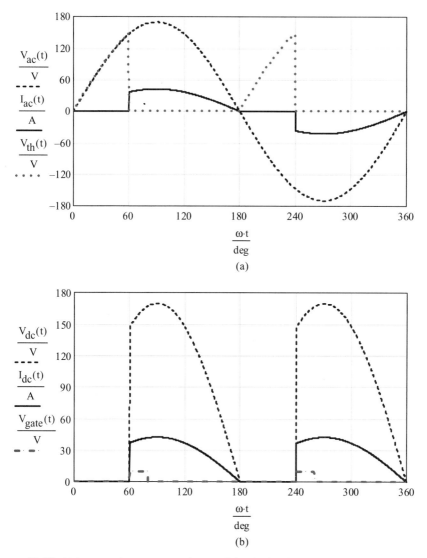

Figure 11.32. Voltage and current waveforms of the single-phase thyristor-controlled bridge rectifier. (a) AC voltage, current, and thyristor voltage; (b) DC voltage, current, and gate pulse.

The shape of the ac current is similar to the dc current. The only difference is the direction. Because of the bridge connection, the dc current is positive in both cycles. Simultaneously, the ac current has the same waveform, but the current direction is positive in the positive cycle and negative in the negative cycle. When a thyristor is not conducting, the voltage across it is the ac voltage. An open circuit represents the nonconducting thyristor, while a short circuit describes a conducting thyristor.

Figure 11.32b also demonstrates that the dc current and voltage waveforms are identical in each half-cycle; consequently, the dc current and voltage values are calculated in Mathcad (see later) using the first half-cycle values. For the ac current equation, the entire first cycle is considered because the direction of the current is different in the positive and negative cycles.

11.4.2.2. Single-Phase Bridge Rectifier with Resistive Load.
The single-phase rectifier operation is analyzed using a numerical example. It is assumed that an ideal bridge-connected rectifier supplies a resistance (see Fig. 11.31). The source impedance is neglected. The ac source voltage and frequency and the load resistance are:

$$V_{rms} := 120 \text{ V} \quad f := 60 \text{ Hz} \quad R_{load} := 4 \text{ }\Omega.$$

The durations of the half- and full cycles for 60 Hz operation are:

$$T_{half} := \frac{1}{2 \cdot f} = 8.33 \text{ ms} \quad T_{cycle} := \frac{1}{f} = 16.67 \text{ ms}.$$

The time is variable, but we select a particular time instant for verification of the equations:

$$\omega := 2 \cdot \pi \cdot f \quad t := \frac{60 \text{ deg}}{\omega} = 2.78 \text{ ms}.$$

The ac voltage equation is:

$$V_{ac}(t) := \sqrt{2} \cdot V_{rms} \cdot \sin(\omega \cdot t) \quad V_{ac}(t) = 147.0 \text{ V}$$
$$\sqrt{2} \cdot V_{rms} = 169.7 \text{ V}.$$

The thyristor pairs are fired by an α delay angle, and the dc voltage is always positive because the thyristor pair conducts only in the positive direction. These two criteria are implemented using an *if* statement in Mathcad. The first condition states that the dc voltage is zero between an angle of zero and α, and the second asserts that the dc voltage cannot be negative. The selected delay angle (α) and dc voltage equations are:

$$V_{dc}(t) = \begin{cases} 0 & 0 < \omega t < \alpha \\ |V_{ac}(t)| & \alpha < \omega t < \pi \end{cases}.$$

This can be accomplished in Mathcad using:

$$\alpha := 60 \text{ deg}$$
$$V_{dc}(t) := \text{if}(\omega \cdot t < \alpha, 0 \text{ V}, |V_{ac}(t)|) \quad V_{dc}(t) = 147.0 \text{ V}.$$

RECTIFIERS

The dc voltage is a periodic function, which can be implemented using:

$$V_{dc}(t) := \text{if}\left(\omega \cdot t < \pi, V_{dc}(t), V_{dc}\left(t - \frac{\pi}{\omega}\right)\right) \quad V_{dc}(t) = 147.0 \text{ V}.$$

The dc current is calculated using Ohm's law. Accordingly, the dc current is equal with the dc voltage divided by the load resistance. The thyristor assures that this current is always positive:

$$I_{dc}(t) := \frac{V_{dc}(t)}{R_{load}} \quad I_{dc}(t) = 36.74 \text{ A}.$$

The ac current is the same as the dc current but its direction in the negative cycle is negative. Again, this is expressed with an *if* statement:

$$I_{ac}(t) := \text{if}\left(t < T_{half}, I_{dc}(t), -I_{dc}(t)\right) \quad I_{ac}(t) = 36.74 \text{ A}.$$

Applying the loop voltage equation, the voltage across each thyristor is equal to the ac source voltage when the thyristor is not conducting and equal to zero when it is conducting:

$$V_{th}(t) := \text{if}\left(\omega \cdot t < \alpha, |V_{ac}(t)|, 0\right) \quad V_{th}\left(\frac{30 \text{ deg}}{\omega}\right) = 84.85 \text{ V}.$$

The resultant ac and dc waveforms of the bridge rectifier are shown in Figure 11.32. According to Figure 11.32a, the maximum thyristor voltage is the peak source voltage, when the delay angle is more than 90°. This peak value is used to select the reverse voltage rating of the thyristor.

The average of the current over a full cycle is needed for the selection of the rated current of the thyristor. The average thyristor current is half of the dc current, because in the positive cycle, Th_1 and Th_2 conduct, and in the negative cycle, Th_3 and Th_4 conduct. Typically, a safety factor of 50–100% is used for the circuit design.

Evaluation of the rectifier performance requires the average dc voltage and current over a full cycle. These averages can be calculated by integrating the instantaneous dc value for the first half-cycle and dividing the integral by the half-cycle period. The average dc voltage is:

$$V_{dc\_ave} := \frac{1}{T_{half}} \cdot \int_0^{T_{half}} V_{dc}(t) dt \quad V_{dc\_ave} = 81.0 \text{ V}.$$

A general expression can be derived for the average dc voltage by performing the integration symbolically or analytically:

$$\frac{1}{T_{half}} \cdot \int_{\frac{\alpha}{\omega}}^{T_{half}} \sqrt{2} \cdot V_{rms} \cdot \sin(\omega \cdot t) dt \rightarrow \frac{\sqrt{2} \cdot V_{rms} \cdot (\cos(T_{half} \cdot \omega) - \cos(\alpha))}{T_{half} \cdot \omega}.$$

The substitution of $T_{half} \, \omega = \pi$ into the equation permits simplification of the expression. The simplified formula is:

$$V_{dc\_average}(\alpha) := \frac{\sqrt{2} \cdot V_{rms}}{\pi} \cdot (1 + \cos(\alpha)) \quad V_{dc\_average}(\alpha) = 81.0 \text{ V}.$$

This equation can only be used for resistive loads.

The average dc current can be calculated either by Ohm's law or by integration:

$$I_{dc\_average}(\alpha) := \frac{V_{dc\_average}(\alpha)}{R_{load}} \quad I_{dc\_average}(\alpha) = 20.26 \text{ A}$$

$$I_{dc\_ave} := \frac{1}{T_{half}} \int_0^{T_{half}} I_{dc}(t) \, dt \quad I_{dc\_ave} = 20.26 \text{ A}.$$

The average thyristor current is half of the average dc current:

$$I_{th\_ave} := \frac{I_{dc\_ave}}{2} \quad I_{th\_ave} = 10.13 \text{ A}.$$

Determination of the input ac power requires the calculation of the rms current. The rms current is computed by integrating the ac current for the half-cycle:

$$I_{ac\_rms} := \sqrt{\frac{1}{T_{half}} \cdot \int_0^{T_{half}} I_{ac}(t)^2 \, dt} \quad I_{ac\_rms} = 26.91 \cdot \text{A}.$$

We use Ohm's law ($I_{ac} = V_{ac}/R_{load}$) to substitute for the current and integrate from α to 180° to determine a general formula for the rms ac current. Using the Mathcad symbolic integration, we obtain:

$$\sqrt{\frac{1}{R_{load}^2 \cdot T_{half}} \cdot \int_{\frac{\alpha}{\omega}}^{T_{half}} \left(\sqrt{2} \cdot V_{rms} \cdot \sin(\omega \cdot t)\right)^2 dt}$$

$$\rightarrow \sqrt{\frac{V_{rms}^2 \cdot (\sin(2 \cdot \alpha) - 2 \cdot \alpha - \sin(2 \cdot T_{half} \cdot \omega) + 2 \cdot T_{half} \cdot \omega)}{2 \cdot R_{load}^2 \cdot T_{half} \cdot \omega}}.$$

This expression is simplified by using the relation $\sin(2\theta) = 2\sin(\theta)\cos(\theta)$ twice and by substituting $T_{half} \, \omega = \pi$ into the equation. The simplified formula is:

$$I_{rms\_ac}(\alpha) := \frac{V_{rms}}{R_{load}} \cdot \sqrt{\frac{1}{\omega \cdot T_{half}} \cdot (\pi - \alpha + \cos(\alpha) \cdot \sin(\alpha))}$$

$$I_{rms\_ac}(\alpha) = 26.91 \text{ A}.$$

The magnitude of the ac complex power (i.e., the apparent power) is the product of the rms current and voltage magnitudes:

$$S_{ac} := I_{ac\_rms} \cdot V_{rms} \quad S_{ac} = 3.23 \text{ kV} \cdot \text{A}.$$

The average power is the integral of the instantaneous power (product of current and voltage) for a full cycle divided by the cycle period:

$$P_{ac} := \frac{1}{T_{cycle}} \cdot \int_0^{T_{cycle}} V_{ac}(t) \cdot I_{ac}(t) dt \quad P_{ac} = 2.9 \cdot \text{kW}.$$

Figure 11.32 shows the delay angle α of 60° between the ac voltage and current. The graph reveals that the ac current wave is not sinusoidal. This distorted wave generates undesired harmonics in the ac system. Assessment of the harmonic generation requires Fourier analysis of the ac current. The frequency spectrum of the ac current is found by computing the Fourier series coefficients (a_n and b_n) using Equation (11.23):

$$n := 0 .. 51$$

$$I_{a_n} := \frac{2}{T_{cycle}} \cdot \int_{0s}^{T_{cycle}} I_{ac}(t) \cdot \cos(n \cdot \omega \cdot t) dt$$

$$I_a^T = \begin{array}{|c|c|c|c|c|c|c|c|c|} \hline & 0 & 1 & 2 & 3 & 4 & 5 & 6 & 7 \\ \hline 0 & 0 & 0 & -10.13 & 0 & 5.06 & 0 & 5.06 & 0 & \ldots \\ \hline \end{array} \text{ A}$$

$$I_{b_n} := \frac{2}{T_{cycle}} \cdot \int_{0s}^{T_{cycle}} I_{ac}(t) \cdot \sin(n \cdot \omega \cdot t) dt$$

$$I_b^T = \begin{array}{|c|c|c|c|c|c|} \hline & 0 & 1 & 2 & 3 & 4 & 5 \\ \hline 0 & 0 & 0 & 34.13 & 0 & -8.77 & 0 & \ldots \\ \hline \end{array} \text{ A}$$

The rms value of each harmonic component of the ac current is:

$$I_{harm_n} := \frac{\sqrt{(I_{a_n})^2 + (I_{b_n})^2}}{\sqrt{2}} \quad I_{harm_0} := \left|\frac{I_{a_0}}{2}\right|$$

$$I_{harm}^T = \begin{array}{|c|c|c|c|c|c|c|c|c|c|} \hline & 0 & 1 & 2 & 3 & 4 & 5 & 6 & 7 & 8 \\ \hline 0 & 0 & 0 & 25.18 & 0 & 7.16 & 0 & 4.13 & 0 & 2.07 & \ldots \\ \hline \end{array} \text{ A}$$

The obtained harmonic current magnitudes are plotted in Figure 11.33 relative to the fundamental frequency (first harmonic) of 60 Hz. The amplitudes of the dc and even harmonic frequency components are zero. The graph shows that the amplitudes of the current harmonics decrease with increasing harmonic number (or frequency). The frequency (f_n) of a particular harmonic (n) is calculated by multiplying the harmonic number n by 60 Hz, that is, $f_n = n$ 60 Hz.

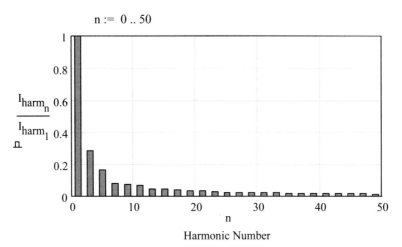

Figure 11.33. Frequency spectrum of the ac current in a single-phase, thyristor-controlled bridge rectifier with a delay angle of 60°.

Using Equation (11.24), the effective (rms) value of the current using the first 51 terms of the Fourier series is:

$$\sqrt{\sum_n (I_{harm_n})^2} = 26.86 \text{ A}.$$

This result of 26.86 A is close to the $I_{rms\_ac}(\alpha)$ value of 26.91 found earlier.

The effect of harmonics is judged using the *total harmonic distortion* (THD) factor. The formula for computing the THD is:

$$\text{THD} := \sqrt{\sum_{n=2}^{51} \left(\frac{I_{harm_n}}{I_{harm_1}}\right)^2} \quad \text{THD} = 37.2 \cdot \%.$$

Historically, only the first 51 harmonics are considered. The power company limits the THD to 10–15%. The THD value previously obtained is unacceptably high. Filters are needed to reduce the harmonic content. For a resistive load, the harmonic content (and THD) depend on the delay angle, because α changes the current waveform.

EXAMPLE 11.4: Analysis of battery charger operation

In this example, a single-phase rectifier charges a battery. The battery is represented by its Thévenin equivalent circuit: a dc voltage source and a resistance connected in series. The circuit diagram is drawn in Figure 11.34.

RECTIFIERS

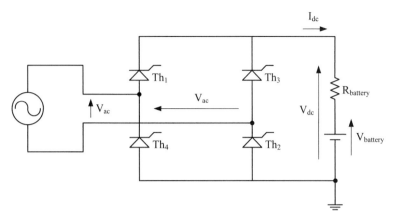

Figure 11.34. Battery charger utilizing a single-phase rectifier.

The ac source rms voltage, battery dc voltage and the battery internal resistance are:

$$V_{rms} := 50 \text{ V} \quad V_{battery} := 48 \text{ V} \quad R_{battery} := 0.4 \, \Omega.$$

The firing angle and the time are variable, but we select initial values for verifying the equations:

$$\alpha := 60 \text{ deg} \quad \omega := 2 \cdot \pi \cdot 60 \text{ Hz} \quad t := \frac{80 \text{ deg}}{\omega}.$$

The ac supply voltage is:

$$V_{ac}(t) := \sqrt{2} \cdot V_{rms} \cdot \sin(\omega \cdot t) \quad V_{ac}(t) = 69.6 \text{ V}.$$

The thyristors affect the dc voltage across the Thévenin equivalent of the battery in two ways. First, the dc current must be positive because of the thyristor, which prevents current drain from the battery when the battery voltage is greater than the dc voltage. Second, the thyristors must be fired before current can flow. Using an *if* statement in combination with these two requirements, the dc voltage is:

$$V_{dc}(t, \alpha) := \text{if}\left[(\omega \cdot t < \alpha) \vee (|V_{ac}(t)| < V_{battery}), V_{battery}, |V_{ac}(t)|\right]$$
$$V_{dc}(t, \alpha) = 69.6 \text{ V},$$

where the symbol "∨" denotes the Boolean *or* operation. Figure 11.35 plots the dc voltage waveform, which is a pulsating periodic function. The modulus operator, *mod(x,y)*, can be used to extend the previous expression for all time. In general, the modulus operator returns the remainder upon dividing *x* by *y*. The remainder of the

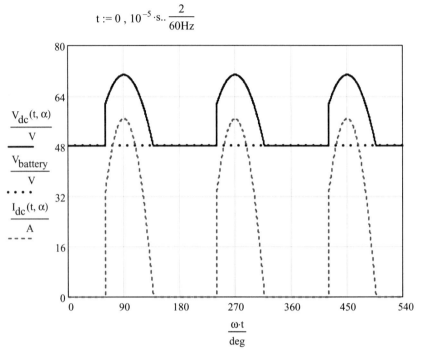

Figure 11.35. Battery charger dc current and voltages.

$\omega t/\pi$ ratio is always $\omega t - n\pi$, where n is the number of half-cycles. Accordingly, in the first half-cycle the remainder is ωt, while for the second half-cycle it is $\omega t - \pi$, and so on. The replacement of $\omega t < \alpha$ in the previous expression with $\mathrm{mod}(\omega t, \pi) < \alpha$ generates the correct periodic function for $t > 0$:

$$V_{dc}(t, \alpha) := \mathrm{if}\left[(\mathrm{mod}(\omega \cdot t, \pi) < \alpha) \vee (|V_{ac}(t)| < V_{battery}), V_{battery}, |V_{ac}(t)|\right].$$

The dc (battery) current can be calculated using Ohm's law. The charging current is equal to the voltage difference divided by the battery resistance. The voltage difference is the dc voltage minus the battery voltage:

$$I_{dc}(t, \alpha) := \frac{V_{dc}(t, \alpha) - V_{battery}}{R_{battery}} \quad I_{dc}(t, \alpha) = 54.1 \text{ A}.$$

The obtained dc current is not constant, but rather is a positive pulsating current. The ac current is the same as the dc current but its polarity in the negative cycle is opposite:

$$I_{ac}(t, \alpha) := \mathrm{if}\,(\mathrm{mod}(\omega \cdot t, 2 \cdot \pi) < \pi, I_{dc}(t, \alpha), -I_{dc}(t, \alpha)).$$

The average dc voltage and current are:

$$V_{dc\_ave}(\alpha) := \frac{\omega}{\pi} \cdot \int_{0s}^{\frac{\pi}{\omega}} V_{dc}(t,\alpha)\,dt \quad V_{dc\_ave}(\alpha) = 55.2 \text{ V}$$

$$I_{dc\_ave}(\alpha) := \frac{\omega}{\pi} \cdot \int_{0s}^{\frac{\pi}{\omega}} I_{dc}(t,\alpha)\,dt \quad I_{dc\_ave}(\alpha) = 18.0 \text{ A}.$$

The dc current and voltage are plotted in Figure 11.35. The graph shows that the current flows only if the dc voltage is larger than the battery voltage. In this case, the current flows from about 43° to 137°, and that also is the range in which current can be controlled. Outside this range, the gate control is ineffectual. Consequently, the delay angle control is effective when the ac voltage is larger than the battery voltage. This is demonstrated in Figure 11.36a. This graph reveals that the current cannot be controlled until a delay angle (α) of 43° is reached.

The real power transfer from the ac supply to the battery is also a function of the delay angle. The average power is the integral of the instantaneous power for a half-cycle divided by the half-cycle period:

$$P_{ac}(\alpha) := \frac{\omega}{\pi} \cdot \int_{0s}^{\frac{\pi}{\omega}} I_{ac}(t,\alpha) \cdot V_{ac}(t)\,dt \quad P_{ac}(\alpha) = 1.2 \text{ kW}.$$

The power transfer as a function of delay angle (α) is plotted in Figure 11.36b. The ac voltage and current are graphed in Figure 11.37. The plot discloses that the control produces a distorted current waveform which may have high harmonic content.

11.4.2.3. Single-Phase Bridge Rectifier with Inductive Load.

The bridge rectifier with a resistive load produces discontinuous, pulsating dc current. This can adversely affect the operation of a dc motor controlled by the bridge. An inductance connected in series with the load resistance can reduce or even eliminate the pulsating current. Figure 11.38 shows the connection diagram of a bridge rectifier with an inductive load.

The effect of the inductance connected in series with the load resistance is investigated using PSpice simulations. Figure 11.39a,b provides typical simulation results with small inductance and medium inductance, respectively. The dc current and voltage at the 4 Ω resistive load are plotted along with the ac source voltage and current. The graphs demonstrate that the inductance lengthens the conduction period beyond the ac voltage zero crossing. However, the dc current remains discontinuous. If the inductance is large, the conduction becomes continuous, as displayed in Figure 11.39c.

We will investigate the very large inductance case when the dc current is constant. In this situation, each thyristor pair conducts 180° (i.e., an entire half-cycle). The firing delay regulates the start of the conduction. The delay shifts the conduction period along the ac voltage signal.

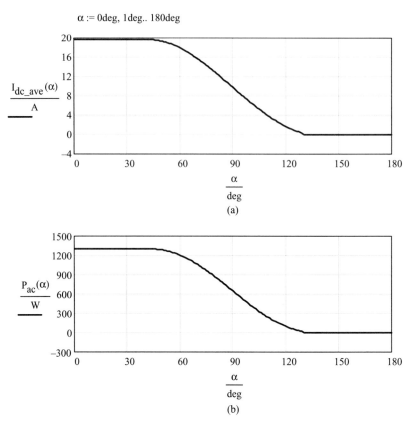

Figure 11.36. Effect of thyristor firing delay on battery charging. (a) Charging current versus delay angle; (b) average power transfer versus delay angle.

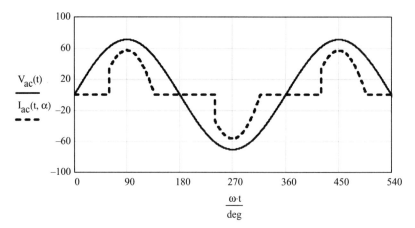

Figure 11.37. Battery charger ac current and voltage.

RECTIFIERS

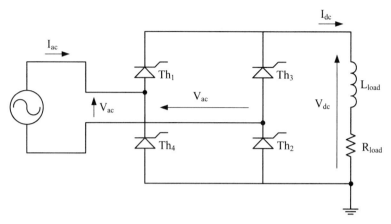

Figure 11.38. Single-phase bridge rectifier with inductive load.

The operation of this circuit is analyzed using a numerical example. The circuit data are:

$$V_{rms} := 120 \text{ V} \quad I_{dc} := 40 \text{ A} \quad f := 60 \text{ Hz} \quad \omega := 2 \cdot \pi \cdot f.$$

The ac supply voltage is:

$$V_{ac}(t) := \sqrt{2} V_{rms} \cdot \sin(\omega \cdot t).$$

The full and half cycle times are:

$$T_{cycle} := \frac{1}{f} \quad T_{half} := \frac{1}{2 \cdot f}.$$

The validity of the equations is tested by using:

$$\alpha := 60 \text{ deg} \quad t := \frac{T_{half}}{3} = 2.78 \text{ ms}.$$

If the thyristor pairs conduct for the half-cycle and the delay angle is α, the dc voltage is:

$$V_{dc}(t, \alpha) := \text{if } (\alpha \leq \omega \cdot t \leq 180 \cdot \text{deg} + \alpha, V_{ac}(t), -V_{ac}(t))$$
$$V_{dc}(t, \alpha) = 147.0 \text{ V}.$$

This equation uses an *if* statement to establish that between α and α + 180°, thyristor Th$_1$ and Th$_2$ conduct for a half-cycle; consequently, the ac voltage is connected to the

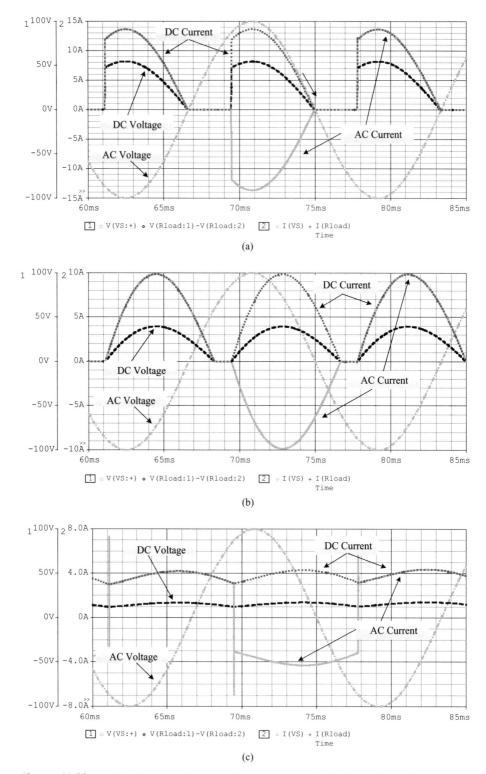

Figure 11.39. Load inductance effect on dc current through the load. (a) Small inductance (10 μH); (b) medium inductance (15 mH); (c) large inductance (200 mH) causes continuous conduction of dc current.

RECTIFIERS

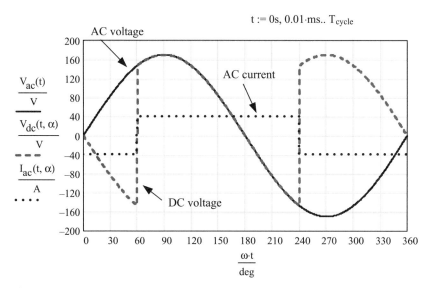

Figure 11.40. AC voltage and current, and dc voltage if the delay angle (α) is 60°.

load in the positive direction. During the next half-cycle, Th$_3$ and Th$_4$ conduct; consequently, the ac voltage is connected to the load in negative direction (see Fig. 11.38).

The ac current is positive when Th$_1$ and Th$_2$ conduct, and negative when Th$_3$ and Th$_4$ conduct, as shown in Figure 11.31. The ac current is:

$$I_{ac}(t, \alpha) := \text{if } (\alpha \le \omega \cdot t \le 180 \text{ deg} + \alpha, I_{dc}, -I_{dc})$$

$$I_{ac}(t, 90 \text{ deg}) = -40 \text{ A}.$$

The ac voltage, dc voltage and ac current waveforms are plotted in Figure 11.40 for a delay angle of 60°. The graph shows the square-shaped ac current and segmented sinusoidal dc voltage.

The nonconducting thyristor voltages are equal with the load voltage. Their maximum value is the peak ac supply voltage.

The average dc voltage is:

$$V_{dc\_ave}(\alpha) := \frac{1}{T_{half}} \cdot \int_{\frac{\alpha}{\omega}}^{T_{half}+\frac{\alpha}{\omega}} V_{dc}(t, \alpha) \, dt \quad V_{dc\_ave}(\alpha) = 54.0 \text{ V}.$$

An analytical expression for the average dc voltage is derived by integrating the ac voltage between α and α + 180°, and then dividing by the half-period duration. Using the Mathcad symbolic integration, we obtain:

$$\frac{1}{T_{half}} \cdot \int_{\frac{\alpha}{\omega}}^{T_{half}+\frac{\alpha}{\omega}} \sqrt{2} \cdot V_{rms} \cdot \sin(\omega \cdot t) \, dt \to -\frac{\sqrt{2} \cdot V_{rms} \cdot (\cos(\alpha + T_{half} \cdot \omega) - \cos(\alpha))}{T_{half} \cdot \omega},$$

where

$$\cos(\alpha + T_{half} \cdot \omega) = \cos(\alpha) \cdot \cos(T_{half} \cdot \omega) - \sin(\alpha) \cdot \sin(T_{half} \cdot \omega)$$
$$T_{half} \cdot \omega = \pi \quad \cos(T_{half} \cdot \omega) = -1 \quad \sin(T_{half} \cdot \omega) = 0.$$

The substitution of these quantities simplifies the formula. The reduced equation is:

$$V_{dc\_ave}(\alpha) := \frac{2 \cdot \sqrt{2} \cdot V_{rms}}{\pi} \cdot \cos(\alpha) \quad V_{dc\_ave}(\alpha) = 54.0 \text{ V}.$$

The dc power is the product of average dc voltage and the constant dc current:

$$P_{dc}(\alpha) := I_{dc} \cdot V_{dc\_ave}(\alpha) \quad P_{dc}(\alpha) = 2.16 \text{ kW}.$$

Figure 11.41 shows the dc power variation with the delay angle. It can be seen that the dc power becomes zero at a delay angle of 90°. A delay angle larger than 90° produces negative dc power. This means that the power flows from the dc side to the ac side. Obviously, this requires a source (battery) at the dc side. Without a dc source, the power remains zero if the delay angle is larger than 90°. Without a dc source, the current cannot be maintained constant beyond a 90° delay angle.

By definition, the rms value of the square-shaped ac current is equal with the dc current:

$$I_{ac\_rms} := I_{dc} \quad I_{ac\_rms} = 40 \text{ A}.$$

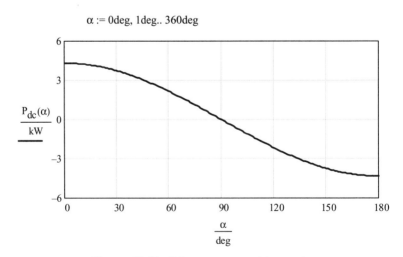

Figure 11.41. DC power versus delay angle.

The harmonic content of the ac current is calculated using Fourier analysis. The first 51 harmonics are calculated for the determination of the total harmonic distortion (THD) factor:

$$N := 51 \quad n := 0 .. N$$

$$I_{a_n} := \frac{1}{T_{half}} \cdot \int_0^{T_{cycle}} I_{ac}(t, \alpha) \cdot \cos(n \cdot \omega \cdot t) \, dt$$

$$I_{b_n} := \frac{1}{T_{half}} \cdot \int_0^{T_{cycle}} I_{ac}(t, \alpha) \cdot \sin(n \cdot \omega \cdot t) \, dt$$

$$I_{h_n} = \frac{\sqrt{(I_{a_n})^2 + (I_{b_n})^2}}{\sqrt{2}} \quad I_{h_0} := \left| \frac{I_{a_0}}{2} \right|$$

$$I_h^T = \begin{array}{|c|c|c|c|c|c|c|c|c|} \hline & 0 & 1 & 2 & 3 & 4 & 5 & 6 & 7 & 8 \\ \hline 0 & 0 & 36.01 & 0 & 12 & 0 & 7.2 & 0 & 5.14 & \ldots \\ \hline \end{array} A$$

The rms value of the first 51 harmonics computed below is close to the rms current of 40 A:

$$\sqrt{\sum_{n=1}^{N} (I_{h_n})^2} = 39.84 \text{ A}.$$

The THD factor is:

$$\text{THD} := \sqrt{\sum_{n=2}^{N} \left(\frac{I_{h_n}}{I_{h_1}} \right)^2} \quad \text{THD} = 47.3\%.$$

In the case of an inductive load, the THD is independent of the delay angle, because the ac current waveform is always a square shape. The obtained THD value is unacceptably large, which requires filters at the ac side of a large rectifier.

The average ac power is the integral of the instantaneous power for a cycle:

$$P_{ave}(\alpha) := \frac{1}{T_{cycle}} \cdot \int_{0 \cdot s}^{T_{cycle}} I_{ac}(t, \alpha) \cdot V_{ac}(t) \, dt$$

$$P_{ave}(\alpha) = 2.16 \cdot \text{kW}.$$

This matches the P_{dc} value computed earlier. The apparent (complex) power and power factor are:

$$S := V_{rms} \cdot I_{ac\_rms} = 4.8 \text{ kV} \cdot \text{A} \quad pf := \frac{P_{ave}(\alpha)}{S} = 0.45.$$

The power factor is very poor. The desired value is above 0.8. Typically, large rectifiers use capacitor banks together with harmonic filters for power factor compensation.

11.4.3. Firing and Snubber Circuits

The operation of the semiconductor switches requires square-shaped firing pulses. A firing circuit generates these pulses. The different semiconductor devices use different firing circuits. The detailed analysis of the firing circuits is outside the scope of this book. However, the concept of the firing circuit is presented using a thyristor firing circuit as an example.

Gate Firing Circuit Concept. The thyristor requires a synchronized, delayed, short firing pulse in each cycle. Large numbers of analog and digital circuits have been developed for generating the delayed firing pulse. A block diagram of a simple auxiliary circuit is drawn in Figure 11.42. This schematic depicts the concept of the generation of the delayed firing pulse for a bridge converter. A transformer reduces the ac supply voltage. This reduced voltage provides a sinusoidal reference signal. The reference signal supplies the "zero crossing detector," which converts the sinusoidal reference signal to a square wave to mark the voltage zero crossing (Fig. 11.43a). The delay angle control integrates the square wave signal. This results in a periodic sawtooth wave, which is compared with the control voltage. The control voltage is a dc voltage that is

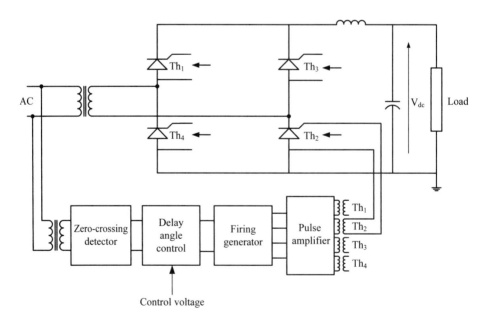

Figure 11.42. Gate firing circuit for a bridge converter.

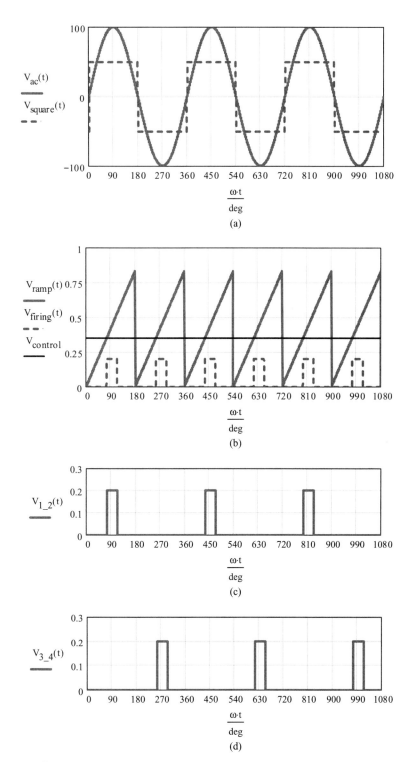

Figure 11.43. Example for thyristor firing signal generation. (a) Reference sinusoidal signal and its square wave conversion; (b) signals to generate delayed firing pulses; (c) and (d) firing signals to the thyristors conducting in the positive and negative directions, respectively.

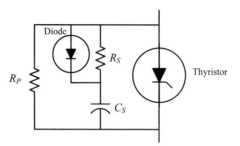

Figure 11.44. Thyristor with snubber circuit.

selected to accomplish the motor speed control within the desired range. A comparator produces a firing impulse when the two signals are equal. Accordingly, the delay angle is controlled by the dc control voltage (Fig. 11.43b). The firing generator separates the firing signals for thyristors Th_1 and Th_2 and thyristors Th_3 and Th_4, as graphed in Figure 11.43c,d, respectively. The firing signal drives a pulse amplifier that supplies the thyristor gates through insulating pulse transformers, as shown in Figure 11.42. In the more advanced circuits, the pulse transformers are replaced by optical couplers or light-fired thyristors are used.

Snubber Circuit. The thyristor turn on and turn off generates voltage transients that can endanger the device. This problem is enhanced by the series connection of several devices. The transients are controlled by snubber circuits connected in parallel with each device. A snubber circuit absorbs the shock of the voltage transients. A typical snubber circuit is presented in Figure 11.44. The circuit consists of a series resistor–capacitor (RC) network and a diode shunting the resistance R_S. R_P is used to equalize the voltage distribution when the thyristors are not conducting. Before the thyristor turns on or when the thyristor is turned off, the current charges the capacitor through the diode and limits dV/dt. When the thyristor turns on, the capacitor discharges through the resistance (R_S), which limits di/dt. The analysis of the snubber circuit operation is beyond the goal of this book.

11.4.4. Three-Phase Rectifiers

Three-phase rectifiers are used for large systems. A three-phase rectifier can be built by adding a third leg to the bridge converter. The three-phase rectifier circuit produces smoother dc voltages. Typically, a large inductance is used to smooth the dc voltage. This results in constant dc current.

Figure 11.45 shows a sketch of the connection diagram of a typical thyristor-controlled three-phase rectifier. In this circuit, the conduction time is 60° and the thyristors are fired in a preselected sequence. The firing sequence is Th_5–Th_6, Th_6–Th_1, Th_1–Th_2, Th_2–Th_3, Th_3–Th_4, Th_4–Th_5. Figure 11.45 illustrates that the firing of any of the thyristor pairs connects the ac line-to-line voltage to the inductive dc load. Consequently, the output dc voltage will be close to the peak value of the line-to-line voltage without firing control. Typically, without gate control the average dc voltage is 1.35 $V_{\text{line-to-line}}$.

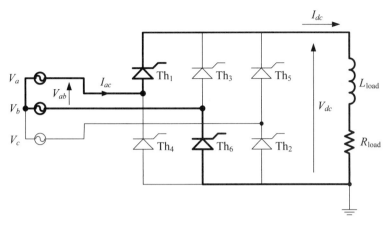

Figure 11.45. Three-phase thyristor-controlled rectifier.

The average dc voltage is controlled by the firing delay. Figure 11.46 shows a plot of the waveforms. The graph clearly shows that firing delay cuts out a part of the dc voltage wave, which reduces the average dc voltage. The dc voltage will be zero when the delay angle is 90°. Beyond 90°, the power flows from the dc to the ac, if a battery or other dc power source is available at the dc side.

11.5. INVERTERS

An *inverter* converts dc voltage and current to an ac voltage and current. The frequency of the ac voltage can be regulated by the inverter operating frequency. Typical power applications are:

1. AC motor drives,
2. solar power conversion to 60 Hz,
3. wind power conversion,
4. fuel cells, and
5. high-voltage dc transmission.

This section discusses the principles of the voltage source inverter with PWM and the line-commutated thyristor-controlled inverter.

The basic concept of inverter operation is illustrated in Figure 11.47. The major components of the circuit are:

- a dc power source (V_{dc}), such as a battery;
- the load, for example a resistance (R_{load}) at the ac side; and
- the bridge circuit with the switches.

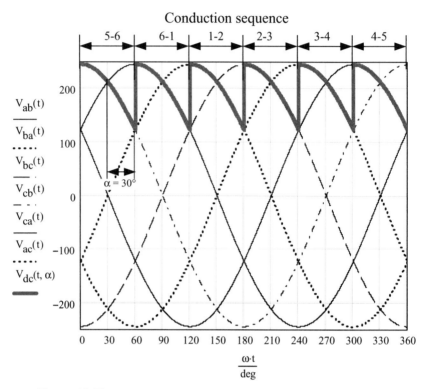

Figure 11.46. Waveforms of three-phase rectifier with 30° firing delay.

First, switches S_1 and S_2 close and connect the dc source to the load, as depicted in Figure 11.47a. The load voltage (V_{ac}) will be equal to the dc voltage. After a selected time, S_1 and S_2 open and disconnect the dc source, such that the load voltage becomes zero.

This is followed by the closing of switches S_3 and S_4, which connect the dc source to the load in the reverse polarity (see Fig. 11.47b). The load voltage will be the negative dc voltage. After a selected time, S_3 and S_4 open and disconnect the dc source, and the load voltage returns to zero.

The repetition of this switching sequence generates the square-shaped gate signal displayed in Figure 11.47c. This is a distorted ac voltage. The waveform can be improved by using filters, which eliminate the higher harmonics, and results in an acceptable sinusoidal voltage.

The following analysis illustrates that the frequency of the generated ac voltage depends on the switching frequency, and the rms value of the ac voltage can be regulated by the on-time duration of the switches.

If the pulse duration is T_{on} and the cycle repetition period is T_{cyc}, then the ac voltage frequency is:

$$f_{ac} = \frac{1}{T_{cyc}}. \qquad (11.30)$$

INVERTERS

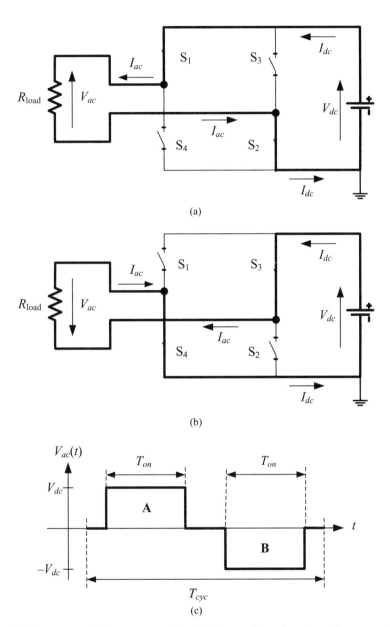

Figure 11.47. Concept of inverter operation. (a) Current flow when S_1 and S_2 are conducting; (b) current flow when S_3 and S_4 are conducting; (c) bridge operation generated voltage waveform.

Carrying out the integration over a half cycle, the rms value of the ac voltage is:

$$V_{\text{rms\_ac}} = \sqrt{\frac{2}{T_{\text{cyc}}} \int_0^{T_{on}} V_{dc}^2 \, dt} = \sqrt{2} V_{dc} \sqrt{\frac{T_{on}}{T_{\text{cyc}}}}. \tag{11.31}$$

The positive and negative cycles must be identical to avoid dc bias in the ac signal. The harmonic content and the THD factor can be calculated by the method presented in Section 11.4.2.2.

The performance of the inverter can be improved by dividing the on time into shorter on and off periods. Typically, this type of inverter is built with GTO switches.

The rectifiers and inverters utilize the same circuit. Actually, every rectifier can operate as an inverter and vice versa. For this reason, the bridge circuit is often termed a *converter*.

The inverter described earlier is called the *voltage source inverter*. The voltage source inverter requires switches (e.g., GTO, MOSFET, or IGBT). We will demonstrate later that the thyristor-controlled rectifier, described previously, can also operate as a current source inverter.

11.5.1. Voltage Source Inverter with Pulse Width Modulation

The major disadvantage of the voltage source inverter described earlier is the large amount of harmonic generation. The amplitude of the harmonics can be reduced by using the *pulse-width-modulation* (PWM) technique. The rms value of the ac voltage is controlled by the on time (T_{on}) of the switches, as shown in Figure 11.47. The basic concept of the PWM method is the division of the on time into several on and off periods with varying duration.

The most frequently used PWM technique is *sinusoidal pulse width modulation*. This approach requires a bridge converter with IGBT or MOSFET switches shunted by an antiparallel connected diode. The diode allows current flow in the opposite direction when the switch is open. These freewheeling diodes prevent inductive current interruption and provide protection against transient overvoltage, which may cause reverse breakdown of the IGBT and MOSFET switches. The typical circuit diagram is provided in Figure 11.48.

The switches in this converter are controlled by gate pulses. The gate signal contains several pulses distributed along the half-cycle. The width of each pulse is varied in proportion to the amplitude of a sine wave. A typical PWM waveform is presented in Figure 11.49.

During the positive cycle, S_1 and S_2 are switched by the high-frequency pulse train shown in Figure 11.49. During the negative cycle, the pulse train switches S_3 and S_4. The load inductance integrates the generated pulse train and produces a sinusoidal voltage (V_{ac}) and current wave, as displayed in Figure 11.49.

The control circuit produces the gate pulse train by generation of a triangular carrier wave and a sinusoidal reference signal. The two signals are compared, and when the carrier wave is larger than the reference signal, the gate signal is positive. When the

INVERTERS

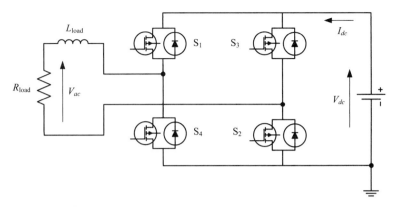

Figure 11.48. Single-phase voltage source converter.

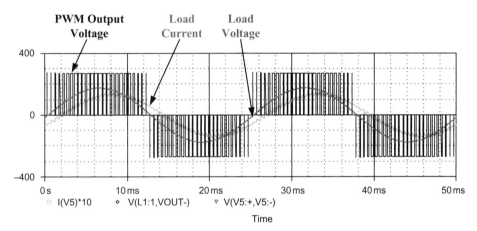

Figure 11.49. Gate pulse input signal and ac voltage and current outputs of a pulse width modulation (PWM) converter.

carrier wave is smaller than the reference signal, the gate signal is zero. This results in a gate pulse with variable width. Figure 11.50a shows the carrier wave and reference sine wave; Figure 11.50b depicts the resulting gate signal with variable width pulses. It has to be noted that several other methods are used for generation of PWM signals. This is a typical example, which demonstrates the technique.

The frequency of the reference sine wave determines the frequency of the generated ac voltage. The amplitude of the ac voltage can be regulated by the variation of the reference signal amplitude. The amplitude of the fundamental component of the ac voltage is:

$$V_{ac} = \frac{V_{control}}{V_{carrier}} V_{dc} = mV_{dc}, \quad (11.32)$$

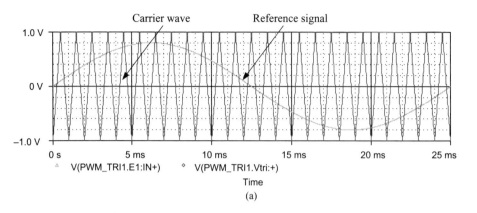

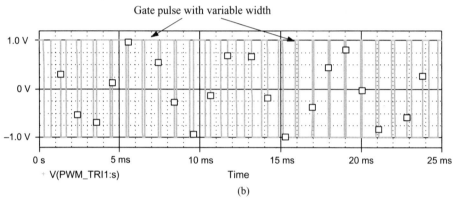

Figure 11.50. Pulse width modulation (PWM) signals. (a) Triangular carrier wave and sinusoidal reference signal; (b) variable-width gate pulse signal.

where

V_{ac} is the amplitude of the ac voltage;
V_{dc} is the dc voltage amplitude;
$V_{control}$ is the control (reference) signal peak voltage;
$V_{carrier}$ is the carrier wave peak voltage; and
m is the modulation index.

The *modulation index* is the ratio of the peak-to-peak ac voltage ($2V_{ac}$) to the dc voltage.

The inverter interrupts the current several times each cycle. The interruption of an inductive current would generate unacceptably high overvoltage. This overvoltage generation is eliminated by providing *freewheeling diodes* connected in parallel with the switches. When the switches open, the current, if inductive, is diverted to the diodes, as shown in Figure 11.51. The diagram exhibits the current path when switches S_1 and S_2 are closed, and switches S_3 and S_4 are open. When switches S_1 and S_2 open (now all

INVERTERS

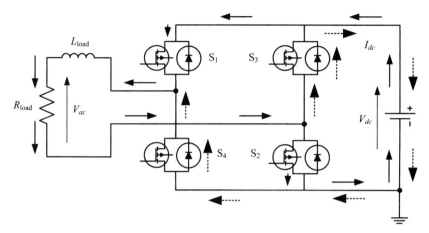

Figure 11.51. Freewheeling diode operation.

switches are open), the current diverts through the diodes of switches S_3 and S_4. This current diversion prevents the interruption of inductive current.

The inductance in the circuit and capacitances connected in parallel (see Fig. 11.42) with the output of the inverter filter out most high-frequency harmonics. The produced ac voltage is sinusoidal. For typical ac motor control, the 60 Hz ac is rectified and the obtained dc voltage supplies the inverter, which in turn produces sinusoidal variable frequency voltage for the motor, as shown in Figure 11.8. It is noted that this inverter circuit can operate as a rectifier.

Furthermore, the addition of a switch pair converts this circuit to a three-phase PWM inverter, which is frequently used for three-phase motor control.

11.5.2. Line-Commutated Thyristor-Controlled Inverter

The rectifier circuit (Fig. 11.30) described in Section 11.4.2 can operate as an inverter. The inverter operation requires both ac and dc sources, and an inductance that maintains the dc current constant or at least assures continuous dc current.

The inverter operation requires a delay angle between 90° and 180°. Figure 11.52 shows graphs of bridge circuit-generated waveforms for 90° and 170° delay angles.

Figure 11.52a shows that the delay angle between the ac voltage and the square-shaped ac current is 90°, and the average dc voltage is zero because the magnitude and duration of the positive and negative segments of the dc voltage are identical. These two facts imply that both the ac and dc power are zero.

Figure 11.52b demonstrates that the delay angle between the ac voltage and the square-shaped ac current is 170°, and the average dc voltage is negative because the

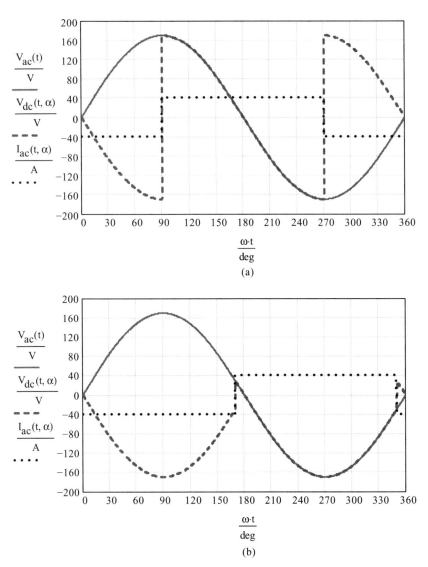

Figure 11.52. Single-phase bridge inverter generated waveforms. (a) Delay angle of 90°; (b) delay angle of 170°.

duration of the positive part of the dc voltage is almost nonexistent (i.e., 10° out of 180°). This implies that power flows from the dc to the ac side.

The practical use of the line-commutated inverter requires the regulation of the dc voltage to maintain constant dc current. Figure 11.53a displays the bridge converter circuit that produces the voltages shown in Figure 11.52. The average value of this dc voltage is V_{dc_inv}. Using the Thévenin equivalent, the converter can be replaced by a dc

INVERTERS

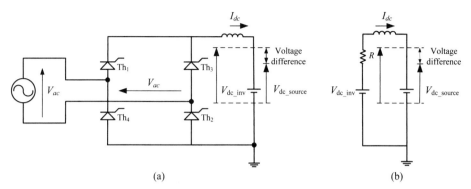

Figure 11.53. Single-phase line-commutated inverter. (a) Circuit diagram; (b) equivalent circuit.

source (V_{dc_source}) and impedance, as shown in Figure 11.53b. The average dc current in this equivalent circuit is:

$$I_{dc} = \frac{V_{dc\_inv} - V_{dc\_source}}{R}. \qquad (11.33)$$

The dc current can be maintained constant by keeping the voltage difference constant. Consequently, if the delay angle increases, V_{dc_inv} is reduced; the maintenance of constant dc current and voltage difference requires the appropriate reduction of the V_{dc_source}. If the firing angle increases beyond 90°, V_{dc_inv} becomes negative. Maintaining constant dc current and voltage difference requires changing the V_{dc_source} polarity. The converter delay angle and the dc source voltage must be controlled simultaneously to maintain a constant voltage difference and dc current. In a practical circuit, the required voltage difference is generally small; consequently, both the inverter-produced voltage and the source voltage must be negative, as shown in Figure 11.54. In this case, the power flows from the dc to the ac circuit, because the current and voltage have the same direction in the dc source (generator) and opposite directions in the inverter (load), as demonstrated in Figure 11.52b.

These inverters are used for driving large synchronous motors, where the motor generates an ac voltage needed for operation. However, this type of inverter is not suitable for induction motor drives and cannot supply a network without an ac source. The operation of a thyristor-controlled, current source type inverter requires availability of ac voltage at the ac side. In the case of a synchronous motor drive, the large motor must generate the variable-frequency ac voltage. The startup of a motor drive requires rotation of the motor, which is achieved by supplying the motor with short duration current pulses which can produce slight rotation and thereby the generation of ac voltage.

A very important application of the line-commutated inverters and rectifiers is HVDC transmission.

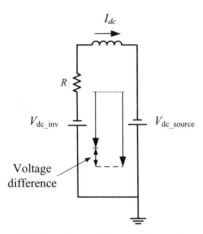

Figure 11.54. Voltage difference during inversion.

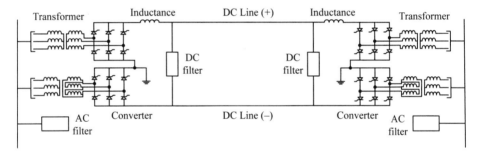

Figure 11.55. Concept of HVDC transmission.

11.5.3. High-Voltage DC Transmission

HVDC lines are used to transport large amounts of energy over a long distance. A representative application is the Pacific DC Intertie, which interconnects the Los Angeles area with Oregon. The voltage of the DC Intertie is ±500 kV and the maximum energy transport is 3100 MW. More than 100 dc transmission systems operate around the world, and one of the oldest and most famous is the cable interconnection between England and France.

Figure 11.55 presents a simplified connection diagram for a HVDC system. The major elements are two converter stations interconnected by a dc transmission line. The converter station can operate in both inverter and rectifier modes, which permits energy transfer in both directions.

Each converter station contains two converters, which are connected in series at the dc side. The series connection node (middle point) is grounded. One of the converters generates the positive, and the other produces the negative dc voltage. The harmonics are filtered at both the ac and dc sides.

Figure 11.56. Valve hall of a dc converter station (courtesy of Siemens, Erlangen, Germany).

A transformer typically supplies each converter. The HVDC system uses two different transformer types at each converter station. A wye–wye transformer supplies one of the converters while the other is connected to a wye–delta transformer. This produces a 30° phase shift between the dc voltage outputs of the two converters. The phase shift produces a smoother dc output voltage.

The regulation of the converter delay angle controls the direction of energy flow. The major advantages of this system are the precise power control, fast power change, and low short-circuit current in the case of a fault. Because of the high price of the electronics, typically this system is only economical for distances longer than 300 miles and power transfer above 1000 MW.

Figure 11.56 shows a picture of the converters in an HVDC station. Each converter contains six high-voltage valves, with several hundred thyristors connected in series. Rounded aluminum electrodes as shown in the photograph shield the valves.

11.6. FLEXIBLE AC TRANSMISSION

Advancements in high-power semiconductor switches and the development of power electronics led to the idea of electronically controlled electrical power system components, which are called FACTS devices. In the last 20 years, several FACTS devices have been developed and tested on the network. This section describes only the major FACTS devices used within the network today.

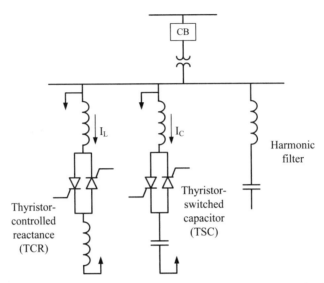

Figure 11.57. One-line diagram of a static VAR compensator (SVC).

11.6.1. Static VAR Compensator

Successful operation of an electrical network requires the availability of variable amounts of reactive power. The SVC is the most frequently used FACTS device. It is designed to provide continuously controlled reactive power. Typically, the SVC regulates injected or consumed reactive power such that the supply voltage remains constant. In the case of heavy inductive load, the SVC injects reactive power to increase the voltage, and at night, when the total load is light, the SVC consumes reactive power to reduce the voltage.

Figure 11.57 shows a single-line diagram of an SVC. The SVC contains thyristor-switched capacitors, thyristor-controlled reactance, and typically a 5th and 7th harmonics tuned filter. Both the thyristor-switched capacitance and the thyristor-controlled reactors are connected in a delta. In most cases, the system is grounded through a grounding transformer.

The thyristor switch contains two thyristors connected antiparallel in order to permit bidirectional current flow. A relatively small inductance is connected in series with the thyristor switches to reduce the inrush current. Because of the high cost of high-voltage thyristor switches, the devices are rated at 12–15 kV and a transformer connects the SVC to the high-voltage network. The SVC is more expensive than switched reactors and capacitors but has the advantage of speed of response when required.

EXAMPLE 11.5: SVC analysis

Figure 11.57 shows an illustration of a typical SVC, which can produce around 150 MVAR capacitive and consume 100 MVAR inductive reactive power. These variables are first defined:

FLEXIBLE AC TRANSMISSION

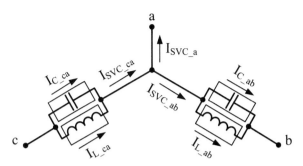

Figure 11.58. Circuit diagram of the SVC components in the legs of the delta nearest to line a.

$$M := 10^6 \quad \Delta Q_C := -150 \, M \cdot V \cdot A \quad \Delta Q_L := 100 \, M \cdot V \cdot A.$$

The supply line-to-line voltage and ac system frequency are taken to be:

$$V_s := 14.5 \, kV \quad f := 60 \, Hz \quad \omega := 2 \cdot \pi \cdot f.$$

The inductor switching-caused transient current and the harmonic filter in Figure 11.57 are neglected in this analysis. The corresponding capacitance and inductance current flows in each phase of the delta for the rated reactive powers are:

$$I_{C\_rms} := \overline{\left(\frac{j \cdot \Delta Q_C}{3 \cdot V_s}\right)} = 3.448 j \cdot kA \quad I_{L\_rms} := \overline{\left(\frac{j \cdot \Delta Q_L}{3 \cdot V_s}\right)} = -2.299 j \cdot kA.$$

Figure 11.58 depicts the SVC components in the legs of the delta connected to node **a**. Normally, the capacitance is switched on as needed, but in this example, the capacitor is always active. The capacitive current in the delta leg between nodes **a** and **b** is:

$$I_{C\_ab}(t) := \sqrt{2} \cdot |I_{C\_rms}| \cdot \cos(\omega \cdot t + 90 \, \deg).$$

The parallel-connected inductor current is regulated by the firing angle delay (α) of the thyristors, which cuts out a part of the inductance current, thereby reducing the rms value of the inductive current:

$$I_{L\_ab}(t, \alpha) := \text{if}\left(\text{mod}(\omega \cdot t, \pi) < \alpha, 0, \sqrt{2} \cdot |I_{L\_rms}| \cdot \cos(\omega \cdot t - 90 \, \deg)\right).$$

The process is demonstrated in Figure 11.59. Figure 11.59a graphs the capacitive current and the thyristor-controlled inductive current for a delay angle (α) of 120°. Figure 11.59b plots the sum of the inductive and capacitive currents in the **ab** leg:

$$I_{SVC\_ab}(t, \alpha) := I_{C\_ab}(t) + I_{L\_ab}(t, \alpha).$$

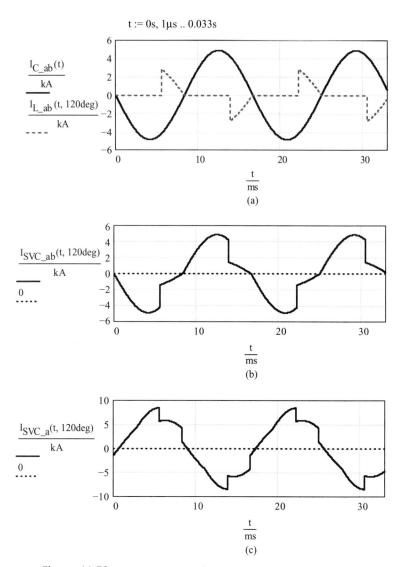

Figure 11.59. Demonstration of SVC current regulation process.

The SVC current in a delta leg is a distorted sine wave with significant harmonics. The thyristor-controlled inductance current cuts out a section of the capacitive current, which reduces the SVC rms capacitive current:

$$T_0 := \frac{1}{f} = 0.017 \text{ s}$$

$$I_{SVCrms} := \sqrt{\frac{1}{T_0} \cdot \int_0^{T_0} I_{SVC\_ab}(t, 120 \text{ deg})^2 \, dt} = 3.134 \cdot \text{kA}.$$

The capacitive and inductive currents in the delta leg **ca** are similarly calculated, but with a −240° (or +120°) phase angle shift:

$$I_{C\_ca}(t) := \sqrt{2} \cdot |I_{C\_rms}| \cdot \cos(\omega \cdot t + 90 \text{ deg} + 120 \text{ deg})$$
$$I_{L\_ca}(t, \alpha) := \text{if} \left(\text{mod}(\omega \cdot t + 120 \text{ deg}, \pi) < \alpha, 0, \sqrt{2} \cdot |I_{L\_rms}| \cdot \cos(\omega \cdot t - 90 \text{ deg} + 120 \text{ deg}) \right)$$
$$I_{SVC\_ca}(t, \alpha) := I_{C\_ca}(t) + I_{L\_ca}(t, \alpha).$$

Referring to Figure 11.58, the current injected by the SVC into node **a** is:

$$I_{SVC\_a}(t, \alpha) := -I_{SVC\_ab}(t, \alpha) + I_{SVC\_ca}(t, \alpha).$$

This overall current is plotted in Figure 11.59c. A visual comparison of the current in a single leg, I_{SVC_ab} in Figure 11.59c, to the overall phase a current reveals that I_{SVC_a} is more sinusoidal like. This qualitative observation can be quantified by performing a spectral analysis, as expounded earlier in this chapter.

SVCs are in service around the world. Figure 11.60 shows a picture of a large SVC unit in Brazil. The capacitor banks and the inductances are outdoors. The thyristor switches are installed in ventilated metal boxes located on the left side of the photograph.

Figure 11.60. SVC installation rated ±250 MVAR in Brazil (courtesy of Siemens, Erlangen, Germany).

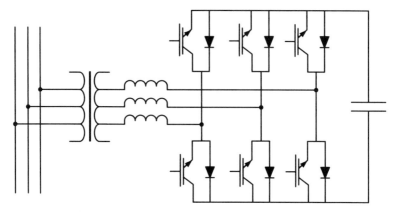

Figure 11.61. STATCOM reactive power generator.

11.6.2. Static Synchronous Compensator

The rapid development of voltage source converters using PWM led to an advanced reactive power generator called the static synchronous compensator (STATCOM). Figure 11.61 exhibits a single-line diagram of a STATCOM reactive power generator. A capacitor bank is connected to the ac system through a voltage source converter, a small reactance, and a coupling transformer. The voltage source converter is built with IGBT switches that are shunted by a diode and the converter operates using the PWM technique.

The converter charges the capacitor to the desired dc voltage and provides a small amount of real power to supply the losses and maintain the dc voltage constant. This is achieved by lagging the converter-generated three-phase voltage by a few degrees behind the ac system voltage.

The converter generates balanced three-phase voltage and provides reactive power. The capacitor does not provide any power; its voltage and charge are maintained constant. The ac network itself supplies the reactive power. The converter interconnects the phases of the network for short durations using the PWM technique in such a way that reactive power is generated. Converter operation analysis is beyond the scope of this book.

11.6.3. Thyristor-Controlled Series Capacitor

The transmission line inductance limits the power transfer in long lines due to system stability. The system stability limit can be increased by connecting a capacitor in series with the line.

The traditional solution is the mechanically switched capacitor. FACTS introduced the thyristor-controlled series capacitor (TCSC), which contains a series-connected capacitor shunted by a thyristor-controlled inductor and protected by a parallel-connected surge arrester or metal oxide varistor (MOV). This has the advantage of fast

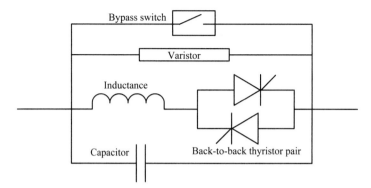

Figure 11.62. Circuit diagram of the TCSC.

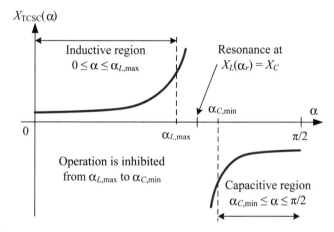

Figure 11.63. Reactance-versus-delay angle characteristic of the thyristor-controlled series capacitor (TCSC).

switching. Also, the device appears capacitive at 60 Hz and inductive at critical subsynchronous frequencies and therefore can be a subsynchronous resonance countermeasure. In addition, the TCSC unit contains a parallel-connected circuit breaker that can bypass the TCSC in the event of a short circuit. Figure 11.62 is an example of a simplified circuit diagram.

The firing of the thyristors connects the inductance in parallel with the capacitor for the thyristor conduction time. When the thyristor is not conducting, the TCSC unit impedance is the capacitance only; in contrast, when the thyristor is conducting, the unit impedance is the parallel-connected capacitance and inductance. This implies that the TCSC has variable impedance. Figure 11.63 shows a graph of the circuit characteristics, which indicate that the TCSC can operate as either an inductance or a capacitance depending on the thyristor delay angle.

Figure 11.64. Thyristor-controlled capacitor (courtesy of ABB).

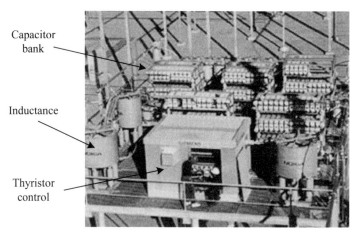

Figure 11.65. TCSC at Kayenta, Arizona (courtesy of Siemens, Erlangen, Germany).

Both Figure 11.64 and Figure 11.65 exhibit a TCSC installation. Because the units are connected in series with the transmission line, all components must be insulated from the ground for the line-to-neutral voltage. This requires that the components are installed on an insulated platform and the unit may be controlled through fiber optic cables. Figure 11.64 displays the post insulator columns, which electrically isolate the metal platform supporting the capacitor bank, reactances, and thyristor valves.

DC-TO-DC CONVERTERS

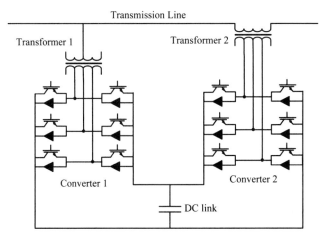

Figure 11.66. Unified power controller.

11.6.4. Unified Power Controller

The power flow in an interconnected ac system depends on the impedance of the parallel-connected lines and typically cannot be controlled. The newly developed "unified power controller" is designed to control independently both active and reactive power in a transmission line.

Figure 11.66 presents a conceptual one-line connection diagram of the Westinghouse developed unified power controller. The diagram reveals that the system is built with two voltage source converters interconnected by a dc link. The first converter is supplied by a transformer connected in shunt to the transmission line. The second converter also supplies a transformer, whose secondary is connected in series to the transmission line. The converters are built with IGBT switches and employ the PWM technique.

Typically, Converter 1 operates as a rectifier and supplies the dc link with real power. Converter 2 is supplied by the dc link and functions as an inverter-injecting voltage, with amplitude and phase angle which can be controlled, into the transmission line. The injected voltage controls both reactive and active power flow in the line. Furthermore, Converter 1 can also inject or absorb reactive power.

The unified power controller is a very flexible device that can provide reactive shunt compensation, series compensation, and phase angle regulation. These units are rather expensive, and presently, there are only three operating worldwide. In contrast, there are hundreds of SVCs and STATCOMs in service.

11.7. DC-TO-DC CONVERTERS

DC-to-DC converters operating at high switching frequencies can be used to either increase or decrease the input voltage. This section introduces both boost and buck converters which produce higher and lower output voltages, respectively, from a given

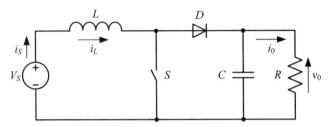

Figure 11.67. Basic boost converter circuit.

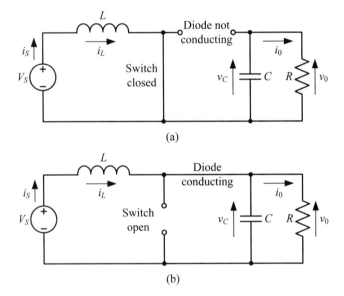

Figure 11.68. Boost converter circuit when switch is (a) closed and (b) open.

dc input voltage. These converters employ diode and resistor–inductor-capacitor (RLC) circuit elements along with high-speed switching to accomplish said functionality. The switching is carried out using semiconductor devices.

11.7.1. Boost Converter

The basic circuit diagram for a boost converter is provided in Figure 11.67. Switch action leads to two different operating periods:

1. *Switch Closed (see Fig. 11.68a)*. During this period, while the switch forms a short circuit, source-driven current increases the energy stored in the inductor, and the diode forms an open circuit, causing energy stored in the capacitor to be discharged through the load.

DC-TO-DC CONVERTERS

2. *Switch Open (see Fig. 11.68b).* When the diode conducts and the energy stored in the inductor is transferred to the capacitor and load, the energy transfer reduces the inductor current and increases the capacitor voltage.

These switching periods are analyzed in the succeeding paragraphs while assuming that the circuit has been operating for a time long enough that startup transients have vanished.

Switch Closed

From time t_0 to t_1 while the switch is closed, the inductor current is related to the dc voltage source via:

$$i_L(t) = i_L(t_0) + \frac{1}{L}\int_{t_0}^{t} V_S \, dt = i_L(t_0) + \frac{V_S}{L}(t - t_0). \quad (11.34)$$

Hence, the inductor current increases linearly from the initial current value to a final current of:

$$i_L(t_1) = i_L(t_0) + \frac{V_S}{L}(t_1 - t_0) = i_L(t_0) + \frac{V_S}{L}T_{on}, \quad (11.35)$$

where T_{on} is the period for which the switch is on (closed).

Switch Open

With the switch open from t_1 to t_2, the current from the inductor now flows through the diode to the capacitor and load. Ignoring the ripple caused by the switching action, the output voltage is approximately constant, that is, $v_0 \approx V_0$. The voltage across the inductor can then be expressed as the difference between the supply voltage and the output voltage, $V_S - V_0$. The inductor current during this operating phase is therefore:

$$i_L(t) = i_L(t_1) + \frac{1}{L}\int_{t_0}^{t}(V_S - V_0) \, dt = i_L(t_1) + \frac{V_S - V_0}{L}(t - t_1). \quad (11.36)$$

During this off state when the switch is open, the inductor current decreases linearly to:

$$i_L(t_2) = i_L(t_1) + \frac{V_S - V_0}{L}(t_2 - t_1) = i_L(t_1) + \frac{V_S - V_0}{L}T_{off}, \quad (11.37)$$

where $T_{off} = t_2 - t_1$.

With continuous periodic operation, the inductor currents at the start and end of the cycle must be equal, that is, $i_L(t_0) = i_L(t_2)$, such that substituting Equation (11.35) into Equation (11.37) yields:

$$i_L(t_2) = i_L(t_0) + \frac{V_S}{L}T_{on} + \frac{V_S - V_0}{L}T_{off},$$
$$V_0 = V_S \frac{T_{on} + T_{off}}{T_{off}} = V_S \frac{T}{T_{off}}.$$
(11.38)

where T is the total cycle period. The *duty cycle* is defined as the fraction of time that the switch is on, that is:

$$D = \frac{T_{on}}{T_{on} + T_{off}} = \frac{T_{on}}{T}.$$
(11.39)

Thus, the output voltage can be expressed in terms of the duty cycle as:

$$V_0 = \frac{V_S}{1-D}.$$
(11.40)

Since the duty cycle ranges from zero to unity, the output voltage is greater than the input voltage ($V_0 > V_S$). Although this result would tend to indicate that any output voltage is possible, another consideration is the average output current. Equating the average input and output powers reveals:

$$I_S V_S = I_0 V_0,$$
$$I_0 = I_S \frac{V_S}{V_0} = I_S(1-D).$$
(11.41)

It stands to reason, and is verified by the previous expression, that the achievable output current is reduced proportionally with the output voltage.

The high-frequency nature of this converter is a consequence of sizing the inductor to reduce the ripple current (I_r), which is the difference between the maximum and minimum inductor current. The inductance is seen to be inversely proportional to the switching frequency ($f = 1/T$):

$$L = \frac{V_S}{I_r}T_{on} = \frac{V_S TD}{I_r}.$$
(11.42)

Generally, the switching signal is operated at frequencies greater than the range of human auditory capability (i.e., above 20 kHz). A higher frequency, consequently, permits the use of a smaller inductance.

While the switch is closed, the capacitor discharges to provide a nearly constant current such that $I_0 \approx V_0/R$. The charge released may be related to the voltage ripple (V_r) via $\Delta Q = CV_r = I_0 T_{on}$. Thus, the capacitor can be sized based on the voltage ripple using:

$$C = \frac{V_0 TD}{V_r R}. \tag{11.43}$$

Larger capacitance then corresponds to a more constant output voltage.

EXAMPLE 11.6: Boost converter analysis

In this example, the ideal voltage and current waveforms are determined for a 100 kHz boost converter designed for 5% voltage ripple and 10% current ripple with a duty time of 55%. The load resistance is taken to be 100 Ω, and the input voltage and current are 12 Vdc and 1 A, respectively. To begin, these data values are established in Mathcad:

$$f := 100 \text{ kHz} \quad D := 0.55$$
$$V_{r\%} := 5\% \quad I_{r\%} := 10\%$$
$$V_S := 12 \text{ V} \quad I_S := 1 \text{ A} \quad R := 100 \text{ } \Omega.$$

The total cycle period as well as the switch-on and -off periods is found from:

$$T := \frac{1}{f} = 10 \text{ } \mu s$$
$$T_{on} := D \cdot T = 5.5 \text{ } \mu s \quad T_{off} := (1-D) \cdot T = 4.5 \text{ } \mu s.$$

The average output voltage is computed from Equation (11.40) and the voltage ripple from its definition:

$$V_0 := \frac{V_S}{1-D} = 26.67 \text{ V}$$
$$V_r := V_{r\%} \cdot V_0 = 1.33 \text{ V}$$

With the average output voltage and the ripple, the minimum and maximum output voltages across the capacitor are readily determined:

$$V_{C\_min} := V_0 - \frac{V_r}{2} = 26.00 \text{ V} \quad V_{C\_max} := V_0 + \frac{V_r}{2} = 27.33 \text{ V}.$$

Using Equation (11.41), the average output current is calculated as:

$$I_0 := I_S \cdot (1-D) = 0.45 \text{ A}.$$

The average inductor current must be equal to the source current, and the inductor ripple current is found from the circuit specification:

$$I_{L\_avg} := I_S = 1 \text{ A}$$
$$I_r := I_{r\%} \cdot I_{L\_avg} = 0.1 \text{ A}.$$

With the ripple known, the minimum and maximum inductor currents are:

$$I_{L\_min} := I_{L\_avg} - \frac{I_r}{2} = 0.95 \cdot \text{A}$$
$$I_{L\_max} := I_{L\_avg} + \frac{I_r}{2} = 1.05 \cdot \text{A}.$$

The design values for the inductor and capacitor are computed using Equation (11.42) and Equation (11.43), respectively:

$$L := \frac{V_S \cdot T \cdot D}{I_r} = 0.66 \text{ mH} \quad C := \frac{V_0 \cdot T \cdot D}{V_r \cdot R} = 1.1 \text{ μF}.$$

Equation (11.35) and Equation (11.37) provide expressions to determine the inductor current when the switch is on and off, respectively. The *mod* function is employed to produce the correct waveform during the continuous cycling operation:

$$i_L(t) := \text{if}\left[\text{mod}(t, T) < T_{on}, I_{L\_min} + \frac{V_S}{L} \cdot \text{mod}(t, T), I_{L\_max} + \frac{V_S - V_0}{L}(\text{mod}(t, T) - T_{on})\right].$$

The inductor current as well as the average supply and average output currents are plotted in Figure 11.69. It can be concluded from the graph that inductor current increases with slope V_S/L while the switch is closed, and then during the off period, the current decreases with a slope of $(V_S - V_0)/L$.

Finally, the capacitor and resistor are in parallel; therefore, the actual load voltage equals the capacitor voltage of:

$$v_C(t) := \text{if}\left[\text{mod}(t, T) < T_{on}, V_{C\_max} - \frac{V_r}{T_{on}} \cdot \text{mod}(t, T), V_{C\_min} + \frac{V_r}{T_{off}}(\text{mod}(t, T) - T_{on})\right].$$

Figure 11.70 provides a graph of the boost circuit voltages. The graph demonstrates that the output voltage is significantly increased over the supply voltage and follows a sawtooth pattern similar to that of the inductor current.

These results are based on the approximations used to derive the defining relationships. Waveforms that account for the full effects of the RLC circuit can be determined using a circuit simulator such as PSpice.

DC-TO-DC CONVERTERS

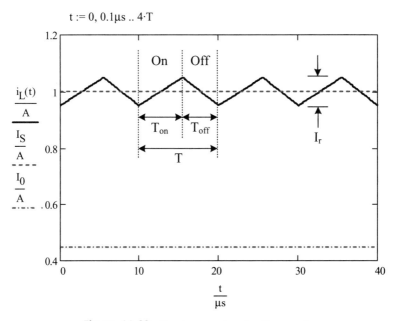

Figure 11.69. Boost converter circuit currents.

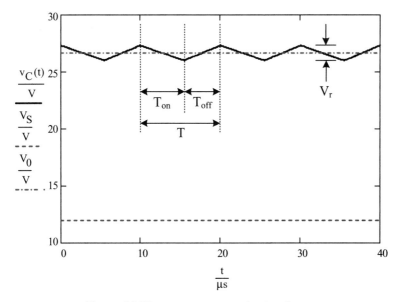

Figure 11.70. Boost converter circuit voltages.

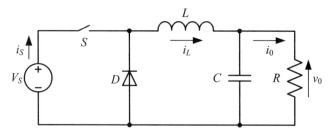

Figure 11.71. Basic buck converter circuit.

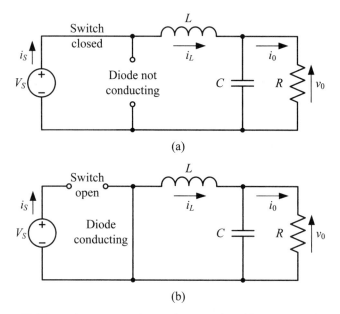

Figure 11.72. Buck converter circuit when switch is (a) closed and (b) open.

11.7.2. Buck Converter

Figure 11.71 illustrates a buck converter circuit used to reduce the input voltage. The operation of the buck converter is akin to that of the boost converter. Again, high-speed switching action leads to dual operating states:

1. *Switch Closed (see Fig. 11.72a).* The switch forms a short circuit, source-driven current that increases the energy stored in the inductor, and the diode forms an open circuit, causing energy to be stored in the capacitor as well.
2. *Switch Open (see Fig. 11.72b).* The source is disconnected from the two storage elements (L and C), and with the diode conducting, energy stored in the inductor and capacitor is transferred to the load.

Switch Closed

From t_0 to t_1 while the switch is closed, the diode is reversed biased and the output voltage is approximately constant ($v_0 \approx V_0$). Similar to the off time for the boost converter, the inductor voltage can be estimated from the difference between the supply voltage and the output voltage, $V_S - V_0$. At the end of this operating stage, the inductor current is:

$$i_L(t_1) = i_L(t_0) + \frac{1}{L}\int_{t_0}^{t_1}(V_S - V_0)\,dt = i_L(t_0) + \frac{V_S - V_0}{L}T_{on}, \quad (11.44)$$

where $T_{on} = t_1 - t_0$ in the case of a buck converter.

Switch Open

The switch is open from t_1 to t_2 and the diode is conducting, thereby causing the inductor voltage to equal the output voltage, but having the opposite polarity. Again, assuming constant output voltage, the inductor current at the end of this time period is:

$$i_L(t_2) = i_L(t_1) + \frac{1}{L}\int_{t_1}^{t_2}(-V_0)\,dt = i_L(t_1) - \frac{V_0}{L}T_{off}, \quad (11.45)$$

where $T_{off} = t_2 - t_1$.

Equating the inductor current at the beginning and end of the overall cycle gives:

$$i_L(t_0) = i_L(t_2),$$
$$\frac{V_S - V_0}{L}T_{on} = \frac{V_0}{L}T_{off}. \quad (11.46)$$

The definition of duty cycle (D) is unchanged from Equation (11.39), allowing the earlier expression to be rewritten as:

$$V_0 = V_S\frac{T_{on}}{T_{off} + T_{on}} = \frac{V_S T_{on}}{T} = V_S D. \quad (11.47)$$

Consequently, the output voltage is reduced compared with the supply voltage. From an energy balance that equates the average input and output powers, the average output current is found to be:

$$I_S V_S = I_0 V_0,$$
$$I_0 = I_S\frac{V_S}{V_0} = \frac{I_S}{D}. \quad (11.48)$$

As should be expected, the reduced voltage output from the buck converter gains a concomitant increase in the output current.

The inductor ripple current (peak-to-peak current difference) can be determined readily from the current change in both Equation (11.44) and Equation (11.45):

$$I_r = \frac{V_S - V_0}{L} T_{on} = \frac{V_0}{L} T_{off}. \qquad (11.49)$$

The earlier formula permits the calculation of the appropriate inductance value:

$$L = \frac{V_0}{I_r} T_{off} = \frac{V_0 T}{I_r}(1-D). \qquad (11.50)$$

The integration of the ripple current over half the period gives the charge that must be filtered by the capacitor, that is, $\Delta Q_r = I_r T/8 = CV_r$, such that the required capacitance is:

$$C = \frac{I_r T}{8 V_r} = \frac{V_0 T^2}{8 L V_r}(1-D). \qquad (11.51)$$

EXAMPLE 11.7: Buck converter circuit parameters

A buck converter operating at 100 kHz and supplied by 5 Vdc is to produce a 3.3 V output. The allowable peak-to-peak voltage and current ripples are ±0.1 V and ±2.5 mA, respectively. Determine: (a) the duty time and (b) the needed inductance and capacitance values. Finally, for a 50 Ω load, plot the actual voltage and current waveforms using circuit analysis software.

To begin, the problem data are defined in Mathcad:

$$V_S := 5 \text{ V} \quad I_S := 1 \text{ A} \quad f := 100 \text{ kHz}$$
$$V_0 := 3.3 \text{ V} \quad V_r := 0.2 \text{ V} \quad I_r := 5 \text{ mA}$$

Using Equation (11.47), the needed duty cycle is found:

$$D := \frac{V_0}{V_S} = 0.66.$$

The cycle period and on time are:

$$T := \frac{1}{f} = 10 \cdot \mu s \quad T_{on} := D \cdot T = 6.6 \ \mu s.$$

APPLICATION EXAMPLES

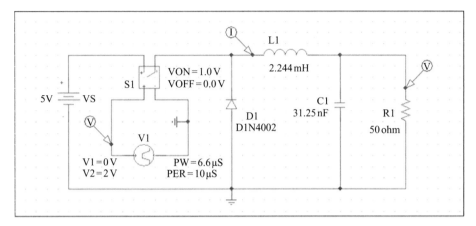

Figure 11.73. Buck converter circuit in PSpice.

The required inductance and capacitance are calculated using Equation (11.50) and Equation (11.51):

$$L := \frac{V_0 \cdot T}{I_r} \cdot (1-D) = 2.244 \cdot mH$$

$$C := \frac{I_r \cdot T}{8 \cdot V_r} = 31.25 \text{ nF}.$$

These circuit parameters were then utilized in a PSpice implementation of the buck converter circuit exhibited in Figure 11.73. Rather than using a transistor, the switch was realized using a switch controlled by a pulse voltage source with a period (PER) equal to T and a pulse width (PW) of T_{on}. A 250 µs transient simulation of the circuit was then performed. Both the voltage output waveform and the switching signal are displayed in Figure 11.74. The graph shows the initializing transient of the circuit, which reaches steady-state behavior at approximately 200 µs. The output does not reach an average of 3.3 V because the circuit parameters are chosen based on an ideal circuit behavior. As should be expected, the output voltage from an actual circuit is not a crisp sawtooth waveform; rather, the output exhibits a certain degree of smoothing.

11.8. APPLICATION EXAMPLES

This section comprises three extended examples, including:

1. PSpice simulation of single-phase bridge converter,
2. DC shunt motor control example, and
3. single-phase induction motor control.

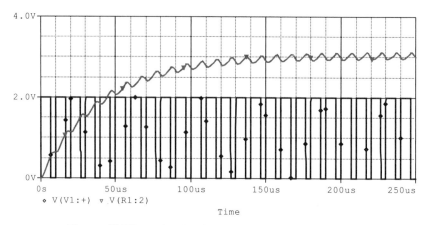

Figure 11.74. PSpice simulation results for buck converter.

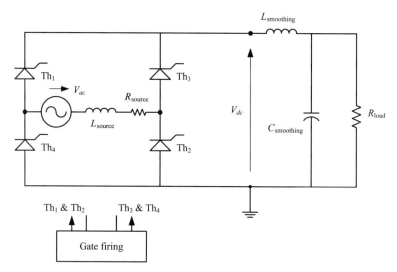

Figure 11.75. Single-phase thyristor-controlled bridge rectifier circuit.

EXAMPLE 11.8: PSpice simulation of single-phase bridge converter

Analytical study of the effects of inductance and capacitance in an inverter circuit is complicated and time consuming; however, computer simulation permits the fast evaluation of converter circuits. For instance, PSpice software can simulate power electronic equipment. PSpice usage is illustrated in this example with a typical circuit.

Figure 11.75 presents the overall circuit of a single-phase thyristor-controlled bridge rectifier. The single-phase bridge converter is supplied by an ac voltage source with a small reactance and resistance connected in series with the source. The converter is built with four thyristors. A gate firing circuit controls the thyristors. The converter has smoothing inductance connected in series, and a smoothing capacitor connected in parallel at the dc side with the resistive load.

PSpice Circuit Model
The corresponding PSpice converter model is provided in Figure 11.76. In this model, the ac voltage source is represented by part VSIN, which provides sinusoidal voltage with amplitude of 100 V at 60 Hz. A 10 µH inductance (LS) and a 1 Ω resistance (RS) are connected in series with the voltage source (VS).

PSpice Thyristor Model
PSpice does not include real power devices, like a thyristor, in its component library. Figure 11.77 shows a developed PSpice model for simulating a thyristor. The main components are a diode (D1a) and a voltage-controlled switch (S1) connected in series. The diode assures that the current flows in only one direction (toward the thyristor cathode), and the switch represents the gate-initiated, delayed turn on of the thyristor. PSpice has several diode models; we selected a frequently used diode designated D1N4002.

Figure 11.77 reveals that the voltage-controlled switch (S1) has four terminals: two control terminals and two main terminals. The main switch terminals are connected in series with a diode (D1a). The negative control terminal is grounded, while the positive terminal is supplied by the control voltage. When the control voltage is applied, the switch turns on.

A thyristor conducts until the anode-to-cathode current direction reverses independently from the gate signal. This behavior is modeled by a current-controlled voltage source (H1), which supplies voltage to the positive terminal of the voltage-controlled switch (S1) until the current flows in the reverse direction. An RC filter and a diode (D1b) are connected in series with the current-controlled voltage source to block the feedback from the positive terminal of the voltage-controlled switch.

In this model, the negative terminal of each voltage-controlled switch and various other points are grounded. PSpice has several different grounds. For this simulation, "0/Design Cache" ground should be used.

Thyristor Firing Control Circuit
The control circuit, titled "Gate Firing Circuit" on Figure 11.76, activates the thyristors. This circuit is supplied by a sinusoidal voltage source (V1). The phase angle of the voltage is varied to control the delay angle. The negative and positive half-periods of the sine wave are separated by using diodes D10 and D11, respectively. The conditional blocks "if(V(%IN)>1,50,0)" convert the half-sine waves to square-shaped pulses. The square-shaped pulse controls the voltage-controlled switches. A pulse from terminal V12 controls switches S1 and S2; that gate pulse is generated by the positive cycle of the sine wave. Switches S3 and S4 are controlled by a pulse from terminal V34; that gate pulse is produced by the negative cycle of the sine wave. In order to simplify the overall circuit, Th_3 and Th_2 are driven by the thyristor internal signals Vp and Vn originating from Th_1 and Th_4, respectively.

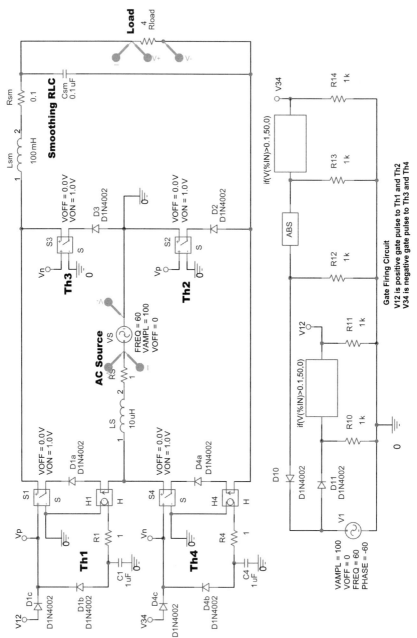

Figure 11.76. PSpice model for single-phase thyristor-controlled bridge converter.

APPLICATION EXAMPLES

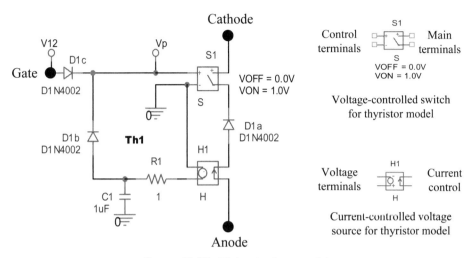

Figure 11.77. PSpice thyristor model.

Circuit Simulation

This PSpice model is used to study the operation of the bridge converter. The phase angle of the control sine wave (V1) determines the delay angle. The load, smoothing inductance, smoothing capacitor, and source impedance are the most frequently varied parameters. The dc current and voltage outputs from the bridge converter are taken at Rload while the ac current and voltage are for the source VS. A current probe placed beside Rload measures the dc current output from the bridge converter. A probe placed at VS measures the ac current from the source. The ac voltage is measured across the sinusoidal voltage source by differential voltage probes. The dc voltage across the load is also measured using a differential probe.

The developed circuit is analyzed using transient analysis simulations. In this mode, the results are the time variation of the selected voltages and currents. However, the software can calculate the rms and average values of each quantity as well. In addition, the frequency spectrum can be determined by FFT analysis. To perform these calculations while in the Probe result window, double-clicking on the quantity (e.g., current through resistor Rload, I(Rload)) opens a list, where the appropriate operation (average, rms, etc.) can be selected. The calculated value will be displayed as a new trace. The values obtained from the simulation can be used for circuit design.

Effect of Smoothing Inductance
For the study of the effect of smoothing reactance, the smoothing inductance (Lsm) was varied from 10 μH to 100 mH, and the generated waveforms were observed. As expected from the results of Figure 11.39, when the reactance was small (a few mH), the dc current was not continuous. But an increase of the inductance increased the duration of the pulsating dc current. At about an inductance of 50 mH, the current

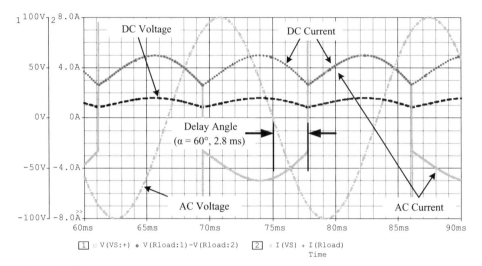

Figure 11.78. Bridge rectifier-generated waveforms, $\alpha = 60°$, smoothing inductance 100 mH, smoothing capacitance 0.1 µF ($C_{sm} = 0.1$ µF, $L_{sm} = 100$ mH).

became continuous. The reader is asked to build the described model and verify this observation. Figure 11.78 shows a plot of a typical run with 100 mH inductance. This graph shows the sinusoidal supply voltage, the ac current, which is a square wave with superimposed sine wave, and the continuous but not smooth dc current.

Effect of Smoothing Capacitance
The converter operation can be improved by connecting a capacitor (Csm) in parallel with the load at the dc side to smooth the dc voltage and current. The capacitor value in the PSpice model was varied, while the smoothing inductance (Lsm) was reduced to 10 µH in the circuit drawn in Figure 11.76. It was observed that an increased capacitance value results in the smoothing of the dc voltage and current. The originally pulsating current becomes continuous, and the ripples are decreased as the capacitance increases.

A typical result from these simulations is displayed in Figure 11.79, with a smoothing capacitance of 2700 µF. The graph shows that pulsating ac current flows when the thyristor is fired, and the ac voltage is higher than the capacitance (dc) voltage; the firing point is marked on the figure with "Conduction starts." During this time, the current charges the capacitor. The peak value of the ac current is around 21 A, but the pulse duration is only about 4.8 ms. The conduction stops when the ac voltage becomes less than the capacitor voltage. This point is marked on the figure by "Conduction stops."

The dc current is continuous with superimposed ripples. The average dc current is near 8 A, and the ripple peak-to-peak amplitude is approximately 3 A.

APPLICATION EXAMPLES

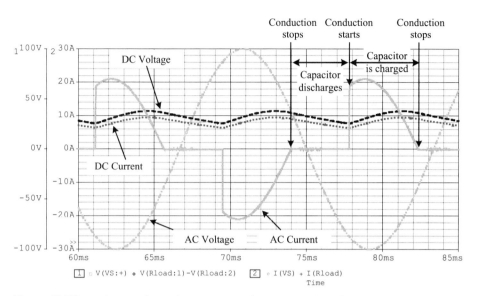

Figure 11.79. Bridge rectifier with resistance and capacitance load ($C_{sm} = 2700\ \mu F$, $L_{sm} = 10\ \mu H$).

Analysis of Thyristor Commutation

In a single-phase bridge circuit, thyristor commutation occurs when the conducting thyristor pair is turned off by firing the nonconducting thyristor pair. In this case, a reverse current is generated, which turns off the conducting thyristor pair. If the source impedance is negligible, the firing of the thyristor produces practically instantaneous commutation. The source impedance delays the commutation time, which is the time needed to turn off the conducting thyristor pair and simultaneously turn on the nonconducting pair. This is illustrated using PSpice simulation.

In the inverter circuit of Figure 11.76, the smoothing capacitor (Csm) is reduced back to 0.1 µF, the smoothing inductance (Lsm) is returned to 100 mH, and the source inductance (LS) is increased from 10 µH to 10 mH. Figure 11.80 displays the simulation results. The graphs demonstrate that an increase of the source inductance reduces both the ac and dc currents. In addition, larger source inductance increases the commutation time. In Figure 11.80b, the commutation time is negligible for a source inductance of 10 µH; simultaneously in Figure 11.80a, the commutation time is significantly longer—about 0.3 ms for a source inductance of 10 mH.

EXAMPLE 11.9: DC shunt motor control

With an understanding of dc motor control and controlled rectifier operation, a practical example is presented for the regulation of dc motor speed using delay angle control. A single-phase bridge converter controls a dc shunt motor as depicted in Figure 11.81. The supply ac voltage is 220 V. The objective of this example is the determination of the delay angle of the rectifier when the motor load and speed are given.

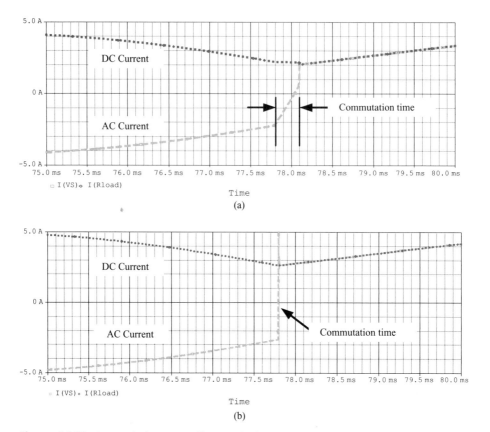

Figure 11.80. Source inductance effect on bridge operation ($C_{sm} = 0.1\ \mu F$, $L_{sm} = 100\ mH$). (a) Source inductance of 10 mH; (b) source inductance of 10 μH.

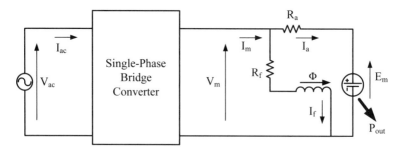

Figure 11.81. Shunt motor controlled by bridge converter.

APPLICATION EXAMPLES

The motor power and voltage ratings and the field and armature (rotor) resistances are:

$$P_{rated} := 50 \text{ hp} \quad V_{rated} := 250 \text{ V}$$
$$R_f := 115 \; \Omega \quad R_a := 0.25 \; \Omega.$$

The dc motor is to supply an 8 hp load at a motor speed of 550 rpm:

$$P_{load} := 8 \text{ hp} \quad rpm := \frac{1}{60 \text{ min}} \quad n_m := 550 \text{ rpm}.$$

The motor no-load voltage and current are measured to permit the calculation of the motor constant. The measured values at the rated motor voltage are:

$$n_{no\_load} := 1000 \text{ rpm} \quad I_{m\_no\_load} := 7.2 \text{ A} \quad V_{m\_no\_load} := V_{rated}.$$

The ac circuit voltage that supplies the rectifier is $V_{ac} := 220 \text{ V}$.

The major steps of this calculation are:

1. determine the motor constant (K_m);
2. compute the motor voltage (V_{mot}) needed to serve the load; and
3. calculate the rectifier delay angle (α).

Motor Constant

The first step of the analysis is the motor constant calculation. This is a shunt motor; therefore, the field current in the no-load condition is:

$$I_{f\_no\_load} := \frac{V_{m\_no\_load}}{R_f} = 2.17 \text{ A}.$$

The armature current is the motor current less the field current:

$$I_{a\_no\_load} := I_{m\_no\_load} - I_{f\_no\_load} = 5.03 \text{ A}.$$

Neglecting the brush voltage drop, the rotor loop voltage equation is:

$$E_{m\_no\_load} := V_{m\_no\_load} - I_{a\_no\_load} \cdot R_a = 248.7 \text{ V}.$$

The motor constant is calculated using the induced voltage equation of Chapter 10:

$$K_m := \frac{E_{m\_no\_load}}{I_{f\_no\_load} \cdot 2 \cdot \pi \cdot n_{no\_load}} = 65.6 \; \frac{V \cdot s}{A}.$$

Motor Voltage

The motor must deliver 8 hp at a speed of 550 rpm. This requires a reduction of the motor voltage from its rated value. The required reduced voltage is calculated using the shunt motor equations of the previous chapter. The field (stator) current and induced voltage equations at the reduced voltage are:

$$I_f = \frac{V_m}{R_f} \qquad E_m = K_m \cdot I_f \cdot 2 \cdot \pi \cdot n_m.$$

The substitution of the field current into the induced voltage equation yields:

$$E_m = K_m \cdot \left(\frac{V_m}{R_f}\right) \cdot 2 \cdot \pi \cdot n_m.$$

The induced voltage computed from the rotor loop and the output motor power are:

$$E_m = V_m - I_a \cdot R_a \qquad P_{out} = E_m \cdot I_a.$$

The armature load current is calculated from the output power formula and is substituted into the induced voltage equation. This results in:

$$E_m = V_m - \frac{P_{out}}{E_m} \cdot R_a.$$

The combination of the two induced voltage equations produces an expression that can be solved for the motor voltage:

$$V_m - \frac{P_{out}}{K_m \cdot \left(\frac{V_m}{R_f}\right) \cdot 2 \cdot \pi \cdot n_m} \cdot R_a = K_m \cdot \left(\frac{V_m}{R_f}\right) \cdot 2 \cdot \pi \cdot n_m.$$

The equation is solved using the Mathcad *root* equation solver. A guess value for the motor voltage is $V_{mot} := 100$ V. The reduced motor voltage for $P_{out} := P_{load} = 8$ hp is found from:

$$V_m := \mathrm{root}\left(V_{mot} - \frac{P_{out}}{K_m \cdot \frac{V_{mot}}{R_f} \cdot 2 \cdot \pi \cdot n_m} \cdot R_a - K_m \cdot \frac{V_{mot}}{R_f} \cdot 2 \cdot \pi \cdot n_m, V_{mot}\right),$$

where the numerical result is $V_m = 77.6$ V.

Using the reduced motor voltage, the field current at load conditions is:

$$I_f := \frac{V_m}{R_f} \quad I_f = 0.675 \text{ A}.$$

The corresponding induced voltage at load is:

$$E_m := K_m \cdot I_f \cdot 2 \cdot \pi \cdot n_m \quad E_m = 42.5 \text{ V}.$$

The rotor (armature) current is:

$$I_a := \frac{P_{out}}{E_m} \quad I_a = 140.5 \text{ A}.$$

The overall motor current at load conditions is:

$$I_m := I_a + I_f \quad I_m = 141.2 \text{ A}.$$

A critical consideration is the possible overload of the motor, because the reduced voltage increases the motor current. The load current has to be less than the rated current. The rated current for this motor is:

$$I_{rated} := \frac{P_{rated}}{V_{rated}} \quad I_{rated} = 149.1 \text{ A}.$$

The calculated operating condition is acceptable, because the actual load current is less than the rated current.

Rectifier Delay Angle

The rectifier delay angle (α) is calculated from the average dc current using the formula developed in Section 11.4.2.3, specifically:

$$V_{dc\_ave}(\alpha) = \frac{2\sqrt{2} \cdot V_{ac}}{\pi} \cdot \cos(\alpha).$$

Rearranging the previous formula, the delay angle is:

$$\alpha := \text{acos}\left(\frac{\pi \cdot V_m}{2\sqrt{2} \cdot V_{ac}}\right) \quad \alpha = 66.9 \text{ deg}.$$

One of the critical problems with the rectifier operation is the ac power factor. We assume that the rectifier has large smoothing inductance, which assures more or less constant dc current.

The input power to the motor is equal with the rectifier output dc power:

$$P_{dc\_in} := I_m \cdot V_m \quad P_{dc\_in} = 10.95 \text{ kW}.$$

Furthermore, the dc power is equal with the ac supply power, assuming the rectifier losses are neglected:

$$P_{ac} := P_{dc\_in} \quad P_{ac} = 10.95 \text{ kW}.$$

We assume that the ac current waveform is a square wave; consequently, the rms value of the ac current is equal with the dc current magnitude:

$$I_{ac\_rms} := I_m \quad I_{ac\_rms} = 141.2 \text{ A}.$$

With these values, the apparent ac power is:

$$S_{ac} := I_{ac\_rms} \cdot V_{ac} \quad S_{ac} = 31.1 \text{ kV} \cdot \text{A}.$$

The power factor of the ac supply is:

$$pf_{ac} := \frac{P_{ac}}{S_{ac}} \quad pf_{ac} = 0.35.$$

The obtained power factor is very low. This rectifier requires capacitors to improve the power factor.

EXAMPLE 11.10: Single-phase induction motor control

The objective of this example is to illustrate control of induction motor speed by regulating the frequency and voltage output from a PWM inverter. A single-phase PWM bridge converter operates in an inverter mode and supplies a single-phase 60 Hz induction motor. The six-pole, 20 hp motor data are:

$$P_{rated} := 20 \text{ hp} \quad V_{rated} := 480 \text{ V} \quad pole := 6$$
$$R_{sta} := 0.65 \text{ }\Omega \quad X_{sta} := 1.20 \text{ }\Omega \quad f_{sys} := 60 \text{ Hz}$$
$$R_{rot} := 0.39 \text{ }\Omega \quad X_{rot} := 1.30 \text{ }\Omega$$
$$R_c := 59 \text{ }\Omega \quad X_m := 28 \text{ }\Omega.$$

The dc voltage that supplies the inverter is:

$$V_{dc} := 600 \text{ V}.$$

APPLICATION EXAMPLES

The ac motor is to deliver 10 hp at 900 rpm:

$$P_{mot} := 10 \text{ hp} \quad \text{rpm} := \text{min}^{-1} \quad n_{mot} := 900 \text{ rpm}.$$

In this example, we will:

1. estimate the approximate inverter frequency;
2. calculate the required motor voltage assuming that the motor flux is maintained constant;
3. compute and plot the actual motor output power versus speed; and
4. determine the actual motor speed at the desired power of P_{mot}.

Inverter Frequency

The motor speed is controlled by the frequency of the inverter. From Chapter 9, the synchronous speed is:

$$n_{syn} = \frac{f_{sys}}{\frac{pole}{2}}.$$

The required supply frequency from the inverter is computed using the synchronous speed equation previously discussed. The approximate value of the inverter frequency is calculated assuming that the motor synchronous speed is equal with the required speed. The result is:

$$n_{syn} := n_{mot} \quad f_{sup} := n_{syn} \cdot \frac{pole}{2} \quad f_{sup} = 45 \text{ Hz}.$$

Motor Voltage

In order to maintain the flux constant, the motor supply voltage is decreased proportionally with the frequency reduction:

$$V_{mot} := V_{rated} \cdot \frac{f_{sup}}{f_{sys}} \quad V_{mot} = 360 \text{ V}.$$

The required modulation index is computed from Equation (11.32), while noting that the motor voltage is an rms value. The result is:

$$modi := \frac{\sqrt{2} \cdot V_{mot}}{V_{dc}} \quad modi = 0.849.$$

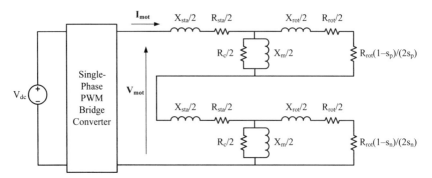

Figure 11.82. Single-phase bridge converter supplying single-phase motor.

Motor Power

The motor output power-versus-speed curve is calculated using the single-phase motor equivalent circuit of Figure 11.82. The motor speed is the variable quantity. The validity of the equations will be tested using $n_m := 850$ rpm.

Because the synchronous speed is equal with the desired motor speed, the positive and negative slips are:

$$s_p(n_m) := \frac{n_{syn} - n_m}{n_{syn}} \qquad s_n(n_m) := \frac{n_{syn} + n_m}{n_{syn}}.$$

The positive and negative rotor impedances are:

$$Z_{rot\_p}(n_m) := \frac{R_{rot}}{2s_p(n_m)} + j \cdot \frac{X_{rot}}{2} \cdot \frac{f_{sup}}{f_{sys}} \qquad Z_{rot\_p}(n_m) = (3.51 + 0.49j)\,\Omega$$

$$Z_{rot\_n}(n_m) := \frac{R_{rot}}{2s_n(n_m)} + j \cdot \frac{X_{rot}}{2} \cdot \frac{f_{sup}}{f_{sys}} \qquad Z_{rot\_n}(n_m) = (0.10 + 0.49j)\,\Omega$$

The parallel magnetizing and rotor impedances are combined:

$$Z_{rot\_m\_p}(n_m) := \left(\frac{1}{\frac{R_c}{2}} + \frac{1}{j \cdot \frac{X_m}{2} \cdot \frac{f_{sup}}{f_{sys}}} + \frac{1}{Z_{rot\_p}(n_m)} \right)^{-1}$$

$$Z_{rot\_m\_n}(n_m) := \left(\frac{1}{\frac{R_c}{2}} + \frac{1}{j \cdot \frac{X_m}{2} \cdot \frac{f_{sup}}{f_{sys}}} + \frac{1}{Z_{rot\_n}(n_m)} \right)^{-1}.$$

$$Z_{rot\_m\_p}(n_m) = (2.70 + 1.15j)\,\Omega$$
$$Z_{rot\_m\_n}(n_m) = (0.10 + 0.46j)\,\Omega$$

APPLICATION EXAMPLES   771

The positive and negative motor impedances are:

$$Z_{mot\_p}(n_m) := \frac{R_{sta}}{2} + j \cdot \frac{X_{sta}}{2} \cdot \frac{f_{sup}}{f_{sys}} + Z_{rot\_m\_p}(n_m)$$

$$Z_{mot\_n}(n_m) := \frac{R_{sta}}{2} + j \cdot \frac{X_{sta}}{2} \cdot \frac{f_{sup}}{f_{sys}} + Z_{rot\_m\_n}(n_m)$$

$$Z_{mot\_p}(n_m) = (3.02 + 1.60j)\ \Omega$$

$$Z_{mot\_n}(n_m) = (0.42 + 0.91j)\ \Omega.$$

The motor current is:

$$I_{mot}(n_m) := \frac{V_{mot}}{Z_{mot\_p}(n_m) + Z_{mot\_n}(n_m)} \quad I_{mot}(n_m) = (68.1 - 49.8j)\ A.$$

The input complex power is:

$$S_{mot}(n_m) := \overline{I_{mot}(n_m)} \cdot V_{mot} \quad S_{mot}(n_m) = (24.5 + 17.9j)\ kV \cdot A.$$

The positive and negative rotor currents are calculated by dividing the voltage on the combined rotor and magnetizing impedance by the rotor impedance. The voltage on the combined rotor and magnetizing impedance is the product of the motor current and the combined impedance. This application of current divisions yields positive and negative rotor currents of:

$$I_{rot\_p}(n_m) := \frac{I_{mot}(n_m) \cdot Z_{rot\_m\_p}(n_m)}{Z_{rot\_p}(n_m)} \quad I_{rot\_p}(n_m) = (65.2 - 24.9j)\ A$$

$$I_{rot\_n}(n_m) := \frac{I_{mot}(n_m) \cdot Z_{rot\_m\_n}(n_m)}{Z_{rot\_n}(n_m)} \quad I_{rot\_n}(n_m) = (64.6 - 47.8j)\ A.$$

The positive and negative output powers are:

$$P_{out\_p}(n_m) := (|I_{rot\_p}(n_m)|)^2 \cdot \frac{R_{rot}}{2} \cdot \frac{1 - s_p(n_m)}{s_p(n_m)} \quad P_{out\_p}(n_m) = 21.7\ hp$$

$$P_{out\_n}(n_m) := (|I_{rot\_n}(n_m)|)^2 \cdot \frac{R_{rot}}{2} \cdot \frac{1 - s_n(n_m)}{s_n(n_m)} \quad P_{out\_n}(n_m) = -0.8\ hp.$$

The total motor output power is:

$$P_{out}(n_m) := P_{out\_p}(n_m) + P_{out\_n}(n_m) \quad P_{out}(n_m) = 20.9\ hp.$$

Figure 11.83 shows a plot of the output power versus motor speed. The graph shows the motor output power–speed curve and the 20 hp power rating. The intersection of

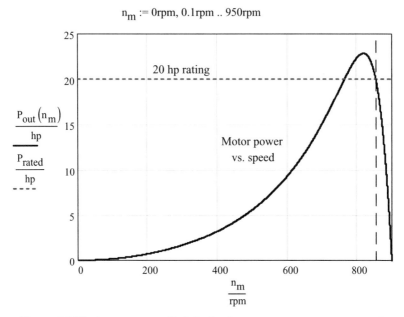

Figure 11.83. Inverter-controlled single-phase motor power versus speed.

the two curves occurs around 850 rpm. The exact value is determined by solving the following equation using the Mathcad *root* equation solver:

$$P_{out}(n_m) - P_{rated} = 0.$$

Using a guess value of $n_m := 850$ rpm, the solution is:

$$n_{rate} := \text{root}(P_{out}(n_m) - P_{rated}, n_m) \quad n_{rate} = 855 \text{ rpm}.$$

Motor at Load

The motor is to be operated at a power of 10 hp. To find the actual operating speed and whether it is at the desired value of 900 rpm, the *root* equation solver is again used:

$$n_{load} := \text{root}(P_{out}(n_m) - P_{mot}, n_m) \quad n_{load} = 884 \text{ rpm}.$$

The results reveal that the motor speed is close to the desired 900 rpm. At this point, we may be satisfied with the present speed, which is subject to the earlier assumption that the motor synchronous speed is equal with the required speed.

Alternatively, we can continue this analysis to find the exact conditions (i.e., supply voltage and frequency, and motor slip) that perfectly match the desired motor power and speed. This could be accomplished using the Mathcad *Find* equation solver and a long list of equations in the solver block, or using an iterative (trial-and-error) approach.

11.9. EXERCISES

1. Discuss the concept of induction motor speed control.
2. Why must the magnetic flux be maintained constant when the frequency of the supply voltage is reduced?
3. Discuss the concept of dc motor speed control.
4. What is the thyristor? Explain its use.
5. Compare and contrast the use of an IGBT, GTO, and MOSFET.
6. Explain the operation of a bridge converter. Draw the current paths and discuss the firing sequence.
7. Draw the connection diagram and the voltage and current waveforms of a thyristor-controlled bridge rectifier in the case of a resistive load.
8. Draw the connection diagram and the voltage and current waveforms of a thyristor-controlled bridge rectifier when the rectifier charges a battery.
9. Draw the connection diagram and the voltage and current waveforms of a thyristor-controlled bridge rectifier in the case of an inductive load.
10. Explain how the average dc voltage and current are calculated for a thyristor-controlled rectifier with a resistive load.
11. Explain how the average dc voltage and current are computed for a thyristor-controlled rectifier with a large inductive load.
12. Explain how the average dc voltage and current are calculated for a thyristor-controlled rectifier when it charges a battery.
13. Explain how to calculate the total harmonic distortion (THD) factor of the ac current generated by a rectifier with an inductive load.
14. Explain the effect of smoothing inductance on the rectifier current. Draw the dc current and voltage waveforms.
15. Explain the effect of the capacitor on the rectifier operation. Draw the dc voltage and current waveforms.
16. Draw the circuit diagram of a three-phase rectifier and explain its operation.
17. Draw the circuit diagram and waveforms for a voltage source inverter.
18. What are the advantages of PWM?
19. What is the role of freewheeling shunt diodes in an inverter circuit?
20. Explain the operation of a sinusoidal PWM inverter.
21. Discuss the operation of a line-commutated, thyristor-controlled inverter.
22. What is an HVDC system? Draw the conceptual circuit block diagram.

11.10. PROBLEMS

Problem 11.1

In this chapter, a sinusoidal ac voltage source was utilized rather than a cosine. Determine the Fourier coefficients for an ac cosinusoidal voltage (i.e., $x(t) = V_M\cos(\omega_0 t)$) using (a) an analytical solution and (b) a numerical FFT analysis. (c) Compare the results for the cosine to those for the pure sine wave.

Problem 11.2

In a manner similar and for the same conditions as Example 11.2, compute and plot the magnitude of the current harmonics from dc to the 30th harmonic for a full-wave rectifier.

Problem 11.3

Use PSpice to implement a full-wave bridge rectifier circuit with a 50 Hz voltage source magnitude of 5 V and a load resistance of 2 kΩ. Plot the voltage waveforms for the source and resistor, and the current waveforms for two different diodes on adjacent legs of the bridge. Then, perform an FFT analysis on the four waveforms.

Problem 11.4

A single-phase thyristor-controlled rectifier supplies the lighting in a theater. The light control requires the reduction of the average dc voltage to 70% of the rated voltage when the lights are dimmed. Calculate the required delay angle. The 60 Hz ac input is 220 V, and the dc output rating is 120 V. The lights normally draw 2 kW at the rated conditions.

Problem 11.5

Calculate the rms current and THD factor of the ac current at 70% load of the rectifier analyzed in the previous problem. Assume that the lighting load can be represented by a resistance, which is calculated from the full-load data.

(a) Calculate the equivalent load resistance. (b) Derive equations for the dc and ac voltage and current. Plot the waveforms to verify proper operation. (c) Perform Fourier analysis on the ac current and calculate the THD at a delay angle corresponding to the 70% load (calculated in the previous problem).

Problem 11.6

For the single-phase bridge rectifier with a resistive load of Section 11.4.2.2, calculate and plot both (a) the THD and (b) the rms current versus delay angle (α). Use Equation (11.24) to compute the effective ac current.

Problem 11.7

Graph the ideal output voltage and inductor current waveforms for a 70 kHz buck converter designed for 5% voltage ripple and 10% current ripple with a duty time of 60%. Use a load resistance of 80 Ω, and the source voltage and current are 9 Vdc and 250 mA, respectively.

Problem 11.8

A boost converter operating at 80 kHz and supplied by 6 Vdc is targeted to produce a 10 V output. The allowable peak-to-peak voltage and current ripples are ±0.2 V and ±75 mA, respectively. Determine: (a) the duty time and (b) the required inductor and capacitor values. (c) Finally, for a 75 Ω load, plot the actual voltage and current waveforms using circuit analysis software.

Problem 11.9

A rectifier charges a battery bank in a substation. The bank rated dc voltage is 48 V. The required charging current is 25 A. The available ac supply is 120 V. The internal resistance of the battery is 2.5 Ω.

(a) Analyze the operating conditions of the charger. Plot the ac and dc voltage and current, and determine the feasibility of delay angle control. (b) Calculate the delay angle needed to maintain the 25 A charging current. (c) Calculate the power and power factor at the ac side.

Problem 11.10

A bridge rectifier supplies a constant current load through a dc feeder, whose resistance is 1.1 Ω. The constant current is achieved by using a large inductance connected in series with the load. The dc voltage and current are 300 V and 45 A, respectively.

(a) Calculate the dc voltage and power at the rectifier terminal. (b) Calculate the ac voltage needed to supply this load (open-ended problem). (c) Calculate the THD and base component (60 Hz) of the ac current.

Problem 11.11

A line-commutated inverter is used in an uninterruptible power supply (UPS). The dc voltage of the battery in the UPS is 400 V. The UPS supplies a 60-Hz electric network with a cogenerator that produces a voltage of 480 V. The inverter has a large inductance, which assures constant dc power. The dc current through the inverter is 150 A; the average dc input voltage for the inverter is 432 V when its output voltage is 480 V.

(a) Calculate the delay angle if the ac and dc voltages are the given values. (b) Derive the equations for ac and dc voltage and current, and plot the waveforms for this operating condition. (c) Calculate the power transported to the ac side and determine the power factor.

Problem 11.12

A single-phase bridge rectifier controls a dc series motor. A 460 V ac circuit supplies the rectifier. The dc motor is rated at 20 hp and 300 V. The motor field and armature winding resistances are both 0.25 Ω.

The no-load speed and current are 1000 rpm and 7.2 A, respectively. The ac circuit voltage that supplies the rectifier is 460 V.

(a) Calculate the motor constant. (b) Calculate the motor voltage needed for delivering 8 hp at 550 rpm. (c) Calculate the rectifier delay angle.

Problem 11.13

A PWM inverter controls the speed of a three-phase induction motor. The 60 Hz, four-pole motor is rated at 17 hp and 440 V. The motor circuit data per phase are:

	Resistance (Ω)	Reactance (Ω)
Stator (series)	0.55	1.3
Rotor (series)	0.35	1.3
Magnetizing (parallel)	25	50

The motor has to deliver 15 hp at 1100 rpm.

(a) Calculate the approximate inverter frequency. (b) Calculate the required motor voltage assuming that the motor flux is maintained constant. (c) Calculate and plot the actual motor output power versus speed. (d) Calculate the actual motor speed at the desired power.

Problem 11.14

A single-phase thyristor-controlled inverter supplies a 25 Ω resistive load. The ac supply voltage and frequency are 220 V and 60 Hz, respectively. Simulate this inverter with varying smoothing inductance using the PSpice circuit of Example 11.8 without a smoothing capacitance. (a) Plot the ac and dc voltages and the dc current when the delay angle is 50° and the smoothing inductance is 10 μH. Determine the average value of the dc current. (b) Modify the smoothing inductance value and determine the inductance that causes continuous conduction. Determine the average value of the dc current.

Problem 11.15

A single-phase thyristor-controlled inverter supplies a resistive load of 20 Ω. The ripples on the dc current are reduced by connecting a large capacitor in parallel with the load. Smoothing inductance is not used to improve the conduction period (i.e., Lsm = 0). The ac supply voltage and frequency are 110 V and 60 Hz, respectively. Simulate this inverter using the PSpice circuit of Example 11.8. (a) Plot the ac and dc voltages and the dc current if the delay angle is 60° and the smoothing capacitor is omitted. What is the average value of the dc current? (b) Determine the capacitance value needed to reduce the voltage ripple to 25 V. What is the average value of the dc current with this capacitor?

APPENDIX A
INTRODUCTION TO MATHCAD

Mathcad® is an all-purpose mathematical software that is designed to solve engineering problems. A major advantage of the program is that the equations appear in a format similar to that found in a book. Simultaneously, if the values of the variables are given, the computer immediately calculates the numerical values. Further, the engineering dimensions (units) of variables and equations are automatically incorporated and evaluated. Table A.1 lists many of the Mathcad built-in dimensions and units. Mathcad can plot functions, which permits trend analysis and finding maximums and minimums, and the program solves equations having complex values and solutions. Calculations with complex numbers are very commonplace in electrical engineering. In this book, we use the basic features of this rich software, which has numerous functions and covers a wide variety of mathematical/technical problems, from matrix algebra to symbolic operations and basic programming.

A.1. WORKSHEET AND TOOLBARS

Figure A.1 displays the Mathcad worksheet and available toolbars. The worksheet is the workspace in which we place text, equations, and graphs. The toolbars give the operators, functions, Greek letters, plots, and so on, which can be inserted in the worksheet by clicking buttons on the toolbar. As an example, we click on the integral symbol [∫] in the **Calculus** toolbar, and this results in the integration function being placed on the worksheet where the cross hair (+) is located (as seen in Fig. A.1). The available toolbars are the following:

- **Calculator** – common arithmetic operators
- **Graph** – various 2 and 3 dimensional plot types and graph tools
- **Matrix** – matrix and vector operators
- **Evaluation** – equal signs for evaluation and definition
- **Calculus** – derivatives, integrals, limits, and iterated sums and products
- **Boolean** – comparative and logical operators for Boolean expression
- **Programming** – programming constructs
- **Greek** – Greek letters
- **Symbolic** – symbolic keywords

Electrical Energy Conversion and Transport: An Interactive Computer-Based Approach, Second Edition. George G. Karady and Keith E. Holbert.
© 2013 Institute of Electrical and Electronics Engineers, Inc. Published 2013 by John Wiley & Sons, Inc.

TABLE A.1. Mathcad Built-In Dimensions and Units

Acceleration		Force		Permittivity	
g	gravity	N	newton	ε_0	epsilon0
Activity		lbf	pound force	**Potential (Voltage)**	
Bq	becquerel	kgf	kilogram force	V	volt
Angular		**Frequency**		**Power**	
deg	degree	Hz	hertz	W	watt
rad	radian	rpm	revolutions per	hp	horsepower
rev	revolution		minute	**Pressure**	
s	steradian	**Inductance**		atm	atmospheres
Area		H	henry	bar	bar
acre	acre	**Length**		Pa	pascal
barn	barns	in	inch	psi	pounds per
hectare	hectare	ft	feet		square inch
Capacitance		mi	mile	torr	torr
F	farad	m	meter	**Resistance**	
Charge		yd	yard	Ω	ohm
C	coulomb	**Magnetic Field Strength**		**Substance**	
Conductance		Oe	oersted	mol	mole
mho	mho	**Magnetic Flux**		**Temperature**	
S	siemens	Wb	weber	K	kelvin
Current		**Magnetic Flux Density**		R	rankine
A	amp	G	gauss	**Time**	
Dose		T	tesla	s	second
Gy	gray	**Mass**		min	minute
Sv	sievert	gm	gram	hr	hour
Energy		kg	kilogram	day	day
Btu	international Btu	oz	ounce	yr	year
cal	calorie	lb	pound mass	**Velocity**	
erg	erg	ton	ton	c	Speed of light
J	joule	tonne	metric ton	**Volume**	
		Permeability		gal	gallon
		μ_0	mu0	L	liter

Btu, British thermal unit.

Mathcad is a large program, which cannot be fully described in this brief introduction. In this appendix, the basic features that are employed in the textbook are primarily discussed. The reader is encouraged to study the extensive Mathcad **Help** files. The solutions presented in the book do not require the professional version of the program—the student version is sufficient.

WORKSHEET AND TOOLBARS

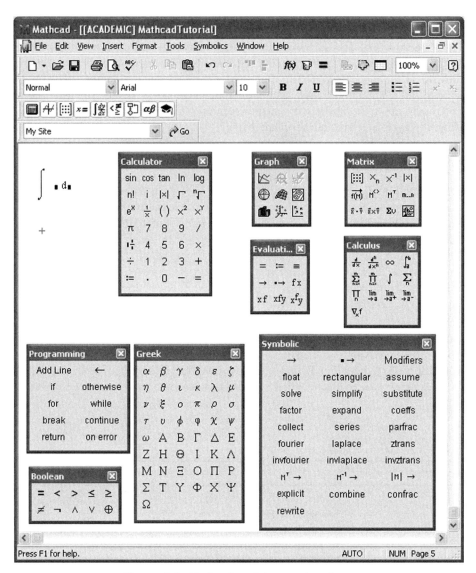

Figure A.1. Mathcad worksheet and toolbars.

A.1.1. Text Regions

The Mathcad workspace is divided into text and math regions. A text region is created by simply typing the desired text in the selected area. The starting point is selected by clicking on the worksheet. The font and the size of the letters may be specified. In addition, the common cut, copy, and paste commands are available. The text editing features are similar to word processor editing features. Table A.2 presents some common Mathcad keystrokes and their resultant actions.

A.1.2. Calculations

A calculation begins with the definition of the variables. As an example, let us define a complex impedance with a 5 Ω resistance and a 3 Ω reactance connected in series. Type "Z" followed by a colon [:] or click on [:=] in the **Calculator** toolbar. The result is Z := ■. The next step is to type the numeral five [5] in the empty *placeholder* (■). This is followed by the definition of the unit. Just type "ohm" or click on the Omega symbol [Ω] in the **Greek** letters toolbar. The result is Z := 5 Ω. The inclusion of the 3 Ω reactance proceeds with typing the plus sign [+], followed by entering the complex number j ($= \sqrt{-1}$). To enter the imaginary number, first type the numeral one [1] and then the letter "j" [j] (or "i") after the number without a space between the two characters to yield: Z := 5 Ω + j. Then enter the multiplication symbol [*], followed by typing the numeral three [3] and the ohm unit again. The result is Z := 5 Ω + j · 3 ohm. Note that we have intentionally mixed the ohm symbol (Ω) and the word "ohm" in the previous expression to illustrate that Mathcad can use either, and it is not confused by

TABLE A.2. Common Mathcad Editing Keystrokes

Keystroke	Action
Enter	Inserts blank line. In text region, begins a new paragraph.
@	Creates X–Y plot.
Ctrl+A	In text region, selects all the text in the text region.
	In a blank spot, selects all regions in the worksheet.
Ctrl+D	Deletes region.
Ctrl+E	Opens the Insert Function dialog box.
Ctrl+G	Toggles between Greek and Roman character.
Ctrl+M	Inserts matrix or vector template.
Ctrl+U	Opens the Insert Unit dialog box.
Ctrl+Z	Undo last edit.
Ctrl+Enter	Inserts a page break. In text region, sets the width of the text region.
	In math region, inserts addition operator with line break.
Ctrl+Shift+Enter	In a region, moves the cursor out of and below the region.
Ctrl+Shift+A	Inserts math in text region.
Ctrl+Shift+K	Allows typing characters that usually insert operators (e.g., [).
Ctrl+Shift+P	Inserts (Greek letter) pi, whose value is 3.1415927...

the varied nomenclature. A more compact expression is formed by grouping the resistance and reactance with parentheses, keeping the "3" with the imaginary number, and stating the ohm unit only at the end, that is, Z := (5 + 3j)Ω.

Variables can be distinguished by using subscripts like Z_2. After typing a variable name, to create a subscript for the variable, type a period [.] and then the subscript designation. As an example, let us define Z_2 as a resistance of 10 Ω. We type [Z], then a period [.], followed by the subscript numeral two [2], and then the colon [:], which is followed by ten [10] and the dimension ohm [Ω]. The result is $Z_2 := 10$ Ω.

For the demonstration of a simple calculation, we assume that the two impedances previously defined (Z and Z_2) are connected in parallel. To combine the two impedances into an equivalent impedance, we type an equation using the keyboard plus [+], minus [−], divide [/], multiply [*], or power [^] signs as necessary. We know that the equivalent impedance of two parallel impedances is their product divided by their sum. The sequence of typing the equation for the parallel connection of two impedances is given here: type [Z], type [.], type [parallel], type colon [:], type [Z], type [*], type [Z], type [.], type [2], type [/], type [Z], type [+], type [Z], type [.], and type [2]. The result is:

$$Z_{parallel} := Z \cdot \frac{Z_2}{Z+Z_2}.$$

The numerical result of the calculation can be displayed by (a) augmenting the earlier expression with an equal sign,

$$Z_{parallel} := Z \cdot \frac{Z_2}{Z+Z_2} = (3.59 + 1.282j) \; \Omega,$$

or (b) typing $Z_{parallel}$, as described previously, and pressing the equal sign [=] on the keyboard. The latter result is $Z_{parallel} = (3.59 + 1.282j)\Omega$.

Note the use of the letter "j" to denote the imaginary number in the previously mentioned complex result. Electrical engineers prefer to use a lowercase "j" to denote the imaginary number rather than a lowercase "i" to avoid confusion with current $i(t)$. However, Mathcad defaults to using the letter "i" for the imaginary number. This can be changed to "j" in the **Display Options** tab of the **Format Result** dialog box, as viewed in Figure A.2.

The units can be changed by clicking on the equation, which results in a *placeholder* (■) on the right-hand side:

$$Z_{parallel} = 3.59 + 1.282j \; \Omega \; \blacksquare.$$

We enter the desired unit in the placeholder. Here we input kΩ as an example, and the result is:

$$Z_{parallel} = \left(3.59 \times 10^{-3} + 1.282j \times 10^{-3}\right) k\Omega.$$

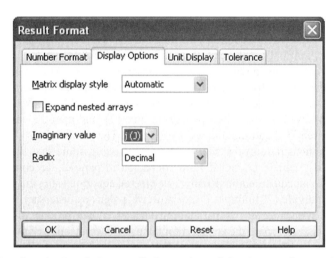

Figure A.2. Using the **Result Format** display options dialog box to change the imaginary number to display as "j."

TABLE A.3. Common Mathcad Operator Keystrokes

Keystroke	Operator
^	Exponentiation
"	Complex conjugate
'	Matched pair of parentheses
|	Magnitude or determinant
!	Factorial
,	Separates arguments in a function; separates expressions to be plotted on the same axis; precedes second number in range
;	Precedes last number in range
\	Square root
Ctrl+\	nth root
Ctrl+Enter	Addition with line break

Electrical engineering circuit analysis frequently requires calculating the absolute value and phase angle of a complex number. The absolute value or magnitude can be calculated by clicking on the **Calculator** toolbar and selecting the |■| operator. The location of the absolute value operator can be seen in the **Calculator** toolbar exhibited in Figure A.1. The result for the parallel impedance is:

$$|Z_{\text{parallel}}| = 3.812 \, \Omega.$$

Other Mathcad operators and their shortcut keystrokes are given in Table A.3.

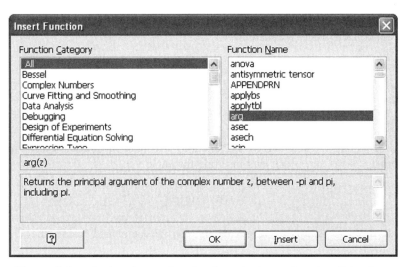

Figure A.3. Selecting the *arg* function from the **Insert Function** dialog box.

A.2. FUNCTIONS

Mathcad has a large number of included functions, which can be displayed by clicking the **Insert** dropdown menu and selecting **Functions**. This results in the display of the list of functions. Figure A.3 shows the display.

As an example, the phase angle of the impedance is calculated with the *arg* function. This function can be obtained by clicking on *arg* in the function list presented in Figure A.3. The result is arg(■). The next step is typing the argument [$Z_{parallel}$] in the placeholder, followed by the equal sign. By default, the result is the angle in radians arg($Z_{parallel}$) = 0.343. The unit can easily be changed to degrees by typing "deg" in the placeholder:

$$\arg(Z_{parallel}) = 19.654 \cdot \deg.$$

We can use the other functions in a similar manner. As an example, we calculate the real and imaginary components of the $Z_{parallel}$ impedance with the Mathcad *Re* and *Im* functions, respectively. Both functions are located in the **Function** list. The calculations are:

$$\text{Re}(\blacksquare) \quad \text{Re}(Z_{parallel}) = 3.59 \, \Omega$$
$$\text{Im}(\blacksquare) \quad \text{Im}(Z_{parallel}) = 1.282 \, \Omega.$$

In some book examples, the value of a current, voltage, or impedance is specified by the absolute value and the phase angle. For example, suppose the absolute value of a line current is 20 A, and the phase angle is 30°. The complex value of the current can be calculated using Euler's equation:

$$I_a := 20 \text{ A} \quad \alpha := 30 \text{ deg}$$
$$I_{complex} := I_a \cdot \cos(\alpha) + j \cdot I_a \cdot \sin(\alpha)$$
$$I_{complex} = (17.321 + 10j) \text{ A}.$$

In these equations, the sine and cosine functions are obtained from the **Function** table or by typing "sin()" or "cos()" and inserting the angle via:

$$\sin(\blacksquare) \quad \sin(\alpha).$$

Alternatively, the complex current value can be computed using the complex exponential:

$$I_{complex} := I_a \cdot e^{j\alpha} \quad I_{complex} = (17.321 + 10j) \text{ A}.$$

In addition, all three of these functions (sine, cosine, and exponential) can be selected from the **Calculator** toolbar, by clicking on the function. A list of common functions is given in Table A.4.

A.2.1. Repetitive Calculations

Repetitive evaluation of an expression is frequently desired in engineering calculations. For the repeated calculation of an expression, Mathcad uses a *range variable*. A typical case is the computation of the values of a sine wave between $\alpha = 0°$ and $90°$ with steps of $30°$.

In this case, α is defined as a *range variable* by clicking on alpha [α] in the **Greek** letters toolbar or using the shortcut: type [a] and press Ctrl+G on the keyboard. This is

TABLE A.4. Common Mathcad Functions

Constants			Exponential
%	Percentage (* 0.01)	exp, e	Exponential (e^)
e	Value of e (15 decimals)	ln	Natural logarithm (base e)
π	Value of pi (15 decimals)	log	Logarithm base 10
Complex			Trigonometric
arg	Argument (phase angle)	sin, cos, tan	Trigonometric
Im	Imaginary part	asin, etc.	Inverse trigonometric
Re	Real part	sinh, etc.	Hyperbolic
| |	Absolute value	asinh, etc.	Inverse hyperbolic

followed by a colon [:] and typing the starting value [0deg]; after this, type a comma [,] and then the next value [30deg], followed by a semicolon [;] (or select [n. . .m] from the Matrix toolbar), finally entering the end value of [90deg]. The intermediate and final results are:

$$\alpha := \blacksquare \quad \alpha := 0 \text{ deg}, \blacksquare \quad \alpha := 0 \text{ deg}, 30 \text{ deg} .. \blacksquare$$
$$\alpha := 0 \text{ deg}, 30 \text{ deg} .. 90 \text{ deg}.$$

The sine wave can be evaluated and displayed by typing "sin(α) =," which yields:

$$\sin(\alpha) = \begin{array}{|c|} \hline 0 \\ \hline 0.5 \\ \hline 0.866 \\ \hline 1 \\ \hline \end{array}.$$

A.2.2. Defining a Function

A user-defined function can be specified by first typing the chosen name of the function. For example, to define an impedance (Z) as a function of frequency (f), the typing sequence is: type [Z], type a set of parentheses [( )], and insert [f] in the parenthesis followed by a colon [:], which yields: Z(f) := $\blacksquare$.

The application of a function within another function is demonstrated using a numerical example. The impedance (Z) of a resistance, capacitance, and inductance connected in series depends on the cyclic frequency (f). An impedance (Z) is generally expressed in terms of the angular frequency (ω). The cyclic frequency (f) is directly proportional to the angular frequency (ω), which we establish using a functional representation of $\omega(f) := 2 \cdot \pi \cdot f$.

The values of the resistance, capacitance, and inductance are taken to be:

$$R_o := 200 \, \Omega \quad L_o := 1 \, H \quad C_o := 1 \, \mu F.$$

The impedance frequency function, using the previously described method for the equation entry, is then formed:

$$Z_o(f) := R_o + j \cdot \omega(f) \cdot L_o + \frac{1}{j \cdot \omega(f) \cdot C_o}.$$

These equations can evaluate the impedance value at any frequency. To test the use of the equations, we select a frequency of 100 Hz: f := 100 Hz.

The resultant impedance can be obtained by typing "Z.o(f) =":

$$Z_o(f) = (200 - 963.231j) \, \Omega.$$

The number of significant digits displayed can be altered using the **Number Format** tab of the **Format Result** dialog box to yield $Z_o(f) = (200.0 - 963.2j)\Omega$.

A.2.3. Plotting a Function

Mathcad has the capability to create different plots ranging from a simple X–Y graph to three-dimensional vector plots. In this appendix, simple X–Y plotting is demonstrated with a numerical example. In particular, we will plot the absolute value of the resistor–capacitor–inductor (RLC) circuit impedance from the previous example as a function of the frequency. The equations are:

$$R_o := 200\,\Omega \qquad L_o := 1\,H \quad C_o := 1\,\mu F$$

$$\omega(f) := 2\cdot \pi \cdot f \qquad Z_o(f) := R_o + j\cdot \omega(f)\cdot L_o + \frac{1}{j\cdot \omega(f)\cdot C_o}$$

$$Z_{abs}(f) := |Z_o(f)| \quad \theta(f) := \arg(Z_o(f)).$$

Here we calculate both the absolute value and the phase angle of the impedance. If the frequency is 100 Hz ($f_{test} := 100\,Hz$), the results are:

$$Z_{abs}(f_{test}) = 983.775\,\Omega \quad \theta(f_{test}) = -78.27\cdot \deg.$$

For plotting the impedance magnitude, the frequency is declared as a range variable from 100 to 250 Hz in steps of 1 Hz:

$$f := 100\,Hz, 101\,Hz\,..\,250\,Hz.$$

An X–Y plot is created by clicking on the **Graph**, **X-Y Plot** in the **Insert** menu, or the X–Y plot can be selected from the **Graph** toolbar. The display initially shows:

The next step is establishing the *x*-axis and *y*-axis variables by typing "f" on the placeholder at the horizontal axis and "$Z_{abs}(f)$" on the vertical axis placeholder. To properly denote the units on a graph, we find it useful to divide each axis variable by its desired unit. Specifically, for the *x*-axis, as "f" is typed, we divide by "Hz," and for the *y*-axis, we divide the impedance by ohms [Ω]. The graph now takes the form given in Figure A.4.

FUNCTIONS

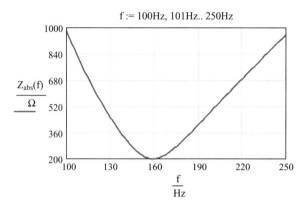

Figure A.4. Mathcad X–Y plot of impedance versus frequency.

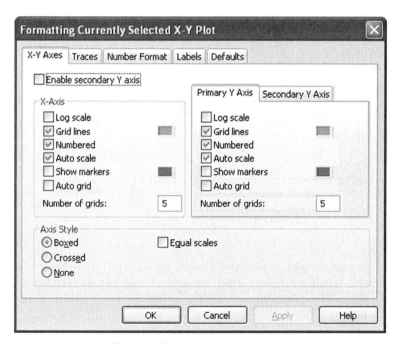

Figure A.5. Plot formatting menu.

The graph can be formatted by double-clicking on the plot area. This opens up the formatting menu seen in Figure A.5. The range of the axis, the color, type, and width of the traces can be changed and the axis can be labeled. In fact, we have selected grid lines for both the x- and y-axes of the plot of Figure A.4, and we set the number of grid lines for each axis to five. The reader should experiment with this tool to learn its features.

A.2.4. Minimum and Maximum Function Values

The maximum and minimum values of a function can be determined by the *Maximize* and *Minimize* functions, respectively. These functions can be found in the **Insert Function** list or by typing:

Maximize (function name, variable)
Minimize (function name, variable)

The procedure for using these functions is as follows:

1. define the function to minimize or maximize,
2. define a *guess value* for the solution,
3. type the word *Given* to start the *solve block*,
4. enter any constraints on the solution using Boolean operators, and
5. enter the *Maximize* or *Minimize* function with the arguments as previously provided.

The application of these functions is demonstrated by finding the minimum value of the $Z_{abs}(f)$ function introduced earlier. Mathcad requires the selection of an initial *guess value* for the minimum (or maximum, if *Maximize* is used). Using the graph provided in Figure A.4, a reasonable guess of f = 160 Hz for the minimum is selected:

Define the function: $Z_{abs}(f) := \left| R_o + j \cdot \omega(f) \cdot L_o + \dfrac{1}{j \cdot \omega(f) \cdot C_o} \right|$

Select a guess value: $f := 160 \text{ Hz}$

$f_{min} := \text{Minimize}(Z_{abs}, f) \quad f_{min} = 159.155 \cdot \text{Hz}$

Note that we have omitted steps 3 and 4 in the previous procedure because there are no constraints, which makes a formal *solve block* unnecessary. Having found the minimum, f_{min} can then be used as a variable to find the actual minimum impedance magnitude using:

$Z_{abs\_min} := Z_{abs}(f_{min}) \quad Z_{abs\_min} = 200 \, \Omega.$

A.3. EQUATION SOLVERS

Mathcad has two equation solvers, which can solve both real and complex-valued equations. The use of both equation solvers is demonstrated in the succeeding sections with numerical examples.

A.3.1. Root Equation Solver

The *root* function can solve a single equation. As an example, suppose we want to determine the frequency where $Z_{abs}(f) = 400\ \Omega$. The formula for $Z_{abs}(f)$ is:

$$Z_{abs}(f) := \left| R_o + j \cdot \omega(f) \cdot L_o + \frac{1}{j \cdot \omega(f) \cdot C_o} \right|.$$

The desired quantity is the frequency where this expression equals 400 Ω. Therefore, the equation that we want to solve is:

$$Z_{abs}(f) - 400\ \Omega = 0,$$

where the Boolean equal sign is employed here. The frequency can be computed by the *root(equation, unknown)* function. The first step is the selection of a guess value. Using the plot of Figure A.4, we can see that there are two solutions: one is around 130 Hz and the other is near 190 Hz. We select a guess value of 100 Hz:

$$f := 100\ Hz.$$

This gives the first root:

$$\text{root}(Z_{abs}(f) - 400\ \Omega, f) = 133.958 \cdot Hz.$$

The selection of a second guess value of 200 Hz yields the second root:

$$f := 200\ Hz \quad \text{root}(Z_{abs}(f) - 400\ \Omega, f) = 189.091 \cdot Hz.$$

The root function can find complex roots, if they exist; however, in such a case, the user should select a complex guess value.

A.3.2. Find Equation Solver

The *Find* equation solver is suitable to solve one equation or a system of equations. This approach is demonstrated by solving the same equation that was previously solved with the *root* function. The procedure for applying the *Find* equation solver is as follows:

1. Select a guess value.
2. Type the word *Given* to start the *solve block*.

3. Using Boolean operators, enter any constraint(s), such as f > 0, if we are searching for positive roots only.
4. Type the equations but replace the default equal sign (:=) with the bold equal sign (=) from the Boolean toolbar. The shortcut for the Boolean equal is to press [Ctrl] and [=] on the keyboard.
5. Type *Find()* and place the unknown variable in the parentheses.

The method is demonstrated on the example previously solved with the *root* function.

The guess value (seed) is first entered; we select a frequency of 110 Hz:

$$f := 110 \text{ Hz}.$$

The *solve block* is:

$$\text{Given}$$
$$f > 150 \text{ Hz}$$
$$Z_{abs}(f) - 400 \, \Omega = 0$$
$$\text{Find}(f) = 189.091 \cdot \text{Hz}.$$

The obtained solution is identical to the second value from the *root* solver.

In the earlier solve block, we defined f such that we are looking for any root larger than 150 Hz. If this constraint is eliminated, the *Find* solver yields the first root for the same guess value of 110 Hz, specifically:

$$\text{Given}$$
$$Z_{abs}(f) - 400 \, \Omega = 0$$
$$\text{Find}(f) = 133.958 \cdot \text{Hz}.$$

If the *Find* function is to yield an array of results with different units, those results must be assigned to an explicit vector of variable names to avoid mixed units in the matrices.

A.4. VECTORS AND MATRICES

To define a vector or matrix, first, type the name of the matrix or vector (e.g., A), followed by a colon [:]. A vector or matrix template is then inserted by clicking the matrix icon in the **Matrix** toolbar or select **Matrix** in the **Insert** menu. Either method opens the dialog box displayed in Figure A.6. The number of rows and columns is then specified. Clicking on **OK** or **Insert** places an empty matrix or vector template to the right of the equal sign:

VECTORS AND MATRICES

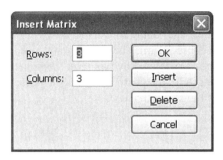

Figure A.6. Matrix or vector insertion dialog box.

$$A := \begin{pmatrix} \blacksquare & \blacksquare & \blacksquare \\ \blacksquare & \blacksquare & \blacksquare \\ \blacksquare & \blacksquare & \blacksquare \end{pmatrix}.$$

The values of the matrix elements have to be inserted manually by moving to each placeholder and entering a number or variable as desired. A typical example matrix is:

$$A := \begin{pmatrix} 5 & 4 & 1 \\ -3 & 7 & 3 \\ -4 & 5 & 2 \end{pmatrix}.$$

Each element can be accessed (or queried) by typing A and then a left square bracket "[," entering the row and column numbers of the element as subscripts separated by a comma, and followed by an equal sign [=]. This method is illustrated as follows:

$$A_{\blacksquare} = \quad A_{0,0} = \quad A_{1,2} = 4.$$

The typical matrix operations are listed in the **Matrix** toolbar (see Fig. A.1). The operations include inversion, dot and scalar product, summation of a vector's elements, and so on. Consult the Mathcad **Help** for details.

The following equation demonstrates matrix inversion. First, click x^{-1} in the **Matrix** toolbar or type in the sequence [x ^ −1]. Then enter A in the placeholder and press = on the keyboard:

$$\blacksquare^{-1} \quad A^{-1} \quad A^{-1} = \begin{pmatrix} 0.062 & 0.188 & -0.313 \\ 0.375 & -0.875 & 1.125 \\ -0.813 & 2.563 & -2.938 \end{pmatrix}.$$

The use of vectors and matrices can accelerate or simplify calculations. As an example, we will compute the distances between three conductors and a selected point. These distances are needed for the determination of the electric field at a distant point from a transmission line. The coordinates of the conductors are: A phase (−11 m, 21 m), B phase (0 m, 27 m), and C phase (11 m, 21 m). The distant point coordinate is (100 m, 1 m).

First, the *x*-coordinates are defined by typing in the sequence [x [{index}], where the indices are 1 for phase A, 2 for phase B, and 3 for phase C. The same method is used for the definition of the *y*-coordinates:

$$x_1 := -11 \text{ m} \quad x_2 := 0 \text{ m} \quad x_3 := 11 \text{ m}$$
$$y_1 := 21 \text{ m} \quad y_2 := 27 \text{ m} \quad y_3 := 21 \text{ m}.$$

The coordinates of the distant point are:

$$x_{dist} := 100 \text{ m} \quad y_{dist} := 1 \text{ m},$$

where we typed in the sequence [x . dist : {value} m].

The calculation of the distances using vectors is now illustrated. These distances can be described by a vector **d**. This vector has three elements, which are indexed by the range variable i. Type in sequence [i : 1; 3], which yields i := 1 .. 3.

Next, let us establish a general formula for computing the distance between an arbitrary conductor and the distant point. To begin the vector equation for the distance, type [d], then a left square bracket "[," followed by [i], and then a colon [:]. Next, the square root template $\sqrt{\blacksquare}$ is entered from the **Calculator** toolbar or by pressing the forward slash key [\]. Under the square root (i.e., in the placeholder), we type [(x [i → − x . dist →)], where "→" represents the right arrow key on the keyboard. This followed by [^ 2] to square the first term. A similar method is used to type the second part of the equation:

$$d_i := \sqrt{(x_i - x_{dist})^2 + (y_i - y_{dist})^2}.$$

The result is obtained by typing [d =]:

$$d = \begin{pmatrix} 0 \\ 112.787 \\ 103.325 \\ 91.22 \end{pmatrix} \text{m}.$$

We note that the resultant vector has four distances, but we should only have three values. This is because Mathcad may default to an array origin of zero. If necessary, the user can change the *array origin* from zero to unity using the **Worksheet Options** dialog box from the **Tools** menu, as shown in Figure A.7. After making that change, we now have only the three distances between the conductors and the distant point:

VECTORS AND MATRICES

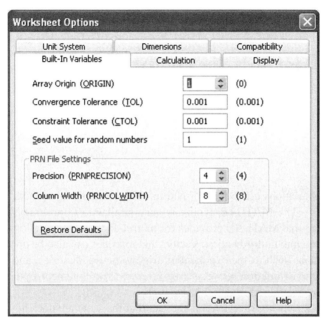

Figure A.7. **Worksheet Options** dialog box to change array origin index to unity (1).

$$d = \begin{pmatrix} 112.787 \\ 103.325 \\ 91.22 \end{pmatrix} m.$$

This appendix has provided a short introduction to Mathcad. The reader should use the **Help** menu to learn additional details of the software use. The student version of Mathcad is adequate to solve the problems in this book.

APPENDIX B
INTRODUCTION TO MATLAB

MATLAB® calculations can be performed in either an interactive mode or using programs (M-files). The MATLAB syntax is very similar to C programming, the main difference being that MATLAB provides for matrix and vector operations such as addition, subtraction, and multiplication. Vectors and matrices can also be operated on using standard functions such as the exponential, trigonometric identities, and so on. Important to note is the distinction between matrix versus array (element-by-element) operations, which are explained in detail later.

A drawback to using MATLAB is the fact that the software does not automatically incorporate the dimensions and units. Experience has shown that the best way to avoid unit mistakes is to keep all variables in the fundamental (International System of Units, SI) units, so that the results will also correspondingly be in the base units.

B.1. DESKTOP TOOLS

Starting MATLAB opens several possible windows. There are three primary windows:

1. MATLAB **Desktop** (main window),
2. **Help** browser, and
3. **Editor/Debugger** for M-files.

Along the left side of the main **Desktop** window (see Fig. B.1), the current folder (i.e., the present working directory path) is listed. Within the **Desktop**, there may be several subwindows (tools), including:

- *Command Window.* For variable assignments, calculations, and input of commands;
- *Launch Pad.* Lists the different MATLAB tools and programs available to the user;

Electrical Energy Conversion and Transport: An Interactive Computer-Based Approach, Second Edition.
George G. Karady and Keith E. Holbert.
© 2013 Institute of Electrical and Electronics Engineers, Inc. Published 2013 by John Wiley & Sons, Inc.

DESKTOP TOOLS 795

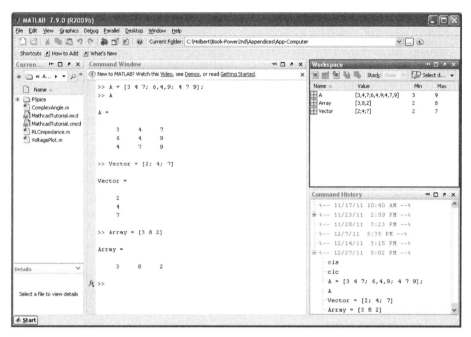

Figure B.1. MATLAB Desktop main window with tiles (subwindows) of the Current Folder, Command Window, Workspace, and Command History (from left to right).

- *Workspace.* Lists the variables in memory;
- *Current Folder.* Lists the programs and other files in the working directory path;
- *Command History.* Shows the calculations and commands recently performed.

The most important window is the **Command Window**, which is used for performing immediate calculations in an interactive mode and entering commands such as executing an M-file.

We will initially demonstrate the use of **Command Window** for direct calculations in the succeeding sections. Performing calculations in the **Command Window** is useful when a few quick results are needed. Generally, the better practice is to use the M-files for program development. The M-files are text files (i.e., a program) in which the MATLAB statements and functions for the calculations are entered. The M-file is saved in the current directory. The calculations can then be performed (i.e., the program can be executed) by entering the name of the M-file in the **Command Window** or simply clicking the green right arrow ("Run") button in the **Editor** window. In the event of a programming error, MATLAB identifies the error type and its location by line and column number of the M-file.

B.2. OPERATORS, VARIABLES, AND FUNCTIONS

Common operators used for calculations include:

Addition	+	Subtraction	–	Multiplication	*
Division	/	Left division	\	Power	^

Useful built-in constants include:

Pi	3.1415..... (π)
i or j	imaginary unit ($\sqrt{-1}$)
Inf	Infinity (∞)
NaN	Not_a_number

Good programming practices recommend the selection of a variable name that is self-evident of its purpose and the quantity that it represents. For example, rather than picking a simple variable name of "a" for the transformer turns ratio, a better choice is "tr," and an even better selection is "Tratio." In addition, the manner and ordering of the variables and operators in a MATLAB formula can help to insure that the equation has been properly entered. As a simple example, consider the calculation of the single-phase equivalent magnetizing resistance (R_m) from the three-phase real power (P_T) and line voltage (V_{line}) using:

$$R_m = \frac{\left(V_{line}/\sqrt{3}\right)^2}{P_T/3}.$$

There are several possible (and equivalent) formulations of the earlier expression as shown later. Some renditions are more cryptic than others, but the last one is the easiest to check for errors. Although the first statement is a simplification of the earlier formula and is quick to type, it obscures the underlying analysis within the equation, and should be avoided. The use of the half-power operation (^0.5) in the second example is not as crisp as using the *sqrt()* function:

```
Rmag = Vline^2 / Ptotal;
Rmag = 3*(Vline/3^0.5)^2/Ptotal;
Rmag = Vline^2/sqrt(3)^2*3/Ptotal;
Rmag = Vline*Vline/3/(Ptotal/3);
Rmag = (Vline/sqrt(3))^2 / (Ptotal/3);
```

MATLAB allows the use of complex variables and functions. MATLAB variable and function names are case sensitive. Of importance to electrical engineering calculations are the following complex functions:

Function	Computes
abs	Absolute value or magnitude
angle	Four-quadrant phase angle
conj	Complex conjugate
exp	Exponential (e)
sqrt	Square root

MATLAB contains all the traditional trigonometric functions, for which MATLAB computes all angles in radians. A second version of the (inverse) trigonometric functions uses degrees; for example, *asind*() computes the inverse sine in degrees, whereas *asin*() determines the arcsine in radians. Using the *atan2*() function is preferred over *atan*(), since *atan2*() calculates the resultant angle over the full $-\pi$ to π range, which clearly identifies the proper quadrant.

The use of MATLAB for simple calculations is first demonstrated in the next section using the **Command Window**. If we append a statement or variable with a semicolon (;), this will prevent the printing of the variable value or calculation result back to the screen.

B.3. VECTORS AND MATRICES

A matrix or vector can be created by typing the elements of a row (i.e., the column entries) separated by commas or blanks. The semicolon is used to separate the rows. Square brackets are used to delineate the entire list of elements. A typical example for a 3×3 matrix is:

```
>> A = [3 4 7; 6,4,9; 4 7 9];
>> A
A =
     3     4     7
     6     4     9
     4     7     9
```

An example for a column vector is:

```
>> Vector = [2; 4; 7]
Vector =
     2
     4
     7
```

Note that by using the semicolon (;) at the end of the definition of the A matrix, the matrix was not printed (echoed) upon its entry. But with the vector, the semicolon was omitted, which resulted in the defined vector being echoed back to the **Command Window**. An example definition for a row vector is:

```
>> Array = [3 8 2]
Array =
     3     8     2
```

Matrix elements can be addressed using the name of the matrix along with the row and column number in parenthesis, that is, *matrix(row, column)*. As an example, A(1,2) = 4. The colon operator can also be used as a row and/or column range, for example, A(2,2:3) denotes row 2, columns 2 through 3 of matrix A.

The mathematical operators listed earlier can also be used with matrices. As an example, the inverse of matrix A is:

```
>> A^-1
ans =
    -0.9310    0.4483    0.2759
    -0.6207   -0.0345    0.5172
     0.8966   -0.1724   -0.4138
```

The same result can be accomplished using the matrix inversion function, which is *inv*(A).

The matrix operators also include additional operators, which are the *element-by-element* equivalent operators that are identified by the period that precedes the normal mathematical operators:

Element-by element multiplication	.*
Element-by element division	./
Element-by element left division	.\
Element-by element power	.^

The difference between the element-by-element and normal math operations on matrices is illustrated. We first establish two row vectors: Z1 and Z2. The element-by-element multiplication of the two vectors yields:

```
>> Z1 = [1,2,3];   Z2 = [4,5,6];
>> Z1 .* Z2
ans =
     4    10    18
```

Matrix multiplication calculates the product of the variables. In the case of Z1 and Z2, we cannot multiply the two vectors without transposing one, and only one, of the vectors. The scalar product of two vectors requires transforming one of the vectors to a column vector by using the (complex conjugate) transpose operator ('), as seen in the following:

```
>> Z1 * Z2'
ans =
    32
```

Transposing the other vector and then multiplying the two arrays creates an entirely different result—a 3 × 3 matrix:

```
>> Z1' * Z2
ans =
     4     5     6
     8    10    12
    12    15    18
```

B.4. COLON OPERATOR

The colon (:) operator is used to generate an array or series of numbers. The syntax is *variable = start : step : end*, where the default *step* is one if omitted. Alternatively, the *linspace* or *logspace* function may be used. As an example, the integers inclusively between 1 and 10 can be generated by:

```
>> 1:10
ans =
     1     2     3     4     5     6     7     8     9    10
```

Nonunity spacing can be obtained by specifying the *step*. Consider a negative *step* of six:

```
>> 100: -6: 40
ans =
   100    94    88    82    76    70    64    58    52    46    40
```

An example using real-valued numbers is:

```
>> 0: pi/3 : 2*pi
ans =
     0    1.0472    2.0944    3.1416    4.1888    5.2360    6.2832
```

B.5. REPEATED EVALUATION OF AN EQUATION

Many calculations in this book require the repeated evaluation of a formula and plotting of the results. The repetitive evaluation of an equation is demonstrated by calculating the absolute value of the impedance of a resistance, inductance, and capacitance connected in series as a function of the frequency. The circuit component values are resistance 25 Ω, inductance 0.1 H, and capacitance 0.1 μF. The equation for the series connected network impedance magnitude in the frequency domain is:

$$|Z_{series}| = \left|25 + j\omega 0.1 + \frac{1}{j\omega 0.1 \cdot 10^{-6}}\right| \quad \omega = 2\pi f.$$

The first step is the generation of a vector covering the selected frequency range. Let us select a frequency range between 2 and 10 kHz, with steps of 2 kHz. This results in five cyclic frequency values (freq). MATLAB creates this vector using the colon operator:

```
>> freq = 2e3 : 2e3 : 1e4     % Hz
freq =
    2000        4000        6000        8000        10000
```

Comment text can be inserted using a percent sign followed with the text, that is, "% comment text." These comments may occupy an entire line by themselves or be at the end of a line, as previously demonstrated. The definition of the frequency array (freq) is followed by the computation of the angular frequency (w) and impedance (Zseries). Note the use of the element-by-element operator in the following assignment statements:

```
>> w = 2*pi .* freq;
>> Zseries = abs(25 + 1j .* w*0.1 + 1 ./ (1j .*w*1e-7))
Zseries =
    1.0e+003 *
    0.4615      2.1155      3.5047      4.8277      6.1241
```

Note that the first line of the answer is a multiplier for all the subsequent values of Zseries. For plotting the results, a larger frequency range and smaller steps are needed.

B.6. PLOTTING

A two-dimensional graph can be produced using the function *plot(xvar, yvar)*, where *xvar* is the *x*-axis vector (e.g., time) and *yvar* is the *y*-axis array (e.g., voltage or power). The *x*- and *y*-axes can be labeled using the *xlabel()* and *ylabel()* functions, respectively, with the axis title given as a string in the function argument. The *title()* function provides the capability to label the entire plot. Often the number conversion to string function of *num2str()* is useful in writing specific numerical results into the title of the plot for reference.

As a simple example, we can plot an alternating current (ac) voltage at 60 Hz using the following code, which first creates a time vector and then computes the cosinusoidal voltage. The resultant graph is seen in Figure B.2.

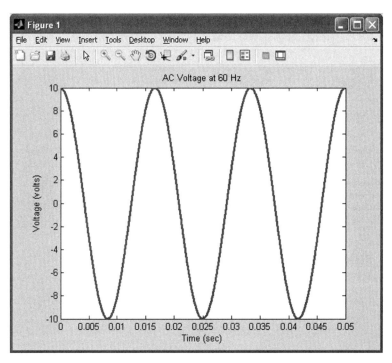

Figure B.2. MATLAB plot of 60 Hz ac voltage waveform for three cycles.

```
% VoltagePlot.m
clear all

freq = 60;    % Hz
omega = 2*pi*freq;
time = 0 : 1/(freq*500) : 3/freq;
acvoltage = 10*cos(omega*time);

plot(time, acvoltage, 'LineWidth',2.5);
xlabel('Time (sec)');
ylabel('Voltage (volts)');
title(['AC Voltage at ',num2str(freq),' Hz']);
```

Similarly, we can analyze the behavior of a series connected resistor–inductor-capacitor (RLC) network as a function of frequency. We will use the earlier RLC circuit example; however, for this analysis, the frequency range is 1–2 kHz, with steps of 0.1 Hz. This simple program can be prepared as an M-file. Figure B.3 displays the M-file of the "RLC circuit impedance" program in the **Editor/Debugger**. The M-file coding is:

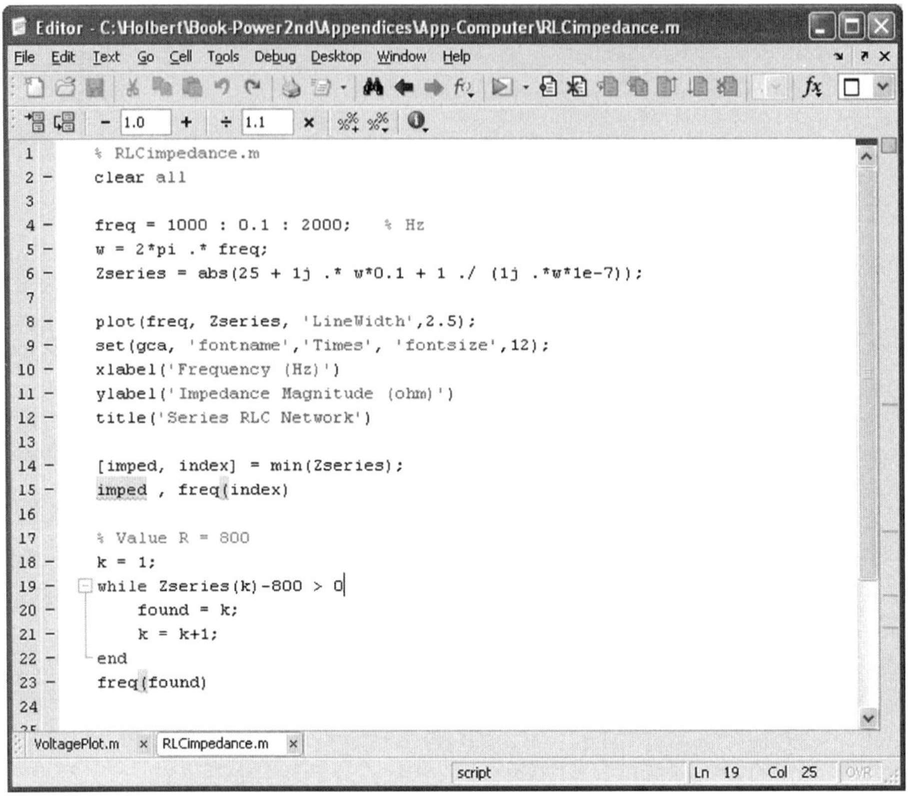

Figure B.3. MATLAB Editor/Debugger window.

```
% RLCimpedance.m
clear all
freq = 1000 : 0.1 : 2000;    % Hz
w = 2*pi .* freq;
Zseries = abs(25 + 1j .* w*0.1 + 1 ./ (1j .*w*1e-7));
plot(freq, Zseries, 'LineWidth',2.5);
set(gca, 'fontname','Times', 'fontsize',12);
xlabel('Frequency (Hz)')
ylabel('Impedance Magnitude (ohm)')
title('Series RLC Network')
```

The result of this calculation is presented in Figure B.4. It can be seen that the resonance occurs at around 1.6 kHz, when the impedance is minimum (25 Ω).

The minimum impedance value can be found with the *min* function, which can also identify the exact array element where the minimum occurs, and therefore in this example, the particular frequency:

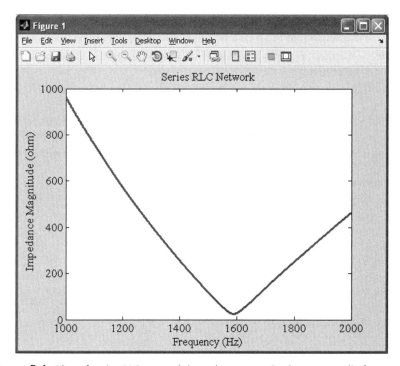

Figure B.4. Plot of series RLC network impedance magnitude versus cyclic frequency.

```
[imped, index] = min(Zseries);
imped , freq(index)
imped =
   25.0001
ans =
  1.5915e+003
```

For multiple curves on a single graph, the *legend* function should be used to properly discriminate between the curves. Furthermore, the line style, marker symbol, and color can be utilized to distinguish the plots. MATLAB figures are readily copied into other applications such as word processors for preparing reports.

B.7. BASIC PROGRAMMING

The programming statements used for elementary MATLAB programming include:

if evaluates a logical statement and executes the following statements when the expressions are true

switch selects a group of statements (a *case*) based on the value of a variable *for* evaluates a group of statements (a loop) using an index counter from a starting value to an ending value with an optional increment

while executes a group of statements (loop) under the control of a logical condition

end denotes the boundary of the previous statement blocks and loops

A practical problem is to find the particular frequency that corresponds to a given impedance value. As an example, we find the frequency when the RLC impedance magnitude is 800 Ω. This can be determined by appending the RLC program developed earlier. The coding for the frequency determination is:

```
% Value R = 800
k = 1;
while Zseries(k)-800 > 0
    found = k;
    k = k+1;
end
freq(found)
```

The result of the calculation is:

```
ans =
  1.0777e+003
```

This answer indicates that the impedance magnitude is 800 Ω at a frequency of 1.0777 kHz. Of course, the frequency spacing in the original definition of freq directly affects the accuracy of the answer, that is, finer spacing leads to better resolution.

MATLAB is a very powerful program, which is suitable to perform the most complicated analysis. A significant advantage of the program is the large number of available functions, which contain dedicated algorithms for performing various analyses. In this introduction, we summarize the major features needed to understand the examples presented in this book. The reader should consult the extensive MATLAB help facility to learn further details about the use of the software.

APPENDIX C
FUNDAMENTAL UNITS AND CONSTANTS

C.1. FUNDAMENTAL UNITS

At least four unit systems are in use today. The *coherent unit systems* utilize a set of *base* quantities from which all *derived* quantities are obtained by multiplication or division without the introduction of numerical factors. The traditional English units, which continue to see use in the United States, do not form a coherent system. The coherent unit systems include SI units and the MKS and cgs systems. The cgs system is based on three basic quantities for length, mass, and time using the **c**entimeter, **g**ram and **s**econd, respectively. In the cgs system, force and energy are quantified with the derived units of dyne (g cm/s^2) and erg (g cm^2/s^2), respectively. The gauss (G) and oersted (Oe) are cgs units for magnetic flux density and magnetic field strength, respectively. MKS is a system of units for mechanics based on the three fundamental quantities of length, mass, and time using the **m**eter, **k**ilogram, and **s**econd, respectively. The International System of Units (SI units) is a coherent system based on the seven basic quantities, and matching units, specified in Table C.1. The derived SI units and their corresponding constituent base units (dimensions) are given in Table C.2, where the angular measures of the radian and steradian are sometimes referred to as *supplementary* units. For volume, the liter (L) is actually not a derived unit since it requires a numeric multiplier, that is, 1 L = 10^{-3} m^3. The SI prefixes incorporated with both base and derived units are provided in Table C.3. Corresponding units between the SI, cgs, and English systems are presented in Table C.4. Some selected conversions between the various units are given in Table C.5. Table C.6 lists the physical quantities common to power engineering as well as the symbol and unit utilized for each.

Electrical Energy Conversion and Transport: An Interactive Computer-Based Approach, Second Edition.
George G. Karady and Keith E. Holbert.
© 2013 Institute of Electrical and Electronics Engineers, Inc. Published 2013 by John Wiley & Sons, Inc.

TABLE C.1. Base SI Quantities and Units

Base Physical Quantity	Base Unit	Unit Symbol
Length	meter	m
Mass	kilogram	kg
Time	second	s
Electric current	ampere	A
Thermodynamic temperature	Kelvin	K
Amount of substance	mole	mol
Luminous intensity	candela	cd

TABLE C.2. Derived SI Quantities and Units

Quantity	Derived Unit	Unit Symbol	Unit Dimension(s)
Frequency	hertz	Hz	$1/s$
Force	newton	N	$kg \times m/s^2 = J/m$
Pressure	pascal	Pa	$N/m^2 = kg/(m \times s^2)$
Energy, Work, Heat	joule	J	$N \times m = kg \times m^2/s^2$
Power	watt	W	$J/s = kg \times m^2/s^3$
Electric charge	coulomb	C	$A \times s$
Electric potential	volt	V	$J/C = kg \times m^2/s^3/A = W/A$
Resistance	ohm	Ω	$V/A = kg \times m^2/s^3/A^2$
Conductance	siemens	S	$1/\Omega = A/V$
Capacitance	farad	F	$C/V = A^2 \times s^4/kg/m^2$
Magnetic flux density	tesla	T	$V \times s/m^2 = kg/A/s^2 = Wb/m^2$
Magnetic flux	weber	Wb	$V \times s = kg \times m^2/s^2/A$
Inductance	henry	H	$V \times s/A = kg \times m^2/s^2/A^2 = Wb/A$
Plane angle	radian	rad	m/m
Solid angle	steradian	sr	m^2/m^2

TABLE C.3. SI Prefixes

Prefix	Factor	Symbol	Prefix	Factor	Symbol
milli	10^{-3}	m	kilo	10^3	k
micro	10^{-6}	μ	Mega	10^6	M
nano	10^{-9}	n	Giga	10^9	G
pico	10^{-12}	p	Tera	10^{12}	T
femto	10^{-15}	f	Peta	10^{15}	P
atto	10^{-18}	a	Exa	10^{18}	E

FUNDAMENTAL UNITS

TABLE C.4. Corresponding Units in the SI, cgs, and English Unit Systems

Quantity	SI	cgs	English
Length	meter, m	centimeter, cm	inch, in; foot, ft
Mass	kilogram, kg	gram, g	pound mass, lb or lb_m
Time	second, s	second, sec	second, sec
Temperature	kelvin, K	degree kelvin, °K; degree Celsius, °C	degree Fahrenheit, °F
Force	newton, N	dyne	pound, lb or lb_f
Pressure	pascal, Pa	dyne/cm^2	pound/inch^2, psi
Energy	joule, J	erg	foot pound, ft lb
Power	watt, W	erg/sec	foot pound/sec, ft lb/sec
Heat	joule, J	calorie, cal	British thermal unit, Btu

TABLE C.5. Unit Conversion Factors

Dimension	Selected Equalities
Length	1 m = 39.37 inches = 3.281 ft
	1 inch = 2.54 cm
	1 ft = 12 inches = 30.48 cm
	1 yard = 3 ft = 0.9144 m
	1 mile = 5280 ft = 1.609 km
Time	1 year = 365.25 days = 8766 hours
Mass	1 lb_m = 16 oz = 0.4536 kg
	1 ton = 2000 lb_m = 907.2 kg
	1 tonne (metric ton) = 1000 kg
Force	1 N = 10^5 dyne = 0.2248 lb_f
	1 lb_f = 4.448 N
Torque	1 N m = 10^7 dyne cm = 0.7376 lb_f ft
Energy	1 J = 10^7 erg = 0.7376 ft lb_f = 9.480 × 10^{-4} Btu = 2.778 × 10^{-7} kWh
	1 Btu = 252 cal = 1055 J
	1 erg = 10^{-7} J
	1 kWh = 3412.3 Btu = 3.6 × 10^6 J
Power	1 W = 0.7376 ft lb_f/s = 1.341 × 10^{-3} hp
	1 Btu/h = 2.93 × 10^{-4} kW
	1 hp = 2545 Btu/h = 0.7457 kW
Magnetic	1 T = 10^4 G = 1 Wb/m^2
	1 Oe = 79.577472 ampere-turns/meter

TABLE C.6. Symbols and Units for Various Quantities

Quantity	Quantity Symbol	Unit(s)
Fundamental		
Length	ℓ	meter (m)
Mass	m	kilogram (kg)
Time	t	second (s)
Current	I	ampere (A)
Mechanical		
Force	F	newton (N)
Torque	T	newton-meter (N m)
Angular displacement	θ	radian (rad)
Velocity	v	meter/second (m/s)
Angular velocity	ω	radian/second (rad/s)
Electrical		
Charge	q	coulomb (C)
Electric potential	V	volt (V)
Electric field strength	**E**	volt/meter (V/m)
Energy	W	joule (J)
Power	P	watt (W)
Electric flux density	**D**	coulomb/meter^2 (C/m^2)
Electric flux	ψ	coulomb (C)
Resistance	R	ohm (Ω)
Conductance (G = 1/R)	G	siemens (S); mho
Resistivity	ρ	ohm-meter (Ω m)
Conductivity ($\sigma = 1/\rho$)	σ	1/(ohm-meter)
Reactance	X	ohm (Ω)
Susceptance (B = 1/X)	B	siemens (S); mho
Impedance	**Z**	ohm (Ω)
Admittance (**Y** = 1/**Z**)	**Y**	siemens (S); mho
Capacitance	C	farad (F)
Permittivity (**D** = ε **E**)	ε	farad/meter (F/m)
Magnetic		
Magnetomotive force ($F_m = \oint \mathbf{H}_s d\mathbf{s}$)	F_m	ampere-turn
Magnetic field strength	**H**	oersted (Oe); ampere-turn/meter
Magnetic flux	Φ	weber (Wb)
Magnetic flux density	**B**	tesla (T); gauss (G)
Inductance	L	henry (H)
Permeability (**B** = μ **H**)	μ	henry/meter (H/m)

C.2. FUNDAMENTAL PHYSICAL CONSTANTS

Constant	Symbol	Value
Speed of light (in vacuum)	c	2.997925×10^8 m/s
Elementary (electron) charge	e	$1.60217646 \times 10^{-19}$ C
Planck's constant	h	6.62608×10^{-34} J s
Boltzmann constant	k	1.38065×10^{-23} J/K
Permittivity of vacuum	ε_0	$10^{-9}/(36\pi) = 8.842 \times 10^{-12}$ F/m
Permeability of vacuum	μ_0	$4\pi \times 10^{-7} = 1.257 \times 10^{-6}$ H/m

APPENDIX D
INTRODUCTION TO PSPICE

Some examples in this book utilize PSpice®, which is a member of the Simulation Program with Integrated Circuit Emphasis (SPICE) family of circuit simulators. The original SPICE program was developed in FORTRAN at the University of California, Berkeley, in the early 1970s. The computation engine of SPICE employs nodal analysis. PSpice, an acronym for Personal SPICE, was marketed and sold by MicroSim Corporation until they were acquired by OrCAD, which is now owned by Cadence Design Systems, Inc.

D.1. OBTAINING AND INSTALLING PSPICE

Cadence provides a *free* demonstration version of PSpice that can be used to solve circuit problems of limited size and complexity. At the writing of this book, the OrCAD *Designer Lite* program and documentation can be downloaded from http://www.cadence.com/products/orcad/pages/downloads.aspx.

The software is about 500 MB and the documentation is 17 MB.

There are presently two choices of graphical circuit editors, either **Schematics** or **Capture**. You must select which of the schematic circuit editors that you will use. **Schematics** is limited to placing and saving a maximum of 50 parts on the design. **Capture** can view and create larger designs, but cannot save a design that contains more than 60 parts. **Capture**, which is intended for printed circuit board (PCB) design, also has the ability to import circuits created by **Schematics**. **Schematics** is probably a little easier to use. Both circuit editors produce an ASCII **netlist** file, which is a listing of the circuit components, their values, and the nodes to which each component is connected.

Electrical Energy Conversion and Transport: An Interactive Computer-Based Approach, Second Edition. George G. Karady and Keith E. Holbert.
© 2013 Institute of Electrical and Electronics Engineers, Inc. Published 2013 by John Wiley & Sons, Inc.

D.2. USING PSPICE

Utilizing PSpice consists of three basic phases:

1. The circuit is created and saved using one of the schematic editors.
2. The simulation parameters are chosen and the circuit is simulated.
3. The circuit simulation results are plotted and analyzed (in **Probe**).

These stages are illustrated with an example in the succeeding sections. PSpice has significant capabilities but this appendix only addresses those most closely tied to the material in this book.

D.2.1. Creating a Circuit

The circuit is drawn in a schematic editor. To draw the circuit, we get, place, and arrange parts, which are subsequently wired together. The component values are then specified. PSpice uses International System of Units (SI)-type prefixes for scaling the component values—it is important to note from the following table that "M" stands for milli and "MEG" denotes mega:

T	tera	10^{12}	M	milli	10^{-3}
G	giga	10^9	U	micro	10^{-6}
MEG	mega	10^6	N	nano	10^{-9}
K	kilo	10^3	P	pico	10^{-12}

We can rename a part and select how much information about the part is displayed on the schematic. Nodes can also be named and annotations can be added to the circuit.

When getting and placing a part, it may be necessary to select the library. Among the provided libraries, the following are of interest here:

Library	Components within the Library (not an All-Inclusive List)
analog	Capacitor (C), inductor (L), resistor (R), voltage- and current-controlled sources (E, F, G, and H), voltage-controlled switch (S), transformer (XFRM_LINEAR)
eval	diode (D), transistor, silicon-controlled rectifiers (SCR), timed switches, metal–oxide–semiconductor field-effect transistor (MOSFET), operational amplifier (op-amp), digital
source	Independent sources including direct current (dc) and alternating current (ac) voltage (VDC and VAC), dc and ac current (IDC and IAC), exponential decay (VSIN and IEXP), transient sinusoidal (VSIN and ISIN), pulse (VPULSE and IPULSE), and piecewise linear sources (VPWL and IPWL)

Every circuit must have at least one ground node, which in **Schematics** is found in the *port* library, and in **Capture** is an item within the Place menu option.

A simple circuit consisting of two ac sources connected through a transmission line with a capacitance in the center of the line is created here. Each part is obtained

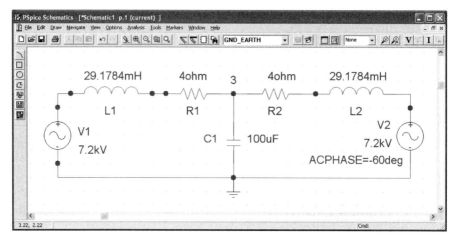

Figure D.1. Circuit in PSpice Schematics graphical editor.

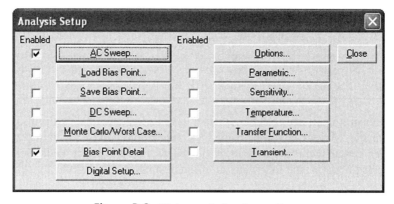

Figure D.2. PSpice analysis setup options.

and placed on the circuit schematic; Ctrl+R may be used to rotate the part as needed. Then, we connect the parts with wire, as shown in Figure D.1. VAC is used to represent an AC voltage source in PSpice; both the magnitude (ACMAG) and phase (ACPHASE) value can be set for the source, but the frequency is not set as a source attribute.

D.2.2. Simulating a Circuit

As viewed in Figure D.2, PSpice has several different simulation types available, including DC, AC sweep, transient, and Monte Carlo. For each analysis type, various options must be selected. The default analysis temperature is 300 K. For the transient analysis, we specify the period over which the simulation is performed.

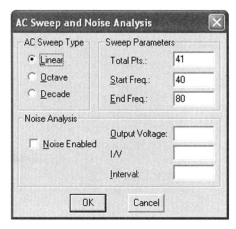

Figure D.3. PSpice AC sweep analysis simulation options.

PSpice can be used to analyze a circuit at a single frequency (e.g., 60 Hz) or for an entire range of frequencies (variable). For the AC sweep analysis, the start and stop frequency are specified along with the number of frequencies at which the results are computed and whether they are linearly or logarithmically spaced. If a single frequency is desired, then select a linear sweep with one point using the same start and end frequency. Because AC analysis simulates the sinusoidal steady-state behavior of the circuit, any nonlinear elements, such as diodes, are replaced by their small-signal models.

For the circuit created earlier, we conduct an AC sweep simulation from 40 to 80 Hz with 41 frequency points linearly separated (see Figure D.3). Once the simulation parameters are entered, the simulation is executed. Had a transient analysis been desired for the earlier circuit, we would have needed to use VSIN sources instead of the VAC sources.

D.2.3. Analyzing Simulation Results

With the simulation run, the results can now be displayed and/or printed in the **Probe** window. Assuming the circuit has no flaws, you will always get an answer, but you must judge its validity. Plots are created in the **Probe** output window exhibited in Figure D.4. Traces and axes are added using the listing provided, as seen in Figure D.5.

For the circuit simulated earlier, the AC sweep analysis results for the voltage at node 3 and the current through the capacitor are plotted in Figure D.4. The voltage and current are placed on different y-axes. The current and voltage magnitudes at 60 Hz are annotated on the plot by first moving the cursor to the 60 Hz position and then using the Mark Label tool.

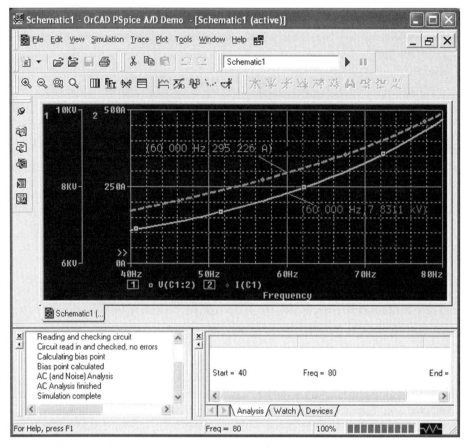

Figure D.4. PSpice Probe output window for plotting simulation results.

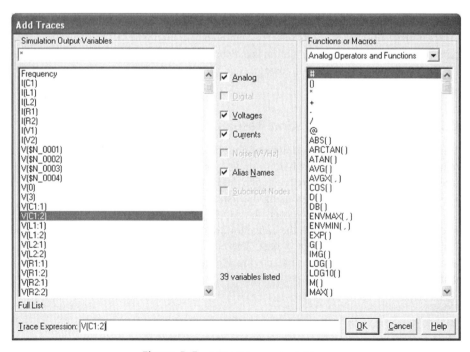

Figure D.5. Add traces menu option.

PROBLEM SOLUTION KEY

Answers to selected problems (e.g., does not include graphs, diagrams, and proofs).

CHAPTER 1: ELECTRIC POWER SYSTEM

1.1. (a) *Short circuit solely on transmission line T2* opens CBA 2 and 3. Supply S feeds T1 through CBA 4, Bus 1 and CBA 1, and serves T3 directly through CBA 5. (b) *Short circuit solely on transmission line T3* activates CBA 6 and 5. Supply S feeds T1 through CBA 4, Bus 1 and CBA 1, and serves T2 through CB4, Bus 1, and CBA 1 and 2

1.2. (a) *CBA 4 fails open.* T3 is supplied directly through CBA 5; T2 is supplied through CBA 5, 6, and 3; and T1 is supplied through CBA 5, 6, 3, and 2. It is evident that the circuit breakers should be sized to pass all three load currents. In this case, the full supply current passes through CBA 5. (b) *CBA 4 fails closed.* If CBA 4 cannot be opened, then a short circuit on Bus 1 cannot be locally isolated since the faulty CBA 4 directly connects the source to the shorted bus.

1.3. CBA 1, 2, 4, and 5 closed; CBA 3 and 6 open.

CHAPTER 2: ELECTRIC GENERATING STATIONS

2.1. 543 days
2.2. (a) $CF_1 = 70.8\%$, $CF_2 = 87.5\%$; (b) $O\&M_2 = 1.2$ ¢/kWh
2.3. $Q_{coal} = 1125$ MWt, $Q_{nuclear} = 2030$ MWt
2.4. $20.5 million
2.5. $\eta_{orig} = 66.3\%$, $\eta_{new} = 92.2\%$
2.6. $V_{out} = 18$ V, $I_{out} = 3$ A, $P_{out} = 54$ W
2.7. two derivations
2.8. (a) plot; (b) $P_{max} = 3.6$ W, $V_{Pmax} = 0.5$ V

Electrical Energy Conversion and Transport: An Interactive Computer-Based Approach, Second Edition.
George G. Karady and Keith E. Holbert.
© 2013 Institute of Electrical and Electronics Engineers, Inc. Published 2013 by John Wiley & Sons, Inc.

2.9. $\eta_{\text{trough}} = 40.2\%$, $\eta_{\text{ptower}} = 75.4\%$
2.10. $\eta = 47.5\%$
2.11. (a) $n_C = 1.53 \times 10^{18}$ atoms; (b) $m_C = 30.5$ µg; (c) $m_C/m_{\text{U-235}} = 2.5 \times 10^6$
2.12. $A = 0.028$ km^2
2.13. $m_{\text{SO2}} = 4.21 \times 10^7$ kg/year, $m_{\text{ash}} = 5.26 \times 10^7$ kg/year
2.14. $P = 3.82$ MW
2.15. (a) $E_{\text{in}} = 1.6 \times 10^5$ MWh, $E_{\text{out}} = 1.22 \times 10^5$ MWh; (b) $\eta_{\text{ta}} = 76\%$

CHAPTER 3: SINGLE-PHASE CIRCUITS

3.1. (a) $T_{50} = 20$ ms; (b) $T_{60} = 16.7$ ms
3.2. (a) $V_M = 169.7$ V; (b) V_{rms}: {114–126 Vrms}
3.3. proof
3.4. phasor diagram
3.5. (a) $\mathbf{Z}_{50} = 50 + j29.7$ Ω; (b) $\mathbf{Z}_{60} = 50 + j82.4$ Ω
3.6. (a) $\mathbf{Z}_{\text{eq}} = 32.6 + j38.2$ Ω; (b) $\mathbf{Y}_{\text{eq}} = 0.013 - j0.015$ S
3.7. $C_{\text{cap}} = 82.1$ µF
3.8. $\mathbf{V}_1$ ($P_1 = -1.7$ MW) operates as generator and $\mathbf{V}_2$ ($P_2 = 1.5$ MW) serves as load
3.9. (a) $P_1 = 1.1$ kW, $P_2 = 789$ W, $P_3 = 302$ W, $P_4 = 139$ W; (b) $Q_1 = -1.76$ kVA, $Q_2 = 394$ VA, $Q_3 = -302$ VA, $Q_4 = 93$ VA
3.10. (a) $R = 64.3$ Ω, $X = 76.6$ Ω; (b) $P = 311$ W, $Q = 371$ VA; (c) $pf = 0.643$ leading
3.11. (a) $P_1 = 512$ W (absorbed), $P_2 = -1.45$ kW (supplied); (b) $Q_1 = 1.76$ kVA (absorbed), $Q_2 = -1.65$ kVA (supplied); (c) $P_{Z1} = 53$ W (absorbed), $Q_{Z2} = 26.5$ VA (absorbed), $Q_{Z3} = -132$ VA (supplied), $P_{Z4} = 883$ W (absorbed); (d) P_{feeder} (180°) = 17.1 kW
3.12. (a) $Q_{\text{cap}} = -332$ kVA; (b) $pf = 0.908$ (lagging)
3.13. (a) diagram; (b) $Q_{\text{cap}} = -46.4$ kVA, $C_{\text{cap}} = 2543$ µF; (c) graph; (d) $C_{\text{pf1}} = 2137$ µF

CHAPTER 4: THREE-PHASE CIRCUITS

4.1. proof
4.2. (a) $I_Y = 12.9$ A; (b) $I_{\Delta\text{phase}} = 7.47$ A, $I_{\Delta\text{line}} = 12.9$ A
4.3. $\mathbf{I}_a = 75.1$ A $\angle-35.5°$, $\mathbf{I}_b = 82.8$ A $\angle-177°$, $\mathbf{I}_c = 52$ A $\angle 65.7°$
4.4. $\mathbf{I}_a = 7.2$ A $\angle-11.9°$, $\mathbf{I}_b = 7.6$ A $\angle-149°$, $\mathbf{I}_c = 5.5$ A $\angle 95.5°$
4.5. (a) $\mathbf{Z}_{1Y} = 5.33 - j2.58$ Ω, $\mathbf{Z}_{2\Delta} = 19.6 - j12.1$ Ω; (b) $I_{\text{load}} = 72.8$ A; (c) $V_{\text{LL\_bus}} = 597$ V; (d) $P_{\text{bus}} = 70$ kW, $Q_{\text{bus}} = 27.5$ kVA
4.6. (a) $\mathbf{I}_{\text{line}} = 7.11$ A $\angle-40.7°$, $\mathbf{V}_{\text{load\_ll}} = 205$ V $\angle 29.3°$; (b) $\mathbf{V}_{\text{load\_ll}} = 215$ V $\angle 26.3°$
4.7. (a) diagram; (b) $\mathbf{Z}_{\text{load1}} = 25 + j23.4$ Ω, $\mathbf{Z}_{\text{load2}} = 21.9 + j7.8$ Ω; (c) $I_{\text{gen}} = 45.3$ A $= 97.99\%$

4.8. (a) $I_{ab} = 33.3$ A, $I_{bc} = 48$ A, $I_{ca} = 80$ A, $I_a = 84.6$ A, $I_b = 23.4$, $I_c = 69.7$ A; (b) Phases **a** and **c** are overloaded

4.9. (a) $P_3 = 13.3$ kW, $|S_3| = 15.1$ kVA, $Q_3 = -7.29$ kVA; (b) $pf_3 = 0.876$ (leading); (c) $R_3 = 0.905$ Ω

4.10. proof

CHAPTER 5: TRANSMISSION LINES AND CABLES

5.1. $L = 0.487$ mH/mile, $X = 0.184$ Ω/mile

5.2. $C = 24.6$ nF/mile, which is negligible

5.3. $R = 0.119$ Ω/mile, $L = 2.14$ mH/mile, $C = 13.9$ nF/mile

5.4. $R = 0.0342$ Ω/km, $L = 0.927$ mH/mile, $C = 12.2$ nF/km

5.5. (a) $\gamma_{line} = 2.4 \times 10^{-5} + 1.19 \times 10^{-3}$ j/km, $Z_{surge} = 316 - 6.3$j Ω; (b) $Z_{ser} = 4.3 + 110.7$j Ω, $Z_{par} = 0.75 - 1759$j Ω

5.6. $V_{sup} = 14.8$ kV, Reg = 7.0%

5.7. $V_{load} = 12.9$ kV, Reg = 7.4%

5.8. $\delta = 43.1°$

5.9. (a) $R = 0.0113$ Ω/km, $X_L = 0.324$ Ω/km; (b) $C = 13.2$ μF/km, $Y_C = 4.96$ μS/km; (c) diagram; (d) $I_{load} = 563$ A, $V_{load} = 164$ kV

5.10. (a) $Z_{line} = 10.6 + j20.7$ Ω; (b) (1) $|V_{LN}| = 13.1$ kV, $|I| = 257$ A, $P = 7.11$ MW, $Q = 7.21$ MVAR (ind); (2) $|V_{LN}| = 9.66$ kV, $|I| = 240$ A, $P = 6.84$ MW, $Q = 1.31$ MVAR (ind); (3) $|V_{LN}| = 10.93$ kV, $|I| = 219$ A, $P = 6.53$ MW, $Q = 2.97$ MVAR (ind); (c) plot

5.11. (a) $R = 7.15$ Ω, $X_L = 96.6$ Ω, $Y_C = 0.698$j mS; (b) diagram; (c) $Z_{Th} = 8.38 + j119$ Ω, $V_{Th} = 315$ kV; (d) plot; (e) $I_{short} = 2.65$ kA

5.12. (a) $R = 2.97$ Ω, $X_L = 63.8$ Ω, $Y_C = 0.988$j mS, $X_{supply1} = 4.41$ Ω, $X_{supply2} = 2.78$ Ω; (b) diagram; (c) $P_{maxXfr} = 779$ MW at $\delta_{max} = 92.4°$

5.13. (a) $R = 4.63$ Ω, $L = 0.148$ H, $C = 1.61$ μF; (b) diagram; (c) $|V_{LL\_source}| = 437$ kV, $S_{sup} = 888 + j827$ MVA; (d) $|S_{10\%}| = 433$ MVA

5.14. $H_{max} = 2.62$ A/m

CHAPTER 6: ELECTROMECHANICAL ENERGY CONVERSION

6.1. $\Phi_{IronError} = 0.015\%$, $B_{IronError} = 0.014\%$, $H_{IronError} = 0.006\%$, $H_{GapError} = 0.014\%$

6.2. (a) $H_{core} \cong 100$ A/m, $\mu_r = 9950$; (b) $\Phi_{core} = 2.45$ mWb, $I_{coil} = 0.224$ A; (c) $L_{coil} = 3.06$ H; (d) $I_{coil2} = 28.6$ A

6.3. (a) $\Phi_{gap} = 1.68$ mWb; (b) $H_{gap} = 540$ kA/m, $B_{gap} = 0.679$ T; (c) $L_{coil} = 112$ mH; (d) $I_{coil} = 4.66$ A; (e) $L_b = 108$ mH

6.4. (a) drawing; (b) $H_{dc} = 9.55$ kA/m, $B_{dc} = 0.012$ T; (c) $F_{dc} = 540$ N

6.5. (a) drawing; (b) $F_{a\_max} = 58$ N, $F_{b\_max} = 72$ N, $F_{c\_max} = 58$ N; (c) phase B in middle is most stressed

6.6. $F_{middle,max} = 315$ N, $F_{side,max} = 157.5$ N

6.7. Relations/plots where $B(g_{min}) = 0.335$ T, $B(g_{max}) = 0.077$ T and $F(g_{min}) = 112$ N, $F(g_{max}) = 5.95$ N

6.8. $V_{coil} = 67$ V

6.9. (a) $B_{gap} = 0.377$ T; (b) plot

6.10. (a) $I(gap_{max}) = 2.8$ A; (b) $F_{max} = 78.5$ N; (c) plot, iron saturation does not occur

6.11. (a) $\ell_{Pmag} = 0.331$ cm; (b) $I_{Vcoil} = 1.105$ A

6.12. $F_{mag} = 1026$ N

6.13. (a) $B_{gap} = 0.77$ T; (b) $E_{coil} = 1.16$ kV

CHAPTER 7: TRANSFORMERS

7.1. $V_{sup} = 7.64$ V

7.2. $V_{load} = 226$ V

7.3. $I_{SC} = 2.88$ kA $\angle -90°$

7.4. (a) $\mathbf{I_{AB}} = 65.1$ A $\angle -36.9°$, $\mathbf{I_{BC}} = 20.8$ A $\angle -161.4°$, $\mathbf{I_{CA}} = 122.4$ A $\angle 88.2°$; (b) $\mathbf{I_a} = 4.3$ A $\angle -36.9°$, $\mathbf{I_b} = 1.4$ A $\angle -161.4°$, $\mathbf{I_c} = 8.2$ A $\angle 88.2°$; (c) $V_a = 8.0$ kV, $V_b = 7.3$ kV, $V_c = 4.9$ kV, $V_{ab} = 1.5$ kV, $V_{bc} = 2.5$ kV, $V_{ca} = 3.2$ kV; (d) $\mathbf{I_{GND}} = 5.6$ A $\angle 64.7°$

7.5. $I_{SC} = 681$ A

7.6. $V_{source} = 230$ V

7.7. (a) $Z_{high} = 242$ Ω, $Z_{low} = 4.5$ Ω; (b) $I_{load} = 250$ A, $V_{sup} = 110$ kV

7.8. (a) $\mathbf{Z_{in}} = 0.646 + j1.684$ Ω; (b) $\mathbf{I_S} = 66.5$ A $\angle -35.5°$, $\mathbf{V_{sup}} = 2.9$ kV $\angle 1.3°$, $P_{in} = 154$ kW, $pf_{in} = 0.8$ (lagging); (c) $P_{5\%} = 216$ kW

7.9. (a) Reg = 9.9%; (b) plot; (c) $\varepsilon_{max}(pf = 0.8$ lag$) = 92.9\%$, $\varepsilon_{max}(pf = 0.9$ lead$) = 94.1\%$

7.10. (a) $R_{e,s} = 0.013$ Ω, $X_{e,s} = 0.087$ Ω, $R_{c,p} = 2.6$ kΩ, $X_{m,p} = 700$ Ω; (b) and (c) plots; (d) explanation

7.11. (a) $R_e = 0.26$ Ω, $X_e = 0.83$ Ω, $R_c = 220$ Ω, $X_m = 22.4$ Ω; (b) $\eta_{0.6lag,max} = 98.8\%$, $\eta_{0.8lag,max} = 99.1\%$, $\eta_{0.9lag,max} = 99.2\%$, $\eta_{0.6lead,max} = 98.9\%$, $\eta_{0.8lead,max} = 99.2\%$, $\eta_{0.9lead,max} = 99.3\%$; (c) $V_{lag} = 219$ V, $V_{lead} = 221$ V

7.12. (a) diagram with wye-connected primary and delta-connected secondary; (b) $\mathbf{I_{pri,ph,a}} = 196$ A $\angle -31.8°$, $\mathbf{I_{sec,ph,ab}} = 1.63$ kA $\angle -31.8°$, $\mathbf{I_{sec,ll,a}} = 2.83$ kA $\angle -1.8°$; (c) $\mathbf{V_{pri,ph,a}} = 20$ kV $\angle 0°$, $\mathbf{V_{pri,ll,ab}} = 34.6$ kV $\angle 30°$, $\mathbf{V_{sec,ll,ab}} = 2.4$ kV $\angle 0°$

7.13. (a) $\eta = 97.82\%$; (b) $\eta = 97.81\%$ at 85.7% load

7.14. (a) diagram; (b) $X_{tr} = 41$ Ω, $R_{ct} = 6.85$ Ω, $X_{mt} = 5.1$ Ω; (c) $\mathbf{I_{mag}} = 17.6$ A $\angle -53.1°$; (d) $\mathbf{I_{load}} = 126$ A $\angle -33.3°$, $\mathbf{V_{load}} = 120$ V $\angle -3.6°$

PROBLEM SOLUTION KEY

CHAPTER 8: SYNCHRONOUS MACHINES

8.1. 1200 rpm
8.2. 750 rpm
8.3. proof that $\Sigma\,S_k = 0$ at each node
8.4. $E_{gen} = 23.4 + j14.1$ kV, $S_{gen} = 125 + j268$ MVA
8.5. $I_{sc} = 5.87$ kA
8.6. $E_{gen} = 35$ kV
8.7. $\delta = 41.9°$
8.8. $P_{net} = 373$ MVA, $Q_{net} = -1.49$ MVA
8.9. (a) $pf = 0.814$; (b) $E_{LL} = 19$ kV
8.10. (a) $E_{rload} = 51.9$ kV, $E_{sc} = 1.27$ kV; (b) $I_{sc}(0\text{ MW}) = 97.7$ A, $I_{sc}(P_{load}) = 4.0$ kA
8.11. (a) $n_s = 3600$ rpm, $\omega_m = 377$ r/s; (b) $\Phi_{dc} = 2.5$ Wb, $B_{dc} = 0.056$ T, $H_{dc} = 44.5$ kA/m; (c) $I_{dc} = 16.3$ A; (d) $X_{syn} = 2.6$ kΩ; (e) diagram
8.12. (a) $Z_{line} = 3.15 + j22.5\ \Omega$, $X_g = 1.53\ \Omega$, $X_{xfmr} = 36.1\ \Omega$; (b) $E_{gen\_ll} = 41.6$ kV $\angle 63.9°$; (c) $\delta_{350} = 50.3°$
8.13. (a) $X_g = 0.722\ \Omega$, $X_{xfmr} = 9.68\ \Omega$, $Z_{line} = 8.4 + j60\ \Omega$, $X_c = 1.77$ kΩ; (b) $E_{g\_ph} = 24$ kV $\angle 33.9°$; (c) $C_{950} = 206\ \mu$F
8.14. For $0°$, TD $= 1/60$ s and $I_{peak} \cong 15$ kA
8.15. For PW $= 40$ ms, $I_{short} \cong 1.7$ kA and $V_{CB,peak} \cong 1.7$ MV

CHAPTER 9: INDUCTION MACHINES

9.1. (a) 16 slots/phase; (b) $30°$ between phases
9.2. $S_{motor} = 19.4 + j17.7$ kVA, $I_{motor} = 31.6$ A
9.3. $C_{cap} = 155\ \mu$F
9.4. $I_{MotorStart} = 356$ A
9.5. $I_{MotorIn} = 167$ A
9.6. $V_{sup} = 141$ V, $I_{motor} = 163$ A
9.7. (a) $n_s = 3600$ rpm; (b) two poles; (c) $f_{rotor} = 1.5$ Hz; (d) $I_{rate} = 314$ A
9.8. (a) diagram; (b) $n_m = 878$ rpm; (c) $I_{in}(2.5\%) = 18.7$ A $\angle -31.1°$, $pf_{in}(2.5\%) = 0.856$ (lagging); (d) $n_{T150} = 827$ rpm; (e) $\eta_{max} = 81.8\%$, $n_{\eta max} = 867$ rpm
9.9. (a) $n_m = 1692$ rpm; (b) $P_{AirGap} = 40.1$ kW, $T_{dev} = 213$ N·m, $T_{load} = 210$ N·m
9.10. (a) $R_{sta} = 1.4\ \Omega$, $X_{sta} = 131\ \Omega = X_{rot\_t}$, $R_c = 1225\ \Omega$, $X_m = 247\ \Omega$, $R_{rot\_t} = 1.865\ \Omega$; (b) diagram; (c) $T_{out}(10\%) = 3.93$ N·m, $P_{out}(10\%) = 666$ W, $\varepsilon(10\%) = 46.3\%$
9.11. (a) $P_{out} = 6$ hp; (b) $P_{ag} = 5.37$ kW; (c) $n = 3265$ rpm; (d) $T_{out} = 13.1$ N·m
9.12. (a) $I_m = 33.6$ A, $pf = 0.784$ (lagging), $V_t = 434$ V; (b) $P_{in} = 20.3$ kW, $Q_{in} = 16.1$ kVA; (c) $P_{ag} = 18.7$ kW; (d) $P_{dev} = 17.8$ kW, $T_{dev} = 99.3$ N·m; (e) $P_{out} = 23/65$ hp, $T_{out} = 98.5$ N·m; (f) $\varepsilon = 87\%$

9.13. (a) diagrams; (b) $s = 6.7\%$, $pf = 0.922$ (lagging); (c) $T_{out} = 118$ N·m; (d) $\varepsilon = 81.9\%$; (e) $I_{start} = 126$ A, $T_{start} = 76$ N·m; (f) $T_{max} = 175$ N·m, $s_{Tmax} = 18.5\%$

9.14. (a) $R_{sta} = 0.182$ Ω, $X_{sta} = 1.008$ Ω $= X_{rot\_t}$, $R_c = 271$ Ω, $X_m = 75.7$ Ω, $R_{rot\_t} = 0.274$ Ω; (b) $I_{mot}(n_m) = 28.2$ A, $I_{rot}(n_m) = 26.7$ A; (c) $T_{out}(n_m) = 112$ N·m; (d) $n_{op} = 1681$ rpm

9.15. (a) $I_{m\_rated} = 5.18$ A; (b) $n_{rated} = 1105$ rpm; (c) $n_{pf0.6} = 1138$ rpm; (d) $n_{m0.5} = 1141$ rpm; (e) $T_{max} = 3.8$ N·m, $n_{Tmax} = 899$ rpm

9.16. (a) $I_{sta} = 5.45$ A, $I_{for} = 1.62$ A, $I_{rev} = 5.08$ A; (b) $P_{out} = 107$ W, $P_{in} = 474$ W; (c) $\varepsilon = 22.5\%$

CHAPTER 10: DC MACHINES

10.1. proof
10.2. $P_{out} = 3.88$ hp
10.3. $P_{out} = 1.64$ hp
10.4. $K_m = 0.633$ V·s/A
10.5. $P_{in} = 0.986$ hp
10.6. $n_{mot} = 649$ rpm
10.7. $n_{mot} = 86$ rpm
10.8. $V_f = 258$ V
10.9. $V_{sup} = 108$ V, $T_{start} = 1126$ N·m; motor can start the pump
10.10. (a) $K_m = 0.837$ V·s/A; (b) $n_{nl} = 1367$ rpm; (c) $I_{99\%} = 24.7$ A
10.11. (a) $K_m = 0.06$ V·s/A; (b) n = 858 rpm, $T = 96.5$ N·m; (c) $I = 18.2$ A, n = 1938 rpm
10.12. (a) diagram; (b) $K_m = 0.651$ V·s/A; (c) $T_{start} = 3175$ rpm, n = 1241 rpm
10.13. (a) $I_f = 2.14$ A, $I_a = 347$ A, $I_{load} = 345$ A; (b) $V_t(0) = 277$ V, $V_t(P) = 246$ V; (c) $Reg = 12.6\%$; (d) $P_{5\%} = 40.6$ kW
10.14. (a) $K_m = 7.86$ V·s/A; (b) $I_{load} = 149$ A, $E_a = 393$ V, n = 672 rpm, (c) $P_{in} = 56.8$ kW, $P_{out} = 75.8$ hp, $\varepsilon = 99.6\%$
10.15. derivation

CHAPTER 11: INTRODUCTION TO POWER ELECTRONICS AND MOTOR CONTROL

11.1. (a) and (b) $a_1 = V_M$, otherwise $a_n = 0$ b_n; (c) a_1 (cosine) = b_1 (sine)
11.2. $I_{fw}(0) = 21.6$ A, $I_{fw}(1) = 0$, $I_{fw}(2) = 14.4$ A, $I_{fw}(3) = 0$, $I_{fw}(4) = 2.88$ A, $I_{fw}(5) = 0$, $I_{fw}(6) = 14.4$ A, ...
11.3. PSpice solution
11.4. $\alpha_{70\%} = 98.7°$
11.5. (a) $R = 7.2$ Ω; (b) plots; (c) THD $= 63.6\%$

11.6. (a) THD increases with delay angle; (b) I_{rms} decreases with α
11.7. $v_C(t) = 5.4 \pm 0.14$ V, $i_L(t) = 250 \pm 42$ mA
11.8. (a) $T_{on} = 5$ μs; (b) $L = 0.2$ mH, $C = 1.67$ μF; (c) PSpice plots
11.9. (a) delay angle control is feasible; (b) $\alpha = 32.2°$; (c) $P_{ac} = 3.66$ kW, $pf_{ac} = 0.97$
11.10. (a) $V_{dc} = 350$ V, $P_{dc} = 15.7$ kW; (b) $V_{ac\_rms} = 480$ V if $\alpha_r = 36.0°$; (c) THD = 47.3%; $I_{h1} = 40.5$ A
11.11. (a) $\alpha = 178.5°$; (b) derivation and plot; (c) $P_{xfr} = 64.8$ kW, $pf = 0.90$
11.12. (a) $K_m = 0.393$ V·s/A; (b) $V_{mot} = 376$ V; (c) $\alpha = 24.9°$
11.13. (a) $f_{inv} = 36.7$ Hz; (b) $V_{mot} = 269$ V, (c) plot; (d) $n_{mot} = 988$ rpm
11.14. (a) $I_{avg,load} \cong 4$ A for $L_{sm} = 10$ μH; (b) $I_{avg,load} \cong 3$ A for $L_{sm} = 200$ mH
11.15. (a) $I_{avg,load} \cong 2.25$ A; (b) $I_{avg,load} \cong 3.4$ A for $C_{sm} = 700$ μF

BIBLIOGRAPHY

Bergseth, F.R., and Venkata, S.S., *Introduction to Electric Energy Devices*, Prentice-Hall, Englewood Cliffs, NJ, 1987.

Burke, J.J., and Lawrence, D.J., *IEEE Trans. Power Apparatus and Systems*, Vol. 103, Jan. 1984, pp. 1–6.

Cathey, J.J., *Electric Machines: Analysis and Design Applying MATLAB*, McGraw-Hill, New York, 2001.

Chapman, C.R., *Electromechanical Energy Conversion*, Blaisdell Publishing, New York, 1965.

Chapman, S.J., *Electric Machinery Fundamentals*, 3rd ed., McGraw-Hill, Burr Ridge, IL, 1999.

Culp, A.W., *Principles of Energy Conversion*, 2nd ed., McGraw-Hill, New York, 1991.

Del Toro, V., *Electric Machines and Power Systems*, Prentice-Hall, Englewood Cliffs, NJ, 1985.

El-Wakil, M.M., *Powerplant Technology*, McGraw-Hill, New York, 1984.

Fitzgerald, A.E., Kingsley, C., Jr., and Umans, S.D., *Electric Machinery*, 4th ed., McGraw-Hill, Burr Ridge, IL, 1983.

Gonen, T., *Electrical Machines*, Power International Press, Carmichael, CA, 1998.

Heck, C., *Magnetic Materials and Their Applications*, Butterworth, London, 1974.

Hubert, C.I., *Electrical Machines*, Macmillan, Columbus, OH, 1991.

IEEE Standard 112-1996, Standard Test Procedure for Poly-phase Induction Motors and Generators, IEEE, Piscataway, NJ, 1996.

IEEE Standard 113-1985, Guide on Test Procedures for DC Machines, IEEE, Piscataway, NJ, 1985.

Jaeger, R.C., and Blalock, T.N., *Microelectronic Circuit Design*, 2nd ed., McGraw-Hill, New York, 2004.

Kosow, I.L., *Electric Machinery and Transformers*, Prentice-Hall, Englewood Cliffs, NJ, 1972.

McPherson, G., *An Introduction to Electrical Machines and Transformers*, John Wiley & Sons, New York, 1981.

Mohan, N., Undeland, T.M., and Robbins, W.P., *Power Electronics*, John Wiley & Sons, New York, 1995.

National Electrical Manufacturers Association, Motors and Generators, Publication No. MGI-1993, NEMA, Washington, DC, 1993.

Slemon, G.R., and Straughen, A., *Electric Machines*, Addison-Wesley, Reading, MA, 1980.

Vithayathil, J., *Power Electronics: Principles and Applications*, McGraw-Hill, New York, 1995.

Electrical Energy Conversion and Transport: An Interactive Computer-Based Approach, Second Edition. George G. Karady and Keith E. Holbert.
© 2013 Institute of Electrical and Electronics Engineers, Inc. Published 2013 by John Wiley & Sons, Inc.

Weisman, J., and Eckart, L.E., *Modern Power Plant Engineering*, Prentice-Hall, Englewood Cliffs, NJ, 1985.

Werninck, E.H. (ed.), *Electric Motor Handbook*, McGraw-Hill, London, 1978.

Wildi, T., *Electric Machines, Drives, and Power Systems*, 5th ed., Prentice-Hall, Inc., Englewood Cliffs, NJ, 2000.

Yamayee, Z.A., and Bala, J.L., Jr., *Electromechanical Energy Devices and Power Systems*, John Wiley & Sons, New York, 2001.

Zorbas, D., *Electric Machines*, West Publishing Company, St. Paul, MN, 1989.

INDEX

acid rain 40
actuator 313, 350, 353
admittance 97, 101
alnico 357
alternator 472
Ampere's law 225, 315, 327, 332, 506
arcing 618
armature reaction 623
ash 41
audio transformer 371
availability factor 78

base load 72, 77, 81
Betz efficiency 63, 75
B-H curve 314, 324–325, 328
  modeling 329–330
biomass 77
black start 603
blackout 20
boiling water reactor 55–56
breadth factor 495
breakdown voltage 350, 686
brownout 135
bundled conductors 218
  equivalent radius 273
bushings 379–381

capacitance 344
capacitance-to-neutral 273
capacity factor 63, 78

capital cost 33, 56, 66, 78, 80
carbon dioxide 39, 40, 85
Carnot 32, 70, 73, 77
cgs units 807
circuit breaker 4, 10–17
  assembly 11, 13
  dead-tank 11, 12
  failure analysis 15–17
  live-tank 10, 12
cogeneration 49
coherent unit systems 805
coil energy 323–324, 348
coil inductance 321
combined cycle 48, 49
commutation 709, 763
compensation capacitor 104, 105, 112, 189
complex numbers 97, 98
complex exponential 91, 113
complex spatial vectors 233–234
condenser 37
conductance 97–99
conductor table 226–227
conductors. *See also* transmission lines
  ACSR 216, 226–227
  bundled 218
  insulated 223
  shield 208–211, 216
  stranded 216
consumer service drop 22–24
  main breaker 24

Electrical Energy Conversion and Transport: An Interactive Computer-Based Approach, Second Edition.
George G. Karady and Keith E. Holbert.
© 2013 Institute of Electrical and Electronics Engineers, Inc. Published 2013 by John Wiley & Sons, Inc.

INDEX **825**

contingency 8
converter, 732. *See also* inverters and rectifiers
  boost 748–751
  buck 754–756
cooling pond 37
cooling tower 37, 38
corona 256
creosote 210
critical clearing time 20
cross-field theory 594
cross-linked polyethylene (PEX) 223
current
  definition 90
  ground 155, 410, 449
  restraint 507
current chopping 535
current division 118

damper, vibration 213
dc machines 616
  armature 617
  brushless 365
  commutator 617–620
  compound connection 629
  efficiency 635, 642
  equivalent circuit 625–628
  excitation methods 616–617, 628–629
  flux 622–623, 625
  generator 623–625
  induced voltage 625–626, 630
  interpoles 618
  machine constant 626
  mechanical output power 634, 640, 648
  motor 620–623
  motor control 651–652, 674–678
  operating point 634
  operating principle 620–625
  rotor 618–620
  separately excited 627, 630–636, 651
  series connection 629, 645–651
  shunt connection 628, 637–645, 651
  speed 628
  starting current 633, 639–640
  stator 618–619
  supply power 630
  terminal voltage 626
  torque 630, 635–636, 642–643, 650–651
  windings 617

dead-end tower 212
delta connection 166, 169
  generators 162, 163
  loads 150, 163, 168, 169, 170
delta-wye transformation
  generators 162
  impedances 150–151
diesel generator 76, 79
differential current 508–509
diode 685–687
disconnect switch 12, 14
distributed power generation 66
distribution system 6, 20–24
  overhead line 210
  pole 212
  substations 6, 10
  radial 21, 22
double revolving field theory 592–593
downcomer 36
duty cycle 750

economics 77–81
economizer 36
eddy current 315, 377
electric field
  charged particle generated 249
  conductor generated 250–252
  definition 249
  force on charge 336–337
  human exposure 257
  intensity 250
  lines 249
  transmission line generated 262–263
electric flux density 249
electric shock 256–257
electrical motor. *See* dc motors; induction motors
electricity cost 78–81
electromagnetic forces 347–349
electromechanical system 343–347
electrostatic force 336
electrostatic precipitator 39, 40, 337
energy harvesting 368
energy sources 31
energy storage 62, 63, 70, 79
English units 807
environmental impact 85
EPDM. *See* ethylene propylene diene monomer

equivalent pi 281
equivalent radius 273
ethylene propylene diene monomer (EPDM) 222
even function 699

FACTS 7, 673, 739–747
Faraday's law 317, 320, 505–506, 548–549
fast Fourier transform (FFT) 700
Ferranti effect 131
ferrite 325
ferromagnetic 314
FFT. *See* fast Fourier transform
firing circuits 726, 759–760
flashover 212, 221, 285, 302
Fleming's left-hand rule 621–622
Fleming's right-hand rule 623–624
fly ash 39, 41
fossil fuels 34, 35
Fourier series 701
   base component 490–491, 495, 707
   trigonometric 699–700
four-wire system 24, 155–158, 175
freewheeling diodes 734–735
frequency scaling 680
fringing fields 347
fuel cell 77
fuel injection 35–36
fundamental frequency 699
fuse cutout 23, 285–286

gate turn-off thyristor (GTO) 692–693
gauss (unit) 228, 805
Gauss' law 249
generator
   load 472
   parameters 488, 529
   protection 507–511
   reactive power regulation 472–473
   voltage 3, 375
geometric mean distance 247
geometric mean radius 244, 248, 274
geothermal 72, 73
greenhouse effect 85
ground current 155, 410, 449
ground fault 9, 182, 447, 507, 529, 608
ground rod 209

grounding 184, 208, 216, 410, 709
GTO. *See* gate turn-off thyristor

harmonic frequencies 699
health concerns 237–238
heat rate 32
heating value 34, 79, 80
high voltage dc 5, 207, 738–739
   HVDC link 7
horsepower 557
hot stick 23
hydroelectric 56–62, 79
   low-head 59
   high-head 60
hysteresis 314–315, 352

IGBT. *See* insulated gate bipolar transistor
image conductors 257–259, 262
impedance 96, 109, 140
   matrix 185
   parallel connection 101, 102
   series connection 100
   triangle 97
induction machines 541
   blocked rotor test 572–573
   control 678–684
   developed power 556, 559
   doubly fed 606–607
   efficiency 557, 564
   electrical power 556
   energy balance 557
   equivalent circuit 553–556
   force generation 552
   generator 603
   induced voltage 550
   leakage reactance 554, 556
   mechanical power 556–557, 559
   motor 548
   motor control 678–684, 735
   no-load test 570–571
   operating point 569
   operating principle 547
   parameter measurement 570
   rotor 546–547
   single-phase 542, 591
   slip 551, 559
   speed 551, 554, 565
   squirrel cage rotor 363, 546
   starting current 562

INDEX **827**

stator 543–545
three-phase 542, 547
windings 543–545
wound rotor 546–547
instrument transformer 13
insulated gate bipolar transistor (IGBT) 695–696
insulator string 219–221
insulators
  ball and socket 218–220
  cap and pin 218–220
  composite 221, 222
  flashover 221
  porcelain post 221
interconnection 6,7
inverters 729–732. *See also* rectifiers
  line-commutated
    thyristor-controlled 735–737
  voltage source 732–735
iron core 325, 377
  B-H (magnetization) curve 324
  neglecting 328–329, 348, 350
  saturation 324, 351, 494
iron damage 152

Kaplan turbine 59–60
Kirchhoff's current law 117, 118
Kirchhoff's voltage law 123

lamination 315
lightning 302
line faults 529
line losses 1, 18, 387
line transposition 247
line voltage 4
load forecasting 84
loads 109–113, 140
loop analysis 124
Lorentz force 337, 549, 621
loudspeaker 361

magnetic circuit 314, 489
  air gap 325
  toroidal 316
magnetic energy 323
magnetic field
  conductor moving in 338
  current-carrying conductor generated 225–228, 230

  Earth's 225
  energy content 229
  force on charge 337
  force generation 337–338
  human exposure 237–238
  resultant 237
magnetic field intensity 228
magnetic flux 229
  inside conductor 241–243
  outside conductor 240–241
  total conductor generated 243–244
magnetic flux density 228
magnetic path length 315
magnetic switch 347, 350, 353
magnetization curve. *See* B-H curve
magnetizing current 382, 391, 548
magnetizing inductance, 391–392. *See also* coil inductance
Mathcad 777
  arg() 783
  complex conjugate 129
  cspline 330
  editing keystrokes 782
  equation solvers 788–790
  Find() 122, 789–790
  function definition 785
  functions 784
  if() 677
  imaginary number representation 120, 781
  matrices 790–793
  Maximize() 788
  Minimize() 788
  modulus 717, 752
  or 717
  plot units 135, 786
  plotting 786–787
  range 784–785
  root() 134, 789
  toolbars 777, 779
  units 778, 781
  vectors 790–793
MATLAB 9, 794
  angle() 797
  colon operator 799
  constants 796
  element-by-element operations 798
  functions 797
  matrices 797–799

MATLAB (cont'd)
  operators 796
  plotting 800–803
  programming 803–804
  vectors 797–799
Maxwell's equations 225, 338
microelectromechanical 365–368
microphone 361–362
MKS units 805
modulation index 734
MOSFET 693–695
motor 170
  brushless 363–365
  permanent magnet 363
  stepper 341
  synchronous 363, 737

natural gas 31, 41, 49
neodymium-iron-boron 357, 363
nodal analysis 118
North American Electric Reliability
  Corporation (NERC) 7
north pole 340
Norton's theorem 127

ocean power 73
  current 75
  thermal 76
  tidal 74
  wave 75
odd function 699
oersted (unit) 318
Ohm's law 96
one-line diagram 8, 101
open circuit voltage 127, 131, 292
operation security 484
overcurrent 23, 282, 285, 447, 509, 609
overvoltage 13, 131, 216, 304, 448, 681

parallel, definition 100
passive circuit elements 98, 344
passive sign convention 90, 131, 140
peak power 77
per unit system 177–182
  base values 177, 180
permanent magnet 356–357, 360–365
permeability 314, 316, 325
  of free space 228
  relative 316, 319

permittivity 252
  of free space 250
PEX. See cross-linked polyethylene
phase sequence 146, 185, 273, 466
phase shift 92, 95
phasor
  conjugate 93
  definition 91
  diagram 95, 147, 154, 166
  mathematical operations 97, 98
  measurement unit 27
photovoltaic 66–70
pi (Π) circuit 275, 281, 291
piezoelectricity 366
piezoresistance 366
pitch 458, 494
polyphase 145
positive phase sequence. See phase sequence
positive sequence capacitance 271
positive sequence inductance 245
potential coefficients 260
power
  active 93
  apparent 93, 111
  average 92, 93
  complex 93, 113, 140, 146
  flow 18, 20, 203
  instantaneous 92, 146
  quadrature 93
  reactive 93, 110, 324
  real 93
power angle 18, 144, 471–472, 515
power factor 111–115
  angle 111, 140
  improvement 105, 128
power plants
  combined cycle 48, 49
  fossil 34–49
  geothermal 72, 73
  hydroelectric 56–62
  nuclear 49–56
  solar 66–72
power system 89
  voltages 4
power triangle 110
pressurized water reactor 53–55
protection
  differential 449, 507–510, 609
  distance 288

generator 507–509
motor 608–610
overcurrent 23, 285, 447, 509–510
transformer 447
transmission line 282
PSpice 810
 circuit editors 810
 examples 198, 302, 304, 528, 707, 757, 758
 libraries 811
 prefixes 811
 pulse source 532
 VSIN 304, 707, 813
pulse-width-modulation (PWM) 695–696, 732–734
pumped storage 62, 63
PWM. *See* pulse-width-modulation

radio interference 208, 256
reactance 96
reactive capability curve 473
rectifiers. *See also* inverters
 bridge 697
 delay angle 709, 712, 716, 719
 diode 697
 full-wave 697–698, 701
 half-wave 697–698, 700
 line commutated 709
 single-phase 709
 three-phase 728–729
 thyristor-controlled 709–711
regenerative breaking 681
relay 13, 47, 282, 287, 373, 447, 507
reserve capacity 78
residential electrical connection 24–26, 96
restraint current 507–509
reverse breakdown current 686
right hand rule 225, 227, 315, 337–338, 382
ripple 698, 750
root mean square 4
 definition 91
 Fourier series 699
 sinusoidal waveform 91
rotor 45, 462–463, 546–547, 618–620

safety 96, 156, 168, 375, 410
SAIFI 20–21
samarium cobalt 357–359

SCR. *See* silicon-controlled rectifier
scrubber 36
sequential components 182
service panel 23–26
SF_6 10, 17
shaded pole motor 601
sheet steel 314
shield conductor 208–211, 216
short circuit 20
 current 132, 303
shunt capacitor 119
silicon-controlled rectifier (SCR) 687
single-line diagram 101
SI units 805–808
slack bus 20
smart grid 26–28
smoothing capacitance 762
smoothing inductance 761
snubber circuit 728
solar cell 66–69
solar thermal plant 70–72
solenoid 340, 353
south pole 340
space potential 250
spinning reserve 78
split-phase motor 601
STATCOM 744
static stability 474–475, 479
static VAR compensator 740
stator 43, 44, 458–460, 543–545, 618–619
steady-state stability limit 299
step motor 341–343
substations 9
subsynchronous resonance 745
sub-transmission lines 209, 211
supply short circuit duty 311
supporting tower 213
surge arrester 13, 14
susceptance 98, 99
switchgear 4
 gas insulated 17–19
switching overvoltage 13, 534–535
switching surge 47
switchyard 35, 54
syllabus xvii
symbols 808
symmetrical components 182–188
synchronization 473–474
synchronous condenser 8, 472

synchronous machines
  armature, *see* stator
  armature flux 468
  armature inductance/reactance 469, 488, 496, 506
  armature reaction 470
  brushless excitation 464–465
  cooling 458
  equivalent circuit 470
  excitation 462–463
  hydro 46
  induced voltage 488
  neutral point 152, 510
  permanent magnet 363
  reactive capability 473
  rotor 458–461
  round rotor 456
  salient pole 46, 60, 457, 461–463
  short circuit 508, 528
  slip rings 463–464
  stator 458–460
  synchronous reactance 470, 506
  synchronous speed 468
  terminal voltage 467, 469–470
  transient impedance 529
system efficiency 102, 297–298
system frequency 91, 456
system protection 20, 169
system voltage 18

television interference 208, 256
tension tower 212, 213
tesla (unit) 229
thermal efficiency 32
thermal energy storage 70
thermal pollution 37, 49
Thévenin equivalent 127
  inverter 736–737
  motor 558
  network 128, 199, 292, 298–299
  transformer 385
three-phase circuits
  advantages 145
  balanced systems 146
  power measurement 174–177
three-wire system 155, 160, 175
thyristor 687, 689–694
torque 557
  starting 568, 599

total harmonic distortion 716
transducer 313, 356
transformer 324–325, 375, 377–378
  circuit parameters 394
  complex power 383
  cooling 378
  core resistance 392, 402
  core type 382
  current 13, 286
  delta-delta 420
  delta-wye 418–420
  distribution 24–25
  dry type 379
  equivalent circuit 393
  generator step-up 507
  ground level 25
  grounding 410, 447
  ideal 382–384
  impedance transfer 386
  inrush current 447–448
  instrument 13
  leakage reactance 391, 393, 432
  oil cooled 380
  open circuit test 401–404, 429–432
  parallel connection 404–405
  phase shift 421
  potential 13, 47
  power cube 697
  power plant 44, 47
  protection 447–450
  rated current 395
  rating 408
  real 391–394
  shell type 382
  short circuit test 400–401, 429, 431–433
  single-phase 381–382
  single-phase equivalent 386–387, 417–418
  step-down 6, 23–25, 96
  step-up 44, 507
  three-phase 408–410, 421
  turns ratio 376, 383–384
  voltage polarity 377
  windings 376–377
  wye-delta connected 415–417
  wye-wye connected 410–411
transient stability 20
transmission line parameters 273
  capacitance 271–273
  inductance 239, 245–248

long 277–280
resistance 224–227
transmission lines
  conductor arrangements 219
  conductors 216–218
  high voltage dc 207
  line voltage 4
  models 274–275
  overhead 209–211
  protection 282–285
  right-of-way 209, 257
  sag 212, 213
  single-phase equivalent 291
  siting 208
  span 212, 213
  splicing 216
  towers 208–210, 215, 266
  vibration damper 213
transmission system 4–6
  faults 529
transposition 247
turbine
  combustion 47
  Francis 60, 61
  hydraulic 56
  Kaplan 59, 60
  Pelton 60
  steam 41–43
  wind 63
turbine-generator 30
turns ratio 376, 383–384

underground cable 23, 213–215, 223
underwater cable 6, 73

unified power controller 747
unit conversions 807
unit dimensions 806

VAR (unit) 93, 170
varistor 744
voice coil 360
voltage
  definition 90
  extra-high 208
  integrated from electric field
    250
  line-to-ground 250
  line-to-line 152, 154, 163, 250
  line-to-neutral 152, 154, 163
  load 156
  open circuit 127, 131, 292
  polarity convention 90
  sinusoidal 91
  standards 4, 208
  ultra-high 4
voltage division 123
voltage regulation 116, 135, 139, 193, 297

water-energy nexus 38, 85
wattmeter 174
weather shed 221–222
Wheatstone bridge 366
wind farm 63–66
winding factor 495
wye connection 155, 169, 186
  generators 151, 152, 155, 156
  loads 150, 168, 169, 170

IEEE Press Series on Power Engineering

1. *Principles of Electric Machines with Power Electronic Applications, Second Edition*
M. E. El-Hawary

2. *Pulse Width Modulation for Power Converters: Principles and Practice*
D. Grahame Holmes and Thomas Lipo

3. *Analysis of Electric Machinery and Drive Systems, Second Edition*
Paul C. Krause, Oleg Wasynczuk, and Scott D. Sudhoff

4. *Risk Assessment of Power Systems: Models, Methods, and Applications*
Wenyuan Li

5. *Optimization Principles: Practical Applications to the Operations of Markets of the Electric Power Industry*
Narayan S. Rau

6. *Electric Economics: Regulation and Deregulation*
Geoffrey Rothwell and Tomas Gomez

7. *Electric Power Systems: Analysis and Control*
Fabio Saccomanno

8. *Electrical Insulation for Rotating Machines: Design, Evaluation, Aging, Testing, and Repair*
Greg Stone, Edward A. Boulter, Ian Culbert, and Hussein Dhirani

9. *Signal Processing of Power Quality Disturbances*
Math H. J. Bollen and Irene Y. H. Gu

10. *Instantaneous Power Theory and Applications to Power Conditioning*
Hirofumi Akagi, Edson H. Watanabe, and Mauricio Aredes

11. *Maintaining Mission Critical Systems in a 24/7 Environment, Second Edition*
Peter M. Curtis

12. *Elements of Tidal-Electric Engineering*
Robert H. Clark

13. *Handbook of Large Turbo-Generator Operation Maintenance, Second Edition*
Geoff Klempner and Isidor Kerszenbaum

14. *Introduction to Electrical Power Systems*
Mohamed E. El-Hawary

15. *Modeling and Control of Fuel Cells: Disturbed Generation Applications*
M. Hashem Nehrir and Caisheng Wang

16. *Power Distribution System Reliability: Practical Methods and Applications*
Ali A. Chowdhury and Don O. Koval

17. *Introduction to FACTS Controllers: Theory, Modeling, and Applications*
Kalyan K. Sen and Mey Ling Sen

18. *Economic Market Design and Planning for Electric Power Systems*
James Momoh and Lamine Mili

19. *Operation and Control of Electric Energy Processing Systems*
James Momoh and Lamine Mili

20. *Restructured Electric Power Systems: Analysis of Electricity Markets with Equilibrium Models*
Xiao-Ping Zhang

21. *An Introduction to Wavelet Modulated Inverters*
S.A. Saleh and M. Azizur Rahman

22. *Probabilistic Transmission System Planning*
Wenyuan Li

23. *Control of Electric Machine Drive Systems*
Seung-Ki Sul

24. *High Voltage and Electrical Insulation Engineering*
Ravindra Arora and Wolfgang Mosch

25. *Practical Lighting Design with LEDs*
Ron Lenk and Carol Lenk

26. *Electricity Power Generation: The Changing Dimensions*
Digambar M. Tagare

27. *Electric Distribution Systems*
Abdelhay A. Sallam and Om P. Malik

28. *Maintaining Mission Critical Systems in a 24/7 Environment, Second Edition*
Peter M. Curtis

29. *Power Conversion and Control of Wind Energy Systems*
Bin Wu, Yongqiang Lang, Navid Zargan, and Samir Kouro

30. *Integration of Distributed Generation in the Power System*
Math Bollen and Fainan Hassan

31. *High Voltage Protection for Telecommunications*
Steven W. Blume

32. *Doubly Fed Induction Machine: Modeling and Control for Wind Energy Generation*
Gonzalo Abad, Jesús Lopéz, Miguel Rodríguez, Luis Marroyo, and Grzegorz Iwanski

33. *Smart Grid: Fundamentals of Design and Analysis*
James Momoh

34. *Electromechanical Motion Devices, Second Edition*
Paul Krause, Oleg Wasynczuk, and Steven Pekarek

35. *Arc Flash Hazard and Analysis and Mitigation*
J. C. Das

36. *Electrical Energy Conversion and Transport: An Interactive Computer-Based Approach, Second Edition*
George G. Karady and Keith E. Holbert

37. *Analysis of Electric Machinery and Drive Systems, Third Edition*
Paul Krause, Oleg Wasynczuk, Scott Sudhoff, and Steven Pekarek